INTRODUCTION TO LINEAR ALGEBRA

Third Edition

GILBERT STRANG
Massachusetts Institute of Technology

WELLESLEY-CAMBRIDGE PRESS
Box 812060 Wellesley MA 02482

Printed in the United States of America 9 8 7 6 5 4 3

ISBN 0-9614088-9-8 **ISBN 0-9614088-4-7**

QA184.S78 2003 512'.5 93-14092

Other texts from Wellesley-Cambridge Press

Wavelets and Filter Banks, Gilbert Strang and Truong Nguyen,
ISBN 0-9614088-7-1.

Linear Algebra, Geodesy, and GPS, Gilbert Strang and Kai Borre,
ISBN 0-9614088-6-3.

Introduction to Applied Mathematics, Gilbert Strang,
ISBN 0-9614088-0-4.

An Analysis of the Finite Element Method, Gilbert Strang and George Fix,
ISBN 0-9614088-8-X.

Calculus, Gilbert Strang,
ISBN 0-9614088-2-0.

Wellesley-Cambridge Press
Box 812060
Wellesley MA 02482 USA
www.wellesleycambridge.com

gs@math.mit.edu
math.mit.edu/~gs
phone/fax (781) 431-8488

LATEX text preparation by Cordula Robinson and Brett Coonley, Massachusetts Institute of Technology

LATEX assembly and book design by Amy Hendrickson, TEXnology Inc., **www.texnology.com**

A Solutions Manual is available to instructors by email from the publisher.
Course material including syllabus and Teaching Codes and exams and videotaped lectures for this course are available on the linear algebra web site: **web.mit.edu/18.06/www**

Linear Algebra is included in MIT's OpenCourseWare site **ocw.mit.edu** with videos of the full course 18.06.

TABLE OF CONTENTS

PREFACE

This preface expresses some personal thoughts. It is my chance to write about how linear algebra can be taught and learned. If we teach pure abstraction, or settle for cookbook formulas, we miss the best part. This course has come a long way, in living up to what it can be.

It may be helpful to mention the web pages connected to this book. So many messages come back with suggestions and encouragement, and I hope that professors and students will make free use of everything. You can directly access **web.mit.edu/18.06/www**, which is continually updated for the MIT course that is taught every semester. Linear Algebra is also on the OpenCourseWare site **ocw.mit.edu**, where 18.06 became exceptional by including videos (which you definitely don't have to watch ...). I can briefly indicate part of what is available now:

1. Lecture schedule and current homeworks and exams with solutions

2. The goals of the course and conceptual questions

3. Interactive Java demos for eigenvalues and least squares and more

4. A table of eigenvalue/eigenvector information (see page 362)

5. ***Glossary***: A Dictionary for Linear Algebra

6. Linear Algebra Teaching Codes and MATLAB problems

7. Videos of the full course (taught in a real classroom).

These web pages are a resource for professors and students worldwide. My goal is to make this book as useful as possible, with all the course material I can provide.

After this preface, the book will speak for itself. You will see the spirit right away. The goal is to show the beauty of linear algebra, and its value. The emphasis is on understanding—*I try to explain rather than to deduce*. This is a book about real mathematics, not endless drill. I am constantly working with examples (create a matrix, find its nullspace, add another column, see what changes, ***ask for help***!). The textbook has to help too, in teaching what students need. The effort is absolutely rewarding, and fortunately this subject is *not too hard*.

The New Edition

A major addition to the book is the large number of Worked Examples, section by section. Their purpose is to connect the text directly to the homework problems. The complete solution to a vector equation $Ax = b$ is $x_{\text{particular}} + x_{\text{nullspace}}$—and the steps are explained as clearly as I can. The Worked Example **3.4 A** converts this explanation into action by taking every step in the solution (starting with the test for solvability). I hope these model examples will bring the content of each section into focus (see **5.1 A** and **5.2 B** on determinants). The "Pascal matrices" are a neat link from the amazing properties of Pascal's triangle to linear algebra.

The book contains new problems of all kinds—more basic practice, applications throughout science and engineering and management, and just fun with matrices. Northwest and southeast matrices wander into Problem 2.4.39. *Google* appears in Chapter 6. Please look at the last exercise in Section 1.1. I hope the problems are a strong point of this book—the newest one is about the six 3 by 3 permutation matrices: What are their determinants and pivots and traces and eigenvalues?

The Glossary is also new, in the book and on the web. I believe students will find it helpful. In addition to defining the important terms of linear algebra, there was also a chance to include many of the key facts for quick reference.

Fortunately, the need for linear algebra is widely recognized. *This subject is absolutely as important as calculus.* I don't concede anything, when I look at how mathematics is used. There is even a light-hearted essay called "*Too Much Calculus*" on the web page. The century of data has begun! So many applications are discrete rather than continuous, digital rather than analog. The truth is that vectors and matrices have become the language to know.

The Linear Algebra Course

The equation $Ax = b$ uses that language right away. The matrix A times any vector x is *a combination of the columns of* A. The equation is asking for ***a combination that produces*** b. Our solution comes at three levels and they are all important:

1. ***Direct solution*** by forward elimination and back substitution.

2. ***Matrix solution*** $x = A^{-1}b$ by inverting the matrix.

3. ***Vector space solution*** by looking at the column space and nullspace of A.

And there is another possibility: $Ax = b$ *may have no solution.* Elimination may lead to $0 = 1$. The matrix approach may fail to find A^{-1}. The vector space approach can look at all combinations Ax of the columns, but b might be outside that column space. Part of mathematics is understanding when $Ax = b$ is solvable, and what to do when it is not (the least squares solution uses $A^{\mathrm{T}}A$ in Chapter 4).

Another part is learning to visualize vectors. A vector v with two components is not hard. Its components v_1 and v_2 tell how far to go across and up—we draw an arrow. A second vector w may be perpendicular to v (and Chapter 1 tells when). If those vectors have six components, we can't draw them but our imagination keeps trying. In six-dimensional space, we can test quickly for a right angle. It is easy to visualize $2v$ (twice as far) and $-w$ (opposite to w). We can *almost* see a combination like $2v - w$.

Most important is the effort to imagine ***all the combinations*** $cv+dw$. They fill a "two-dimensional plane" inside the six-dimensional space. As I write these words, I am not at all sure that I can see this subspace. But linear algebra works easily with vectors and matrices of any size. If we have currents on six edges, or prices for six products, or just position and velocity of an airplane, we are dealing with six dimensions. For image processing or web searches (or the human genome), six might change to a million. It is still linear algebra, and linear combinations still hold the key.

Structure of the Textbook

Already in this preface, you can see the style of the book and its goal. The style is informal but the goal is absolutely serious. Linear algebra is great mathematics, and I certainly hope that each professor who teaches this course will learn something new. The author always does. The student will notice how the applications reinforce the ideas. I hope you will see how this book moves forward, *gradually and steadily*.

I want to note six points about the organization of the book:

1. Chapter 1 provides a brief introduction to vectors and dot products. If the class has met them before, the course can begin with Chapter 2. That chapter solves n by n systems $Ax = b$, and prepares for the whole course.
2. I now use the *reduced row echelon form* more than before. The MATLAB command rref(A) produces bases for the row space and column space. Better than that, reducing the combined matrix $\begin{bmatrix} A & I \end{bmatrix}$ produces total information about all four of the fundamental subspaces.
3. Those four subspaces are an excellent way to learn about linear independence and bases and dimension. They go to the heart of the matrix, and they are genuinely the key to applications. I hate just making up vector spaces when so many important ones come naturally. If the class sees plenty of examples, independence is virtually understood in advance: A has independent columns when $x = 0$ is the only solution to $Ax = 0$.
4. Section 6.1 introduces eigenvalues for ***2 by 2 matrices***. Many courses want to see eigenvalues early. It is absolutely possible to go directly from Chapter 3 to Section 6.1. The determinant is easy for a 2 by 2 matrix, and eigshow on the web captures graphically the moment when $Ax = \lambda x$.
5. Every section in Chapters 1 to 7 ends with a highlighted ***Review of the Key Ideas***. The reader can recapture the main points by going carefully through this review.
6. Chapter 8 (*Applications*) has a new section on ***Matrices in Engineering***.

When software is available (and time to use it), I see two possible approaches. One is to carry out instantly the steps of testing linear independence, orthogonalizing by Gram-Schmidt, and solving $Ax = b$ and $Ax = \lambda x$. The Teaching Codes follow the steps described in class—MATLAB and Maple and Mathematica compute a little differently. All can be used (*optionally*) with this book. The other approach is to experiment on bigger problems—like finding the largest determinant of a ± 1 matrix, or the average size of a pivot. The time to compute $A^{-1}b$ is measured by tic; inv(A) * b; toc. Choose A = rand(1000) and compare with tic; A/b; toc by direct elimination.

A one-semester course that moves steadily will reach eigenvalues. The key idea is to diagonalize A by its eigenvector matrix S. When that succeeds, the eigenvalues appear in $S^{-1}AS$. For symmetric matrices we can choose $S^{-1} = S^{\mathrm{T}}$. When A is rectangular we need $U^{\mathrm{T}}AV$ (U comes from eigenvectors of AA^{T} and V from $A^{\mathrm{T}}A$). Chapters 1 to 6 are the heart of a basic course in linear algebra—*theory plus applications*. The beauty of this subject is in the way those come together.

May I end with this thought for professors. You might feel that the direction is right, and wonder if your students are ready. ***Just give them a chance***! Literally thousands of students have written to me, frequently with suggestions and surprisingly often with thanks. They know when the course has a purpose, because the professor and the book are on their side. Linear algebra is a fantastic subject, enjoy it.

Acknowledgements

This book owes a big debt to readers everywhere. Thousands of students and colleagues have been involved in every step. I have not forgotten the warm welcome for the first sentence written 30 years ago: "I believe that the teaching of linear algebra has become too abstract." A less formal approach is now widely accepted as the right choice for the basic course. And this course has steadily improved—the homework problems, the lectures, the Worked Examples, even the Web. I really hope you see that linear algebra is not some optional elective, it is *needed*. The first step in all subjects is linear!

I owe a particular debt to friends who offered suggestions and corrections and ideas. David Arnold in California and Mike Kerckhove in Virginia teach this course *well*. Per-Olof Persson created MATLAB codes for the experiments, as Cleve Moler and Steven Lee did earlier for the Teaching Codes. And the Pascal matrix examples, in the book and on the Web, owe a lot to Alan Edelman (and a little to Pascal). It is just a pleasure to work with friends.

My deepest thanks of all go to Cordula Robinson and Brett Coonley. They created the LaTeX pages that you see. Day after day, new words and examples have gone back and forth across the hall. After 2000 problems (and 3000 attempted solutions) this expression of my gratitude to them is almost the last sentence, of work they have beautifully done.

Amy Hendrickson of **texnology.com** produced the book itself, and you will recognize the quality of her ideas. My favorites are the clear boxes that highlight key points. The quilt on the front cover was created by Chris Curtis (it appears in *Great American Quilts: Book 5*, by Oxmoor House). Those houses show nine linear transformations of the plane. (At least they are linear in Figure 7.1, possibly superlinear in the quilt.) Tracy Baldwin has succeeded again to combine art and color and mathematics, in her fourth neat cover for Wellesley-Cambridge Press.

May I dedicate this book to grandchildren who are very precious: Roger, Sophie, Kathryn, Alexander, Scott, Jack, William, Caroline, and Elizabeth. I hope you might take linear algebra one day. Especially I hope you like it. The author is proud of you.

1

INTRODUCTION TO VECTORS

The heart of linear algebra is in two operations—both with vectors. We add vectors to get $\boldsymbol{v} + \boldsymbol{w}$. We multiply by numbers c and d to get $c\boldsymbol{v}$ and $d\boldsymbol{w}$. Combining those two operations (adding $c\boldsymbol{v}$ to $d\boldsymbol{w}$) gives the *linear combination* $c\boldsymbol{v} + d\boldsymbol{w}$.

Linear combinations are all-important in this subject! Sometimes we want one particular combination, a specific choice of c and d that produces a desired $c\boldsymbol{v} + d\boldsymbol{w}$. Other times we want to visualize *all the combinations* (coming from all c and d). The vectors $c\boldsymbol{v}$ lie along a line. The combinations $c\boldsymbol{v} + d\boldsymbol{w}$ normally fill a two-dimensional plane. (I have to say "two-dimensional" because linear algebra allows higher-dimensional planes.) From four vectors $\boldsymbol{u}, \boldsymbol{v}, \boldsymbol{w}, \boldsymbol{z}$ in four-dimensional space, their combinations are likely to fill the whole space.

Chapter 1 explains these central ideas, on which everything builds. We start with two-dimensional vectors and three-dimensional vectors, which are reasonable to draw. Then we move into higher dimensions. The really impressive feature of linear algebra is how smoothly it takes that step into n-dimensional space. Your mental picture stays completely correct, even if drawing a ten-dimensional vector is impossible.

This is where the book is going (into n-dimensional space), and the first steps are the operations in Sections 1.1 and 1.2:

1.1 ***Vector addition*** $\boldsymbol{v} + \boldsymbol{w}$ ***and linear combinations*** $c\boldsymbol{v} + d\boldsymbol{w}$.

1.2 ***The dot product*** $v \cdot w$ ***and the length*** $\|\boldsymbol{v}\| = \sqrt{\boldsymbol{v} \cdot \boldsymbol{v}}$.

VECTORS AND LINEAR COMBINATIONS ■ 1.1

"You can't add apples and oranges." In a strange way, this is the reason for vectors! If we keep the number of apples separate from the number of oranges, we have a pair of numbers. That pair is a ***two-dimensional vector*** $\boldsymbol{v}$, with "components" v_1 and v_2:

$$\boldsymbol{v} = \begin{bmatrix} v_1 \\ v_2 \end{bmatrix} \qquad \begin{aligned} v_1 &= \text{number of apples} \\ v_2 &= \text{number of oranges.} \end{aligned}$$

We write $\boldsymbol{v}$ as a ***column vector***. The main point so far is to have a single letter $\boldsymbol{v}$ (in ***boldface italic***) for this pair of numbers v_1 and v_2 (in *lightface italic*).

Even if we don't add v_1 to v_2, we do *add vectors*. The first components of $\boldsymbol{v}$ and $\boldsymbol{w}$ stay separate from the second components:

VECTOR ADDITION $$\boldsymbol{v} = \begin{bmatrix} v_1 \\ v_2 \end{bmatrix} \quad \text{and} \quad \boldsymbol{w} = \begin{bmatrix} w_1 \\ w_2 \end{bmatrix} \quad \text{add to} \quad \boldsymbol{v}+\boldsymbol{w} = \begin{bmatrix} v_1 + w_1 \\ v_2 + w_2 \end{bmatrix}.$$

You see the reason. We want to add apples to apples. Subtraction of vectors follows the same idea: *The components of* $\boldsymbol{v} - \boldsymbol{w}$ *are* $v_1 - w_1$ *and* _____.

The other basic operation is *scalar multiplication*. Vectors can be multiplied by 2 or by -1 or by any number c. There are two ways to double a vector. One way is to add $\boldsymbol{v} + \boldsymbol{v}$. The other way (the usual way) is to multiply each component by 2:

SCALAR MULTIPLICATION $$2\boldsymbol{v} = \begin{bmatrix} 2v_1 \\ 2v_2 \end{bmatrix} \quad \text{and} \quad -\boldsymbol{v} = \begin{bmatrix} -v_1 \\ -v_2 \end{bmatrix}.$$

The components of $c\boldsymbol{v}$ are cv_1 and cv_2. The number c is called a "scalar".

Notice that the sum of $-\boldsymbol{v}$ and $\boldsymbol{v}$ is the zero vector. This is $\mathbf{0}$, which is not the same as the number zero! The vector $\mathbf{0}$ has components 0 and 0. Forgive me for hammering away at the difference between a vector and its components. Linear algebra is built on these operations $\boldsymbol{v} + \boldsymbol{w}$ and $c\boldsymbol{v}$—***adding vectors and multiplying by scalars***.

The order of addition makes no difference: $\boldsymbol{v} + \boldsymbol{w}$ equals $\boldsymbol{w} + \boldsymbol{v}$. Check that by algebra: The first component is $v_1 + w_1$ which equals $w_1 + v_1$. Check also by an example:

$$\boldsymbol{v}+\boldsymbol{w} = \begin{bmatrix} 1 \\ 5 \end{bmatrix} + \begin{bmatrix} 3 \\ 3 \end{bmatrix} = \begin{bmatrix} 4 \\ 8 \end{bmatrix} = \begin{bmatrix} 3 \\ 3 \end{bmatrix} + \begin{bmatrix} 1 \\ 5 \end{bmatrix} = \boldsymbol{w}+\boldsymbol{v}.$$

Linear Combinations

By combining these operations, we now form "*linear combinations*" of $\boldsymbol{v}$ and $\boldsymbol{w}$. Multiply $\boldsymbol{v}$ by c and multiply $\boldsymbol{w}$ by d; then add $c\boldsymbol{v} + d\boldsymbol{w}$.

DEFINITION ***The sum of*** $c\boldsymbol{v}$ ***and*** $d\boldsymbol{w}$ ***is a*** *linear combination* ***of*** $\boldsymbol{v}$ ***and*** $\boldsymbol{w}$.

Four special linear combinations are: sum, difference, zero, and a scalar multiple $c\boldsymbol{v}$:

$$\begin{aligned} 1\boldsymbol{v} + 1\boldsymbol{w} &= \text{sum of vectors in Figure 1.1} \\ 1\boldsymbol{v} - 1\boldsymbol{w} &= \text{difference of vectors in Figure 1.1} \\ 0\boldsymbol{v} + 0\boldsymbol{w} &= \textit{zero vector} \\ c\boldsymbol{v} + 0\boldsymbol{w} &= \text{vector } c\boldsymbol{v} \text{ in the direction of } \boldsymbol{v} \end{aligned}$$

The zero vector is always a possible combination (when the coefficients are zero). Every time we see a "space" of vectors, that zero vector will be included. It is this big view, taking all the combinations of $\boldsymbol{v}$ and $\boldsymbol{w}$, that makes the subject work.

The figures show how you can visualize vectors. For algebra, we just need the components (like 4 and 2). In the plane, that vector $\boldsymbol{v}$ is represented by an arrow. The arrow goes $v_1 = 4$ units to the right and $v_2 = 2$ units up. It ends at the point whose x, y coordinates are $4, 2$. This point is another representation of the vector—so we have three ways to describe $\boldsymbol{v}$, by an *arrow* or a *point* or a *pair of numbers*.

Using arrows, you can see how to visualize the sum $\boldsymbol{v}+\boldsymbol{w}$:

Vector addition (head to tail) ***At the end of $\boldsymbol{v}$, place the start of $\boldsymbol{w}$.***

We travel along $\boldsymbol{v}$ and then along $\boldsymbol{w}$. Or we take the shortcut along $\boldsymbol{v}+\boldsymbol{w}$. We could also go along $\boldsymbol{w}$ and then $\boldsymbol{v}$. In other words, $\boldsymbol{w}+\boldsymbol{v}$ gives the same answer as $\boldsymbol{v}+\boldsymbol{w}$. These are different ways along the parallelogram (in this example it is a rectangle). The endpoint in Figure 1.1 is the diagonal $\boldsymbol{v}+\boldsymbol{w}$ which is also $\boldsymbol{w}+\boldsymbol{v}$.

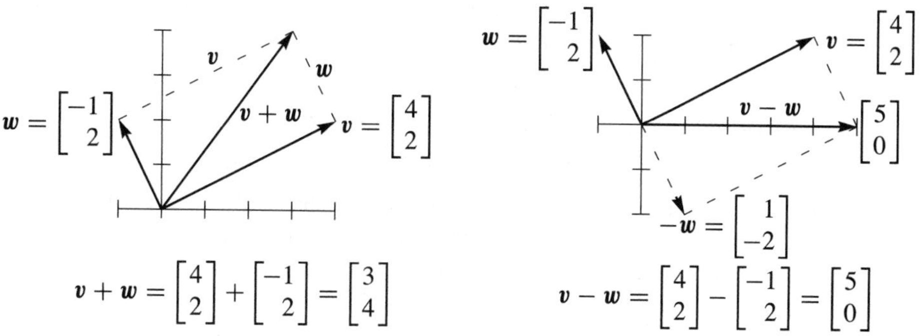

Figure 1.1 Vector addition $\boldsymbol{v}+\boldsymbol{w}$ produces the diagonal of a parallelogram. The linear combination on the right is $\boldsymbol{v}-\boldsymbol{w}$.

The zero vector has $v_1 = 0$ and $v_2 = 0$. It is too short to draw a decent arrow, but you know that $\boldsymbol{v}+\mathbf{0} = \boldsymbol{v}$. For $2\boldsymbol{v}$ we double the length of the arrow. We reverse its direction for $-\boldsymbol{v}$. This reversing gives the subtraction on the right side of Figure 1.1.

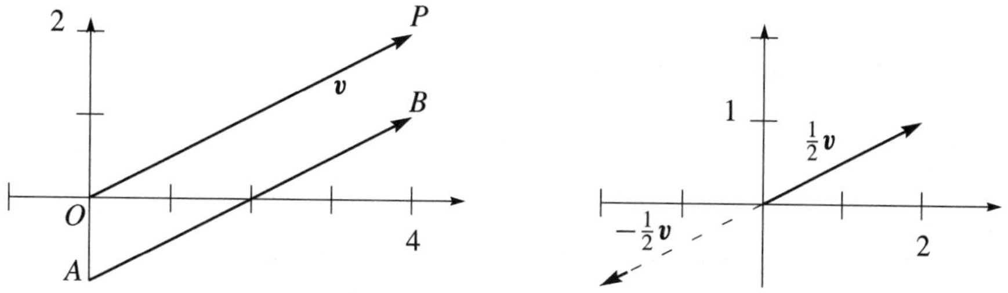

Figure 1.2 The arrow usually starts at the origin $(0,0)$; $c\boldsymbol{v}$ is always parallel to $\boldsymbol{v}$.

Vectors in Three Dimensions

A vector with two components corresponds to a point in the xy plane. The components of $\boldsymbol{v}$ are the coordinates of the point: $x = v_1$ and $y = v_2$. The arrow ends at this point (v_1, v_2), when it starts from $(0, 0)$. Now we allow vectors to have three components (v_1, v_2, v_3). The xy plane is replaced by three-dimensional space.

Here are typical vectors (still column vectors but with three components):

$$\boldsymbol{v} = \begin{bmatrix} 1 \\ 2 \\ 2 \end{bmatrix} \quad \text{and} \quad \boldsymbol{w} = \begin{bmatrix} 2 \\ 3 \\ -1 \end{bmatrix} \quad \text{and} \quad \boldsymbol{v} + \boldsymbol{w} = \begin{bmatrix} 3 \\ 5 \\ 1 \end{bmatrix}.$$

The vector $\boldsymbol{v}$ corresponds to an arrow in 3-space. Usually the arrow starts at the origin, where the xyz axes meet and the coordinates are $(0, 0, 0)$. The arrow ends at the point with coordinates v_1, v_2, v_3. There is a perfect match between the ***column vector*** and the ***arrow from the origin*** and the ***point where the arrow ends***.

From now on $\quad \boldsymbol{v} = \begin{bmatrix} 1 \\ 2 \\ 2 \end{bmatrix} \quad$ ***is also written as*** $\quad \boldsymbol{v} = (1, 2, 2)$.

The reason for the row form (in parentheses) is to save space. But $\boldsymbol{v} = (1, 2, 2)$ is not a row vector! It is in actuality a column vector, just temporarily lying down. The row vector $[\,1 \;\; 2 \;\; 2\,]$ is absolutely different, even though it has the same three components. It is the "transpose" of the column $\boldsymbol{v}$.

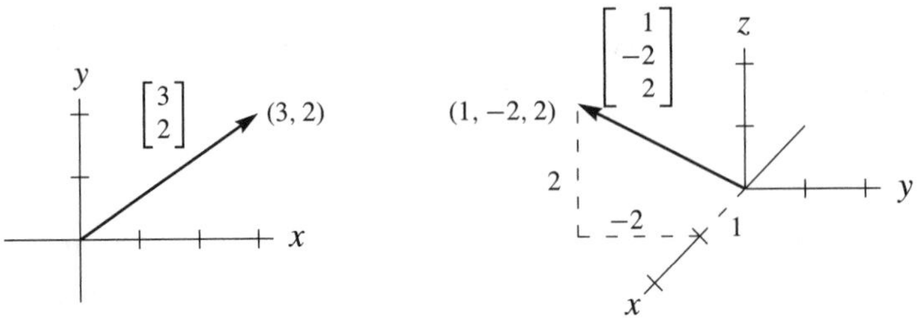

Figure 1.3 Vectors $\begin{bmatrix} x \\ y \end{bmatrix}$ and $\begin{bmatrix} x \\ y \\ z \end{bmatrix}$ correspond to points (x, y) and (x, y, z).

In three dimensions, $\boldsymbol{v} + \boldsymbol{w}$ is still done a component at a time. The sum has components $v_1 + w_1$ and $v_2 + w_2$ and $v_3 + w_3$. You see how to add vectors in 4 or 5 or n dimensions. When $\boldsymbol{w}$ starts at the end of $\boldsymbol{v}$, the third side is $\boldsymbol{v} + \boldsymbol{w}$. The other way around the parallelogram is $\boldsymbol{w} + \boldsymbol{v}$. Question: Do the four sides all lie in the same plane? *Yes.* And the sum $\boldsymbol{v} + \boldsymbol{w} - \boldsymbol{v} - \boldsymbol{w}$ goes completely around to produce ______.

A typical linear combination of three vectors in three dimensions is $\boldsymbol{u} + 4\boldsymbol{v} - 2\boldsymbol{w}$:

$$\textbf{Linear combination} \qquad \begin{bmatrix} 1 \\ 0 \\ 3 \end{bmatrix} + 4 \begin{bmatrix} 1 \\ 2 \\ 1 \end{bmatrix} - 2 \begin{bmatrix} 2 \\ 3 \\ -1 \end{bmatrix} = \begin{bmatrix} 1 \\ 2 \\ 9 \end{bmatrix}.$$

The Important Questions

For one vector $\boldsymbol{u}$, the only linear combinations are the multiples $c\boldsymbol{u}$. For two vectors, the combinations are $c\boldsymbol{u}+d\boldsymbol{v}$. For three vectors, the combinations are $c\boldsymbol{u}+d\boldsymbol{v}+e\boldsymbol{w}$. Will you take the big step from *one* linear combination to *all* linear combinations? Every c and d and e are allowed. Suppose the vectors $\boldsymbol{u}, \boldsymbol{v}, \boldsymbol{w}$ are in three-dimensional space:

1 What is the picture of all combinations $c\boldsymbol{u}$?

2 What is the picture of all combinations $c\boldsymbol{u}+d\boldsymbol{v}$?

3 What is the picture of all combinations $c\boldsymbol{u}+d\boldsymbol{v}+e\boldsymbol{w}$?

The answers depend on the particular vectors $\boldsymbol{u}$, $\boldsymbol{v}$, and $\boldsymbol{w}$. If they were all zero vectors (a very extreme case), then every combination would be zero. If they are typical nonzero vectors (components chosen at random), here are the three answers. This is the key to our subject:

1 The combinations $c\boldsymbol{u}$ fill a ***line***.

2 The combinations $c\boldsymbol{u}+d\boldsymbol{v}$ fill a ***plane***.

3 The combinations $c\boldsymbol{u}+d\boldsymbol{v}+e\boldsymbol{w}$ fill ***three-dimensional space***.

The line is infinitely long, in the direction of $\boldsymbol{u}$ (forward and backward, going through the zero vector). It is the plane of all $c\boldsymbol{u}+d\boldsymbol{v}$ (combining two lines) that I especially ask you to think about.

Adding all $c\boldsymbol{u}$ on one line to all $d\boldsymbol{v}$ on the other line fills in the plane in Figure 1.4.

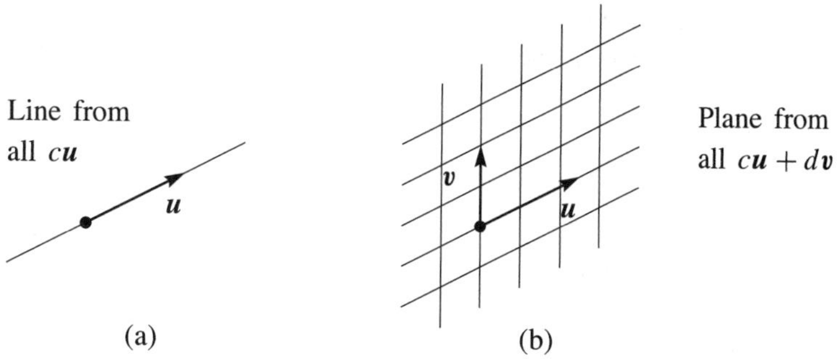

Figure 1.4 (a) The line through $\boldsymbol{u}$. (b) The plane containing the lines through $\boldsymbol{u}$ and $\boldsymbol{v}$.

When we include a third vector $\boldsymbol{w}$, the multiples $e\boldsymbol{w}$ give a third line. Suppose that line is not in the plane of $\boldsymbol{u}$ and $\boldsymbol{v}$. Then combining all $e\boldsymbol{w}$ with all $c\boldsymbol{u}+d\boldsymbol{v}$ fills up the whole three-dimensional space.

This is the typical situation! Line, then plane, then space. But other possibilities exist. When $\boldsymbol{w}$ happens to be $c\boldsymbol{u}+d\boldsymbol{v}$, the third vector is in the plane of the first two. The combinations of $\boldsymbol{u}, \boldsymbol{v}, \boldsymbol{w}$ will not go outside that $\boldsymbol{uv}$ plane. We do not get the full three-dimensional space. Please think about the special cases in Problem 1.

■ REVIEW OF THE KEY IDEAS ■

1. A vector $\boldsymbol{v}$ in two-dimensional space has two components v_1 and v_2.

2. $\boldsymbol{v}+\boldsymbol{w} = (v_1+w_1,\ v_2+w_2)$ and $c\boldsymbol{v} = (cv_1, cv_2)$ are executed a component at a time.

3. A linear combination of $\boldsymbol{u}$ and $\boldsymbol{v}$ and $\boldsymbol{w}$ is $c\boldsymbol{u}+d\boldsymbol{v}+e\boldsymbol{w}$.

4. Take all linear combinations of $\boldsymbol{u}$, or $\boldsymbol{u}$ and $\boldsymbol{v}$, or $\boldsymbol{u}$ and $\boldsymbol{v}$ and $\boldsymbol{w}$. In three dimensions, those combinations typically fill a line, a plane, and the whole space.

■ WORKED EXAMPLES ■

1.1 A Describe all the linear combinations of $\boldsymbol{v} = (1,1,0)$ and $\boldsymbol{w} = (0,1,1)$. Find a vector that is *not* a combination of $\boldsymbol{v}$ and $\boldsymbol{w}$.

Solution These are vectors in three-dimensional space $\mathbf{R}^3$. Their combinations $c\boldsymbol{v}+d\boldsymbol{w}$ fill a *plane* in $\mathbf{R}^3$. The vectors in that plane allow any c and d:

$$c\boldsymbol{v}+d\boldsymbol{w} = c\begin{bmatrix}1\\1\\0\end{bmatrix} + d\begin{bmatrix}0\\1\\1\end{bmatrix} = \begin{bmatrix}c\\c+d\\d\end{bmatrix}.$$

Four particular vectors in that plane are $(0,0,0)$ and $(2,3,1)$ and $(5,7,2)$ and $(\sqrt{2}, 0, -\sqrt{2})$. The second component is always the sum of the first and third components. The vector $(1,1,1)$ is *not* in the plane.

Another description of this plane through $(0,0,0)$ is to know a vector *perpendicular* to the plane. In this case $\boldsymbol{n} = (1,-1,1)$ is perpendicular, as Section 1.2 will confirm by testing dot products: $\boldsymbol{v}\cdot\boldsymbol{n} = 0$ and $\boldsymbol{w}\cdot\boldsymbol{n} = 0$.

1.1 B For $\boldsymbol{v} = (1,0)$ and $\boldsymbol{w} = (0,1)$, describe all the points $c\boldsymbol{v}$ and all the combinations $c\boldsymbol{v}+d\boldsymbol{w}$ with any d and **(1)** *whole numbers* c **(2)** *nonnegative* $c \geq 0$.

Solution

(1) The vectors $c\boldsymbol{v} = (c, 0)$ with whole numbers c are equally spaced points along the x axis (the direction of $\boldsymbol{v}$). They include $(-2, 0)$, $(-1, 0)$, $(0, 0)$, $(1, 0)$, $(2, 0)$. Adding all vectors $d\boldsymbol{w} = (0, d)$ puts a full line in the y direction through those points. We have infinitely many *parallel lines* from $c\boldsymbol{v} + d\boldsymbol{w} =$ (*whole number, any number*). These are vertical lines in the xy plane, through equally spaced points on the x axis.

(2) The vectors $c\boldsymbol{v}$ with $c \geq 0$ fill a "half-line". It is the *positive* x axis, starting at $(0, 0)$ where $c = 0$. It includes $(\pi, 0)$ but not $(-\pi, 0)$. Adding all vectors $d\boldsymbol{w}$ puts a full line in the y direction crossing every point on that half-line. Now we have a *half-plane*. It is the right half of the xy plane, where $x \geq 0$.

Problem Set 1.1

Problems 1–9 are about addition of vectors and linear combinations.

1 Describe geometrically (as a line, plane, ...) all linear combinations of

$$\text{(a)}\ \begin{bmatrix}1\\1\\1\end{bmatrix} \text{ and } \begin{bmatrix}4\\4\\4\end{bmatrix} \quad \text{(b)}\ \begin{bmatrix}1\\0\\0\end{bmatrix} \text{ and } \begin{bmatrix}0\\1\\1\end{bmatrix} \quad \text{(c)}\ \begin{bmatrix}1\\0\\0\end{bmatrix} \text{ and } \begin{bmatrix}0\\1\\1\end{bmatrix} \text{ and } \begin{bmatrix}1\\1\\1\end{bmatrix}$$

2 Draw the vectors $\boldsymbol{v} = \begin{bmatrix}4\\1\end{bmatrix}$ and $\boldsymbol{w} = \begin{bmatrix}-2\\2\end{bmatrix}$ and $\boldsymbol{v} + \boldsymbol{w}$ and $\boldsymbol{v} - \boldsymbol{w}$ in a single xy plane.

3 If $\boldsymbol{v} + \boldsymbol{w} = \begin{bmatrix}3\\1\end{bmatrix}$ and $\boldsymbol{v} - \boldsymbol{w} = \begin{bmatrix}1\\3\end{bmatrix}$, compute and draw $\boldsymbol{v}$ and $\boldsymbol{w}$.

4 From $\boldsymbol{v} = \begin{bmatrix}2\\1\end{bmatrix}$ and $\boldsymbol{w} = \begin{bmatrix}1\\2\end{bmatrix}$, find the components of $3\boldsymbol{v} + \boldsymbol{w}$ and $\boldsymbol{v} - 3\boldsymbol{w}$ and $c\boldsymbol{v} + d\boldsymbol{w}$.

5 Compute $\boldsymbol{u} + \boldsymbol{v}$ and $\boldsymbol{u} + \boldsymbol{v} + \boldsymbol{w}$ and $2\boldsymbol{u} + 2\boldsymbol{v} + \boldsymbol{w}$ when

$$\boldsymbol{u} = \begin{bmatrix}1\\2\\3\end{bmatrix}, \quad \boldsymbol{v} = \begin{bmatrix}-3\\1\\-2\end{bmatrix}, \quad \boldsymbol{w} = \begin{bmatrix}2\\-3\\-1\end{bmatrix}.$$

6 Every combination of $\boldsymbol{v} = (1, -2, 1)$ and $\boldsymbol{w} = (0, 1, -1)$ has components that add to ______. Find c and d so that $c\boldsymbol{v} + d\boldsymbol{w} = (4, 2, -6)$.

7 In the xy plane mark all nine of these linear combinations:

$$c\begin{bmatrix}3\\1\end{bmatrix} + d\begin{bmatrix}0\\1\end{bmatrix} \quad \text{with} \quad c = 0, 1, 2 \quad \text{and} \quad d = 0, 1, 2.$$

8 The parallelogram in Figure 1.1 has diagonal $\boldsymbol{v}+\boldsymbol{w}$. What is its other diagonal? What is the sum of the two diagonals? Draw that vector sum.

9 If three corners of a parallelogram are $(1,1)$, $(4,2)$, and $(1,3)$, what are all the possible fourth corners? Draw two of them.

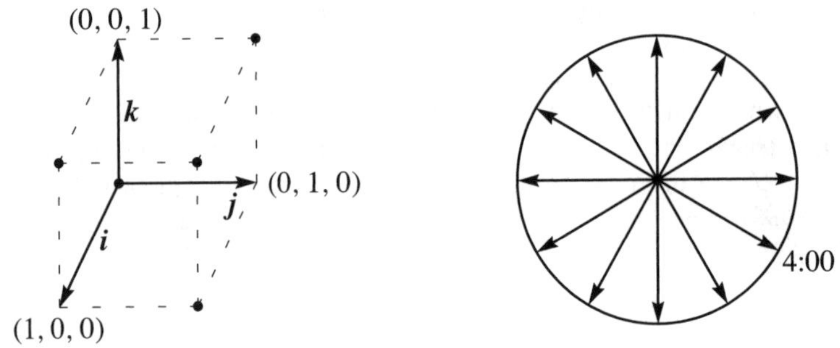

Figure 1.5 Unit cube from $\boldsymbol{i}, \boldsymbol{j}, \boldsymbol{k}$; twelve clock vectors.

Problems 10–14 are about special vectors on cubes and clocks.

10 Copy the cube and draw the vector sum of $\boldsymbol{i}=(1,0,0)$ and $\boldsymbol{j}=(0,1,0)$ and $\boldsymbol{k}=(0,0,1)$. The addition $\boldsymbol{i}+\boldsymbol{j}$ yields the diagonal of ______.

11 Four corners of the cube are $(0,0,0)$, $(1,0,0)$, $(0,1,0)$, $(0,0,1)$. What are the other four corners? Find the coordinates of the center point of the cube. The center points of the six faces are ______.

12 How many corners does a cube have in 4 dimensions? How many faces? How many edges? A typical corner is $(0,0,1,0)$.

13 (a) What is the sum $\boldsymbol{V}$ of the twelve vectors that go from the center of a clock to the hours 1:00, 2:00, ..., 12:00?

(b) If the vector to 4:00 is removed, find the sum of the eleven remaining vectors.

(c) What is the unit vector to 1:00?

14 Suppose the twelve vectors start from 6:00 at the bottom instead of $(0,0)$ at the center. The vector to 12:00 is doubled to $2\boldsymbol{j}=(0,2)$. Add the new twelve vectors.

Problems 15–19 go further with linear combinations of $\boldsymbol{v}$ and $\boldsymbol{w}$ (Figure 1.6)

15 The figure shows $\frac{1}{2}\boldsymbol{v}+\frac{1}{2}\boldsymbol{w}$. Mark the points $\frac{3}{4}\boldsymbol{v}+\frac{1}{4}\boldsymbol{w}$ and $\frac{1}{4}\boldsymbol{v}+\frac{1}{4}\boldsymbol{w}$ and $\boldsymbol{v}+\boldsymbol{w}$.

16 Mark the point $-\boldsymbol{v}+2\boldsymbol{w}$ and any other combination $c\boldsymbol{v}+d\boldsymbol{w}$ with $c+d=1$. Draw the line of all combinations that have $c+d=1$.

17 Locate $\frac{1}{3}\boldsymbol{v}+\frac{1}{3}\boldsymbol{w}$ and $\frac{2}{3}\boldsymbol{v}+\frac{2}{3}\boldsymbol{w}$. The combinations $c\boldsymbol{v}+c\boldsymbol{w}$ fill out what line? Restricted by $c \geq 0$ those combinations with $c=d$ fill out what half line?

18 Restricted by $0 \leq c \leq 1$ and $0 \leq d \leq 1$, shade in all combinations $c\boldsymbol{v}+d\boldsymbol{w}$.

19 Restricted only by $c \geq 0$ and $d \geq 0$ draw the "cone" of all combinations $c\boldsymbol{v}+d\boldsymbol{w}$.

Problems 20–27 deal with $\boldsymbol{u}, \boldsymbol{v}, \boldsymbol{w}$ in three-dimensional space (see Figure 1.6).

20 Locate $\frac{1}{3}\boldsymbol{u}+\frac{1}{3}\boldsymbol{v}+\frac{1}{3}\boldsymbol{w}$ and $\frac{1}{2}\boldsymbol{u}+\frac{1}{2}\boldsymbol{w}$ in the dashed triangle. Challenge problem: Under what restrictions on c, d, e, will the combinations $c\boldsymbol{u}+d\boldsymbol{v}+e\boldsymbol{w}$ fill in the dashed triangle?

21 The three sides of the dashed triangle are $\boldsymbol{v}-\boldsymbol{u}$ and $\boldsymbol{w}-\boldsymbol{v}$ and $\boldsymbol{u}-\boldsymbol{w}$. Their sum is ______. Draw the head-to-tail addition around a plane triangle of $(3,1)$ plus $(-1,1)$ plus $(-2,-2)$.

22 Shade in the pyramid of combinations $c\boldsymbol{u}+d\boldsymbol{v}+e\boldsymbol{w}$ with $c \geq 0$, $d \geq 0$, $e \geq 0$ and $c+d+e \leq 1$. Mark the vector $\frac{1}{2}(\boldsymbol{u}+\boldsymbol{v}+\boldsymbol{w})$ as inside or outside this pyramid.

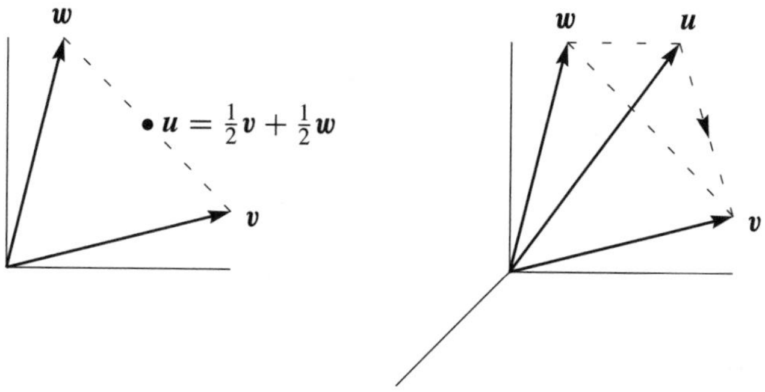

Figure 1.6 Problems **15–19** in a plane Problems **20–27** in 3-dimensional space

23 If you look at *all* combinations of those $\boldsymbol{u}$, $\boldsymbol{v}$, and $\boldsymbol{w}$, is there any vector that can't be produced from $c\boldsymbol{u}+d\boldsymbol{v}+e\boldsymbol{w}$?

24 Which vectors are combinations of $\boldsymbol{u}$ and $\boldsymbol{v}$, and *also* combinations of $\boldsymbol{v}$ and $\boldsymbol{w}$?

25 Draw vectors $\boldsymbol{u}$, $\boldsymbol{v}$, $\boldsymbol{w}$ so that their combinations $c\boldsymbol{u}+d\boldsymbol{v}+e\boldsymbol{w}$ fill only a line. Draw vectors $\boldsymbol{u}$, $\boldsymbol{v}$, $\boldsymbol{w}$ so that their combinations $c\boldsymbol{u}+d\boldsymbol{v}+e\boldsymbol{w}$ fill only a plane.

26 What combination of the vectors $\begin{bmatrix}1\\2\end{bmatrix}$ and $\begin{bmatrix}3\\1\end{bmatrix}$ produces $\begin{bmatrix}14\\8\end{bmatrix}$? Express this question as two equations for the coefficients c and d in the linear combination.

27 *Review Question.* In xyz space, where is the plane of all linear combinations of $\boldsymbol{i}=(1,0,0)$ and $\boldsymbol{j}=(0,1,0)$?

28 If (a, b) is a multiple of (c, d) with $abcd \neq 0$, *show that (a, c) is a multiple of (b, d)*. This is surprisingly important; call it a challenge question. You could use numbers first to see how a, b, c, d are related. The question will lead to:

If $A = \begin{bmatrix} a & b \\ c & d \end{bmatrix}$ has dependent rows then it has dependent columns.

And eventually: If $AB = \begin{bmatrix} 1 & 0 \\ 0 & 1 \end{bmatrix}$ then $BA = \begin{bmatrix} 1 & 0 \\ 0 & 1 \end{bmatrix}$. That looks so simple...

LENGTHS AND DOT PRODUCTS ■ 1.2

The first section mentioned multiplication of vectors, but it backed off. Now we go forward to define the *"dot product"* of $\boldsymbol{v}$ and $\boldsymbol{w}$. This multiplication involves the separate products $v_1 w_1$ and $v_2 w_2$, but it doesn't stop there. Those two numbers are added to produce the single number $\boldsymbol{v} \cdot \boldsymbol{w}$.

DEFINITION The ***dot product*** or ***inner product*** of $\boldsymbol{v} = (v_1, v_2)$ and $\boldsymbol{w} = (w_1, w_2)$ is the number

$$\boldsymbol{v} \cdot \boldsymbol{w} = v_1 w_1 + v_2 w_2. \tag{1}$$

Example 1 The vectors $\boldsymbol{v} = (4, 2)$ and $\boldsymbol{w} = (-1, 2)$ have a *zero* dot product:

$$\begin{bmatrix} 4 \\ 2 \end{bmatrix} \cdot \begin{bmatrix} -1 \\ 2 \end{bmatrix} = -4 + 4 = 0.$$

In mathematics, zero is always a special number. For dot products, it means that *these two vectors are perpendicular*. The angle between them is 90°. When we drew them in Figure 1.1, we saw a rectangle (not just any parallelogram). The clearest example of perpendicular vectors is $\boldsymbol{i} = (1, 0)$ along the x axis and $\boldsymbol{j} = (0, 1)$ up the y axis. Again the dot product is $\boldsymbol{i} \cdot \boldsymbol{j} = 0 + 0 = 0$. Those vectors $\boldsymbol{i}$ and $\boldsymbol{j}$ form a right angle.

The dot product of $\boldsymbol{v} = (1, 2)$ and $\boldsymbol{w} = (2, 1)$ is 4. *Please check this.* Soon that will reveal the angle between $\boldsymbol{v}$ and $\boldsymbol{w}$ (not 90°).

Example 2 Put a weight of 4 at the point $x = -1$ and a weight of 2 at the point $x = 2$. The x axis will balance on the center point $x = 0$ (like a see-saw). The weights balance because the dot product is $(4)(-1) + (2)(2) = 0$.

This example is typical of engineering and science. The vector of weights is $(w_1, w_2) = (4, 2)$. The vector of distances from the center is $(v_1, v_2) = (-1, 2)$. The weights times the distances, $w_1 v_1$ and $w_2 v_2$, give the "moments". The equation for the see-saw to balance is $w_1 v_1 + w_2 v_2 = 0$.

The dot product $\boldsymbol{w} \cdot \boldsymbol{v}$ equals $\boldsymbol{v} \cdot \boldsymbol{w}$. The order of $\boldsymbol{v}$ and $\boldsymbol{w}$ makes no difference.

Example 3 Dot products enter in economics and business. We have three products to buy and sell. Their prices are (p_1, p_2, p_3) for each unit—this is the "price vector" $\boldsymbol{p}$.

The quantities we buy or sell are (q_1, q_2, q_3)—positive when we sell, negative when we buy. Selling q_1 units of the first product at the price p_1 brings in $q_1 p_1$. The total income is the dot product $\boldsymbol{q} \cdot \boldsymbol{p}$:

$$\textbf{\textit{Income}} = (q_1, q_2, q_3) \cdot (p_1, p_2, p_3) = q_1 p_1 + q_2 p_2 + q_3 p_3.$$

A zero dot product means that "the books balance." Total sales equal total purchases if $\boldsymbol{q} \cdot \boldsymbol{p} = 0$. Then $\boldsymbol{p}$ is perpendicular to $\boldsymbol{q}$ (in three-dimensional space). With three products, ***the vectors are three-dimensional***. A supermarket goes quickly into high dimensions.

Small note: Spreadsheets have become essential in management. They compute linear combinations and dot products. What you see on the screen is a matrix.

Main point To compute the dot product $\boldsymbol{v} \cdot \boldsymbol{w}$, multiply each v_i times w_i. Then add.

Lengths and Unit Vectors

An important case is the dot product of a vector *with itself*. In this case $\boldsymbol{v} = \boldsymbol{w}$. When the vector is $\boldsymbol{v} = (1, 2, 3)$, the dot product with itself is $\boldsymbol{v} \cdot \boldsymbol{v} = 14$:

$$\boldsymbol{v} \cdot \boldsymbol{v} = \begin{bmatrix} 1 \\ 2 \\ 3 \end{bmatrix} \cdot \begin{bmatrix} 1 \\ 2 \\ 3 \end{bmatrix} = 1 + 4 + 9 = 14.$$

The answer is not zero because $\boldsymbol{v}$ is not perpendicular to itself. Instead of a 90° angle between vectors we have 0°. The dot product $\boldsymbol{v} \cdot \boldsymbol{v}$ gives the *length of* $\boldsymbol{v}$ *squared.*

DEFINITION The ***length*** (or ***norm***) of a vector $\boldsymbol{v}$ is the square root of $\boldsymbol{v} \cdot \boldsymbol{v}$:

$$\text{length} = \|\boldsymbol{v}\| = \sqrt{\boldsymbol{v} \cdot \boldsymbol{v}}.$$

In two dimensions the length is $\sqrt{v_1^2 + v_2^2}$. In three dimensions it is $\sqrt{v_1^2 + v_2^2 + v_3^2}$. By the calculation above, the length of $\boldsymbol{v} = (1, 2, 3)$ is $\|\boldsymbol{v}\| = \sqrt{14}$.

We can explain this definition. $\|\boldsymbol{v}\|$ is just the ordinary length of the arrow that represents the vector. In two dimensions, the arrow is in a plane. If the components are 1 and 2, the arrow is the third side of a right triangle (Figure 1.7). The formula $a^2 + b^2 = c^2$, which connects the three sides, is $1^2 + 2^2 = \|\boldsymbol{v}\|^2$.

For the length of $\boldsymbol{v} = (1, 2, 3)$, we used the right triangle formula twice. The vector $(1, 2, 0)$ in the base has length $\sqrt{5}$. This base vector is perpendicular to $(0, 0, 3)$ that goes straight up. So the diagonal of the box has length $\|\boldsymbol{v}\| = \sqrt{5 + 9} = \sqrt{14}$.

The length of a four-dimensional vector would be $\sqrt{v_1^2 + v_2^2 + v_3^2 + v_4^2}$. Thus $(1, 1, 1, 1)$ has length $\sqrt{1^2 + 1^2 + 1^2 + 1^2} = 2$. This is the diagonal through a unit cube in four-dimensional space. The diagonal in n dimensions has length $\sqrt{n}$.

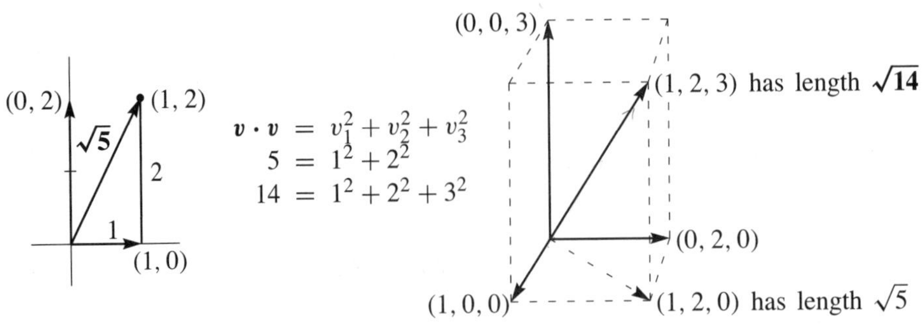

Figure 1.7 The length $\sqrt{v \cdot v}$ of two-dimensional and three-dimensional vectors.

The word "unit" is always indicating that some measurement equals "one." The unit price is the price for one item. A unit cube has sides of length one. A unit circle is a circle with radius one. Now we define the idea of a "unit vector."

DEFINITION A *unit vector* ***u is a vector whose length equals one***. Then $u \cdot u = 1$.

An example in four dimensions is $u = (\frac{1}{2}, \frac{1}{2}, \frac{1}{2}, \frac{1}{2})$. Then $u \cdot u$ is $\frac{1}{4} + \frac{1}{4} + \frac{1}{4} + \frac{1}{4} = 1$. We divided $v = (1, 1, 1, 1)$ by its length $\|v\| = 2$ to get this unit vector.

Example 4 The standard unit vectors along the x and y axes are written i and j. In the xy plane, the unit vector that makes an angle "theta" with the x axis is $(\cos\theta, \sin\theta)$:

$$\textbf{Unit vectors} \quad i = \begin{bmatrix} 1 \\ 0 \end{bmatrix} \quad \text{and} \quad j = \begin{bmatrix} 0 \\ 1 \end{bmatrix} \quad \text{and} \quad u = \begin{bmatrix} \cos\theta \\ \sin\theta \end{bmatrix}.$$

When $\theta = 0$, the horizontal vector u is i. When $\theta = 90°$ (or $\frac{\pi}{2}$ radians), the vertical vector is j. At any angle, the components $\cos\theta$ and $\sin\theta$ produce $u \cdot u = 1$ because $\cos^2\theta + \sin^2\theta = 1$. These vectors reach out to the unit circle in Figure 1.8. Thus $\cos\theta$ and $\sin\theta$ are simply the coordinates of that point at angle θ on the unit circle.
In three dimensions, the unit vectors along the axes are i, j, and k. Their components are $(1, 0, 0)$ and $(0, 1, 0)$ and $(0, 0, 1)$. Notice how every three-dimensional vector is a linear combination of i, j, and k. The vector $v = (2, 2, 1)$ is equal to $2i + 2j + k$. Its length is $\sqrt{2^2 + 2^2 + 1^2}$. This is the square root of 9, so $\|v\| = 3$.

Since $(2, 2, 1)$ has length 3, the vector $(\frac{2}{3}, \frac{2}{3}, \frac{1}{3})$ has length 1. Check that $u \cdot u = \frac{4}{9} + \frac{4}{9} + \frac{1}{9} = 1$. To create a unit vector, just divide v by its length $\|v\|$.

1A Unit vectors ***Divide any nonzero vector v by its length***. Then $u = v/\|v\|$ is a unit vector in the same direction as v.

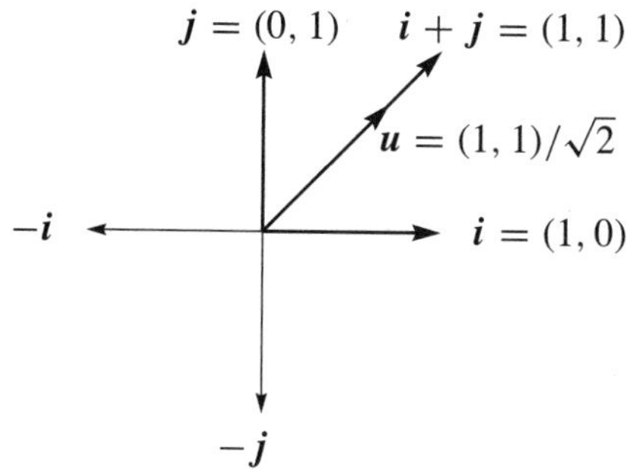

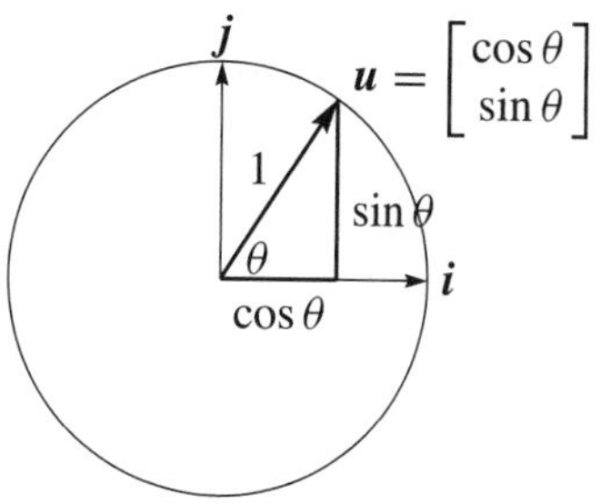

Figure 1.8 The coordinate vectors i and j. The unit vector u at angle 45° (left) and the unit vector $(\cos\theta, \sin\theta)$ at angle θ.

The Angle Between Two Vectors

We stated that perpendicular vectors have $v \cdot w = 0$. The dot product is zero when the angle is 90°. To explain this, we have to connect angles to dot products. Then we show how $v \cdot w$ finds the angle between any two nonzero vectors v and w.

1B Right angles ***The dot product is $v \cdot w = 0$ when v is perpendicular to w.***

Proof When v and w are perpendicular, they form two sides of a right triangle. The third side is $v - w$ (the hypotenuse going across in Figure 1.7). The *Pythagoras Law* for the sides of a right triangle is $a^2 + b^2 = c^2$:

$$\textbf{Perpendicular vectors} \quad \|v\|^2 + \|w\|^2 = \|v - w\|^2 \tag{2}$$

Writing out the formulas for those lengths in two dimensions, this equation is

$$\left(v_1^2 + v_2^2\right) + \left(w_1^2 + w_2^2\right) = (v_1 - w_1)^2 + (v_2 - w_2)^2. \tag{3}$$

The right side begins with $v_1^2 - 2v_1w_1 + w_1^2$. Then v_1^2 and w_1^2 are on both sides of the equation and they cancel, leaving $-2v_1w_1$. Similarly v_2^2 and w_2^2 cancel, leaving $-2v_2w_2$. (In three dimensions there would also be $-2v_3w_3$.) The last step is to divide by -2:

$$0 = -2v_1w_1 - 2v_2w_2 \quad \text{which leads to} \quad v_1w_1 + v_2w_2 = 0. \tag{4}$$

Conclusion Right angles produce $v \cdot w = 0$. We have proved **Theorem 1B**. The dot product is zero when the angle is $\theta = 90°$. Then $\cos\theta = 0$. The zero vector $v = \mathbf{0}$ is perpendicular to every vector w because $\mathbf{0} \cdot w$ is always zero.

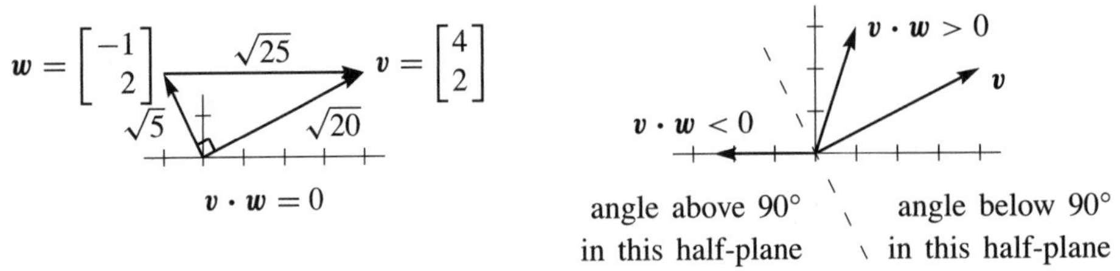

Figure 1.9 Perpendicular vectors have $\boldsymbol{v} \cdot \boldsymbol{w} = 0$. The angle is below 90° when $\boldsymbol{v} \cdot \boldsymbol{w} > 0$.

Now suppose $\boldsymbol{v} \cdot \boldsymbol{w}$ is not zero. It may be positive, it may be negative. The sign of $\boldsymbol{v} \cdot \boldsymbol{w}$ immediately tells whether we are below or above a right angle. The angle is less than 90° when $\boldsymbol{v} \cdot \boldsymbol{w}$ is positive. The angle is above 90° when $\boldsymbol{v} \cdot \boldsymbol{w}$ is negative. Figure 1.9 shows a typical vector $\boldsymbol{v} = (3, 1)$. The angle with $\boldsymbol{w} = (1, 3)$ is less than 90°.

The borderline is where vectors are perpendicular to $\boldsymbol{v}$. On that dividing line between plus and minus, where we find $\boldsymbol{w} = (1, -3)$, the dot product is zero.

The next page takes one more step, to find the exact angle θ. This is not necessary for linear algebra—you could stop here! Once we have matrices and linear equations, we won't come back to θ. But while we are on the subject of angles, this is the place for the formula.

Start with unit vectors $\boldsymbol{u}$ and $\boldsymbol{U}$. The sign of $\boldsymbol{u} \cdot \boldsymbol{U}$ tells whether $\theta < 90°$ or $\theta > 90°$. Because the vectors have length 1, we learn more than that. ***The dot product*** $\boldsymbol{u} \cdot \boldsymbol{U}$ ***is the cosine of*** θ. This is true in any number of dimensions.

1C If $\boldsymbol{u}$ and $\boldsymbol{U}$ are unit vectors then $\boldsymbol{u} \cdot \boldsymbol{U} = \cos\theta$. Certainly $|\boldsymbol{u} \cdot \boldsymbol{U}| \le 1$.

Remember that $\cos\theta$ is never greater than 1. It is never less than -1. ***The dot product of unit vectors is between*** -1 ***and*** 1.

Figure 1.10 shows this clearly when the vectors are $\boldsymbol{u} = (\cos\theta, \sin\theta)$ and $\boldsymbol{i} = (1, 0)$. The dot product is $\boldsymbol{u} \cdot \boldsymbol{i} = \cos\theta$. That is the cosine of the angle between them.

After rotation through any angle α, these are still unit vectors. Call the vectors $\boldsymbol{u} = (\cos\beta, \sin\beta)$ and $\boldsymbol{U} = (\cos\alpha, \sin\alpha)$. Their dot product is $\cos\alpha\cos\beta + \sin\alpha\sin\beta$. From trigonometry this is the same as $\cos(\beta - \alpha)$. Since $\beta - \alpha$ equals θ (no change in the angle between them) we have reached the formula $\boldsymbol{u} \cdot \boldsymbol{U} = \cos\theta$.

Problem 26 proves $|\boldsymbol{u} \cdot \boldsymbol{U}| \le 1$ directly, without mentioning angles. The inequality and the cosine formula $\boldsymbol{u} \cdot \boldsymbol{U} = \cos\theta$ are always true for unit vectors.

What if $\boldsymbol{v}$ *and* $\boldsymbol{w}$ *are not unit vectors*? Divide by their lengths to get $\boldsymbol{u} = \boldsymbol{v}/\|\boldsymbol{v}\|$ and $\boldsymbol{U} = \boldsymbol{w}/\|\boldsymbol{w}\|$. Then the dot product of those unit vectors $\boldsymbol{u}$ and $\boldsymbol{U}$ gives $\cos\theta$.

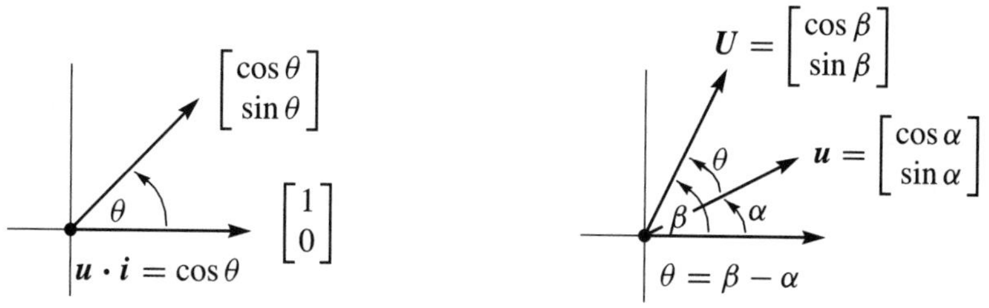

Figure 1.10 The dot product of unit vectors is the cosine of the angle θ.

Whatever the angle, this dot product of $v/\|v\|$ with $w/\|w\|$ never exceeds one. That is the ***"Schwarz inequality"*** for dot products—or more correctly the Cauchy-Schwarz-Buniakowsky inequality. It was found in France and Germany and Russia (and maybe elsewhere—it is the most important inequality in mathematics). With the division by $\|v\|\,\|w\|$ from rescaling to unit vectors, we have $\cos\theta$:

1D (a) **COSINE FORMULA** If v and w are nonzero vectors then $\dfrac{v \cdot w}{\|v\|\,\|w\|} = \cos\theta$.

(b) **SCHWARZ INEQUALITY** If v and w are any vectors then $|v \cdot w| \le \|v\|\,\|w\|$.

Example 5 Find $\cos\theta$ for $v = \left[\begin{smallmatrix}3\\1\end{smallmatrix}\right]$ and $w = \left[\begin{smallmatrix}1\\3\end{smallmatrix}\right]$ in Figure 1.9b.

Solution The dot product is $v \cdot w = 6$. Both v and w have length $\sqrt{10}$. The cosine is

$$\cos\theta = \frac{v \cdot w}{\|v\|\,\|w\|} = \frac{6}{\sqrt{10}\sqrt{10}} = \frac{3}{5}.$$

The angle is below 90° because $v \cdot w = 6$ is positive. By the Schwarz inequality, $\|v\|\,\|w\| = 10$ is larger than $v \cdot w = 6$.

Example 6 The dot product of $v = (a, b)$ and $w = (b, a)$ is $2ab$. Both lengths are $\sqrt{a^2 + b^2}$. The Schwarz inequality says that $2ab \le a^2 + b^2$.

Reason The difference between $a^2 + b^2$ and $2ab$ can never be negative:

$$a^2 + b^2 - 2ab = (a - b)^2 \ge 0.$$

This is more famous if we write $x = a^2$ and $y = b^2$. Then the "geometric mean" $\sqrt{xy}$ is not larger than the "arithmetic mean," which is the average $\frac{1}{2}(x + y)$:

$$ab \le \frac{a^2 + b^2}{2} \quad \text{becomes} \quad \sqrt{xy} \le \frac{x + y}{2}.$$

Notes on Computing

Write the components of $\boldsymbol{v}$ as $v(1), \ldots, v(N)$ and similarly for $\boldsymbol{w}$. In FORTRAN, the sum $\boldsymbol{v}+\boldsymbol{w}$ requires a loop to add components separately. The dot product also loops to add the separate $v(i)w(i)$:

```
   DO 10 I = 1,N                DO 10 I = 1,N
10 VPLUSW(I) = v(I)+w(I)     10 VDOTW = VDOTW + V(I) * W(I)
```

MATLAB *works directly with whole vectors, not their components.* No loop is needed. When $\boldsymbol{v}$ and $\boldsymbol{w}$ have been defined, $\boldsymbol{v}+\boldsymbol{w}$ is immediately understood. It is printed unless the line ends in a semicolon. Input $\boldsymbol{v}$ and $\boldsymbol{w}$ as rows—the prime $'$ at the end transposes them to columns. The combination $2\boldsymbol{v}+3\boldsymbol{w}$ uses $*$ for multiplication.

$$\boldsymbol{v} = [2 \quad 3 \quad 4]' \quad ; \quad \boldsymbol{w} = [1 \quad 1 \quad 1]' \quad ; \quad \boldsymbol{u} = 2 * \boldsymbol{v} + 3 * \boldsymbol{w}$$

The dot product $\boldsymbol{v}\cdot\boldsymbol{w}$ is usually seen as ***a row times a column*** **(*with no dot*)**:

$$\text{Instead of} \quad \begin{bmatrix}1\\2\end{bmatrix}\cdot\begin{bmatrix}3\\4\end{bmatrix} \quad \text{we more often see} \quad [1 \;\; 2]\begin{bmatrix}3\\4\end{bmatrix} \quad \text{or} \quad \boldsymbol{v}' * \boldsymbol{w}$$

The length of $\boldsymbol{v}$ is already known to MATLAB as norm$(\boldsymbol{v})$. We could define it ourselves as sqrt$(\boldsymbol{v}' * \boldsymbol{v})$, using the square root function—also known. The cosine we have to define ourselves! Then the angle (in radians) comes from the *arc cosine* (acos) function:

$$\begin{aligned} &\text{cosine} = \boldsymbol{v}' * \boldsymbol{w}/(\text{norm}\,(\boldsymbol{v}) * \text{norm}\,(\boldsymbol{w}));\\ &\text{angle} \;\;= \text{acos}\,(\text{cosine}) \end{aligned}$$

An M-file would create a new function **cosine** $(\boldsymbol{v}, \boldsymbol{w})$ for future use. (Quite a few M-files have been created especially for this book. They are listed at the end.)

■ REVIEW OF THE KEY IDEAS ■

1. The dot product $\boldsymbol{v}\cdot\boldsymbol{w}$ multiplies each component v_i by w_i and adds the $v_i w_i$.
2. The length $\|\boldsymbol{v}\|$ is the square root of $\boldsymbol{v}\cdot\boldsymbol{v}$.
3. The vector $\boldsymbol{v}/\|\boldsymbol{v}\|$ is a ***unit vector***. Its length is 1.
4. The dot product is $\boldsymbol{v}\cdot\boldsymbol{w}=0$ when $\boldsymbol{v}$ and $\boldsymbol{w}$ are perpendicular.
5. The cosine of θ (the angle between any nonzero $\boldsymbol{v}$ and $\boldsymbol{w}$) never exceeds 1:

$$\cos\theta = \frac{\boldsymbol{v}\cdot\boldsymbol{w}}{\|\boldsymbol{v}\|\|\boldsymbol{w}\|} \qquad \textbf{\textit{Schwarz inequality}} \quad |\boldsymbol{v}\cdot\boldsymbol{w}| \le \|\boldsymbol{v}\|\,\|\boldsymbol{w}\|.$$

■ WORKED EXAMPLES ■

1.2 A For the vectors $\boldsymbol{v} = (3,4)$ and $\boldsymbol{w} = (4,3)$ test the Schwarz inequality on $\boldsymbol{v} \cdot \boldsymbol{w}$ and the triangle inequality on $\|\boldsymbol{v}+\boldsymbol{w}\|$. Find $\cos\theta$ for the angle between $\boldsymbol{v}$ and $\boldsymbol{w}$. When will we have *equality* $|\boldsymbol{v} \cdot \boldsymbol{w}| = \|\boldsymbol{v}\|\,\|\boldsymbol{w}\|$ and $\|\boldsymbol{v}+\boldsymbol{w}\| = \|\boldsymbol{v}\| + \|\boldsymbol{w}\|$?

Solution The dot product is $\boldsymbol{v} \cdot \boldsymbol{w} = (3)(4) + (4)(3) = 24$. The length of $\boldsymbol{v}$ is $\|\boldsymbol{v}\| = \sqrt{9+16} = 5$ and also $\|\boldsymbol{w}\| = 5$. The sum $\boldsymbol{v}+\boldsymbol{w} = (7,7)$ has length $\|\boldsymbol{v}+\boldsymbol{w}\| = 7\sqrt{2} \approx 9.9$.

Schwarz inequality	$\lvert\boldsymbol{v} \cdot \boldsymbol{w}\rvert \le \|\boldsymbol{v}\|\,\|\boldsymbol{w}\|$	is $24 < 25$.
Triangle inequality	$\|\boldsymbol{v}+\boldsymbol{w}\| \le \|\boldsymbol{v}\| + \|\boldsymbol{w}\|$	is $7\sqrt{2} < 10$.
Cosine of angle	$\cos\theta = \frac{24}{25}$	(Thin angle!)

If one vector is a multiple of the other as in $\boldsymbol{w} = -2\boldsymbol{v}$, then the angle is $0°$ or $180°$ and $|\cos\theta| = 1$ and $|\boldsymbol{v} \cdot \boldsymbol{w}|$ *equals* $\|\boldsymbol{v}\|\,\|\boldsymbol{w}\|$. If the angle is $0°$, as in $\boldsymbol{w} = 2\boldsymbol{v}$, then $\|\boldsymbol{v}+\boldsymbol{w}\| = \|\boldsymbol{v}\| + \|\boldsymbol{w}\|$. The triangle is flat.

1.2 B Find a unit vector $\boldsymbol{u}$ in the direction of $\boldsymbol{v} = (3,4)$. Find a unit vector $\boldsymbol{U}$ perpendicular to $\boldsymbol{u}$. How many possibilities for $\boldsymbol{U}$?

Solution For a unit vector $\boldsymbol{u}$, divide $\boldsymbol{v}$ by its length $\|\boldsymbol{v}\| = 5$. For a perpendicular vector $\boldsymbol{V}$ we can choose $(-4,3)$ since the dot product $\boldsymbol{v} \cdot \boldsymbol{V}$ is $(3)(-4) + (4)(3) = 0$. For a *unit* vector $\boldsymbol{U}$, divide $\boldsymbol{V}$ by its length $\|\boldsymbol{V}\|$:

$$\boldsymbol{u} = \frac{\boldsymbol{v}}{\|\boldsymbol{v}\|} = \frac{(3,4)}{5} = \left(\frac{3}{5}, \frac{4}{5}\right) \qquad \boldsymbol{U} = \frac{\boldsymbol{V}}{\|\boldsymbol{V}\|} = \frac{(-4,3)}{5} = \left(-\frac{4}{5}, \frac{3}{5}\right)$$

The only other perpendicular unit vector would be $-\boldsymbol{U} = (\frac{4}{5}, -\frac{3}{5})$.

Problem Set 1.2

1 Calculate the dot products $\boldsymbol{u} \cdot \boldsymbol{v}$ and $\boldsymbol{u} \cdot \boldsymbol{w}$ and $\boldsymbol{v} \cdot \boldsymbol{w}$ and $\boldsymbol{w} \cdot \boldsymbol{v}$:

$$\boldsymbol{u} = \begin{bmatrix} -.6 \\ .8 \end{bmatrix} \qquad \boldsymbol{v} = \begin{bmatrix} 3 \\ 4 \end{bmatrix} \qquad \boldsymbol{w} = \begin{bmatrix} 4 \\ 3 \end{bmatrix}.$$

2 Compute the lengths $\|\boldsymbol{u}\|$ and $\|\boldsymbol{v}\|$ and $\|\boldsymbol{w}\|$ of those vectors. Check the Schwarz inequalities $|\boldsymbol{u} \cdot \boldsymbol{v}| \le \|\boldsymbol{u}\|\,\|\boldsymbol{v}\|$ and $|\boldsymbol{v} \cdot \boldsymbol{w}| \le \|\boldsymbol{v}\|\,\|\boldsymbol{w}\|$.

3 Find unit vectors in the directions of $\boldsymbol{v}$ and $\boldsymbol{w}$ in Problem 1, and the cosine of the angle θ. Choose vectors that make $0°$, $90°$, and $180°$ angles with $\boldsymbol{w}$.

4 Find unit vectors $\boldsymbol{u}_1$ and $\boldsymbol{u}_2$ in the directions of $\boldsymbol{v} = (3,1)$ and $\boldsymbol{w} = (2,1,2)$. Find unit vectors $\boldsymbol{U}_1$ and $\boldsymbol{U}_2$ that are perpendicular to $\boldsymbol{u}_1$ and $\boldsymbol{u}_2$.

5 For any *unit* vectors $\boldsymbol{v}$ and $\boldsymbol{w}$, find the dot products (actual numbers) of

(a) $\boldsymbol{v}$ and $-\boldsymbol{v}$ (b) $\boldsymbol{v}+\boldsymbol{w}$ and $\boldsymbol{v}-\boldsymbol{w}$ (c) $\boldsymbol{v}-2\boldsymbol{w}$ and $\boldsymbol{v}+2\boldsymbol{w}$

6 Find the angle θ (from its cosine) between

(a) $\boldsymbol{v}=\begin{bmatrix}1\\\sqrt{3}\end{bmatrix}$ and $\boldsymbol{w}=\begin{bmatrix}1\\0\end{bmatrix}$ (b) $\boldsymbol{v}=\begin{bmatrix}2\\2\\-1\end{bmatrix}$ and $\boldsymbol{w}=\begin{bmatrix}2\\-1\\2\end{bmatrix}$

(c) $\boldsymbol{v}=\begin{bmatrix}1\\\sqrt{3}\end{bmatrix}$ and $\boldsymbol{w}=\begin{bmatrix}-1\\\sqrt{3}\end{bmatrix}$ (d) $\boldsymbol{v}=\begin{bmatrix}3\\1\end{bmatrix}$ and $\boldsymbol{w}=\begin{bmatrix}-1\\-2\end{bmatrix}$.

7 (a) Describe every vector $\boldsymbol{w}=(w_1,w_2)$ that is perpendicular to $\boldsymbol{v}=(2,-1)$.
(b) The vectors that are perpendicular to $\boldsymbol{V}=(1,1,1)$ lie on a ____.
(c) The vectors that are perpendicular to $(1,1,1)$ and $(1,2,3)$ lie on a ____.

8 True or false (give a reason if true or a counterexample if false):

(a) If $\boldsymbol{u}$ is perpendicular (in three dimensions) to $\boldsymbol{v}$ and $\boldsymbol{w}$, then $\boldsymbol{v}$ and $\boldsymbol{w}$ are parallel.
(b) If $\boldsymbol{u}$ is perpendicular to $\boldsymbol{v}$ and $\boldsymbol{w}$, then $\boldsymbol{u}$ is perpendicular to $\boldsymbol{v}+2\boldsymbol{w}$.
(c) If $\boldsymbol{u}$ and $\boldsymbol{v}$ are perpendicular unit vectors then $\|\boldsymbol{u}-\boldsymbol{v}\|=\sqrt{2}$.

9 The slopes of the arrows from $(0,0)$ to (v_1,v_2) and (w_1,w_2) are v_2/v_1 and w_2/w_1. If the product v_2w_2/v_1w_1 of those slopes is -1, show that $\boldsymbol{v}\cdot\boldsymbol{w}=0$ and the vectors are perpendicular.

10 Draw arrows from $(0,0)$ to the points $\boldsymbol{v}=(1,2)$ and $\boldsymbol{w}=(-2,1)$. Multiply their slopes. That answer is a signal that $\boldsymbol{v}\cdot\boldsymbol{w}=0$ and the arrows are ____.

11 If $\boldsymbol{v}\cdot\boldsymbol{w}$ is negative, what does this say about the angle between $\boldsymbol{v}$ and $\boldsymbol{w}$? Draw a 2-dimensional vector $\boldsymbol{v}$ (an arrow), and show where to find all $\boldsymbol{w}$'s with $\boldsymbol{v}\cdot\boldsymbol{w}<0$.

12 With $\boldsymbol{v}=(1,1)$ and $\boldsymbol{w}=(1,5)$ choose a number c so that $\boldsymbol{w}-c\boldsymbol{v}$ is perpendicular to $\boldsymbol{v}$. Then find the formula that gives this number c for any nonzero $\boldsymbol{v}$ and $\boldsymbol{w}$.

13 Find two vectors $\boldsymbol{v}$ and $\boldsymbol{w}$ that are perpendicular to $(1,0,1)$ and to each other.

14 Find three vectors $\boldsymbol{u},\boldsymbol{v},\boldsymbol{w}$ that are perpendicular to $(1,1,1,1)$ and to each other.

15 The geometric mean of $x=2$ and $y=8$ is $\sqrt{xy}=4$. The arithmetic mean is larger: $\frac{1}{2}(x+y)=$ ____. This came in Example 6 from the Schwarz inequality for $\boldsymbol{v}=(\sqrt{2},\sqrt{8})$ and $\boldsymbol{w}=(\sqrt{8},\sqrt{2})$. Find $\cos\theta$ for this $\boldsymbol{v}$ and $\boldsymbol{w}$.

16 How long is the vector $\boldsymbol{v}=(1,1,\ldots,1)$ in 9 dimensions? Find a unit vector $\boldsymbol{u}$ in the same direction as $\boldsymbol{v}$ and a vector $\boldsymbol{w}$ that is perpendicular to $\boldsymbol{v}$.

17 What are the cosines of the angles α,β,θ between the vector $(1,0,-1)$ and the unit vectors $\boldsymbol{i},\boldsymbol{j},\boldsymbol{k}$ along the axes? Check the formula $\cos^2\alpha+\cos^2\beta+\cos^2\theta=1$.

Problems 18–31 lead to the main facts about lengths and angles in triangles.

18 The parallelogram with sides $\boldsymbol{v} = (4, 2)$ and $\boldsymbol{w} = (-1, 2)$ is a rectangle. Check the Pythagoras formula $a^2 + b^2 = c^2$ which is for ***right triangles only***:

$$(\text{length of } \boldsymbol{v})^2 + (\text{length of } \boldsymbol{w})^2 = (\text{length of } \boldsymbol{v} + \boldsymbol{w})^2.$$

19 In this 90° case, $a^2 + b^2 = c^2$ also works for $\boldsymbol{v} - \boldsymbol{w}$:

$$(\text{length of } \boldsymbol{v})^2 + (\text{length of } \boldsymbol{w})^2 = (\text{length of } \boldsymbol{v} - \boldsymbol{w})^2.$$

Give an example of $\boldsymbol{v}$ and $\boldsymbol{w}$ (not at right angles) for which this equation fails.

20 (Rules for dot products) These equations are simple but useful:
(1) $\boldsymbol{v} \cdot \boldsymbol{w} = \boldsymbol{w} \cdot \boldsymbol{v}$ **(2)** $\boldsymbol{u} \cdot (\boldsymbol{v} + \boldsymbol{w}) = \boldsymbol{u} \cdot \boldsymbol{v} + \boldsymbol{u} \cdot \boldsymbol{w}$ **(3)** $(c\boldsymbol{v}) \cdot \boldsymbol{w} = c(\boldsymbol{v} \cdot \boldsymbol{w})$
Use **(1)** and **(2)** with $\boldsymbol{u} = \boldsymbol{v} + \boldsymbol{w}$ to prove $\|\boldsymbol{v} + \boldsymbol{w}\|^2 = \boldsymbol{v} \cdot \boldsymbol{v} + 2\boldsymbol{v} \cdot \boldsymbol{w} + \boldsymbol{w} \cdot \boldsymbol{w}$.

21 The ***triangle inequality*** says: (length of $\boldsymbol{v} + \boldsymbol{w}$) $\le$ (length of $\boldsymbol{v}$) + (length of $\boldsymbol{w}$). Problem 20 found $\|\boldsymbol{v} + \boldsymbol{w}\|^2 = \|\boldsymbol{v}\|^2 + 2\boldsymbol{v} \cdot \boldsymbol{w} + \|\boldsymbol{w}\|^2$. Use the Schwarz inequality $\boldsymbol{v} \cdot \boldsymbol{w} \le \|\boldsymbol{v}\|\,\|\boldsymbol{w}\|$ to turn this into the triangle inequality:

$$\|\boldsymbol{v} + \boldsymbol{w}\|^2 \le (\|\boldsymbol{v}\| + \|\boldsymbol{w}\|)^2 \quad \text{or} \quad \|\boldsymbol{v} + \boldsymbol{w}\| \le \|\boldsymbol{v}\| + \|\boldsymbol{w}\|.$$

22 A right triangle in three dimensions still obeys $\|\boldsymbol{v}\|^2 + \|\boldsymbol{w}\|^2 = \|\boldsymbol{v} + \boldsymbol{w}\|^2$. Show how this leads in Problem 20 to $v_1 w_1 + v_2 w_2 + v_3 w_3 = 0$.

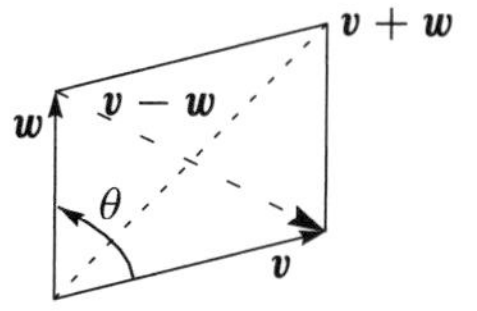

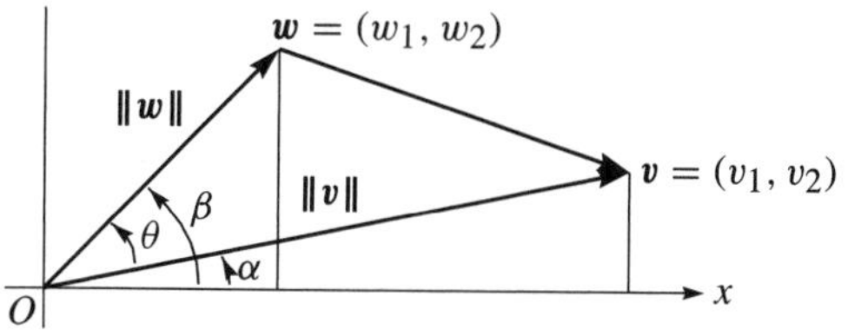

23 The figure shows that $\cos\alpha = v_1/\|\boldsymbol{v}\|$ and $\sin\alpha = v_2/\|\boldsymbol{v}\|$. Similarly $\cos\beta$ is _____ and $\sin\beta$ is _____. The angle θ is $\beta - \alpha$. Substitute into the formula $\cos\beta\cos\alpha + \sin\beta\sin\alpha$ for $\cos(\beta - \alpha)$ to find $\cos\theta = \boldsymbol{v} \cdot \boldsymbol{w}/\|\boldsymbol{v}\|\,\|\boldsymbol{w}\|$.

24 With $\boldsymbol{v}$ and $\boldsymbol{w}$ at angle θ, the "Law of Cosines" comes from $(\boldsymbol{v} - \boldsymbol{w}) \cdot (\boldsymbol{v} - \boldsymbol{w})$:

$$\|\boldsymbol{v} - \boldsymbol{w}\|^2 = \|\boldsymbol{v}\|^2 - 2\|\boldsymbol{v}\|\,\|\boldsymbol{w}\|\cos\theta + \|\boldsymbol{w}\|^2.$$

If $\theta < 90^\circ$ show that $\|\boldsymbol{v}\|^2 + \|\boldsymbol{w}\|^2$ is larger than $\|\boldsymbol{v} - \boldsymbol{w}\|^2$ (the third side).

25 The Schwarz inequality $|\boldsymbol{v} \cdot \boldsymbol{w}| \le \|\boldsymbol{v}\|\,\|\boldsymbol{w}\|$ by algebra instead of trigonometry:

(a) Multiply out both sides of $(v_1 w_1 + v_2 w_2)^2 \le (v_1^2 + v_2^2)(w_1^2 + w_2^2)$.

(b) Show that the difference between those sides equals $(v_1 w_2 - v_2 w_1)^2$. This cannot be negative since it is a square—so the inequality is true.

26 One-line proof of the Schwarz inequality $|\boldsymbol{u} \cdot \boldsymbol{U}| \leq 1$ for unit vectors:

$$|\boldsymbol{u} \cdot \boldsymbol{U}| \leq |u_1|\,|U_1| + |u_2|\,|U_2| \leq \frac{u_1^2 + U_1^2}{2} + \frac{u_2^2 + U_2^2}{2} = \frac{1+1}{2} = 1.$$

Put $(u_1, u_2) = (.6, .8)$ and $(U_1, U_2) = (.8, .6)$ in that whole line and find $\cos\theta$.

27 Why is $|\cos\theta|$ never greater than 1 in the first place?

28 Pick any numbers that add to $x + y + z = 0$. Find the angle between your vector $\boldsymbol{v} = (x, y, z)$ and the vector $\boldsymbol{w} = (z, x, y)$. Challenge question: Explain why $\boldsymbol{v} \cdot \boldsymbol{w}/\|\boldsymbol{v}\|\|\boldsymbol{w}\|$ is always $-\frac{1}{2}$.

29 (*Recommended*) If $||\boldsymbol{v}|| = 5$ and $||\boldsymbol{w}|| = 3$, what are the smallest and largest values of $||\boldsymbol{v} - \boldsymbol{w}||$? What are the smallest and largest values of $\boldsymbol{v} \cdot \boldsymbol{w}$?

30 If $\boldsymbol{v} = (1, 2)$ draw all vectors $\boldsymbol{w} = (x, y)$ in the xy plane with $\boldsymbol{v} \cdot \boldsymbol{w} = 5$. Which is the shortest $\boldsymbol{w}$?

31 Can three vectors in the xy plane have $\boldsymbol{u} \cdot \boldsymbol{v} < 0$ and $\boldsymbol{v} \cdot \boldsymbol{w} < 0$ and $\boldsymbol{u} \cdot \boldsymbol{w} < 0$? I don't know how many vectors in xyz space can have all negative dot products. (Four of those vectors in the plane would be impossible...).

2

SOLVING LINEAR EQUATIONS

VECTORS AND LINEAR EQUATIONS ■ 2.1

The central problem of linear algebra is to solve a system of equations. Those equations are linear, which means that the unknowns are only multiplied by numbers—we never see x times y. Our first example of a linear system is certainly not big. It has two equations in two unknowns. But you will see how far it leads:

$$\begin{array}{rcrcr} x & - & 2y & = & 1 \\ 3x & + & 2y & = & 11 \end{array} \qquad (1)$$

We begin *a row at a time*. The first equation $x-2y=1$ produces a straight line in the xy plane. The point $x=1, y=0$ is on the line because it solves that equation. The point $x=3, y=1$ is also on the line because $3-2=1$. If we choose $x=101$ we find $y=50$. The slope of this particular line is $\frac{1}{2}$ (y increases by 50 when x changes by 100). But slopes are important in calculus and this is linear algebra!

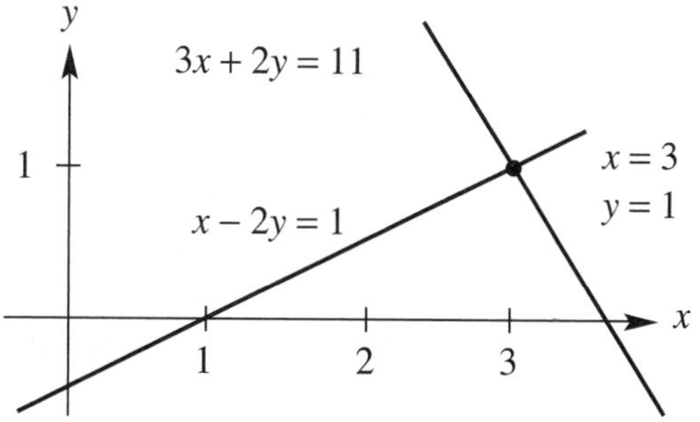

Figure 2.1 *Row picture*: The point (3, 1) where the lines meet is the solution.

Figure 2.1 shows that line $x-2y=1$. The second line in this "row picture" comes from the second equation $3x+2y=11$. You can't miss the intersection point where the two lines meet. *The point* $x = 3$, $y = 1$ *lies on both lines.* That point solves both equations at once. This is the solution to our system of linear equations.

R ***The row picture shows two lines meeting at a single point.***

Turn now to the column picture. I want to recognize the linear system as a "vector equation". Instead of numbers we need to see *vectors.* If you separate the original system into its columns instead of its rows, you get

$$x\begin{bmatrix} 1 \\ 3 \end{bmatrix} + y\begin{bmatrix} -2 \\ 2 \end{bmatrix} = \begin{bmatrix} 1 \\ 11 \end{bmatrix} = \boldsymbol{b}. \qquad (2)$$

This has two column vectors on the left side. The problem is *to find the combination of those vectors that equals the vector on the right.* We are multiplying the first column by x and the second column by y, and adding. With the right choices $x = 3$ and $y = 1$, this produces $3(\textbf{\textit{column 1}}) + 1(\textbf{\textit{column 2}}) = \boldsymbol{b}$.

C ***The column picture combines the column vectors on the left side to produce the vector b on the right side.***

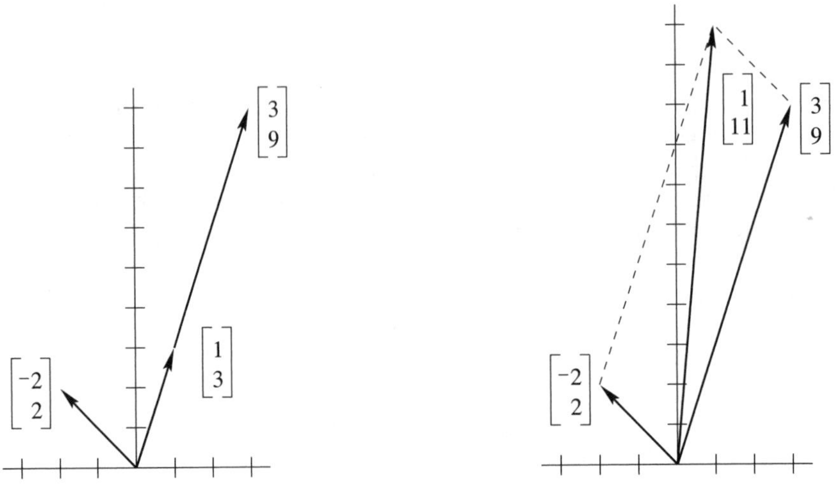

Figure 2.2 *Column picture*: A combination of columns produces the right side (1,11).

Figure 2.2 is the "column picture" of two equations in two unknowns. The first part shows the two separate columns, and that first column multiplied by 3. This multiplication by a *scalar* (a number) is one of the two basic operations in linear algebra:

$$\textbf{Scalar multiplication} \qquad 3\begin{bmatrix} 1 \\ 3 \end{bmatrix} = \begin{bmatrix} 3 \\ 9 \end{bmatrix}.$$

If the components of a vector $\boldsymbol{v}$ are v_1 and v_2, then $c\boldsymbol{v}$ has components cv_1 and cv_2.

The other basic operation is *vector addition*. We add the first components and the second components separately. The vector sum is (1, 11) as desired:

$$\textbf{Vector addition} \qquad \begin{bmatrix} 3 \\ 9 \end{bmatrix} + \begin{bmatrix} -2 \\ 2 \end{bmatrix} = \begin{bmatrix} 1 \\ 11 \end{bmatrix}.$$

The graph in Figure 2.2 shows a parallelogram. The sum (1, 11) is along the diagonal:

$$\textbf{\textit{The sides are}} \begin{bmatrix} 3 \\ 9 \end{bmatrix} \textbf{\textit{ and }} \begin{bmatrix} -2 \\ 2 \end{bmatrix}. \quad \textbf{\textit{The diagonal sum is}} \begin{bmatrix} 3-2 \\ 9+2 \end{bmatrix} = \begin{bmatrix} 1 \\ 11 \end{bmatrix}.$$

We have multiplied the original columns by $x = 3$ and $y = 1$. That combination produces the vector $\boldsymbol{b} = (1, 11)$ on the right side of the linear equations.

To repeat: The left side of the vector equation is a ***linear combination*** of the columns. The problem is to find the right coefficients $x = 3$ and $y = 1$. We are combining scalar multiplication and vector addition into one step. That step is crucially important, because it contains both of the basic operations:

$$\textbf{Linear combination} \qquad 3\begin{bmatrix} 1 \\ 3 \end{bmatrix} + \begin{bmatrix} -2 \\ 2 \end{bmatrix} = \begin{bmatrix} 1 \\ 11 \end{bmatrix}.$$

Of course the solution $x = 3$, $y = 1$ is the same as in the row picture. I don't know which picture you prefer! I suspect that the two intersecting lines are more familiar at first. You may like the row picture better, but only for one day. My own preference is to combine column vectors. It is a lot easier to see a combination of four vectors in four-dimensional space, than to visualize how four hyperplanes might possibly meet at a point. (*Even one hyperplane is hard enough. . .*)

The ***coefficient matrix*** on the left side of the equations is the 2 by 2 matrix A:

$$\textbf{Coefficient matrix} \qquad A = \begin{bmatrix} 1 & -2 \\ 3 & 2 \end{bmatrix}.$$

This is very typical of linear algebra, to look at a matrix by rows and by columns. Its rows give the row picture and its columns give the column picture. Same numbers, different pictures, same equations. We write those equations as a matrix problem $A\boldsymbol{x} = \boldsymbol{b}$:

$$\textbf{Matrix equation} \qquad \begin{bmatrix} 1 & -2 \\ 3 & 2 \end{bmatrix}\begin{bmatrix} x \\ y \end{bmatrix} = \begin{bmatrix} 1 \\ 11 \end{bmatrix}.$$

The row picture deals with the two rows of A. The column picture combines the columns. The numbers $x = 3$ and $y = 1$ go into the solution vector $\boldsymbol{x}$. Then

$$A\boldsymbol{x} = \boldsymbol{b} \quad \text{is} \quad \begin{bmatrix} 1 & -2 \\ 3 & 2 \end{bmatrix}\begin{bmatrix} 3 \\ 1 \end{bmatrix} = \begin{bmatrix} 1 \\ 11 \end{bmatrix}.$$

Three Equations in Three Unknowns

The three unknowns are x, y, z. The linear equations $A\boldsymbol{x} = \boldsymbol{b}$ are

$$\begin{aligned} x &+ 2y + 3z = 6 \\ 2x &+ 5y + 2z = 4 \\ 6x &- 3y + z = 2 \end{aligned} \qquad (3)$$

We look for numbers x, y, z that solve all three equations at once. Those desired numbers might or might not exist. For this system, they do exist. When the number of unknowns matches the number of equations, there is *usually* one solution. Before solving the problem, we visualize it both ways:

R ***The row picture shows three planes meeting at a single point.***

C ***The column picture combines three columns to produce the vector*** $(6, 4, 2)$.

In the row picture, each equation is a *plane* in three-dimensional space. The first plane comes from the first equation $x + 2y + 3z = 6$. That plane crosses the x and y and z axes at the points $(6, 0, 0)$ and $(0, 3, 0)$ and $(0, 0, 2)$. Those three points solve the equation and they determine the whole plane.

The vector $(x, y, z) = (0, 0, 0)$ does not solve $x + 2y + 3z = 6$. Therefore the plane in Figure 2.3 does not contain the origin.

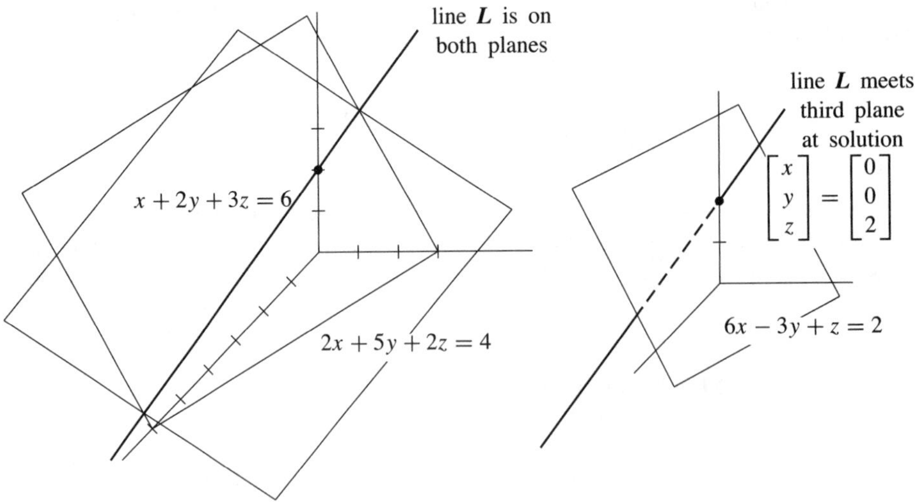

Figure 2.3 Row picture of three equations: Three planes meet at a point.

The plane $x + 2y + 3z = 0$ does pass through the origin, and it is parallel to $x + 2y + 3z = 6$. When the right side increases to 6, the plane moves away from the origin.

The second plane is given by the second equation $2x + 5y + 2z = 4$. *It intersects the first plane in a line* $\boldsymbol{L}$. The usual result of two equations in three unknowns is a line $\boldsymbol{L}$ of solutions.

The third equation gives a third plane. It cuts the line $\boldsymbol{L}$ at a single point. That point lies on all three planes and it solves all three equations. It is harder to draw this triple intersection point than to imagine it. The three planes meet at the solution (which we haven't found yet). The column form shows immediately why $z = 2$!

The column picture starts with the vector form of the equations:

$$x \begin{bmatrix} 1 \\ 2 \\ 6 \end{bmatrix} + y \begin{bmatrix} 2 \\ 5 \\ -3 \end{bmatrix} + z \begin{bmatrix} 3 \\ 2 \\ 1 \end{bmatrix} = \begin{bmatrix} 6 \\ 4 \\ 2 \end{bmatrix}. \tag{4}$$

The unknown numbers x, y, z are the coefficients in this linear combination. We want to multiply the three column vectors by the correct numbers x, y, z to produce $\boldsymbol{b} = (6, 4, 2)$.

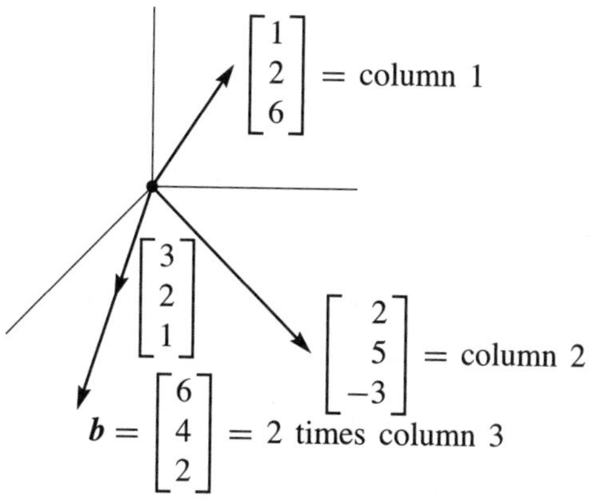

Figure 2.4 *Column picture*: $(x, y, z) = (0, 0, 2)$ because $2(3, 2, 1) = (6, 4, 2) = \boldsymbol{b}$.

Figure 2.4 shows this column picture. Linear combinations of those columns can produce any vector $\boldsymbol{b}$! The combination that produces $\boldsymbol{b} = (6, 4, 2)$ is just 2 times the third column. *The coefficients we need are* $x = 0$, $y = 0$, *and* $z = 2$. This is also the intersection point of the three planes in the row picture. It solves the system:

$$\textbf{Correct combination} \quad 0 \begin{bmatrix} 1 \\ 2 \\ 6 \end{bmatrix} + 0 \begin{bmatrix} 2 \\ 5 \\ -3 \end{bmatrix} + 2 \begin{bmatrix} 3 \\ 2 \\ 1 \end{bmatrix} = \begin{bmatrix} 6 \\ 4 \\ 2 \end{bmatrix}.$$

The Matrix Form of the Equations

We have three rows in the row picture and three columns in the column picture (plus the right side). The three rows and three columns contain nine numbers. *These nine numbers fill a* 3 *by* 3 *matrix.* The "coefficient matrix" has the rows and columns that have so far been kept separate:

$$\textbf{\textit{The coefficient matrix is}} \quad A = \begin{bmatrix} 1 & 2 & 3 \\ 2 & 5 & 2 \\ 6 & -3 & 1 \end{bmatrix}.$$

The capital letter A stands for all nine coefficients (in this square array). The letter $\boldsymbol{b}$ denotes the column vector with components $6, 4, 2$. The unknown $\boldsymbol{x}$ is also a column vector, with components x, y, z. (We use boldface because it is a vector, $\boldsymbol{x}$ because it is unknown.) By rows the equations were (3), by columns they were (4), and now by matrices they are (5). The shorthand is $A\boldsymbol{x} = \boldsymbol{b}$:

$$\textbf{\textit{Matrix equation}} \qquad \begin{bmatrix} 1 & 2 & 3 \\ 2 & 5 & 2 \\ 6 & -3 & 1 \end{bmatrix} \begin{bmatrix} x \\ y \\ z \end{bmatrix} = \begin{bmatrix} 6 \\ 4 \\ 2 \end{bmatrix}. \tag{5}$$

We multiply the matrix A times the unknown vector $\boldsymbol{x}$ to get the right side $\boldsymbol{b}$.

Basic question: What does it mean to "multiply A times $\boldsymbol{x}$"? We can multiply by rows or by columns. Either way, $A\boldsymbol{x} = \boldsymbol{b}$ must be a correct representation of the three equations. You do the same nine multiplications either way.

Multiplication by rows $\quad A\boldsymbol{x}$ comes from *dot products*, each row times the column $\boldsymbol{x}$:

$$A\boldsymbol{x} = \begin{bmatrix} (\textbf{\textit{row 1}}) \cdot \boldsymbol{x} \\ (\textbf{\textit{row 2}}) \cdot \boldsymbol{x} \\ (\textbf{\textit{row 3}}) \cdot \boldsymbol{x} \end{bmatrix}. \tag{6}$$

Multiplication by columns $\quad A\boldsymbol{x}$ is a ***combination of column vectors***:

$$A\boldsymbol{x} = x\ (\textbf{\textit{column 1}}) + y\ (\textbf{\textit{column 2}}) + z\ (\textbf{\textit{column 3}}). \tag{7}$$

When we substitute the solution $\boldsymbol{x} = (0, 0, 2)$, the multiplication $A\boldsymbol{x}$ produces $\boldsymbol{b}$:

$$\begin{bmatrix} 1 & 2 & 3 \\ 2 & 5 & 2 \\ 6 & -3 & 1 \end{bmatrix} \begin{bmatrix} 0 \\ 0 \\ 2 \end{bmatrix} = \textbf{\textit{2 times column 3}} = \begin{bmatrix} 6 \\ 4 \\ 2 \end{bmatrix}.$$

The first dot product in row multiplication is $(1, 2, 3) \cdot (0, 0, 2) = 6$. The other dot products are 4 and 2. Multiplication by columns is simply 2 times column 3.

This book sees $A\boldsymbol{x}$ as a combination of the columns of A.

Example 1 Here are 3 by 3 matrices A and I, with three ones and six zeros:

$$A\boldsymbol{x} = \begin{bmatrix} 1 & 0 & 0 \\ 1 & 0 & 0 \\ 1 & 0 & 0 \end{bmatrix} \begin{bmatrix} 4 \\ 5 \\ 6 \end{bmatrix} = \begin{bmatrix} 4 \\ 4 \\ 4 \end{bmatrix} \qquad I\boldsymbol{x} = \begin{bmatrix} 1 & 0 & 0 \\ 0 & 1 & 0 \\ 0 & 0 & 1 \end{bmatrix} \begin{bmatrix} 4 \\ 5 \\ 6 \end{bmatrix} = \begin{bmatrix} 4 \\ 5 \\ 6 \end{bmatrix}$$

If you are a row person, the product of every row $(1, 0, 0)$ with $(4, 5, 6)$ is 4. If you are a column person, the linear combination is 4 times the first column $(1, 1, 1)$. In that matrix A, the second and third columns are zero vectors.

The example with $I\boldsymbol{x}$ deserves a careful look, because the matrix I is special. It has ones on the "main diagonal". Off that diagonal, all the entries are zeros. *Whatever vector this matrix multiplies, that vector is not changed.* This is like multiplication by 1, but for matrices and vectors. The exceptional matrix in this example is the 3 by 3 ***identity matrix***:

$$I = \begin{bmatrix} 1 & 0 & 0 \\ 0 & 1 & 0 \\ 0 & 0 & 1 \end{bmatrix} \quad \text{always yields the multiplication} \quad I\boldsymbol{x} = \boldsymbol{x}\,.$$

Matrix Notation

The first row of a 2 by 2 matrix contains a_{11} and a_{12}. The second row contains a_{21} and a_{22}. The first index gives the row number, so that a_{ij} is an entry in row i. The second index j gives the column number. But those subscripts are not convenient on a keyboard! Instead of a_{ij} it is easier to type $A(i, j)$. ***The entry*** $a_{57} = A(5, 7)$ ***would be in row 5, column 7.***

$$A = \begin{bmatrix} a_{11} & a_{12} \\ a_{21} & a_{22} \end{bmatrix} = \begin{bmatrix} A(1,1) & A(1,2) \\ A(2,1) & A(2,2) \end{bmatrix}.$$

For an m by n matrix, the row index i goes from 1 to m. The column index j stops at n. There are mn entries in the matrix. A square matrix (order n) has n^2 entries.

Multiplication in MATLAB

I want to express A and $\boldsymbol{x}$ and their product $A\boldsymbol{x}$ using MATLAB commands. This is a first step in learning that language. I begin by defining the matrix A and the vector $\boldsymbol{x}$. This vector is a 3 by 1 matrix, with three rows and one column. Enter matrices a row at a time, and use a semicolon to signal the end of a row:

$$A = [1\ \ 2\ \ 3;\ \ 2\ \ 5\ \ 2;\ \ 6\ -3\ \ 1]$$
$$\boldsymbol{x} = [0;\ 0;\ 2]$$

Here are three ways to multiply Ax in MATLAB. In reality, $A * \boldsymbol{x}$ is the way to do it. MATLAB is a high level language, and it works with matrices:

$$\textit{Matrix multiplication} \quad \boldsymbol{b} = A * \boldsymbol{x}$$

We can also pick out the first row of A (as a smaller matrix!). The notation for that 1 by 3 submatrix is $A(1, :)$. **Here the colon symbol keeps all columns of row 1**:

$$\textit{Row at a time} \quad \boldsymbol{b} = [\, A(1, :) * \boldsymbol{x}\,;\ A(2, :) * \boldsymbol{x}\,;\ A(3, :) * \boldsymbol{x}\,]$$

Those are dot products, row times column, 1 by 3 matrix times 3 by 1 matrix.

The other way to multiply uses the columns of A. The first column is the 3 by 1 submatrix $A(:, 1)$. Now the colon symbol : is keeping all rows of column 1. This column multiplies $x(1)$ and the other columns multiply $x(2)$ and $x(3)$:

$$\textit{Column at a time} \quad \boldsymbol{b} = A(:, 1) * x(1) + A(:, 2) * x(2) + A(:, 3) * x(3)$$

I think that matrices are stored by columns. Then multiplying a column at a time will be a little faster. So $A * \boldsymbol{x}$ is actually executed by columns.

You can see the same choice in a FORTRAN-type structure, which operates on single entries of A and $\boldsymbol{x}$. This lower level language needs an outer and inner "DO loop". When the outer loop uses the row number I, multiplication is a row at a time. The inner loop $J = 1, 3$ goes along each row I.

When the outer loop uses J, multiplication is a column at a time. I will do that in MATLAB , which needs two more lines "end" "end" to close "for I" and "for J":

FORTRAN by rows

```
DO 10  I = 1, 3
DO 10  J = 1, 3
10  B(I) = B(I) + A(I, J) * X(J)
```

MATLAB **by columns**

```
for J = 1 : 3
for I = 1 : 3
b(I) = b(I) + A(I, J) * x(J)
```

Notice that MATLAB is sensitive to upper case versus lower case (capital letters and small letters). If the matrix is A then its entries are $A(I, J)$ not $a(I, J)$.

I think you will prefer the higher level $A * \boldsymbol{x}$. FORTRAN won't appear again in this book. *Maple* and *Mathematica* and graphing calculators also operate at the higher level. Multiplication is $A.\ x$ in *Mathematica*. It is **multiply**(A, x); or **evalm**$(A\&{*}x)$; in *Maple*. Those languages allow symbolic entries $a, b, x, \ldots$ and not only real numbers. Like MATLAB's Symbolic Toolbox, they give the symbolic answer.

■ REVIEW OF THE KEY IDEAS ■

1. The basic operations on vectors are multiplication $c\boldsymbol{v}$ and vector addition $\boldsymbol{v} + \boldsymbol{w}$.

2. Together those operations give linear combinations $c\boldsymbol{v} + d\boldsymbol{w}$.

3. Matrix-vector multiplication $A\boldsymbol{x}$ can be executed by rows (dot products). But it should be understood as a combination of the columns of A!

4. Column picture: $A\boldsymbol{x} = \boldsymbol{b}$ asks for a combination of columns to produce $\boldsymbol{b}$.

5. Row picture: Each equation in $A\boldsymbol{x} = \boldsymbol{b}$ gives a line ($n = 2$) or a plane ($n = 3$) or a "hyperplane" ($n > 3$). They intersect at the solution or solutions.

■ WORKED EXAMPLES ■

2.1 A Describe the column picture of these three equations. Solve by careful inspection of the columns (instead of elimination):

$$\begin{aligned} x + 3y + 2z &= -3 \\ 2x + 2y + 2z &= -2 \\ 3x + 5y + 4z &= -5 \end{aligned} \qquad \text{which is } A\boldsymbol{x} = \boldsymbol{b}: \quad \begin{bmatrix} 1 & 3 & 2 \\ 2 & 2 & 2 \\ 3 & 5 & 4 \end{bmatrix} \begin{bmatrix} x \\ y \\ z \end{bmatrix} = \begin{bmatrix} -3 \\ -2 \\ -5 \end{bmatrix}.$$

Solution The column picture asks for a linear combination that produces $\boldsymbol{b}$ from the three columns of A. In this example $\boldsymbol{b}$ is *minus the second column.* So the solution is $x = 0, y = -1, z = 0$. To show that $(0, -1, 0)$ is the *only* solution we have to know that "A is invertible" and "the columns are independent" and "the determinant isn't zero". Those words are not yet defined but the test comes from elimination: We need (and we find!) a full set of three nonzero pivots.

If the right side changes to $\boldsymbol{b} = (4, 4, 8) =$ sum of the first two columns, then the right combination has $x = 1, y = 1, z = 0$. The solution becomes $\boldsymbol{x} = (1, 1, 0)$.

2.1 B This system has *no solution*, because the three planes in the row picture don't pass through a point. No combination of the three columns produces $\boldsymbol{b}$:

$$\begin{aligned} x + 3y + 5z &= 4 \\ x + 2y - 3z &= 5 \\ 2x + 5y + 2z &= 8 \end{aligned} \qquad \begin{bmatrix} 1 & 3 & 5 \\ 1 & 2 & -3 \\ 2 & 5 & 2 \end{bmatrix} \begin{bmatrix} x \\ y \\ z \end{bmatrix} = \begin{bmatrix} 4 \\ 5 \\ 8 \end{bmatrix} = \boldsymbol{b}$$

(1) Multiply the equations by $1, 1, -1$ and add to show that these planes don't meet at a point. Are any two of the planes parallel? What are the equations of planes parallel to $x + 3y + 5z = 4$?

(2) Take the dot product of each column (and also $\boldsymbol{b}$) with $\boldsymbol{y} = (1, 1, -1)$. How do those dot products show that the system has no solution?

(3) Find three right side vectors $\boldsymbol{b}^*$ and $\boldsymbol{b}^{**}$ and $\boldsymbol{b}^{***}$ that *do* allow solutions.

Solution

(1) Multiplying the equations by $1, 1, -1$ and adding gives

$$\begin{array}{rl} x+3y+5z=4 & \\ x+2y-3z=5 & \\ -[2x+5y+2z=8] & \\ \hline 0x+0y+0z=1 & \textit{No Solution} \end{array}$$

The planes don't meet at any point, but no two planes are parallel. For a plane parallel to $x+3y+5z=4$, just change the "4". The parallel plane $x+3y+5z=0$ goes through the origin $(0,0,0)$. And the equation multiplied by any nonzero constant still gives the same plane, as in $2x+6y+10z=8$.

(2) The dot product of each column with $\boldsymbol{y}=(1,1,-1)$ is *zero*. On the right side, $\boldsymbol{y}\cdot\boldsymbol{b}=(1,1,-1)\cdot(4,5,8)=1$ is *not zero*. So a solution is impossible. (If a combination of columns could produce $\boldsymbol{b}$, take dot products with $\boldsymbol{y}$. Then a combination of zeros would produce 1.)

(3) There is a solution when $\boldsymbol{b}$ is a combination of the columns. These three examples $\boldsymbol{b}^*, \boldsymbol{b}^{**}, \boldsymbol{b}^{***}$ have solutions $\boldsymbol{x}^*=(1,0,0)$ and $\boldsymbol{x}^{**}=(1,1,1)$ and $\boldsymbol{x}^{***}=(0,0,0)$:

$$\boldsymbol{b}^*=\begin{bmatrix}1\\1\\2\end{bmatrix}=\text{first column} \qquad \boldsymbol{b}^{**}=\begin{bmatrix}9\\0\\9\end{bmatrix}=\text{sum of columns} \qquad \boldsymbol{b}^{***}=\begin{bmatrix}0\\0\\0\end{bmatrix}.$$

Problem Set 2.1

Problems 1–9 are about the row and column pictures of $A\boldsymbol{x}=\boldsymbol{b}$.

1 With $A=I$ (the identity matrix) draw the planes in the row picture. Three sides of a box meet at the solution $\boldsymbol{x}=(x,y,z)=(2,3,4)$:

$$\begin{aligned}1x+0y+0z&=2\\0x+1y+0z&=3\\0x+0y+1z&=4\end{aligned} \quad\text{or}\quad \begin{bmatrix}1&0&0\\0&1&0\\0&0&1\end{bmatrix}\begin{bmatrix}x\\y\\z\end{bmatrix}=\begin{bmatrix}2\\3\\4\end{bmatrix}.$$

2 Draw the vectors in the column picture of Problem 1. Two times column 1 plus three times column 2 plus four times column 3 equals the right side $\boldsymbol{b}$.

3 If the equations in Problem 1 are multiplied by $2, 3, 4$ they become $\widehat{A\boldsymbol{x}}=\widehat{\boldsymbol{b}}$:

$$\begin{aligned}2x+0y+0z&=4\\0x+3y+0z&=9\\0x+0y+4z&=16\end{aligned} \quad\text{or}\quad \begin{bmatrix}2&0&0\\0&3&0\\0&0&4\end{bmatrix}\begin{bmatrix}x\\y\\z\end{bmatrix}=\begin{bmatrix}4\\9\\16\end{bmatrix}$$

Why is the row picture the same? Is the solution $\widehat{\boldsymbol{x}}$ the same as $\boldsymbol{x}$? What is changed in the column picture—the columns or the right combination to give $\widehat{\boldsymbol{b}}$?

4 If equation 1 is added to equation 2, which of these are changed: the planes in the row picture, the column picture, the coefficient matrix, the solution? The new equations in Problem 1 would be $x = 2$, $x + y = 5$, $z = 4$.

5 Find a point with $z = 2$ on the intersection line of the planes $x + y + 3z = 6$ and $x - y + z = 4$. Find the point with $z = 0$ and a third point halfway between.

6 The first of these equations plus the second equals the third:

$$\begin{aligned} x + y + z &= 2 \\ x + 2y + z &= 3 \\ 2x + 3y + 2z &= 5. \end{aligned}$$

The first two planes meet along a line. The third plane contains that line, because if x, y, z satisfy the first two equations then they also _____. The equations have infinitely many solutions (the whole line **L**). Find three solutions on **L**.

7 Move the third plane in Problem 6 to a parallel plane $2x + 3y + 2z = 9$. Now the three equations have no solution—*why not*? The first two planes meet along the line **L**, but the third plane doesn't _____ that line.

8 In Problem 6 the columns are $(1, 1, 2)$ and $(1, 2, 3)$ and $(1, 1, 2)$. This is a "singular case" because the third column is _____. Find two combinations of the columns that give $\boldsymbol{b} = (2, 3, 5)$. This is only possible for $\boldsymbol{b} = (4, 6, c)$ if $c =$ _____.

9 Normally 4 "planes" in 4-dimensional space meet at a _____. Normally 4 column vectors in 4-dimensional space can combine to produce $\boldsymbol{b}$. What combination of $(1, 0, 0, 0)$, $(1, 1, 0, 0)$, $(1, 1, 1, 0)$, $(1, 1, 1, 1)$ produces $\boldsymbol{b} = (3, 3, 3, 2)$? What 4 equations for x, y, z, t are you solving?

Problems 10–15 are about multiplying matrices and vectors.

10 Compute each $A\boldsymbol{x}$ by dot products of the rows with the column vector:

$$\text{(a)} \quad \begin{bmatrix} 1 & 2 & 4 \\ -2 & 3 & 1 \\ -4 & 1 & 2 \end{bmatrix} \begin{bmatrix} 2 \\ 2 \\ 3 \end{bmatrix} \qquad \text{(b)} \quad \begin{bmatrix} 2 & 1 & 0 & 0 \\ 1 & 2 & 1 & 0 \\ 0 & 1 & 2 & 1 \\ 0 & 0 & 1 & 2 \end{bmatrix} \begin{bmatrix} 1 \\ 1 \\ 1 \\ 2 \end{bmatrix}$$

11 Compute each $A\boldsymbol{x}$ in Problem 10 as a combination of the columns:

$$\text{10(a) becomes} \quad A\boldsymbol{x} = 2 \begin{bmatrix} 1 \\ -2 \\ -4 \end{bmatrix} + 2 \begin{bmatrix} 2 \\ 3 \\ 1 \end{bmatrix} + 3 \begin{bmatrix} 4 \\ 1 \\ 2 \end{bmatrix} = \begin{bmatrix} \\ \\ \\ \end{bmatrix}.$$

How many separate multiplications for $A\boldsymbol{x}$, when the matrix is "3 by 3"?

12 Find the two components of Ax by rows or by columns:

$$\begin{bmatrix} 2 & 3 \\ 5 & 1 \end{bmatrix}\begin{bmatrix} 4 \\ 2 \end{bmatrix} \quad \text{and} \quad \begin{bmatrix} 3 & 6 \\ 6 & 12 \end{bmatrix}\begin{bmatrix} 2 \\ -1 \end{bmatrix} \quad \text{and} \quad \begin{bmatrix} 1 & 2 & 4 \\ 2 & 0 & 1 \end{bmatrix}\begin{bmatrix} 3 \\ 1 \\ 1 \end{bmatrix}.$$

13 Multiply A times x to find three components of Ax:

$$\begin{bmatrix} 0 & 0 & 1 \\ 0 & 1 & 0 \\ 1 & 0 & 0 \end{bmatrix}\begin{bmatrix} x \\ y \\ z \end{bmatrix} \quad \text{and} \quad \begin{bmatrix} 2 & 1 & 3 \\ 1 & 2 & 3 \\ 3 & 3 & 6 \end{bmatrix}\begin{bmatrix} 1 \\ 1 \\ -1 \end{bmatrix} \quad \text{and} \quad \begin{bmatrix} 2 & 1 \\ 1 & 2 \\ 3 & 3 \end{bmatrix}\begin{bmatrix} 1 \\ 1 \end{bmatrix}.$$

14 (a) A matrix with m rows and n columns multiplies a vector with _____ components to produce a vector with _____ components.

(b) The planes from the m equations $Ax = b$ are in _____ -dimensional space. The combination of the columns of A is in _____ -dimensional space.

15 Write $2x + 3y + z + 5t = 8$ as a matrix A (how many rows?) multiplying the column vector $\boldsymbol{x} = (x, y, z, t)$ to produce $\boldsymbol{b}$. The solutions $\boldsymbol{x}$ fill a plane or "hyperplane" in 4-dimensional space. *The plane is 3-dimensional with no 4D volume.*

Problems 16–23 ask for matrices that act in special ways on vectors.

16 (a) What is the 2 by 2 identity matrix? I times $\begin{bmatrix} x \\ y \end{bmatrix}$ equals $\begin{bmatrix} x \\ y \end{bmatrix}$.

(b) What is the 2 by 2 exchange matrix? P times $\begin{bmatrix} x \\ y \end{bmatrix}$ equals $\begin{bmatrix} y \\ x \end{bmatrix}$.

17 (a) What 2 by 2 matrix R rotates every vector by 90°? R times $\begin{bmatrix} x \\ y \end{bmatrix}$ is $\begin{bmatrix} y \\ -x \end{bmatrix}$.

(b) What 2 by 2 matrix rotates every vector by 180°?

18 Find the matrix P that multiplies (x, y, z) to give (y, z, x). Find the matrix Q that multiplies (y, z, x) to bring back (x, y, z).

19 What 2 by 2 matrix E subtracts the first component from the second component? What 3 by 3 matrix does the same?

$$E\begin{bmatrix} 3 \\ 5 \end{bmatrix} = \begin{bmatrix} 3 \\ 2 \end{bmatrix} \quad \text{and} \quad E\begin{bmatrix} 3 \\ 5 \\ 7 \end{bmatrix} = \begin{bmatrix} 3 \\ 2 \\ 7 \end{bmatrix}.$$

20 What 3 by 3 matrix E multiplies (x, y, z) to give $(x, y, z+x)$? What matrix E^{-1} multiplies (x, y, z) to give $(x, y, z-x)$? If you multiply $(3, 4, 5)$ by E and then multiply by E^{-1}, the two results are (_____) and (_____).

21 What 2 by 2 matrix P_1 projects the vector (x, y) onto the x axis to produce $(x, 0)$? What matrix P_2 projects onto the y axis to produce $(0, y)$? If you multiply $(5, 7)$ by P_1 and then multiply by P_2, you get (_____) and (_____).

22 What 2 by 2 matrix R rotates every vector through 45°? The vector $(1,0)$ goes to $(\sqrt{2}/2, \sqrt{2}/2)$. The vector $(0,1)$ goes to $(-\sqrt{2}/2, \sqrt{2}/2)$. Those determine the matrix. Draw these particular vectors in the xy plane and find R.

23 Write the dot product of $(1,4,5)$ and (x,y,z) as a matrix multiplication $A\boldsymbol{x}$. The matrix A has one row. The solutions to $A\boldsymbol{x} = \boldsymbol{0}$ lie on a ____ perpendicular to the vector ____. The columns of A are only in ____-dimensional space.

24 In MATLAB notation, write the commands that define this matrix A and the column vectors $\boldsymbol{x}$ and $\boldsymbol{b}$. What command would test whether or not $A\boldsymbol{x} = \boldsymbol{b}$?

$$A = \begin{bmatrix} 1 & 2 \\ 3 & 4 \end{bmatrix} \qquad \boldsymbol{x} = \begin{bmatrix} 5 \\ -2 \end{bmatrix} \qquad \boldsymbol{b} = \begin{bmatrix} 1 \\ 7 \end{bmatrix}$$

25 The MATLAB commands A = eye(3) and v = [3 : 5]′ produce the 3 by 3 identity matrix and the column vector $(3,4,5)$. What are the outputs from A ∗ v and v′ ∗ v? (Computer not needed!) If you ask for v ∗ A, what happens?

26 If you multiply the 4 by 4 all-ones matrix A = ones(4,4) and the column v = ones(4,1), what is A ∗ v? (Computer not needed.) If you multiply B = eye(4) + ones(4,4) times w = zeros(4,1) + 2 ∗ ones(4,1), what is B ∗ w?

Questions 27–29 are a review of the row and column pictures.

27 Draw the two pictures in two planes for the equations $x - 2y = 0$, $x + y = 6$.

28 For two linear equations in three unknowns x, y, z, the row picture will show (2 or 3) (lines or planes) in (2 or 3)-dimensional space. The column picture is in (2 or 3)-dimensional space. The solutions normally lie on a ____.

29 For four linear equations in two unknowns x and y, the row picture shows four ____. The column picture is in ____-dimensional space. The equations have no solution unless the vector on the right side is a combination of ____.

30 Start with the vector $\boldsymbol{u}_0 = (1,0)$. Multiply again and again by the same "Markov matrix" A below. The next three vectors are $\boldsymbol{u}_1$, $\boldsymbol{u}_2$, $\boldsymbol{u}_3$:

$$\boldsymbol{u}_1 = \begin{bmatrix} .8 & .3 \\ .2 & .7 \end{bmatrix} \begin{bmatrix} 1 \\ 0 \end{bmatrix} = \begin{bmatrix} .8 \\ .2 \end{bmatrix} \qquad \boldsymbol{u}_2 = A\boldsymbol{u}_1 = \underline{\quad\quad} \qquad \boldsymbol{u}_3 = A\boldsymbol{u}_2 = \underline{\quad\quad}.$$

What property do you notice for all four vectors $\boldsymbol{u}_0$, $\boldsymbol{u}_1$, $\boldsymbol{u}_2$, $\boldsymbol{u}_3$?

31 With a computer, continue from $\boldsymbol{u}_0 = (1, 0)$ to $\boldsymbol{u}_7$, and from $\boldsymbol{v}_0 = (0, 1)$ to $\boldsymbol{v}_7$. What do you notice about $\boldsymbol{u}_7$ and $\boldsymbol{v}_7$? Here are two MATLAB codes, one with while and one with for. They plot $\boldsymbol{u}_0$ to $\boldsymbol{u}_7$—you can use other languages:

```
u = [1 ; 0]; A = [.8 .3 ; .2 .7];
x = u; k = [0 : 7];
while size(x,2) <= 7
   u = A*u; x = [x u];
end
plot(k, x)
```

```
u = [1 ; 0]; A = [.8 .3 ; .2 .7];
x = u; k = [0 : 7];
for j=1 : 7
   u = A*u; x = [x u];
end
plot(k, x)
```

32 The $\boldsymbol{u}$'s and $\boldsymbol{v}$'s in Problem 31 are approaching a steady state vector $\boldsymbol{s}$. Guess that vector and check that $A\boldsymbol{s} = \boldsymbol{s}$. If you start with $\boldsymbol{s}$, you stay with $\boldsymbol{s}$.

33 This MATLAB code allows you to input $\boldsymbol{x}_0$ with a mouse click, by ginput. With $t = 1$, A rotates vectors by *theta*. The plot will show $A\boldsymbol{x}_0, A^2\boldsymbol{x}_0, \ldots$ going around a circle ($t > 1$ will spiral out and $t < 1$ will spiral in). You can change *theta* and the stop at j=10. We plan to put this code on **web.mit.edu/18.06/www**:

```
theta = 15*pi/180; t = 1.0;
A = t*[cos(theta) -sin(theta) ; sin(theta) cos(theta)];
disp('Click to select starting point')
[x1 , x2] = ginput(1); x = [x1 ; x2];
for j=1:10
   x = [x A*x( : , end)];
end
plot(x(1,:), x(2,:), 'o')
hold off
```

34 Invent a 3 by 3 **magic matrix** M_3 with entries $1, 2, \ldots, 9$. All rows and columns and diagonals add to 15. The first row could be 8, 3, 4. What is M_3 times $(1, 1, 1)$? What is M_4 times $(1, 1, 1, 1)$ if this magic matrix has entries $1, \ldots, 16$?

THE IDEA OF ELIMINATION ■ 2.2

This chapter explains a systematic way to solve linear equations. The method is called ***"elimination"***, and you can see it immediately in our 2 by 2 example. Before elimination, x and y appear in both equations. After elimination, the first unknown x has disappeared from the second equation:

$$\textbf{Before}\quad \begin{aligned} x - 2y &= 1 \\ 3x + 2y &= 11 \end{aligned} \qquad \textbf{After}\quad \begin{aligned} x - 2y &= 1 \\ 8y &= 8 \end{aligned} \qquad \begin{aligned} &(\textit{multiply by 3 and subtract}) \\ &(x \textit{ has been eliminated}) \end{aligned}$$

The last equation $8y = 8$ instantly gives $y = 1$. Substituting for y in the first equation leaves $x - 2 = 1$. Therefore $x = 3$ and the solution $(x, y) = (3, 1)$ is complete.

Elimination produces an ***upper triangular system***—this is the goal. The nonzero coefficients $1, -2, 8$ form a triangle. The last equation $8y = 8$ reveals $y = 1$, and we go up the triangle to x. This quick process is called ***back substitution***. It is used for upper triangular systems of any size, after forward elimination is complete.

Important point: The original equations have the same solution $x = 3$ and $y = 1$. Figure 2.5 repeats this original system as a pair of lines, intersecting at the solution point $(3, 1)$. After elimination, the lines still meet at the same point! One line is horizontal because its equation $8y = 8$ does not contain x.

How did we get from the first pair of lines to the second pair? We subtracted 3 times the first equation from the second equation. The step that eliminates x from equation 2 is the fundamental operation in this chapter. We use it so often that we look at it closely:

To eliminate x: Subtract a multiple of equation 1 from equation 2.

Three times $x - 2y = 1$ gives $3x - 6y = 3$. When this is subtracted from $3x + 2y = 11$, the right side becomes 8. The main point is that $3x$ cancels $3x$. What remains on the left side is $2y - (-6y)$ or $8y$, and x is eliminated.

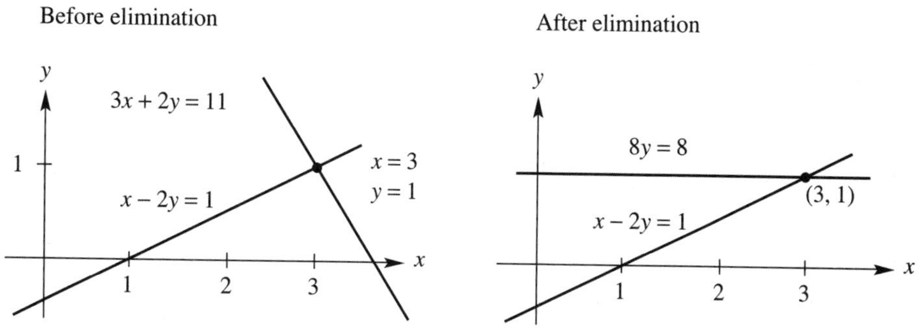

Figure 2.5 Two lines meet at the solution. So does the new line $8y = 8$.

Ask yourself how that multiplier $\ell = 3$ was found. The first equation contains x. ***The first pivot is*** 1 (the coefficient of x). The second equation contains $3x$, so the first equation was multiplied by 3. Then subtraction $3x - 3x$ produced the zero.

You will see the multiplier rule if we change the first equation to $4x - 8y = 4$. (Same straight line but the first pivot becomes 4.) The correct multiplier is now $\ell = \frac{3}{4}$. *To find the multiplier, divide the coefficient "3" to be eliminated by the pivot "4":*

$4x - 8y = 4$	**Multiply equation 1 by $\frac{3}{4}$**	$4x - 8y = 4$
$3x + 2y = 11$	**Subtract from equation 2**	$8y = 8.$

The final system is triangular and the last equation still gives $y = 1$. Back substitution produces $4x - 8 = 4$ and $4x = 12$ and $x = 3$. We changed the numbers but not the lines or the solution. ***Divide by the pivot to find that multiplier $\ell = \frac{3}{4}$:***

Pivot	=	***first nonzero in the row that does the elimination***
Multiplier	=	***(entry to eliminate) divided by (pivot)*** $= \frac{3}{4}$.

The new second equation starts with the second pivot, which is 8. We would use it to eliminate y from the third equation if there were one. *To solve n equations we want n pivots.* ***The pivots are on the diagonal of the triangle after elimination.***

You could have solved those equations for x and y without reading this book. It is an extremely humble problem, but we stay with it a little longer. Even for a 2 by 2 system, elimination might break down and we have to see how. By understanding the possible breakdown (when we can't find a full set of pivots), you will understand the whole process of elimination.

Breakdown of Elimination

Normally, elimination produces the pivots that take us to the solution. But failure is possible. At some point, the method might ask us to *divide by zero*. We can't do it. The process has to stop. There might be a way to adjust and continue—or failure may be unavoidable. Example 1 fails with no solution. Example 2 fails with too many solutions. Example 3 succeeds by exchanging the equations.

Example 1 ***Permanent failure with no solution.*** Elimination makes this clear:

$x - 2y = 1$	Subtract 3 times	$x - 2y = 1$
$3x - 6y = 11$	eqn. 1 from eqn. 2	$\mathbf{0}y = 8.$

The last equation is $0y = 8$. There is *no* solution. Normally we divide the right side 8 by the second pivot, but *this system has no second pivot.* ***(Zero is never allowed as a pivot!)*** The row and column pictures of this 2 by 2 system show that failure was unavoidable. If there is no solution, elimination must certainly have trouble.

The row picture in Figure 2.6 shows parallel lines—which never meet. A solution must lie on both lines. With no meeting point, the equations have no solution.

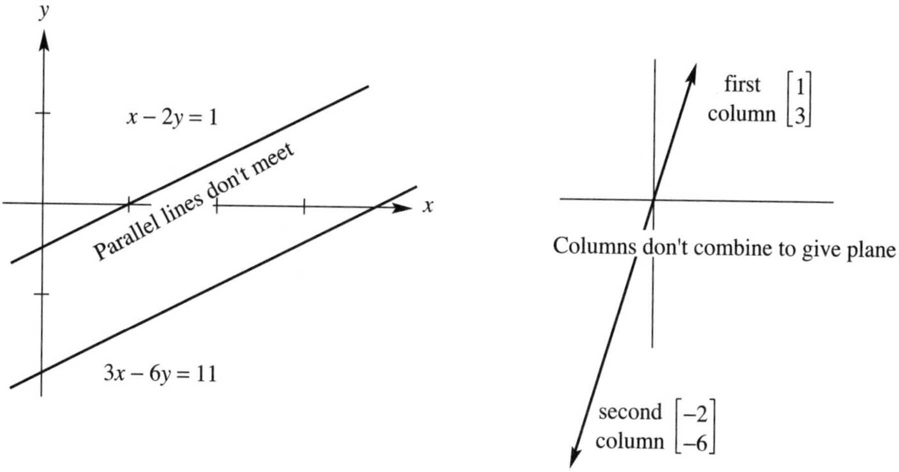

Figure 2.6 Row picture and column picture for Example 1: ***no solution***.

The column picture shows the two columns $(1, 3)$ and $(-2, -6)$ in the same direction. *All combinations of the columns lie along a line.* But the column from the right side is in a different direction $(1, 11)$. No combination of the columns can produce this right side—therefore no solution.

When we change the right side to $(1, 3)$, failure shows as a whole line of solutions. Instead of no solution there are infinitely many:

Example 2 ***Permanent failure with infinitely many solutions:***

$$\begin{aligned} x - 2y &= 1 \\ 3x - 6y &= 3 \end{aligned} \qquad \begin{array}{c}\text{Subtract 3 times}\\ \text{eqn. 1 from eqn. 2}\end{array} \qquad \begin{aligned} x - 2y &= 1 \\ \mathbf{0}y &= \mathbf{0}. \end{aligned}$$

Every y satisfies $0y = 0$. There is really only one equation $x - 2y = 1$. The unknown y is ***"free"***. After y is freely chosen, x is determined as $x = 1 + 2y$.

In the row picture, the parallel lines have become the same line. Every point on that line satisfies both equations. We have a whole line of solutions.

In the column picture, the right side $(1, 3)$ is now the same as the first column. So we can choose $x = 1$ and $y = 0$. We can also choose $x = 0$ and $y = -\frac{1}{2}$; the second column times $-\frac{1}{2}$ equals the right side. There are infinitely many other solutions. Every (x, y) that solves the row problem also solves the column problem.

Elimination can go wrong in a third way—but this time it can be fixed. *Suppose the first pivot position contains zero.* We refuse to allow zero as a pivot. When the first equation has no term involving x, we can exchange it with an equation below:

Example 3 ***Temporary failure but a row exchange produces two pivots***:

$$\begin{aligned} 0x + 2y &= 4 \\ 3x - 2y &= 5 \end{aligned} \qquad \begin{array}{c}\text{Exchange the}\\ \text{two equations}\end{array} \qquad \begin{aligned} 3x - 2y &= 5 \\ 2y &= 4. \end{aligned}$$

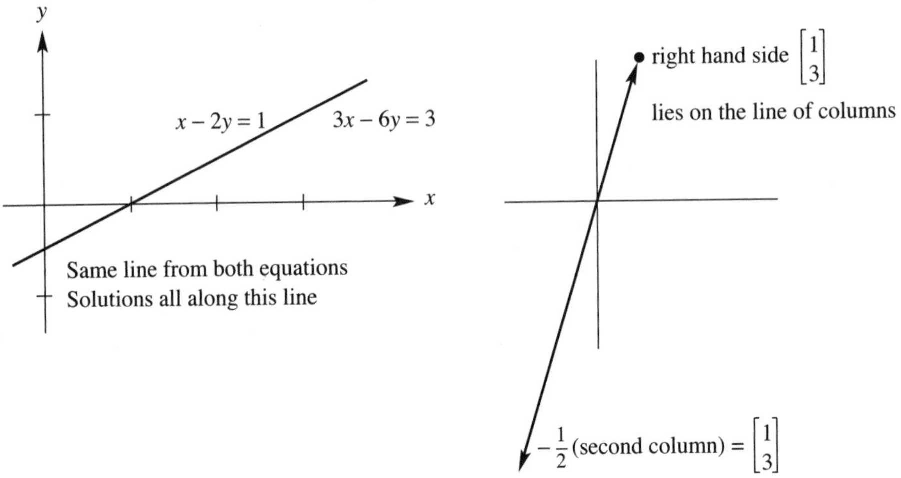

Figure 2.7 Row and column pictures for Example 2: *infinitely many solutions.*

The new system is already triangular. This small example is ready for back substitution. The last equation gives $y = 2$, and then the first equation gives $x = 3$. The row picture is normal (two intersecting lines). The column picture is also normal (column vectors not in the same direction). The pivots 3 and 2 are normal—but an exchange was required to put the rows in a good order.

Examples 1 and 2 are ***singular***—there is no second pivot. Example 3 is ***nonsingular***—there is a full set of pivots and exactly one solution. Singular equations have no solution or infinitely many solutions. Pivots must be nonzero because we have to divide by them.

Three Equations in Three Unknowns

To understand Gaussian elimination, you have to go beyond 2 by 2 systems. Three by three is enough to see the pattern. For now the matrices are square—an equal number of rows and columns. Here is a 3 by 3 system, specially constructed so that all steps lead to whole numbers and not fractions:

$$\begin{aligned} \mathbf{2}x + 4y - 2z &= 2 \\ 4x + 9y - 3z &= 8 \\ -2x - 3y + 7z &= 10 \end{aligned} \tag{1}$$

What are the steps? The first pivot is the boldface 2 (upper left). Below that pivot we want to create zeros. The first multiplier is the ratio $4/2 = 2$. Multiply the pivot equation by $\ell_{21} = 2$ and subtract. Subtraction removes the $4x$ from the second equation:

Step 1 Subtract 2 times equation 1 from equation 2.
We also eliminate $-2x$ from equation 3—still using the first pivot. The quick way is to add equation 1 to equation 3. Then $2x$ cancels $-2x$. We do exactly that, but the rule in this book is to *subtract rather than add.* The systematic pattern has multiplier $\ell_{31} = -2/2 = -1$. Subtracting -1 times an equation is the same as adding:

Step 2 Subtract -1 times equation 1 from equation 3.
The two new equations involve only y and z. The second pivot (boldface) is 1:

$$\begin{aligned} \mathbf{1}y + 1z &= 4 \\ 1y + 5z &= 12 \end{aligned}$$

We have reached a 2 *by* 2 *system.* The final step eliminates y to make it 1 by 1:

Step 3 Subtract equation 2_{new} from 3_{new}. The multiplier is 1. Then $4z = 8$.

The original system $A\boldsymbol{x} = \boldsymbol{b}$ has been converted into a triangular system $U\boldsymbol{x} = \boldsymbol{c}$:

$$\begin{aligned} 2x + 4y - 2z &= 2 \\ 4x + 9y - 3z &= 8 \\ -2x - 3y + 7z &= 10 \end{aligned} \quad \text{has become} \quad \begin{aligned} \mathbf{2}x + 4y - 2z &= 2 \\ \mathbf{1}y + 1z &= 4 \\ \mathbf{4}z &= 8. \end{aligned} \tag{2}$$

The goal is achieved—forward elimination is complete. ***Notice the pivots* 2,1,4 *along the diagonal*.** Those pivots 1 and 4 were hidden in the original system! Elimination brought them out. This triangle is ready for back substitution, which is quick:

$(4z = 8 \text{ gives } z = 2) \quad (y + z = 4 \text{ gives } y = 2) \quad (\text{equation 1 gives } x = -1)$

The solution is $(x, y, z) = (-1, 2, 2)$. The row picture has three planes from three equations. All the planes go through this solution. The original planes are sloping, but the last plane $4z = 8$ after elimination is horizontal.

The column picture shows a combination of column vectors producing the right side $\boldsymbol{b}$. The coefficients in that combination $A\boldsymbol{x}$ are $-1, 2, 2$ (the solution):

$$(-1)\begin{bmatrix} 2 \\ 4 \\ -2 \end{bmatrix} + 2\begin{bmatrix} 4 \\ 9 \\ -3 \end{bmatrix} + 2\begin{bmatrix} -2 \\ -3 \\ 7 \end{bmatrix} \text{ equals } \begin{bmatrix} 2 \\ 8 \\ 10 \end{bmatrix}. \tag{3}$$

The numbers x, y, z multiply columns 1, 2, 3 in the original system $A\boldsymbol{x} = \boldsymbol{b}$ and also in the triangular system $U\boldsymbol{x} = \boldsymbol{c}$.

For a 4 by 4 problem, or an n by n problem, elimination proceeds the same way. Here is the whole idea of forward elimination, column by column:

Column 1. ***Use the first equation to create zeros below the first pivot.***

Column 2. ***Use the new equation 2 to create zeros below the second pivot.***

Columns 3 to n. ***Keep going to find the other pivots and the triangular U.***

After column 2 we have $\begin{bmatrix} \boldsymbol{x} & x & x & x \\ 0 & \boldsymbol{x} & x & x \\ 0 & 0 & x & x \\ 0 & 0 & x & x \end{bmatrix}$. We want $\begin{bmatrix} \boldsymbol{x} & x & x & x \\ & \boldsymbol{x} & x & x \\ & & \boldsymbol{x} & x \\ & & & \boldsymbol{x} \end{bmatrix}$. (4)

The result of forward elimination is an upper triangular system. It is nonsingular if there is a full set of n pivots (never zero!). *Question*: Which x could be changed to boldface $\boldsymbol{x}$ because the pivot is known? Here is a final example to show the original $A\boldsymbol{x} = \boldsymbol{b}$, the triangular system $U\boldsymbol{x} = \boldsymbol{c}$, and the solution from back substitution:

$$\begin{aligned} x + y + z &= 6 \\ x + 2y + 2z &= 9 \\ x + 2y + 3z &= 10 \end{aligned} \qquad \begin{aligned} x + y + z &= 6 \\ y + z &= 3 \\ z &= 1 \end{aligned} \qquad \begin{bmatrix} x \\ y \\ z \end{bmatrix} = \begin{bmatrix} 3 \\ 2 \\ 1 \end{bmatrix}$$

All multipliers are 1. All pivots are 1. All planes meet at the solution (3, 2, 1). The columns combine with coefficients 3, 2, 1 to give $\boldsymbol{b} = (6, 9, 10)$ and $\boldsymbol{c} = (6, 3, 1)$.

■ REVIEW OF THE KEY IDEAS ■

1. A linear system becomes upper triangular after elimination.

2. The upper triangular system is solved by back substitution (starting at the bottom).

3. Elimination subtracts ℓ_{ij} times equation j from equation i, to make the (i, j) entry zero.

4. The multiplier is $\ell_{ij} = \dfrac{\text{entry to eliminate in row } i}{\text{pivot in row } j}$. Pivots can not be zero!

5. A zero in the pivot position can be repaired if there is a nonzero below it.

6. When breakdown is permanent, the system has no solution or infinitely many.

■ WORKED EXAMPLES ■

2.2 A When elimination is applied to this matrix A, what are the first and second pivots? What is the multiplier ℓ_{21} in the first step (ℓ_{21} times row 1 is subtracted from row 2)? What entry in the 2, 2 position (instead of 9) would force an exchange of rows 2 and 3? Why is the multiplier $\ell_{31} = 0$, subtracting 0 times row 1 from row 3?

$$A = \begin{bmatrix} 3 & 1 & 0 \\ 6 & 9 & 2 \\ 0 & 1 & 5 \end{bmatrix}$$

Solution The first pivot is 3. The multiplier ℓ_{21} is $\frac{6}{3} = 2$. When 2 times row 1 is subtracted from row 2, the second pivot is revealed as 7. If we reduce the entry "9" to "2", that drop of 7 in the $(2, 2)$ position would force a row exchange. (The second row would start with $6, 2$ which is an exact multiple of $3, 1$ in the first row. Zero will appear in the second pivot position.) The multiplier ℓ_{31} is zero because $a_{31} = 0$. A zero at the start of a row needs no elimination.

2.2 B Use elimination to reach upper triangular matrices U. Solve by back substitution or explain why this is impossible. What are the pivots (never zero)? Exchange equations when necessary. The only difference is the $-x$ in equation (3).

$$\begin{aligned} x + y + z &= 7 \\ x + y - z &= 5 \\ x - y + z &= 3 \end{aligned} \qquad \begin{aligned} x + y + z &= 7 \\ x + y - z &= 5 \\ -x - y + z &= 3 \end{aligned}$$

Solution For the first system, subtract equation 1 from equations 2 and 3 (the multipliers are $\ell_{21} = 1$ and $\ell_{31} = 1$). The 2, 2 entry becomes zero, so exchange equations:

$$\begin{aligned} x + y + z &= 7 \\ \mathbf{0}y - 2z &= -2 \\ -2y + 0z &= -4 \end{aligned} \quad \text{exchanges into} \quad \begin{aligned} x + y + z &= 7 \\ \mathbf{-2}y + 0z &= -4 \\ -2z &= -2 \end{aligned}$$

Then back substitution gives $z = 1$ and $y = 2$ and $x = 4$. The pivots are $1, -2, -2$.

For the second system, subtract equation 1 from equation 2 as before. Add equation 1 to equation 3. This leaves zero in the 2, 2 entry *and below*:

$$\begin{aligned} x + y + z &= 7 \\ \mathbf{0}y - 2z &= -2 \\ \mathbf{0}y + 2z &= 10 \end{aligned} \qquad \begin{array}{l} \text{There is } \textit{no pivot} \text{ in column 2.} \\ \text{A further elimination step gives } 0z = 8 \\ \text{The three planes don't meet!} \end{array}$$

Plane 1 meets plane 2 in a line. Plane 1 meets plane 3 in a parallel line. *No solution.*

If we change the "3" in the original third equation to "-5" then elimination would leave $2z = 2$ instead of $2z = 10$. Now $z = 1$ would be consistent—we have moved the third plane. Substituting $z = 1$ in the first equation leaves $x + y = 6$. There are infinitely many solutions! *The three planes now meet along a whole line.*

Problem Set 2.2

Problems 1–10 are about elimination on 2 by 2 systems.

1 What multiple ℓ of equation 1 should be subtracted from equation 2?

$$\begin{aligned} 2x + 3y &= 1 \\ 10x + 9y &= 11. \end{aligned}$$

After this elimination step, write down the upper triangular system and circle the two pivots. The numbers 1 and 11 have no influence on those pivots.

2 Solve the triangular system of Problem 1 by back substitution, y before x. Verify that x times $(2, 10)$ plus y times $(3, 9)$ equals $(1, 11)$. If the right side changes to $(4, 44)$, what is the new solution?

3 What multiple of equation 1 should be *subtracted* from equation 2?

$$\begin{aligned} 2x - 4y &= 6 \\ -x + 5y &= 0. \end{aligned}$$

After this elimination step, solve the triangular system. If the right side changes to $(-6, 0)$, what is the new solution?

4 What multiple ℓ of equation 1 should be subtracted from equation 2?

$$\begin{aligned} ax + by &= f \\ cx + dy &= g. \end{aligned}$$

The first pivot is a (assumed nonzero). Elimination produces what formula for the second pivot? What is y? The second pivot is missing when $ad = bc$.

5 Choose a right side which gives no solution and another right side which gives infinitely many solutions. What are two of those solutions?

$$\begin{aligned} 3x + 2y &= 10 \\ 6x + 4y &= \end{aligned}$$

6 Choose a coefficient b that makes this system singular. Then choose a right side g that makes it solvable. Find two solutions in that singular case.

$$\begin{aligned} 2x + by &= 16 \\ 4x + 8y &= g. \end{aligned}$$

7 For which numbers a does elimination break down (1) permanently (2) temporarily?

$$\begin{aligned} ax + 3y &= -3 \\ 4x + 6y &= 6. \end{aligned}$$

Solve for x and y after fixing the second breakdown by a row exchange.

8 For which three numbers k does elimination break down? Which is fixed by a row exchange? In each case, is the number of solutions 0 or 1 or ∞?

$$\begin{aligned} kx + 3y &= 6 \\ 3x + ky &= -6. \end{aligned}$$

9 What test on b_1 and b_2 decides whether these two equations allow a solution? How many solutions will they have? Draw the column picture.

$$\begin{aligned} 3x - 2y &= b_1 \\ 6x - 4y &= b_2. \end{aligned}$$

10 In the xy plane, draw the lines $x + y = 5$ and $x + 2y = 6$ and the equation $y =$ ____ that comes from elimination. The line $5x - 4y = c$ will go through the solution of these equations if $c =$ ____.

Problems 11–20 study elimination on 3 by 3 systems (and possible failure).

11 Reduce this system to upper triangular form by two row operations:

$$\begin{aligned} 2x + 3y + z &= 8 \\ 4x + 7y + 5z &= 20 \\ -2y + 2z &= 0. \end{aligned}$$

Circle the pivots. Solve by back substitution for z, y, x.

12 Apply elimination (circle the pivots) and back substitution to solve

$$\begin{aligned} 2x - 3y \qquad &= 3 \\ 4x - 5y + z &= 7 \\ 2x - y - 3z &= 5. \end{aligned}$$

List the three row operations: Subtract ____ times row ____ from row ____.

13 Which number d forces a row exchange, and what is the triangular system (not singular) for that d? Which d makes this system singular (no third pivot)?

$$\begin{aligned} 2x + 5y + z &= 0 \\ 4x + dy + z &= 2 \\ y - z &= 3. \end{aligned}$$

14 Which number b leads later to a row exchange? Which b leads to a missing pivot? In that singular case find a nonzero solution x, y, z.

$$\begin{aligned} x + by \qquad &= 0 \\ x - 2y - z &= 0 \\ y + z &= 0. \end{aligned}$$

15 (a) Construct a 3 by 3 system that needs two row exchanges to reach a triangular form and a solution.

(b) Construct a 3 by 3 system that needs a row exchange to keep going, but breaks down later.

16 If rows 1 and 2 are the same, how far can you get with elimination (allowing row exchange)? If columns 1 and 2 are the same, which pivot is missing?

$$\begin{aligned} 2x - y + z &= 0 \\ 2x - y + z &= 0 \\ 4x + y + z &= 2 \end{aligned} \qquad \begin{aligned} 2x + 2y + z &= 0 \\ 4x + 4y + z &= 0 \\ 6x + 6y + z &= 2. \end{aligned}$$

17 Construct a 3 by 3 example that has 9 different coefficients on the left side, but rows 2 and 3 become zero in elimination. How many solutions to your system with $\boldsymbol{b} = (1, 10, 100)$ and how many with $\boldsymbol{b} = (0, 0, 0)$?

18 Which number q makes this system singular and which right side t gives it infinitely many solutions? Find the solution that has $z = 1$.

$$\begin{aligned} x + 4y - 2z &= 1 \\ x + 7y - 6z &= 6 \\ 3y + qz &= t. \end{aligned}$$

19 (Recommended) It is impossible for a system of linear equations to have exactly two solutions. *Explain why.*

(a) If (x, y, z) and (X, Y, Z) are two solutions, what is another one?

(b) If 25 planes meet at two points, where else do they meet?

20 Three planes can fail to have an intersection point, when no two planes are parallel. The system is singular if row 3 of A is a ______ of the first two rows. Find a third equation that can't be solved if $x + y + z = 0$ and $x - 2y - z = 1$.

Problems 21–23 move up to 4 by 4 and n by n.

21 Find the pivots and the solution for these four equations:

$$\begin{aligned} 2x + y \qquad\qquad &= 0 \\ x + 2y + z \qquad &= 0 \\ y + 2z + t &= 0 \\ z + 2t &= 5. \end{aligned}$$

22 This system has the same pivots and right side as Problem 21. How is the solution different (if it is)?

$$\begin{aligned} 2x - y \qquad\qquad &= 0 \\ -x + 2y - z \qquad &= 0 \\ - y + 2z - t &= 0 \\ - z + 2t &= 5. \end{aligned}$$

23 If you extend Problems 21–22 following the $1, 2, 1$ pattern or the $-1, 2, -1$ pattern, what is the fifth pivot? What is the nth pivot?

24 If elimination leads to these equations, find three possible original matrices A:

$$\begin{aligned} x + y + z &= 0 \\ y + z &= 0 \\ 3z &= 0. \end{aligned}$$

25 For which two numbers a will elimination fail on $A = \begin{bmatrix} a & 2 \\ a & a \end{bmatrix}$?

26 For which three numbers a will elimination fail to give three pivots?

$$A = \begin{bmatrix} a & 2 & 3 \\ a & a & 4 \\ a & a & a \end{bmatrix}.$$

27 Look for a matrix that has row sums 4 and 8, and column sums 2 and s:

$$\text{Matrix } = \begin{bmatrix} a & b \\ c & d \end{bmatrix} \qquad \begin{matrix} a + b = 4 & a + c = 2 \\ c + d = 8 & b + d = s \end{matrix}$$

The four equations are solvable only if $s =$ ______. Then find two different matrices that have the correct row and column sums. *Extra credit*: Write down the 4 by 4 system $A\boldsymbol{x} = \boldsymbol{b}$ with $\boldsymbol{x} = (a, b, c, d)$ and make A triangular by elimination.

28 Elimination in the usual order gives what pivot matrix and what solution to this "lower triangular" system? We are really solving by *forward substitution*:

$$\begin{aligned} 3x \qquad\quad &= 3 \\ 6x + 2y \quad &= 8 \\ 9x - 2y + z &= 9. \end{aligned}$$

29 Create a MATLAB command A(2, :) = ... for the new row 2, to subtract 3 times row 1 from the existing row 2 if the matrix A is already known.

30 Find experimentally the average first and second and third pivot sizes (use the absolute value) in MATLAB's $A = \textbf{rand}(3, 3)$. The average of $\textbf{abs}(A(1, 1))$ should be 0.5 but I don't know the others.

ELIMINATION USING MATRICES ■ 2.3

We now combine two ideas—elimination and matrices. The goal is to express all the steps of elimination (and the final result) in the clearest possible way. In a 3 by 3 example, elimination could be described in words. For larger systems, a long list of steps would be hopeless. You will see how to subtract a multiple of one row from another row—*using matrices.*

The matrix form of a linear system is $A\boldsymbol{x} = \boldsymbol{b}$. Here are $\boldsymbol{b}, \boldsymbol{x}$, and A:

1 The vector of right sides is $\boldsymbol{b}$.

2 The vector of unknowns is $\boldsymbol{x}$. (The unknowns change to $x_1, x_2, x_3, \ldots$ because we run out of letters before we run out of numbers.)

3 The coefficient matrix is A. In this chapter A is square.

The example in the previous section has the beautifully short form $A\boldsymbol{x} = \boldsymbol{b}$:

$$\begin{aligned} 2x_1 + 4x_2 - 2x_3 &= 2 \\ 4x_1 + 9x_2 - 3x_3 &= 8 \\ -2x_1 - 3x_2 + 7x_3 &= 10 \end{aligned} \quad \text{is the same as} \quad \begin{bmatrix} 2 & 4 & -2 \\ 4 & 9 & -3 \\ -2 & -3 & 7 \end{bmatrix} \begin{bmatrix} x_1 \\ x_2 \\ x_3 \end{bmatrix} = \begin{bmatrix} 2 \\ 8 \\ 10 \end{bmatrix}. \tag{1}$$

The nine numbers on the left go into the matrix A. That matrix not only sits beside $\boldsymbol{x}$, it *multiplies* $\boldsymbol{x}$. The rule for "A times $\boldsymbol{x}$" is exactly chosen to yield the three equations.

Review of A times x. A matrix times a vector gives a vector. The matrix is square when the number of equations (three) matches the number of unknowns (three). Our matrix is 3 by 3. A general square matrix is n by n. Then the vector $\boldsymbol{x}$ is in n-dimensional space. This example is in 3-dimensional space:

$$\textbf{\textit{The unknown is}} \quad \boldsymbol{x} = \begin{bmatrix} x_1 \\ x_2 \\ x_3 \end{bmatrix} \quad \textbf{\textit{and the solution is}} \quad \boldsymbol{x} = \begin{bmatrix} -1 \\ 2 \\ 2 \end{bmatrix}.$$

Key point: $A\boldsymbol{x} = \boldsymbol{b}$ represents the row form and also the column form of the equations. We can multiply by taking a column of A at a time:

$$A\boldsymbol{x} = (-1)\begin{bmatrix} 2 \\ 4 \\ -2 \end{bmatrix} + 2\begin{bmatrix} 4 \\ 9 \\ -3 \end{bmatrix} + 2\begin{bmatrix} -2 \\ -3 \\ 7 \end{bmatrix} = \begin{bmatrix} 2 \\ 8 \\ 10 \end{bmatrix}. \tag{2}$$

This rule is used so often that we express it once more for emphasis.

> **2A** ***The product** $A\boldsymbol{x}$ **is a combination of the columns of** A.* Components of $\boldsymbol{x}$ multiply columns: $A\boldsymbol{x} = x_1$ times (column 1) $+ \cdots + x_n$ times (column n).

One point to repeat about matrix notation: The entry in row 1, column 1 (the top left corner) is called a_{11}. The entry in row 1, column 3 is a_{13}. The entry in row 3, column 1 is a_{31}. (Row number comes before column number.) The word "entry" for a matrix corresponds to the word "component" for a vector. General rule: ***The entry in row** i**, column** j **of the matrix** A **is** a_{ij}.*

Example 1 This matrix has $a_{ij} = 2i + j$. Then $a_{11} = 3$. Also $a_{12} = 4$ and $a_{21} = 5$. Here is Ax with numbers and letters:

$$\begin{bmatrix} 3 & 4 \\ 5 & 6 \end{bmatrix}\begin{bmatrix} 2 \\ 1 \end{bmatrix} = \begin{bmatrix} 3 \cdot 2 + 4 \cdot 1 \\ 5 \cdot 2 + 6 \cdot 1 \end{bmatrix} \qquad \begin{bmatrix} a_{11} & a_{12} \\ a_{21} & a_{22} \end{bmatrix}\begin{bmatrix} x_1 \\ x_2 \end{bmatrix} = \begin{bmatrix} a_{11}x_1 + a_{12}x_2 \\ a_{21}x_1 + a_{22}x_2 \end{bmatrix}.$$

The first component of Ax is $6 + 4 = 10$. That is the product of the row $[3 \quad 4]$ with the column (2, 1). ***A row times a column gives a dot product!***

The ith component of Ax involves row i, which is $[\, a_{i1} \quad a_{i2} \quad \cdots \quad a_{in} \,]$. The short formula for its dot product with $\boldsymbol{x}$ uses "sigma notation":

2B The ith component of Ax is $a_{i1}x_1 + a_{i2}x_2 + \cdots + a_{in}x_n$. This is $\displaystyle\sum_{j=1}^{n} a_{ij}x_j$.

The sigma symbol $\sum$ is an instruction to add. Start with $j = 1$ and stop with $j = n$. Start the sum with $a_{i1}x_1$ and stop with $a_{in}x_n$.[1]

The Matrix Form of One Elimination Step

$Ax = \boldsymbol{b}$ is a convenient form for the original equation. What about the elimination steps? The first step in this example subtracts 2 times the first equation from the second equation. On the right side, 2 times the first component of $\boldsymbol{b}$ is subtracted from the second component:

$$\boldsymbol{b} = \begin{bmatrix} 2 \\ 8 \\ 10 \end{bmatrix} \quad \text{changes to} \quad \boldsymbol{b}_{\text{new}} = \begin{bmatrix} 2 \\ 4 \\ 10 \end{bmatrix}.$$

We want to do that subtraction with a matrix! The same result $\boldsymbol{b}_{\text{new}} = E\boldsymbol{b}$ is achieved when we multiply an "elimination matrix" E times $\boldsymbol{b}$. It subtracts $2b_1$ from b_2:

$$\textit{The elimination matrix is} \quad E = \begin{bmatrix} 1 & 0 & 0 \\ -2 & 1 & 0 \\ 0 & 0 & 1 \end{bmatrix}.$$

Multiplication by E subtracts 2 times row 1 from row 2. Rows 1 and 3 stay the same:

$$\begin{bmatrix} 1 & 0 & 0 \\ -2 & 1 & 0 \\ 0 & 0 & 1 \end{bmatrix}\begin{bmatrix} 2 \\ 8 \\ 10 \end{bmatrix} = \begin{bmatrix} 2 \\ 4 \\ 10 \end{bmatrix} \qquad \begin{bmatrix} 1 & 0 & 0 \\ -2 & 1 & 0 \\ 0 & 0 & 1 \end{bmatrix}\begin{bmatrix} b_1 \\ b_2 \\ b_3 \end{bmatrix} = \begin{bmatrix} b_1 \\ b_2 - 2b_1 \\ b_3 \end{bmatrix}$$

Notice how $b_1 = 2$ and $b_3 = 10$ stay the same. The first and third rows of E are the first and third rows of the identity matrix I. The new second component is the number 4 that appeared after the elimination step. This is $b_2 - 2b_1$.

---

[1]Einstein shortened this even more by omitting the $\sum$. The repeated j in $a_{ij}x_j$ automatically meant addition. He also wrote the sum as $a_i^j x_j$. Not being Einstein, we include the $\sum$.

It is easy to describe the "elementary matrices" or "elimination matrices" like E. Start with the identity matrix I. *Change one of its zeros to the multiplier* $-\ell$:

2C The ***identity matrix*** has 1's on the diagonal and otherwise 0's. Then $I\boldsymbol{b} = \boldsymbol{b}$. The ***elementary matrix or elimination matrix*** E_{ij} that subtracts a multiple ℓ of row j from row i has the extra nonzero entry $-\ell$ in the i, j position.

Example 2

$$\textbf{Identity} \quad I = \begin{bmatrix} 1 & 0 & 0 \\ 0 & 1 & 0 \\ 0 & 0 & 1 \end{bmatrix} \qquad \textbf{Elimination} \quad E_{31} = \begin{bmatrix} 1 & 0 & 0 \\ 0 & 1 & 0 \\ -\ell & 0 & 1 \end{bmatrix}.$$

When you multiply I times $\boldsymbol{b}$, you get $\boldsymbol{b}$. But E_{31} subtracts ℓ times the first component from the third component. With $\ell = 4$ we get $9 - 4 = 5$:

$$I\boldsymbol{b} = \begin{bmatrix} 1 & 0 & 0 \\ 0 & 1 & 0 \\ 0 & 0 & 1 \end{bmatrix}\begin{bmatrix} 1 \\ 3 \\ 9 \end{bmatrix} = \begin{bmatrix} 1 \\ 3 \\ 9 \end{bmatrix} \quad \text{and} \quad E\boldsymbol{b} = \begin{bmatrix} 1 & 0 & 0 \\ 0 & 1 & 0 \\ -4 & 0 & 1 \end{bmatrix}\begin{bmatrix} 1 \\ 3 \\ 9 \end{bmatrix} = \begin{bmatrix} 1 \\ 3 \\ 5 \end{bmatrix}.$$

What about the left side of $A\boldsymbol{x} = \boldsymbol{b}$? The multiplier $\ell = 4$ was chosen to produce a zero, by subtracting 4 times the pivot. $\boldsymbol{E_{31}}$ ***creates a zero in the*** $\mathbf{(3, 1)}$ ***position.***

The notation fits this purpose. Start with A. Apply E's to produce zeros below the pivots (the first E is E_{21}). End with a triangular U. We now look in detail at those steps.

First a small point. The vector $\boldsymbol{x}$ stays the same. The solution is not changed by elimination. (That may be more than a small point.) It is the coefficient matrix that is changed! When we start with $A\boldsymbol{x} = \boldsymbol{b}$ and multiply by E, the result is $EA\boldsymbol{x} = E\boldsymbol{b}$. The new matrix EA is the result of multiplying E times A.

Matrix Multiplication

The big question is: ***How do we multiply two matrices?*** When the first matrix is E (an elimination matrix), there is already an important clue. We know A, and we know what it becomes after the elimination step. To keep everything right, we hope and expect that EA is

$$\begin{bmatrix} 1 & 0 & 0 \\ -2 & 1 & 0 \\ 0 & 0 & 1 \end{bmatrix}\begin{bmatrix} 2 & 4 & -2 \\ 4 & 9 & -3 \\ -2 & -3 & 7 \end{bmatrix} = \begin{bmatrix} 2 & 4 & -2 \\ 0 & 1 & 1 \\ -2 & -3 & 7 \end{bmatrix} \quad \textit{(with the zero)}.$$

This step does not change rows 1 and 3 of A. Those rows are unchanged in EA—only row 2 is different. *Twice the first row has been subtracted from the second row.* Matrix multiplication agrees with elimination—and the new system of equations is $EA\boldsymbol{x} = E\boldsymbol{b}$.

$EA\boldsymbol{x}$ is simple but it involves a subtle idea. Multiplying both sides of the original equation gives $E(A\boldsymbol{x}) = E\boldsymbol{b}$. With our proposed multiplication of matrices, this is also

$(EA)\boldsymbol{x} = E\boldsymbol{b}$. The first was E times $A\boldsymbol{x}$, the second is EA times $\boldsymbol{x}$. They are the same! The parentheses are not needed. We just write $EA\boldsymbol{x} = E\boldsymbol{b}$.

When multiplying ABC, you can do BC first or you can do AB first. This is the point of an "associative law" like $3 \times (4 \times 5) = (3 \times 4) \times 5$. We multiply 3 times 20, or we multiply 12 times 5. Both answers are 60. That law seems so obvious that it is hard to imagine it could be false. But the "commutative law" $3 \times 4 = 4 \times 3$ looks even more obvious. For matrices, EA is different from AE.

2D ASSOCIATIVE LAW	$A(BC) = (AB)C$
NOT COMMUTATIVE LAW	Often $AB \neq BA$.

There is another requirement on matrix multiplication. Suppose B has only one column (this column is $\boldsymbol{b}$). The matrix-matrix law for EB should be consistent with the old matrix-vector law for $E\boldsymbol{b}$. Even more, we should be able to *multiply matrices a column at a time*:

If B has several columns $\boldsymbol{b}_1, \boldsymbol{b}_2, \boldsymbol{b}_3$, then EB has columns $E\boldsymbol{b}_1, E\boldsymbol{b}_2, E\boldsymbol{b}_3$.

This holds true for the matrix multiplication above (where the matrix is A instead of B). If you multiply column 1 of A by E, you get column 1 of EA:

$$\begin{bmatrix} 1 & 0 & 0 \\ -2 & 1 & 0 \\ 0 & 0 & 1 \end{bmatrix} \begin{bmatrix} 2 \\ 4 \\ -2 \end{bmatrix} = \begin{bmatrix} 2 \\ 0 \\ -2 \end{bmatrix} \quad \text{and } E(\text{column } j \text{ of } A) = \text{column } j \text{ of } EA.$$

This requirement deals with columns, while elimination deals with rows. The next section describes each individual entry of the product. The beauty of matrix multiplication is that all three approaches (rows, columns, whole matrices) come out right.

The Matrix P_{ij} for a Row Exchange

To subtract row j from row i we use E_{ij}. To exchange or "permute" those rows we use another matrix P_{ij}. Row exchanges are needed when zero is in the pivot position. Lower down that pivot column may be a nonzero. By exchanging the two rows, we have a pivot (never zero!) and elimination goes forward.

What matrix P_{23} exchanges row 2 with row 3? We can find it by exchanging rows of the identity matrix I:

$$\textbf{Permutation matrix} \qquad P_{23} = \begin{bmatrix} 1 & 0 & 0 \\ 0 & 0 & 1 \\ 0 & 1 & 0 \end{bmatrix}.$$

This is a ***row exchange matrix***. Multiplying by P_{23} exchanges components 2 and 3 of any column vector. Therefore it also exchanges rows 2 and 3 of any matrix:

$$\begin{bmatrix} 1 & 0 & 0 \\ 0 & 0 & 1 \\ 0 & 1 & 0 \end{bmatrix} \begin{bmatrix} 1 \\ 3 \\ 5 \end{bmatrix} = \begin{bmatrix} 1 \\ 5 \\ 3 \end{bmatrix} \quad \text{and} \quad \begin{bmatrix} 1 & 0 & 0 \\ 0 & 0 & 1 \\ 0 & 1 & 0 \end{bmatrix} \begin{bmatrix} 2 & 4 & 1 \\ 0 & 0 & 3 \\ 0 & 6 & 5 \end{bmatrix} = \begin{bmatrix} 2 & 4 & 1 \\ 0 & 6 & 5 \\ 0 & 0 & 3 \end{bmatrix}.$$

On the right, P_{23} is doing what it was created for. With zero in the second pivot position and "6" below it, the exchange puts 6 into the pivot.

Matrices *act*. They don't just sit there. We will soon meet other permutation matrices, which can change the order of several rows. Rows 1, 2, 3 can be moved to 3, 1, 2. Our P_{23} is one particular permutation matrix—it exchanges rows 2 and 3.

2E Row Exchange Matrix P_{ij} is the identity matrix with rows i and j reversed. When P_{ij} multiplies a matrix A, it exchanges rows i and j of A.

$$\textbf{\textit{To exchange equations 1 and 3 multiply by}} \quad P_{13} = \begin{bmatrix} \mathbf{0} & \mathbf{0} & \mathbf{1} \\ \mathbf{0} & \mathbf{1} & \mathbf{0} \\ \mathbf{1} & \mathbf{0} & \mathbf{0} \end{bmatrix}.$$

Usually row exchanges are not required. The odds are good that elimination uses only the E_{ij}. But the P_{ij} are ready if needed, to move a pivot up to the diagonal.

The Augmented Matrix

This book eventually goes far beyond elimination. Matrices have all kinds of practical applications, in which they are multiplied. Our best starting point was a square E times a square A, because we met this in elimination—and we know what answer to expect for EA. The next step is to allow a *rectangular matrix*. It still comes from our original equations, but now it includes the right side $\boldsymbol{b}$.

Key idea: Elimination does the same row operations to A and to $\boldsymbol{b}$. ***We can include b as an extra column and follow it through elimination***. The matrix A is enlarged or "augmented" by the extra column $\boldsymbol{b}$:

$$\textbf{\textit{Augmented matrix}} \quad \begin{bmatrix} A & \boldsymbol{b} \end{bmatrix} = \begin{bmatrix} 2 & 4 & -2 & \mathbf{2} \\ 4 & 9 & -3 & \mathbf{8} \\ -2 & -3 & 7 & \mathbf{10} \end{bmatrix}.$$

Elimination acts on whole rows of this matrix. The left side and right side are both multiplied by E, to subtract 2 times equation 1 from equation 2. With $\begin{bmatrix} A & \boldsymbol{b} \end{bmatrix}$ those steps happen together:

$$\begin{bmatrix} 1 & 0 & 0 \\ -2 & 1 & 0 \\ 0 & 0 & 1 \end{bmatrix} \begin{bmatrix} 2 & 4 & -2 & \mathbf{2} \\ 4 & 9 & -3 & \mathbf{8} \\ -2 & -3 & 7 & \mathbf{10} \end{bmatrix} = \begin{bmatrix} 2 & 4 & -2 & \mathbf{2} \\ 0 & 1 & 1 & \mathbf{4} \\ -2 & -3 & 7 & \mathbf{10} \end{bmatrix}.$$

The new second row contains $0, 1, 1, 4$. The new second equation is $x_2 + x_3 = 4$. Matrix multiplication works by rows and at the same time by columns:

R (by rows): Each row of E acts on $[\,A \;\; \boldsymbol{b}\,]$ to give a row of $[\,EA \;\; E\boldsymbol{b}\,]$.

C (by columns): E acts on each column of $[\,A \;\; \boldsymbol{b}\,]$ to give a column of $[\,EA \;\; E\boldsymbol{b}\,]$.

Notice again that word "acts." This is essential. Matrices do something! The matrix A acts on $\boldsymbol{x}$ to produce $\boldsymbol{b}$. The matrix E operates on A to give EA. The whole process of elimination is a sequence of row operations, alias matrix multiplications. A goes to $E_{21}A$ which goes to $E_{31}E_{21}A$. Finally $E_{32}E_{31}E_{21}A$ is a triangular matrix.

The right side is included in the augmented matrix. The end result is a triangular system of equations. We stop for exercises on multiplication by E, before writing down the rules for all matrix multiplications (including block multiplication).

■ REVIEW OF THE KEY IDEAS ■

1. $A\boldsymbol{x} = x_1$ times column $1 + \cdots + x_n$ times column n. And $(A\boldsymbol{x})_i = \sum_{j=1}^{n} a_{ij}x_j$.
2. Identity matrix $= I$, elimination matrix $= E_{ij}$, exchange matrix $= P_{ij}$.
3. Multiplying $A\boldsymbol{x} = \boldsymbol{b}$ by E_{21} subtracts a multiple ℓ_{21} of equation 1 from equation 2. The number $-\ell_{21}$ is the $(2,1)$ entry of the elimination matrix E_{21}.
4. For the augmented matrix $\begin{bmatrix} A & \boldsymbol{b} \end{bmatrix}$, that elimination step gives $\begin{bmatrix} E_{21}A & E_{21}\boldsymbol{b} \end{bmatrix}$.
5. When A multiplies any matrix B, it multiplies each column of B separately.

■ WORKED EXAMPLES ■

2.3 A What 3 by 3 matrix E_{21} subtracts 4 times row 1 from row 2? What matrix P_{32} exchanges row 2 and row 3? If you multiply A on the *right* instead of the left, describe the results AE_{21} and AP_{32}.

Solution By doing those operations on the identity matrix I, we find

$$E_{21} = \begin{bmatrix} 1 & 0 & 0 \\ -4 & 1 & 0 \\ 0 & 0 & 1 \end{bmatrix} \quad \text{and} \quad P_{32} = \begin{bmatrix} 1 & 0 & 0 \\ 0 & 0 & 1 \\ 0 & 1 & 0 \end{bmatrix}.$$

Multiplying by E_{21} on the right side will subtract 4 times **column 2** from **column 1**. Multiplying by P_{32} on the right will exchange **columns 2** and **3**.

2.3 B Write down the augmented matrix $[A \;\; \boldsymbol{b}]$ with an extra column:

$$\begin{aligned} x + 2y + 2z &= 1 \\ 4x + 8y + 9z &= 3 \\ 3y + 2z &= 1 \end{aligned}$$

Apply E_{21} and then P_{32} to reach a triangular system. Solve by back substitution. What combined matrix $P_{32}\,E_{21}$ will do both steps at once?

Solution The augmented matrix and the result of using E_{21} are

$$[A \quad \boldsymbol{b}] = \begin{bmatrix} 1 & 2 & 2 & 1 \\ 4 & 8 & 9 & 3 \\ 0 & 3 & 2 & 1 \end{bmatrix} \quad \text{and} \quad E_{21}[A \quad \boldsymbol{b}] = \begin{bmatrix} 1 & 2 & 2 & 1 \\ 0 & 0 & 1 & -1 \\ 0 & 3 & 2 & 1 \end{bmatrix}$$

P_{32} exchanges equation 2 and 3. Back substitution produces (x, y, z):

$$P_{32}\,E_{21}[A \quad \boldsymbol{b}] = \begin{bmatrix} 1 & 2 & 2 & 1 \\ 0 & 3 & 2 & 1 \\ 0 & 0 & 1 & -1 \end{bmatrix} \quad \text{and} \quad \begin{bmatrix} x \\ y \\ z \end{bmatrix} = \begin{bmatrix} 1 \\ 1 \\ -1 \end{bmatrix}$$

For the matrix $P_{32}\,E_{21}$ that does both steps at once, apply P_{32} to E_{21}!

$$P_{32}\,E_{21} = \text{exchange the rows of } E_{21} = \begin{bmatrix} 1 & 0 & 0 \\ 0 & 0 & 1 \\ -4 & 1 & 0 \end{bmatrix}.$$

2.3 C Multiply these matrices in two ways: first, rows of A times columns of B to find each entry of AB, and second, columns of A times rows of B to produce two matrices that add to AB. How many separate ordinary multiplications are needed?

$$AB = \begin{bmatrix} 3 & 4 \\ 1 & 5 \\ 2 & 0 \end{bmatrix} \begin{bmatrix} 2 & 4 \\ 1 & 1 \end{bmatrix} = (3 \text{ by } 2)(2 \text{ by } 2)$$

Solution Rows of A times columns of B are dot products of vectors:

$$(\text{row } 1) \cdot (\text{column } 1) = \begin{bmatrix} 3 & 4 \end{bmatrix} \begin{bmatrix} 2 \\ 1 \end{bmatrix} = 10 \qquad \text{is the } (1,1) \text{ entry of } AB$$

$$(\text{row } 2) \cdot (\text{column } 1) = \begin{bmatrix} 1 & 5 \end{bmatrix} \begin{bmatrix} 2 \\ 1 \end{bmatrix} = 7 \qquad \text{is the } (2,1) \text{ entry of } AB$$

The first columns of AB are $(10, 7, 4)$ and $(16, 9, 8)$. We need 6 dot products, 2 multiplications each, 12 in all $(3 \cdot 2 \cdot 2)$. The same AB comes from *columns of A times rows of B*:

$$AB = \begin{bmatrix} 3 \\ 1 \\ 2 \end{bmatrix} \begin{bmatrix} 2 & 4 \end{bmatrix} + \begin{bmatrix} 4 \\ 5 \\ 0 \end{bmatrix} \begin{bmatrix} 1 & 1 \end{bmatrix} = \begin{bmatrix} 6 & 12 \\ 2 & 4 \\ 4 & 8 \end{bmatrix} + \begin{bmatrix} 4 & 4 \\ 5 & 5 \\ 0 & 0 \end{bmatrix} = \begin{bmatrix} \mathbf{10} & \mathbf{16} \\ \mathbf{7} & \mathbf{9} \\ \mathbf{4} & \mathbf{8} \end{bmatrix}.$$

Problem Set 2.3

Problems 1–15 are about elimination matrices.

1 Write down the 3 by 3 matrices that produce these elimination steps:

(a) E_{21} subtracts 5 times row 1 from row 2.

(b) E_{32} subtracts -7 times row 2 from row 3.

(c) P exchanges rows 1 and 2, then rows 2 and 3.

2 In Problem 1, applying E_{21} and then E_{32} to the column $\boldsymbol{b} = (1, 0, 0)$ gives $E_{32}E_{21}\boldsymbol{b} =$ ____. Applying E_{32} before E_{21} gives $E_{21}E_{32}\boldsymbol{b} =$ ____. When E_{32} comes first, row ____ feels no effect from row ____.

3 Which three matrices E_{21}, E_{31}, E_{32} put A into triangular form U?

$$A = \begin{bmatrix} 1 & 1 & 0 \\ 4 & 6 & 1 \\ -2 & 2 & 0 \end{bmatrix} \quad \text{and} \quad E_{32}E_{31}E_{21}A = U.$$

Multiply those E's to get one matrix M that does elimination: $MA = U$.

4 Include $\boldsymbol{b} = (1, 0, 0)$ as a fourth column in Problem 3 to produce $[\,A \;\; \boldsymbol{b}\,]$. Carry out the elimination steps on this augmented matrix to solve $A\boldsymbol{x} = \boldsymbol{b}$.

5 Suppose $a_{33} = 7$ and the third pivot is 5. If you change a_{33} to 11, the third pivot is ____. If you change a_{33} to ____, there is no third pivot.

6 If every column of A is a multiple of $(1, 1, 1)$, then $A\boldsymbol{x}$ is always a multiple of $(1, 1, 1)$. Do a 3 by 3 example. How many pivots are produced by elimination?

7 Suppose E_{31} subtracts 7 times row 1 from row 3. To reverse that step you should ____ 7 times row ____ to row ____. This "inverse matrix" is $R_{31} =$ ____.

8 Suppose E_{31} subtracts 7 times row 1 from row 3. What matrix R_{31} is changed into I? Then $E_{31}R_{31} = I$ where Problem 7 has $R_{31}E_{31} = I$. Both are true!

9 (a) E_{21} subtracts row 1 from row 2 and then P_{23} exchanges rows 2 and 3. What matrix $M = P_{23}E_{21}$ does both steps at once?

(b) P_{23} exchanges rows 2 and 3 and then E_{31} subtracts row 1 from row 3. What matrix $M = E_{31}P_{23}$ does both steps at once? Explain why the M's are the same but the E's are different.

10 (a) What 3 by 3 matrix E_{13} will add row 3 to row 1?

(b) What matrix adds row 1 to row 3 and *at the same time* row 3 to row 1?

(c) What matrix adds row 1 to row 3 and *then* adds row 3 to row 1?

11 Create a matrix that has $a_{11} = a_{22} = a_{33} = 1$ but elimination produces two negative pivots without row exchanges. (The first pivot is 1.)

12 Multiply these matrices:

$$\begin{bmatrix} 0 & 0 & 1 \\ 0 & 1 & 0 \\ 1 & 0 & 0 \end{bmatrix} \begin{bmatrix} 1 & 2 & 3 \\ 4 & 5 & 6 \\ 7 & 8 & 9 \end{bmatrix} \begin{bmatrix} 0 & 0 & 1 \\ 0 & 1 & 0 \\ 1 & 0 & 0 \end{bmatrix} \qquad \begin{bmatrix} 1 & 0 & 0 \\ -1 & 1 & 0 \\ -1 & 0 & 1 \end{bmatrix} \begin{bmatrix} 1 & 2 & 3 \\ 1 & 3 & 1 \\ 1 & 4 & 0 \end{bmatrix}.$$

13 Explain these facts. If the third column of B is all zero, the third column of EB is all zero (for any E). If the third *row* of B is all zero, the third row of EB might *not* be zero.

14 This 4 by 4 matrix will need elimination matrices E_{21} and E_{32} and E_{43}. What are those matrices?

$$A = \begin{bmatrix} 2 & -1 & 0 & 0 \\ -1 & 2 & -1 & 0 \\ 0 & -1 & 2 & -1 \\ 0 & 0 & -1 & 2 \end{bmatrix}.$$

15 Write down the 3 by 3 matrix that has $a_{ij} = 2i - 3j$. This matrix has $a_{32} = 0$, but elimination still needs E_{32} to produce a zero in the 3, 2 position. Which previous step destroys the original zero and what is E_{32}?

Problems 16–23 are about creating and multiplying matrices.

16 Write these ancient problems in a 2 by 2 matrix form $A\boldsymbol{x} = \boldsymbol{b}$ and solve them:

(a) X is twice as old as Y and their ages add to 33.

(b) $(x, y) = (2, 5)$ and $(3, 7)$ lie on the line $y = mx + c$. Find m and c.

17 The parabola $y = a + bx + cx^2$ goes through the points $(x, y) = (1, 4)$ and $(2, 8)$ and $(3, 14)$. Find and solve a matrix equation for the unknowns (a, b, c).

18 Multiply these matrices in the orders EF and FE and E^2:

$$E = \begin{bmatrix} 1 & 0 & 0 \\ a & 1 & 0 \\ b & 0 & 1 \end{bmatrix} \qquad F = \begin{bmatrix} 1 & 0 & 0 \\ 0 & 1 & 0 \\ 0 & c & 1 \end{bmatrix}.$$

Also compute $E^2 = EE$ and $F^3 = FFF$.

19 Multiply these row exchange matrices in the orders PQ and QP and P^2:

$$P = \begin{bmatrix} 0 & 1 & 0 \\ 1 & 0 & 0 \\ 0 & 0 & 1 \end{bmatrix} \quad \text{and} \quad Q = \begin{bmatrix} 0 & 0 & 1 \\ 0 & 1 & 0 \\ 1 & 0 & 0 \end{bmatrix}.$$

Find four matrices whose squares are $M^2 = I$.

20 (a) Suppose all columns of B are the same. Then all columns of EB are the same, because each one is E times ______.

(b) Suppose all rows of B are $[\,1 \;\; 2 \;\; 4\,]$. Show by example that all rows of EB are *not* $[\,1 \;\; 2 \;\; 4\,]$. It is true that those rows are ______.

21 If E adds row 1 to row 2 and F adds row 2 to row 1, does EF equal FE?

22 The entries of A and $\boldsymbol{x}$ are a_{ij} and x_j. So the first component of $A\boldsymbol{x}$ is $\sum a_{1j}x_j = a_{11}x_1 + \cdots + a_{1n}x_n$. If E_{21} subtracts row 1 from row 2, write a formula for

(a) the third component of $A\boldsymbol{x}$

(b) the (2, 1) entry of $E_{21}A$

(c) the (2, 1) entry of $E_{21}(E_{21}A)$

(d) the first component of $EA\boldsymbol{x}$.

23 The elimination matrix $E = \begin{bmatrix} 1 & 0 \\ -2 & 1 \end{bmatrix}$ subtracts 2 times row 1 of A from row 2 of A. The result is EA. What is the effect of $E(EA)$? In the opposite order AE, we are subtracting 2 times ______ of A from ______. (Do examples.)

Problems 24–29 include the column $\boldsymbol{b}$ in the augmented matrix $[\,A \;\; \boldsymbol{b}\,]$.

24 Apply elimination to the 2 by 3 augmented matrix $[\,A \;\; \boldsymbol{b}\,]$. What is the triangular system $U\boldsymbol{x} = \boldsymbol{c}$? What is the solution $\boldsymbol{x}$?

$$A\boldsymbol{x} = \begin{bmatrix} 2 & 3 \\ 4 & 1 \end{bmatrix} \begin{bmatrix} x_1 \\ x_2 \end{bmatrix} = \begin{bmatrix} 1 \\ 17 \end{bmatrix}.$$

25 Apply elimination to the 3 by 4 augmented matrix $[\,A \;\; \boldsymbol{b}\,]$. How do you know this system has no solution? Change the last number 6 so there *is* a solution.

$$A\boldsymbol{x} = \begin{bmatrix} 1 & 2 & 3 \\ 2 & 3 & 4 \\ 3 & 5 & 7 \end{bmatrix} \begin{bmatrix} x \\ y \\ z \end{bmatrix} = \begin{bmatrix} 1 \\ 2 \\ 6 \end{bmatrix}.$$

26 The equations $A\boldsymbol{x} = \boldsymbol{b}$ and $A\boldsymbol{x}^* = \boldsymbol{b}^*$ have the same matrix A. What double augmented matrix should you use in elimination to solve both equations at once?

Solve both of these equations by working on a 2 by 4 matrix:

$$\begin{bmatrix} 1 & 4 \\ 2 & 7 \end{bmatrix} \begin{bmatrix} x \\ y \end{bmatrix} = \begin{bmatrix} 1 \\ 0 \end{bmatrix} \quad \text{and} \quad \begin{bmatrix} 1 & 4 \\ 2 & 7 \end{bmatrix} \begin{bmatrix} u \\ v \end{bmatrix} = \begin{bmatrix} 0 \\ 1 \end{bmatrix}.$$

27 Choose the numbers a, b, c, d in this augmented matrix so that there is (a) no solution (b) infinitely many solutions.

$$\begin{bmatrix} A & \boldsymbol{b} \end{bmatrix} = \begin{bmatrix} 1 & 2 & 3 & a \\ 0 & 4 & 5 & b \\ 0 & 0 & d & c \end{bmatrix}$$

Which of the numbers a, b, c, or d have no effect on the solvability?

28 If $AB = I$ and $BC = I$ use the associative law to prove $A = C$.

29 Choose two matrices $M = \begin{bmatrix} a & b \\ c & d \end{bmatrix}$ with $\det M = ad - bc = 1$ and with a, b, c, d positive integers. Prove that every such matrix M either has

$$\text{EITHER} \quad \text{row } 1 \le \text{row } 2 \quad \text{OR} \quad \text{row } 2 \le \text{row } 1 .$$

Subtraction makes $\begin{bmatrix} 1 & 0 \\ -1 & 1 \end{bmatrix} M$ or $\begin{bmatrix} 1 & -1 \\ 0 & 1 \end{bmatrix} M$ nonnegative but smaller than M. If you continue and reach I, write your M's as products of the inverses $\begin{bmatrix} 1 & 0 \\ 1 & 1 \end{bmatrix}$ and $\begin{bmatrix} 1 & 1 \\ 0 & 1 \end{bmatrix}$.

30 Find the triangular matrix E that reduces "*Pascal's matrix*" to a smaller Pascal:

$$E \begin{bmatrix} 1 & 0 & 0 & 0 \\ 1 & 1 & 0 & 0 \\ 1 & 2 & 1 & 0 \\ 1 & 3 & 3 & 1 \end{bmatrix} = \begin{bmatrix} 1 & 0 & 0 & 0 \\ 0 & 1 & 0 & 0 \\ 0 & 1 & 1 & 0 \\ 0 & 1 & 2 & 1 \end{bmatrix}.$$

Challenge question: Which M (from several E's) reduces Pascal all the way to I?

RULES FOR MATRIX OPERATIONS ■ 2.4

I will start with basic facts. A matrix is a rectangular array of numbers or "entries." When A has m rows and n columns, it is an "m by n" matrix. Matrices can be added if their shapes are the same. They can be multiplied by any constant c. Here are examples of $A + B$ and $2A$, for 3 by 2 matrices:

$$\begin{bmatrix} 1 & 2 \\ 3 & 4 \\ 0 & 0 \end{bmatrix} + \begin{bmatrix} 2 & 2 \\ 4 & 4 \\ 9 & 9 \end{bmatrix} = \begin{bmatrix} 3 & 4 \\ 7 & 8 \\ 9 & 9 \end{bmatrix} \quad \text{and} \quad 2 \begin{bmatrix} 1 & 2 \\ 3 & 4 \\ 0 & 0 \end{bmatrix} = \begin{bmatrix} 2 & 4 \\ 6 & 8 \\ 0 & 0 \end{bmatrix}.$$

Matrices are added exactly as vectors are—one entry at a time. We could even regard a column vector as a matrix with only one column (so $n = 1$). The matrix $-A$ comes from multiplication by $c = -1$ (reversing all the signs). Adding A to $-A$ leaves the *zero matrix*, with all entries zero.

The 3 by 2 zero matrix is different from the 2 by 3 zero matrix. Even zero has a shape (several shapes) for matrices. All this is only common sense.

The entry in row i and column j is called a_{ij} ***or*** $A(i, j)$. The n entries along the first row are $a_{11}, a_{12}, \ldots, a_{1n}$. The lower left entry in the matrix is a_{m1} and the lower right is a_{mn}. The row number i goes from 1 to m. The column number j goes from 1 to n.

Matrix addition is easy. The serious question is ***matrix multiplication.*** When can we multiply A times B, and what is the product AB? We cannot multiply when A and B are 3 by 2. They don't pass the following test:

To multiply AB: ***If A has n columns, B must have n rows.***

If A has two columns, B must have two rows. When A is 3 by 2, the matrix B can be 2 by 1 (a vector) or 2 by 2 (square) or 2 by 20. Every column of B is ready to be multiplied by A. Then AB is 3 by 1 (a vector) or 3 by 2 or 3 by 20.

Suppose A is m by n and B is n by p. We can multiply. The product AB is m by p.

$$\begin{bmatrix} \textbf{m rows} \\ n\ columns \end{bmatrix} \begin{bmatrix} n\ rows \\ \textbf{p columns} \end{bmatrix} = \begin{bmatrix} \textbf{m rows} \\ \textbf{p columns} \end{bmatrix}.$$

A row times a column is an extreme case. Then 1 by n multiplies n by 1. The result is 1 by 1. That single number is the "dot product."

In every case AB is filled with dot products. For the top corner, the (1, 1) entry of AB is (row 1 of A) · (column 1 of B). To multiply matrices, take all these dot products: (***each row of*** A) · (***each column of*** B).

2F ***The entry in row i and column j of AB is*** (row i of A) · (column j of B) .

Figure 2.8 picks out the second row ($i = 2$) of a 4 by 5 matrix A. It picks out the third column ($j = 3$) of a 5 by 6 matrix B. Their dot product goes into row 2 and column 3 of AB. The matrix AB has *as many rows as* A (4 rows), and *as many columns as* B.

$$\begin{bmatrix} * & & \\ a_{i1} \ a_{i2} & \cdots & a_{i5} \\ * & & \\ * & & \end{bmatrix} \begin{bmatrix} * & * & b_{1j} & * & * & * \\ & & b_{2j} & & & \\ & & \vdots & & & \\ & & b_{5j} & & & \end{bmatrix} = \begin{bmatrix} & & * & & & \\ * & * & (AB)_{ij} & * & * & * \\ & & * & & & \\ & & * & & & \end{bmatrix}$$

A is 4 by 5 $\qquad$ B is 5 by 6 $\qquad$ AB is 4 by 6

Figure 2.8 Here $i = 2$ and $j = 3$. Then $(AB)_{23}$ is (row 2) · (column 3) $= \Sigma a_{2k} b_{k3}$.

Example 1 Square matrices can be multiplied if and only if they have the same size:

$$\begin{bmatrix} 1 & 1 \\ 2 & -1 \end{bmatrix} \begin{bmatrix} 2 & 2 \\ 3 & 4 \end{bmatrix} = \begin{bmatrix} 5 & 6 \\ 1 & 0 \end{bmatrix}.$$

The first dot product is $1 \cdot 2 + 1 \cdot 3 = 5$. Three more dot products give 6, 1, and 0. Each dot product requires two multiplications—thus eight in all.

If A and B are n by n, so is AB. It contains n^2 dot products, row of A times column of B. Each dot product needs n multiplications, so ***the computation of*** AB ***uses*** n^3 ***separate multiplications***. For $n = 100$ we multiply a million times. For $n = 2$ we have $n^3 = 8$.

Mathematicians thought until recently that AB absolutely needed $2^3 = 8$ multiplications. Then somebody found a way to do it with 7 (and extra additions). By breaking n by n matrices into 2 by 2 blocks, this idea also reduced the count for large matrices. Instead of n^3 it went below $n^{2.8}$, and the exponent keeps falling.[1] The best

[1]Maybe the exponent won't stop falling before 2. No number in between looks special.

at this moment is $n^{2.376}$. But the algorithm is so awkward that scientific computing is done the regular way: n^2 dot products in AB, and n multiplications for each one.

Example 2 Suppose A is a row vector (1 by 3) and B is a column vector (3 by 1). Then AB is 1 by 1 (only one entry, the dot product). On the other hand B times A (***a column times a row***) is a full 3 by 3 matrix. This multiplication is allowed!

$$\textbf{\textit{Column times row:}} \quad \begin{bmatrix} 0 \\ 1 \\ 2 \end{bmatrix} \begin{bmatrix} 1 & 2 & 3 \end{bmatrix} = \begin{bmatrix} 0 & 0 & 0 \\ 1 & 2 & 3 \\ 2 & 4 & 6 \end{bmatrix}.$$

A row times a column is an *"inner"* product—that is another name for dot product. A column times a row is an *"outer"* product. These are extreme cases of matrix multiplication, with very thin matrices. They follow the rule for shapes in multiplication: (n by 1) times (1 by n). The product of column times row is n by n.

Example 3 will show how to multiply AB ***using columns times rows.***

Rows and Columns of AB

In the big picture, A multiplies each column of B. The result is a column of AB. In that column, we are combining the columns of A. ***Each column of AB is a combination of the columns of A.*** That is the column picture of matrix multiplication:

$$\textbf{\textit{Column of}}\ AB \text{ is } (\textit{matrix } A)\ \textbf{\textit{times}}\ (\textit{column of } B).$$

The row picture is reversed. Each row of A multiplies the whole matrix B. The result is a row of AB. It is a combination of the rows of B:

$$\begin{bmatrix} \text{row } i \text{ of } A \end{bmatrix} \begin{bmatrix} 1 & 2 & 3 \\ 4 & 5 & 6 \\ 7 & 8 & 9 \end{bmatrix} = \begin{bmatrix} \text{row } i \text{ of } AB \end{bmatrix}.$$

We see row operations in elimination (E times A). We see columns in A times $\boldsymbol{x}$. The "row-column picture" has the dot products of rows with columns. Believe it or not, ***there is also a "column-row picture."*** Not everybody knows that columns $1, \ldots, n$ of A multiply rows $1, \ldots, n$ of B and add up to the same answer AB.

The Laws for Matrix Operations

May I put on record six laws that matrices do obey, while emphasizing an equation they don't obey? The matrices can be square or rectangular, and the laws involving $A + B$ are all simple and all obeyed. Here are three addition laws:

$$\begin{aligned} A + B &= B + A && \text{(commutative law)} \\ c(A + B) &= cA + cB && \text{(distributive law)} \\ A + (B + C) &= (A + B) + C && \text{(associative law).} \end{aligned}$$

Three more laws hold for multiplication, but $AB = BA$ is not one of them:

$$
\begin{aligned}
AB &\neq BA && \text{(the commutative “law” is \textit{usually broken})}\\
C(A+B) &= CA + CB && \text{(distributive law from the left)}\\
(A+B)C &= AC + BC && \text{(distributive law from the right)}\\
A(BC) &= (AB)C && \text{(associative law for } ABC\text{) (\textbf{\textit{parentheses not needed}}).}
\end{aligned}
$$

When A and B are not square, AB is a different size from BA. These matrices can't be equal—even if both multiplications are allowed. For square matrices, almost any example shows that AB is different from BA:

$$AB = \begin{bmatrix} 0 & 0 \\ 1 & 0 \end{bmatrix}\begin{bmatrix} 0 & 1 \\ 0 & 0 \end{bmatrix} = \begin{bmatrix} 0 & 0 \\ 0 & 1 \end{bmatrix} \quad \text{but} \quad BA = \begin{bmatrix} 0 & 1 \\ 0 & 0 \end{bmatrix}\begin{bmatrix} 0 & 0 \\ 1 & 0 \end{bmatrix} = \begin{bmatrix} 1 & 0 \\ 0 & 0 \end{bmatrix}.$$

It is true that $AI = IA$. All square matrices commute with I and also with cI. Only these matrices cI commute with all other matrices.

The law $A(B+C) = AB + AC$ is proved a column at a time. Start with $A(\boldsymbol{b} + \boldsymbol{c}) = A\boldsymbol{b} + A\boldsymbol{c}$ for the first column. That is the key to everything—***linearity***. Say no more.

The law $A(BC) = (AB)C$ ***means that you can multiply*** BC ***first or*** AB ***first.*** The direct proof is sort of awkward (Problem 16) but this law is extremely useful. We highlighted it above; it is the key to the way we multiply matrices.

Look at the special case when $A = B = C =$ square matrix. Then (A *times* A^2) $=$ (A^2 *times* A). The product in either order is A^3. The matrix powers A^p follow the same rules as numbers:

$$A^p = AAA \cdots A \ (p \text{ factors}) \qquad (A^p)(A^q) = A^{p+q} \qquad (A^p)^q = A^{pq}.$$

Those are the ordinary laws for exponents. A^3 times A^4 is A^7 (seven factors). A^3 to the fourth power is A^{12} (twelve A's). When p and q are zero or negative these rules still hold, provided A has a "-1 power"—which is the *inverse matrix* A^{-1}. Then $A^0 = I$ is the identity matrix (no factors).

For a number, a^{-1} is $1/a$. For a matrix, the inverse is written A^{-1}. (It is *never* I/A, except this is allowed in MATLAB.) Every number has an inverse except $a = 0$. To decide when A has an inverse is a central problem in linear algebra. Section 2.5 will start on the answer. This section is a Bill of Rights for matrices, to say when A and B can be multiplied and how.

Block Matrices and Block Multiplication

We have to say one more thing about matrices. They can be cut into ***blocks*** (which are smaller matrices). This often happens naturally. Here is a 4 by 6 matrix broken into blocks of size 2 by 2—and each block is just I:

$$A = \left[\begin{array}{cc|cc|cc} 1 & 0 & 1 & 0 & 1 & 0 \\ 0 & 1 & 0 & 1 & 0 & 1 \\ \hline 1 & 0 & 1 & 0 & 1 & 0 \\ 0 & 1 & 0 & 1 & 0 & 1 \end{array}\right] = \begin{bmatrix} I & I & I \\ I & I & I \end{bmatrix}.$$

If B is also 4 by 6 and its block sizes match the block sizes in A, you can add $A + B$ *a block at a time*.

We have seen block matrices before. The right side vector $\boldsymbol{b}$ was placed next to A in the "augmented matrix." Then $[\, A \quad \boldsymbol{b} \,]$ has two blocks of different sizes. Multiplying by an elimination matrix gave $[\, EA \quad E\boldsymbol{b} \,]$. No problem to multiply blocks times blocks, when their shapes permit:

2G Block multiplication If the cuts between columns of A match the cuts between rows of B, then block multiplication of AB is allowed:

$$\begin{bmatrix} A_{11} & A_{12} \\ A_{21} & A_{22} \end{bmatrix} \begin{bmatrix} B_{11} & \cdots \\ B_{21} & \cdots \end{bmatrix} = \begin{bmatrix} A_{11}B_{11} + A_{12}B_{21} & \cdots \\ A_{21}B_{11} + A_{22}B_{21} & \cdots \end{bmatrix}. \tag{1}$$

This equation is the same as if the blocks were numbers (which are 1 by 1 blocks). We are careful to keep A's in front of B's, because BA can be different. The cuts between rows of A give cuts between rows of AB. Any column cuts in B are also column cuts in AB.

Main point When matrices split into blocks, it is often simpler to see how they act. The block matrix of I's above is much clearer than the original 4 by 6 matrix A.

Example 3 **(Important special case)** Let the blocks of A be its n columns. Let the blocks of B be its n rows. Then block multiplication AB adds up ***columns times rows***:

$$AB = \begin{bmatrix} | & & | \\ \boldsymbol{a}_1 & \cdots & \boldsymbol{a}_n \\ | & & | \end{bmatrix} \begin{bmatrix} - & \boldsymbol{b}_1 & - \\ & \vdots & \\ - & \boldsymbol{b}_n & - \end{bmatrix} = \begin{bmatrix} \boldsymbol{a}_1\boldsymbol{b}_1 + \cdots + \boldsymbol{a}_n\boldsymbol{b}_n \end{bmatrix}. \tag{2}$$

This is another way to multiply matrices! Compare it with the usual rows times columns. Row 1 of A times column 1 of B gave the $(1, 1)$ entry in AB. Now *column* 1 of A

times *row* 1 of B gives a full matrix—not just a single number. Look at this example:

$$\begin{bmatrix}1 & 4\\1 & 5\end{bmatrix}\begin{bmatrix}3 & 2\\1 & 0\end{bmatrix} = \begin{bmatrix}1\\1\end{bmatrix}\begin{bmatrix}3 & 2\end{bmatrix}+\begin{bmatrix}4\\5\end{bmatrix}\begin{bmatrix}1 & 0\end{bmatrix}$$

$$= \begin{bmatrix}3 & 2\\3 & 2\end{bmatrix}+\begin{bmatrix}4 & 0\\5 & 0\end{bmatrix}. \qquad (3)$$

We stop there so you can see columns multiplying rows. If a 2 by 1 matrix (a column) multiplies a 1 by 2 matrix (a row), the result is 2 by 2. That is what we found. Dot products are "inner products," these are "outer products."

When you add the two matrices at the end of equation (3), you get the correct answer AB. In the top left corner the answer is $3 + 4 = 7$. This agrees with the row-column dot product of $(1, 4)$ with $(3, 1)$.

Summary The usual way, rows times columns, gives four dot products (8 multiplications). The new way, columns times rows, gives two full matrices (8 multiplications). The eight multiplications, and also the four additions, are all the same. You just execute them in a different order.

Example 4 (Elimination by blocks) Suppose the first column of A contains $1, 3, 4$. To change 3 and 4 to 0 and 0, multiply the pivot row by 3 and 4 and subtract. Those row operations are really multiplications by elimination matrices E_{21} and E_{31}:

$$E_{21} = \begin{bmatrix}1 & 0 & 0\\-3 & 1 & 0\\0 & 0 & 1\end{bmatrix} \quad\text{and}\quad E_{31} = \begin{bmatrix}1 & 0 & 0\\0 & 1 & 0\\-4 & 0 & 1\end{bmatrix}.$$

The "block idea" is to do both eliminations with one matrix E. That matrix clears out the whole first column of A below the pivot $a = 2$:

$$E = \begin{bmatrix}1 & 0 & 0\\-3 & 1 & 0\\-4 & 0 & 1\end{bmatrix} \quad\text{multiplies}\quad \begin{bmatrix}\mathbf{1} & x & x\\\mathbf{3} & x & x\\\mathbf{4} & x & x\end{bmatrix} \quad\text{to give}\quad EA = \begin{bmatrix}\mathbf{1} & x & x\\\mathbf{0} & x & x\\\mathbf{0} & x & x\end{bmatrix}.$$

Block multiplication gives a formula for EA. The matrix A has four blocks a, $\boldsymbol{b}$, $\boldsymbol{c}$, D: the pivot, the rest of row 1, the rest of column 1, and the rest of the matrix. Watch how E multiplies A by blocks:

$$EA = \left[\begin{array}{c|c}1 & \mathbf{0}\\\hline -\boldsymbol{c}/a & I\end{array}\right]\left[\begin{array}{c|c}a & \boldsymbol{b}\\\hline \boldsymbol{c} & D\end{array}\right] = \left[\begin{array}{c|c}a & \boldsymbol{b}\\\hline \mathbf{0} & D - \boldsymbol{c}\boldsymbol{b}/a\end{array}\right]. \qquad (4)$$

Elimination multiplies the first row $[\,a \;\; \boldsymbol{b}\,]$ by $\boldsymbol{c}/a$. It subtracts from $\boldsymbol{c}$ to get zeros in the first column. It subtracts from D to get $D - \boldsymbol{c}\boldsymbol{b}/a$. This is ordinary elimination, a column at a time—written in blocks.

■ REVIEW OF THE KEY IDEAS ■

1. The (i, j) entry of AB is (row i of A) · (column j of B).
2. An m by n matrix times an n by p matrix uses mnp separate multiplications.
3. A times BC equals AB times C (surprisingly important).
4. AB is also the sum of these matrices: (column j of A) times (row j of B).
5. Block multiplication is allowed when the block shapes match correctly.

■ WORKED EXAMPLES ■

2.4 A Put yourself in the position of the author! I want to show you matrix multiplications that are *special*, but mostly I am stuck with small matrices. There is one terrific family of **Pascal matrices**, and they come in all sizes, and above all they have real meaning. I think 4 by 4 is a good size to show some of their amazing patterns.

Here is the lower triangular Pascal matrix L. Its entries come from "*Pascal's triangle*". I will multiply L times the **ones** vector, and the **powers** vector:

$$\textbf{Pascal matrix} \quad \begin{bmatrix} 1 & & & \\ 1 & 1 & & \\ 1 & 2 & 1 & \\ 1 & 3 & 3 & 1 \end{bmatrix} \begin{bmatrix} \mathbf{1} \\ \mathbf{1} \\ \mathbf{1} \\ \mathbf{1} \end{bmatrix} = \begin{bmatrix} \mathbf{1} \\ \mathbf{2} \\ \mathbf{4} \\ \mathbf{8} \end{bmatrix} \qquad \begin{bmatrix} 1 & & & \\ 1 & 1 & & \\ 1 & 2 & 1 & \\ 1 & 3 & 3 & 1 \end{bmatrix} \begin{bmatrix} \mathbf{1} \\ \boldsymbol{x} \\ \boldsymbol{x}^2 \\ \boldsymbol{x}^3 \end{bmatrix} = \begin{bmatrix} \mathbf{1} \\ \mathbf{1+x} \\ \mathbf{(1+x)^2} \\ \mathbf{(1+x)^3} \end{bmatrix}.$$

Each row of L leads to the next row: *Add an entry to the one on its left to get the entry below.* In symbols $\ell_{i\,j} + \ell_{i\,j-1} = \ell_{i+1\,j}$. The numbers after $1, 3, 3, 1$ would be $1, 4, 6, 4, 1$. Pascal lived in the 1600's, long before matrices, but his triangle fits perfectly into L.

Multiplying by **ones** is the same as adding up each row, to get powers of 2. In fact **powers** = **ones** when $x = 1$. By writing out the last rows of L times **powers**, you see the entries of L as the "binomial coefficients" that are so essential to gamblers:

$$\mathbf{1} + \mathbf{2}x + \mathbf{1}x^2 = (1+x)^2 \qquad \mathbf{1} + \mathbf{3}x + \mathbf{3}x^2 + \mathbf{1}x^3 = (1+x)^3$$

The number "3" counts the ways to get Heads once and Tails twice in three coin flips: HTT and THT and TTH. The other "3" counts the ways to get Heads twice: HHT and HTH and THH. Those are examples of "i choose j" = the number of ways to get j heads in i coin flips. That number is exactly ℓ_{ij}, if we start counting rows and columns of L at $i = 0$ and $j = 0$ (and remember $0! = 1$):

$$\ell_{ij} = \binom{i}{j} = \boldsymbol{i \text{ choose } j} = \frac{i!}{j!\,(i-j)!} \qquad \binom{4}{2} = \frac{4!}{2!\,2!} = 6$$

There are six ways to choose two aces out of four aces. We will see Pascal's triangle and these matrices again. Here are the questions I want to ask now:

1. What is $H = L^2$? This is the "hypercube matrix".
2. Multiply H times **ones** and **powers**.
3. The last row of H is $8, 12, 6, 1$. A cube has 8 corners, 12 edges, 6 faces, 1 box. *What would the next row of H tell about a hypercube in* **4D**?

Solution Multiply L times L to get the hypercube matrix $H = L^2$:

$$\begin{bmatrix} 1 & & & \\ 1 & 1 & & \\ 1 & 2 & 1 & \\ 1 & 3 & 3 & 1 \end{bmatrix}\begin{bmatrix} 1 & & & \\ 1 & 1 & & \\ 1 & 2 & 1 & \\ 1 & 3 & 3 & 1 \end{bmatrix} = \begin{bmatrix} \mathbf{1} & & & \\ \mathbf{2} & \mathbf{1} & & \\ \mathbf{4} & \mathbf{4} & \mathbf{1} & \\ \mathbf{8} & \mathbf{12} & \mathbf{6} & \mathbf{1} \end{bmatrix} = H.$$

Now multiply H times the vectors of **ones** and **powers**:

$$\begin{bmatrix} 1 & & & \\ 2 & 1 & & \\ 4 & 4 & 1 & \\ 8 & 12 & 6 & 1 \end{bmatrix}\begin{bmatrix} 1 \\ 1 \\ 1 \\ 1 \end{bmatrix} = \begin{bmatrix} \mathbf{1} \\ \mathbf{3} \\ \mathbf{9} \\ \mathbf{27} \end{bmatrix} \qquad \begin{bmatrix} 1 & & & \\ 2 & 1 & & \\ 4 & 4 & 1 & \\ 8 & 12 & 6 & 1 \end{bmatrix}\begin{bmatrix} 1 \\ x \\ x^2 \\ x^3 \end{bmatrix} = \begin{bmatrix} \mathbf{1} \\ \mathbf{2+x} \\ \mathbf{(2+x)^2} \\ \mathbf{(2+x)^3} \end{bmatrix}$$

If $x = 1$ we get the powers of 3. If $x = 0$ we get powers of 2 (where do $1, 2, 4, 8$ appear in H?). Where L changed x to $1+x$, applying L again changes $1+x$ to $2+x$.

How do the rows of H count corners and edges and faces of a cube? A square in **2D** has 4 corners, 4 edges, 1 face. Add one dimension at a time:

Connect two squares to get a **3D** *cube. Connect two cubes to get a* **4D** *hypercube.*

The cube has 8 corners and 12 edges: 4 edges in each square and 4 between the squares. The cube has 6 faces: 1 in each square and 4 faces between the squares. This row $8, 12, 6, 1$ of H will lead to the next row (one more dimension) by $\mathbf{2}h_{i\,j} + h_{i\,j-1} = h_{i+1\,j}$.

Can you see this in four dimensions? The hypercube has 16 corners, no problem. It has 12 edges from one cube, 12 from the other cube, 8 that connect corners between those cubes: total $2 \times 12 + 8 = 32$ edges. It has 6 faces from each separate cube and 12 more from connecting pairs of edges: total $2 \times 6 + 12 = 24$ faces. It has one box from each cube and 6 more from connecting pairs of faces: total $2 \times 1 + 6 = 8$ boxes. And sure enough, the next row of H is $16, 32, 24, 8, 1$.

2.4 B For these matrices, when does $AB = BA$? When does $BC = CB$? When does A times BC equal AB times C? Give the conditions on their entries p, q, r, z:

$$A = \begin{bmatrix} p & 0 \\ q & r \end{bmatrix} \qquad B = \begin{bmatrix} 1 & 1 \\ 0 & 1 \end{bmatrix} \qquad C = \begin{bmatrix} 0 & z \\ 0 & 0 \end{bmatrix}$$

If $p, q, r, 1, z$ are 4 by 4 blocks instead of numbers, do the answers change?

Solution First of all, A times BC *always* equals AB times C. We don't need parentheses in $A(BC) = (AB)C = ABC$. But we do need to keep the matrices in this order A, B, C. Compare AB with BA:

$$AB = \begin{bmatrix} p & p \\ q & q+r \end{bmatrix} \qquad BA = \begin{bmatrix} p+q & r \\ q & r \end{bmatrix}.$$

We only have $AB = BA$ if $q = 0$ and $p = r$. Now compare BC with CB:

$$BC = \begin{bmatrix} 0 & z \\ 0 & 0 \end{bmatrix} \qquad CB = \begin{bmatrix} 0 & z \\ 0 & 0 \end{bmatrix}.$$

B and C happen to commute. One explanation is that the diagonal part of B is I, which commutes with all 2 by 2 matrices. The off-diagonal part of B looks exactly like C (except for a scalar factor z) and every matrix commutes with itself.

When p, q, r, z are 4 by 4 blocks and 1 changes to the 4 by 4 identity matrix, all these products remain correct. So the answers are the same. (If the I's in B were changed to blocks t, t, t, then BC would have the block tz and CB would have the block zt. Those would normally be different—the order is important in block multiplication.)

2.4 C A **directed graph** starts with n nodes. There are n^2 possible edges—each edge leaves one of the n nodes and enters one of the n nodes (possibly itself). The n by n **adjacency matrix** has $a_{ij} = 1$ when an edge leaves node i and enters node j; if no edge then $a_{ij} = 0$. Here are two directed graphs and their adjacency matrices:

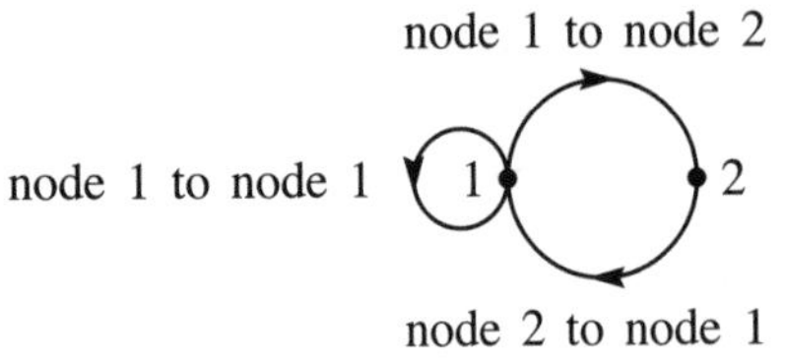

$$A = \begin{bmatrix} 1 & 1 \\ 1 & 0 \end{bmatrix}$$

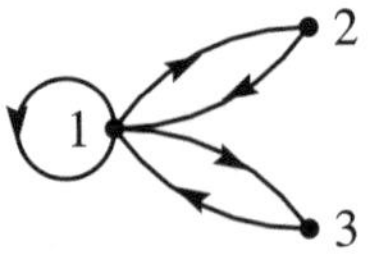

$$\begin{bmatrix} 1 & 1 & 1 \\ 1 & 0 & 0 \\ 1 & 0 & 0 \end{bmatrix}$$

The $\boldsymbol{i, j}$ ***entry of*** $\boldsymbol{A^2}$ ***is*** $\boldsymbol{a_{i1}a_{1j} + \cdots + a_{in}a_{nj}}$. Why does that sum count the *two-step paths* from i to any node to j? The i, j entry of A^k counts k-step paths:

$$\begin{bmatrix} 1 & 1 \\ 1 & 0 \end{bmatrix}^2 = \begin{bmatrix} 2 & 1 \\ 1 & 1 \end{bmatrix} \quad \begin{matrix}\text{counts the paths} \\ \text{with two edges}\end{matrix} \quad \begin{bmatrix} \text{1 to 2 to 1, 1 to 1 to 1} & \text{1 to 1 to 2} \\ \text{2 to 1 to 1} & \text{2 to 1 to 2} \end{bmatrix}$$

List all of the 3-step paths between each pair of nodes and compare with A^3. When A^k has ***no zeros***, that number k is the ***diameter*** of the graph—the number of edges needed to connect the most distant pair of nodes. What is the diameter of the second graph?

Solution The number $a_{ik}a_{kj}$ will be "1" if there is an edge from node i to k and an edge from k to j. This is a 2-step path. The number $a_{ik}a_{kj}$ will be "0" if either of those edges (i to k, k to j) is missing. So the sum of $a_{ik}a_{kj}$ is the number of 2-step paths leaving i and entering j. Matrix multiplication is just right for this count.

The 3-step paths are counted by A^3; we look at paths to node 2:

$$A^3 = \begin{bmatrix} 3 & \mathbf{2} \\ 2 & \mathbf{1} \end{bmatrix} \quad \text{counts the paths with three steps} \quad \begin{bmatrix} \cdots & 1 \text{ to } 1 \text{ to } 1 \text{ to } 2,\ 1 \text{ to } 2 \text{ to } 1 \text{ to } 2 \\ \cdots & 2 \text{ to } 1 \text{ to } 1 \text{ to } 2 \end{bmatrix}$$

These A^k contain the Fibonacci numbers $0, 1, 1, 2, 3, 5, 8, 13, \ldots$ coming in Section 6.2. Fibonacci's rule $F_{k+2} = F_{k+1} + F_k$ (as in $13 = 8 + 5$) shows up in $(A)(A^k) = A^{k+1}$:

$$\begin{bmatrix} 1 & 1 \\ 1 & 0 \end{bmatrix} \begin{bmatrix} F_{k+1} & F_k \\ F_k & F_{k-1} \end{bmatrix} = \begin{bmatrix} F_{k+2} & F_{k+1} \\ F_{k+1} & F_k \end{bmatrix} = A^{k+1}.$$

There are 13 six-step paths from node 1 to node 1, but I can't find them all.

A^k also counts words. A path like 1 to 1 to 2 to 1 corresponds to the number 1121 or the word **aaba**. The number 2 (the letter **b**) is not allowed to repeat because the graph has no edge from node 2 to node 2. The i, j entry of A^k counts the allowed numbers (or words) of length $k + 1$ that start with the ith letter and end with the jth.

The second graph also has diameter 2; *A^2 has no zeros.*

Problem Set 2.4

Problems 1–17 are about the laws of matrix multiplication.

1 A is 3 by 5, B is 5 by 3, C is 5 by 1, and D is 3 by 1. *All entries are* 1. Which of these matrix operations are allowed, and what are the results?

$$BA \qquad AB \qquad ABD \qquad DBA \qquad A(B + C).$$

2 What rows or columns or matrices do you multiply to find

(a) the third column of AB?

(b) the first row of AB?

(c) the entry in row 3, column 4 of AB?

(d) the entry in row 1, column 1 of CDE?

3 Add AB to AC and compare with $A(B + C)$:

$$A = \begin{bmatrix} 1 & 5 \\ 2 & 3 \end{bmatrix} \quad \text{and} \quad B = \begin{bmatrix} 0 & 2 \\ 0 & 1 \end{bmatrix} \quad \text{and} \quad C = \begin{bmatrix} 3 & 1 \\ 0 & 0 \end{bmatrix}.$$

4 In Problem 3, multiply A times BC. Then multiply AB times C.

5 Compute A^2 and A^3. Make a prediction for A^5 and A^n:

$$A = \begin{bmatrix} 1 & b \\ 0 & 1 \end{bmatrix} \quad \text{and} \quad A = \begin{bmatrix} 2 & 2 \\ 0 & 0 \end{bmatrix}.$$

6 Show that $(A+B)^2$ is different from $A^2+2AB+B^2$, when

$$A=\begin{bmatrix}1&2\\0&0\end{bmatrix} \quad\text{and}\quad B=\begin{bmatrix}1&0\\3&0\end{bmatrix}.$$

Write down the correct rule for $(A+B)(A+B)=A^2+$ _____ $+B^2$.

7 True or false. Give a specific example when false:

(a) If columns 1 and 3 of B are the same, so are columns 1 and 3 of AB.

(b) If rows 1 and 3 of B are the same, so are rows 1 and 3 of AB.

(c) If rows 1 and 3 of A are the same, so are rows 1 and 3 of ABC.

(d) $(AB)^2=A^2B^2$.

8 How is each row of DA and EA related to the rows of A, when

$$D=\begin{bmatrix}3&0\\0&5\end{bmatrix} \quad\text{and}\quad E=\begin{bmatrix}0&1\\0&1\end{bmatrix} \quad\text{and}\quad A=\begin{bmatrix}a&b\\c&d\end{bmatrix}?$$

How is each column of AD and AE related to the columns of A?

9 Row 1 of A is added to row 2. This gives EA below. Then column 1 of EA is added to column 2 to produce $(EA)F$:

$$EA=\begin{bmatrix}1&0\\1&1\end{bmatrix}\begin{bmatrix}a&b\\c&d\end{bmatrix}=\begin{bmatrix}a&b\\a+c&b+d\end{bmatrix}$$

$$\text{and}\quad (EA)F=(EA)\begin{bmatrix}1&1\\0&1\end{bmatrix}=\begin{bmatrix}a&a+b\\a+c&a+c+b+d\end{bmatrix}.$$

(a) Do those steps in the opposite order. First add column 1 of A to column 2 by AF, then add row 1 of AF to row 2 by $E(AF)$.

(b) Compare with $(EA)F$. What law is obeyed by matrix multiplication?

10 Row 1 of A is again added to row 2 to produce EA. Then F adds row 2 of EA to row 1. The result is $F(EA)$:

$$F(EA)=\begin{bmatrix}1&1\\0&1\end{bmatrix}\begin{bmatrix}a&b\\a+c&b+d\end{bmatrix}=\begin{bmatrix}2a+c&2b+d\\a+c&b+d\end{bmatrix}.$$

(a) Do those steps in the opposite order: first add row 2 to row 1 by FA, then add row 1 of FA to row 2.

(b) What law is or is not obeyed by matrix multiplication?

11 (3 by 3 matrices) Choose the only B so that for every matrix A

(a) $BA=4A$

(b) $BA=4B$

(c) BA has rows 1 and 3 of A reversed and row 2 unchanged

(d) All rows of BA are the same as row 1 of A.

12 Suppose $AB = BA$ and $AC = CA$ for these two particular matrices B and C:

$$A = \begin{bmatrix} a & b \\ c & d \end{bmatrix} \quad \text{commutes with} \quad B = \begin{bmatrix} 1 & 0 \\ 0 & 0 \end{bmatrix} \quad \text{and} \quad C = \begin{bmatrix} 0 & 1 \\ 0 & 0 \end{bmatrix}.$$

Prove that $a = d$ and $b = c = 0$. Then A is a multiple of I. The only matrices that commute with B and C and all other 2 by 2 matrices are $A =$ multiple of I.

13 Which of the following matrices are guaranteed to equal $(A - B)^2$: $A^2 - B^2$, $(B - A)^2$, $A^2 - 2AB + B^2$, $A(A - B) - B(A - B)$, $A^2 - AB - BA + B^2$?

14 True or false:

(a) If A^2 is defined then A is necessarily square.

(b) If AB and BA are defined then A and B are square.

(c) If AB and BA are defined then AB and BA are square.

(d) If $AB = B$ then $A = I$.

15 If A is m by n, how many separate multiplications are involved when

(a) A multiplies a vector $\boldsymbol{x}$ with n components?

(b) A multiplies an n by p matrix B?

(c) A multiplies itself to produce A^2? Here $m = n$.

16 To prove that $(AB)C = A(BC)$, use the column vectors $\boldsymbol{b}_1, \ldots, \boldsymbol{b}_n$ of B. First suppose that C has only one column $\boldsymbol{c}$ with entries $c_1, \ldots, c_n$:

AB has columns $A\boldsymbol{b}_1, \ldots, A\boldsymbol{b}_n$ and $B\boldsymbol{c}$ has one column $c_1\boldsymbol{b}_1 + \cdots + c_n\boldsymbol{b}_n$.

Then $(AB)\boldsymbol{c} = c_1 A\boldsymbol{b}_1 + \cdots + c_n A\boldsymbol{b}_n$ equals $A(c_1\boldsymbol{b}_1 + \cdots + c_n\boldsymbol{b}_n) = A(B\boldsymbol{c})$.

Linearity gives equality of those two sums, and $(AB)\boldsymbol{c} = A(B\boldsymbol{c})$. The same is true for all other ______ of C. Therefore $(AB)C = A(BC)$.

17 For $A = \begin{bmatrix} 2 & -1 \\ 3 & -2 \end{bmatrix}$ and $B = \begin{bmatrix} 1 & 0 & 4 \\ 1 & 0 & 6 \end{bmatrix}$, compute these answers *and nothing more*:

(a) column 2 of AB

(b) row 2 of AB

(c) row 2 of $AA = A^2$

(d) row 2 of $AAA = A^3$.

Problems 18–20 use a_{ij} for the entry in row i, column j of A.

18 Write down the 3 by 3 matrix A whose entries are

(a) a_{ij} = minimum of i and j

(b) $a_{ij} = (-1)^{i+j}$

(c) $a_{ij} = i/j$.

19 What words would you use to describe each of these classes of matrices? Give a 3 by 3 example in each class. Which matrix belongs to all four classes?

(a) $a_{ij} = 0$ if $i \neq j$

(b) $a_{ij} = 0$ if $i < j$

(c) $a_{ij} = a_{ji}$

(d) $a_{ij} = a_{1j}$.

20 The entries of A are a_{ij}. Assuming that zeros don't appear, what is

(a) the first pivot?

(b) the multiplier ℓ_{31} of row 1 to be subtracted from row 3?

(c) the new entry that replaces a_{32} after that subtraction?

(d) the second pivot?

Problems 21–25 involve powers of A.

21 Compute A^2, A^3, A^4 and also $A\boldsymbol{v}$, $A^2\boldsymbol{v}$, $A^3\boldsymbol{v}$, $A^4\boldsymbol{v}$ for

$$A = \begin{bmatrix} 0 & 2 & 0 & 0 \\ 0 & 0 & 2 & 0 \\ 0 & 0 & 0 & 2 \\ 0 & 0 & 0 & 0 \end{bmatrix} \quad \text{and} \quad \boldsymbol{v} = \begin{bmatrix} x \\ y \\ z \\ t \end{bmatrix}.$$

22 Find all the powers $A^2, A^3, \ldots$ and $AB, (AB)^2, \ldots$ for

$$A = \begin{bmatrix} .5 & .5 \\ .5 & .5 \end{bmatrix} \quad \text{and} \quad B = \begin{bmatrix} 1 & 0 \\ 0 & -1 \end{bmatrix}.$$

23 By trial and error find real nonzero 2 by 2 matrices such that

$$A^2 = -I \qquad BC = 0 \qquad DE = -ED \text{ (not allowing } DE = 0\text{)}.$$

24 (a) Find a nonzero matrix A for which $A^2 = 0$.

(b) Find a matrix that has $A^2 \neq 0$ but $A^3 = 0$.

25 By experiment with $n = 2$ and $n = 3$ predict A^n for

$$A_1 = \begin{bmatrix} 2 & 1 \\ 0 & 1 \end{bmatrix} \quad \text{and} \quad A_2 = \begin{bmatrix} 1 & 1 \\ 1 & 1 \end{bmatrix} \quad \text{and} \quad A_3 = \begin{bmatrix} a & b \\ 0 & 0 \end{bmatrix}.$$

Problems 26–34 use column-row multiplication and block multiplication.

26 Multiply AB using columns times rows:

$$AB = \begin{bmatrix} 1 & 0 \\ 2 & 4 \\ 2 & 1 \end{bmatrix} \begin{bmatrix} 3 & 3 & 0 \\ 1 & 2 & 1 \end{bmatrix} = \begin{bmatrix} 1 \\ 2 \\ 2 \end{bmatrix} \begin{bmatrix} 3 & 3 & 0 \end{bmatrix} + \underline{\quad\quad} = \underline{\quad\quad}.$$

27 The product of upper triangular matrices is always upper triangular:

$$AB = \begin{bmatrix} x & x & x \\ 0 & x & x \\ 0 & 0 & x \end{bmatrix} \begin{bmatrix} x & x & x \\ 0 & x & x \\ 0 & 0 & x \end{bmatrix} = \begin{bmatrix} & & \\ 0 & & \\ 0 & 0 & \end{bmatrix}.$$

Row times column is dot product (Row 2 of A) $\cdot$ (column 1 of B) $= 0$. Which other dot products give zeros?

Column times row is full matrix Draw x's and 0's in (column 2 of A) times (row 2 of B) and in (column 3 of A) times (row 3 of B).

28 Draw the cuts in A (2 by 3) and B (3 by 4) and AB to show how each of the four multiplication rules is really a block multiplication:

(1) Matrix A times columns of B.

(2) Rows of A times matrix B.

(3) Rows of A times columns of B.

(4) Columns of A times rows of B.

29 Draw cuts in A and $\boldsymbol{x}$ to multiply $A\boldsymbol{x}$ a column at a time: x_1(column 1) $+ \cdots$.

30 Which matrices E_{21} and E_{31} produce zeros in the (2, 1) and (3, 1) positions of $E_{21}A$ and $E_{31}A$?

$$A = \begin{bmatrix} 2 & 1 & 0 \\ -2 & 0 & 1 \\ 8 & 5 & 3 \end{bmatrix}.$$

Find the single matrix $E = E_{31}E_{21}$ that produces both zeros at once. Multiply EA.

31 Block multiplication says in the text that column 1 is eliminated by

$$EA = \begin{bmatrix} 1 & \mathbf{0} \\ -\boldsymbol{c}/a & I \end{bmatrix} \begin{bmatrix} a & \boldsymbol{b} \\ \boldsymbol{c} & D \end{bmatrix} = \begin{bmatrix} a & \boldsymbol{b} \\ \mathbf{0} & D - \boldsymbol{cb}/a \end{bmatrix}.$$

In Problem 30, what are $\boldsymbol{c}$ and D and what is $D - \boldsymbol{cb}/a$?

32 With $i^2 = -1$, the product of $(A+iB)$ and $(\boldsymbol{x}+i\boldsymbol{y})$ is $A\boldsymbol{x}+iB\boldsymbol{x}+iA\boldsymbol{y}-B\boldsymbol{y}$. Use blocks to separate the real part without i from the imaginary part that multiplies i:

$$\begin{bmatrix} A & -B \\ ? & ? \end{bmatrix}\begin{bmatrix} \boldsymbol{x} \\ \boldsymbol{y} \end{bmatrix} = \begin{bmatrix} A\boldsymbol{x} - B\boldsymbol{y} \\ ? \end{bmatrix} \begin{matrix} \text{real part} \\ \text{imaginary part} \end{matrix}$$

33 Suppose you solve $A\boldsymbol{x} = \boldsymbol{b}$ for three special right sides $\boldsymbol{b}$:

$$A\boldsymbol{x}_1 = \begin{bmatrix} 1 \\ 0 \\ 0 \end{bmatrix} \quad \text{and} \quad A\boldsymbol{x}_2 = \begin{bmatrix} 0 \\ 1 \\ 0 \end{bmatrix} \quad \text{and} \quad A\boldsymbol{x}_3 = \begin{bmatrix} 0 \\ 0 \\ 1 \end{bmatrix}.$$

If the three solutions $\boldsymbol{x}_1, \boldsymbol{x}_2, \boldsymbol{x}_3$ are the columns of a matrix X, what is A times X?

34 If the three solutions in Question 33 are $\boldsymbol{x}_1 = (1, 1, 1)$ and $\boldsymbol{x}_2 = (0, 1, 1)$ and $\boldsymbol{x}_3 = (0, 0, 1)$, solve $A\boldsymbol{x} = \boldsymbol{b}$ when $\boldsymbol{b} = (3, 5, 8)$. Challenge problem: What is A?

35 *Elimination for a 2 by 2 block matrix*: When you multiply the first block row by CA^{-1} and subtract from the second row, what is the "*Schur complement*" S that appears?

$$\begin{bmatrix} I & 0 \\ -CA^{-1} & I \end{bmatrix}\begin{bmatrix} A & B \\ C & D \end{bmatrix} = \begin{bmatrix} A & B \\ 0 & S \end{bmatrix}.$$

36 Find all matrices $A = \left[\begin{smallmatrix} a & b \\ c & d \end{smallmatrix}\right]$ that satisfy $A\left[\begin{smallmatrix} 1 & 1 \\ 1 & 1 \end{smallmatrix}\right] = \left[\begin{smallmatrix} 1 & 1 \\ 1 & 1 \end{smallmatrix}\right]A$.

37 Suppose a "circle graph" has 5 nodes connected (in both directions) by edges around a circle. What is its adjacency matrix from Worked Example **2.4 C**? What are A^2 and A^3 and the diameter of this graph?

38 If 5 edges in Question 37 go in one direction only, from nodes 1, 2, 3, 4, 5 to 2, 3, 4, 5, 1, what are A and A^2 and the diameter of this one-way circle?

39 If you multiply a *northwest matrix* A and a *southeast matrix* B, what type of matrices are AB and BA? "Northwest" and "southeast" mean zeros below and above the antidiagonal going from $(1, n)$ to $(n, 1)$.

INVERSE MATRICES ■ 2.5

Suppose A is a square matrix. We look for an "***inverse matrix***" A^{-1} of the same size, such that A^{-1} *times* A *equals* I. Whatever A does, A^{-1} undoes. Their product is the identity matrix—which does nothing. But A^{-1} might not exist.

What a matrix mostly does is to multiply a vector $\boldsymbol{x}$. Multiplying $A\boldsymbol{x} = \boldsymbol{b}$ by A^{-1} gives $A^{-1}A\boldsymbol{x} = A^{-1}\boldsymbol{b}$. The left side is just $\boldsymbol{x}$! The product $A^{-1}A$ is like multiplying by a number and then dividing by that number. An ordinary number has an inverse if it is not zero—matrices are more complicated and more interesting. The matrix A^{-1} is called "A inverse."

DEFINITION The matrix A is ***invertible*** if there exists a matrix A^{-1} such that

$$A^{-1}A = I \quad \text{and} \quad AA^{-1} = I. \tag{1}$$

Not all matrices have inverses. This is the first question we ask about a square matrix: Is A invertible? We don't mean that we immediately calculate A^{-1}. In most problems we never compute it! Here are six "notes" about A^{-1}.

Note 1 ***The inverse exists if and only if elimination produces n pivots*** (row exchanges allowed). Elimination solves $A\boldsymbol{x} = \boldsymbol{b}$ without explicitly using A^{-1}.

Note 2 The matrix A cannot have two different inverses. Suppose $BA = I$ and also $AC = I$. Then $B = C$, according to this "proof by parentheses":

$$B(AC) = (BA)C \quad \text{gives} \quad BI = IC \quad \text{or} \quad B = C. \tag{2}$$

This shows that a *left-inverse* B (multiplying from the left) and a *right-inverse* C (multiplying A from the right to give $AC = I$) must be the *same matrix*.

Note 3 If A is invertible, the one and only solution to $A\boldsymbol{x} = \boldsymbol{b}$ is $\boldsymbol{x} = A^{-1}\boldsymbol{b}$:

$$\textbf{\textit{Multiply}} \quad A\boldsymbol{x} = \boldsymbol{b} \quad \textbf{\textit{by}} \quad A^{-1}. \quad \textbf{\textit{Then}} \quad \boldsymbol{x} = A^{-1}A\boldsymbol{x} = A^{-1}\boldsymbol{b}.$$

Note 4 (Important) ***Suppose there is a nonzero vector x such that*** $A\boldsymbol{x} = \boldsymbol{0}$. *Then A cannot have an inverse.* No matrix can bring $\boldsymbol{0}$ back to $\boldsymbol{x}$.

If A is invertible, then $A\boldsymbol{x} = \boldsymbol{0}$ can only have the zero solution $\boldsymbol{x} = \boldsymbol{0}$.

Note 5 A 2 by 2 matrix is invertible if and only if $ad - bc$ is not zero:

$$\textbf{2 by 2 Inverse:} \quad \begin{bmatrix} a & b \\ c & d \end{bmatrix}^{-1} = \frac{1}{ad - bc}\begin{bmatrix} d & -b \\ -c & a \end{bmatrix}. \tag{3}$$

This number $ad - bc$ is the *determinant* of A. A matrix is invertible if its determinant is not zero (Chapter 5). The test for n pivots is usually decided before the determinant appears.

Note 6 A diagonal matrix has an inverse provided no diagonal entries are zero:

$$\text{If}\quad A = \begin{bmatrix} d_1 & & \\ & \ddots & \\ & & d_n \end{bmatrix} \quad \text{then} \quad A^{-1} = \begin{bmatrix} 1/d_1 & & \\ & \ddots & \\ & & 1/d_n \end{bmatrix}.$$

Example 1 The 2 by 2 matrix $A = \left[\begin{smallmatrix} 1 & 2 \\ 1 & 2 \end{smallmatrix}\right]$ is not invertible. It fails the test in Note 5, because $ad - bc$ equals $2 - 2 = 0$. It fails the test in Note 3, because $A\boldsymbol{x} = \mathbf{0}$ when $\boldsymbol{x} = (2, -1)$. It fails to have two pivots as required by Note 1. Elimination turns the second row of A into a zero row.

The Inverse of a Product AB

For two nonzero numbers a and b, the sum $a + b$ might or might not be invertible. The numbers $a = 3$ and $b = -3$ have inverses $\frac{1}{3}$ and $-\frac{1}{3}$. Their sum $a + b = 0$ has no inverse. But the product $ab = -9$ does have an inverse, which is $\frac{1}{3}$ times $-\frac{1}{3}$.

For two matrices A and B, the situation is similar. It is hard to say much about the invertibility of $A + B$. But the *product* AB has an inverse, whenever the factors A and B are separately invertible (and the same size). The important point is that A^{-1} and B^{-1} come in ***reverse order***:

2H If A and B are invertible then so is AB. The inverse of a product AB is

$$(AB)^{-1} = B^{-1}A^{-1}. \tag{4}$$

To see why the order is reversed, multiply AB times $B^{-1}A^{-1}$. The inside step is $BB^{-1} = I$:

$$(AB)(B^{-1}A^{-1}) = AIA^{-1} = AA^{-1} = I.$$

We moved parentheses to multiply BB^{-1} first. Similarly $B^{-1}A^{-1}$ times AB equals I. This illustrates a basic rule of mathematics: Inverses come in reverse order. It is also common sense: If you put on socks and then shoes, the first to be taken off are the _____. The same idea applies to three or more matrices:

$$\textbf{Reverse order} \quad (ABC)^{-1} = C^{-1}B^{-1}A^{-1}. \tag{5}$$

Example 2 ***Inverse of an Elimination Matrix.*** If E subtracts 5 times row 1 from row 2, then E^{-1} *adds* 5 times row 1 to row 2:

$$E = \begin{bmatrix} 1 & 0 & 0 \\ -5 & 1 & 0 \\ 0 & 0 & 1 \end{bmatrix} \quad \text{and} \quad E^{-1} = \begin{bmatrix} 1 & 0 & 0 \\ 5 & 1 & 0 \\ 0 & 0 & 1 \end{bmatrix}.$$

Multiply EE^{-1} to get the identity matrix I. Also multiply $E^{-1}E$ to get I. We are adding and subtracting the same 5 times row 1. Whether we add and then subtract (this is EE^{-1}) or subtract and then add (this is $E^{-1}E$), we are back at the start.

For square matrices, an inverse on one side is automatically an inverse on the other side. If $AB = I$ then automatically $BA = I$. In that case B is A^{-1}. This is very useful to know but we are not ready to prove it.

Example 3 Suppose F subtracts 4 times row 2 from row 3, and F^{-1} adds it back:

$$F = \begin{bmatrix} 1 & 0 & 0 \\ 0 & 1 & 0 \\ 0 & -4 & 1 \end{bmatrix} \quad \text{and} \quad F^{-1} = \begin{bmatrix} 1 & 0 & 0 \\ 0 & 1 & 0 \\ 0 & 4 & 1 \end{bmatrix}.$$

Now multiply F by the matrix E in Example 2 to find FE. Also multiply E^{-1} times F^{-1} to find $(FE)^{-1}$. Notice the orders FE and $E^{-1}F^{-1}$!

$$FE = \begin{bmatrix} 1 & 0 & 0 \\ -5 & 1 & 0 \\ 20 & -4 & 1 \end{bmatrix} \quad \text{is inverted by} \quad E^{-1}F^{-1} = \begin{bmatrix} 1 & 0 & 0 \\ 5 & 1 & 0 \\ 0 & 4 & 1 \end{bmatrix}. \tag{6}$$

The result is strange but correct. The product FE contains "20" but its inverse doesn't. E subtracts 5 times row 1 from row 2. Then F subtracts 4 times the *new* row 2 (changed by row 1) from row 3. ***In this order FE, row 3 feels an effect from row 1.***

In the order $E^{-1}F^{-1}$, that effect does not happen. First F^{-1} adds 4 times row 2 to row 3. After that, E^{-1} adds 5 times row 1 to row 2. There is no 20, because row 3 doesn't change again. ***In this order, row 3 feels no effect from row 1.***

> For elimination with normal order FE, ***the product of inverses $E^{-1}F^{-1}$ is quick. The multipliers fall into place below the diagonal of 1's.***

This special property of $E^{-1}F^{-1}$ and $E^{-1}F^{-1}G^{-1}$ will be useful in the next section. We will explain it again, more completely. In this section our job is A^{-1}, and we expect some serious work to compute it. Here is a way to organize that computation.

Calculating A^{-1} by Gauss-Jordan Elimination

I hinted that A^{-1} might not be explicitly needed. The equation $A\boldsymbol{x} = \boldsymbol{b}$ is solved by $\boldsymbol{x} = A^{-1}\boldsymbol{b}$. But it is not necessary or efficient to compute A^{-1} and multiply it times $\boldsymbol{b}$. *Elimination goes directly to* $\boldsymbol{x}$. Elimination is also the way to calculate A^{-1}, as we now show. The Gauss-Jordan idea is to solve $AA^{-1} = I$, *finding each column of* A^{-1}.

A multiplies the first column of A^{-1} (call that $\boldsymbol{x}_1$) to give the first column of I (call that $\boldsymbol{e}_1$). This is our equation $A\boldsymbol{x}_1 = \boldsymbol{e}_1 = (1, 0, 0)$. Each of the columns $\boldsymbol{x}_1$, $\boldsymbol{x}_2$, $\boldsymbol{x}_3$ of A^{-1} is multiplied by A to produce a column of I:

$$AA^{-1} = A\begin{bmatrix} \boldsymbol{x}_1 & \boldsymbol{x}_2 & \boldsymbol{x}_3 \end{bmatrix} = \begin{bmatrix} \boldsymbol{e}_1 & \boldsymbol{e}_2 & \boldsymbol{e}_3 \end{bmatrix} = I. \tag{7}$$

To invert a 3 by 3 matrix A, we have to solve three systems of equations: $A\boldsymbol{x}_1 = \boldsymbol{e}_1$ and $A\boldsymbol{x}_2 = \boldsymbol{e}_2 = (0, 1, 0)$ and $A\boldsymbol{x}_3 = \boldsymbol{e}_3 = (0, 0, 1)$. This already shows why computing A^{-1} is expensive. We must solve n equations for its n columns. To solve $A\boldsymbol{x} = \boldsymbol{b}$ without A^{-1}, we deal only with *one* column.

In defense of A^{-1}, we want to say that its cost is not n times the cost of one system $A\boldsymbol{x} = \boldsymbol{b}$. Surprisingly, the cost for n columns is only multiplied by 3. This saving is because the n equations $A\boldsymbol{x}_i = \boldsymbol{e}_i$ all involve the same matrix A. Working with the right sides is relatively cheap, because elimination only has to be done once on A. The complete A^{-1} needs n^3 elimination steps, where a single $\boldsymbol{x}$ needs $n^3/3$. The next section calculates these costs.

The ***Gauss-Jordan method*** computes A^{-1} by solving all n equations together. Usually the "augmented matrix" has one extra column $\boldsymbol{b}$, from the right side of the equations. Now we have three right sides $\boldsymbol{e}_1, \boldsymbol{e}_2, \boldsymbol{e}_3$ (when A is 3 by 3). They are the columns of I, so the augmented matrix is really the block matrix $[\,A \quad I\,]$. Here is a worked-out example when A has 2's on the main diagonal and -1's next to the 2's:

$$\begin{bmatrix} A & \boldsymbol{e}_1 & \boldsymbol{e}_2 & \boldsymbol{e}_3 \end{bmatrix} = \begin{bmatrix} 2 & -1 & 0 & 1 & 0 & 0 \\ -1 & 2 & -1 & 0 & 1 & 0 \\ 0 & -1 & 2 & 0 & 0 & 1 \end{bmatrix} \quad \textbf{Start Gauss-Jordan}$$

$$\rightarrow \begin{bmatrix} 2 & -1 & 0 & 1 & 0 & 0 \\ \mathbf{0} & \mathbf{\frac{3}{2}} & \mathbf{-1} & \mathbf{\frac{1}{2}} & \mathbf{1} & \mathbf{0} \\ 0 & -1 & 2 & 0 & 0 & 1 \end{bmatrix} \quad (\tfrac{1}{2}\ \textit{row}\ 1 + \textit{row}\ 2)$$

$$\rightarrow \begin{bmatrix} 2 & -1 & 0 & 1 & 0 & 0 \\ 0 & \frac{3}{2} & -1 & \frac{1}{2} & 1 & 0 \\ \mathbf{0} & \mathbf{0} & \mathbf{\frac{4}{3}} & \mathbf{\frac{1}{3}} & \mathbf{\frac{2}{3}} & \mathbf{1} \end{bmatrix} \quad (\tfrac{2}{3}\ \textit{row}\ 2 + \textit{row}\ 3)$$

We are now halfway. The matrix in the first three columns is U (upper triangular). The pivots $2, \frac{3}{2}, \frac{4}{3}$ are on its diagonal. Gauss would finish by back substitution. The contribution of Jordan is *to continue with elimination*! He goes all the way to the ***"reduced echelon form"***. Rows are added to rows above them, to produce ***zeros above the pivots***:

$$\rightarrow \begin{bmatrix} 2 & -1 & 0 & 1 & 0 & 0 \\ \mathbf{0} & \mathbf{\frac{3}{2}} & \mathbf{0} & \mathbf{\frac{3}{4}} & \mathbf{\frac{3}{2}} & \mathbf{\frac{3}{4}} \\ 0 & 0 & \frac{4}{3} & \frac{1}{3} & \frac{2}{3} & 1 \end{bmatrix} \quad (\tfrac{3}{4}\ \textit{row}\ 3 + \textit{row}\ 2)$$

$$\rightarrow \begin{bmatrix} \mathbf{2} & \mathbf{0} & \mathbf{0} & \mathbf{\frac{3}{2}} & \mathbf{1} & \mathbf{\frac{1}{2}} \\ 0 & \frac{3}{2} & 0 & \frac{3}{4} & \frac{3}{2} & \frac{3}{4} \\ 0 & 0 & \frac{4}{3} & \frac{1}{3} & \frac{2}{3} & 1 \end{bmatrix} \quad (\tfrac{2}{3}\ \textit{row}\ 2 + \textit{row}\ 1)$$

The last Gauss-Jordan step is to divide each row by its pivot. The new pivots are 1. We have reached I in the first half of the matrix, because A is invertible. ***The three columns of*** A^{-1} ***are in the second half of*** $[\, I \;\; A^{-1} \,]$:

$$\begin{matrix} \text{(divide by 2)} \\ \text{(divide by } \frac{3}{2}) \\ \text{(divide by } \frac{4}{3}) \end{matrix} \quad \left[\begin{array}{ccc|ccc} 1 & 0 & 0 & \frac{3}{4} & \frac{1}{2} & \frac{1}{4} \\ 0 & 1 & 0 & \frac{1}{2} & 1 & \frac{1}{2} \\ 0 & 0 & 1 & \frac{1}{4} & \frac{1}{2} & \frac{3}{4} \end{array}\right] = \left[\, I \;\; x_1 \;\; x_2 \;\; x_3 \,\right].$$

Starting from the 3 by 6 matrix $[\, A \;\; I \,]$, we ended with $[\, I \;\; A^{-1} \,]$. Here is the whole Gauss-Jordan process on one line:

$$\textbf{\textit{Multiply}} \;\; [\, \boldsymbol{A} \;\; \boldsymbol{I} \,] \;\; \textbf{\textit{by}} \;\; \boldsymbol{A^{-1}} \;\; \textbf{\textit{to get}} \;\; [\boldsymbol{I} \;\; \boldsymbol{A^{-1}}].$$

The elimination steps gradually create the inverse matrix. For large matrices, we probably don't want A^{-1} at all. But for small matrices, it can be very worthwhile to know the inverse. We add three observations about this particular A^{-1} because it is an important example. We introduce the words *symmetric*, *tridiagonal*, and *determinant* (Chapter 5):

1. A is ***symmetric*** across its main diagonal. So is A^{-1}.

2. A is ***tridiagonal*** (only three nonzero diagonals). But A^{-1} is a full matrix with no zeros. That is another reason we don't often compute A^{-1}.

3. The product of pivots is $2(\frac{3}{2})(\frac{4}{3}) = 4$. This number 4 is the ***determinant*** of A.

$$\boldsymbol{A^{-1}} \textbf{\textit{ involves division by the determinant}} \qquad A^{-1} = \frac{1}{4}\begin{bmatrix} 3 & 2 & 1 \\ 2 & 4 & 2 \\ 1 & 2 & 3 \end{bmatrix}. \tag{8}$$

Example 4 Find A^{-1} by Gauss-Jordan elimination starting from $A = \left[\begin{smallmatrix} 2 & 3 \\ 4 & 7 \end{smallmatrix}\right]$. There are two row operations and then a division to put 1's in the pivots:

$$[\, A \;\; I \,] = \begin{bmatrix} \mathbf{2} & \mathbf{3} & 1 & 0 \\ \mathbf{4} & \mathbf{7} & 0 & 1 \end{bmatrix} \to \begin{bmatrix} 2 & 3 & 1 & 0 \\ \mathbf{0} & \mathbf{1} & \mathbf{-2} & \mathbf{1} \end{bmatrix}$$

$$\to \begin{bmatrix} \mathbf{2} & \mathbf{0} & \mathbf{7} & \mathbf{-3} \\ 0 & 1 & -2 & 1 \end{bmatrix} \to \begin{bmatrix} 1 & 0 & \frac{7}{2} & -\frac{3}{2} \\ 0 & 1 & -\mathbf{2} & \mathbf{1} \end{bmatrix} = [\, I \;\; \boldsymbol{A^{-1}} \,].$$

The reduced echelon form of $[\, A \;\; I \,]$ is $[\, I \;\; A^{-1} \,]$. This A^{-1} involves division by the determinant $2 \cdot 7 - 3 \cdot 4 = 2$. The code for $X =$ inverse(A) has three important lines!

```
I = eye (n, n);            % Define the identity matrix
R = rref ([A I]);          % Eliminate on the augmented matrix
X = R(:, n + 1 : n + n)    % Pick A^-1 from the last n columns of R
```

A must be invertible, or elimination will not reduce it (in the left half of R) to I.

Singular versus Invertible

We come back to the central question. Which matrices have inverses? The start of this section proposed the pivot test: A^{-1} ***exists exactly when A has a full set of n pivots.*** (Row exchanges allowed.) Now we can prove that by Gauss-Jordan elimination:

1. With n pivots, elimination solves all the equations $A\boldsymbol{x}_i = \boldsymbol{e}_i$. The columns $\boldsymbol{x}_i$ go into A^{-1}. Then $AA^{-1} = I$ and A^{-1} is at least a ***right-inverse***.

2. Elimination is really a sequence of multiplications by E's and P's and D^{-1}:

$$(D^{-1} \cdots E \cdots P \cdots E)A = I. \tag{9}$$

D^{-1} divides by the pivots. The matrices E produce zeros below and above the pivots. P will exchange rows if needed (see Section 2.7). The product matrix in equation (9) is evidently a ***left-inverse***. With n pivots we reach $A^{-1}A = I$.

The right-inverse equals the left-inverse. That was Note 2 in this section. So a square matrix with a full set of pivots will always have a two-sided inverse.

Reasoning in reverse will now show that A must have n pivots if $AC = I$. Then we deduce that C is also a left-inverse. Here is one route to those conclusions:

1. If A doesn't have n pivots, elimination will lead to a *zero row*.
2. Those elimination steps are taken by an invertible M. *So a row of MA is zero.*
3. If $AC = I$ then $MAC = M$. The zero row of MA, times C, gives a zero row of M.
4. The invertible matrix M can't have a zero row! A *must* have n pivots if $AC = I$.
5. Then equation (9) displays the left inverse in $BA = I$, and Note 2 proves $B = C$.

That argument took five steps, but the outcome is short and important.

2I A complete test for invertibility of a square matrix A comes from elimination. A^{-1} ***exists (and Gauss-Jordan finds it) exactly when A has n pivots***. The full argument shows more:

$$\text{If} \quad AC = I \quad \text{then} \quad CA = I \quad \text{and} \quad C = A^{-1} \quad !$$

Example 5 If L is lower triangular with 1's on the diagonal, so is L^{-1}.

Use the Gauss-Jordan method to construct L^{-1}. Start by subtracting multiples of pivot rows from rows *below*. Normally this gets us halfway to the inverse, but for L it gets us all the way. L^{-1} appears on the right when I appears on the left:

$$\begin{bmatrix} L & I \end{bmatrix} = \begin{bmatrix} \mathbf{1} & \mathbf{0} & \mathbf{0} & 1 & 0 & 0 \\ \mathbf{3} & \mathbf{1} & \mathbf{0} & 0 & 1 & 0 \\ \mathbf{4} & \mathbf{5} & \mathbf{1} & 0 & 0 & 1 \end{bmatrix}$$

$$\begin{matrix} \\ \to \\ \to \end{matrix} \begin{bmatrix} 1 & 0 & 0 & 1 & 0 & 0 \\ 0 & 1 & 0 & -3 & 1 & 0 \\ 0 & 5 & 1 & -4 & 0 & 1 \end{bmatrix} \quad \begin{matrix} \\ \text{(3 times row 1 from row 2)} \\ \text{(4 times row 1 from row 3)} \end{matrix}$$

$$\begin{matrix} \\ \\ \to \end{matrix} \begin{bmatrix} 1 & 0 & 0 & \mathbf{1} & \mathbf{0} & \mathbf{0} \\ 0 & 1 & 0 & \mathbf{-3} & \mathbf{1} & \mathbf{0} \\ 0 & 0 & 1 & \mathbf{11} & \mathbf{-5} & \mathbf{1} \end{bmatrix} = \begin{bmatrix} I & \mathbf{L^{-1}} \end{bmatrix}.$$

When L goes to I by elimination, I goes to L^{-1}. In other words, the product of elimination matrices $E_{32}E_{31}E_{21}$ is L^{-1}. All pivots are 1's (a full set). L^{-1} is lower triangular. The strange entry "11" in L^{-1} does not appear in $E_{21}^{-1}E_{31}^{-1}E_{32}^{-1} = L$.

■ REVIEW OF THE KEY IDEAS ■

1. The inverse matrix gives $AA^{-1} = I$ and $A^{-1}A = I$.
2. A is invertible if and only if it has n pivots (row exchanges allowed).
3. If $A\boldsymbol{x} = \mathbf{0}$ for a nonzero vector $\boldsymbol{x}$, then A has no inverse.
4. The inverse of AB is the reverse product $B^{-1}A^{-1}$.
5. The Gauss-Jordan method solves $AA^{-1} = I$ to find the n columns of A^{-1}. The augmented matrix $\begin{bmatrix} A & I \end{bmatrix}$ is row-reduced to $\begin{bmatrix} I & A^{-1} \end{bmatrix}$.

■ WORKED EXAMPLES ■

2.5 A Three of these matrices are invertible, and three are singular. Find the inverse when it exists. Give reasons for noninvertibility (zero determinant, too few pivots, nonzero solution to $A\boldsymbol{x} = \mathbf{0}$) for the other three, in that order. The matrices A, B, C, D, E, F are

$$\begin{bmatrix} 4 & 3 \\ 8 & 6 \end{bmatrix} \begin{bmatrix} 4 & 3 \\ 8 & 7 \end{bmatrix} \begin{bmatrix} 6 & 6 \\ 6 & 0 \end{bmatrix} \begin{bmatrix} 6 & 6 \\ 6 & 6 \end{bmatrix} \begin{bmatrix} 1 & 0 & 0 \\ 1 & 1 & 0 \\ 1 & 1 & 1 \end{bmatrix} \begin{bmatrix} 1 & 1 & 1 \\ 1 & 1 & 0 \\ 1 & 1 & 1 \end{bmatrix}$$

Solution

$$B^{-1} = \frac{1}{4}\begin{bmatrix} 7 & -3 \\ -8 & 4 \end{bmatrix} \quad C^{-1} = \frac{1}{36}\begin{bmatrix} 0 & 6 \\ 6 & -6 \end{bmatrix} \quad E^{-1} = \begin{bmatrix} 1 & 0 & 0 \\ -1 & 1 & 0 \\ 0 & -1 & 1 \end{bmatrix}$$

A is not invertible because its determinant is $4 \cdot 6 - 3 \cdot 8 = 24 - 24 = 0$. D is not invertible because there is only one pivot; the second row becomes zero when the first row is subtracted. F is not invertible because a combination of the columns (the second column minus the first column) is zero—in other words $F\boldsymbol{x} = \mathbf{0}$ has the solution $\boldsymbol{x} = (-1, 1, 0)$.

Of course all three reasons for noninvertibility would apply to each of A, D, F.

2.5 B Apply the Gauss-Jordan method to find the inverse of this triangular "Pascal matrix" A = abs(pascal(4,1)). You see **Pascal's triangle**—adding each entry to the entry on its left gives the entry below. The entries are "binomial coefficients":

$$\textbf{Triangular Pascal matrix} \quad A = \begin{bmatrix} \mathbf{1} & 0 & 0 & 0 \\ \mathbf{1} & \mathbf{1} & 0 & 0 \\ \mathbf{1} & \mathbf{2} & \mathbf{1} & 0 \\ \mathbf{1} & \mathbf{3} & \mathbf{3} & \mathbf{1} \end{bmatrix}.$$

Solution Gauss-Jordan starts with $[A \;\; I]$ and produces zeros by subtracting row 1:

$$[A \;\; I] = \left[\begin{array}{cccc|cccc} \mathbf{1} & 0 & 0 & 0 & 1 & 0 & 0 & 0 \\ \mathbf{1} & \mathbf{1} & 0 & 0 & 0 & 1 & 0 & 0 \\ \mathbf{1} & \mathbf{2} & \mathbf{1} & 0 & 0 & 0 & 1 & 0 \\ \mathbf{1} & \mathbf{3} & \mathbf{3} & \mathbf{1} & 0 & 0 & 0 & 1 \end{array}\right] \to \left[\begin{array}{cccc|cccc} 1 & 0 & 0 & 0 & 1 & 0 & 0 & 0 \\ 0 & 1 & 0 & 0 & -1 & 1 & 0 & 0 \\ 0 & 2 & 1 & 0 & -1 & 0 & 1 & 0 \\ 0 & 3 & 3 & 1 & -1 & 0 & 0 & 1 \end{array}\right].$$

The next stage creates zeros below the second pivot, using multipliers 2 and 3. Then the last stage subtracts 3 times the new row 3 from the new row 4:

$$\to \left[\begin{array}{cccc|cccc} 1 & 0 & 0 & 0 & 1 & 0 & 0 & 0 \\ 0 & 1 & 0 & 0 & -1 & 1 & 0 & 0 \\ 0 & 0 & 1 & 0 & 1 & -2 & 1 & 0 \\ 0 & 0 & 3 & 1 & 2 & -3 & 0 & 1 \end{array}\right] \to \left[\begin{array}{cccc|cccc} 1 & 0 & 0 & 0 & \mathbf{1} & 0 & 0 & 0 \\ 0 & 1 & 0 & 0 & \mathbf{-1} & \mathbf{1} & 0 & 0 \\ 0 & 0 & 1 & 0 & \mathbf{1} & \mathbf{-2} & \mathbf{1} & 0 \\ 0 & 0 & 0 & 1 & \mathbf{-1} & \mathbf{3} & \mathbf{-3} & \mathbf{1} \end{array}\right] = [I \;\; \boldsymbol{A^{-1}}].$$

All the pivots were 1! So we didn't need to divide rows by pivots to get I. The inverse matrix A^{-1} looks like A itself, except odd-numbered diagonals are multiplied by -1.

Please notice that 4 by 4 matrix A^{-1}, we will see Pascal matrices again. The same pattern continues to n by n Pascal matrices—the inverse has "alternating diagonals".

Problem Set 2.5

1 Find the inverses (directly or from the 2 by 2 formula) of A, B, C:

$$A = \begin{bmatrix} 0 & 3 \\ 4 & 0 \end{bmatrix} \quad \text{and} \quad B = \begin{bmatrix} 2 & 0 \\ 4 & 2 \end{bmatrix} \quad \text{and} \quad C = \begin{bmatrix} 3 & 4 \\ 5 & 7 \end{bmatrix}.$$

2 For these "permutation matrices" find P^{-1} by trial and error (with 1's and 0's):

$$P = \begin{bmatrix} 0 & 0 & 1 \\ 0 & 1 & 0 \\ 1 & 0 & 0 \end{bmatrix} \quad \text{and} \quad P = \begin{bmatrix} 0 & 1 & 0 \\ 0 & 0 & 1 \\ 1 & 0 & 0 \end{bmatrix}.$$

3 Solve for the columns of $A^{-1} = \left[\begin{smallmatrix} x & t \\ y & z \end{smallmatrix}\right]$:

$$\begin{bmatrix} 10 & 20 \\ 20 & 50 \end{bmatrix}\begin{bmatrix} x \\ y \end{bmatrix} = \begin{bmatrix} 1 \\ 0 \end{bmatrix} \quad \text{and} \quad \begin{bmatrix} 10 & 20 \\ 20 & 50 \end{bmatrix}\begin{bmatrix} t \\ z \end{bmatrix} = \begin{bmatrix} 0 \\ 1 \end{bmatrix}.$$

4 Show that $\left[\begin{smallmatrix} 1 & 2 \\ 3 & 6 \end{smallmatrix}\right]$ has no inverse by trying to solve for the column (x, y):

$$\begin{bmatrix} 1 & 2 \\ 3 & 6 \end{bmatrix}\begin{bmatrix} x & t \\ y & z \end{bmatrix} = \begin{bmatrix} 1 & 0 \\ 0 & 1 \end{bmatrix} \quad \text{must include} \quad \begin{bmatrix} 1 & 2 \\ 3 & 6 \end{bmatrix}\begin{bmatrix} x \\ y \end{bmatrix} = \begin{bmatrix} 1 \\ 0 \end{bmatrix}.$$

5 Find an upper triangular U (not diagonal) with $U^2 = I$ and $U = U^{-1}$.

6 (a) If A is invertible and $AB = AC$, prove quickly that $B = C$.

(b) If $A = \left[\begin{smallmatrix} 1 & 1 \\ 1 & 1 \end{smallmatrix}\right]$, find two matrices $B \neq C$ such that $AB = AC$.

7 (Important) If A has row 1 + row 2 = row 3, show that A is not invertible:

(a) Explain why $A\boldsymbol{x} = (1, 0, 0)$ cannot have a solution.

(b) Which right sides (b_1, b_2, b_3) might allow a solution to $A\boldsymbol{x} = \boldsymbol{b}$?

(c) What happens to row 3 in elimination?

8 If A has column 1 + column 2 = column 3, show that A is not invertible:

(a) Find a nonzero solution $\boldsymbol{x}$ to $A\boldsymbol{x} = \boldsymbol{0}$. The matrix is 3 by 3.

(b) Elimination keeps column 1 + column 2 = column 3. Explain why there is no third pivot.

9 Suppose A is invertible and you exchange its first two rows to reach B. Is the new matrix B invertible and how would you find B^{-1} from A^{-1}?

10 Find the inverses (in any legal way) of

$$A = \begin{bmatrix} 0 & 0 & 0 & 2 \\ 0 & 0 & 3 & 0 \\ 0 & 4 & 0 & 0 \\ 5 & 0 & 0 & 0 \end{bmatrix} \quad \text{and} \quad B = \begin{bmatrix} 3 & 2 & 0 & 0 \\ 4 & 3 & 0 & 0 \\ 0 & 0 & 6 & 5 \\ 0 & 0 & 7 & 6 \end{bmatrix}.$$

11 (a) Find invertible matrices A and B such that $A + B$ is not invertible.

(b) Find singular matrices A and B such that $A + B$ is invertible.

12 If the product $C = AB$ is invertible (A and B are square), then A itself is invertible. Find a formula for A^{-1} that involves C^{-1} and B.

13 If the product $M = ABC$ of three square matrices is invertible, then B is invertible. (So are A and C.) Find a formula for B^{-1} that involves M^{-1} and A and C.

14 If you add row 1 of A to row 2 to get B, how do you find B^{-1} from A^{-1}?

$$\text{Notice the order.}\quad \text{The inverse of}\quad B = \begin{bmatrix} 1 & 0 \\ 1 & 1 \end{bmatrix}\begin{bmatrix} & A & \end{bmatrix} \quad \text{is} \quad \underline{\qquad}.$$

15 Prove that a matrix with a column of zeros cannot have an inverse.

16 Multiply $\begin{bmatrix} a & b \\ c & d \end{bmatrix}$ times $\begin{bmatrix} d & -b \\ -c & a \end{bmatrix}$. What is the inverse of each matrix if $ad \neq bc$?

17 (a) What matrix E has the same effect as these three steps? Subtract row 1 from row 2, subtract row 1 from row 3, then subtract row 2 from row 3.

(b) What single matrix L has the same effect as these three reverse steps? Add row 2 to row 3, add row 1 to row 3, then add row 1 to row 2.

18 If B is the inverse of A^2, show that AB is the inverse of A.

19 Find the numbers a and b that give the inverse of 5 ∗ eye(4) − ones(4,4):

$$\begin{bmatrix} 4 & -1 & -1 & -1 \\ -1 & 4 & -1 & -1 \\ -1 & -1 & 4 & -1 \\ -1 & -1 & -1 & 4 \end{bmatrix}^{-1} = \begin{bmatrix} a & b & b & b \\ b & a & b & b \\ b & b & a & b \\ b & b & b & a \end{bmatrix}.$$

What are a and b in the inverse of 6 ∗ eye(5) − ones(5,5)?

20 Show that A = 4 ∗ eye(4) − ones(4,4) is *not* invertible: Multiply A ∗ ones(4,1).

21 There are sixteen 2 by 2 matrices whose entries are 1's and 0's. How many of them are invertible?

Questions 22–28 are about the Gauss-Jordan method for calculating A^{-1}.

22 Change I into A^{-1} as you reduce A to I (by row operations):

$$\begin{bmatrix} A & I \end{bmatrix} = \begin{bmatrix} 1 & 3 & 1 & 0 \\ 2 & 7 & 0 & 1 \end{bmatrix} \quad \text{and} \quad \begin{bmatrix} A & I \end{bmatrix} = \begin{bmatrix} 1 & 4 & 1 & 0 \\ 3 & 9 & 0 & 1 \end{bmatrix}$$

23 Follow the 3 by 3 text example but with plus signs in A. Eliminate above and below the pivots to reduce $[\,A \;\; I\,]$ to $[\,I \;\; A^{-1}\,]$:

$$\begin{bmatrix} A & I \end{bmatrix} = \begin{bmatrix} 2 & 1 & 0 & 1 & 0 & 0 \\ 1 & 2 & 1 & 0 & 1 & 0 \\ 0 & 1 & 2 & 0 & 0 & 1 \end{bmatrix}.$$

24 Use Gauss-Jordan elimination on $[\,A \quad I\,]$ to solve $AA^{-1} = I$:

$$\begin{bmatrix} 1 & a & b \\ 0 & 1 & c \\ 0 & 0 & 1 \end{bmatrix} \begin{bmatrix} \boldsymbol{x}_1 & \boldsymbol{x}_2 & \boldsymbol{x}_3 \end{bmatrix} = \begin{bmatrix} 1 & 0 & 0 \\ 0 & 1 & 0 \\ 0 & 0 & 1 \end{bmatrix}.$$

25 Find A^{-1} and B^{-1} (*if they exist*) by elimination on $[\,A \quad I\,]$ and $[\,B \quad I\,]$:

$$A = \begin{bmatrix} 2 & 1 & 1 \\ 1 & 2 & 1 \\ 1 & 1 & 2 \end{bmatrix} \quad \text{and} \quad B = \begin{bmatrix} 2 & -1 & -1 \\ -1 & 2 & -1 \\ -1 & -1 & 2 \end{bmatrix}.$$

26 What three matrices E_{21} and E_{12} and D^{-1} reduce $A = \left[\begin{smallmatrix} 1 & 2 \\ 2 & 6 \end{smallmatrix}\right]$ to the identity matrix? Multiply $D^{-1}E_{12}E_{21}$ to find A^{-1}.

27 Invert these matrices A by the Gauss-Jordan method starting with $[\,A \quad I\,]$:

$$A = \begin{bmatrix} 1 & 0 & 0 \\ 2 & 1 & 3 \\ 0 & 0 & 1 \end{bmatrix} \quad \text{and} \quad A = \begin{bmatrix} 1 & 1 & 1 \\ 1 & 2 & 2 \\ 1 & 2 & 3 \end{bmatrix}.$$

28 Exchange rows and continue with Gauss-Jordan to find A^{-1}:

$$[\,A \quad I\,] = \begin{bmatrix} 0 & 2 & 1 & 0 \\ 2 & 2 & 0 & 1 \end{bmatrix}.$$

29 True or false (with a counterexample if false and a reason if true):

(a) A 4 by 4 matrix with a row of zeros is not invertible.

(b) A matrix with 1's down the main diagonal is invertible.

(c) If A is invertible then A^{-1} is invertible.

(d) If A is invertible then A^2 is invertible.

30 For which three numbers c is this matrix not invertible, and why not?

$$A = \begin{bmatrix} 2 & c & c \\ c & c & c \\ 8 & 7 & c \end{bmatrix}.$$

31 Prove that A is invertible if $a \neq 0$ and $a \neq b$ (find the pivots or A^{-1}):

$$A = \begin{bmatrix} a & b & b \\ a & a & b \\ a & a & a \end{bmatrix}.$$

32 This matrix has a remarkable inverse. Find A^{-1} by elimination on $[\,A \quad I\,]$. Extend to a 5 by 5 "alternating matrix" and guess its inverse; then multiply to confirm.

$$A = \begin{bmatrix} 1 & -1 & 1 & -1 \\ 0 & 1 & -1 & 1 \\ 0 & 0 & 1 & -1 \\ 0 & 0 & 0 & 1 \end{bmatrix}.$$

33 Use the 4 by 4 inverse in Question 32 to solve $A\boldsymbol{x} = (1, 1, 1, 1)$.

34 Suppose P and Q have the same rows as I but in any order. Show that $P - Q$ is singular by solving $(P - Q)\boldsymbol{x} = \boldsymbol{0}$.

35 Find and check the inverses (assuming they exist) of these block matrices:

$$\begin{bmatrix} I & 0 \\ C & I \end{bmatrix} \quad \begin{bmatrix} A & 0 \\ C & D \end{bmatrix} \quad \begin{bmatrix} 0 & I \\ I & D \end{bmatrix}.$$

36 If an invertible matrix A commutes with C (this means $AC = CA$) show that A^{-1} commutes with C. If also B commutes with C, show that AB commutes with C. Translation: If $AC = CA$ and $BC = CB$ then $(AB)C = C(AB)$.

37 Could a 4 by 4 matrix A be invertible if every row contains the numbers $0, 1, 2, 3$ in some order? What if every row of B contains $0, 1, 2, -3$ in some order?

38 In the worked example **2.5 B**, the triangular Pascal matrix A has an inverse with "alternating diagonals". Check that this A^{-1} is DAD, where the diagonal matrix D has alternating entries $1, -1, 1, -1$. Then $ADAD = I$, so what is the inverse of $AD =$ pascal (4,1)?

39 The Hilbert matrices have $H_{ij} = 1/(i + j - 1)$. Ask MATLAB for the exact 6 by 6 inverse invhilb(6). Then ask for inv(hilb(6)). How can these be different, when the computer never makes mistakes?

40 Use inv(S) to invert MATLAB's 4 by 4 symmetric matrix S = pascal(4). Create Pascal's lower triangular A = abs(pascal(4,1)) and test inv(S) = inv(A′) * inv(A).

41 If $A =$ ones(4,4) and $\boldsymbol{b} =$ rand(4,1), how does MATLAB tell you that $A\boldsymbol{x} = \boldsymbol{b}$ has no solution? If $\boldsymbol{b} =$ ones(4,1), which solution to $A\boldsymbol{x} = \boldsymbol{b}$ is found by $A\backslash\boldsymbol{b}$?

42 If $AC = I$ and $AC^* = I$ (all square matrices) use **2I** to prove that $C = C^*$.

43 Direct multiplication gives $MM^{-1} = I$, and I would recommend doing #3. M^{-1} shows the change in A^{-1} (useful to know) when a matrix is subtracted from A:

1	$M = I - \boldsymbol{u}\boldsymbol{v}$	and	$M^{-1} = I + \boldsymbol{u}\boldsymbol{v}/(1 - \boldsymbol{v}\boldsymbol{u})$
2	$M = A - \boldsymbol{u}\boldsymbol{v}$	and	$M^{-1} = A^{-1} + A^{-1}\boldsymbol{u}\boldsymbol{v}A^{-1}/(1 - \boldsymbol{v}A^{-1}\boldsymbol{u})$
3	$M = I - UV$	and	$M^{-1} = I_n + U(I_m - VU)^{-1}V$
4	$M = A - UW^{-1}V$	and	$M^{-1} = A^{-1} + A^{-1}U(W - VA^{-1}U)^{-1}VA^{-1}$

The Woodbury-Morrison formula **4** is the "matrix inversion lemma" in engineering. The four identities come from the 1, 1 block when inverting these matrices ($\boldsymbol{v}$ is 1 by n, $\boldsymbol{u}$ is n by 1, V is m by n, U is n by m, $m \le n$):

$$\begin{bmatrix} I & \boldsymbol{u} \\ \boldsymbol{v} & 1 \end{bmatrix} \qquad \begin{bmatrix} A & \boldsymbol{u} \\ \boldsymbol{v} & 1 \end{bmatrix} \qquad \begin{bmatrix} I_n & U \\ V & I_m \end{bmatrix} \qquad \begin{bmatrix} A & U \\ V & W \end{bmatrix}$$

ELIMINATION = FACTORIZATION: $A = LU$ ■ 2.6

Students often say that mathematics courses are too theoretical. Well, not this section. It is almost purely practical. The goal is to describe Gaussian elimination in the most useful way. Many key ideas of linear algebra, when you look at them closely, are really *factorizations* of a matrix. The original matrix A becomes the product of two or three special matrices. The first factorization—also the most important in practice—comes now from elimination. ***The factors are triangular matrices. The factorization that comes from elimination is $A = LU$.***

We already know U, the upper triangular matrix with the pivots on its diagonal. The elimination steps take A to U. We will show how reversing those steps (taking U back to A) is achieved by a lower triangular L. ***The entries of L are exactly the multipliers ℓ_{ij}***—which multiplied row j when it was subtracted from row i.

Start with a 2 by 2 example. The matrix A contains 2, 1, 6, 8. The number to eliminate is 6. ***Subtract 3 times row* 1 *from row* 2.** That step is E_{21} in the forward direction. The return step from U to A is $L = E_{21}^{-1}$ (an addition using $+3$):

$$\textit{Forward from } A \textit{ to } U: \quad E_{21}A = \begin{bmatrix} 1 & 0 \\ -3 & 1 \end{bmatrix}\begin{bmatrix} 2 & 1 \\ 6 & 8 \end{bmatrix} = \begin{bmatrix} 2 & 1 \\ 0 & 5 \end{bmatrix} = U$$

$$\textit{Back from } U \textit{ to } A: \quad E_{21}^{-1}U = \begin{bmatrix} 1 & 0 \\ 3 & 1 \end{bmatrix}\begin{bmatrix} 2 & 1 \\ 0 & 5 \end{bmatrix} = \begin{bmatrix} 2 & 1 \\ 6 & 8 \end{bmatrix} = A.$$

The second line is our factorization. Instead of $E_{21}^{-1}U = A$ we write $LU = A$. Move now to larger matrices with many E's. ***Then L will include all their inverses.***

Each step from A to U multiplies by a matrix E_{ij} to produce zero in the (i, j) position. To keep this clear, we stay with the most frequent case—***when no row exchanges are involved.*** If A is 3 by 3, we multiply by E_{21} and E_{31} and E_{32}. The multipliers ℓ_{ij} produce zeros in the (2, 1) and (3, 1) and (3, 2) positions—all below the diagonal. Elimination ends with the upper triangular U.

Now move those E's onto the other side, *where their inverses multiply U*:

$$(E_{32}E_{31}E_{21})A = U \quad \textbf{becomes} \quad A = (E_{21}^{-1}E_{31}^{-1}E_{32}^{-1})U \quad \textbf{which is} \quad A = LU. \qquad (1)$$

The inverses go in opposite order, as they must. That product of three inverses is L. ***We have reached $A = LU$.*** Now we stop to understand it.

Explanation and Examples

First point: Every inverse matrix E_{ij}^{-1} is *lower triangular*. Its off-diagonal entry is ℓ_{ij}, to undo the subtraction with $-\ell_{ij}$. The main diagonals of E and E^{-1} contain 1's. Our example above had $\ell_{21} = 3$ and $E = \begin{bmatrix} 1 & 0 \\ -3 & 1 \end{bmatrix}$ and $E^{-1} = \begin{bmatrix} 1 & 0 \\ 3 & 1 \end{bmatrix}$.

Second point: Equation (1) shows a lower triangular matrix (the product of E_{ij}) multiplying A. It also shows a lower triangular matrix (the product of E_{ij}^{-1}) multiplying U to bring back A. ***This product of inverses is L.***

One reason for working with the inverses is that we want to factor A, not U. The "inverse form" gives $A = LU$. The second reason is that we get something extra, almost more than we deserve. This is the third point, showing that L is exactly right.

Third point: Each multiplier ℓ_{ij} goes directly into its i, j position—*unchanged*—in the product of inverses which is L. Usually matrix multiplication will mix up all the numbers. Here that doesn't happen. The order is right for the inverse matrices, to keep the ℓ's unchanged. The reason is given below in equation (3).

Since each E^{-1} has 1's down its diagonal, the final good point is that L does too.

2J ($A = LU$) ***This is elimination without row exchanges.*** The upper triangular U has the pivots on its diagonal. The lower triangular L has all 1's on its diagonal. ***The multipliers ℓ_{ij} are below the diagonal of L.***

Example 1 The matrix A has 1, 2, 1 on its diagonals. Elimination subtracts $\frac{1}{2}$ times row 1 from row 2. The last step subtracts $\frac{2}{3}$ times row 2 from row 3. The lower triangular L has $\ell_{21} = \frac{1}{2}$ and $\ell_{32} = \frac{2}{3}$. Multiplying LU produces A:

$$A = \begin{bmatrix} 2 & 1 & 0 \\ 1 & 2 & 1 \\ 0 & 1 & 2 \end{bmatrix} = \begin{bmatrix} 1 & 0 & 0 \\ \frac{1}{2} & 1 & 0 \\ 0 & \frac{2}{3} & 1 \end{bmatrix} \begin{bmatrix} 2 & 1 & 0 \\ 0 & \frac{3}{2} & 1 \\ 0 & 0 & \frac{4}{3} \end{bmatrix} = LU.$$

The (3, 1) multiplier is zero because the (3, 1) entry in A is zero. No operation needed.

Example 2 Change the top left entry from 2 to 1. The pivots all become 1. The multipliers are all 1. That pattern continues when A is 4 by 4:

$$A = \begin{bmatrix} \mathbf{1} & 1 & 0 & 0 \\ 1 & 2 & 1 & 0 \\ 0 & 1 & 2 & 1 \\ 0 & 0 & 1 & 2 \end{bmatrix} = \begin{bmatrix} 1 & & & \\ 1 & 1 & & \\ 0 & 1 & 1 & \\ 0 & 0 & 1 & 1 \end{bmatrix} \begin{bmatrix} 1 & 1 & 0 & 0 \\ & 1 & 1 & 0 \\ & & 1 & 1 \\ & & & 1 \end{bmatrix}.$$

These LU examples are showing something extra, which is very important in practice. Assume no row exchanges. When can we predict *zeros* in L and U?

When a row of A starts with zeros, so does that row of L.

When a column of A starts with zeros, so does that column of U.

If a row starts with zero, we don't need an elimination step. L has a zero, which saves computer time. Similarly, zeros at the *start* of a column survive into U. But please realize: Zeros in the *middle* of a matrix are likely to be filled in, while elimination sweeps forward. We now explain why L has the multipliers ℓ_{ij} in position, with no mix-up.

The key reason why A equals LU: Ask yourself about the pivot rows that are subtracted from lower rows. Are they the original rows of A? *No*, elimination probably changed them. Are they rows of U? *Yes*, the pivot rows never change again. When computing the third row of U, we subtract multiples of earlier rows of U (*not rows of* A!):

$$\text{Row 3 of } U = (\text{Row 3 of } A) - \ell_{31}(\text{Row 1 of } U) - \ell_{32}(\text{Row 2 of } U). \qquad (2)$$

Rewrite this equation to see that the row $[\,\ell_{31} \;\; \ell_{32} \;\; 1\,]$ is multiplying U:

$$(\text{Row 3 of } A) = \ell_{31}(\text{Row 1 of } U) + \ell_{32}(\text{Row 2 of } U) + 1(\text{Row 3 of } U). \qquad (3)$$

This is exactly row 3 of $A = LU$. All rows look like this, whatever the size of A. With no row exchanges, we have $A = LU$.

Remark The LU factorization is "unsymmetric" because U has the pivots on its diagonal where L has 1's. This is easy to change. ***Divide U by a diagonal matrix D that contains the pivots.*** That leaves a new matrix with 1's on the diagonal:

$$\text{Split } U \text{ into } \begin{bmatrix} d_1 & & & \\ & d_2 & & \\ & & \ddots & \\ & & & d_n \end{bmatrix} \begin{bmatrix} 1 & u_{12}/d_1 & u_{13}/d_1 & \cdot \\ & 1 & u_{23}/d_2 & \cdot \\ & & \ddots & \vdots \\ & & & 1 \end{bmatrix}.$$

It is convenient (but a little confusing) to keep the same letter U for this new upper triangular matrix. It has 1's on the diagonal (like L). Instead of the normal LU, the new form has D in the middle: ***Lower triangular L times diagonal D times upper triangular U.***

The triangular factorization can be written $A = LU$ or $A = LDU$.

Whenever you see LDU, it is understood that U has 1's on the diagonal. *Each row is divided by its first nonzero entry—the pivot.* Then L and U are treated evenly in LDU:

$$\begin{bmatrix} 1 & 0 \\ 3 & 1 \end{bmatrix} \begin{bmatrix} 2 & 8 \\ 0 & 5 \end{bmatrix} \quad \text{splits further into} \quad \begin{bmatrix} 1 & 0 \\ 3 & 1 \end{bmatrix} \begin{bmatrix} 2 & \\ & 5 \end{bmatrix} \begin{bmatrix} 1 & 4 \\ 0 & 1 \end{bmatrix}. \qquad (4)$$

The pivots 2 and 5 went into D. Dividing the rows by 2 and 5 left the rows $[\,1 \;\; 4\,]$ and $[\,0 \;\; 1\,]$ in the new U. The multiplier 3 is still in L.

My own lectures sometimes stop at this point. The next paragraphs show how elimination codes are organized, and how long they take. If MATLAB (or any software) is available, I strongly recommend the last problems 32 to 35. You can measure the computing time by just counting the seconds!

One Square System = Two Triangular Systems

The matrix L contains our memory of Gaussian elimination. It holds the numbers that multiplied the pivot rows, before subtracting them from lower rows. When do we need this record and how do we use it?

We need L as soon as there is a *right side* $\boldsymbol{b}$. The factors L and U were completely decided by the left side (the matrix A). On the right side of $A\boldsymbol{x} = \boldsymbol{b}$, we use ***Solve***:

1 ***Factor*** (into L and U, by forward elimination on A)

2 ***Solve*** (forward elimination on $\boldsymbol{b}$ using L, then back substitution using U).

Earlier, we worked on $\boldsymbol{b}$ while we were working on A. No problem with that—just augment A by an extra column $\boldsymbol{b}$. But most computer codes keep the two sides separate. The memory of forward elimination is held in L and U, at no extra cost in storage. Then we process $\boldsymbol{b}$ whenever we want to. The User's Guide to LINPACK remarks that "This situation is so common and the savings are so important that no provision has been made for solving a single system with just one subroutine."

How does ***Solve*** work on $\boldsymbol{b}$? First, apply forward elimination to the right side (the multipliers are stored in L, use them now). This changes $\boldsymbol{b}$ to a new right side $\boldsymbol{c}$—*we are really solving* $L\boldsymbol{c} = \boldsymbol{b}$. Then back substitution solves $U\boldsymbol{x} = \boldsymbol{c}$ as always. The original system $A\boldsymbol{x} = \boldsymbol{b}$ is factored into ***two triangular systems***:

$$\textbf{\textit{Solve}} \quad L\boldsymbol{c} = \boldsymbol{b} \quad \textbf{\textit{and then solve}} \quad U\boldsymbol{x} = \boldsymbol{c}\,. \tag{5}$$

To see that $\boldsymbol{x}$ is correct, multiply $U\boldsymbol{x} = \boldsymbol{c}$ by L. Then $LU\boldsymbol{x} = L\boldsymbol{c}$ is just $A\boldsymbol{x} = \boldsymbol{b}$.

To emphasize: There is *nothing new* about those steps. This is exactly what we have done all along. We were really solving the triangular system $L\boldsymbol{c} = \boldsymbol{b}$ as elimination went forward. Then back substitution produced $\boldsymbol{x}$. An example shows it all.

Example 3 Forward elimination on $A\boldsymbol{x} = \boldsymbol{b}$ ends at $U\boldsymbol{x} = \boldsymbol{c}$:

$$\begin{aligned} u + 2v &= 5 \\ 4u + 9v &= 21 \end{aligned} \qquad \text{becomes} \qquad \begin{aligned} u + 2v &= 5 \\ v &= 1\,. \end{aligned}$$

The multiplier was 4, which is saved in L. The right side used it to find $\boldsymbol{c}$:

$$L\boldsymbol{c} = \boldsymbol{b} \text{ The lower triangular system} \quad \begin{bmatrix} 1 & 0 \\ 4 & 1 \end{bmatrix} \begin{bmatrix} \boldsymbol{c} \end{bmatrix} = \begin{bmatrix} 5 \\ 21 \end{bmatrix} \quad \text{gives} \quad \boldsymbol{c} = \begin{bmatrix} 5 \\ 1 \end{bmatrix}.$$

$$U\boldsymbol{x} = \boldsymbol{c} \text{ The upper triangular system} \quad \begin{bmatrix} 1 & 2 \\ 0 & 1 \end{bmatrix} \begin{bmatrix} \boldsymbol{x} \end{bmatrix} = \begin{bmatrix} 5 \\ 1 \end{bmatrix} \quad \text{gives} \quad \boldsymbol{x} = \begin{bmatrix} 3 \\ 1 \end{bmatrix}.$$

It is satisfying that L and U can take the n^2 storage locations that originally held A. The ℓ's go below the diagonal. The whole discussion is only looking to see what elimination actually did.

The Cost of Elimination

A very practical question is cost—or computing time. Can we solve 1000 equations on a PC? What if $n = 10,000$? Large systems come up all the time in scientific computing, where a three-dimensional problem can easily lead to a million unknowns. We can let the calculation run overnight, but we can't leave it for 100 years.

The first stage of elimination, on column 1, produces zeros below the first pivot. To find each new entry below the pivot row requires one multiplication and one subtraction. *We will count this first stage as n^2 multiplications and n^2 subtractions.* It is actually less, $n^2 - n$, because row 1 does not change.

The next stage clears out the second column below the second pivot. The working matrix is now of size $n - 1$. Estimate this stage by $(n-1)^2$ multiplications and subtractions. The matrices are getting smaller as elimination goes forward. The rough count to reach U is the sum of squares $n^2 + (n-1)^2 + \cdots + 2^2 + 1^2$.

There is an exact formula $\frac{1}{3}n(n+\frac{1}{2})(n+1)$ for this sum of squares. When n is large, the $\frac{1}{2}$ and the 1 are not important. *The number that matters is* $\frac{1}{3}n^3$. The sum of squares is like the integral of x^2! The integral from 0 to n is $\frac{1}{3}n^3$:

Elimination on A requires about $\frac{1}{3}n^3$ multiplications and $\frac{1}{3}n^3$ subtractions.

What about the right side $\boldsymbol{b}$? Going forward, we subtract multiples of b_1 from the lower components $b_2, \ldots, b_n$. This is $n-1$ steps. The second stage takes only $n-2$ steps, because b_1 is not involved. The last stage of forward elimination takes one step.

Now start back substitution. Computing x_n uses one step (divide by the last pivot). The next unknown uses two steps. When we reach x_1 it will require n steps ($n - 1$ substitutions of the other unknowns, then division by the first pivot). The total count on the right side, from $\boldsymbol{b}$ to $\boldsymbol{c}$ to $\boldsymbol{x}$—*forward to the bottom and back to the top*—is exactly n^2:

$$[(n-1)+(n-2)+\cdots+1] \; + \; [1+2+\cdots+(n-1)+n] = n^2. \tag{6}$$

To see that sum, pair off $(n-1)$ with 1 and $(n-2)$ with 2. The pairings leave n terms, each equal to n. That makes n^2. The right side costs a lot less than the left side!

Each right side needs n^2 multiplications and n^2 subtractions.

Here are the MATLAB codes to factor A into LU and to solve $Ax = \boldsymbol{b}$. The program **slu** stops right away if a number smaller than the tolerance "*tol*" appears in a pivot position. Later the program **plu** will look down the column for a pivot, to execute a row exchange and continue solving. These Teaching Codes are on **web.mit.edu/18.06/www**.

```
function [L, U] = slu(A)
%      Square LU factorization with no row exchanges!
[n, n] = size(A);    tol = 1.e - 6;
for k = 1 : n
  if abs(A(k, k)) < tol
  end        % Cannot proceed without a row exchange: stop
  L(k, k) = 1;
  for i = k + 1 : n   % Multipliers for column k are put into L
    L(i, k) = A(i, k)/A(k, k);
    for j = k + 1 : n   % Elimination beyond row k and column k
      A(i, j) = A(i, j) - L(i, k) * A(k, j);    % Matrix still called A
    end
  end
  for j = k : n
    U(k, j) = A(k, j);        % row k is settled, now name it U
  end
end
```

```
function x = slv(A, b)
%      Solve Ax = b using L and U from slu(A). No row exchanges!
[L, U] = slu(A);
for k = 1 : n
  for j = 1 : k - 1
    s = s + L(k, j) * c(j);
  end
  c(k) = b(k) - s;    % Forward elimination to solve Lc = b
end
for k = n : -1 : 1   % Going backwards from x(n) to x(1)
  for j = k + 1 : n   % Back substitution
    t = t + U(k, j) * x(j);
  end
  x(k) = (c(k) - t)/U(k, k);    % Divide by pivot
end
x = x';    % Transpose to column vector
```

How long does it take to solve $Ax = b$? For a random matrix of order $n = 1000$, we tried the MATLAB command tic; $A\backslash b$; toc. The time on my PC was 3 seconds. For $n = 2000$ the time was 20 seconds, which is approaching the n^3 rule. The time is multiplied by about 8 when n is multiplied by 2.

According to this n^3 rule, matrices that are 10 times as large (order 10,000) will take thousands of seconds. Matrices of order 100,000 will take millions of seconds.

This is too expensive without a supercomputer, but remember that these matrices are full. Most matrices in practice are sparse (many zero entries). In that case $A = LU$ is much faster. For tridiagonal matrices of order 10,000, storing only the nonzeros, solving $Ax = b$ is a breeze.

■ REVIEW OF THE KEY IDEAS ■

1. Gaussian elimination (with no row exchanges) factors A into L times U.
2. The lower triangular L contains the numbers that multiply pivot rows, going from A to U. The product LU adds those rows back to recover A.
3. On the right side we solve $Lc = b$ (forward) and $Ux = c$ (backwards).
4. There are $\frac{1}{3}(n^3 - n)$ multiplications and subtractions on the left side.
5. There are n^2 multiplications and subtractions on the right side.

■ WORKED EXAMPLES ■

2.6 A The lower triangular Pascal matrix P_L was in the worked example **2.5 B**. (It contains the "*Pascal triangle*" and Gauss-Jordan found its inverse.) This problem connects P_L to the *symmetric* Pascal matrix P_S and the upper triangular P_U. The symmetric P_S has Pascal's triangle tilted, so each entry is the sum of the entry above and the entry to the left. The n by n symmetric P_S is pascal(n) in MATLAB.

Problem: *Establish the amazing lower-upper factorization* $P_S = P_L P_U$:

$$\text{pascal(4)} = \begin{bmatrix} 1 & 1 & 1 & 1 \\ 1 & 2 & 3 & 4 \\ 1 & 3 & 6 & 10 \\ 1 & 4 & 10 & 20 \end{bmatrix} = \begin{bmatrix} 1 & 0 & 0 & 0 \\ 1 & 1 & 0 & 0 \\ 1 & 2 & 1 & 0 \\ 1 & 3 & 3 & 1 \end{bmatrix} \begin{bmatrix} 1 & 1 & 1 & 1 \\ 0 & 1 & 2 & 3 \\ 0 & 0 & 1 & 3 \\ 0 & 0 & 0 & 1 \end{bmatrix} = P_L P_U.$$

Then predict and check the next row and column for 5 by 5 Pascal matrices.

Solution You could multiply $P_L P_U$ to get P_S. Better to start with the symmetric P_S and reach the upper triangular P_U by elimination:

$$P_S = \begin{bmatrix} 1 & 1 & 1 & 1 \\ 1 & 2 & 3 & 4 \\ 1 & 3 & 6 & 10 \\ 1 & 4 & 10 & 20 \end{bmatrix} \to \begin{bmatrix} 1 & 1 & 1 & 1 \\ 0 & 1 & 2 & 3 \\ 0 & 2 & 5 & 9 \\ 0 & 3 & 9 & 19 \end{bmatrix} \to \begin{bmatrix} 1 & 1 & 1 & 1 \\ 0 & 1 & 2 & 3 \\ 0 & 0 & 1 & 3 \\ 0 & 0 & 3 & 10 \end{bmatrix} \to \begin{bmatrix} 1 & 1 & 1 & 1 \\ 0 & 1 & 2 & 3 \\ 0 & 0 & 1 & 3 \\ 0 & 0 & 0 & 1 \end{bmatrix} = P_U.$$

The multipliers ℓ_{ij} that entered these steps go perfectly into P_L. Then $P_S = P_L P_U$ is a particularly neat example of $A = LU$. *Notice that every pivot is* 1! The pivots are on the diagonal of P_U. The next section will show how symmetry produces a special relationship between the triangular L and U. You see P_U as the "transpose" of P_L.

You might expect the MATLAB command lu(pascal(4)) to produce these factors P_L and P_U. That doesn't happen because the **lu** subroutine chooses the largest available pivot in each column (it will exchange rows so the second pivot is 3). But a different command chol factors without row exchanges. Then $[L, U] =$ chol(pascal(4)) produces the triangular Pascal matrices as L and U. Try it.

In the 5 by 5 case the new fifth rows do maintain $P_S = P_L P_U$:

Next Row $\quad 1 \quad 5 \quad 15 \quad 35 \quad 70 \quad$ for $P_S \qquad 1 \quad 4 \quad 6 \quad 4 \quad 1 \quad$ for P_L

I will only check that this fifth row of P_L times the (same) fifth column of P_U gives $1^2 + 4^2 + 6^2 + 4^2 + 1^2 = 70$ in the fifth row of P_S. The full proof of $P_S = P_L P_U$ is quite fascinating—this factorization can be reached in at least four different ways. I am going to put these proofs on the course web page **web.mit.edu/18.06/www**, which is also available through MIT's *OpenCourseWare* at **ocw.mit.edu**.

These Pascal matrices P_S, P_L, P_U have so many remarkable properties—we will see them again. You could locate them using the Index at the end of the book.

2.6 B The problem is: *Solve* $P_S \boldsymbol{x} = \boldsymbol{b} = (1, 0, 0, 0)$. This special right side means that $\boldsymbol{x}$ will be the first column of P_S^{-1}. That is Gauss-Jordan, matching the columns of $P_S P_S^{-1} = I$. We already know the triangular P_L and P_U from **2.6 A**, so we solve

$$P_L \boldsymbol{c} = \boldsymbol{b} \text{ (forward substitution)} \qquad P_U \boldsymbol{x} = \boldsymbol{c} \text{ (back substitution)}.$$

Use MATLAB to find the full inverse matrix P_S^{-1}.

Solution The lower triangular system $P_L \boldsymbol{c} = \boldsymbol{b}$ is solved *top to bottom*:

$$\begin{aligned} c_1 \qquad\qquad\quad &= 1 \\ c_1 + c_2 \qquad\quad &= 0 \\ c_1 + 2c_2 + c_3 \quad &= 0 \\ c_1 + 3c_2 + 3c_3 + c_4 &= 0 \end{aligned} \qquad \text{gives} \qquad \begin{aligned} c_1 &= +1 \\ c_2 &= -1 \\ c_3 &= +1 \\ c_4 &= -1 \end{aligned}$$

Forward elimination is multiplication by P_L^{-1}. It produces the upper triangular system $P_U \boldsymbol{x} = \boldsymbol{c}$. The solution $\boldsymbol{x}$ comes as always by back substitution, *bottom to top*:

$$\begin{aligned} x_1 + x_2 + x_3 + x_4 &= 1 \\ x_2 + 2x_3 + 3x_4 &= -1 \\ x_3 + 3x_4 &= 1 \\ x_4 &= -1 \end{aligned} \qquad \text{gives} \qquad \begin{aligned} x_1 &= +4 \\ x_2 &= -6 \\ x_3 &= +4 \\ x_4 &= -1 \end{aligned}$$

The complete inverse matrix P_S^{-1} has that $\boldsymbol{x}$ in its first column:

$$\text{inv(pascal(4))} = \begin{bmatrix} 4 & -6 & 4 & -1 \\ -6 & 14 & -11 & 3 \\ 4 & -11 & 10 & -3 \\ -1 & 3 & -3 & 1 \end{bmatrix}.$$

Problem Set 2.6

Problems 1–14 compute the factorization $A = LU$ (and also $A = LDU$).

1 (Important) Forward elimination changes $\begin{bmatrix} 1 & 1 \\ 1 & 2 \end{bmatrix}\boldsymbol{x} = \boldsymbol{b}$ to a triangular $\begin{bmatrix} 1 & 1 \\ 0 & 1 \end{bmatrix}\boldsymbol{x} = \boldsymbol{c}$:

$$\begin{aligned} x + y &= 5 \\ x + 2y &= 7 \end{aligned} \quad \longrightarrow \quad \begin{aligned} x + y &= 5 \\ y &= 2 \end{aligned} \qquad \begin{bmatrix} 1 & 1 & \mathbf{5} \\ 1 & 2 & \mathbf{7} \end{bmatrix} \longrightarrow \begin{bmatrix} 1 & 1 & \mathbf{5} \\ 0 & 1 & \mathbf{2} \end{bmatrix}$$

That step subtracted $\ell_{21} =$ _____ times row 1 from row 2. The reverse step *adds* ℓ_{21} times row 1 to row 2. The matrix for that reverse step is $L =$ _____. Multiply this L times the triangular system $\begin{bmatrix} 1 & 1 \\ 0 & 1 \end{bmatrix}\boldsymbol{x} = \begin{bmatrix} 5 \\ 2 \end{bmatrix}$ to get _____ = _____. In letters, L multiplies $U\boldsymbol{x} = \boldsymbol{c}$ to give _____.

2 (Move to 3 by 3) Forward elimination changes $A\boldsymbol{x} = \boldsymbol{b}$ to a triangular $U\boldsymbol{x} = \boldsymbol{c}$:

$$\begin{aligned} x + y + z &= 5 \\ x + 2y + 3z &= 7 \\ x + 3y + 6z &= 11 \end{aligned} \qquad \begin{aligned} x + y + z &= 5 \\ y + 2z &= 2 \\ 2y + 5z &= 6 \end{aligned} \qquad \begin{aligned} x + y + z &= 5 \\ y + 2z &= 2 \\ z &= 2 \end{aligned}$$

The equation $z = 2$ in $U\boldsymbol{x} = \boldsymbol{c}$ comes from the original $x + 3y + 6z = 11$ in $A\boldsymbol{x} = \boldsymbol{b}$ by subtracting $\ell_{31} =$ _____ times equation 1 and $\ell_{32} =$ _____ times the *final* equation 2. Reverse that to recover $[1\ \ 3\ \ 6\ \ 11]$ in A and $\boldsymbol{b}$ from the final $[1\ \ 1\ \ 1\ \ 5]$ and $[0\ \ 1\ \ 2\ \ 2]$ and $[0\ \ 0\ \ 1\ \ 2]$ in U and $\boldsymbol{c}$:

$$\text{Row 3 of } \begin{bmatrix} A & \boldsymbol{b} \end{bmatrix} = (\ell_{31} \text{ Row } 1 + \ell_{32} \text{ Row } 2 + 1 \text{ Row } 3) \text{ of } \begin{bmatrix} U & \boldsymbol{c} \end{bmatrix}.$$

In matrix notation this is multiplication by L. So $A = LU$ and $\boldsymbol{b} = L\boldsymbol{c}$.

3 Write down the 2 by 2 triangular systems $L\boldsymbol{c} = \boldsymbol{b}$ and $U\boldsymbol{x} = \boldsymbol{c}$ from Problem 1. Check that $\boldsymbol{c} = (5, 2)$ solves the first one. Find $\boldsymbol{x}$ that solves the second one.

4 What are the 3 by 3 triangular systems $L\boldsymbol{c} = \boldsymbol{b}$ and $U\boldsymbol{x} = \boldsymbol{c}$ from Problem 2? Check that $\boldsymbol{c} = (5, 2, 2)$ solves the first one. Which $\boldsymbol{x}$ solves the second one?

5 What matrix E puts A into triangular form $EA = U$? Multiply by $E^{-1} = L$ to factor A into LU:

$$A = \begin{bmatrix} 2 & 1 & 0 \\ 0 & 4 & 2 \\ 6 & 3 & 5 \end{bmatrix}.$$

6 What two elimination matrices E_{21} and E_{32} put A into upper triangular form $E_{32}E_{21}A = U$? Multiply by E_{32}^{-1} and E_{21}^{-1} to factor A into $LU = E_{21}^{-1}E_{32}^{-1}U$:

$$A = \begin{bmatrix} 1 & 1 & 1 \\ 2 & 4 & 5 \\ 0 & 4 & 0 \end{bmatrix}.$$

7 What three elimination matrices E_{21}, E_{31}, E_{32} put A into upper triangular form $E_{32}E_{31}E_{21}A = U$? Multiply by E_{32}^{-1}, E_{31}^{-1} and E_{21}^{-1} to factor A into LU where $L = E_{21}^{-1}E_{31}^{-1}E_{32}^{-1}$. Find L and U:

$$A = \begin{bmatrix} 1 & 0 & 1 \\ 2 & 2 & 2 \\ 3 & 4 & 5 \end{bmatrix}.$$

8 Suppose A is already lower triangular with 1's on the diagonal. Then $U = I$!

$$A = L = \begin{bmatrix} 1 & 0 & 0 \\ a & 1 & 0 \\ b & c & 1 \end{bmatrix}.$$

The elimination matrices E_{21}, E_{31}, E_{32} contain $-a$ then $-b$ then $-c$.

(a) Multiply $E_{32}E_{31}E_{21}$ to find the single matrix E that produces $EA = I$.

(b) Multiply $E_{21}^{-1}E_{31}^{-1}E_{32}^{-1}$ to bring back L (nicer than E).

9 When zero appears in a pivot position, $A = LU$ *is not possible*! (We are requiring nonzero pivots in U.) Show directly why these are both impossible:

$$\begin{bmatrix} 0 & 1 \\ 2 & 3 \end{bmatrix} = \begin{bmatrix} 1 & 0 \\ l & 1 \end{bmatrix} \begin{bmatrix} d & e \\ 0 & f \end{bmatrix} \qquad \begin{bmatrix} 1 & 1 & 0 \\ 1 & 1 & 2 \\ 1 & 2 & 1 \end{bmatrix} = \begin{bmatrix} 1 & & \\ l & 1 & \\ m & n & 1 \end{bmatrix} \begin{bmatrix} d & e & g \\ & f & h \\ & & i \end{bmatrix}.$$

This difficulty is fixed by a row exchange. That needs a "permutation" P.

10 Which number c leads to zero in the second pivot position? A row exchange is needed and $A = LU$ is not possible. Which c produces zero in the third pivot position? Then a row exchange can't help and elimination fails:

$$A = \begin{bmatrix} 1 & c & 0 \\ 2 & 4 & 1 \\ 3 & 5 & 1 \end{bmatrix}.$$

11 What are L and D for this matrix A? What is U in $A = LU$ and what is the new U in $A = LDU$?

$$A = \begin{bmatrix} 2 & 4 & 8 \\ 0 & 3 & 9 \\ 0 & 0 & 7 \end{bmatrix}.$$

12 A and B are symmetric across the diagonal (because $4 = 4$). Find their triple factorizations LDU and say how U is related to L for these symmetric matrices:

$$A = \begin{bmatrix} 2 & 4 \\ 4 & 11 \end{bmatrix} \quad \text{and} \quad B = \begin{bmatrix} 1 & 4 & 0 \\ 4 & 12 & 4 \\ 0 & 4 & 0 \end{bmatrix}.$$

13 (Recommended) Compute L and U for the symmetric matrix

$$A = \begin{bmatrix} a & a & a & a \\ a & b & b & b \\ a & b & c & c \\ a & b & c & d \end{bmatrix}.$$

Find four conditions on a, b, c, d to get $A = LU$ with four pivots.

14 Find L and U for the nonsymmetric matrix

$$A = \begin{bmatrix} a & r & r & r \\ a & b & s & s \\ a & b & c & t \\ a & b & c & d \end{bmatrix}.$$

Find the four conditions on a, b, c, d, r, s, t to get $A = LU$ with four pivots.

Problems 15-16 use L and U (without needing A) to solve $Ax = b$.

15 Solve the triangular system $L\boldsymbol{c} = \boldsymbol{b}$ to find $\boldsymbol{c}$. Then solve $U\boldsymbol{x} = \boldsymbol{c}$ to find $\boldsymbol{x}$:

$$L = \begin{bmatrix} 1 & 0 \\ 4 & 1 \end{bmatrix} \quad \text{and} \quad U = \begin{bmatrix} 2 & 4 \\ 0 & 1 \end{bmatrix} \quad \text{and} \quad \boldsymbol{b} = \begin{bmatrix} 2 \\ 11 \end{bmatrix}.$$

For safety find $A = LU$ and solve $A\boldsymbol{x} = \boldsymbol{b}$ as usual. Circle $\boldsymbol{c}$ when you see it.

16 Solve $L\boldsymbol{c} = \boldsymbol{b}$ to find $\boldsymbol{c}$. Then solve $U\boldsymbol{x} = \boldsymbol{c}$ to find $\boldsymbol{x}$. What was A?

$$L = \begin{bmatrix} 1 & 0 & 0 \\ 1 & 1 & 0 \\ 1 & 1 & 1 \end{bmatrix} \quad \text{and} \quad U = \begin{bmatrix} 1 & 1 & 1 \\ 0 & 1 & 1 \\ 0 & 0 & 1 \end{bmatrix} \quad \text{and} \quad \boldsymbol{b} = \begin{bmatrix} 4 \\ 5 \\ 6 \end{bmatrix}.$$

17 (a) When you apply the usual elimination steps to L, what matrix do you reach?

$$L = \begin{bmatrix} 1 & 0 & 0 \\ l_{21} & 1 & 0 \\ l_{31} & l_{32} & 1 \end{bmatrix}.$$

(b) When you apply the same steps to I, what matrix do you get?

(c) When you apply the same steps to LU, what matrix do you get?

18 If $A = LDU$ and also $A = L_1D_1U_1$ with all factors invertible, then $L = L_1$ and $D = D_1$ and $U = U_1$. *"The factors are unique."*

Derive the equation $L_1^{-1}LD = D_1U_1U^{-1}$. Are the two sides triangular or diagonal? Deduce $L = L_1$ and $U = U_1$ (they all have diagonal 1's). Then $D = D_1$.

19 *Tridiagonal matrices* have zero entries except on the main diagonal and the two adjacent diagonals. Factor these into $A = LU$ and $A = LDL^{\mathrm{T}}$:

$$A = \begin{bmatrix} 1 & 1 & 0 \\ 1 & 2 & 1 \\ 0 & 1 & 2 \end{bmatrix} \quad \text{and} \quad A = \begin{bmatrix} a & a & 0 \\ a & a+b & b \\ 0 & b & b+c \end{bmatrix}.$$

20 When T is tridiagonal, its L and U factors have only two nonzero diagonals. How would you take advantage of the zeros in T in a computer code for Gaussian elimination? Find L and U.

$$T = \begin{bmatrix} 1 & 2 & 0 & 0 \\ 2 & 3 & 1 & 0 \\ 0 & 1 & 2 & 3 \\ 0 & 0 & 3 & 4 \end{bmatrix}.$$

21 If A and B have nonzeros in the positions marked by x, which zeros (marked by 0) are still zero in their factors L and U?

$$A = \begin{bmatrix} x & x & x & x \\ x & x & x & 0 \\ 0 & x & x & x \\ 0 & 0 & x & x \end{bmatrix} \quad \text{and} \quad B = \begin{bmatrix} x & x & x & 0 \\ x & x & 0 & x \\ x & 0 & x & x \\ 0 & x & x & x \end{bmatrix}.$$

22 After elimination has produced zeros below the first pivot, put x's to show which blank entries are known in the final L and U:

$$\begin{bmatrix} x & x & x \\ x & x & x \\ x & x & x \end{bmatrix} = \begin{bmatrix} 1 & 0 & 0 \\ & 1 & 0 \\ & & 1 \end{bmatrix} \begin{bmatrix} & & \\ 0 & & \\ 0 & & \end{bmatrix}.$$

23 Suppose you eliminate upwards (almost unheard of). Use the last row to produce zeros in the last column (the pivot is 1). Then use the second row to produce zero above the second pivot. Find the factors in $A = UL(!)$:

$$A = \begin{bmatrix} 5 & 3 & 1 \\ 3 & 3 & 1 \\ 1 & 1 & 1 \end{bmatrix}.$$

24 Collins uses elimination in both directions, meeting at the center. Substitution goes out from the center. After eliminating both 2's in A, one from above and one from below, what 4 by 4 matrix is left? Solve $A\boldsymbol{x} = \boldsymbol{b}$ his way.

$$A = \begin{bmatrix} 1 & 1 & 0 & 0 \\ 2 & 1 & 1 & 0 \\ 0 & 1 & 3 & 2 \\ 0 & 0 & 1 & 1 \end{bmatrix} \quad \text{and} \quad \boldsymbol{b} = \begin{bmatrix} 5 \\ 8 \\ 8 \\ 2 \end{bmatrix}.$$

25 (Important) If A has pivots $2, 7, 6$ with no row exchanges, what are the pivots for the upper left 2 by 2 submatrix B (without row 3 and column 3)? Explain why.

26 Starting from a 3 by 3 matrix A with pivots $2, 7, 6$, add a fourth row and column to produce M. What are the first three pivots for M, and why? What fourth row and column are sure to produce 9 as the fourth pivot?

27 Use chol(pascal(5)) to find the triangular Pascal factors as in Worked Example **2.6 A**. Show how row exchanges in $[\,L, U\,] =$ lu(pascal(5)) spoil Pascal's pattern!

28 (Careful review) For which numbers c is $A = LU$ impossible—with three pivots?

$$A = \begin{bmatrix} 1 & 2 & 0 \\ 3 & c & 1 \\ 0 & 1 & 1 \end{bmatrix}$$

29 Change the program **slu**(A) into **sldu**(A), so that it produces L, D, and U. Put L, D, U into the n^2 storage locations that held the original A. The extra storage used for L is not required.

30 Explain in words why $x(k)$ is $(c(k) - t)/U(k,k)$ at the end of **slv**(A, b).

31 Write a program that multiplies a two-diagonal L times a two-diagonal U. Don't loop from 1 to n when you know there are zeros! L times U should undo **slu**.

32 I just learned MATLAB's tic–toc command, which measures computing time. Previously I counted seconds until the answer appeared, which required very large problems—now A = rand(1000) and $\boldsymbol{b}$ = rand(1000,1) is large enough. How much faster is tic; $A\backslash\boldsymbol{b}$; toc for elimination than tic; inv$(A) * \boldsymbol{b}$; toc which computes A^{-1}?

33 Compare tic; inv(A); toc for A = rand(500) and A = rand(1000). The n^3 operation count says that doubling n should multiply computing time by 8.

34 I = eye(1000); A = rand(1000); B = triu(A); produces a random *triangular* matrix B. Compare the times for inv(B) and $B \setminus I$. Backslash is engineered to use the zeros in B, while inv uses the zeros in I when reducing $[\,B \;\; I\,]$ by Gauss-Jordan. (Compare also with inv(A) and $A \setminus I$ for the full matrix A.)

35 Estimate the time difference for each new right side $\boldsymbol{b}$ when $n = 800$. Create A = rand(800) and $\boldsymbol{b}$ = rand(800,1) and B = rand(800,9). Compare tic; $A\backslash\boldsymbol{b}$; toc and tic; $A\backslash B$; toc (which solves for 9 right sides).

36 Show that L^{-1} has entries j/i on and below its main diagonal:

$$L = \begin{bmatrix} 1 & 0 & 0 & 0 \\ -\frac{1}{2} & 1 & 0 & 0 \\ 0 & -\frac{2}{3} & 1 & 0 \\ 0 & 0 & -\frac{3}{4} & 1 \end{bmatrix} \quad \text{and} \quad L^{-1} = \begin{bmatrix} 1 & 0 & 0 & 0 \\ \frac{1}{2} & 1 & 0 & 0 \\ \frac{1}{3} & \frac{2}{3} & 1 & 0 \\ \frac{1}{4} & \frac{2}{4} & \frac{3}{4} & 1 \end{bmatrix}.$$

I think this pattern continues for $L = \text{eye}(5) - \text{diag}(1{:}5)\backslash\text{diag}(1{:}4, -1)$ and $\text{inv}(L)$.

TRANSPOSES AND PERMUTATIONS ■ 2.7

We need one more matrix, and fortunately it is much simpler than the inverse. It is the ***"transpose"*** of A, which is denoted by A^{T}. *The columns of A^{T} are the rows of A.*

When A is an m by n matrix, the transpose is n by m:

$$\text{If} \quad A = \begin{bmatrix} 1 & 2 & 3 \\ 0 & 0 & 4 \end{bmatrix} \quad \text{then} \quad A^{\mathrm{T}} = \begin{bmatrix} 1 & 0 \\ 2 & 0 \\ 3 & 4 \end{bmatrix}.$$

You can write the rows of A into the columns of A^{T}. Or you can write the columns of A into the rows of A^{T}. The matrix "flips over" its main diagonal. The entry in row i, column j of A^{T} comes from row j, column i of the original A:

$$(A^{\mathrm{T}})_{ij} = A_{ji}.$$

The transpose of a lower triangular matrix is upper triangular. (But the inverse is still lower triangular.) The transpose of A^{T} is A.

Note MATLAB's symbol for the transpose of A is A'. Typing [1 2 3] gives a row vector and the column vector is $\boldsymbol{v} = [\,1 \;\; 2 \;\; 3\,]'$. To enter a matrix M with second column $\boldsymbol{w} = [\,4 \;\, 5 \;\, 6\,]'$ you could define $M = [\,\boldsymbol{v} \;\; \boldsymbol{w}\,]$. Quicker to enter by rows and then transpose the whole matrix: $M = [\,1 \;\; 2 \;\; 3; \;\; 4 \;\; 5 \;\; 6\,]'$.

The rules for transposes are very direct. We can transpose $A+B$ to get $(A+B)^{\mathrm{T}}$. Or we can transpose A and B separately, and then add $A^{\mathrm{T}} + B^{\mathrm{T}}$—same result. The serious questions are about the transpose of a product AB and an inverse A^{-1}:

$$\text{The transpose of} \quad A+B \quad \text{is} \quad A^{\mathrm{T}} + B^{\mathrm{T}}. \tag{1}$$

$$\text{The transpose of} \quad AB \quad \text{is} \quad (AB)^{\mathrm{T}} = B^{\mathrm{T}} A^{\mathrm{T}}. \tag{2}$$

$$\text{The transpose of} \quad A^{-1} \quad \text{is} \quad (A^{-1})^{\mathrm{T}} = (A^{\mathrm{T}})^{-1}. \tag{3}$$

Notice especially how $B^{\mathrm T}A^{\mathrm T}$ comes in reverse order. For inverses, this reverse order was quick to check: $B^{-1}A^{-1}$ times AB produces I. To understand $(AB)^{\mathrm T} = B^{\mathrm T}A^{\mathrm T}$, start with $(A\boldsymbol{x})^{\mathrm T} = \boldsymbol{x}^{\mathrm T}A^{\mathrm T}$:

$$A\boldsymbol{x} \textbf{ combines the columns of } A \textbf{ while } \boldsymbol{x}^{\mathrm T}A^{\mathrm T} \textbf{ combines the rows of } A^{\mathrm T}.$$

It is the same combination of the same vectors! In A they are columns, in $A^{\mathrm T}$ they are rows. So the transpose of the column $A\boldsymbol{x}$ is the row $\boldsymbol{x}^{\mathrm T}A^{\mathrm T}$. That fits our formula $(A\boldsymbol{x})^{\mathrm T} = \boldsymbol{x}^{\mathrm T}A^{\mathrm T}$. Now we can prove the formula for $(AB)^{\mathrm T}$.

When $B = [\boldsymbol{x}_1 \ \boldsymbol{x}_2]$ has two columns, apply the same idea to each column. The columns of AB are $A\boldsymbol{x}_1$ and $A\boldsymbol{x}_2$. Their transposes are the rows of $B^{\mathrm T}A^{\mathrm T}$:

$$\text{Transposing } AB = \begin{bmatrix} A\boldsymbol{x}_1 & A\boldsymbol{x}_2 & \cdots \end{bmatrix} \text{ gives } \begin{bmatrix} \boldsymbol{x}_1^{\mathrm T}A^{\mathrm T} \\ \boldsymbol{x}_2^{\mathrm T}A^{\mathrm T} \\ \vdots \end{bmatrix} \text{ which is } B^{\mathrm T}A^{\mathrm T}. \quad (4)$$

The right answer $B^{\mathrm T}A^{\mathrm T}$ comes out a row at a time. There is also a "***transparent proof***" by looking through the page at the end of the problem set. Here are numbers!

$$AB = \begin{bmatrix} 1 & 0 \\ 1 & 1 \end{bmatrix}\begin{bmatrix} 5 & 0 \\ 4 & 1 \end{bmatrix} = \begin{bmatrix} 5 & 0 \\ 9 & 1 \end{bmatrix} \quad \text{and} \quad B^{\mathrm T}A^{\mathrm T} = \begin{bmatrix} 5 & 4 \\ 0 & 1 \end{bmatrix}\begin{bmatrix} 1 & 1 \\ 0 & 1 \end{bmatrix} = \begin{bmatrix} 5 & 9 \\ 0 & 1 \end{bmatrix}.$$

The reverse order rule extends to three or more factors: $(ABC)^{\mathrm T}$ equals $C^{\mathrm T}B^{\mathrm T}A^{\mathrm T}$.

$$\textbf{\textit{If}}\ \ A = LDU \ \ \textbf{\textit{then}}\ \ A^{\mathrm T} = U^{\mathrm T}D^{\mathrm T}L^{\mathrm T}. \quad \textbf{\textit{The pivot matrix has}}\ \ D = D^{\mathrm T}.$$

Now apply this product rule to both sides of $A^{-1}A = I$. On one side, $I^{\mathrm T}$ is I. We confirm the rule that $(A^{-1})^{\mathrm T}$ is the inverse of $A^{\mathrm T}$:

$$A^{-1}A = I \quad \text{is transposed to} \quad A^{\mathrm T}(A^{-1})^{\mathrm T} = I. \qquad (5)$$

Similarly $AA^{-1} = I$ leads to $(A^{-1})^{\mathrm T}A^{\mathrm T} = I$. We can invert the transpose or transpose the inverse. Notice especially: $A^{\mathrm T}$ ***is invertible exactly when*** A ***is invertible.***

Example 1 The inverse of $A = \begin{bmatrix} 1 & 0 \\ 6 & 1 \end{bmatrix}$ is $A^{-1} = \begin{bmatrix} 1 & 0 \\ -6 & 1 \end{bmatrix}$. The transpose is $A^{\mathrm T} = \begin{bmatrix} 1 & 6 \\ 0 & 1 \end{bmatrix}$.

$$(A^{-1})^{\mathrm T} \ \ \textbf{\textit{and}} \ \ (A^{\mathrm T})^{-1} \ \ \textbf{\textit{are both equal to}} \ \ \begin{bmatrix} 1 & -6 \\ 0 & 1 \end{bmatrix}.$$

Before leaving these rules, we call attention to dot products. The following statement looks extremely simple, but it actually contains the deep purpose for the transpose. For any vectors $\boldsymbol{x}$ and $\boldsymbol{y}$,

$$(A\boldsymbol{x})^{\mathrm T}\boldsymbol{y} \quad \text{equals} \quad \boldsymbol{x}^{\mathrm T}A^{\mathrm T}\boldsymbol{y} \quad \text{equals} \quad \boldsymbol{x}^{\mathrm T}(A^{\mathrm T}\boldsymbol{y}). \qquad (6)$$

When A moves from one side of a dot product to the other side, it becomes $A^{\mathrm T}$.

Here are two quick applications to electrical engineering and mechanical engineering (with more in Chapter 8). The same A and A^{T} appear in both applications.

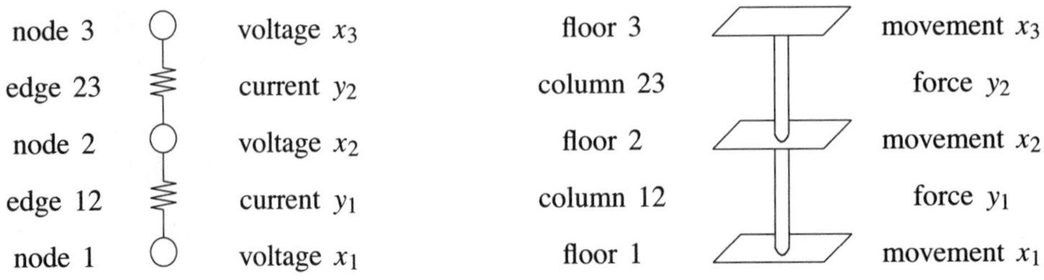

Figure 2.9 A line of resistors and a structure, both governed by A and A^{T}.

Electrical Networks The vector $\boldsymbol{x} = (x_1, x_2, x_3)$ gives voltages at 3 nodes, and $A\boldsymbol{x}$ gives the voltage differences across 2 edges. The "difference matrix" A is 2 by 3:

$$A\boldsymbol{x} = \begin{bmatrix} 1 & -1 & 0 \\ 0 & 1 & -1 \end{bmatrix} \begin{bmatrix} x_1 \\ x_2 \\ x_3 \end{bmatrix} = \begin{bmatrix} x_1 - x_2 \\ x_2 - x_3 \end{bmatrix} = \text{voltage differences.}$$

The vector $\boldsymbol{y} = (y_1, y_2)$ gives currents on those edges (node 1 to 2, and node 2 to 3). Look how $A^{\mathrm{T}}\boldsymbol{y}$ finds the total currents leaving each node in Kirchhoff's Current Law:

$$A^{\mathrm{T}}\boldsymbol{y} = \begin{bmatrix} 1 & 0 \\ -1 & 1 \\ 0 & -1 \end{bmatrix} \begin{bmatrix} y_1 \\ y_2 \end{bmatrix} = \begin{bmatrix} y_1 \\ y_2 - y_1 \\ -y_2 \end{bmatrix} = \begin{bmatrix} \text{current leaving node 1} \\ \textbf{out minus in} \text{ at node 2} \\ \text{current leaving node 3} \end{bmatrix}.$$

Section 8.2 studies networks in detail. Here we look at the energy $\boldsymbol{x}^{\mathrm{T}}A^{\mathrm{T}}\boldsymbol{y}$ lost as heat:

$$\textit{Energy } (\textbf{voltages } \boldsymbol{x}) \cdot (\textbf{inputs } A^{\mathrm{T}}\boldsymbol{y}) = \textit{Heat loss } (\textbf{voltage drops } A\boldsymbol{x}) \cdot (\textbf{currents } \boldsymbol{y}).$$

Forces on a Structure The vector $\boldsymbol{x} = (x_1, x_2, x_3)$ gives the movement of each floor under the weight of the floors above. The matrix A takes differences of the x's to give the strains $A\boldsymbol{x}$, the movements between floors:

$$A\boldsymbol{x} = \begin{bmatrix} 1 & -1 & 0 \\ 0 & 1 & -1 \end{bmatrix} \begin{bmatrix} x_1 \\ x_2 \\ x_3 \end{bmatrix} = \begin{bmatrix} x_1 - x_2 \\ x_2 - x_3 \end{bmatrix} = \begin{bmatrix} \text{movement between 1 and 2} \\ \text{movement between 2 and 3} \end{bmatrix}.$$

The vector $\boldsymbol{y} = (y_1, y_2)$ gives the stresses (internal forces from the columns that resist the movement and save the structure). Then $A^{\mathrm{T}}\boldsymbol{y}$ gives the forces that balance the weight:

$$A^{\mathrm{T}}\boldsymbol{y} = \begin{bmatrix} 1 & 0 \\ -1 & 1 \\ 0 & -1 \end{bmatrix} \begin{bmatrix} y_1 \\ y_2 \end{bmatrix} = \begin{bmatrix} y_1 \\ y_2 - y_1 \\ -y_2 \end{bmatrix} \quad \text{balances} \quad \begin{bmatrix} \text{weight of floor 1} \\ \text{weight of floor 2} \\ \text{weight of floor 3} \end{bmatrix}$$

In resistors, the relation of $\boldsymbol{y}$ to $A\boldsymbol{x}$ is Ohm's Law (current proportional to voltage difference). For elastic structures this is Hooke's Law (stress proportional to strain). The catastrophe on September 11 came when the fires in the World Trade Center weakened the steel columns. Hooke's Law eventually failed. The internal forces couldn't balance the weight of the tower. After the first columns buckled, the columns below couldn't take the extra weight.

For a linearly elastic structure, the work balance equation is $(A\boldsymbol{x})^{\mathrm{T}}\boldsymbol{y} = \boldsymbol{x}^{\mathrm{T}}(A^{\mathrm{T}}\boldsymbol{y})$:

Internal work (**strain** $A\boldsymbol{x}$) $\cdot$ (**stress** $\boldsymbol{y}$) $=$ *External work* (**movement** $\boldsymbol{x}$) $\cdot$ (**force** $A^{\mathrm{T}}\boldsymbol{y}$).

Symmetric Matrices

For a *symmetric* matrix—these are the most important matrices—transposing A to A^{T} produces no change. Then $A^{\mathrm{T}} = A$. The matrix is symmetric across the main diagonal. A symmetric matrix is necessarily square. Its (j,i) and (i,j) entries are equal.

DEFINITION A ***symmetric matrix*** has $A^{\mathrm{T}} = A$. This means that $a_{ji} = a_{ij}$.

Example 2 $$A = \begin{bmatrix} 1 & 2 \\ 2 & 5 \end{bmatrix} = A^{\mathrm{T}} \quad \text{and} \quad D = \begin{bmatrix} 1 & 0 \\ 0 & 10 \end{bmatrix} = D^{\mathrm{T}}.$$

A is symmetric because of the 2's on opposite sides of the diagonal. The rows agree with the columns. In D those 2's are zeros. Every diagonal matrix is symmetric.

The inverse of a symmetric matrix is also symmetric. (We have to add: "*If* A *is invertible.*") The transpose of A^{-1} is $(A^{-1})^{\mathrm{T}} = (A^{\mathrm{T}})^{-1} = A^{-1}$, so A^{-1} is symmetric:

$$A^{-1} = \begin{bmatrix} 5 & -2 \\ -2 & 1 \end{bmatrix} \quad \text{and} \quad D^{-1} = \begin{bmatrix} 1 & 0 \\ 0 & 0.1 \end{bmatrix}.$$

Now we show that ***multiplying any matrix R by*** $\boldsymbol{R}^{\mathrm{T}}$ ***gives a symmetric matrix.***

Symmetric Products $R^{\mathrm{T}}R$ and RR^{T} and LDL^{T}

Choose any matrix R, probably rectangular. Multiply R^{T} times R. Then the product $R^{\mathrm{T}}R$ is automatically a square symmetric matrix:

$$\textit{The transpose of } \boldsymbol{R}^{\mathrm{T}}\boldsymbol{R} \textit{ is } \boldsymbol{R}^{\mathrm{T}}(\boldsymbol{R}^{\mathrm{T}})^{\mathrm{T}} \textit{ which is } \boldsymbol{R}^{\mathrm{T}}\boldsymbol{R}. \tag{7}$$

That is a quick proof of symmetry for $R^{\mathrm{T}}R$. We could also look at the (i,j) entry of $R^{\mathrm{T}}R$. It is the dot product of row i of R^{T} (column i of R) with column j of R. The (j,i) entry is the same dot product, column j with column i. So $R^{\mathrm{T}}R$ is symmetric.

The matrix RR^{T} is also symmetric. (The shapes of R and R^{T} allow multiplication.) But RR^{T} is a different matrix from $R^{\mathrm{T}}R$. In our experience, most scientific problems that start with a rectangular matrix R end up with $R^{\mathrm{T}}R$ or RR^{T} or both.

Example 3 $R = \begin{bmatrix} 1 & 2 \end{bmatrix}$ and $R^{\mathrm{T}} = \begin{bmatrix} 1 \\ 2 \end{bmatrix}$ produce $R^{\mathrm{T}}R = \begin{bmatrix} 1 & 2 \\ 2 & 4 \end{bmatrix}$ and $RR^{\mathrm{T}} = [\,5\,]$.

The product $R^{\mathrm{T}}R$ is n by n. In the opposite order, RR^{T} is m by m. Even if $m = n$, it is not very likely that $R^{\mathrm{T}}R = RR^{\mathrm{T}}$. Equality can happen, but it is abnormal.

When elimination is applied to a symmetric matrix, $A^{\mathrm{T}} = A$ is an advantage. The smaller matrices stay symmetric as elimination proceeds, and we can work with half the matrix! It is true that the upper triangular U cannot be symmetric. ***The symmetry is in LDU.*** Remember how the diagonal matrix D of pivots can be divided out, to leave 1's on the diagonal of both L and U:

$$\begin{bmatrix} 1 & 2 \\ 2 & 7 \end{bmatrix} = \begin{bmatrix} 1 & 0 \\ 2 & 1 \end{bmatrix} \begin{bmatrix} 1 & 2 \\ 0 & 3 \end{bmatrix} \qquad (LU \text{ misses the symmetry})$$

$$= \begin{bmatrix} 1 & 0 \\ 2 & 1 \end{bmatrix} \begin{bmatrix} 1 & 0 \\ 0 & 3 \end{bmatrix} \begin{bmatrix} 1 & 2 \\ 0 & 1 \end{bmatrix} \qquad \begin{matrix} (LDU \text{ captures the symmetry}) \\ \text{Now } U \textit{ is the transpose of } L. \end{matrix}$$

When A is symmetric, the usual form $A = LDU$ becomes $A = LDL^{\mathrm{T}}$. The final U (with 1's on the diagonal) is the transpose of L (also with 1's on the diagonal). The diagonal D—the matrix of pivots—is symmetric by itself.

2K If $A = A^{\mathrm{T}}$ can be factored into LDU with no row exchanges, then $U = L^{\mathrm{T}}$. ***The symmetric factorization of a symmetric matrix is*** $A = LDL^{\mathrm{T}}$.

Notice that the transpose of LDL^{T} is automatically $(L^{\mathrm{T}})^{\mathrm{T}}D^{\mathrm{T}}L^{\mathrm{T}}$ which is LDL^{T} again. The work of elimination is cut in half, from $n^3/3$ multiplications to $n^3/6$. The storage is also cut essentially in half. We only keep L and D, not U.

Permutation Matrices

The transpose plays a special role for a *permutation matrix*. This matrix P has a single "1" in every row and every column. Then P^{T} is also a permutation matrix—maybe the same or maybe different. Any product P_1P_2 is again a permutation matrix. We now create every P from the identity matrix, by reordering the rows of I.

The simplest permutation matrix is $P = I$ (*no exchanges*). The next simplest are the row exchanges P_{ij}. Those are constructed by exchanging two rows i and j of I. Other permutations reorder more rows. By doing all possible row exchanges to I, we get all possible permutation matrices:

DEFINITION ***A permutation matrix P has the rows of I in any order.***

Example 4 There are six 3 by 3 permutation matrices. Here they are without the zeros:

$$I = \begin{bmatrix} 1 & & \\ & 1 & \\ & & 1 \end{bmatrix} \quad P_{21} = \begin{bmatrix} & 1 & \\ 1 & & \\ & & 1 \end{bmatrix} \quad P_{32}P_{21} = \begin{bmatrix} & 1 & \\ & & 1 \\ 1 & & \end{bmatrix}$$

$$P_{31} = \begin{bmatrix} & & 1 \\ & 1 & \\ 1 & & \end{bmatrix} \quad P_{32} = \begin{bmatrix} 1 & & \\ & & 1 \\ & 1 & \end{bmatrix} \quad P_{21}P_{32} = \begin{bmatrix} & & 1 \\ 1 & & \\ & 1 & \end{bmatrix}.$$

There are $n!$ permutation matrices of order n. The symbol $n!$ means "n factorial," the product of the numbers $(1)(2)\cdots(n)$. Thus $3! = (1)(2)(3)$ which is 6. There will be 24 permutation matrices of order $n = 4$. And 120 permutations of order 5.

There are only two permutation matrices of order 2, namely $\left[\begin{smallmatrix} 1 & 0 \\ 0 & 1 \end{smallmatrix}\right]$ and $\left[\begin{smallmatrix} 0 & 1 \\ 1 & 0 \end{smallmatrix}\right]$.

Important: P^{-1} is also a permutation matrix. Among the six 3 by 3 P's displayed above, the four matrices on the left are their own inverses. The two matrices on the right are inverses of each other. In all cases, a single row exchange is its own inverse. If we repeat the exchange we are back to I. But for $P_{32}P_{21}$, the inverses go in opposite order (of course). The inverse is $P_{21}P_{32}$.

More important: ***P^{-1} is always the same as P^{T}.*** The two matrices on the right are transposes—and inverses—of each other. When we multiply PP^{T}, the "1" in the first row of P hits the "1" in the first column of P^{T} (since the first row of P is the first column of P^{T}). It misses the ones in all the other columns. So $PP^{\mathrm{T}} = I$.

Another proof of $P^{\mathrm{T}} = P^{-1}$ looks at P as a product of row exchanges. A row exchange is its own transpose and its own inverse. P^{T} and P^{-1} both come from the product of row exchanges in the opposite order. So P^{T} and P^{-1} are the same.
Symmetric matrices led to $A = LDL^{\mathrm{T}}$. Now permutations lead to $PA = LU$.

The LU Factorization with Row Exchanges

We sure hope you remember $A = LU$. It started with $A = (E_{21}^{-1} \cdots E_{ij}^{-1} \cdots)U$. Every elimination step was carried out by an E_{ij} and it was inverted by E_{ij}^{-1}. Those inverses were compressed into one matrix L, bringing U back to A. The lower triangular L has 1's on the diagonal, and the result is $A = LU$.

This is a great factorization, but it doesn't always work! Sometimes row exchanges are needed to produce pivots. Then $A = (E^{-1} \cdots P^{-1} \cdots E^{-1} \cdots P^{-1} \cdots)U$. Every row exchange is carried out by a P_{ij} and inverted by that P_{ij}. We now compress those row exchanges into a *single permutation matrix* P. This gives a factorization for every invertible matrix A—which we naturally want.

The main question is where to collect the P_{ij}'s. There are two good possibilities—do all the exchanges before elimination, or do them after the E_{ij}'s. The first way gives $PA = LU$. The second way has a permutation matrix P_1 in the middle.

1. The row exchanges can be done *in advance*. Their product P puts the rows of A in the right order, so that no exchanges are needed for PA. ***Then $PA = LU$.***

2. If we hold row exchanges until *after elimination*, the pivot rows are in a strange order. P_1 puts them in the correct triangular order in U_1. ***Then*** $\boldsymbol{A = L_1 P_1 U_1}$.

$PA = LU$ is constantly used in almost all computing (and always in MATLAB). ***We will concentrate on this form*** $PA = LU$. The factorization $A = L_1 P_1 U_1$ might be more elegant. If we mention both, it is because the difference is not well known. Probably you will not spend a long time on either one. Please don't. The most important case has $P = I$, when A equals LU with no exchanges.

For this matrix A, exchange rows 1 and 2 to put the first pivot in its usual place. Then go through elimination on PA:

$$\underset{A}{\begin{bmatrix} 0 & 1 & 1 \\ 1 & 2 & 1 \\ 2 & 7 & 9 \end{bmatrix}} \rightarrow \underset{PA}{\begin{bmatrix} 1 & 2 & 1 \\ 0 & 1 & 1 \\ 2 & 7 & 9 \end{bmatrix}} \rightarrow \underset{\ell_{31}=2}{\begin{bmatrix} 1 & 2 & 1 \\ 0 & 1 & 1 \\ 0 & 3 & 7 \end{bmatrix}} \rightarrow \underset{\ell_{32}=3}{\begin{bmatrix} 1 & 2 & 1 \\ 0 & 1 & 1 \\ 0 & 0 & 4 \end{bmatrix}}.$$

The matrix PA is in good order, and it factors as usual into LU:

$$PA = \begin{bmatrix} 1 & 0 & 0 \\ 0 & 1 & 0 \\ 2 & 3 & 1 \end{bmatrix} \begin{bmatrix} 1 & 2 & 1 \\ 0 & 1 & 1 \\ 0 & 0 & 4 \end{bmatrix} = LU. \tag{8}$$

We started with A and ended with U. *The only requirement is invertibility of* A.

2L If A is invertible, a permutation P will put its rows in the right order to factor $\boldsymbol{PA = LU}$. There must be a full set of pivots, after row exchanges.

In the MATLAB code, $A([r\ k],:) = A([k\ r],:)$ exchanges row k with row r below it (where the kth pivot has been found). Then we update L and P and the sign of P:

$$\begin{aligned} &A([r\ k],:) = A([k\ r],:);\\ &L([r\ k],1:k-1) = L([k\ r],1:k-1);\\ &P([r\ k],:) = P([k\ r],:);\\ &\text{sign} \ = -\text{sign} \end{aligned}$$

The "**sign**" of P tells whether the number of row exchanges is even (sign $= +1$) or odd (sign $= -1$). At the start, P is I and sign $= +1$. When there is a row exchange, the sign is reversed. The final value of sign is the **determinant of** P and it does not depend on the order of the row exchanges.

For PA we get back to the familiar LU. This is the usual factorization. In reality, MATLAB might not use the first available pivot. Mathematically we can accept a small pivot—anything but zero. It is better if the computer looks down the column for the largest pivot. (Section 9.1 explains why this "***partial pivoting***" reduces the roundoff error.) P may contain row exchanges that are not algebraically necessary. Still $PA = LU$.

Our advice is to understand permutations but let MATLAB do the computing. Calculations of $A = LU$ are enough to do by hand, without P. The Teaching Code **splu** (A) factors $PA = LU$ and **splv** (A, b) solves $Ax = b$ for any invertible A. The program **splu** stops if no pivot can be found in column k. That fact is printed.

■ REVIEW OF THE KEY IDEAS ■

1. The transpose puts the rows of A into the columns of A^{T}. Then $(A^{\mathrm{T}})_{ij} = A_{ji}$.
2. The transpose of AB is $B^{\mathrm{T}}A^{\mathrm{T}}$. The transpose of A^{-1} is the inverse of A^{T}.
3. The dot product $(A\boldsymbol{x})^{\mathrm{T}}\boldsymbol{y}$ equals the dot product $\boldsymbol{x}^{\mathrm{T}}(A^{\mathrm{T}}\boldsymbol{y})$.
4. When A is symmetric ($A^{\mathrm{T}} = A$), its LDU factorization is symmetric: $A = LDL^{\mathrm{T}}$.
5. A permutation matrix P has a 1 in each row and column, and $P^{\mathrm{T}} = P^{-1}$.
6. If A is invertible then a permutation P will reorder its rows for $PA = LU$.

■ WORKED EXAMPLES ■

2.7 A Applying the permutation P to the rows of A destroys its symmetry:

$$P = \begin{bmatrix} 0 & 1 & 0 \\ 0 & 0 & 1 \\ 1 & 0 & 0 \end{bmatrix} \qquad A = \begin{bmatrix} 1 & 4 & 5 \\ 4 & 2 & 6 \\ 5 & 6 & 3 \end{bmatrix} \qquad PA = \begin{bmatrix} 4 & 2 & 6 \\ 5 & 6 & 3 \\ 1 & 4 & 5 \end{bmatrix}$$

What permutation matrix Q applied to the columns of PA will recover symmetry in PAQ? The numbers $1, 2, 3$ must come back to the main diagonal (not necessarily in order). How is Q related to P, when symmetry is saved by PAQ?

Solution To recover symmetry and put "2" on the diagonal, column 2 of PA must move to column 1. Column 3 of PA (containing "3") must move to column 2. Then the "1" moves to the $3, 3$ position. The matrix that permutes columns is Q:

$$PA = \begin{bmatrix} 4 & 2 & 6 \\ 5 & 6 & 3 \\ 1 & 4 & 5 \end{bmatrix} \qquad Q = \begin{bmatrix} 0 & 0 & 1 \\ 1 & 0 & 0 \\ 0 & 1 & 0 \end{bmatrix} \qquad PAQ = \begin{bmatrix} 2 & 6 & 4 \\ 6 & 3 & 5 \\ 4 & 5 & 1 \end{bmatrix} \text{ is symmetric.}$$

The matrix Q is P^{T}. This choice always recovers symmetry, because PAP^{T} is guaranteed to be symmetric. (Its transpose is again PAP^{T}.) *The matrix Q is also P^{-1}, because the inverse of every permutation matrix is its transpose.*

If we look only at the main diagonal D of A, we are finding that PDP^{T} is guaranteed diagonal. When P moves row 1 down to row 3, P^{T} on the right will move column 1 to column 3. The $(1,1)$ entry moves down to $(3,1)$ and over to $(3,3)$.

2.7 B Find the symmetric factorization $A = LDL^{\mathrm{T}}$ for the matrix A above. Is A invertible? Find also the $PQ = LU$ factorization for Q, which needs row exchanges.

Solution To factor A into LDL^{T} we eliminate below the pivots:

$$A = \begin{bmatrix} 1 & 4 & 5 \\ 4 & 2 & 6 \\ 5 & 6 & 3 \end{bmatrix} \longrightarrow \begin{bmatrix} 1 & 4 & 5 \\ 0 & -14 & -14 \\ 0 & -14 & -22 \end{bmatrix} \longrightarrow \begin{bmatrix} 1 & 4 & 5 \\ 0 & -14 & -14 \\ 0 & 0 & -8 \end{bmatrix} = U.$$

The multipliers were $\ell_{21} = 4$ and $\ell_{31} = 5$ and $\ell_{32} = 1$. The pivots $1, -14, -8$ go into D. When we divide the rows of U by those pivots, L^{T} should appear:

$$A = LDL^{\mathrm{T}} = \begin{bmatrix} 1 & 0 & 0 \\ 4 & 1 & 0 \\ 5 & 1 & 1 \end{bmatrix} \begin{bmatrix} 1 & & \\ & -14 & \\ & & -8 \end{bmatrix} \begin{bmatrix} 1 & 4 & 5 \\ 0 & 1 & 1 \\ 0 & 0 & 1 \end{bmatrix}.$$

This matrix A is invertible because *it has three pivots.* Its inverse is $(L^{\mathrm{T}})^{-1}D^{-1}L^{-1}$ and it is also symmetric. The numbers 14 and 8 will turn up in the denominators of A^{-1}. The "determinant" of A is the product of the pivots $(1)(-14)(-8) = 112$.

The matrix Q is certainly invertible. But elimination needs two row exchanges:

$$Q = \begin{bmatrix} 0 & 0 & 1 \\ 1 & 0 & 0 \\ 0 & 1 & 0 \end{bmatrix} \underset{1 \leftrightarrow 2}{\xrightarrow{\text{rows}}} \begin{bmatrix} 1 & 0 & 0 \\ 0 & 0 & 1 \\ 0 & 1 & 0 \end{bmatrix} \underset{2 \leftrightarrow 3}{\xrightarrow{\text{rows}}} \begin{bmatrix} 1 & 0 & 0 \\ 0 & 1 & 0 \\ 0 & 0 & 1 \end{bmatrix} = I.$$

Then $L = I$ and $U = I$ are the LU factors. We only need the permutation P that put the rows of Q into their right order in I. Well, P must be Q^{-1}. It is the same P as above! We could find it as a product of the two row exchanges, $1 \leftrightarrow 2$ and $2 \leftrightarrow 3$:

$$P = P_{23}P_{12} = \begin{bmatrix} 1 & 0 & 0 \\ 0 & 0 & 1 \\ 0 & 1 & 0 \end{bmatrix} \begin{bmatrix} 0 & 1 & 0 \\ 1 & 0 & 0 \\ 0 & 0 & 1 \end{bmatrix} = \begin{bmatrix} 0 & 1 & 0 \\ 0 & 0 & 1 \\ 1 & 0 & 0 \end{bmatrix} \quad \text{reorders } Q \text{ into } PQ = I.$$

Problem Set 2.7

Questions 1–7 are about the rules for transpose matrices.

1 Find A^{T} and A^{-1} and $(A^{-1})^{\mathrm{T}}$ and $(A^{\mathrm{T}})^{-1}$ for

$$A = \begin{bmatrix} 1 & 0 \\ 9 & 3 \end{bmatrix} \quad \text{and also} \quad A = \begin{bmatrix} 1 & c \\ c & 0 \end{bmatrix}.$$

2 Verify that $(AB)^{\mathrm{T}}$ equals $B^{\mathrm{T}}A^{\mathrm{T}}$ but those are different from $A^{\mathrm{T}}B^{\mathrm{T}}$:

$$A = \begin{bmatrix} 1 & 0 \\ 2 & 1 \end{bmatrix} \qquad B = \begin{bmatrix} 1 & 3 \\ 0 & 1 \end{bmatrix} \qquad AB = \begin{bmatrix} 1 & 3 \\ 2 & 7 \end{bmatrix}.$$

In case $AB = BA$ (not generally true!) how do you prove that $B^{\mathrm{T}}A^{\mathrm{T}} = A^{\mathrm{T}}B^{\mathrm{T}}$?

3 (a) The matrix $((AB)^{-1})^{\mathrm{T}}$ comes from $(A^{-1})^{\mathrm{T}}$ and $(B^{-1})^{\mathrm{T}}$. *In what order?*

(b) If U is upper triangular then $(U^{-1})^{\mathrm{T}}$ is _____ triangular.

4 Show that $A^2 = 0$ is possible but $A^{\mathrm{T}}A = 0$ is not possible (unless $A =$ zero matrix).

5 (a) The row vector $\boldsymbol{x}^{\mathrm{T}}$ times A times the column $\boldsymbol{y}$ produces what number?

$$\boldsymbol{x}^{\mathrm{T}}A\boldsymbol{y} = \begin{bmatrix} 0 & 1 \end{bmatrix} \begin{bmatrix} 1 & 2 & 3 \\ 4 & 5 & 6 \end{bmatrix} \begin{bmatrix} 0 \\ 1 \\ 0 \end{bmatrix} = \_\_\_\_ .$$

(b) This is the row $\boldsymbol{x}^{\mathrm{T}}A =$ _____ times the column $\boldsymbol{y} = (0, 1, 0)$.

(c) This is the row $\boldsymbol{x}^{\mathrm{T}} = [0 \ \ 1]$ times the column $A\boldsymbol{y} =$ _____.

6 When you transpose a block matrix $M = \left[\begin{smallmatrix} \mathbf{A} & \mathbf{B} \\ \mathbf{C} & \mathbf{D} \end{smallmatrix}\right]$ the result is $M^{\mathrm{T}} =$ _____. Test it. Under what conditions on A, B, C, D is the block matrix symmetric?

7 True or false:

(a) The block matrix $\left[\begin{smallmatrix} \mathbf{0} & \mathbf{A} \\ \mathbf{A} & \mathbf{0} \end{smallmatrix}\right]$ is automatically symmetric.

(b) If A and B are symmetric then their product AB is symmetric.

(c) If A is not symmetric then A^{-1} is not symmetric.

(d) When A, B, C are symmetric, the transpose of ABC is CBA.

Questions 8–15 are about permutation matrices.

8 Why are there $n!$ permutation matrices of order n?

9 If P_1 and P_2 are permutation matrices, so is P_1P_2. This still has the rows of I in some order. Give examples with $P_1P_2 \neq P_2P_1$ and $P_3P_4 = P_4P_3$.

10 There are 12 "*even*" permutations of $(1, 2, 3, 4)$, with an *even number of exchanges*. Two of them are $(1, 2, 3, 4)$ with no exchanges and $(4, 3, 2, 1)$ with two exchanges. List the other ten. Instead of writing each 4 by 4 matrix, use the numbers 4, 3, 2, 1 to give the position of the 1 in each row.

11 (Try this question) Which permutation makes PA upper triangular? Which permutations make P_1AP_2 lower triangular? ***Multiplying*** A ***on the right by*** P_2 ***exchanges the*** _____ ***of*** A.

$$A = \begin{bmatrix} 0 & 0 & 6 \\ 1 & 2 & 3 \\ 0 & 4 & 5 \end{bmatrix}.$$

12 Explain why the dot product of $\boldsymbol{x}$ and $\boldsymbol{y}$ equals the dot product of $P\boldsymbol{x}$ and $P\boldsymbol{y}$. Then from $(P\boldsymbol{x})^{\mathrm{T}}(P\boldsymbol{y}) = \boldsymbol{x}^{\mathrm{T}}\boldsymbol{y}$ deduce that $P^{\mathrm{T}}P = I$ for any permutation. With $\boldsymbol{x} = (1, 2, 3)$ and $\boldsymbol{y} = (1, 4, 2)$ choose P to show that $P\boldsymbol{x} \cdot \boldsymbol{y}$ is not always equal to $\boldsymbol{x} \cdot P\boldsymbol{y}$.

13 Find a 3 by 3 permutation matrix with $P^3 = I$ (but not $P = I$). Find a 4 by 4 permutation $\widehat{P}$ with $\widehat{P}^4 \neq I$.

14 If you take powers of a permutation matrix, why is some P^k eventually equal to I?

Find a 5 by 5 permutation P so that the smallest power to equal I is P^6. (This is a challenge question. Combine a 2 by 2 block with a 3 by 3 block.)

15 Row exchange matrices are symmetric: $P^{\mathrm{T}} = P$. Then $P^{\mathrm{T}}P = I$ becomes $P^2 = I$. Some other permutation matrices are also symmetric.

(a) If P sends row 1 to row 4, then P^{T} sends row _____ to row _____. When $P^{\mathrm{T}} = P$ the row exchanges come in pairs with no overlap.

(b) Find a 4 by 4 example with $P^{\mathrm{T}} = P$ that moves all four rows.

Questions 16–21 are about symmetric matrices and their factorizations.

16 If $A = A^{\mathrm{T}}$ and $B = B^{\mathrm{T}}$, which of these matrices are certainly symmetric?

(a) $A^2 - B^2$

(b) $(A + B)(A - B)$

(c) ABA

(d) $ABAB$.

17 Find 2 by 2 symmetric matrices $A = A^{\mathrm{T}}$ with these properties:

(a) A is not invertible.

(b) A is invertible but cannot be factored into LU (row exchanges needed).

(c) A can be factored into LDL^{T} but not into LL^{T} (because of negative D).

18 (a) How many entries of A can be chosen independently, if $A = A^{\mathrm{T}}$ is 5 by 5?

(b) How do L and D (still 5 by 5) give the same number of choices?

(c) How many entries can be chosen if A is *skew-symmetric*? ($A^{\mathrm{T}} = -A$).

19 Suppose R is rectangular (m by n) and A is symmetric (m by m).

(a) Transpose $R^{\mathrm{T}}AR$ to show its symmetry. What shape is this matrix?

(b) Show why $R^{\mathrm{T}}R$ has no negative numbers on its diagonal.

20 Factor these symmetric matrices into $A = LDL^{\mathrm{T}}$. The pivot matrix D is diagonal:

$$A = \begin{bmatrix} 1 & 3 \\ 3 & 2 \end{bmatrix} \quad \text{and} \quad A = \begin{bmatrix} 1 & b \\ b & c \end{bmatrix} \quad \text{and} \quad A = \begin{bmatrix} 2 & -1 & 0 \\ -1 & 2 & -1 \\ 0 & -1 & 2 \end{bmatrix}.$$

21 After elimination clears out column 1 below the first pivot, find the symmetric 2 by 2 matrix that appears in the lower right corner:

$$A = \begin{bmatrix} 2 & 4 & 8 \\ 4 & 3 & 9 \\ 8 & 9 & 0 \end{bmatrix} \quad \text{and} \quad A = \begin{bmatrix} 1 & b & c \\ b & d & e \\ c & e & f \end{bmatrix}.$$

Questions 22–30 are about the factorizations $PA = LU$ and $A = L_1P_1U_1$.

22 Find the $PA = LU$ factorizations (and check them) for

$$A = \begin{bmatrix} 0 & 1 & 1 \\ 1 & 0 & 1 \\ 2 & 3 & 4 \end{bmatrix} \quad \text{and} \quad A = \begin{bmatrix} 1 & 2 & 0 \\ 2 & 4 & 1 \\ 1 & 1 & 1 \end{bmatrix}.$$

23 Find a 3 by 3 permutation matrix (call it A) that needs two row exchanges to reach the end of elimination. For this matrix, what are its factors P, L, and U?

24 Factor the following matrix into $PA = LU$. Factor it also into $A = L_1P_1U_1$ (hold the exchange of row 3 until 3 times row 1 is subtracted from row 2):

$$A = \begin{bmatrix} 0 & 1 & 2 \\ 0 & 3 & 8 \\ 2 & 1 & 1 \end{bmatrix}.$$

25 Write out P after each step of the MATLAB code splu, when

$$A = \begin{bmatrix} 0 & 1 \\ 2 & 3 \end{bmatrix} \quad \text{and} \quad A = \begin{bmatrix} 0 & 0 & 1 \\ 2 & 3 & 4 \\ 0 & 5 & 6 \end{bmatrix}.$$

26 Write out P and L after each step of the code splu when

$$A = \begin{bmatrix} 0 & 1 & 2 \\ 1 & 1 & 0 \\ 2 & 5 & 4 \end{bmatrix}.$$

27 Extend the MATLAB code splu to a code spldu which factors PA into LDU.

28 What is the matrix L_1 in $A = L_1P_1U_1$?

$$A = \begin{bmatrix} 1 & 1 & 1 \\ 1 & 1 & 3 \\ 2 & 5 & 8 \end{bmatrix} \to \begin{bmatrix} 1 & 1 & 1 \\ 0 & 0 & 2 \\ 0 & 3 & 6 \end{bmatrix} = P_1U_1 = \begin{bmatrix} 1 & 0 & 0 \\ 0 & 0 & 1 \\ 0 & 1 & 0 \end{bmatrix} \begin{bmatrix} 1 & 1 & 1 \\ 0 & 3 & 6 \\ 0 & 0 & 2 \end{bmatrix}.$$

29 Prove that the identity matrix cannot be the product of three row exchanges (or five). It can be the product of two exchanges (or four).

30 (a) Choose E_{21} to remove the 3 below the first pivot. Then multiply $E_{21}AE_{21}^{\mathrm{T}}$ to remove both 3's:

$$A = \begin{bmatrix} 1 & 3 & 0 \\ 3 & 11 & 4 \\ 0 & 4 & 9 \end{bmatrix} \quad \text{is going toward} \quad D = \begin{bmatrix} 1 & 0 & 0 \\ 0 & 2 & 0 \\ 0 & 0 & 1 \end{bmatrix}.$$

(b) Choose E_{32} to remove the 4 below the second pivot. Then A is reduced to D by $E_{32}E_{21}AE_{21}^{\mathrm{T}}E_{32}^{\mathrm{T}} = D$. Invert the E's to find L in $A = LDL^{\mathrm{T}}$.

The next questions are about applications of the identity $(Ax)^{\mathrm{T}}y = x^{\mathrm{T}}(A^{\mathrm{T}}y)$.

31 Wires go between Boston, Chicago, and Seattle. Those cities are at voltages x_B, x_C, x_S. With unit resistances between cities, the currents between cities are in $\boldsymbol{y}$:

$$\boldsymbol{y} = A\boldsymbol{x} \quad \text{is} \quad \begin{bmatrix} y_{BC} \\ y_{CS} \\ y_{BS} \end{bmatrix} = \begin{bmatrix} 1 & -1 & 0 \\ 0 & 1 & -1 \\ 1 & 0 & -1 \end{bmatrix} \begin{bmatrix} x_B \\ x_C \\ x_S \end{bmatrix}.$$

(a) Find the total currents $A^{\mathrm{T}}\boldsymbol{y}$ out of the three cities.

(b) Verify that $(A\boldsymbol{x})^{\mathrm{T}}\boldsymbol{y}$ agrees with $\boldsymbol{x}^{\mathrm{T}}(A^{\mathrm{T}}\boldsymbol{y})$—six terms in both.

32 Producing x_1 trucks and x_2 planes needs $x_1 + 50x_2$ tons of steel, $40x_1 + 1000x_2$ pounds of rubber, and $2x_1 + 50x_2$ months of labor. If the unit costs y_1, y_2, y_3 are \$700 per ton, \$3 per pound, and \$3000 per month, what are the values of one truck and one plane? Those are the components of $A^{\mathrm{T}}\boldsymbol{y}$.

33 $A\boldsymbol{x}$ gives the amounts of steel, rubber, and labor to produce $\boldsymbol{x}$ in Problem 32. Find A. Then $A\boldsymbol{x} \cdot \boldsymbol{y}$ is the _____ of inputs while $\boldsymbol{x} \cdot A^{\mathrm{T}}\boldsymbol{y}$ is the value of _____.

34 The matrix P that multiplies (x, y, z) to give (z, x, y) is also a rotation matrix. Find P and P^3. The rotation axis $\boldsymbol{a} = (1, 1, 1)$ doesn't move, it equals $P\boldsymbol{a}$. What is the angle of rotation from $\boldsymbol{v} = (2, 3, -5)$ to $P\boldsymbol{v} = (-5, 2, 3)$?

35 Write $A = \begin{bmatrix} 1 & 2 \\ 4 & 9 \end{bmatrix}$ as the product EH of an elementary row operation matrix E and a symmetric matrix H.

36 Here is a new factorization of A into *triangular times symmetric*:

$$\text{Start from } A = LDU. \text{ Then } A = L(U^{\mathrm{T}})^{-1} \text{ times } U^{\mathrm{T}}DU.$$

Why is $L(U^{\mathrm{T}})^{-1}$ triangular? Its diagonal is all 1's. Why is $U^{\mathrm{T}}DU$ symmetric?

37 A *group* of matrices includes AB and A^{-1} if it includes A and B. "Products and inverses stay in the group." Which of these sets are groups? Lower triangular matrices L with 1's on the diagonal, symmetric matrices S, positive matrices M, diagonal invertible matrices D, permutation matrices P, matrices with $Q^{\mathrm{T}} = Q^{-1}$. Invent two more matrix groups.

38 If every row of a 4 by 4 matrix contains the numbers $0, 1, 2, 3$ in some order, can the matrix be symmetric?

39 Prove that no reordering of rows and reordering of columns can transpose a typical matrix.

40 A square ***northwest matrix*** B is zero in the southeast corner, below the antidiagonal that connects $(1, n)$ to $(n, 1)$. Will B^{T} and B^2 be northwest matrices? Will B^{-1} be northwest or southeast? What is the shape of BC = ***northwest times southeast***? OK to combine permutations with the usual L and U (southwest and northeast).

41 If P has 1's on the antidiagonal from $(1, n)$ to $(n, 1)$, describe PAP.

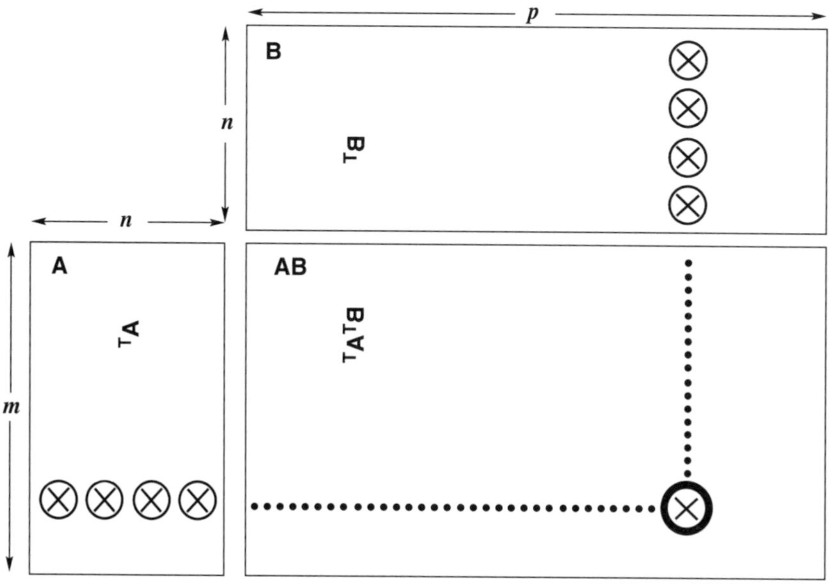

* **Transparent proof that $(AB)^{\mathrm{T}} = B^{\mathrm{T}}A^{\mathrm{T}}$.** Matrices can be transposed by looking through the page from the other side. Hold up to the light and practice with B. Its column with four entries $\otimes$ becomes a row, when you look from the back and the symbol B^{T} is upright.

The three matrices are in position for matrix multiplication: the row of A times the column of B gives the entry in AB. Looking from the reverse side, the row of B^{T} times the column of A^{T} gives the correct entry in $B^{\mathrm{T}}A^{\mathrm{T}} = (AB)^{\mathrm{T}}$.

3

VECTOR SPACES AND SUBSPACES

SPACES OF VECTORS ■ 3.1

To a newcomer, matrix calculations involve a lot of numbers. To you, they involve vectors. The columns of Ax and AB are linear combinations of n vectors—the columns of A. This chapter moves from numbers and vectors to a third level of understanding (the highest level). Instead of individual columns, we look at "spaces" of vectors. Without seeing ***vector spaces*** and especially their ***subspaces***, you haven't understood everything about $Ax = b$.

Since this chapter goes a little deeper, it may seem a little harder. That is natural. We are looking inside the calculations, to find the mathematics. The author's job is to make it clear. These pages go to the heart of linear algebra.

We begin with the most important vector spaces. They are denoted by $\mathbf{R}^1$, $\mathbf{R}^2$, $\mathbf{R}^3$, $\mathbf{R}^4$, ... Each space $\mathbf{R}^n$ consists of a whole collection of vectors. $\mathbf{R}^5$ contains all column vectors with five components. This is called "5-dimensional space."

DEFINITION ***The space $\mathbf{R}^n$ consists of all column vectors v with n components.***

The components of v are real numbers, which is the reason for the letter $\mathbf{R}$. A vector whose n components are complex numbers lies in the space $\mathbf{C}^n$.

The vector space $\mathbf{R}^2$ is represented by the usual xy plane. Each vector v in $\mathbf{R}^2$ has two components. The word "space", asks us to think of all those vectors—the whole plane. Each vector gives the x and y coordinates of a point in the plane.

Similarly the vectors in $\mathbf{R}^3$ correspond to points (x, y, z) in three-dimensional space. The one-dimensional space $\mathbf{R}^1$ is a line (like the x axis). As before, we print vectors as a column between brackets, or along a line using commas and parentheses:

$$\begin{bmatrix} 4 \\ 0 \\ 1 \end{bmatrix} \text{ is in } \mathbf{R}^3, \quad (1, 1, 0, 1, 1) \text{ is in } \mathbf{R}^5, \quad \begin{bmatrix} 1+i \\ 1-i \end{bmatrix} \text{ is in } \mathbf{C}^2.$$

The great thing about linear algebra is that it deals easily with five-dimensional space. We don't draw the vectors, we just need the five numbers (or n numbers). To multiply $\boldsymbol{v}$ by 7, multiply every component by 7. Here 7 is a "scalar." To add vectors in $\mathbf{R}^5$, add them a component at a time. The two essential vector operations go on *inside the vector space*:

We can add any vectors in $\mathbf{R}^n$, and we can multiply any vector by any scalar.

"Inside the vector space" means that ***the result stays in the space.*** If $\boldsymbol{v}$ is the vector in $\mathbf{R}^4$ with components $1, 0, 0, 1$, then $2\boldsymbol{v}$ is the vector in $\mathbf{R}^4$ with components $2, 0, 0, 2$. (In this case 2 is the scalar.) A whole series of properties can be verified in $\mathbf{R}^n$. The commutative law is $\boldsymbol{v}+\boldsymbol{w}=\boldsymbol{w}+\boldsymbol{v}$; the distributive law is $c(\boldsymbol{v}+\boldsymbol{w})=c\boldsymbol{v}+c\boldsymbol{w}$. There is a unique "zero vector" satisfying $\mathbf{0}+\boldsymbol{v}=\boldsymbol{v}$. Those are three of the eight conditions listed at the start of the problem set.

These eight conditions are required of every vector space. There are vectors other than column vectors, and vector spaces other than $\mathbf{R}^n$, and they have to obey the eight reasonable rules.

A real vector space is a set of "vectors" *together with rules for vector addition and for multiplication by real numbers.* The addition and the multiplication must produce vectors that are in the space. And the eight conditions must be satisfied (which is usually no problem). Here are three vector spaces other than $\mathbf{R}^n$:

M The vector space of ***all real 2 by 2 matrices.***
F The vector space of ***all real functions*** $f(x)$.
Z The vector space that consists only of a ***zero vector.***

In **M** the "vectors" are really matrices. In **F** the vectors are functions. In **Z** the only addition is $\mathbf{0}+\mathbf{0}=\mathbf{0}$. In each case we can add: matrices to matrices, functions to functions, zero vector to zero vector. We can multiply a matrix by 4 or a function by 4 or the zero vector by 4. The result is still in **M** or **F** or **Z**. The eight conditions are all easily checked.

The space **Z** is zero-dimensional (by any reasonable definition of dimension). It is the smallest possible vector space. We hesitate to call it $\mathbf{R}^0$, which means no components—you might think there was no vector. The vector space **Z** contains exactly *one vector* (zero). No space can do without that zero vector. Each space has its own zero vector—the zero matrix, the zero function, the vector $(0, 0, 0)$ in $\mathbf{R}^3$.

Subspaces

At different times, we will ask you to think of matrices and functions as vectors. But at all times, the vectors that we need most are ordinary column vectors. They are vectors with n components—but *maybe not all* of the vectors with n components. There are important vector spaces ***inside*** $\mathbf{R}^n$.

Start with the usual three-dimensional space $\mathbf{R}^3$. Choose a plane through the origin $(0, 0, 0)$. ***That plane is a vector space in its own right.*** If we add two vectors

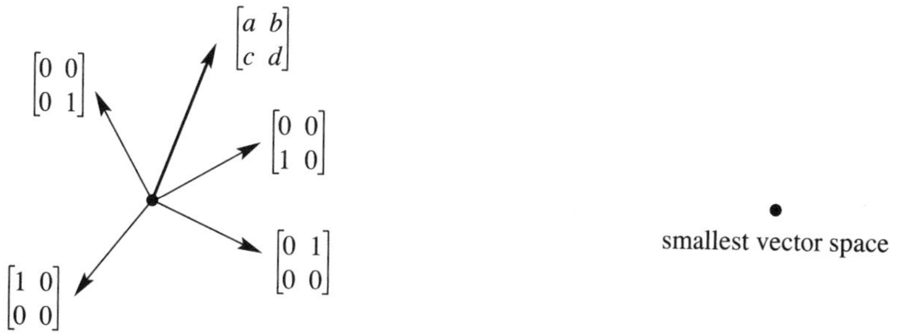

Figure 3.1 "Four-dimensional" matrix space **M**. The "zero-dimensional" space **Z**.

in the plane, their sum is in the plane. If we multiply an in-plane vector by 2 or -5, it is still in the plane. The plane is not $\mathbf{R}^2$ (even if it looks like $\mathbf{R}^2$). The vectors have three components and they belong to $\mathbf{R}^3$. The plane is a vector space inside $\mathbf{R}^3$.

This illustrates one of the most fundamental ideas in linear algebra. The plane is a ***subspace*** of the full vector space $\mathbf{R}^3$.

DEFINITION A ***subspace*** of a vector space is a set of vectors (including $\mathbf{0}$) that satisfies two requirements: ***If v and w are vectors in the subspace and c is any scalar, then*** (**i**) $v + w$ is in the subspace and (**ii**) cv is in the subspace.

In other words, the set of vectors is "closed" under addition $v + w$ and multiplication cv (and cw). Those operations leave us in the subspace. We can also subtract, because $-w$ is in the subspace and its sum with v is $v - w$. In short, ***all linear combinations stay in the subspace.***

All these operations follow the rules of the host space, so the eight required conditions are automatic. We just have to check the requirements (**i**) and (**ii**) for a subspace.

First fact: ***Every subspace contains the zero vector.*** The plane in $\mathbf{R}^3$ has to go through $(0, 0, 0)$. We mention this separately, for extra emphasis, but it follows directly from rule (**ii**). Choose $c = 0$, and the rule requires $0v$ to be in the subspace.

Planes that don't contain the origin fail those tests. When v is on such a plane, $-v$ and $0v$ are *not* on the plane. A plane that misses the origin is not a subspace.

Lines through the origin are also subspaces. When we multiply by 5, or add two vectors on the line, we stay on the line. But the line must go through $(0, 0, 0)$.

Another subspace is all of $\mathbf{R}^3$. The whole space is a subspace (*of itself*). Here is a list of all the possible subspaces of $\mathbf{R}^3$:

(**L**) Any line through $(0, 0, 0)$ ($\mathbf{R}^3$) The whole space
(**P**) Any plane through $(0, 0, 0)$ (**Z**) The single vector $(0, 0, 0)$

If we try to keep only *part* of a plane or line, the requirements for a subspace don't hold. Look at these examples in $\mathbf{R}^2$.

Example 1 Keep only the vectors (x, y) whose components are positive or zero (this is a quarter-plane). The vector $(2, 3)$ is included but $(-2, -3)$ is not. So rule **(ii)** is violated when we try to multiply by $c = -1$. ***The quarter-plane is not a subspace.***

Example 2 Include also the vectors whose components are both negative. Now we have two quarter-planes. Requirement **(ii)** is satisfied; we can multiply by any c. But rule **(i)** now fails. The sum of $\boldsymbol{v} = (2, 3)$ and $\boldsymbol{w} = (-3, -2)$ is $(-1, 1)$, which is outside the quarter-planes. ***Two quarter-planes don't make a subspace.***

Rules **(i)** and **(ii)** involve vector addition $\boldsymbol{v} + \boldsymbol{w}$ and multiplication by scalars like c and d. The rules can be combined into a single requirement—*the rule for subspaces*:

A subspace containing v and w must contain all linear combinations $c\boldsymbol{v} + d\boldsymbol{w}$.

Example 3 Inside the vector space $\mathbf{M}$ of all 2 by 2 matrices, here are two subspaces:

$$(\mathbf{U}) \text{ All upper triangular matrices } \begin{bmatrix} a & b \\ 0 & d \end{bmatrix} \quad (\mathbf{D}) \text{ All diagonal matrices } \begin{bmatrix} a & 0 \\ 0 & d \end{bmatrix}.$$

Add any two matrices in $\mathbf{U}$, and the sum is in $\mathbf{U}$. Add diagonal matrices, and the sum is diagonal. In this case $\mathbf{D}$ is also a subspace of $\mathbf{U}$! Of course the zero matrix is in these subspaces, when a, b, and d all equal zero.

To find a smaller subspace of diagonal matrices, we could require $a = d$. The matrices are multiples of the identity matrix I. The sum $2I + 3I$ is in this subspace, and so is 3 times $4I$. It is a "line of matrices" inside $\mathbf{M}$ and $\mathbf{U}$ and $\mathbf{D}$.

Is the matrix I a subspace by itself? Certainly not. Only the zero matrix is. Your mind will invent more subspaces of 2 by 2 matrices—write them down for Problem 5.

The Column Space of A

The most important subspaces are tied directly to a matrix A. We are trying to solve $A\boldsymbol{x} = \boldsymbol{b}$. If A is not invertible, the system is solvable for some $\boldsymbol{b}$ and not solvable for other $\boldsymbol{b}$. We want to describe the good right sides $\boldsymbol{b}$—the vectors that *can* be written as A times some vector $\boldsymbol{x}$.

Remember that $A\boldsymbol{x}$ is a combination of the columns of A. To get every possible $\boldsymbol{b}$, we use every possible $\boldsymbol{x}$. So start with the columns of A, and ***take all their linear combinations. This produces the column space of A.*** It is a vector space made up of column vectors—not just the n columns of A, but all their combinations $A\boldsymbol{x}$.

DEFINITION The ***column space*** consists of ***all linear combinations of the columns.*** The combinations are all possible vectors $A\boldsymbol{x}$. They fill the column space $\boldsymbol{C}(A)$.

This column space is crucial to the whole book, and here is why. ***To solve*** $A\boldsymbol{x} = \boldsymbol{b}$ ***is to express b as a combination of the columns.*** The right side $\boldsymbol{b}$ has to be *in the column space* produced by A on the left side, or no solution!

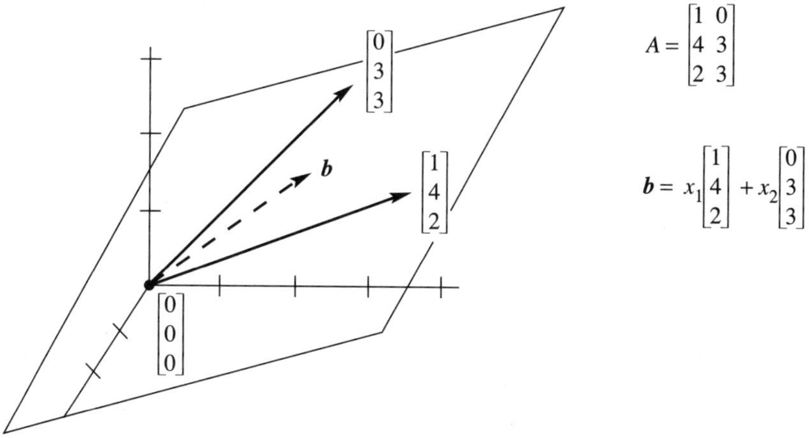

Figure 3.2 The column space $C(A)$ is a plane containing the two columns. $Ax = b$ is solvable when b is on that plane. Then b is a combination of the columns.

3A The system $Ax = b$ is solvable if and only if b is in the column space of A.

When b is in the column space, it is a combination of the columns. The coefficients in that combination give us a solution x to the system $Ax = b$.

Suppose A is an m by n matrix. Its columns have m components (not n). So the columns belong to $\mathbf{R}^m$. ***The column space of A is a subspace of*** $\mathbf{R}^m$ ***(not*** $\mathbf{R}^n$***).*** The set of all column combinations Ax satisfies rules (**i**) and (**ii**) for a subspace: When we add linear combinations or multiply by scalars, we still produce combinations of the columns. The word "subspace" is justified *by taking all linear combinations.*

Here is a 3 by 2 matrix A, whose column space is a subspace of $\mathbf{R}^3$. It is a plane.

Example 4

$$Ax \quad \text{is} \quad \begin{bmatrix}1 & 0\\4 & 3\\2 & 3\end{bmatrix}\begin{bmatrix}x_1\\x_2\end{bmatrix} \quad \text{which is} \quad x_1\begin{bmatrix}1\\4\\2\end{bmatrix} + x_2\begin{bmatrix}0\\3\\3\end{bmatrix}.$$

The column space consists of all combinations of the two columns—any x_1 times the first column plus any x_2 times the second column. *Those combinations fill up a plane in* $\mathbf{R}^3$ (Figure 3.2). If the right side b lies on that plane, then it is one of the combinations and (x_1, x_2) is a solution to $Ax = b$. The plane has zero thickness, so it is more likely that b is not in the column space. Then there is no solution to our 3 equations in 2 unknowns.

Of course $(0, 0, 0)$ is in the column space. The plane passes through the origin. There is certainly a solution to $Ax = \mathbf{0}$. That solution, always available, is $x =$ _____.

To repeat, the attainable right sides $\boldsymbol{b}$ are exactly the vectors in the column space. One possibility is the first column itself—take $x_1 = 1$ and $x_2 = 0$. Another combination is the second column—take $x_1 = 0$ and $x_2 = 1$. The new level of understanding is to see *all* combinations—the whole subspace is generated by those two columns.

Notation The column space of A is denoted by $\boldsymbol{C}(A)$. Start with the columns and take all their linear combinations. We might get the whole $\mathbf{R}^m$ or only a subspace.

Example 5 Describe the column spaces (they are subspaces of $\mathbf{R}^2$) for

$$I = \begin{bmatrix} 1 & 0 \\ 0 & 1 \end{bmatrix} \quad \text{and} \quad A = \begin{bmatrix} 1 & 2 \\ 2 & 4 \end{bmatrix} \quad \text{and} \quad B = \begin{bmatrix} 1 & 2 & 3 \\ 0 & 0 & 4 \end{bmatrix}.$$

Solution The column space of I is the *whole space* $\mathbf{R}^2$. Every vector is a combination of the columns of I. In vector space language, $\boldsymbol{C}(I)$ is $\mathbf{R}^2$.

The column space of A is only a line. The second column $(2, 4)$ is a multiple of the first column $(1, 2)$. Those vectors are different, but our eye is on vector ***spaces***. The column space contains $(1, 2)$ and $(2, 4)$ and all other vectors $(c, 2c)$ along that line. The equation $A\boldsymbol{x} = \boldsymbol{b}$ is only solvable when $\boldsymbol{b}$ is on the line.

The third matrix (with three columns) places no restriction on $\boldsymbol{b}$. The column space $\boldsymbol{C}(B)$ is all of $\mathbf{R}^2$. Every $\boldsymbol{b}$ is attainable. The vector $\boldsymbol{b} = (5, 4)$ is column 2 plus column 3, so $\boldsymbol{x}$ can be $(0, 1, 1)$. The same vector $(5, 4)$ is also 2(column 1) + column 3, so another possible $\boldsymbol{x}$ is $(2, 0, 1)$. This matrix has the same column space as I—any $\boldsymbol{b}$ is allowed. But now $\boldsymbol{x}$ has extra components and there are more solutions.

The next section creates another vector space, to describe all the solutions of $A\boldsymbol{x} = \boldsymbol{0}$. This section created the column space, to describe all the attainable right sides $\boldsymbol{b}$.

■ REVIEW OF THE KEY IDEAS ■

1. $\mathbf{R}^n$ contains all column vectors with n real components.
2. $\mathbf{M}$ (2 by 2 matrices) and $\mathbf{F}$ (functions) and $\mathbf{Z}$ (zero vector alone) are vector spaces.
3. A subspace containing $\boldsymbol{v}$ and $\boldsymbol{w}$ must contain all their combinations $c\boldsymbol{v} + d\boldsymbol{w}$.
4. The combinations of the columns of A form the ***column space*** $\boldsymbol{C}(A)$.
5. $A\boldsymbol{x} = \boldsymbol{b}$ has a solution exactly when $\boldsymbol{b}$ is in the column space of A.

■ WORKED EXAMPLES ■

3.1 A We are given three different vectors $\boldsymbol{b}_1, \boldsymbol{b}_2, \boldsymbol{b}_3$. Construct a matrix so that the equations $A\boldsymbol{x} = \boldsymbol{b}_1$ and $A\boldsymbol{x} = \boldsymbol{b}_2$ are solvable, but $A\boldsymbol{x} = \boldsymbol{b}_3$ is not solvable. How can you decide if this is possible? How could you construct A?

Solution We want to have $\boldsymbol{b}_1$ and $\boldsymbol{b}_2$ in the column space of A. Then $A\boldsymbol{x} = \boldsymbol{b}_1$ and $A\boldsymbol{x} = \boldsymbol{b}_2$ will be solvable. *The quickest way is to make $\boldsymbol{b}_1$ and $\boldsymbol{b}_2$ the two columns of* A. Then the solutions are $\boldsymbol{x} = (1, 0)$ and $\boldsymbol{x} = (0, 1)$.

Also, we don't want $A\boldsymbol{x} = \boldsymbol{b}_3$ to be solvable. So don't make the column space any larger! Keeping only the columns of $\boldsymbol{b}_1$ and $\boldsymbol{b}_2$, the question is:

$$\text{Is } A\boldsymbol{x} = \begin{bmatrix} \boldsymbol{b}_1 & \boldsymbol{b}_2 \end{bmatrix} \begin{bmatrix} x_1 \\ x_2 \end{bmatrix} = \boldsymbol{b}_3 \text{ solvable?} \qquad \text{Is } \boldsymbol{b}_3 \text{ a combination of } \boldsymbol{b}_1 \text{ and } \boldsymbol{b}_2\text{?}$$

If the answer is *no*, we have the desired matrix A. If the answer is *yes*, then it is *not possible* to construct A. When the column space contains $\boldsymbol{b}_1$ and $\boldsymbol{b}_2$, it will have to contain all their linear combinations. So $\boldsymbol{b}_3$ would necessarily be in that column space and $A\boldsymbol{x} = \boldsymbol{b}_3$ would necessarily be solvable.

3.1 B Describe a subspace **S** of each vector space **V**, and then a subspace **SS** of **S**.

$\mathbf{V}_1$ = all combinations of $(1, 1, 0, 0)$ and $(1, 1, 1, 0)$ and $(1, 1, 1, 1)$
$\mathbf{V}_2$ = all vectors perpendicular to $\boldsymbol{u} = (1, 2, 2, 1)$
$\mathbf{V}_3$ = all symmetric 2 by 2 matrices
$\mathbf{V}_4$ = all solutions to the equation $d^4y/dx^4 = 0$.

Describe each **V** two ways: *All combinations of ...., all solutions of the equations ....*

Solution A subspace **S** of $\mathbf{V}_1$ comes from all combinations of the first two vectors $(1, 1, 0, 0)$ and $(1, 1, 1, 0)$. A subspace **SS** of **S** comes from all multiples $(c, c, 0, 0)$ of the first vector.

A subspace **S** of $\mathbf{V}_2$ comes from all combinations of two vectors $(1, 0, 0, -1)$ and $(0, 1, -1, 0)$ that are perpendicular to $\boldsymbol{u}$. The vector $\boldsymbol{x} = (1, 1, -1, -1)$ is in **S** and all its multiples $c\boldsymbol{x}$ give a subspace **SS**.

The diagonal matrices are a subspace **S** of the symmetric matrices. The multiples cI are a subspace **SS** of the diagonal matrices.

$\mathbf{V}_4$ contains all cubic polynomials $y = a + bx + cx^2 + dx^3$. The quadratic polynomials give a subspace **S**. The linear polynomials are one choice of **SS**. The constants could be **SSS**.

In all four parts we could have chosen **S** = **V** itself, and **SS** = the zero subspace **Z**.

Each **V** can be described as all combinations of .... and as all solutions of ....:

$\mathbf{V}_1$ = all combinations of the 3 vectors = all solutions of $v_1 - v_2 = 0$
$\mathbf{V}_2$ = all combinations of $(1, 0, 0, -1)$, $(0, 1, -1, 0)$, $(2, -1, 0, 0)$
= all solutions of $\boldsymbol{u}^{\mathrm{T}}\boldsymbol{v} = 0$
$\mathbf{V}_3$ = all combinations of $\left[\begin{smallmatrix} 1 & 0 \\ 0 & 0 \end{smallmatrix}\right], \left[\begin{smallmatrix} 0 & 1 \\ 1 & 0 \end{smallmatrix}\right], \left[\begin{smallmatrix} 0 & 0 \\ 0 & 1 \end{smallmatrix}\right]$ = all solutions $\left[\begin{smallmatrix} a & b \\ c & d \end{smallmatrix}\right]$ of $b = c$
$\mathbf{V}_4$ = all combinations of $1, x, x^2, x^3$ = all solutions to $d^4y/dx^4 = 0$.

Problem Set 3.1

The first problems 1–8 are about vector spaces in general. The vectors in those spaces are not necessarily column vectors. In the definition of a *vector space*, vector addition $x+y$ and scalar multiplication cx must obey the following eight rules:

(1) $\boldsymbol{x} + \boldsymbol{y} = \boldsymbol{y} + \boldsymbol{x}$

(2) $\boldsymbol{x} + (\boldsymbol{y} + \boldsymbol{z}) = (\boldsymbol{x} + \boldsymbol{y}) + \boldsymbol{z}$

(3) There is a unique "zero vector" such that $\boldsymbol{x} + \mathbf{0} = \boldsymbol{x}$ for all $\boldsymbol{x}$

(4) For each $\boldsymbol{x}$ there is a unique vector $-\boldsymbol{x}$ such that $\boldsymbol{x} + (-\boldsymbol{x}) = \mathbf{0}$

(5) 1 times $\boldsymbol{x}$ equals $\boldsymbol{x}$

(6) $(c_1c_2)\boldsymbol{x} = c_1(c_2\boldsymbol{x})$

(7) $c(\boldsymbol{x} + \boldsymbol{y}) = c\boldsymbol{x} + c\boldsymbol{y}$

(8) $(c_1 + c_2)\boldsymbol{x} = c_1\boldsymbol{x} + c_2\boldsymbol{x}$.

1 Suppose $(x_1, x_2) + (y_1, y_2)$ is defined to be $(x_1 + y_2, x_2 + y_1)$. With the usual multiplication $c\boldsymbol{x} = (cx_1, cx_2)$, which of the eight conditions are not satisfied?

2 Suppose the multiplication $c\boldsymbol{x}$ is defined to produce $(cx_1, 0)$ instead of (cx_1, cx_2). With the usual addition in $\mathbf{R}^2$, are the eight conditions satisfied?

3 (a) Which rules are broken if we keep only the positive numbers $x > 0$ in $\mathbf{R}^1$? Every c must be allowed. The half-line is not a subspace.

(b) The positive numbers with $\boldsymbol{x} + \boldsymbol{y}$ and $c\boldsymbol{x}$ redefined to equal the usual xy and x^c *do* satisfy the eight rules. Test rule 7 when $c = 3, x = 2, y = 1$. (Then $\boldsymbol{x} + \boldsymbol{y} = 2$ and $c\boldsymbol{x} = 8$.) Which number acts as the "zero vector"?

4 The matrix $A = \begin{bmatrix} 2 & -2 \\ 2 & -2 \end{bmatrix}$ is a "vector" in the space $\mathbf{M}$ of all 2 by 2 matrices. Write down the zero vector in this space, the vector $\frac{1}{2}A$, and the vector $-A$. What matrices are in the smallest subspace containing A?

5 (a) Describe a subspace of $\mathbf{M}$ that contains $A = \begin{bmatrix} 1 & 0 \\ 0 & 0 \end{bmatrix}$ but not $B = \begin{bmatrix} 0 & 0 \\ 0 & -1 \end{bmatrix}$.

(b) If a subspace of $\mathbf{M}$ contains A and B, must it contain I?

(c) Describe a subspace of $\mathbf{M}$ that contains no nonzero diagonal matrices.

6 The functions $\boldsymbol{f}(x) = x^2$ and $\boldsymbol{g}(x) = 5x$ are "vectors" in $\mathbf{F}$. This is the vector space of all real functions. (The functions are defined for $-\infty < x < \infty$.) The combination $3\boldsymbol{f}(x) - 4\boldsymbol{g}(x)$ is the function $\boldsymbol{h}(x) =$ ______.

7 Which rule is broken if multiplying $f(x)$ by c gives the function $f(cx)$? Keep the usual addition $f(x) + g(x)$.

8 If the sum of the "vectors" $f(x)$ and $g(x)$ is defined to be the function $f(g(x))$, then the "zero vector" is $g(x) = x$. Keep the usual scalar multiplication $cf(x)$ and find two rules that are broken.

Questions 9–18 are about the "subspace requirements": $x + y$ and cx (and then all linear combinations $cx + dy$) must stay in the subspace.

9 One requirement can be met while the other fails. Show this by finding

(a) A set of vectors in $\mathbf{R}^2$ for which $\boldsymbol{x} + \boldsymbol{y}$ stays in the set but $\frac{1}{2}\boldsymbol{x}$ may be outside.

(b) A set of vectors in $\mathbf{R}^2$ (other than two quarter-planes) for which every $c\boldsymbol{x}$ stays in the set but $\boldsymbol{x} + \boldsymbol{y}$ may be outside.

10 Which of the following subsets of $\mathbf{R}^3$ are actually subspaces?

(a) The plane of vectors (b_1, b_2, b_3) with $b_1 = b_2$.

(b) The plane of vectors with $b_1 = 1$.

(c) The vectors with $b_1b_2b_3 = 0$.

(d) All linear combinations of $\boldsymbol{v} = (1, 4, 0)$ and $\boldsymbol{w} = (2, 2, 2)$.

(e) All vectors that satisfy $b_1 + b_2 + b_3 = 0$.

(f) All vectors with $b_1 \leq b_2 \leq b_3$.

11 Describe the smallest subspace of the matrix space $\mathbf{M}$ that contains

(a) $\begin{bmatrix} 1 & 0 \\ 0 & 0 \end{bmatrix}$ and $\begin{bmatrix} 0 & 1 \\ 0 & 0 \end{bmatrix}$

(b) $\begin{bmatrix} 1 & 1 \\ 0 & 0 \end{bmatrix}$ (c) $\begin{bmatrix} 1 & 0 \\ 0 & 0 \end{bmatrix}$ and $\begin{bmatrix} 1 & 0 \\ 0 & 1 \end{bmatrix}$.

12 Let P be the plane in $\mathbf{R}^3$ with equation $x + y - 2z = 4$. The origin $(0, 0, 0)$ is not in P! Find two vectors in P and check that their sum is not in P.

13 Let $\mathbf{P}_0$ be the plane through $(0, 0, 0)$ parallel to the previous plane P. What is the equation for $\mathbf{P}_0$? Find two vectors in $\mathbf{P}_0$ and check that their sum is in $\mathbf{P}_0$.

14 The subspaces of $\mathbf{R}^3$ are planes, lines, $\mathbf{R}^3$ itself, or $\mathbf{Z}$ containing only $(0, 0, 0)$.

(a) Describe the three types of subspaces of $\mathbf{R}^2$.

(b) Describe the five types of subspaces of $\mathbf{R}^4$.

15 (a) The intersection of two planes through $(0, 0, 0)$ is probably a ____ but it could be a ____. It can't be $\mathbf{Z}$!

(b) The intersection of a plane through $(0,0,0)$ with a line through $(0,0,0)$ is probably a ____ but it could be a ____ .

(c) If $\mathbf{S}$ and $\mathbf{T}$ are subspaces of $\mathbf{R}^5$, prove that their intersection $\mathbf{S} \cap \mathbf{T}$ (vectors in both subspaces) is a subspace of $\mathbf{R}^5$. *Check the requirements on $\boldsymbol{x}+\boldsymbol{y}$ and $c\boldsymbol{x}$.*

16 Suppose $\mathbf{P}$ is a plane through $(0,0,0)$ and $\mathbf{L}$ is a line through $(0,0,0)$. The smallest vector space containing both $\mathbf{P}$ and $\mathbf{L}$ is either ____ or ____ .

17 (a) Show that the set of *invertible* matrices in $\mathbf{M}$ is not a subspace.

(b) Show that the set of *singular* matrices in $\mathbf{M}$ is not a subspace.

18 True or false (check addition in each case by an example):

(a) The symmetric matrices in $\mathbf{M}$ (with $A^{\mathrm{T}} = A$) form a subspace.

(b) The skew-symmetric matrices in $\mathbf{M}$ (with $A^{\mathrm{T}} = -A$) form a subspace.

(c) The unsymmetric matrices in $\mathbf{M}$ (with $A^{\mathrm{T}} \neq A$) form a subspace.

Questions 19–27 are about column spaces $C(A)$ and the equation $Ax = b$.

19 Describe the column spaces (lines or planes) of these particular matrices:

$$A = \begin{bmatrix} 1 & 2 \\ 0 & 0 \\ 0 & 0 \end{bmatrix} \quad \text{and} \quad B = \begin{bmatrix} 1 & 0 \\ 0 & 2 \\ 0 & 0 \end{bmatrix} \quad \text{and} \quad C = \begin{bmatrix} 1 & 0 \\ 2 & 0 \\ 0 & 0 \end{bmatrix}.$$

20 For which right sides (find a condition on b_1, b_2, b_3) are these systems solvable?

$$\text{(a)} \quad \begin{bmatrix} 1 & 4 & 2 \\ 2 & 8 & 4 \\ -1 & -4 & -2 \end{bmatrix} \begin{bmatrix} x_1 \\ x_2 \\ x_3 \end{bmatrix} = \begin{bmatrix} b_1 \\ b_2 \\ b_3 \end{bmatrix} \qquad \text{(b)} \quad \begin{bmatrix} 1 & 4 \\ 2 & 9 \\ -1 & -4 \end{bmatrix} \begin{bmatrix} x_1 \\ x_2 \end{bmatrix} = \begin{bmatrix} b_1 \\ b_2 \\ b_3 \end{bmatrix}.$$

21 Adding row 1 of A to row 2 produces B. Adding column 1 to column 2 produces C. A combination of the columns of ____ is also a combination of the columns of A. Which two matrices have the same column ____ ?

$$A = \begin{bmatrix} 1 & 2 \\ 2 & 4 \end{bmatrix} \quad \text{and} \quad B = \begin{bmatrix} 1 & 2 \\ 3 & 6 \end{bmatrix} \quad \text{and} \quad C = \begin{bmatrix} 1 & 3 \\ 2 & 6 \end{bmatrix}.$$

22 For which vectors (b_1, b_2, b_3) do these systems have a solution?

$$\begin{bmatrix} 1 & 1 & 1 \\ 0 & 1 & 1 \\ 0 & 0 & 1 \end{bmatrix} \begin{bmatrix} x_1 \\ x_2 \\ x_3 \end{bmatrix} = \begin{bmatrix} b_1 \\ b_2 \\ b_3 \end{bmatrix} \quad \text{and} \quad \begin{bmatrix} 1 & 1 & 1 \\ 0 & 1 & 1 \\ 0 & 0 & 0 \end{bmatrix} \begin{bmatrix} x_1 \\ x_2 \\ x_3 \end{bmatrix} = \begin{bmatrix} b_1 \\ b_2 \\ b_3 \end{bmatrix}$$

$$\text{and} \quad \begin{bmatrix} 1 & 1 & 1 \\ 0 & 0 & 1 \\ 0 & 0 & 1 \end{bmatrix} \begin{bmatrix} x_1 \\ x_2 \\ x_3 \end{bmatrix} = \begin{bmatrix} b_1 \\ b_2 \\ b_3 \end{bmatrix}.$$

23 (Recommended) If we add an extra column $\boldsymbol{b}$ to a matrix A, then the column space gets larger unless ____. Give an example where the column space gets larger and an example where it doesn't. Why is $A\boldsymbol{x} = \boldsymbol{b}$ solvable exactly when the column space *doesn't* get larger—it is the same for A and $\begin{bmatrix} A & \boldsymbol{b} \end{bmatrix}$?

24 The columns of AB are combinations of the columns of A. This means: *The column space of AB is contained in* (possibly equal to) *the column space of A.* Give an example where the column spaces of A and AB are not equal.

25 Suppose $A\boldsymbol{x} = \boldsymbol{b}$ and $A\boldsymbol{y} = \boldsymbol{b}^*$ are both solvable. Then $A\boldsymbol{z} = \boldsymbol{b} + \boldsymbol{b}^*$ is solvable. What is $\boldsymbol{z}$? This translates into: If $\boldsymbol{b}$ and $\boldsymbol{b}^*$ are in the column space $\boldsymbol{C}(A)$, then $\boldsymbol{b} + \boldsymbol{b}^*$ is in $\boldsymbol{C}(A)$.

26 If A is any 5 by 5 invertible matrix, then its column space is ____. Why?

27 True or false (with a counterexample if false):

(a) The vectors $\boldsymbol{b}$ that are not in the column space $\boldsymbol{C}(A)$ form a subspace.

(b) If $\boldsymbol{C}(A)$ contains only the zero vector, then A is the zero matrix.

(c) The column space of $2A$ equals the column space of A.

(d) The column space of $A - I$ equals the column space of A.

28 Construct a 3 by 3 matrix whose column space contains $(1, 1, 0)$ and $(1, 0, 1)$ but not $(1, 1, 1)$. Construct a 3 by 3 matrix whose column space is only a line.

29 If the 9 by 12 system $A\boldsymbol{x} = \boldsymbol{b}$ is solvable for every $\boldsymbol{b}$, then $\boldsymbol{C}(A) =$ ____.

THE NULLSPACE OF A: SOLVING $AX = 0$ ■ 3.2

This section is about the space of solutions to $A\boldsymbol{x} = \mathbf{0}$. The matrix A can be square or rectangular. *One immediate solution is* $\boldsymbol{x} = \mathbf{0}$. For invertible matrices this is the only solution. For other matrices, not invertible, there are nonzero solutions to $A\boldsymbol{x} = \mathbf{0}$. *Each solution* $\boldsymbol{x}$ *belongs to the nullspace of* A.

Elimination will find all solutions and identify this very important subspace.

DEFINITION ***The nullspace of*** A ***consists of all solutions to*** $A\boldsymbol{x} = \mathbf{0}$. These solution vectors $\boldsymbol{x}$ are in $\mathbf{R}^n$. The nullspace containing all solutions is denoted by $\boldsymbol{N}(A)$.

Check that the solution vectors form a subspace. Suppose $\boldsymbol{x}$ and $\boldsymbol{y}$ are in the nullspace (this means $A\boldsymbol{x} = \mathbf{0}$ and $A\boldsymbol{y} = \mathbf{0}$). The rules of matrix multiplication give $A(\boldsymbol{x}+\boldsymbol{y}) = \mathbf{0}+\mathbf{0}$. The rules also give $A(c\boldsymbol{x}) = c\mathbf{0}$. The right sides are still zero. Therefore $\boldsymbol{x}+\boldsymbol{y}$ and $c\boldsymbol{x}$ are also in the nullspace $\boldsymbol{N}(A)$. Since we can add and multiply without leaving the nullspace, it is a subspace.

To repeat: The solution vectors $\boldsymbol{x}$ have n components. They are vectors in $\mathbf{R}^n$, so *the nullspace is a subspace of* $\mathbf{R}^n$. The column space $\boldsymbol{C}(A)$ is a subspace of $\mathbf{R}^m$.

If the right side $\boldsymbol{b}$ is not zero, the solutions of $A\boldsymbol{x} = \boldsymbol{b}$ do *not* form a subspace. The vector $\boldsymbol{x} = \mathbf{0}$ is only a solution if $\boldsymbol{b} = \mathbf{0}$. When the set of solutions does not include $\boldsymbol{x} = \mathbf{0}$, it cannot be a subspace. Section 3.4 will show how the solutions to $A\boldsymbol{x} = \boldsymbol{b}$ (if there are any solutions) are shifted away from the origin by one particular solution.

Example 1 The equation $x + 2y + 3z = 0$ comes from the 1 by 3 matrix $A = [1 \;\; 2 \;\; 3]$. This equation produces a plane through the origin. The plane is a subspace of $\mathbf{R}^3$. *It is the nullspace of* A.

The solutions to $x + 2y + 3z = 6$ also form a plane, but not a subspace.

Example 2 Describe the nullspace of $A = \begin{bmatrix} 1 & 2 \\ 3 & 6 \end{bmatrix}$.

Solution Apply elimination to the linear equations $A\boldsymbol{x} = \mathbf{0}$:

$$\begin{bmatrix} x_1 + 2x_2 = 0 \\ 3x_1 + 6x_2 = 0 \end{bmatrix} \quad \rightarrow \quad \begin{bmatrix} x_1 + 2x_2 = 0 \\ 0 = 0 \end{bmatrix}$$

There is really only one equation. The second equation is the first equation multiplied by 3. In the row picture, the line $x_1 + 2x_2 = 0$ is the same as the line $3x_1 + 6x_2 = 0$. That line is the nullspace $\boldsymbol{N}(A)$.

To describe this line of solutions, here is an efficient way. Choose one point on the line (one "***special solution***"). Then all points on the line are multiples of this one. We choose the second component to be $x_2 = 1$ (a special choice). From the equation $x_1 + 2x_2 = 0$, the first component must be $x_1 = -2$. The special solution is $(-2, 1)$:

$$\text{The nullspace } \boldsymbol{N}(A) \text{ contains all multiples of } \boldsymbol{s} = \begin{bmatrix} -2 \\ 1 \end{bmatrix}.$$

This is the best way to describe the nullspace, by computing special solutions to $A\boldsymbol{x} = \mathbf{0}$. ***The nullspace consists of all combinations of those special solutions.*** This example has one special solution and the nullspace is a line.

The plane $x + 2y + 3z = 0$ in Example 1 had *two* special solutions:

$$\begin{bmatrix} 1 & 2 & 3 \end{bmatrix} \begin{bmatrix} x \\ y \\ z \end{bmatrix} = 0 \text{ has the special solutions } \boldsymbol{s}_1 = \begin{bmatrix} -2 \\ 1 \\ 0 \end{bmatrix} \text{ and } \boldsymbol{s}_2 = \begin{bmatrix} -3 \\ 0 \\ 1 \end{bmatrix}.$$

Those vectors $\boldsymbol{s}_1$ and $\boldsymbol{s}_2$ lie on the plane $x + 2y + 3z = 0$, which is the nullspace of $A = \begin{bmatrix} 1 & 2 & 3 \end{bmatrix}$. All vectors on the plane are combinations of $\boldsymbol{s}_1$ and $\boldsymbol{s}_2$.

Notice what is special about $\boldsymbol{s}_1$ and $\boldsymbol{s}_2$. They have ones and zeros in the last two components. *Those components are "free" and we choose them specially.* Then the first components -2 and -3 are determined by the equation $A\boldsymbol{x} = \mathbf{0}$.

The first column of $A = \begin{bmatrix} 1 & 2 & 3 \end{bmatrix}$ contains the *pivot*, so the first component of $\boldsymbol{x}$ is *not free*. The free components correspond to columns without pivots. This description of special solutions will be completed after one more example.

The special choice (one or zero) is only for the free variables.

Example 3 Describe the nullspaces of these three matrices A, B, C:

$$A = \begin{bmatrix} 1 & 2 \\ 3 & 8 \end{bmatrix} \quad B = \begin{bmatrix} A \\ 2A \end{bmatrix} = \begin{bmatrix} 1 & 2 \\ 3 & 8 \\ 2 & 4 \\ 6 & 16 \end{bmatrix} \quad C = \begin{bmatrix} A & 2A \end{bmatrix} = \begin{bmatrix} 1 & 2 & 2 & 4 \\ 3 & 8 & 6 & 16 \end{bmatrix}.$$

Solution The equation $A\boldsymbol{x} = \mathbf{0}$ has only the zero solution $\boldsymbol{x} = \mathbf{0}$. *The nullspace is* $\mathbf{Z}$. It contains only the single point $\boldsymbol{x} = \mathbf{0}$ in $\mathbf{R}^2$. This comes from elimination:

$$\begin{bmatrix} 1 & 2 \\ 3 & 8 \end{bmatrix} \begin{bmatrix} x_1 \\ x_2 \end{bmatrix} = \begin{bmatrix} 0 \\ 0 \end{bmatrix} \text{ yields } \begin{bmatrix} 1 & 2 \\ 0 & 2 \end{bmatrix} \begin{bmatrix} x_1 \\ x_2 \end{bmatrix} = \begin{bmatrix} 0 \\ 0 \end{bmatrix} \text{ and } \begin{bmatrix} x_1 = 0 \\ x_2 = 0 \end{bmatrix}.$$

A is invertible. There are no special solutions. All columns have pivots.

The rectangular matrix B has the same nullspace $\mathbf{Z}$. The first two equations in $B\boldsymbol{x} = \mathbf{0}$ again require $\boldsymbol{x} = \mathbf{0}$. The last two equations would also force $\boldsymbol{x} = \mathbf{0}$. When we add extra equations, the nullspace certainly cannot become larger. The extra rows impose more conditions on the vectors $\boldsymbol{x}$ in the nullspace.

The rectangular matrix C is different. It has extra columns instead of extra rows. The solution vector $\boldsymbol{x}$ has *four* components. Elimination will produce pivots in the first two columns of C, but the last two columns are "free". They don't have pivots:

$$C = \begin{bmatrix} 1 & 2 & 2 & 4 \\ 3 & 8 & 6 & 16 \end{bmatrix} \text{ becomes } U = \begin{bmatrix} 1 & 2 & 2 & 4 \\ 0 & 2 & 0 & 4 \end{bmatrix}$$

$$\uparrow \; \uparrow \; \uparrow \; \uparrow$$

pivot columns free columns

For the free variables x_3 and x_4, we make special choices of ones and zeros. First $x_3 = 1$, $x_4 = 0$ and second $x_3 = 0$, $x_4 = 1$. The pivot variables x_1 and x_2 are determined by the equation $Ux = \mathbf{0}$. We get two special solutions in the nullspace of C (and also the nullspace of U). The special solutions are:

$$s_1 = \begin{bmatrix} -2 \\ 0 \\ 1 \\ 0 \end{bmatrix} \quad \text{and } s_2 = \begin{bmatrix} 0 \\ -2 \\ 0 \\ 1 \end{bmatrix} \begin{matrix} \leftarrow \text{pivot} \\ \leftarrow \quad \text{variables} \\ \leftarrow \text{free} \\ \leftarrow \quad \text{variables} \end{matrix}$$

One more comment to anticipate what is coming soon. Elimination will not stop at the upper triangular U! We can continue to make this matrix simpler, in two ways:

1. ***Produce zeros above the pivots***, by eliminating upward.
2. ***Produce ones in the pivots***, by dividing the whole row by its pivot.

Those steps don't change the zero vector on the right side of the equation. The nullspace stays the same. This nullspace becomes easiest to see when we reach the ***reduced row echelon form*** R. It has I in the pivot columns:

$$U = \begin{bmatrix} 1 & 2 & 2 & 4 \\ 0 & 2 & 0 & 4 \end{bmatrix} \quad \text{becomes} \quad R = \begin{bmatrix} 1 & 0 & 2 & 0 \\ 0 & 1 & 0 & 2 \end{bmatrix}.$$
$$\uparrow \; \uparrow$$

pivot columns contain I

I subtracted row 2 of U from row 1, and then multiplied row 2 by $\frac{1}{2}$. The original two equations have simplified to $x_1 + 2x_3 = 0$ and $x_2 + 2x_4 = 0$.

The first special solution is still $s_1 = (-2, 0, 1, 0)$, and s_2 is unchanged. Special solutions are much easier to find from the reduced system $Rx = \mathbf{0}$.

Before moving to m by n matrices A and their nullspaces $N(A)$ and special solutions, allow me to repeat one comment. For many matrices, the only solution to $Ax = \mathbf{0}$ is $x = \mathbf{0}$. Their nullspaces $N(A) = \mathbf{Z}$ contain only that one vector. The only combination of the columns that produces $b = \mathbf{0}$ is then the "zero combination" or "trivial combination". The solution is trivial (just $x = \mathbf{0}$) but the idea is not trivial.

This case of a zero nullspace $\mathbf{Z}$ is of the greatest importance. It says that the columns of A are independent. No combination of columns gives the zero vector (except the zero combination). All columns have pivots and no columns are free. You will see this idea of independence again ...

Solving $Ax = 0$ by Elimination

This is important. ***A is rectangular and we still use elimination***. We solve m equations in n unknowns when $b = 0$. After A is simplified by row operations, we read off the solution (or solutions). Remember the two stages in solving $Ax = \mathbf{0}$:

1. Forward elimination from A to a triangular U (or its reduced form R).

2. Back substitution in $U\boldsymbol{x} = \mathbf{0}$ or $R\boldsymbol{x} = \mathbf{0}$ to find $\boldsymbol{x}$.

You will notice a difference in back substitution, when A and U have fewer than n pivots. *We are allowing all matrices in this chapter*, not just the nice ones (which are square matrices with inverses).

Pivots are still nonzero. The columns below the pivots are still zero. But it might happen that a column has no pivot. In that case, don't stop the calculation. ***Go on to the next column***. The first example is a 3 by 4 matrix with two pivots:

$$A = \begin{bmatrix} 1 & 1 & 2 & 3 \\ 2 & 2 & 8 & 10 \\ 3 & 3 & 10 & 13 \end{bmatrix}.$$

Certainly $a_{11} = 1$ is the first pivot. Clear out the 2 and 3 below that pivot:

$$A \to \begin{bmatrix} 1 & 1 & 2 & 3 \\ 0 & 0 & 4 & 4 \\ 0 & 0 & 4 & 4 \end{bmatrix} \qquad \begin{matrix} \\ \text{(subtract } 2 \times \text{ row 1)} \\ \text{(subtract } 3 \times \text{ row 1)} \end{matrix}$$

The second column has a zero in the pivot position. We look below the zero for a nonzero entry, ready to do a row exchange. *The entry below that position is also zero.* Elimination can do nothing with the second column. This signals trouble, which we expect anyway for a rectangular matrix. There is no reason to quit, and we go on to the third column.

The second pivot is 4 (but it is in the third column). Subtracting row 2 from row 3 clears out that column below the pivot. We arrive at

$$\textbf{Triangular } U: \quad U = \begin{bmatrix} \mathbf{1} & 1 & 2 & 3 \\ 0 & 0 & \mathbf{4} & 4 \\ 0 & 0 & 0 & 0 \end{bmatrix} \qquad \begin{matrix} \textit{(only two pivots)} \\ \textit{(the last equation} \\ \textit{became } 0 = 0) \end{matrix}$$

The fourth column also has a zero in the pivot position—but nothing can be done. There is no row below it to exchange, and forward elimination is complete. The matrix has three rows, four columns, and *only two pivots*. The original $A\boldsymbol{x} = \mathbf{0}$ seemed to involve three different equations, but the third equation is the sum of the first two. It is automatically satisfied $(0 = 0)$ when the first two equations are satisfied. Elimination reveals the inner truth about a system of equations. Soon we push on from U to R.

Now comes back substitution, to find all solutions to $U\boldsymbol{x} = \mathbf{0}$. With four unknowns and only two pivots, there are many solutions. The question is how to write them all down. A good method is to separate the ***pivot variables*** from the ***free variables***.

P The ***pivot*** variables are x_1 and x_3, since columns 1 and 3 contain pivots.
F The ***free*** variables are x_2 and x_4, because columns 2 and 4 have no pivots.

The free variables x_2 and x_4 can be given any values whatsoever. Then back substitution finds the pivot variables x_1 and x_3. (In Chapter 2 no variables were free. When A is invertible, all variables are pivot variables.) The simplest choices for the free variables are ones and zeros. Those choices give the *special solutions*.

Special Solutions to $x_1 + x_2 + 2x_3 + 3x_4 = 0$ and $4x_3 + 4x_4 = 0$

- Set $x_2 = 1$ and $x_4 = 0$. By back substitution $x_3 = 0$. Then $x_1 = -1$.
- Set $x_2 = 0$ and $x_4 = 1$. By back substitution $x_3 = -1$. Then $x_1 = -1$.

These special solutions solve $U\boldsymbol{x} = \boldsymbol{0}$ and therefore $A\boldsymbol{x} = \boldsymbol{0}$. They are in the nullspace. The good thing is that *every solution is a combination of the special solutions*.

Complete Solution
$$\boldsymbol{x} = x_2 \underbrace{\begin{bmatrix} -1 \\ 1 \\ 0 \\ 0 \end{bmatrix}}_{\textbf{special}} + x_4 \underbrace{\begin{bmatrix} -1 \\ 0 \\ -1 \\ 1 \end{bmatrix}}_{\textbf{special}} = \underbrace{\begin{bmatrix} -x_2 - x_4 \\ x_2 \\ -x_4 \\ x_4 \end{bmatrix}}_{\textbf{complete}}. \tag{1}$$

Please look again at that answer. It is the main goal of this section. The vector $\boldsymbol{s}_1 = (-1, 1, 0, 0)$ is the special solution when $x_2 = 1$ and $x_4 = 0$. The second special solution has $x_2 = 0$ and $x_4 = 1$. ***All solutions are linear combinations of $\boldsymbol{s}_1$ and $\boldsymbol{s}_2$.*** The special solutions are in the nullspace $\boldsymbol{N}(A)$, and their combinations fill out the whole nullspace.

The MATLAB code **nullbasis** computes these special solutions. They go into the columns of a ***nullspace matrix*** N. The complete solution to $A\boldsymbol{x} = \boldsymbol{0}$ is a combination of those columns. Once we have the special solutions, we have the whole nullspace.

There is a special solution for each free variable. If no variables are free—this means there are n pivots—then the only solution to $U\boldsymbol{x} = \boldsymbol{0}$ and $A\boldsymbol{x} = \boldsymbol{0}$ is the trivial solution $\boldsymbol{x} = \boldsymbol{0}$. All variables are pivot variables. In that case the nullspaces of A and U contain only the zero vector. With no free variables, and pivots in every column, the output from **nullbasis** is an empty matrix.

Example 4 Find the nullspace of $U = \begin{bmatrix} 1 & 5 & 7 \\ 0 & 0 & 9 \end{bmatrix}$.

The second column of U has no pivot. So x_2 is free. The special solution has $x_2 = 1$. Back substitution into $9x_3 = 0$ gives $x_3 = 0$. Then $x_1 + 5x_2 = 0$ or $x_1 = -5$. The solutions to $U\boldsymbol{x} = \boldsymbol{0}$ are multiples of one special solution:

$$\boldsymbol{x} = x_2 \begin{bmatrix} -5 \\ 1 \\ 0 \end{bmatrix}$$

The nullspace of U is a line in $\mathbf{R}^3$.
It contains multiples of the special solution.
One variable is free, and $N = $ **nullbasis** (U) has one column.

In a minute we will continue elimination on U, to get *zeros above the pivots and ones in the pivots.* The 7 is eliminated and the pivot changes from 9 to 1. The final result of this elimination will be the reduced row echelon matrix R:

$$U = \begin{bmatrix} 1 & 5 & 7 \\ 0 & 0 & 9 \end{bmatrix} \text{ reduces to } R = \begin{bmatrix} 1 & 5 & 0 \\ 0 & 0 & 1 \end{bmatrix}.$$

This makes it even clearer that the special solution (column of N) is $s = (-5, 1, 0)$.

Echelon Matrices

Forward elimination goes from A to U. The process starts with an m by n matrix A. It acts by row operations, including row exchanges. It goes on to the next column when no pivot is available in the current column. The m by n "staircase," U is an ***echelon matrix***.

Here is a 4 by 7 echelon matrix with the three pivots highlighted in boldface:

$$U = \begin{bmatrix} \mathbf{x} & x & x & x & x & x & x \\ 0 & \mathbf{x} & x & x & x & x & x \\ 0 & 0 & 0 & 0 & 0 & \mathbf{x} & x \\ 0 & 0 & 0 & 0 & 0 & 0 & 0 \end{bmatrix} \quad \begin{array}{l} \text{Three pivot variables } x_1, x_2, x_6 \\ \text{Four free variables } x_3, x_4, x_5, x_7 \\ \text{Four special solutions in } N(U) \end{array}$$

Question What are the column space and the nullspace for this matrix?

Answer The columns have four components so they lie in $\mathbf{R}^4$. (Not in $\mathbf{R}^3$!) The fourth component of every column is zero. Every combination of the columns—every vector in the column space—has fourth component zero. *The column space* $\boldsymbol{C}(U)$ *consists of all vectors of the form* $(b_1, b_2, b_3, 0)$. For those vectors we can solve $U\boldsymbol{x} = \boldsymbol{b}$ by back substitution. These vectors $\boldsymbol{b}$ are all possible combinations of the seven columns.

The nullspace $\boldsymbol{N}(U)$ is a subspace of $\mathbf{R}^7$. The solutions to $U\boldsymbol{x} = \mathbf{0}$ are all the combinations of the four special solutions—*one for each free variable*:

1. Columns $3, 4, 5, 7$ have no pivots. So the free variables are x_3, x_4, x_5, x_7.
2. Set one free variable to 1 and set the other free variables to zero.
3. Solve $U\boldsymbol{x} = \mathbf{0}$ for the pivot variables x_1, x_2, x_6.
4. This gives one of the four special solutions in the nullspace matrix N.

The nonzero rows of an echelon matrix go down in a staircase pattern. The pivots are the first nonzero entries in those rows. There is a column of zeros below every pivot.

Counting the pivots leads to an extremely important theorem. Suppose A has more columns than rows. ***With*** $n > m$ ***there is at least one free variable.*** The system $A\boldsymbol{x} = \mathbf{0}$ has at least one special solution. This solution is *not zero*!

3B If $Ax = \mathbf{0}$ has more unknowns than equations ($n > m$, more columns than rows), then it has nonzero solutions. There must be free columns, without pivots.

In other words, a short wide matrix ($n > m$) always has nonzero vectors in its nullspace. There must be at least $n-m$ free variables, since the number of pivots cannot exceed m. (The matrix only has m rows, and a row never has two pivots.) Of course a row might have *no* pivot—which means an extra free variable. But here is the point: When there is a free variable, it can be set to 1. Then the equation $Ax = \mathbf{0}$ has a nonzero solution.

To repeat: There are at most m pivots. With $n > m$, the system $Ax = \mathbf{0}$ has a nonzero solution. Actually there are infinitely many solutions, since any multiple cx is also a solution. The nullspace contains at least a line of solutions. With two free variables, there are two special solutions and the nullspace is even larger.

The nullspace is a subspace. Its "dimension" is the number of free variables. This central idea—the ***dimension*** of a subspace—is defined and explained in this chapter.

The Reduced Row Echelon Matrix R

>From an echelon matrix U we can go one more step. Continue with our example

$$U = \begin{bmatrix} 1 & 1 & 2 & 3 \\ 0 & 0 & 4 & 4 \\ 0 & 0 & 0 & 0 \end{bmatrix}.$$

We can divide the second row by 4. Then both pivots equal 1. *We can subtract* 2 *times this new row* $\begin{bmatrix} 0 & 0 & 1 & 1 \end{bmatrix}$ *from the row above*. The ***reduced row echelon matrix R has zeros above the pivots as well as below***:

$$R = \begin{bmatrix} \mathbf{1} & 1 & 0 & 1 \\ 0 & 0 & \mathbf{1} & 1 \\ 0 & 0 & 0 & 0 \end{bmatrix}.$$

R has 1's as pivots. Zeros above pivots come from ***upward elimination***.

If A is invertible, its reduced row echelon form is the identity matrix $R = I$. This is the ultimate in row reduction. Of course the nullspace is then $\mathbf{Z}$.

The zeros in R make it easy to find the special solutions (the same as before):

1. Set $x_2 = 1$ and $x_4 = 0$. Solve $Rx = \mathbf{0}$. Then $x_1 = -1$ and $x_3 = 0$.

Those numbers -1 and 0 are sitting in column 2 of R (with plus signs).

2. Set $x_2 = 0$ and $x_4 = 1$. Solve $Rx = \mathbf{0}$. Then $x_1 = -1$ and $x_3 = -1$.

Those numbers -1 and -1 are sitting in column 4 (with plus signs).

By reversing signs we can read off the special solutions directly from R. The nullspace $N(A) = N(U) = N(R)$ contains all combinations of the special solutions:

$$x = x_2 \begin{bmatrix} -1 \\ 1 \\ 0 \\ 0 \end{bmatrix} + x_4 \begin{bmatrix} -1 \\ 0 \\ -1 \\ 1 \end{bmatrix} = (\textbf{\textit{complete solution of}}\ Ax = \mathbf{0}).$$

The next section of the book moves firmly from U to R. The MATLAB command $[R, pivcol] = \mathbf{rref}(A)$ produces R and also a list of the pivot columns.

■ REVIEW OF THE KEY IDEAS ■

1. The nullspace $N(A)$, a subspace of $\mathbf{R}^n$, contains all solutions to $Ax = \mathbf{0}$.
2. Elimination produces an echelon matrix U, and then a row reduced R, with pivot columns and free columns.
3. Every free column of U or R leads to a special solution. The free variable equals 1 and the other free variables equal 0. Back substitution solves $Ax = \mathbf{0}$.
4. The complete solution to $Ax = \mathbf{0}$ is a combination of the special solutions.
5. If $n > m$ then A has at least one column without pivots, giving a special solution. So there are nonzero vectors x in the nullspace of this rectangular A.

■ WORKED EXAMPLES ■

3.2 A Create a 3 by 4 matrix whose special solutions to $Ax = \mathbf{0}$ are s_1 and s_2:

$$s_1 = \begin{bmatrix} -3 \\ 1 \\ 0 \\ 0 \end{bmatrix} \quad \text{and} \quad s_2 = \begin{bmatrix} -2 \\ 0 \\ -6 \\ 1 \end{bmatrix} \qquad \begin{array}{l} \text{pivot columns 1 and 3} \\ \text{free variables } x_2 \text{ and } x_4 \end{array}$$

You could create the matrix A in row reduced form R. Then describe all possible matrices A with the required nullspace $N(A) =$ all combinations of s_1 and s_2.

Solution The reduced matrix R has pivots $= 1$ in columns 1 and 3. There is no third pivot, so the third row of R is all zeros. The free columns 2 and 4 will be combinations of the pivot columns:

$$R = \begin{bmatrix} 1 & 3 & 0 & 2 \\ 0 & 0 & 1 & 6 \\ 0 & 0 & 0 & 0 \end{bmatrix} \quad \text{has} \quad Rs_1 = \mathbf{0} \quad \text{and} \quad Rs_2 = \mathbf{0}.$$

The entries $3, 2, 6$ are the negatives of $-3, -2, -6$ in the special solutions!

R is only one matrix (one possible A) with the required nullspace. We could do any elementary operations on R—exchange rows, multiply a row by any $c \neq 0$, subtract any multiple of one row from another. R can be multiplied by *any invertible matrix*, without changing the row space and nullspace.

Every 3 by 4 matrix has at least one special solution. *These* A's *have two.*

3.2 B Find the special solutions and describe the *complete solution* to $A\boldsymbol{x} = \mathbf{0}$ for

$$A_1 = 3 \text{ by } 4 \textit{ zero matrix} \qquad A_2 = \begin{bmatrix} 3 & 6 \\ 1 & 2 \end{bmatrix} \qquad A_3 = \begin{bmatrix} A_2 & A_2 \end{bmatrix}$$

Which are the pivot columns? Which are the free variables? What is R in each case?

Solution $A_1\boldsymbol{x} = \mathbf{0}$ has four special solutions. They are the columns $\boldsymbol{s}_1, \boldsymbol{s}_2, \boldsymbol{s}_3, \boldsymbol{s}_4$ of the 4 by 4 identity matrix. The nullspace is all of $\mathbf{R}^4$. The complete solution is any $\boldsymbol{x} = c_1\boldsymbol{s}_1 + c_2\boldsymbol{s}_2 + c_3\boldsymbol{s}_3 + c_4\boldsymbol{s}_4$ in $\mathbf{R}^4$. There are no pivot columns; all variables are free; the reduced R is the same zero matrix as A_1.

$A_2\boldsymbol{x} = \mathbf{0}$ has only one special solution $\boldsymbol{s} = (-2, 1)$. The multiples $\boldsymbol{x} = c\boldsymbol{s}$ give the complete solution. The first column of A_2 is its pivot column, and x_2 is the free variable. The row reduced matrices R_2 for A_2 and R_3 for $A_3 = [\,A_2 \;\; A_2\,]$ have 1's in the pivot:

$$R_2 = \begin{bmatrix} 1 & 2 \\ 0 & 0 \end{bmatrix} \qquad R_3 = \begin{bmatrix} 1 & 2 & 1 & 2 \\ 0 & 0 & 0 & 0 \end{bmatrix}$$

Notice that R_3 has only one pivot column (the first column). All the variables x_2, x_3, x_4 are free. There are three special solutions to $A_3\boldsymbol{x} = \mathbf{0}$ (and also $R_3\boldsymbol{x} = \mathbf{0}$):

$$\boldsymbol{s}_1 = \begin{bmatrix} -2 \\ 1 \\ 0 \\ 0 \end{bmatrix} \quad \boldsymbol{s}_2 = \begin{bmatrix} -1 \\ 0 \\ 1 \\ 0 \end{bmatrix} \quad \boldsymbol{s}_3 = \begin{bmatrix} -2 \\ 0 \\ 0 \\ 1 \end{bmatrix} \quad \textbf{Complete solution } \boldsymbol{x} = c_1\boldsymbol{s}_1 + c_2\boldsymbol{s}_2 + c_3\boldsymbol{s}_3.$$

With r pivots, A has $n - r$ free variables and $A\boldsymbol{x} = \mathbf{0}$ has $n - r$ special solutions.

Problem Set 3.2

Questions 1–4 and 5–8 are about the matrices in Problems 1 and 5.

1 Reduce these matrices to their ordinary echelon forms U:

$$\text{(a)} \quad A = \begin{bmatrix} 1 & 2 & 2 & 4 & 6 \\ 1 & 2 & 3 & 6 & 9 \\ 0 & 0 & 1 & 2 & 3 \end{bmatrix} \qquad \text{(b)} \quad B = \begin{bmatrix} 2 & 4 & 2 \\ 0 & 4 & 4 \\ 0 & 8 & 8 \end{bmatrix}.$$

Which are the free variables and which are the pivot variables?

2 For the matrices in Problem 1, find a special solution for each free variable. (Set the free variable to 1. Set the other free variables to zero.)

3 By combining the special solutions in Problem 2, describe every solution to $Ax = \mathbf{0}$ and $Bx = \mathbf{0}$. The nullspace contains only $x = \mathbf{0}$ when there are no _____ .

4 By further row operations on each U in Problem 1, find the reduced echelon form R. *True or false*: The nullspace of R equals the nullspace of U.

5 By row operations reduce each matrix to its echelon form U. Write down a 2 by 2 lower triangular L such that $B = LU$.

(a) $A = \begin{bmatrix} -1 & 3 & 5 \\ -2 & 6 & 10 \end{bmatrix}$ (b) $B = \begin{bmatrix} -1 & 3 & 5 \\ -2 & 6 & 7 \end{bmatrix}$.

6 Find the special solutions to $Ax = \mathbf{0}$ and $Bx = \mathbf{0}$. For an m by n matrix, the number of pivot variables plus the number of free variables is _____ .

7 In Problem 5, describe the nullspaces of A and B in two ways. Give the equations for the plane or the line, and give all vectors x that satisfy those equations as combinations of the special solutions.

8 Reduce the echelon forms U in Problem 5 to R. For each R draw a box around the identity matrix that is in the pivot rows and pivot columns.

Questions 9–17 are about free variables and pivot variables.

9 True or false (with reason if true or example to show it is false):

(a) A square matrix has no free variables.

(b) An invertible matrix has no free variables.

(c) An m by n matrix has no more than n pivot variables.

(d) An m by n matrix has no more than m pivot variables.

10 Construct 3 by 3 matrices A to satisfy these requirements (if possible):

(a) A has no zero entries but $U = I$.

(b) A has no zero entries but $R = I$.

(c) A has no zero entries but $R = U$.

(d) $A = U = 2R$.

11 Put as many 1's as possible in a 4 by 7 echelon matrix U whose pivot variables are

(a) 2, 4, 5

(b) 1, 3, 6, 7

(c) 4 and 6.

12 Put as many 1's as possible in a 4 by 8 *reduced* echelon matrix R so that the free variables are

(a) 2, 4, 5, 6

(b) 1, 3, 6, 7, 8.

13 Suppose column 4 of a 3 by 5 matrix is all zero. Then x_4 is certainly a ______ variable. The special solution for this variable is the vector $\boldsymbol{x} =$ ______.

14 Suppose the first and last columns of a 3 by 5 matrix are the same (not zero). Then ______ is a free variable. Find the special solution for this variable.

15 Suppose an m by n matrix has r pivots. The number of special solutions is ______. The nullspace contains only $\boldsymbol{x} = \mathbf{0}$ when $r =$ ______. The column space is all of $\mathbf{R}^m$ when $r =$ ______.

16 The nullspace of a 5 by 5 matrix contains only $\boldsymbol{x} = \mathbf{0}$ when the matrix has ______ pivots. The column space is $\mathbf{R}^5$ when there are ______ pivots. Explain why.

17 The equation $x - 3y - z = 0$ determines a plane in $\mathbf{R}^3$. What is the matrix A in this equation? Which are the free variables? The special solutions are $(3, 1, 0)$ and ______.

18 (Recommended) The plane $x - 3y - z = 12$ is parallel to the plane $x - 3y - z = 0$ in Problem 17. One particular point on this plane is $(12, 0, 0)$. All points on the plane have the form (fill in the first components)

$$\begin{bmatrix} x \\ y \\ z \end{bmatrix} = \begin{bmatrix} \\ 0 \\ 0 \end{bmatrix} + y \begin{bmatrix} \\ 1 \\ 0 \end{bmatrix} + z \begin{bmatrix} \\ 0 \\ 1 \end{bmatrix}.$$

19 Prove that U and $A = LU$ have the same nullspace when L is invertible:

$$\text{If } U\boldsymbol{x} = \mathbf{0} \text{ then } LU\boldsymbol{x} = \mathbf{0}. \quad \text{If } LU\boldsymbol{x} = \mathbf{0}, \text{ how do you know } U\boldsymbol{x} = \mathbf{0}?$$

20 Suppose column 1 + column 3 + column 5 = $\mathbf{0}$ in a 4 by 5 matrix with four pivots. Which column is sure to have no pivot (and which variable is free)? What is the special solution? What is the nullspace?

Questions 21–28 ask for matrices (if possible) with specific properties.

21 Construct a matrix whose nullspace consists of all combinations of $(2, 2, 1, 0)$ and $(3, 1, 0, 1)$.

22 Construct a matrix whose nullspace consists of all multiples of $(4, 3, 2, 1)$.

23 Construct a matrix whose column space contains $(1, 1, 5)$ and $(0, 3, 1)$ and whose nullspace contains $(1, 1, 2)$.

24 Construct a matrix whose column space contains $(1, 1, 0)$ and $(0, 1, 1)$ and whose nullspace contains $(1, 0, 1)$ and $(0, 0, 1)$.

25 Construct a matrix whose column space contains $(1, 1, 1)$ and whose nullspace is the line of multiples of $(1, 1, 1, 1)$.

26 Construct a 2 by 2 matrix whose nullspace equals its column space. This is possible.

27 Why does no 3 by 3 matrix have a nullspace that equals its column space?

28 If $AB = 0$ then the column space of B is contained in the _____ of A. Give an example of A and B.

29 The reduced form R of a 3 by 3 matrix with randomly chosen entries is almost sure to be _____. What R is virtually certain if the random A is 4 by 3?

30 Show by example that these three statements are generally *false*:

(a) A and A^{T} have the same nullspace.

(b) A and A^{T} have the same free variables.

(c) If R is the reduced form $\mathbf{rref}(A)$ then R^{T} is $\mathbf{rref}(A^{\mathrm{T}})$.

31 If the nullspace of A consists of all multiples of $\boldsymbol{x} = (2, 1, 0, 1)$, how many pivots appear in U? What is R?

32 If the special solutions to $R\boldsymbol{x} = \mathbf{0}$ are in the columns of these N, go backward to find the nonzero rows of the reduced matrices R:

$$N = \begin{bmatrix} 2 & 3 \\ 1 & 0 \\ 0 & 1 \end{bmatrix} \quad \text{and} \quad N = \begin{bmatrix} 0 \\ 0 \\ 1 \end{bmatrix} \quad \text{and} \quad N = \begin{bmatrix} \quad \end{bmatrix} \quad \text{(empty 3 by 1)}.$$

33 (a) What are the five 2 by 2 reduced echelon matrices R whose entries are all 0's and 1's?

(b) What are the eight 1 by 3 matrices containing only 0's and 1's? Are all eight of them reduced echelon matrices R?

34 Explain why A and $-A$ always have the same reduced echelon form R.

THE RANK AND THE ROW REDUCED FORM ■ 3.3

This section completes the step from A to its reduced row echelon form R. The matrix A is m by n (completely general). The matrix R is also m by n, but each pivot column has only one nonzero entry (the pivot which is always 1). This example is 3 by 5:

$$\textbf{Reduced Row Echelon Form} \qquad R = \begin{bmatrix} \mathbf{1} & 3 & \mathbf{0} & 2 & -1 \\ \mathbf{0} & 0 & \mathbf{1} & 4 & -3 \\ \mathbf{0} & 0 & \mathbf{0} & 0 & 0 \end{bmatrix}.$$

You see zero above the second pivot as well as below. R is the final result of elimination, and MATLAB uses the command **rref**. The Teaching Code **elim** for this book has **rref** built into it. Of course **rref**(R) would give R again!

MATLAB: $[R, pivcol] = \mathbf{rref}(A)$ Teaching Code: $[E, R] = \mathbf{elim}(A)$

The extra output *pivcol* gives the numbers of the pivot columns. They are the same in A and R. The extra output E is the m by m ***elimination matrix*** that puts the original A (whatever it was) into its row reduced form R:

$$EA = R. \tag{1}$$

The square matrix E is the product of elementary matrices E_{ij} and P_{ij} and D^{-1}. Now we allow $j > i$, when E_{ij} subtracts a multiple of row j from row i. P_{ij} exchanges these rows. D^{-1} divides rows by their pivots to produce 1's.

If we want E, we can apply row reduction to the matrix $\begin{bmatrix} A & I \end{bmatrix}$ with $n + m$ columns. All the elementary matrices that multiply A (to produce R) will also multiply I (to produce E). The whole augmented matrix is being multiplied by E:

$$E\,[A \;\; I] \quad = \quad [R \;\; E] \tag{2}$$

This is exactly what "Gauss-Jordan" did in Chapter 2 to compute A^{-1}. ***When A is square and invertible, its reduced row echelon form is*** $R = I$. Then $EA = R$ becomes $EA = I$. In this invertible case, E is A^{-1}. This chapter is going further, to any (rectangular) matrix A and its reduced form R. The matrix E that multiplies A is still square and invertible, but the best it can do is to produce R. The pivot columns are reduced to ones and zeros.

The Rank of a Matrix

The numbers m and n give the size of a matrix—but not necessarily the *true size* of a linear system. An equation like $0 = 0$ should not count. If there are two identical rows in A, the second one disappears in R. Also if row 3 is a combination of rows 1 and 2, then row 3 will become all zeros in R. We don't want to count rows of zeros. ***The true size of A is given by its rank***:

DEFINITION ***The rank of A is the number of pivots. This number is r.***

The matrix R at the start of this section has rank $r = 2$. It has two pivots and two pivot columns. So does the unknown matrix A that produced R. This number $r = 2$ will be crucial to the theory, but its first definition is entirely computational. To execute the command $r =$ **rank** (A), the computer just counts the pivots. When *pivcol* gives a list of the pivot columns, the length of that list is r.

Actually the computer has a hard time to decide whether a small number is really zero. When it subtracts 3 times $\frac{1}{3}$ from 1, does it obtain zero? Our Teaching Codes treat numbers below the tolerance 10^{-6} as zero.

We know right away that $r \le m$ and $r \le n$. The number of pivots can't be greater than the number of rows. It can't be greater than the number of columns. The cases $r = m$ and $r = n$ of "full row rank" and "full column rank" will be especially important. We mention them here and come back to them soon:

- ***A has full row rank if every row has a pivot:*** $r = m$. ***No zero rows in R.***
- ***A has full column rank if every column has a pivot:*** $r = n$. ***No free variables.***

A square invertible matrix has $r = m = n$. Then R is the same as I.

At the other extreme are the matrices of ***rank one***. There is only *one pivot*. When elimination clears out the first column, it clears out all the columns. *Every row is a multiple of the pivot row.* At the same time, every column is a multiple of the pivot column!

$$\textbf{Rank one matrix} \qquad A = \begin{bmatrix} \mathbf{1} & 3 & 10 \\ \mathbf{2} & 6 & 20 \\ \mathbf{3} & 9 & 30 \end{bmatrix} \longrightarrow R = \begin{bmatrix} \mathbf{1} & 3 & 10 \\ 0 & 0 & 0 \\ 0 & 0 & 0 \end{bmatrix}.$$

The column space of a rank one matrix is "one-dimensional". Here all columns are on the line through $\boldsymbol{u} = (1, 2, 3)$. The columns of A are $\boldsymbol{u}$ and $3\boldsymbol{u}$ and $10\boldsymbol{u}$. Put those numbers into the row $\boldsymbol{v}^{\mathrm{T}} = [\,1 \quad 3 \quad 10\,]$ and you have the special rank one form $A = \boldsymbol{u}\boldsymbol{v}^{\mathrm{T}}$:

$$A = \textbf{column times row} = \boldsymbol{u}\boldsymbol{v}^{\mathrm{T}} \qquad \begin{bmatrix} 1 & 3 & 10 \\ 2 & 6 & 20 \\ 3 & 9 & 30 \end{bmatrix} = \begin{bmatrix} \mathbf{1} \\ \mathbf{2} \\ \mathbf{3} \end{bmatrix} [1 \quad 3 \quad 10]. \qquad (3)$$

Example 1 When all rows are multiples of one pivot row, the rank is $r = 1$:

$$\begin{bmatrix} 1 & 3 & 4 \\ 2 & 6 & 8 \end{bmatrix} \text{ and } \begin{bmatrix} 0 & 3 \\ 0 & \mathbf{5} \end{bmatrix} \text{ and } \begin{bmatrix} 5 \\ 2 \end{bmatrix} \text{ and } \begin{bmatrix} 6 \end{bmatrix} \text{ all have rank 1.}$$

The reduced row echelon forms $R =$ **rref** (A) can be checked by eye:

$$R = \begin{bmatrix} 1 & 3 & 4 \\ 0 & 0 & 0 \end{bmatrix} \text{ and } \begin{bmatrix} 0 & 1 \\ 0 & 0 \end{bmatrix} \text{ and } \begin{bmatrix} 1 \\ 0 \end{bmatrix} \text{ and } \begin{bmatrix} 1 \end{bmatrix} \text{ have only one pivot.}$$

Our second definition of rank is coming at a higher level. It deals with entire rows and entire columns—vectors and not just numbers. The matrices A and U and R

have r *independent* rows (the pivot rows). They also have r *independent* columns (the pivot columns). Section 3.5 says what it means for rows or columns to be independent.

A third definition of rank, at the top level of linear algebra, will deal with *spaces* of vectors. The rank r is the "dimension" of the column space. It is also the dimension of the row space. The great thing is that r also reveals the dimension of the nullspace.

The Pivot Columns

The pivot columns of R have 1's in the pivots and 0's everywhere else. The r pivot columns taken together contain an r by r identity matrix I. It sits above $m - r$ rows of zeros. The numbers of the pivot columns are in the list *pivcol*.

The pivot columns of A are probably *not* obvious from A itself. But their column numbers are given by the *same list pivcol*. The r columns of A that eventually have pivots (in U and R) are the pivot columns. The first matrix R in this section is the row reduced echelon form of this matrix A, with $pivcol = (1, 3)$:

Pivot Columns

$$A = \begin{bmatrix} \mathbf{1} & 3 & \mathbf{0} & 2 & -1 \\ \mathbf{0} & 0 & \mathbf{1} & 4 & -3 \\ \mathbf{1} & 3 & \mathbf{1} & 6 & -4 \end{bmatrix} \quad \text{yields } R = \begin{bmatrix} \mathbf{1} & 3 & \mathbf{0} & 2 & -1 \\ \mathbf{0} & 0 & \mathbf{1} & 4 & -3 \\ \mathbf{0} & 0 & \mathbf{0} & 0 & 0 \end{bmatrix}.$$

The column spaces of R and A can be different! All columns of this R end with zeros. E subtracts rows 1 and 2 of A from row 3 (to produce that zero row in R):

$$E = \begin{bmatrix} 1 & 0 & 0 \\ 0 & 1 & 0 \\ -1 & -1 & 1 \end{bmatrix} \quad \text{and} \quad E^{-1} = \begin{bmatrix} 1 & 0 & 0 \\ 0 & 1 & 0 \\ 1 & 1 & 1 \end{bmatrix}.$$

The r pivot columns of A are also the first r columns of E^{-1}. The reason is that each column of A is E^{-1} times a column of R. The r by r identity matrix inside R just picks out the first r columns of E^{-1}.

One more fact about pivot columns. Their definition has been purely computational, based on R. Here is a direct mathematical description of the pivot columns of A:

3C ***The pivot columns are not combinations of earlier columns.*** The free columns ***are*** combinations of earlier columns. These combinations are the special solutions!

A pivot column of R (with 1 in the pivot row) cannot be a combination of earlier columns (with 0's in that row). The same column of A can't be a combination of earlier columns, because $A\boldsymbol{x} = \mathbf{0}$ *exactly when* $R\boldsymbol{x} = \mathbf{0}$. Now we look at the special solution $\boldsymbol{x}$ from each free column.

The Special Solutions

Each special solution to $Ax = \mathbf{0}$ and $Rx = \mathbf{0}$ has one free variable equal to 1. The other free variables are all zero. The solutions come directly from the echelon form R:

$$\textbf{Free columns}\quad \textbf{Free variables}\qquad Rx = \begin{bmatrix} 1 & \mathbf{3} & 0 & \mathbf{2} & \mathbf{-1} \\ 0 & \mathbf{0} & 1 & \mathbf{4} & \mathbf{-3} \\ 0 & \mathbf{0} & 0 & \mathbf{0} & \mathbf{0} \end{bmatrix} \begin{bmatrix} x_1 \\ \boldsymbol{x_2} \\ x_3 \\ \boldsymbol{x_4} \\ \boldsymbol{x_5} \end{bmatrix} = \begin{bmatrix} 0 \\ 0 \\ 0 \end{bmatrix}.$$

The free variables are in boldface. Set the first free variable to $x_2 = 1$ with $x_4 = x_5 = 0$. The equations give the pivot variables $x_1 = -3$ and $x_3 = 0$. This says that column 2 (a free column) is 3 times column 1. The special solution is $s_1 = (-3, 1, 0, 0, 0)$.

The next special solution has $x_4 = 1$. The other free variables are $x_2 = x_5 = 0$. The solution is $s_2 = (-2, 0, -4, 1, 0)$. Notice -2 and -4 in R, with plus signs.

The third special solution has $x_5 = 1$. With $x_2 = 0$ and $x_4 = 0$ we find $s_3 = (1, 0, 3, 0, 1)$. The numbers $x_1 = 1$ and $x_3 = 3$ are in column 5 of R, again with opposite signs. This is a general rule as we soon verify. The nullspace matrix N contains the three special solutions in its columns:

$$\textbf{Nullspace matrix}\qquad N = \begin{bmatrix} \mathbf{-3} & \mathbf{-2} & \mathbf{1} \\ 1 & 0 & 0 \\ \mathbf{0} & \mathbf{-4} & \mathbf{3} \\ 0 & 1 & 0 \\ 0 & 0 & 1 \end{bmatrix} \begin{matrix} \text{not free} \\ \text{free} \\ \text{not free} \\ \text{free} \\ \text{free} \end{matrix}$$

The linear combinations of these three columns give all vectors in the nullspace. This is the complete solution to $Ax = \mathbf{0}$ (and $Rx = \mathbf{0}$). Where R had the identity matrix (2 by 2) in its pivot columns, N has the identity matrix (3 by 3) in its free rows.

There is a special solution for every free variable. Since r columns have pivots, that leaves $n - r$ free variables. This is the key to $Ax = \mathbf{0}$.

3D $Ax = \mathbf{0}$ ***has*** $n - r$ ***free variables and special solutions***: n columns minus r pivot columns. The ***nullspace matrix*** N has $n - r$ columns (the special solutions).

When we introduce the idea of "independent" vectors, we will show that the special solutions are independent. You can see in N that no column is a combination of the other columns. The beautiful thing is that the count is exactly right:

$$Ax = \mathbf{0} \textit{ has } r \textit{ independent equations so } n - r \textit{ independent solutions.}$$

To complete this section, look again at the special solutions. Suppose for simplicity that the first r columns are the pivot columns, and the last $n-r$ columns are free (no pivots). Then the reduced row echelon form looks like

$$R = \begin{bmatrix} \boldsymbol{I} & \boldsymbol{F} \\ \mathbf{0} & \mathbf{0} \end{bmatrix} \begin{matrix} r \text{ pivot rows} \\ m-r \text{ zero rows} \end{matrix} \tag{4}$$

r pivot columns $\quad n-r$ free columns

3E The pivot variables in the $n-r$ special solutions come by changing F to $-F$:

$$\textbf{Nullspace matrix} \quad N = \begin{bmatrix} -\boldsymbol{F} \\ \boldsymbol{I} \end{bmatrix} \begin{matrix} r \text{ pivot variables} \\ n-r \text{ free variables} \end{matrix} \tag{5}$$

Check $RN = 0$. The first block row of RN is (I times $-F$) + (F times I) = zero. The columns of N solve $R\boldsymbol{x} = \mathbf{0}$. When the free part of $R\boldsymbol{x} = \mathbf{0}$ moves to the right side, the left side just holds the identity matrix:

$$I \begin{bmatrix} \text{pivot} \\ \text{variables} \end{bmatrix} = -F \begin{bmatrix} \text{free} \\ \text{variables} \end{bmatrix}. \tag{6}$$

In each special solution, the free variables are a column of I. Then the pivot variables are a column of $-F$. Those special solutions give the nullspace matrix N.

The idea is still true if the pivot columns are mixed in with the free columns. Then I and F are mixed together. You can still see $-F$ in the solutions. Here is an example where $I = [1]$ comes first and $F = \begin{bmatrix} 2 & 3 \end{bmatrix}$ comes last.

Example 2 The special solutions of $R\boldsymbol{x} = x_1 + 2x_2 + 3x_3 = 0$ are the columns of N:

$$R = \begin{bmatrix} 1 & 2 & 3 \end{bmatrix} \qquad N = \begin{bmatrix} -2 & -3 \\ 1 & 0 \\ 0 & 1 \end{bmatrix}.$$

The rank is one. There are $n - r = 3 - 1$ special solutions $(-2, 1, 0)$ and $(-3, 0, 1)$.

Final Note How can I write confidently about R when I don't know which steps MATLAB will take? A could be reduced to R in different ways. Very likely you and Mathematica and Maple would do the elimination differently. The key point is that ***the final matrix R is always the same***. The original A *completely determines the* I *and* F *and zero rows in* R, according to **3C**:

The pivot columns *are not* combinations of earlier columns of A.

The free columns *are* combinations of earlier columns (F tells the combinations).

A small example with rank one will show two E's that produce the correct $EA = R$:

$$A = \begin{bmatrix} 2 & 2 \\ 1 & 1 \end{bmatrix} \quad \text{reduces to} \quad R = \begin{bmatrix} 1 & 1 \\ 0 & 0 \end{bmatrix} \quad \text{and no other } R.$$

You could multiply row 1 of A by $\frac{1}{2}$, and subtract row 1 from row 2:

$$\begin{bmatrix} 1 & 0 \\ -1 & 1 \end{bmatrix} \begin{bmatrix} 1/2 & 0 \\ 0 & 1 \end{bmatrix} = \begin{bmatrix} 1/2 & 0 \\ -1/2 & 1 \end{bmatrix} = E.$$

Or you could exchange rows in A, and then subtract 2 times row 1 from row 2:

$$\begin{bmatrix} 1 & 0 \\ -2 & 1 \end{bmatrix} \begin{bmatrix} 0 & 1 \\ 1 & 0 \end{bmatrix} = \begin{bmatrix} 0 & 1 \\ 1 & -2 \end{bmatrix} = E_{\text{new}}.$$

Multiplication gives $EA = R$ and also $E_{\text{new}}A = R$. ***Different E's but the same R.***

■ REVIEW OF THE KEY IDEAS ■

1. The rank of A is the number of pivots (which are 1's in R).
2. The r pivot columns of A and R are in the same list *pivcol*.
3. Those r pivot columns are not combinations of earlier columns.
4. The $n - r$ free columns are combinations of earlier columns.
5. Those combinations (using $-F$ taken from R) give the $n - r$ special solutions to $Ax = \mathbf{0}$ and $Rx = \mathbf{0}$. They are the $n - r$ columns of the nullspace matrix N.

■ WORKED EXAMPLES ■

3.3 A Factor these rank one matrices into $A = \boldsymbol{u}\boldsymbol{v}^{\mathrm{T}}$ = row times column:

$$A = \begin{bmatrix} 1 & 2 & 3 \\ 2 & 4 & 6 \\ 3 & 6 & 9 \end{bmatrix} \qquad A = \begin{bmatrix} a & b \\ c & d \end{bmatrix} \quad (\text{find } d \text{ from } a^{-1}, b, c)$$

Split this rank two matrix into $\boldsymbol{u}_1\boldsymbol{v}_1^{\mathrm{T}} + \boldsymbol{u}_2\boldsymbol{v}_2^{\mathrm{T}}$ = (3 by 2) times (2 by 4) using E^{-1} and R:

$$A = \begin{bmatrix} 1 & 1 & 0 & 2 \\ 1 & 2 & 0 & 3 \\ 2 & 3 & 0 & 5 \end{bmatrix} = \begin{bmatrix} 1 & 1 & 0 \\ 1 & 2 & 0 \\ 2 & 3 & 1 \end{bmatrix} \begin{bmatrix} 1 & 0 & 0 & 1 \\ 0 & 1 & 0 & 1 \\ 0 & 0 & 0 & 0 \end{bmatrix} = E^{-1}R.$$

Solution For the 3 by 3 matrix A, all rows are multiples of $\boldsymbol{v}^{\mathrm{T}} = [1 \ \ 2 \ \ 3]$. All columns are multiples of the column $\boldsymbol{u} = (1, 2, 3)$. This symmetric matrix has $\boldsymbol{u} = \boldsymbol{v}$ and A is $\boldsymbol{u}\boldsymbol{u}^{\mathrm{T}}$. Every rank one symmetric matrix will have this form or else $-\boldsymbol{u}\boldsymbol{u}^{\mathrm{T}}$.

If the 2 by 2 matrix $\left[\begin{smallmatrix} a & b \\ c & d \end{smallmatrix}\right]$ has rank one, it must be singular. In Chapter 5, its determinant is $ad - bc = 0$. In this chapter, row 2 is a multiple of row 1. That multiple is $\frac{c}{a}$ (the problem assumes $a \neq 0$). Rank one always produces column times row:

$$\begin{bmatrix} a & b \\ c & d \end{bmatrix} = \begin{bmatrix} 1 \\ c/a \end{bmatrix} \begin{bmatrix} a & b \end{bmatrix} = \begin{bmatrix} a & b \\ c & bc/a \end{bmatrix}. \quad \text{So} \quad d = \frac{bc}{a}.$$

The 3 by 4 matrix of rank two is a sum of *two matrices of rank one*. All columns of A are combinations of the pivot columns 1 and 2. All rows are combinations of the nonzero rows of R. The pivot columns are $\boldsymbol{u}_1$ and $\boldsymbol{u}_2$ and those nonzero rows are $\boldsymbol{v}_1^{\mathrm{T}}$ and $\boldsymbol{v}_2^{\mathrm{T}}$. Then A is $\boldsymbol{u}_1\boldsymbol{v}_1^{\mathrm{T}} + \boldsymbol{u}_2\boldsymbol{v}_2^{\mathrm{T}}$, multiplying columns of E^{-1} times rows of R:

$$\begin{bmatrix} 1 & 1 & 0 & 2 \\ 1 & 2 & 0 & 3 \\ 2 & 3 & 0 & 5 \end{bmatrix} = \begin{bmatrix} 1 \\ 1 \\ 2 \end{bmatrix} \begin{bmatrix} 1 & 0 & 0 & 1 \end{bmatrix} + \begin{bmatrix} 1 \\ 2 \\ 3 \end{bmatrix} \begin{bmatrix} 0 & 1 & 0 & 1 \end{bmatrix}$$

3.3 B Find the row reduced form R and the rank r of A (*those depend on* c). Which are the pivot columns of A? Which variables are free? What are the special solutions and the nullspace matrix N (always depending on c)?

$$A = \begin{bmatrix} 1 & 2 & 1 \\ 3 & 6 & 3 \\ 4 & 8 & c \end{bmatrix} \quad \text{and} \quad A = \begin{bmatrix} c & c \\ c & c \end{bmatrix}.$$

Solution The 3 by 3 matrix A has rank $r = 2$ ***except if*** $c = 4$. The pivots are in columns 1 and 3. The second variable x_2 is free. Notice the form of R:

$$c \neq 4 \quad R = \begin{bmatrix} 1 & 2 & 0 \\ 0 & 0 & 1 \\ 0 & 0 & 0 \end{bmatrix} \qquad c = 4 \quad R = \begin{bmatrix} 1 & 2 & 1 \\ 0 & 0 & 0 \\ 0 & 0 & 0 \end{bmatrix}.$$

When $c = 4$, the only pivot is in column 1 (one pivot column). Columns 2 and 3 are multiples of column 1 (so rank $= 1$). The second and third variables are free, producing two special solutions:

$$c \neq 4 \quad \text{Special solution with } x_2 = 1 \text{ goes into} \quad N = \begin{bmatrix} -2 \\ 1 \\ 0 \end{bmatrix}.$$

$$c = 4 \quad \text{Another special solution goes into} \quad N = \begin{bmatrix} -2 & -1 \\ 1 & 0 \\ 0 & 1 \end{bmatrix}.$$

The 2 by 2 matrix $\begin{bmatrix} c & c \\ c & c \end{bmatrix}$ has rank $r = 1$ ***except if*** $c = 0$, when the rank is zero!

$$c \neq 0 \quad R = \begin{bmatrix} 1 & 1 \\ 0 & 0 \end{bmatrix} \qquad\qquad c = 0 \quad R = \begin{bmatrix} 0 & 0 \\ 0 & 0 \end{bmatrix}$$

The first column is the pivot column if $c \neq 0$, and the second variable is free (one special solution in N). The matrix has *no pivot columns* if $c = 0$, and both variables are free:

$$c \neq 0 \quad N = \begin{bmatrix} -1 \\ 1 \end{bmatrix} \qquad\qquad c = 0 \quad N = \begin{bmatrix} 1 & 0 \\ 0 & 1 \end{bmatrix}.$$

Problem Set 3.3

1 Which of these rules gives a correct definition of the *rank* of A?

(a) The number of nonzero rows in R.

(b) The number of columns minus the total number of rows.

(c) The number of columns minus the number of free columns.

(d) The number of 1's in the matrix R.

2 Find the reduced row echelon forms R and the rank of these matrices:

(a) The 3 by 4 matrix of all ones.

(b) The 3 by 4 matrix with $a_{ij} = i + j - 1$.

(c) The 3 by 4 matrix with $a_{ij} = (-1)^j$.

3 Find R for each of these (block) matrices:

$$A = \begin{bmatrix} 0 & 0 & 0 \\ 0 & 0 & 3 \\ 2 & 4 & 6 \end{bmatrix} \quad B = \begin{bmatrix} A & A \end{bmatrix} \quad C = \begin{bmatrix} A & A \\ A & 0 \end{bmatrix}$$

4 Suppose all the pivot variables come *last* instead of first. Describe all four blocks in the reduced echelon form (the block B should be r by r):

$$R = \begin{bmatrix} A & B \\ C & D \end{bmatrix}.$$

What is the nullspace matrix N containing the special solutions?

5 (Silly problem) Describe all 2 by 3 matrices A_1 and A_2, with row echelon forms R_1 and R_2, such that $R_1 + R_2$ is the row echelon form of $A_1 + A_2$. Is is true that $R_1 = A_1$ and $R_2 = A_2$ in this case?

6 If A has r pivot columns, how do you know that A^{T} has r pivot columns? Give a 3 by 3 example for which the column numbers are different.

7 What are the special solutions to $R\boldsymbol{x} = \mathbf{0}$ and $\boldsymbol{y}^{\mathrm{T}}R = \mathbf{0}$ for these R?

$$R = \begin{bmatrix} 1 & 0 & 2 & 3 \\ 0 & 1 & 4 & 5 \\ 0 & 0 & 0 & 0 \end{bmatrix} \qquad R = \begin{bmatrix} 0 & 1 & 2 \\ 0 & 0 & 0 \\ 0 & 0 & 0 \end{bmatrix}$$

Problems 8–11 are about matrices of rank $r = 1$.

8 Fill out these matrices so that they have rank 1:

$$A = \begin{bmatrix} 1 & 2 & 4 \\ 2 & & \\ 4 & & \end{bmatrix} \quad \text{and} \quad B = \begin{bmatrix} 2 & & \\ 1 & & \\ 2 & 6 & -3 \end{bmatrix} \quad \text{and} \quad M = \begin{bmatrix} a & b \\ c & \end{bmatrix}.$$

9 If A is an m by n matrix with $r = 1$, its columns are multiples of one column and its rows are multiples of one row. The column space is a _____ in $\mathbf{R}^m$. The nullspace is a _____ in $\mathbf{R}^n$. Also the column space of A^{T} is a _____ in $\mathbf{R}^n$.

10 Choose vectors $\boldsymbol{u}$ and $\boldsymbol{v}$ so that $A = \boldsymbol{u}\boldsymbol{v}^{\mathrm{T}}$ = column times row:

$$A = \begin{bmatrix} 3 & 6 & 6 \\ 1 & 2 & 2 \\ 4 & 8 & 8 \end{bmatrix} \quad \text{and} \quad A = \begin{bmatrix} 2 & 2 & 6 & 4 \\ -1 & -1 & -3 & -2 \end{bmatrix}.$$

$A = \boldsymbol{u}\boldsymbol{v}^{\mathrm{T}}$ *is the natural form for every matrix that has rank* $r = 1$.

11 If A is a rank one matrix, the second row of U is _____. Do an example.

Problems 12–14 are about r by r invertible matrices inside A.

12 *If A has rank r, then it has an r by r submatrix S that is invertible.* Remove $m - r$ rows and $n - r$ columns to find an invertible submatrix S inside each A (you could keep the pivot rows and pivot columns of A):

$$A = \begin{bmatrix} 1 & 2 & 3 \\ 1 & 2 & 4 \end{bmatrix} \qquad A = \begin{bmatrix} 1 & 2 & 3 \\ 2 & 4 & 6 \end{bmatrix} \qquad A = \begin{bmatrix} 0 & 1 & 0 \\ 0 & 0 & 0 \\ 0 & 0 & 1 \end{bmatrix}.$$

13 Suppose P is the submatrix of A containing only the pivot columns. Explain why this m by r submatrix P has rank r.

14 In Problem 13, we can transpose P and find the r pivot columns of P^{T}. Transposing back, we have an r by r invertible submatrix S inside P:

$$\text{For } A = \begin{bmatrix} 1 & 2 & 3 \\ 2 & 4 & 6 \\ 2 & 4 & 7 \end{bmatrix} \quad \text{find } P \text{ (3 by 2) and then } S \text{ (2 by 2).}$$

Problems 15–20 show that rank(AB) is not greater than rank(A) or rank(B).

15 Find the ranks of AB and AM (rank one matrix times rank one matrix):

$$A = \begin{bmatrix} 1 & 2 \\ 2 & 4 \end{bmatrix} \quad \text{and} \quad B = \begin{bmatrix} 2 & 1 & 4 \\ 3 & 1.5 & 6 \end{bmatrix} \quad \text{and} \quad M = \begin{bmatrix} 1 & b \\ c & bc \end{bmatrix}.$$

16 The rank one matrix $\boldsymbol{u}\boldsymbol{v}^{\mathrm{T}}$ times the rank one matrix $\boldsymbol{w}\boldsymbol{z}^{\mathrm{T}}$ is $\boldsymbol{u}\boldsymbol{z}^{\mathrm{T}}$ times the number _____. This has rank one unless _____ $= 0$.

17 (a) Suppose column j of B is a combination of previous columns of B. Show that column j of AB is the same combination of previous columns of AB. Then **rank(AB) $\le$ rank(B)**, because AB cannot have new pivot columns.

(b) Find A_1 and A_2 so that rank$(A_1B) = 1$ and rank$(A_2B) = 0$ for $B = \left[\begin{smallmatrix} 1 & 1 \\ 1 & 1 \end{smallmatrix}\right]$.

18 Problem 17 proved that rank$(AB) \le$ rank(B). Then the same reasoning gives rank$(B^{\mathrm{T}}A^{\mathrm{T}}) \le$ rank(A^{T}). How do you deduce that **rank(AB) $\le$ rank A**?

19 *(Important)* Suppose A and B are n by n matrices, and $AB = I$. Prove from rank$(AB) \le$ rank(A) that the rank of A is n. So A is invertible and B must be its two-sided inverse (Section 2.5). Therefore $BA = I$ *(which is not so obvious!)*.

20 If A is 2 by 3 and B is 3 by 2 and $AB = I$, show from its rank that $BA \ne I$. Give an example of A and B. For $m < n$, a right inverse is not a left inverse.

21 Suppose A and B have the *same* reduced row echelon form R.

(a) Show that A and B have the same nullspace and the same row space.

(b) We know $E_1A = R$ and $E_2B = R$. So A equals an _____ matrix times B.

22 Every m by n matrix of rank r reduces to (m by r) times (r by n):

$$A = (\text{pivot columns of } A)\ (\text{first } r \text{ rows of } R) = (\mathbf{COL})(\mathbf{ROW})^{\mathrm{T}}.$$

Write the 3 by 5 matrix A at the start of this section as the product of the 3 by 2 matrix from the pivot columns and the 2 by 5 matrix from R.

23 $A = (\text{COL})(\text{ROW})^{\mathrm{T}}$ is a sum of r rank one matrices (multiply columns times rows). Express A and B as the sum of two rank one matrices:

$$A = \begin{bmatrix} 1 & 1 & 0 \\ 1 & 1 & 4 \\ 1 & 1 & 8 \end{bmatrix} \qquad B = \begin{bmatrix} A & A \end{bmatrix}.$$

24 Suppose A is an m by n matrix of rank r. Its reduced echelon form is R. Describe exactly the matrix Z (its shape and all its entries) that comes from *transposing the reduced row echelon form of R'* (prime means transpose):

$$R = \mathbf{rref}(A) \quad \text{and} \quad Z = (\mathbf{rref}(R'))'.$$

25 Instead of transposing R (Problem 24) we could transpose A first. Explain in one line why $Y = Z$:

$$V = \textbf{rref}(A') \quad \text{and} \quad Y = \textbf{rref}(V').$$

26 Answer the same questions as in Worked Example **3.3 B** for

$$A = \begin{bmatrix} 1 & 1 & 2 & 2 \\ 2 & 2 & 4 & 4 \\ 1 & c & 2 & 2 \end{bmatrix} \quad \text{and} \quad A = \begin{bmatrix} 1-c & 2 \\ 0 & 2-c \end{bmatrix}.$$

27 What is the nullspace matrix N (containing the special solutions) for A, B, C?

$$A = \begin{bmatrix} I & I \end{bmatrix} \quad \text{and} \quad B = \begin{bmatrix} I & I \\ 0 & 0 \end{bmatrix} \quad \text{and} \quad C = [\, I \;\; I \;\; I \,].$$

THE COMPLETE SOLUTION TO $AX = B$ ■ 3.4

The last section totally solved $A\boldsymbol{x} = \mathbf{0}$. Elimination converted the problem to $R\boldsymbol{x} = \mathbf{0}$. The free variables were given special values (one and zero). Then the pivot variables were found by back substitution. We paid no attention to the right side $\boldsymbol{b}$ because it started and ended as zero. The solution $\boldsymbol{x}$ was in the nullspace of A.

Now $\boldsymbol{b}$ is not zero. Row operations on the left side must act also on the right side. One way to organize that is to ***add b as an extra column of the matrix***. We keep the same example A as before. But we *"augment"* A with the right side $(b_1, b_2, b_3) = (1, 6, 7)$:

$$\begin{bmatrix} 1 & 3 & 0 & 2 \\ 0 & 0 & 1 & 4 \\ 1 & 3 & 1 & 6 \end{bmatrix} \begin{bmatrix} x_1 \\ x_2 \\ x_3 \\ x_4 \end{bmatrix} = \begin{bmatrix} 1 \\ 6 \\ 7 \end{bmatrix} \quad \begin{matrix} \text{has the} \\ \text{augmented} \\ \text{matrix} \end{matrix} \quad \begin{bmatrix} 1 & 3 & 0 & 2 & \mathbf{1} \\ 0 & 0 & 1 & 4 & \mathbf{6} \\ 1 & 3 & 1 & 6 & \mathbf{7} \end{bmatrix} = \begin{bmatrix} A & \boldsymbol{b} \end{bmatrix}.$$

The augmented matrix is just $\begin{bmatrix} A & \boldsymbol{b} \end{bmatrix}$. When we apply the usual elimination steps to A, we also apply them to $\boldsymbol{b}$. In this example we subtract row 1 from row 3 and then subtract row 2 from row 3. This produces a *complete row of zeros*:

$$\begin{bmatrix} 1 & 3 & 0 & 2 \\ 0 & 0 & 1 & 4 \\ 0 & 0 & 0 & 0 \end{bmatrix} \begin{bmatrix} x_1 \\ x_2 \\ x_3 \\ x_4 \end{bmatrix} = \begin{bmatrix} 1 \\ 6 \\ 0 \end{bmatrix} \quad \begin{matrix} \text{has the} \\ \text{augmented} \\ \text{matrix} \end{matrix} \quad \begin{bmatrix} 1 & 3 & 0 & 2 & \mathbf{1} \\ 0 & 0 & 1 & 4 & \mathbf{6} \\ 0 & 0 & 0 & 0 & \mathbf{0} \end{bmatrix} = \begin{bmatrix} R & \boldsymbol{d} \end{bmatrix}.$$

That very last zero is crucial. It means that the equations can be solved; the third equation has become $0 = 0$. In the original matrix A, the first row plus the second row equals the third row. If the equations are consistent, this must be true on the right side of the equations also! The all-important property on the right side was $1+6=7$.

Here are the same augmented matrices for a general $\boldsymbol{b} = (b_1, b_2, b_3)$:

$$\begin{bmatrix} 1 & 3 & 0 & 2 & \boldsymbol{b_1} \\ 0 & 0 & 1 & 4 & \boldsymbol{b_2} \\ 1 & 3 & 1 & 6 & \boldsymbol{b_3} \end{bmatrix} \longrightarrow \begin{bmatrix} 1 & 3 & 0 & 2 & \boldsymbol{b_1} \\ 0 & 0 & 1 & 4 & \boldsymbol{b_2} \\ 0 & 0 & 0 & 0 & \boldsymbol{b_3 - b_1 - b_2} \end{bmatrix}.$$

Now we get $0 = 0$ in the third equation provided $b_3 - b_1 - b_2 = 0$. This is $b_1 + b_2 = b_3$.

One Particular Solution

Choose the free variables to be $x_2 = x_4 = 0$. Then the equations give the pivot variables $x_1 = 1$ and $x_3 = 6$. They are in the last column $\boldsymbol{d}$ of the reduced augmented matrix. The code $\boldsymbol{x} =$ **partic** $(A, \boldsymbol{b})$ gives this particular solution (call it $\boldsymbol{x}_p$) to $A\boldsymbol{x} = \boldsymbol{b}$. First A and $\boldsymbol{b}$ reduce to R and $\boldsymbol{d}$. Zero rows in R must also be zero in $\boldsymbol{d}$. Then the r pivot variables in $\boldsymbol{x}$ are taken directly from $\boldsymbol{d}$, because the pivot columns in R contain the identity matrix. After row reduction we are just solving $I\boldsymbol{x} = \boldsymbol{d}$.

Notice how we *choose* the free variables (as zero) and *solve* for the pivot variables. After the row reduction to R, those steps are quick. When the free variables are zero, the pivot variables for $\boldsymbol{x}_p$ are in the extra column:

The particular solution solves	$A\boldsymbol{x}_p = \boldsymbol{b}$
The $n - r$ ***special solutions solve***	$A\boldsymbol{x}_n = \boldsymbol{0}$.

In this example the particular solution is $(1, 0, 6, 0)$. The two special (nullspace) solutions to $R\boldsymbol{x} = \boldsymbol{0}$ come from the two free columns of R, by reversing signs of $3, 2$, and 4. ***Please notice how I write the complete solution*** $\boldsymbol{x}_p + \boldsymbol{x}_n$ ***to*** $A\boldsymbol{x} = \boldsymbol{b}$:

$$\textbf{Complete solution:} \quad \boldsymbol{x} = \boldsymbol{x}_p + \boldsymbol{x}_n = \begin{bmatrix} 1 \\ 0 \\ 6 \\ 0 \end{bmatrix} + x_2 \begin{bmatrix} -3 \\ 1 \\ 0 \\ 0 \end{bmatrix} + x_4 \begin{bmatrix} -2 \\ 0 \\ -4 \\ 1 \end{bmatrix}.$$

Question Suppose A is a square invertible matrix, $m = n = r$. What are $\boldsymbol{x}_p$ and $\boldsymbol{x}_n$?

Answer The particular solution is the one and *only* solution $A^{-1}\boldsymbol{b}$. There are no special solutions or free variables. $R = I$ has no zero rows. The only vector in the nullspace is $\boldsymbol{x}_n = \boldsymbol{0}$. The complete solution is $\boldsymbol{x} = \boldsymbol{x}_p + \boldsymbol{x}_n = A^{-1}\boldsymbol{b} + \boldsymbol{0}$.

This was the situation in Chapter 2. We didn't mention the nullspace in that chapter. $\boldsymbol{N}(A)$ contained only the zero vector. Reduction goes from $\begin{bmatrix} A & \boldsymbol{b} \end{bmatrix}$ to $\begin{bmatrix} I & A^{-1}\boldsymbol{b} \end{bmatrix}$. The original $A\boldsymbol{x} = \boldsymbol{b}$ is reduced all the way to $\boldsymbol{x} = A^{-1}\boldsymbol{b}$. This is a special case here, but square invertible matrices are the ones we see most often in practice. So they got their own chapter at the start of the book.

For small examples we can put $\begin{bmatrix} A & \boldsymbol{b} \end{bmatrix}$ into reduced row echelon form. For a large matrix, MATLAB can do it better. Here is a small example with *full column rank*. Both columns have pivots.

Example 1 Find the condition on (b_1, b_2, b_3) for $A\boldsymbol{x} = \boldsymbol{b}$ to be solvable, if

$$A = \begin{bmatrix} 1 & 1 \\ 1 & 2 \\ -2 & -3 \end{bmatrix} \quad \text{and} \quad \boldsymbol{b} = \begin{bmatrix} b_1 \\ b_2 \\ b_3 \end{bmatrix}.$$

This condition puts $\boldsymbol{b}$ in the column space of A. Find the complete $\boldsymbol{x} = \boldsymbol{x}_p + \boldsymbol{x}_n$.

Solution Use the augmented matrix, with its extra column $\boldsymbol{b}$. Elimination subtracts row 1 from row 2, and adds 2 times row 1 to row 3:

$$\begin{bmatrix} 1 & 1 & b_1 \\ 1 & 2 & b_2 \\ -2 & -3 & b_3 \end{bmatrix} \longrightarrow \begin{bmatrix} 1 & 1 & b_1 \\ 0 & 1 & b_2 - b_1 \\ 0 & -1 & b_3 + 2b_1 \end{bmatrix} \rightarrow \begin{bmatrix} 1 & 0 & 2b_1 - b_2 \\ 0 & 1 & b_2 - b_1 \\ 0 & 0 & b_3 + b_1 + b_2 \end{bmatrix}.$$

The last equation is $0 = 0$ provided $b_3 + b_1 + b_2 = 0$. This is the condition to put $\boldsymbol{b}$ in the column space; then the system is solvable. The rows of A add to the zero row. So for consistency (these are equations!) the entries of $\boldsymbol{b}$ must also add to zero.

This example has no free variables and no special solutions. The nullspace solution is $\boldsymbol{x}_n = \boldsymbol{0}$. The (only) particular solution $\boldsymbol{x}_p$ is at the top of the augmented column:

$$\boldsymbol{x} = \boldsymbol{x}_p + \boldsymbol{x}_n = \begin{bmatrix} 2b_1 - b_2 \\ b_2 - b_1 \end{bmatrix} + \begin{bmatrix} 0 \\ 0 \end{bmatrix}.$$

If $b_3 + b_1 + b_2$ is not zero, there is no solution to $A\boldsymbol{x} = \boldsymbol{b}$ ($\boldsymbol{x}_p$ doesn't exist).

This example is typical of the extremely important case when A has *full column rank*. Every column has a pivot. *The rank is* $r = n$. The matrix is tall and thin ($m \geq n$). Row reduction puts I at the top, when A is reduced to R:

$$\textbf{Full column rank} \quad R = \begin{bmatrix} n \text{ by } n \text{ identity matrix} \\ m - n \text{ rows of zeros} \end{bmatrix} = \begin{bmatrix} I \\ 0 \end{bmatrix}. \tag{1}$$

There are no free columns or free variables. The nullspace matrix is empty!

We will collect together the different ways of recognizing this type of matrix.

3F Every matrix A with **full column rank** ($r = n$) has all these properties:

1. All columns of A are pivot columns.

2. There are no free variables or special solutions.

3. The nullspace $N(A)$ contains only the zero vector $\boldsymbol{x} = \boldsymbol{0}$.

4. If $A\boldsymbol{x} = \boldsymbol{b}$ has a solution (it might not) then it has only *one solution*.

In the language of the next section, this A has ***independent columns***. In Chapter 4 we will add one more fact to the list: *The square matrix* $A^{\mathrm{T}}A$ *is invertible.*

In this case the nullspace of A (and R) has shrunk to the zero vector. The solution to $A\boldsymbol{x} = \boldsymbol{b}$ is *unique* (if it exists). There will be $m - n$ (here $3 - 2$) zero rows in R. So there are $m - n$ (here 1 condition) conditions on $\boldsymbol{b}$ in order to have $0 = 0$ in those rows. If $b_3 + b_1 + b_2 = 0$ is satisfied, $A\boldsymbol{x} = \boldsymbol{b}$ has exactly one solution.

The Complete Solution

The other extreme case is full row rank. Now $A\boldsymbol{x} = \boldsymbol{b}$ either has one or infinitely may solutions. In this case A is *short and wide* ($m \le n$). The number of unknowns is at least the number of equations. A matrix has ***full row rank*** if $r = m$. The nullspace of A^{T} shrinks to the zero vector. Every row has a pivot, and here is an example.

Example 2 There are $n = 3$ unknowns but only two equations. The rank is $r = m = 2$:

$$\begin{aligned} x + y + z &= 3 \\ x + 2y - z &= 4 \end{aligned}$$

These are two planes in xyz space. The planes are not parallel so they intersect in a line. This line of solutions is exactly what elimination will find. ***The particular solution will be one point on the line. Adding the nullspace vectors $\boldsymbol{x}_n$ will move us along the line***. Then $\boldsymbol{x} = \boldsymbol{x}_p + \boldsymbol{x}_n$ gives the whole line of solutions.

We find $\boldsymbol{x}_p$ and $\boldsymbol{x}_n$ by elimination. Subtract row 1 from row 2 and then subtract row 2 from row 1:

$$\begin{bmatrix} 1 & 1 & 1 & \mathbf{3} \\ 1 & 2 & -1 & \mathbf{4} \end{bmatrix} \rightarrow \begin{bmatrix} 1 & 1 & 1 & \mathbf{3} \\ 0 & 1 & -2 & \mathbf{1} \end{bmatrix} \rightarrow \begin{bmatrix} 1 & 0 & 3 & \mathbf{2} \\ 0 & 1 & -2 & \mathbf{1} \end{bmatrix} = \begin{bmatrix} R & \boldsymbol{d} \end{bmatrix}.$$

The particular solution has free variable $x_3 = 0$. The special solution has $x_3 = 1$:

$\boldsymbol{x}_{\textbf{particular}}$ comes directly from $\boldsymbol{d}$ the right side: $\boldsymbol{x}_p = (2, 1, 0)$
$\boldsymbol{x}_{\textbf{special}}$ comes from the third column (free column F) of R: $\boldsymbol{s} = (-3, 2, 1)$

It is wise to check that $\boldsymbol{x}_p$ and $\boldsymbol{s}$ satisfy the original equations $A\boldsymbol{x}_p = \boldsymbol{b}$ and $A\boldsymbol{s} = \boldsymbol{0}$:

$$\begin{aligned} 2 + 1 &= 3 \qquad & -3 + 2 + 1 &= 0 \\ 2 + 2 &= 4 \qquad & -3 + 4 - 1 &= 0 \end{aligned}$$

The nullspace solution $\boldsymbol{x}_n$ is any multiple of $\boldsymbol{s}$. It moves along the line of solutions, starting at $\boldsymbol{x}_{\text{particular}}$. ***Please notice again how to write the answer***:

Complete Solution: $$\boldsymbol{x} = \boldsymbol{x}_p + \boldsymbol{x}_n = \begin{bmatrix} 2 \\ 1 \\ 0 \end{bmatrix} + x_3 \begin{bmatrix} -3 \\ 2 \\ 1 \end{bmatrix}.$$

This line is drawn in Figure 3.3. Any point on the line *could* have been chosen as the particular solution; we chose the point with $x_3 = 0$. The particular solution is *not* multiplied by an arbitrary constant! The special solution is, and you understand why.

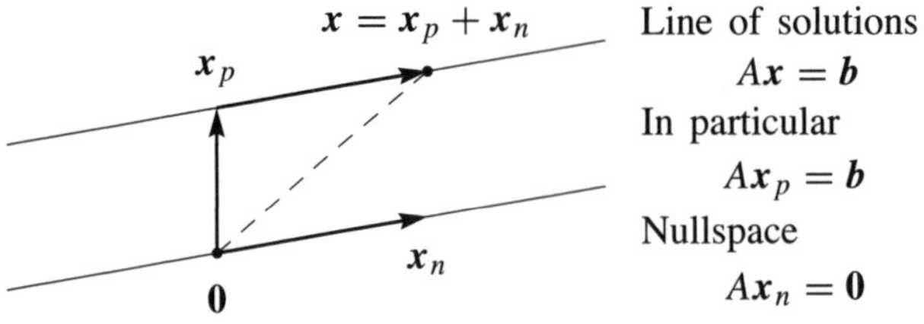

Figure 3.3 The complete solution is *one* particular solution plus *all* nullspace solutions.

Now we summarize this short wide case ($m \le n$) of *full row rank*:

3G Every matrix A with ***full row rank*** ($r = m$) has all these properties:

1. All rows have pivots, and R has no zero rows.

2. $A\boldsymbol{x} = \boldsymbol{b}$ has a solution for every right side $\boldsymbol{b}$.

3. The column space is the whole space $\mathbf{R}^m$.

4. There are $n - r = n - m$ special solutions in the nullspace of A.

In this case with m pivots, the rows are "linearly independent". In other words, the columns of A^{T} are linearly independent. We are more than ready for the definition of linear independence, as soon as we summarize the four possibilities—which depend on the rank. Notice how r, m, n are the critical numbers!

The four possibilities for linear equations depend on the rank r:

$r = m$	and	$r = n$	*Square and invertible*	$A\boldsymbol{x} = \boldsymbol{b}$	has 1 solution
$r = m$	and	$r < n$	*Short and wide*	$A\boldsymbol{x} = \boldsymbol{b}$	has ∞ solutions
$r < m$	and	$r = n$	*Tall and thin*	$A\boldsymbol{x} = \boldsymbol{b}$	has 0 or 1 solution
$r < m$	and	$r < n$	*Unknown shape*	$A\boldsymbol{x} = \boldsymbol{b}$	has 0 or ∞ solutions

The reduced R will fall in the same category as the matrix A. In case the pivot columns happen to come first, we can display these four possibilities for R:

$$R = \begin{bmatrix} I \end{bmatrix} \qquad \begin{bmatrix} I & F \end{bmatrix} \qquad \begin{bmatrix} I \\ 0 \end{bmatrix} \qquad \begin{bmatrix} I & F \\ 0 & 0 \end{bmatrix}$$

$$r = m = n \qquad r = m < n \qquad r = n < m \qquad r < m, r < n$$

Cases 1 and 2 have full row rank $r = m$. Cases 1 and 3 have full column rank $r = n$. Case 4 is the most general in theory and the least common in practice.

Note In the first edition of this textbook, we generally stopped at U before reaching R. Instead of reading the complete solution directly from $R\boldsymbol{x} = \boldsymbol{d}$, we found it by back substitution from $U\boldsymbol{x} = \boldsymbol{c}$. That combination of reduction to U and back substitution for $\boldsymbol{x}$ is slightly faster. Now we prefer the complete reduction: a single "1" in each pivot column. We find that everything is so much clearer in R (and the computer should do the hard work anyway) that we reduce all the way.

■ REVIEW OF THE KEY IDEAS ■

1. The rank r is the number of pivots. The matrix R has $m - r$ zero rows.
2. $A\boldsymbol{x} = \boldsymbol{b}$ is solvable if and only if the last $m - r$ equations reduce to $0 = 0$.
3. One particular solution $\boldsymbol{x}_p$ has all free variables equal to zero.
4. The pivot variables are determined after the free variables are chosen.
5. Full column rank $r = n$ means no free variables: one solution or none.
6. Full row rank $r = m$ means one solution if $m = n$ or infinitely many if $m < n$.

■ WORKED EXAMPLES ■

3.4 A This question connects elimination-pivot columns-back substitution to column space-nullspace-rank-solvability (the full picture). The 3 by 3 matrix A has rank 2:

$$A\boldsymbol{x} = \boldsymbol{b} \quad \text{is} \quad \begin{aligned} x_1 + 2x_2 + 3x_3 + \ \ 5x_4 &= b_1 \\ 2x_1 + 4x_2 + 8x_3 + 12x_4 &= b_2 \\ 3x_1 + 6x_2 + 7x_3 + 13x_4 &= b_3 \end{aligned}$$

1. Reduce $[\,A \ \ \boldsymbol{b}\,]$ to $[\,U \ \ \boldsymbol{c}\,]$, so that $A\boldsymbol{x} = \boldsymbol{b}$ becomes a triangular system $U\boldsymbol{x} = \boldsymbol{c}$.
2. Find the condition on b_1, b_2, b_3 for $A\boldsymbol{x} = \boldsymbol{b}$ to have a solution.
3. Describe the column space of A. Which plane in $\mathbf{R}^3$?
4. Describe the nullspace of A. Which special solutions in $\mathbf{R}^4$?
5. Find a particular solution to $A\boldsymbol{x} = (0, 6, -6)$ and then the complete solution.
6. Reduce $[\,U \ \ \boldsymbol{c}\,]$ to $[\,R \ \ \boldsymbol{d}\,]$: Special solutions from R, particular solution from $\boldsymbol{d}$.

Solution

1. The multipliers in elimination are 2 and 3 and -1. They take $[\,A \ \ \boldsymbol{b}\,]$ into $[\,U \ \ \boldsymbol{c}\,]$.

$$\begin{bmatrix} 1 & 2 & 3 & 5 & \boldsymbol{b_1} \\ 2 & 4 & 8 & 12 & \boldsymbol{b_2} \\ 3 & 6 & 7 & 13 & \boldsymbol{b_3} \end{bmatrix} \to \left[\begin{array}{cccc|c} 1 & 2 & 3 & 5 & \boldsymbol{b_1} \\ 0 & 0 & 2 & 2 & \boldsymbol{b_2 - 2b_1} \\ 0 & 0 & -2 & -2 & \boldsymbol{b_3 - 3b_1} \end{array}\right] \to \left[\begin{array}{cccc|c} 1 & 2 & 3 & 5 & \boldsymbol{b_1} \\ 0 & 0 & 2 & 2 & \boldsymbol{b_2 - 2b_1} \\ 0 & 0 & 0 & 0 & \boldsymbol{b_3 + b_2 - 5b_1} \end{array}\right]$$

2. The last equation shows the solvability condition $b_3 + b_2 - 5b_1 = 0$. Then $0 = 0$.

3. **First description**: The column space is the plane containing all combinations of the pivot columns $(1, 2, 3)$ and $(3, 8, 7)$, since the pivots are in columns 1 and 3. **Second description**: The column space contains all vectors with $b_3 + b_2 - 5b_1 = 0$. That makes $A\boldsymbol{x} = \boldsymbol{b}$ solvable, so $\boldsymbol{b}$ is in the column space. *All columns of A pass this test $b_3 + b_2 - 5b_1 = 0$. This is the equation for the plane in the first description.*

4. The special solutions have free variables $x_2 = 1, x_4 = 0$ and then $x_2 = 0, x_4 = 1$:

Special solutions to $A\boldsymbol{x} = \boldsymbol{0}$
Back substitution in $U\boldsymbol{x} = \boldsymbol{0}$

$$\boldsymbol{s}_1 = \begin{bmatrix} -2 \\ 1 \\ 0 \\ 0 \end{bmatrix} \qquad \boldsymbol{s}_2 = \begin{bmatrix} -2 \\ 0 \\ -1 \\ 1 \end{bmatrix}$$

The nullspace $\boldsymbol{N}(A)$ in $\mathbf{R}^4$ contains all $\boldsymbol{x}_n = c_1\boldsymbol{s}_1 + c_2\boldsymbol{s}_2 = (-2c_1 - 2c_2, c_1, -c_2, c_2)$.

5. One particular solution $\boldsymbol{x}_p$ has free variables = zero. Back substitute in $U\boldsymbol{x} = \boldsymbol{c}$:

Particular solution to $A\boldsymbol{x}_p = (0, 6, -6)$
This vector $\boldsymbol{b}$ satisfies $b_3 + b_2 - 5b_1 = 0$

$$\boldsymbol{x}_p = \begin{bmatrix} -9 \\ 0 \\ 3 \\ 0 \end{bmatrix}$$

The complete solution to $A\boldsymbol{x} = (0, 6, -6)$ is $\boldsymbol{x} = \boldsymbol{x}_p +$ all $\boldsymbol{x}_n$.

6. In the reduced form R, the third column changes from $(3, 2, 0)$ in U to $(0, 1, 0)$. The right side $\boldsymbol{c} = (0, 6, 0)$ now becomes $\boldsymbol{d} = (-9, 3, 0)$ showing -9 and 3 in $\boldsymbol{x}_p$:

$$[\,U \quad \boldsymbol{c}\,] = \begin{bmatrix} 1 & 2 & 3 & 5 & \mathbf{0} \\ 0 & 0 & 2 & 2 & \mathbf{6} \\ 0 & 0 & 0 & 0 & \mathbf{0} \end{bmatrix} \longrightarrow [\,R \quad \boldsymbol{d}\,] = \begin{bmatrix} 1 & 2 & 0 & 2 & \mathbf{-9} \\ 0 & 0 & 1 & 1 & \mathbf{3} \\ 0 & 0 & 0 & 0 & \mathbf{0} \end{bmatrix}$$

3.4 B If you have this information about the solutions to $A\boldsymbol{x} = \boldsymbol{b}$ for a specific $\boldsymbol{b}$, what does that tell you about the shape of A (and A itself)? And possibly about $\boldsymbol{b}$.

1. There is exactly one solution.
2. All solutions to $A\boldsymbol{x} = \boldsymbol{b}$ have the form $\boldsymbol{x} = \begin{bmatrix} 1 \\ 2 \end{bmatrix} + c\begin{bmatrix} 1 \\ 1 \end{bmatrix}$.
3. There are no solutions.
4. All solutions to $A\boldsymbol{x} = \boldsymbol{b}$ have the form $\boldsymbol{x} = \begin{bmatrix} 1 \\ 1 \\ 0 \end{bmatrix} + c\begin{bmatrix} 1 \\ 0 \\ 1 \end{bmatrix}$
5. There are infinitely many solutions.

Solution In case **1**, with exactly one solution, A must have full column rank $r = n$. The nullspace of A contains only the zero vector. Necessarily $m \geq n$.

In case **2**, A must have $n = 2$ columns (and m is arbitrary). With $\begin{bmatrix} 1 \\ 1 \end{bmatrix}$ in the nullspace of A, column 2 is the *negative* of column 1. With $x = \begin{bmatrix} 1 \\ 2 \end{bmatrix}$ as a solution, $\boldsymbol{b} = (\text{column } 1) + 2\,(\text{column } 2) = \text{column } 2$. The columns can't be zero vectors.

In case **3** we only know that $\boldsymbol{b}$ is not in the column space of A. The rank of A must be less than m. I guess we know $\boldsymbol{b} \neq \mathbf{0}$, otherwise $\boldsymbol{x} = \mathbf{0}$ would be a solution.

In case **4**, A must have $n = 3$ columns. With $(1, 0, 1)$ in the nullspace of A, column 3 is the negative of column 1. Column 2 must *not* be a multiple of column 1, or the nullspace would contain another special solution. So the rank of A is $3 - 1 = 2$. Necessarily A has $m \geq 2$ rows. The right side $\boldsymbol{b}$ is column $1 +$ column 2.

In case **5** with infinitely many solutions, the nullspace must contain nonzero vectors. The rank r must be less than n (not full column rank), and $\boldsymbol{b}$ must be in the column space of A. We don't know if *every* $\boldsymbol{b}$ is in the column space, so we don't know if $r = m$.

3.4 C Find the complete solution $\boldsymbol{x} = \boldsymbol{x}_p + \boldsymbol{x}_n$ by forward elimination on $[A \ \ \boldsymbol{b}]$:

$$\begin{bmatrix} 1 & 2 & 1 & 0 \\ 2 & 4 & 4 & 8 \\ 4 & 8 & 6 & 8 \end{bmatrix} \begin{bmatrix} x_1 \\ x_2 \\ x_3 \\ x_4 \end{bmatrix} = \begin{bmatrix} 4 \\ 2 \\ 10 \end{bmatrix}.$$

Find numbers y_1, y_2, y_3 so that y_1 (row 1)$+y_2$ (row 2)$+y_3$ (row 3) $=$ ***zero row***. Check that $\boldsymbol{b} = (4, 2, 10)$ satisfies the condition $y_1 b_1 + y_2 b_2 + y_3 b_3 = 0$. Why is this the condition for the equations to be solvable and $\boldsymbol{b}$ to be in the column space?

Solution Forward elimination on $[A \ \ \boldsymbol{b}]$ produces a zero row in $[U \ \ \boldsymbol{c}]$. The third equation becomes $0 = 0$ and the equations are consistent (and solvable):

$$\begin{bmatrix} 1 & 2 & 1 & 0 & \mathbf{4} \\ 2 & 4 & 4 & 8 & \mathbf{2} \\ 4 & 8 & 6 & 8 & \mathbf{10} \end{bmatrix} \longrightarrow \begin{bmatrix} 1 & 2 & 1 & 0 & \mathbf{4} \\ 0 & 0 & 2 & 8 & \mathbf{-6} \\ 0 & 0 & 2 & 8 & \mathbf{-6} \end{bmatrix} \longrightarrow \begin{bmatrix} 1 & 2 & 1 & 0 & \mathbf{4} \\ 0 & 0 & 2 & 8 & \mathbf{-6} \\ 0 & 0 & 0 & 0 & \mathbf{0} \end{bmatrix}.$$

Columns 1 and 3 contain pivots. The variables x_2 and x_4 are free. If we set those to zero we can solve (back substitution) for the particular solution $\boldsymbol{x}_p = (7, 0, -3, 0)$. We see 7 and -3 again if elimination continues all the way to $[R \ \ \boldsymbol{d}]$:

$$\begin{bmatrix} 1 & 2 & 1 & 0 & \mathbf{4} \\ 0 & 0 & 2 & 8 & \mathbf{-6} \\ 0 & 0 & 0 & 0 & \mathbf{0} \end{bmatrix} \longrightarrow \begin{bmatrix} 1 & 2 & 1 & 0 & \mathbf{4} \\ 0 & 0 & 1 & 4 & \mathbf{-3} \\ 0 & 0 & 0 & 0 & \mathbf{0} \end{bmatrix} \longrightarrow \begin{bmatrix} 1 & 2 & 0 & -4 & \mathbf{7} \\ 0 & 0 & 1 & 4 & \mathbf{-3} \\ 0 & 0 & 0 & 0 & \mathbf{0} \end{bmatrix}.$$

For the nullspace part $\boldsymbol{x}_n$ with $\boldsymbol{b} = \mathbf{0}$, set the free variables x_2, x_4 to $1, 0$ and also $0, 1$:

Special solutions $\quad s_1 = (-2, 1, 0, 0)$ **and** $s_2 = (4, 0, -4, 1)$

Then the complete solution to $A\boldsymbol{x} = \boldsymbol{b}$ (and $R\boldsymbol{x} = \boldsymbol{d}$) is $\boldsymbol{x}_{\text{complete}} = \boldsymbol{x}_p + c_1 s_1 + c_2 s_2$.

The rows of A produced the zero row from 2(row 1) + (row 2) – (row 3) $=$ $(0, 0, 0, 0)$. The same combination for $\boldsymbol{b} = (4, 2, 10)$ gives $2(4) + (2) - (10) = 0$. If a combination of the rows (on the left side) gives the zero row, then the same combination must give zero on the right side. Of course! Otherwise no solution.

Later we will say this again in different words: If every column of A is perpendicular to $\boldsymbol{y} = (2, 1, -1)$, then any combination $\boldsymbol{b}$ of those columns must also be perpendicular to $\boldsymbol{y}$. Otherwise $\boldsymbol{b}$ is not in the column space and $A\boldsymbol{x} = \boldsymbol{b}$ is not solvable.

And again: If $\boldsymbol{y}$ is in the nullspace of A^{T} then $\boldsymbol{y}$ must be perpendicular to every $\boldsymbol{b}$ in the column space. Just looking ahead ...

Problem Set 3.4

1 (Recommended) Execute the six steps of Worked Example **3.4 A** to describe the column space and nullspace of A and the complete solution to $A\boldsymbol{x} = \boldsymbol{b}$:

$$A = \begin{bmatrix} 2 & 4 & 6 & 4 \\ 2 & 5 & 7 & 6 \\ 2 & 3 & 5 & 2 \end{bmatrix} \qquad \boldsymbol{b} = \begin{bmatrix} b_1 \\ b_2 \\ b_3 \end{bmatrix} = \begin{bmatrix} 4 \\ 3 \\ 5 \end{bmatrix}$$

2 Carry out the same six steps for this matrix A with rank one. You will find *two* conditions on b_1, b_2, b_3 for $A\boldsymbol{x} = \boldsymbol{b}$ to be solvable. Together these two conditions put $\boldsymbol{b}$ into the _____ space (two planes give a line):

$$A = \begin{bmatrix} 1 \\ 3 \\ 2 \end{bmatrix} \begin{bmatrix} 2 & 1 & 3 \end{bmatrix} = \begin{bmatrix} 2 & 1 & 3 \\ 6 & 3 & 9 \\ 4 & 2 & 6 \end{bmatrix} \qquad \boldsymbol{b} = \begin{bmatrix} b_1 \\ b_2 \\ b_3 \end{bmatrix} = \begin{bmatrix} 10 \\ 30 \\ 20 \end{bmatrix}$$

Questions 3–15 are about the solution of $A\boldsymbol{x} = \boldsymbol{b}$. Follow the steps in the text to $\boldsymbol{x}_p$ and $\boldsymbol{x}_n$. Use the augmented matrix with last column $\boldsymbol{b}$.

3 Write the complete solution as $\boldsymbol{x}_p$ plus any multiple of $\boldsymbol{s}$ in the nullspace:

$$\begin{aligned} x + 3y + 3z &= 1 \\ 2x + 6y + 9z &= 5 \\ -x - 3y + 3z &= 5. \end{aligned}$$

4 Find the complete solution (also called the *general solution*) to

$$\begin{bmatrix} 1 & 3 & 1 & 2 \\ 2 & 6 & 4 & 8 \\ 0 & 0 & 2 & 4 \end{bmatrix} \begin{bmatrix} x \\ y \\ z \\ t \end{bmatrix} = \begin{bmatrix} 1 \\ 3 \\ 1 \end{bmatrix}.$$

5 Under what condition on b_1, b_2, b_3 is this system solvable? Include $\boldsymbol{b}$ as a fourth column in elimination. Find all solutions when that condition holds:

$$\begin{aligned} x + 2y - 2z &= b_1 \\ 2x + 5y - 4z &= b_2 \\ 4x + 9y - 8z &= b_3. \end{aligned}$$

6 What conditions on b_1, b_2, b_3, b_4 make each system solvable? Find $\boldsymbol{x}$ in that case:

$$\begin{bmatrix} 1 & 2 \\ 2 & 4 \\ 2 & 5 \\ 3 & 9 \end{bmatrix} \begin{bmatrix} x_1 \\ x_2 \end{bmatrix} = \begin{bmatrix} b_1 \\ b_2 \\ b_3 \\ b_4 \end{bmatrix} \qquad \begin{bmatrix} 1 & 2 & 3 \\ 2 & 4 & 6 \\ 2 & 5 & 7 \\ 3 & 9 & 12 \end{bmatrix} \begin{bmatrix} x_1 \\ x_2 \\ x_3 \end{bmatrix} = \begin{bmatrix} b_1 \\ b_2 \\ b_3 \\ b_4 \end{bmatrix}.$$

7 Show by elimination that (b_1, b_2, b_3) is in the column space if $b_3 - 2b_2 + 4b_1 = 0$.

$$A = \begin{bmatrix} 1 & 3 & 1 \\ 3 & 8 & 2 \\ 2 & 4 & 0 \end{bmatrix}.$$

What combination of the rows of A gives the zero row?

8 Which vectors (b_1, b_2, b_3) are in the column space of A? Which combinations of the rows of A give zero?

(a) $A = \begin{bmatrix} 1 & 2 & 1 \\ 2 & 6 & 3 \\ 0 & 2 & 5 \end{bmatrix}$ (b) $A = \begin{bmatrix} 1 & 1 & 1 \\ 1 & 2 & 4 \\ 2 & 4 & 8 \end{bmatrix}$.

9 (a) The Worked Example **3.4 A** reached $[\,U \;\; \boldsymbol{c}\,]$ from $[\,A \;\; \boldsymbol{b}\,]$. Put the multipliers into L and verify that LU equals A and $L\boldsymbol{c}$ equals $\boldsymbol{b}$.

(b) Combine the pivot columns of A with the numbers -9 and 3 in the particular solution $\boldsymbol{x}_p$. What is that linear combination and why?

10 Construct a 2 by 3 system $A\boldsymbol{x} = \boldsymbol{b}$ with particular solution $\boldsymbol{x}_p = (2, 4, 0)$ and homogeneous solution $\boldsymbol{x}_n =$ any multiple of $(1, 1, 1)$.

11 Why can't a 1 by 3 system have $\boldsymbol{x}_p = (2, 4, 0)$ and $\boldsymbol{x}_n =$ any multiple of $(1, 1, 1)$?

12 (a) If $A\boldsymbol{x} = \boldsymbol{b}$ has two solutions $\boldsymbol{x}_1$ and $\boldsymbol{x}_2$, find two solutions to $A\boldsymbol{x} = \boldsymbol{0}$.

(b) Then find another solution to $A\boldsymbol{x} = \boldsymbol{0}$ and another solution to $A\boldsymbol{x} = \boldsymbol{b}$.

13 Explain why these are all false:

(a) The complete solution is any linear combination of $\boldsymbol{x}_p$ and $\boldsymbol{x}_n$.

(b) A system $A\boldsymbol{x} = \boldsymbol{b}$ has at most one particular solution.

(c) The solution $\boldsymbol{x}_p$ with all free variables zero is the shortest solution (minimum length $\|\boldsymbol{x}\|$). Find a 2 by 2 counterexample.

(d) If A is invertible there is no solution $\boldsymbol{x}_n$ in the nullspace.

14 Suppose column 5 of U has no pivot. Then x_5 is a ______ variable. The zero vector (is) (is not) the only solution to $A\boldsymbol{x} = \boldsymbol{0}$. If $A\boldsymbol{x} = \boldsymbol{b}$ has a solution, then it has ______ solutions.

15 Suppose row 3 of U has no pivot. Then that row is ____. The equation $U\boldsymbol{x} = \boldsymbol{c}$ is only solvable provided ____. The equation $A\boldsymbol{x} = \boldsymbol{b}$ (*is*) (*is not*) (*might not be*) solvable.

Questions 16–20 are about matrices of "full rank" $r = m$ or $r = n$.

16 The largest possible rank of a 3 by 5 matrix is ____. Then there is a pivot in every ____ of U and R. The solution to $A\boldsymbol{x} = \boldsymbol{b}$ (*always exists*) (*is unique*). The column space of A is ____. An example is $A =$ ____.

17 The largest possible rank of a 6 by 4 matrix is ____. Then there is a pivot in every ____ of U and R. The solution to $A\boldsymbol{x} = \boldsymbol{b}$ (*always exists*) (*is unique*). The nullspace of A is ____. An example is $A =$ ____.

18 Find by elimination the rank of A and also the rank of A^{T}:

$$A = \begin{bmatrix} 1 & 4 & 0 \\ 2 & 11 & 5 \\ -1 & 2 & 10 \end{bmatrix} \quad \text{and} \quad A = \begin{bmatrix} 1 & 0 & 1 \\ 1 & 1 & 2 \\ 1 & 1 & q \end{bmatrix} \text{ (rank depends on } q).$$

19 Find the rank of A and also of $A^{\mathrm{T}}A$ and also of AA^{T}:

$$A = \begin{bmatrix} 1 & 1 & 5 \\ 1 & 0 & 1 \end{bmatrix} \quad \text{and} \quad A = \begin{bmatrix} 2 & 0 \\ 1 & 1 \\ 1 & 2 \end{bmatrix}.$$

20 Reduce A to its echelon form U. Then find a triangular L so that $A = LU$.

$$A = \begin{bmatrix} 3 & 4 & 1 & 0 \\ 6 & 5 & 2 & 1 \end{bmatrix} \quad \text{and} \quad A = \begin{bmatrix} 1 & 0 & 1 & 0 \\ 2 & 2 & 0 & 3 \\ 0 & 6 & 5 & 4 \end{bmatrix}.$$

21 Find the complete solution in the form $\boldsymbol{x}_p + \boldsymbol{x}_n$ to these full rank systems:

(a) $x + y + z = 4$ (b) $\begin{aligned} x + y + z &= 4 \\ x - y + z &= 4. \end{aligned}$

22 If $A\boldsymbol{x} = \boldsymbol{b}$ has infinitely many solutions, why is it impossible for $A\boldsymbol{x} = \boldsymbol{B}$ (new right side) to have only one solution? Could $A\boldsymbol{x} = \boldsymbol{B}$ have no solution?

23 Choose the number q so that (if possible) the ranks are (a) 1, (b) 2, (c) 3:

$$A = \begin{bmatrix} 6 & 4 & 2 \\ -3 & -2 & -1 \\ 9 & 6 & q \end{bmatrix} \quad \text{and} \quad B = \begin{bmatrix} 3 & 1 & 3 \\ q & 2 & q \end{bmatrix}.$$

24 Give examples of matrices A for which the number of solutions to $A\boldsymbol{x} = \boldsymbol{b}$ is

(a) 0 or 1, depending on $\boldsymbol{b}$

(b) ∞, regardless of $\boldsymbol{b}$

(c) 0 or ∞, depending on $\boldsymbol{b}$

(d) 1, regardless of $\boldsymbol{b}$.

25 Write down all known relations between r and m and n if $A\boldsymbol{x} = \boldsymbol{b}$ has

(a) no solution for some $\boldsymbol{b}$

(b) infinitely many solutions for every $\boldsymbol{b}$

(c) exactly one solution for some $\boldsymbol{b}$, no solution for other $\boldsymbol{b}$

(d) exactly one solution for every $\boldsymbol{b}$.

Questions 26–33 are about Gauss-Jordan elimination (upwards as well as downwards) and the reduced echelon matrix R.

26 Continue elimination from U to R. Divide rows by pivots so the new pivots are all 1. Then produce zeros *above* those pivots to reach R:

$$U = \begin{bmatrix} 2 & 4 & 4 \\ 0 & 3 & 6 \\ 0 & 0 & 0 \end{bmatrix} \quad \text{and} \quad U = \begin{bmatrix} 2 & 4 & 4 \\ 0 & 3 & 6 \\ 0 & 0 & 5 \end{bmatrix}.$$

27 Suppose U is square with n pivots (an invertible matrix). *Explain why* $R = I$.

28 Apply Gauss-Jordan elimination to $U\boldsymbol{x} = \boldsymbol{0}$ and $U\boldsymbol{x} = \boldsymbol{c}$. Reach $R\boldsymbol{x} = \boldsymbol{0}$ and $R\boldsymbol{x} = \boldsymbol{d}$:

$$\begin{bmatrix} U & \boldsymbol{0} \end{bmatrix} = \begin{bmatrix} 1 & 2 & 3 & \mathbf{0} \\ 0 & 0 & 4 & \mathbf{0} \end{bmatrix} \quad \text{and} \quad \begin{bmatrix} U & \boldsymbol{c} \end{bmatrix} = \begin{bmatrix} 1 & 2 & 3 & \mathbf{5} \\ 0 & 0 & 4 & \mathbf{8} \end{bmatrix}.$$

Solve $R\boldsymbol{x} = \boldsymbol{0}$ to find $\boldsymbol{x}_n$ (its free variable is $x_2 = 1$). Solve $R\boldsymbol{x} = \boldsymbol{d}$ to find $\boldsymbol{x}_p$ (its free variable is $x_2 = 0$).

29 Apply Gauss-Jordan elimination to reduce to $R\boldsymbol{x} = \boldsymbol{0}$ and $R\boldsymbol{x} = \boldsymbol{d}$:

$$\begin{bmatrix} U & 0 \end{bmatrix} = \begin{bmatrix} 3 & 0 & 6 & \mathbf{0} \\ 0 & 0 & 2 & \mathbf{0} \\ 0 & 0 & 0 & \mathbf{0} \end{bmatrix} \quad \text{and} \quad \begin{bmatrix} U & \boldsymbol{c} \end{bmatrix} = \begin{bmatrix} 3 & 0 & 6 & \mathbf{9} \\ 0 & 0 & 2 & \mathbf{4} \\ 0 & 0 & 0 & \mathbf{5} \end{bmatrix}.$$

Solve $U\boldsymbol{x} = \boldsymbol{0}$ or $R\boldsymbol{x} = \boldsymbol{0}$ to find $\boldsymbol{x}_n$ (free variable $= 1$). What are the solutions to $R\boldsymbol{x} = \boldsymbol{d}$?

30 Reduce to $U\boldsymbol{x} = \boldsymbol{c}$ (Gaussian elimination) and then $R\boldsymbol{x} = \boldsymbol{d}$ (Gauss-Jordan):

$$A\boldsymbol{x} = \begin{bmatrix} 1 & 0 & 2 & 3 \\ 1 & 3 & 2 & 0 \\ 2 & 0 & 4 & 9 \end{bmatrix} \begin{bmatrix} x_1 \\ x_2 \\ x_3 \\ x_4 \end{bmatrix} = \begin{bmatrix} 2 \\ 5 \\ 10 \end{bmatrix} = \boldsymbol{b}.$$

Find a particular solution $\boldsymbol{x}_p$ and all homogeneous solutions $\boldsymbol{x}_n$.

31 Find matrices A and B with the given property or explain why you can't: The only solution of $A\boldsymbol{x} = \begin{bmatrix} 1 \\ 2 \\ 3 \end{bmatrix}$ is $\boldsymbol{x} = \begin{bmatrix} 0 \\ 1 \end{bmatrix}$. The only solution of $B\boldsymbol{x} = \begin{bmatrix} 0 \\ 1 \end{bmatrix}$ is $\boldsymbol{x} = \begin{bmatrix} 1 \\ 2 \\ 3 \end{bmatrix}$.

32 Find the LU factorization of A and the complete solution to $A\boldsymbol{x} = \boldsymbol{b}$:

$$A = \begin{bmatrix} 1 & 3 & 1 \\ 1 & 2 & 3 \\ 2 & 4 & 6 \\ 1 & 1 & 5 \end{bmatrix} \quad \text{and} \quad \boldsymbol{b} = \begin{bmatrix} 1 \\ 3 \\ 6 \\ 5 \end{bmatrix} \quad \text{and then} \quad \boldsymbol{b} = \begin{bmatrix} 1 \\ 0 \\ 0 \\ 0 \end{bmatrix}.$$

33 The complete solution to $A\boldsymbol{x} = \begin{bmatrix} 1 \\ 3 \end{bmatrix}$ is $\boldsymbol{x} = \begin{bmatrix} 1 \\ 0 \end{bmatrix} + c\begin{bmatrix} 0 \\ 1 \end{bmatrix}$. Find A.

34 Suppose you know that the 3 by 4 matrix A has the vector $\boldsymbol{s} = (2, 3, 1, 0)$ as a basis for its nullspace

(a) What is the *rank* of A and the complete solution to $A\boldsymbol{x} = \boldsymbol{0}$?

(b) What is the exact row reduced echelon form R of A?

INDEPENDENCE, BASIS AND DIMENSION ■ 3.5

This important section is about the true size of a subspace. There are n columns in an m by n matrix, and each column has m components. But the true "dimension" of the column space is not necessarily m or n. The dimension is measured by counting ***independent columns***—and we have to say what that means. We will see that *the true dimension of the column space is the rank* r.

The idea of independence applies to any vectors $\boldsymbol{v}_1, \ldots, \boldsymbol{v}_n$ in any vector space. Most of this section concentrates on the subspaces that we know and use—especially the column space in $\mathbf{R}^m$ and the nullspace in $\mathbf{R}^n$. In the last part we also study "vectors" that are not column vectors. They can be matrices and functions; they can be linearly independent (or not). First come the key examples using column vectors.

The final goal is to understand a ***basis*** for a vector space. A basis contains independent vectors that "span the space". We are at the heart of our subject, and we cannot go on without a basis. The four essential ideas in this section (with first hints at their meaning) are:

1. **Independent vectors** (*not too many*)
2. **Spanning a space** (*not too few*)
3. **Basis for a space** (*not too many or too few*)
4. **Dimension of a space** (*the right number of vectors*).

Linear Independence

Our first definition of independence is not so conventional, but you are ready for it.

DEFINITION The columns of A are *linearly independent* when the only solution to $A\boldsymbol{x} = \mathbf{0}$ is $\boldsymbol{x} = \mathbf{0}$. ***No other combination*** $A\boldsymbol{x}$ ***of the columns gives the zero vector.***

With linearly independent columns, the nullspace $\boldsymbol{N}(A)$ contains only the zero vector. Let me illustrate linear independence (and linear dependence) with three vectors in $\mathbf{R}^3$:

1. If three vectors are *not* in the same plane, they are independent. No combination of $\boldsymbol{v}_1, \boldsymbol{v}_2, \boldsymbol{v}_3$ in Figure 3.4 gives zero except $0\boldsymbol{v}_1 + 0\boldsymbol{v}_2 + 0\boldsymbol{v}_3$.

2. If three vectors $\boldsymbol{w}_1, \boldsymbol{w}_2, \boldsymbol{w}_3$ are *in the same plane*, they are dependent.

This idea of independence applies to 7 vectors in 12-dimensional space. If they are the columns of A, and independent, the nullspace only contains $\boldsymbol{x} = \mathbf{0}$. Now we choose different words to express the same idea. The following definition of independence will apply to any sequence of vectors in any vector space. When the vectors are the columns of A, the two definitions say exactly the same thing.

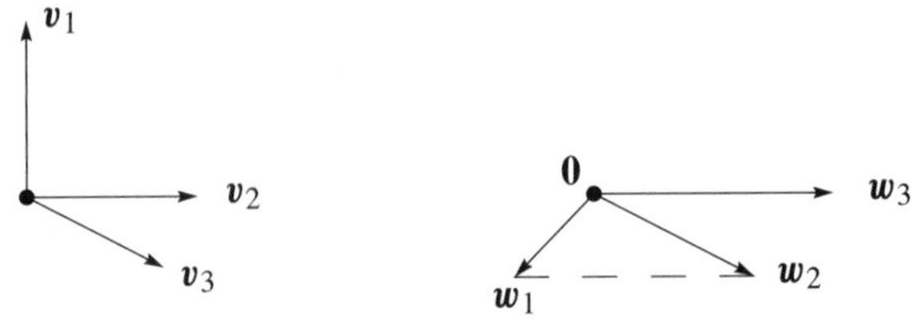

Figure 3.4 Independent vectors v_1, v_2, v_3. Dependent vectors w_1, w_2, w_3. The combination $w_1 - w_2 + w_3$ is $(0, 0, 0)$.

DEFINITION The sequence of vectors $v_1, \ldots, v_n$ is ***linearly independent*** if the only combination that gives the zero vector is $0v_1 + 0v_2 + \cdots + 0v_n$. Thus linear independence means that

$$x_1 v_1 + x_2 v_2 + \cdots + x_n v_n = \mathbf{0} \quad \text{only happens when all } x\text{'s are zero.} \tag{1}$$

If a combination gives $\mathbf{0}$, when the x's are not all zero, the vectors are *dependent.*

Correct language: "The sequence of vectors is linearly independent." Acceptable shortcut: "The vectors are independent." Unacceptable: "The matrix is independent."

A sequence of vectors is either dependent or independent. They can be combined to give the zero vector (with nonzero x's) or they can't. So the key question is: Which combinations of the vectors give zero? We begin with some small examples in $\mathbf{R}^2$:

(a) The vectors $(1, 0)$ and $(0, 1)$ are independent.

(b) The vectors $(1, 1)$ and $(1, 0.00001)$ are independent.

(c) The vectors $(1, 1)$ and $(2, 2)$ are *dependent.*

(d) The vectors $(1, 1)$ and $(0, 0)$ are *dependent.*

Geometrically, $(1, 1)$ and $(2, 2)$ are on a line through the origin. They are not independent. To use the definition, find numbers x_1 and x_2 so that $x_1(1, 1) + x_2(2, 2) = (0, 0)$. This is the same as solving $Ax = \mathbf{0}$:

$$\begin{bmatrix} 1 & 2 \\ 1 & 2 \end{bmatrix} \begin{bmatrix} x_1 \\ x_2 \end{bmatrix} = \begin{bmatrix} 0 \\ 0 \end{bmatrix} \quad \text{for } x_1 = 2 \text{ and } x_2 = -1.$$

The columns are dependent exactly when *there is a nonzero vector in the nullspace.*

If one of the v's is the zero vector, independence has no chance. Why not?

Now move to three vectors in $\mathbf{R}^3$. If one of them is a multiple of another one, these vectors are dependent. But the complete test involves all three vectors at once. We put them in a matrix and try to solve $Ax = \mathbf{0}$.

Example 1 The columns of A are dependent. $A\boldsymbol{x} = \mathbf{0}$ has a nonzero solution:

$$A\boldsymbol{x} = \begin{bmatrix} 1 & 0 & 3 \\ 2 & 1 & 5 \\ 1 & 0 & 3 \end{bmatrix} \begin{bmatrix} -3 \\ 1 \\ 1 \end{bmatrix} \quad \text{is} \quad -3 \begin{bmatrix} 1 \\ 2 \\ 1 \end{bmatrix} + 1 \begin{bmatrix} 0 \\ 1 \\ 0 \end{bmatrix} + 1 \begin{bmatrix} 3 \\ 5 \\ 3 \end{bmatrix} = \begin{bmatrix} 0 \\ 0 \\ 0 \end{bmatrix}.$$

The rank of A is only $r = 2$. *Independent columns would give full column rank* $r = n = 3$.

In that matrix the rows are also dependent. Row 1 minus row 3 is the zero row. For a *square matrix*, we will show that dependent columns imply dependent rows (and vice versa).

Question How do you find that solution to $A\boldsymbol{x} = \mathbf{0}$? The systematic way is elimination.

$$A = \begin{bmatrix} 1 & 0 & 3 \\ 2 & 1 & 5 \\ 1 & 0 & 3 \end{bmatrix} \quad \text{reduces to } R = \begin{bmatrix} 1 & 0 & 3 \\ 0 & 1 & -1 \\ 0 & 0 & 0 \end{bmatrix}.$$

The solution $\boldsymbol{x} = (-3, 1, 1)$ was exactly the special solution. It shows how the free column (column 3) is a combination of the pivot columns. That kills independence!

3H The columns of A are independent exactly when the rank is $r = n$. There are n pivots and no free variables. Only $\boldsymbol{x} = \mathbf{0}$ is in the nullspace.

One case is of special importance because it is clear from the start. Suppose seven columns have five components each ($m = 5$ is less than $n = 7$). Then the columns *must be dependent*. Any seven vectors from $\mathbf{R}^5$ are dependent. The rank of A cannot be larger than 5. There cannot be more than five pivots in five rows. The system $A\boldsymbol{x} = \mathbf{0}$ has at least $7 - 5 = 2$ free variables, so it has nonzero solutions—which means that the columns are dependent.

3I Any set of n vectors in $\mathbf{R}^m$ must be linearly dependent if $n > m$.

The matrix has more columns than rows—it is short and wide. The columns are certainly dependent if $n > m$, because $A\boldsymbol{x} = \mathbf{0}$ has a nonzero solution. The columns might be dependent or might be independent if $n \leq m$. Elimination will reveal the pivot columns. *It is those pivot columns that are independent.*

Note Another way to describe linear independence is this: "*One vector is a combination of the other vectors.*" That sounds clear. Why don't we say this from the start? Our definition was longer: "*Some combination gives the zero vector, other than the trivial combination with every* $x = 0$." We must rule out the easy way to get the zero vector. That trivial combination of zeros gives every author a headache. If one vector is a combination of the others, that vector has coefficient $x = 1$.

The point is, our definition doesn't pick out one particular vector as guilty. All columns of A are treated the same. We look at $Ax = \mathbf{0}$, and it has a nonzero solution or it hasn't. In the end that is better than asking if the last column (or the first, or a column in the middle) is a combination of the others.

Vectors that Span a Subspace

The first subspace in this book was the column space. Starting with columns $v_1, \cdots, v_n$, the subspace was filled out by including all combinations $x_1 v_1 + \cdots + x_n v_n$. *The column space consists of all combinations Ax of the columns.* We now introduce the single word "span" to describe this: The column space is ***spanned*** by the columns.

DEFINITION A set of vectors ***spans*** a space if their linear combinations fill the space.

Example 2 $v_1 = \begin{bmatrix} 1 \\ 0 \end{bmatrix}$ and $v_2 = \begin{bmatrix} 0 \\ 1 \end{bmatrix}$ span the full two-dimensional space $\mathbf{R}^2$.

Example 3 $v_1 = \begin{bmatrix} 1 \\ 0 \end{bmatrix}$, $v_2 = \begin{bmatrix} 0 \\ 1 \end{bmatrix}$, $v_3 = \begin{bmatrix} 4 \\ 7 \end{bmatrix}$ also span the full space $\mathbf{R}^2$.

Example 4 $w_1 = \begin{bmatrix} 1 \\ 1 \end{bmatrix}$ and $w_2 = \begin{bmatrix} -1 \\ -1 \end{bmatrix}$ only span a line in $\mathbf{R}^2$. So does w_1 by itself. So does w_2 by itself.

Think of two vectors coming out from $(0, 0, 0)$ in 3-dimensional space. Generally they span a plane. Your mind fills in that plane by taking linear combinations. Mathematically you know other possibilities: two vectors spanning a line, three vectors spanning all of $\mathbf{R}^3$, three vectors spanning only a plane. It is even possible that three vectors span only a line, or ten vectors span only a plane. They are certainly not independent!

The columns span the column space. Here is a new subspace—*which is spanned by the rows.* ***The combinations of the rows produce the "row space".***

DEFINITION The ***row space*** of a matrix is the subspace of $\mathbf{R}^n$ spanned by the rows.

The rows of an m by n matrix have n components. They are vectors in $\mathbf{R}^n$—or they would be if they were written as column vectors. There is a quick way to fix that: *Transpose the matrix.* Instead of the rows of A, look at the columns of A^{T}. Same numbers, but now in columns.

The row space of A is $C(A^{\mathrm{T}})$. It is the column space of A^{T}. It is a subspace of $\mathbf{R}^n$. The vectors that span it are the columns of A^{T}, which are the rows of A.

Example 5

$$A = \begin{bmatrix} 1 & 4 \\ 2 & 7 \\ 3 & 5 \end{bmatrix} \text{ and } A^{\mathrm{T}} = \begin{bmatrix} 1 & 2 & 3 \\ 4 & 7 & 5 \end{bmatrix}. \text{ Here } m = 3 \text{ and } n = 2.$$

The column space of A is spanned by the two columns of A. It is a plane in $\mathbf{R}^3$. *The row space of A is spanned by the three rows of A* (which are columns of A^{T}). This row space is all of $\mathbf{R}^2$. Remember: The rows are in $\mathbf{R}^n$. The columns are in $\mathbf{R}^m$. Same numbers, different vectors, different spaces.

A Basis for a Vector Space

In the xy plane, a set of independent vectors could be quite small—just one vector. A set that spans the xy plane could be large—three vectors, or four, or infinitely many. One vector won't span the plane. Three vectors won't be independent. A "***basis***" is just right. We want ***enough independent vectors to span the space***.

DEFINITION A ***basis*** for a vector space is a sequence of vectors that has two properties at once:

1. The vectors are ***linearly independent***.
2. The vectors ***span the space***.

This combination of properties is fundamental to linear algebra. Every vector $\boldsymbol{v}$ in the space is a combination of the basis vectors, because they span the space. More than that, the combination that produces $\boldsymbol{v}$ is *unique*, because the basis vectors $\boldsymbol{v}_1, \ldots, \boldsymbol{v}_n$ are independent:

There is one and only one way to write $\boldsymbol{v}$ as a combination of the basis vectors.

Reason: Suppose $\boldsymbol{v} = a_1\boldsymbol{v}_1 + \cdots + a_n\boldsymbol{v}_n$ and also $\boldsymbol{v} = b_1\boldsymbol{v}_1 + \cdots + b_n\boldsymbol{v}_n$. By subtraction $(a_1 - b_1)\boldsymbol{v}_1 + \cdots + (a_n - b_n)\boldsymbol{v}_n$ is the zero vector. From the independence of the $\boldsymbol{v}$'s, each $a_i - b_i = 0$. Hence $a_i = b_i$.

Example 6 The columns of $I = \begin{bmatrix} 1 & 0 \\ 0 & 1 \end{bmatrix}$ produce the "standard basis" for $\mathbf{R}^2$.

$$\text{The basis vectors } \boldsymbol{i} = \begin{bmatrix} 1 \\ 0 \end{bmatrix} \text{ and } \boldsymbol{j} = \begin{bmatrix} 0 \\ 1 \end{bmatrix} \text{ are independent. They span } \mathbf{R}^2.$$

Everybody thinks of this basis first. The vector $\boldsymbol{i}$ goes across and $\boldsymbol{j}$ goes straight up. The columns of the 3 by 3 identity matrix are the standard basis $\boldsymbol{i}, \boldsymbol{j}, \boldsymbol{k}$. The columns of the n by n identity matrix give the "standard basis" for $\mathbf{R}^n$. Now we find other bases.

Example 7 (Important) The columns of *any invertible n by n matrix* give a basis for $\mathbf{R}^n$:

$$A = \begin{bmatrix} 1 & 2 \\ 2 & 5 \end{bmatrix} \quad \text{and} \quad A = \begin{bmatrix} 1 & 0 & 0 \\ 1 & 1 & 0 \\ 1 & 1 & 1 \end{bmatrix} \quad \text{but not} \quad A = \begin{bmatrix} 1 & 2 \\ 2 & 4 \end{bmatrix}.$$

When A is invertible, its columns are independent. The only solution to $Ax = \mathbf{0}$ is $x = \mathbf{0}$. The columns span the whole space $\mathbf{R}^n$—because every vector b is a combination of the columns. $Ax = b$ can always be solved by $x = A^{-1}b$. Do you see how everything comes together for invertible matrices? Here it is in one sentence:

3J The vectors $v_1, \ldots, v_n$ are a ***basis for*** $\mathbf{R}^n$ exactly when they are ***the columns of an n by n invertible matrix***. Thus $\mathbf{R}^n$ has infinitely many different bases.

When any matrix has independent columns, they are a basis for its column space. When the columns are dependent, we keep only the *pivot columns*—the r columns with pivots. They are independent and they span the column space.

3K ***The pivot columns of*** A ***are a basis for its column space.*** The pivot rows of A are a basis for its row space. So are the pivot rows of its echelon form R.

Example 8 This matrix is not invertible. Its columns are not a basis for anything!

$$A = \begin{bmatrix} 2 & 4 \\ 3 & 6 \end{bmatrix} \text{ which reduces to } R = \begin{bmatrix} 1 & 2 \\ 0 & 0 \end{bmatrix}.$$

Column 1 of A is the pivot column. That column alone is a basis for its column space. The second column of A would be a different basis. So would any nonzero multiple of that column. There is no shortage of bases! So we often make a definite choice: the pivot columns.

Notice that the pivot column of this R ends in zero. That column is a basis for the column space of R, but it is not even a member of the column space of A. The column spaces of A and R are different. Their bases are different.

The row space of A is the *same* as the row space of R. It contains $(2, 4)$ and $(1, 2)$ and all other multiples of those vectors. As always, there are infinitely many bases to choose from. I think the most natural choice is to pick the nonzero rows of R (rows with a pivot). So this matrix A with rank one has only one vector in the basis:

$$\text{Basis for the column space: } \begin{bmatrix} 2 \\ 3 \end{bmatrix}. \quad \text{Basis for the row space: } \begin{bmatrix} 1 \\ 2 \end{bmatrix}.$$

The next chapter will come back to these bases for the column space and row space. We are happy first with examples where the situation is clear (and the idea of a basis is still new). The next example is larger but still clear.

Example 9 Find bases for the column and row spaces of a rank two matrix:

$$R = \begin{bmatrix} 1 & 2 & 0 & 3 \\ 0 & 0 & 1 & 4 \\ 0 & 0 & 0 & 0 \end{bmatrix}.$$

Columns 1 and 3 are the pivot columns. They are a basis for the column space (of R!). The vectors in that column space all have the form $\boldsymbol{b} = (x, y, 0)$. The column space of R is the "xy plane" inside the full 3-dimensional xyz space. That plane is not $\mathbf{R}^2$, it is a subspace of $\mathbf{R}^3$. Columns 2 and 3 are a basis for the same column space. So are columns 1 and 4, and also columns 2 and 4. Which pairs of columns of R are *not* a basis for its column space?

The row space of R is a subspace of $\mathbf{R}^4$. The simplest basis for that row space is the two nonzero rows of R. The third row (the zero vector) is in the row space too. But it is not in a *basis* for the row space. The basis vectors must be independent.

Question Given five vectors in $\mathbf{R}^7$, ***how do you find a basis for the space they span?***

First answer Make them the rows of A, and eliminate to find the nonzero rows of R.
Second answer Put the five vectors into the columns of A. Eliminate to find the pivot columns (of A not R!). The program **colbasis** uses the column numbers from *pivcol*.

Could another basis have more vectors, or fewer? This is a crucial question with a good answer. ***All bases for a vector space contain the same*** number ***of vectors. This number is the "dimension" of the space.***

Dimension of a Vector Space

We have to prove what was just stated. There are many choices for the basis vectors, but the *number* of basis vectors doesn't change.

3L If $\boldsymbol{v}_1, \ldots, \boldsymbol{v}_m$ and $\boldsymbol{w}_1, \ldots, \boldsymbol{w}_n$ are both bases for the same vector space, then $m = n$.

Proof Suppose that there are more $\boldsymbol{w}$'s than $\boldsymbol{v}$'s. From $n > m$ we want to reach a contradiction. The $\boldsymbol{v}$'s are a basis, so $\boldsymbol{w}_1$ must be a combination of the $\boldsymbol{v}$'s. If $\boldsymbol{w}_1$ equals $a_{11}\boldsymbol{v}_1 + \cdots + a_{m1}\boldsymbol{v}_m$, this is the first column of a matrix multiplication VA:

$$W = \begin{bmatrix} \\ \boldsymbol{w}_1 \boldsymbol{w}_2 \ldots \boldsymbol{w}_n \\ \\ \end{bmatrix} = \begin{bmatrix} \\ \boldsymbol{v}_1 & \cdots & \boldsymbol{v}_m \\ \\ \end{bmatrix} \begin{bmatrix} a_{11} & \\ \vdots & \\ a_{m1} & \end{bmatrix} = VA.$$

We don't know each a_{ij}, but we know the shape of A (it is m by n). The second vector $\boldsymbol{w}_2$ is also a combination of the $\boldsymbol{v}$'s. The coefficients in that combination fill the second column of A. The key is that A has a row for every $\boldsymbol{v}$ and a column for every $\boldsymbol{w}$. A is a short wide matrix, since $n > m$. ***There is a nonzero solution to*** $A\boldsymbol{x} = \mathbf{0}$. Then $VA\boldsymbol{x} = \mathbf{0}$ which is $W\boldsymbol{x} = \mathbf{0}$. *A combination of the* $\boldsymbol{w}$*'s gives zero*! The $\boldsymbol{w}$'s could not be a basis—so we cannot have $n > m$.

If $m > n$ we exchange the $\boldsymbol{v}$'s and $\boldsymbol{w}$'s and repeat the same steps. The only way to avoid a contradiction is to have $m = n$. This completes the proof that $m = n$.

The number of basis vectors depends on the space—not on a particular basis. The number is the same for every basis, and it tells how many "degrees of freedom" the vector space allows. The dimension of $\mathbf{R}^n$ is n. We now introduce the important word ***dimension*** for other vector spaces too.

DEFINITION The ***dimension of a space*** is the number of vectors in every basis.

This matches our intuition. The line through $\boldsymbol{v} = (1, 5, 2)$ has dimension one. It is a subspace with one vector $\boldsymbol{v}$ in its basis. Perpendicular to that line is the plane $x + 5y + 2z = 0$. This plane has dimension 2. To prove it, we find a basis $(-5, 1, 0)$ and $(-2, 0, 1)$. The dimension is 2 because the basis contains two vectors.

The plane is the nullspace of the matrix $A = \begin{bmatrix} 1 & 5 & 2 \end{bmatrix}$, which has two free variables. Our basis vectors $(-5, 1, 0)$ and $(-2, 0, 1)$ are the "special solutions" to $A\boldsymbol{x} = \mathbf{0}$. The next section shows that the $n - r$ special solutions always give a basis for the nullspace. So $N(A)$ has dimension $n - r$. Here we emphasize only this: All bases for a space contain the same number of vectors.

Note about the language of linear algebra We never say "the rank of a space" or "the dimension of a basis" or "the basis of a matrix". Those terms have no meaning. It is the ***dimension of the column space*** that equals the ***rank of the matrix.***

Bases for Matrix Spaces and Function Spaces

The words "independence" and "basis" and "dimension" are not at all restricted to column vectors. We can ask whether three 3 by 4 matrices A_1, A_2, A_3 are independent. They are members of the space of all 3 by 4 matrices; some combination might give the zero matrix. We can also ask the dimension of that matrix space (it is 12).

In differential equations, the space of solutions to $d^2y/dx^2 = y$ contains functions. One basis is $y = e^x$ and $y = e^{-x}$. Counting the basis functions gives the dimension 2 (for the space of all solutions).

We think matrix spaces and function spaces are optional. Your class can go past this page—no problem. But in some way, you haven't got the ideas of basis and dimension straight until you can apply them to "vectors" other than column vectors.

Matrix spaces The vector space $\mathbf{M}$ contains all 2 by 2 matrices. Its dimension is 4.

$$\textbf{One basis is} \quad A_1, A_2, A_3, A_4 = \begin{bmatrix} 1 & 0 \\ 0 & 0 \end{bmatrix}, \begin{bmatrix} 0 & 1 \\ 0 & 0 \end{bmatrix}, \begin{bmatrix} 0 & 0 \\ 1 & 0 \end{bmatrix}, \begin{bmatrix} 0 & 0 \\ 0 & 1 \end{bmatrix}.$$

Those matrices are linearly independent. We are not looking at their columns, but at the whole matrix. Combinations of those four matrices can produce any matrix in **M**, so they span the space:

$$c_1A_1 + c_2A_2 + c_3A_3 + c_4A_4 = \begin{bmatrix} c_1 & c_2 \\ c_3 & c_4 \end{bmatrix}.$$

This is zero only if the c's are all zero—which proves independence.

The matrices A_1, A_2, A_4 are a basis for a subspace—the upper triangular matrices. Its dimension is 3. A_1 and A_4 are a basis for the diagonal matrices. What is a basis for the symmetric matrices? Keep A_1 and A_4, and throw in $A_2 + A_3$.

To push this further, think about the space of all n by n matrices. For a basis, choose matrices that have only a single nonzero entry (that entry is 1). There are n^2 positions for that 1, so there are n^2 basis matrices:

The dimension of the whole n by n matrix space is n^2.

The dimension of the subspace of *upper triangular* matrices is $\frac{1}{2}n^2 + \frac{1}{2}n$.

The dimension of the subspace of *diagonal* matrices is n.

The dimension of the subspace of *symmetric* matrices is $\frac{1}{2}n^2 + \frac{1}{2}n$.

Function spaces The equations $d^2y/dx^2 = 0$ and $d^2y/dx^2 = -y$ and $d^2y/dx^2 = y$ involve the second derivative. In calculus we solve to find the functions $y(x)$:

$$\begin{array}{ll} y'' = 0 & \text{is solved by any linear function } y = cx + d \\ y'' = -y & \text{is solved by any combination } y = c\sin x + d\cos x \\ y'' = y & \text{is solved by any combination } y = ce^x + de^{-x}. \end{array}$$

The second solution space has two basis functions: $\sin x$ and $\cos x$. The third solution space has basis functions e^x and e^{-x}. The first space has x and 1. It is the "nullspace" of the second derivative! The dimension is 2 in each case (these are second-order equations).

The solutions of $y'' = 2$ don't form a subspace—the right side $b = 2$ is not zero. A particular solution is $y(x) = x^2$. The complete solution is $y(x) = x^2 + cx + d$. All those functions satisfy $y'' = 2$. Notice the particular solution plus any function $cx + d$ in the nullspace. A linear differential equation is like a linear matrix equation $A\boldsymbol{x} = \boldsymbol{b}$. But we solve it by calculus instead of linear algebra.

We end here with the space **Z** that contains only the zero vector. The dimension of this space is *zero. The empty set* (containing no vectors at all) *is a basis.* We can never allow the zero vector into a basis, because then linear independence is lost.

■ REVIEW OF THE KEY IDEAS ■

1. The columns of A are ***independent*** if $\boldsymbol{x} = \mathbf{0}$ is the only solution to $A\boldsymbol{x} = \mathbf{0}$.

2. The vectors $\boldsymbol{v}_1, \ldots, \boldsymbol{v}_r$ ***span*** a space if their combinations fill that space.

3. ***A basis consists of linearly independent vectors that span the space.*** Every vector is a *unique* combination of the basis vectors.

4. All bases for a space have the same number of vectors. This number is the ***dimension*** of the space.

5. The pivot columns are a basis for the column space and the dimension is r.

■ WORKED EXAMPLES ■

3.5 A Start with the vectors $\boldsymbol{v}_1 = (1, 2, 0)$ and $\boldsymbol{v}_2 = (2, 3, 0)$. **(a)** Are they linearly independent? **(b)** Are they a basis for any space? **(c)** What space V do they span? **(d)** What is the dimension of that space? **(e)** Which matrices A have $\boldsymbol{V}$ as their column space? **(f)** Which matrices have $\boldsymbol{V}$ as their nullspace? **(g)** Describe all vectors $\boldsymbol{v}_3$ that complete a basis $\boldsymbol{v}_1, \boldsymbol{v}_2, \boldsymbol{v}_3$ for $\mathbf{R}^3$.

Solution

(a) $\boldsymbol{v}_1$ and $\boldsymbol{v}_2$ are independent—the only combination to give $\mathbf{0}$ is $0\boldsymbol{v}_1 + 0\boldsymbol{v}_2$.

(b) Yes, they are a basis for whatever space $\boldsymbol{V}$ they span.

(c) That space $\boldsymbol{V}$ contains all vectors $(x, y, 0)$. It is the xy plane in $\mathbf{R}^3$.

(d) The dimension of $\boldsymbol{V}$ is 2 since the basis contains two vectors.

(e) This $\boldsymbol{V}$ is the column space of any 3 by n matrix A of rank 2, if every column is a combination of $\boldsymbol{v}_1$ and $\boldsymbol{v}_2$. In particular A could just have columns $\boldsymbol{v}_1$ and $\boldsymbol{v}_2$.

(f) This V is the nullspace of any m by 3 matrix B of rank 1, if every row is a multiple of $(0, 0, 1)$. In particular take $B = [0\ 0\ 1]$. Then $B\boldsymbol{v}_1 = \mathbf{0}$ and $B\boldsymbol{v}_2 = \mathbf{0}$.

(g) Any third vector $\boldsymbol{v}_3 = (a, b, c)$ will complete a basis for $\mathbf{R}^3$ provided $c \neq 0$.

3.5 B Start with three independent vectors $\boldsymbol{w}_1, \boldsymbol{w}_2, \boldsymbol{w}_3$. Take combinations of those vectors to produce $\boldsymbol{v}_1, \boldsymbol{v}_2, \boldsymbol{v}_3$. Write the combinations in matrix form as $V = WM$:

$$\begin{aligned} \boldsymbol{v}_1 &= \boldsymbol{w}_1 + \boldsymbol{w}_2 \\ \boldsymbol{v}_2 &= \boldsymbol{w}_1 + 2\boldsymbol{w}_2 + \boldsymbol{w}_3 \\ \boldsymbol{v}_3 &= \phantom{\boldsymbol{w}_1 +{}} \boldsymbol{w}_2 + c\boldsymbol{w}_3 \end{aligned} \qquad \text{which is} \qquad \begin{bmatrix} \boldsymbol{v}_1 & \boldsymbol{v}_2 & \boldsymbol{v}_3 \end{bmatrix} = \begin{bmatrix} \boldsymbol{w}_1 & \boldsymbol{w}_2 & \boldsymbol{w}_3 \end{bmatrix} \begin{bmatrix} 1 & 1 & 0 \\ 1 & 2 & 1 \\ 0 & 1 & c \end{bmatrix}$$

What is the test on a matrix V to see if its columns are linearly independent? If $c \neq 1$ show that $\boldsymbol{v}_1, \boldsymbol{v}_2, \boldsymbol{v}_3$ are linearly independent. If $c = 1$ show that the $\boldsymbol{v}$'s are linearly *dependent.*

Solution The test on V for independence of its columns was in our first definition: *The nullspace of V must contain only the zero vector.* Then $\boldsymbol{x} = (0, 0, 0)$ is the only combination of the columns that gives $V\boldsymbol{x}$ = zero vector.

In $c = 1$ in our problem, we can see *dependence* in two ways. First, $\boldsymbol{v}_1 + \boldsymbol{v}_3$ will be the same as $\boldsymbol{v}_2$. (If you add $\boldsymbol{w}_1 + \boldsymbol{w}_2$ to $\boldsymbol{w}_2 + \boldsymbol{w}_3$ you get $\boldsymbol{w}_1 + 2\boldsymbol{w}_2 + \boldsymbol{w}_3$ which is $\boldsymbol{v}_2$.) In other words $\boldsymbol{v}_1 - \boldsymbol{v}_2 + \boldsymbol{v}_3 = \mathbf{0}$—which says that the $\boldsymbol{v}$'s are not independent.

The other way is to look at the nullspace of M. If $c = 1$, the vector $\boldsymbol{x} = (1, -1, 1)$ is in that nullspace, and $M\boldsymbol{x} = \mathbf{0}$. Then certainly $WM\boldsymbol{x} = \mathbf{0}$ which is the same as $V\boldsymbol{x} = 0$. So the $\boldsymbol{v}$'s are dependent. This specific $\boldsymbol{x} = (1, -1, 1)$ from the nullspace tells us again that $\boldsymbol{v}_1 - \boldsymbol{v}_2 + \boldsymbol{v}_3 = \mathbf{0}$.

Now suppose $c \neq 1$. Then the matrix M is invertible. So if $\boldsymbol{x}$ is *any nonzero vector* we know that $M\boldsymbol{x}$ is nonzero. Since the $\boldsymbol{w}$'s are given as independent, we further know that $WM\boldsymbol{x}$ is nonzero. Since $V = WM$, this says that $\boldsymbol{x}$ is *not* in the nullspace of V. In other words $\boldsymbol{v}_1, \boldsymbol{v}_2, \boldsymbol{v}_3$ are independent.

The general rule is "independent $\boldsymbol{v}$'s from independent $\boldsymbol{w}$'s when M is invertible". And if these vectors are in $\mathbf{R}^3$, they are not only independent—they are a basis for $\mathbf{R}^3$. "Basis of $\boldsymbol{v}$'s from basis of $\boldsymbol{w}$'s when the change of basis matrix M is invertible."

3.5 C Suppose $\boldsymbol{v}_1, \ldots, \boldsymbol{v}_n$ is a basis for $\mathbf{R}^n$ and the n by n matrix A is invertible. Show that $A\boldsymbol{v}_1, \ldots, A\boldsymbol{v}_n$ is also a basis for $\mathbf{R}^n$.

Solution In *matrix language*: Put the basis vectors $\boldsymbol{v}_1, \ldots, \boldsymbol{v}_n$ in the columns of an invertible(!) matrix V. Then $A\boldsymbol{v}_1, \ldots, A\boldsymbol{v}_n$ are the columns of AV. Since A is invertible, so is AV and its columns give a basis.

In *vector language*: Suppose $c_1A\boldsymbol{v}_1 + \cdots + c_nA\boldsymbol{v}_n = \mathbf{0}$. This is $A\boldsymbol{v} = \mathbf{0}$ with $\boldsymbol{v} = c_1\boldsymbol{v}_1 + \cdots + c_n\boldsymbol{v}_n$. Multiply by A^{-1} to get $\boldsymbol{v} = \mathbf{0}$. By linear independence of the $\boldsymbol{v}$'s, all $c_i = 0$. So the $A\boldsymbol{v}$'s are independent.

To show that the $A\boldsymbol{v}$'s span $\mathbf{R}^n$, solve $c_1A\boldsymbol{v}_1 + \cdots + c_nA\boldsymbol{v}_n = \boldsymbol{b}$ which is the same as $c_1\boldsymbol{v}_1 + \cdots + c_n\boldsymbol{v}_n = A^{-1}\boldsymbol{b}$. Since the $\boldsymbol{v}$'s are a basis, this must be solvable.

Problem Set 3.5

Questions 1–10 are about linear independence and linear dependence.

1 Show that $\boldsymbol{v}_1, \boldsymbol{v}_2, \boldsymbol{v}_3$ are independent but $\boldsymbol{v}_1, \boldsymbol{v}_2, \boldsymbol{v}_3, \boldsymbol{v}_4$ are dependent:

$$\boldsymbol{v}_1 = \begin{bmatrix} 1 \\ 0 \\ 0 \end{bmatrix} \quad \boldsymbol{v}_2 = \begin{bmatrix} 1 \\ 1 \\ 0 \end{bmatrix} \quad \boldsymbol{v}_3 = \begin{bmatrix} 1 \\ 1 \\ 1 \end{bmatrix} \quad \boldsymbol{v}_4 = \begin{bmatrix} 2 \\ 3 \\ 4 \end{bmatrix}.$$

Solve either $c_1\boldsymbol{v}_1 + c_2\boldsymbol{v}_2 + c_3\boldsymbol{v}_3 = \mathbf{0}$ or $A\boldsymbol{x} = \mathbf{0}$. The $\boldsymbol{v}$'s go in the columns of A.

2 (Recommended) Find the largest possible number of independent vectors among

$$v_1 = \begin{bmatrix} 1 \\ -1 \\ 0 \\ 0 \end{bmatrix} \quad v_2 = \begin{bmatrix} 1 \\ 0 \\ -1 \\ 0 \end{bmatrix} \quad v_3 = \begin{bmatrix} 1 \\ 0 \\ 0 \\ -1 \end{bmatrix} \quad v_4 = \begin{bmatrix} 0 \\ 1 \\ -1 \\ 0 \end{bmatrix} \quad v_5 = \begin{bmatrix} 0 \\ 1 \\ 0 \\ -1 \end{bmatrix} \quad v_6 = \begin{bmatrix} 0 \\ 0 \\ 1 \\ -1 \end{bmatrix}.$$

3 Prove that if $a = 0$ or $d = 0$ or $f = 0$ (3 cases), the columns of U are dependent:

$$U = \begin{bmatrix} a & b & c \\ 0 & d & e \\ 0 & 0 & f \end{bmatrix}.$$

4 If a, d, f in Question 3 are all nonzero, show that the only solution to $U\boldsymbol{x} = \boldsymbol{0}$ is $\boldsymbol{x} = \boldsymbol{0}$. Then U has independent columns.

5 Decide the dependence or independence of

(a) the vectors $(1, 3, 2)$ and $(2, 1, 3)$ and $(3, 2, 1)$

(b) the vectors $(1, -3, 2)$ and $(2, 1, -3)$ and $(-3, 2, 1)$.

6 Choose three independent columns of U. Then make two other choices. Do the same for A.

$$U = \begin{bmatrix} 2 & 3 & 4 & 1 \\ 0 & 6 & 7 & 0 \\ 0 & 0 & 0 & 9 \\ 0 & 0 & 0 & 0 \end{bmatrix} \quad \text{and} \quad A = \begin{bmatrix} 2 & 3 & 4 & 1 \\ 0 & 6 & 7 & 0 \\ 0 & 0 & 0 & 9 \\ 4 & 6 & 8 & 2 \end{bmatrix}.$$

7 If $\boldsymbol{w}_1, \boldsymbol{w}_2, \boldsymbol{w}_3$ are independent vectors, show that the differences $\boldsymbol{v}_1 = \boldsymbol{w}_2 - \boldsymbol{w}_3$ and $\boldsymbol{v}_2 = \boldsymbol{w}_1 - \boldsymbol{w}_3$ and $\boldsymbol{v}_3 = \boldsymbol{w}_1 - \boldsymbol{w}_2$ are *dependent*. Find a combination of the $\boldsymbol{v}$'s that gives zero.

8 If $\boldsymbol{w}_1, \boldsymbol{w}_2, \boldsymbol{w}_3$ are independent vectors, show that the sums $\boldsymbol{v}_1 = \boldsymbol{w}_2 + \boldsymbol{w}_3$ and $\boldsymbol{v}_2 = \boldsymbol{w}_1 + \boldsymbol{w}_3$ and $\boldsymbol{v}_3 = \boldsymbol{w}_1 + \boldsymbol{w}_2$ are *independent*. (Write $c_1\boldsymbol{v}_1 + c_2\boldsymbol{v}_2 + c_3\boldsymbol{v}_3 = \boldsymbol{0}$ in terms of the $\boldsymbol{w}$'s. Find and solve equations for the c's.)

9 Suppose $\boldsymbol{v}_1, \boldsymbol{v}_2, \boldsymbol{v}_3, \boldsymbol{v}_4$ are vectors in $\mathbf{R}^3$.

(a) These four vectors are dependent because ______.

(b) The two vectors $\boldsymbol{v}_1$ and $\boldsymbol{v}_2$ will be dependent if ______.

(c) The vectors $\boldsymbol{v}_1$ and $(0, 0, 0)$ are dependent because ______.

10 Find two independent vectors on the plane $x + 2y - 3z - t = 0$ in $\mathbf{R}^4$. Then find three independent vectors. Why not four? This plane is the nullspace of what matrix?

Questions 11–15 are about the space *spanned* by a set of vectors. Take all linear combinations of the vectors.

11 Describe the subspace of $\mathbf{R}^3$ (is it a line or plane or $\mathbf{R}^3$?) spanned by

(a) the two vectors $(1, 1, -1)$ and $(-1, -1, 1)$

(b) the three vectors $(0, 1, 1)$ and $(1, 1, 0)$ and $(0, 0, 0)$

(c) the columns of a 3 by 5 echelon matrix with 2 pivots

(d) all vectors with positive components.

12 The vector $\boldsymbol{b}$ is in the subspace spanned by the columns of A when there is a solution to ______. The vector $\boldsymbol{c}$ is in the row space of A when there is a solution to ______.

True or false: If the zero vector is in the row space, the rows are dependent.

13 Find the dimensions of these 4 spaces. Which two of the spaces are the same? (a) column space of A, (b) column space of U, (c) row space of A, (d) row space of U:

$$A = \begin{bmatrix} 1 & 1 & 0 \\ 1 & 3 & 1 \\ 3 & 1 & -1 \end{bmatrix} \quad \text{and} \quad U = \begin{bmatrix} 1 & 1 & 0 \\ 0 & 2 & 1 \\ 0 & 0 & 0 \end{bmatrix}.$$

14 Choose $\boldsymbol{x} = (x_1, x_2, x_3, x_4)$ in $\mathbf{R}^4$. It has 24 rearrangements like (x_2, x_1, x_3, x_4) and (x_4, x_3, x_1, x_2). Those 24 vectors, including $\boldsymbol{x}$ itself, span a subspace $\mathbf{S}$. Find specific vectors $\boldsymbol{x}$ so that the dimension of $\mathbf{S}$ is: (a) zero, (b) one, (c) three, (d) four.

15 $\boldsymbol{v} + \boldsymbol{w}$ and $\boldsymbol{v} - \boldsymbol{w}$ are combinations of $\boldsymbol{v}$ and $\boldsymbol{w}$. Write $\boldsymbol{v}$ and $\boldsymbol{w}$ as combinations of $\boldsymbol{v} + \boldsymbol{w}$ and $\boldsymbol{v} - \boldsymbol{w}$. The two pairs of vectors ______ the same space. When are they a basis for the same space?

Questions 16–26 are about the requirements for a basis.

16 If $\boldsymbol{v}_1, \ldots, \boldsymbol{v}_n$ are linearly independent, the space they span has dimension ______. These vectors are a ______ for that space. If the vectors are the columns of an m by n matrix, then m is ______ than n.

17 Find a basis for each of these subspaces of $\mathbf{R}^4$:

(a) All vectors whose components are equal.

(b) All vectors whose components add to zero.

(c) All vectors that are perpendicular to $(1, 1, 0, 0)$ and $(1, 0, 1, 1)$.

(d) The column space (in $\mathbf{R}^2$) and nullspace (in $\mathbf{R}^5$) of $U = \left[\begin{smallmatrix} 1 & 0 & 1 & 0 & 1 \\ 0 & 1 & 0 & 1 & 0 \end{smallmatrix}\right]$.

18 Find three different bases for the column space of U above. Then find two different bases for the row space of U.

19 Suppose $\boldsymbol{v}_1, \boldsymbol{v}_2, \ldots, \boldsymbol{v}_6$ are six vectors in $\mathbf{R}^4$.

(a) Those vectors (do)(do not)(might not) span $\mathbf{R}^4$.

(b) Those vectors (are)(are not)(might be) linearly independent.

(c) Any four of those vectors (are)(are not)(might be) a basis for $\mathbf{R}^4$.

20 The columns of A are n vectors from $\mathbf{R}^m$. If they are linearly independent, what is the rank of A? If they span $\mathbf{R}^m$, what is the rank? If they are a basis for $\mathbf{R}^m$, what then?

21 Find a basis for the plane $x - 2y + 3z = 0$ in $\mathbf{R}^3$. Then find a basis for the intersection of that plane with the xy plane. Then find a basis for all vectors perpendicular to the plane.

22 Suppose the columns of a 5 by 5 matrix A are a basis for $\mathbf{R}^5$.

(a) The equation $A\boldsymbol{x} = \mathbf{0}$ has only the solution $\boldsymbol{x} = \mathbf{0}$ because ____ .

(b) If $\boldsymbol{b}$ is in $\mathbf{R}^5$ then $A\boldsymbol{x} = \boldsymbol{b}$ is solvable because ____ .

Conclusion: A is invertible. Its rank is 5.

23 Suppose $\mathbf{S}$ is a 5-dimensional subspace of $\mathbf{R}^6$. True or false:

(a) Every basis for $\mathbf{S}$ can be extended to a basis for $\mathbf{R}^6$ by adding one more vector.

(b) Every basis for $\mathbf{R}^6$ can be reduced to a basis for $\mathbf{S}$ by removing one vector.

24 U comes from A by subtracting row 1 from row 3:

$$A = \begin{bmatrix} 1 & 3 & 2 \\ 0 & 1 & 1 \\ 1 & 3 & 2 \end{bmatrix} \quad \text{and} \quad U = \begin{bmatrix} 1 & 3 & 2 \\ 0 & 1 & 1 \\ 0 & 0 & 0 \end{bmatrix}.$$

Find bases for the two column spaces. Find bases for the two row spaces. Find bases for the two nullspaces.

25 True or false (give a good reason):

(a) If the columns of a matrix are dependent, so are the rows.

(b) The column space of a 2 by 2 matrix is the same as its row space.

(c) The column space of a 2 by 2 matrix has the same dimension as its row space.

(d) The columns of a matrix are a basis for the column space.

26 For which numbers c and d do these matrices have rank 2?

$$A = \begin{bmatrix} 1 & 2 & 5 & 0 & 5 \\ 0 & 0 & c & 2 & 2 \\ 0 & 0 & 0 & d & 2 \end{bmatrix} \quad \text{and} \quad B = \begin{bmatrix} c & d \\ d & c \end{bmatrix}.$$

Questions 27–32 are about spaces where the "vectors" are matrices.

27 Find a basis for each of these subspaces of 3 by 3 matrices:

(a) All diagonal matrices.

(b) All symmetric matrices ($A^{\mathrm{T}} = A$).

(c) All skew-symmetric matrices ($A^{\mathrm{T}} = -A$).

28 Construct six linearly independent 3 by 3 echelon matrices $U_1, \ldots, U_6$.

29 Find a basis for the space of all 2 by 3 matrices whose columns add to zero. Find a basis for the subspace whose rows also add to zero.

30 Show that the six 3 by 3 permutation matrices (Section 2.6) are linearly dependent.

31 What subspace of 3 by 3 matrices is spanned by

(a) all invertible matrices?

(b) all echelon matrices?

(c) the identity matrix?

32 Find a basis for the space of 2 by 3 matrices whose nullspace contains (2, 1, 1).

Questions 33–37 are about spaces where the "vectors" are functions.

33 (a) Find all functions that satisfy $\frac{dy}{dx} = 0$.

(b) Choose a particular function that satisfies $\frac{dy}{dx} = 3$.

(c) Find all functions that satisfy $\frac{dy}{dx} = 3$.

34 The cosine space $\mathbf{F}_3$ contains all combinations $y(x) = A\cos x + B\cos 2x + C\cos 3x$. Find a basis for the subspace with $y(0) = 0$.

35 Find a basis for the space of functions that satisfy

(a) $\frac{dy}{dx} - 2y = 0$

(b) $\frac{dy}{dx} - \frac{y}{x} = 0$.

36 Suppose $y_1(x), y_2(x), y_3(x)$ are three different functions of x. The vector space they span could have dimension 1, 2, or 3. Give an example of y_1, y_2, y_3 to show each possibility.

37 Find a basis for the space of polynomials $p(x)$ of degree ≤ 3. Find a basis for the subspace with $p(1) = 0$.

38 Find a basis for the space **S** of vectors (a, b, c, d) with $a+c+d = 0$ and also for the space **T** with $a+b = 0$ and $c = 2d$. What is the dimension of the intersection $\mathbf{S} \cap \mathbf{T}$?

39 Write the 3 by 3 identity matrix as a combination of the other five permutation matrices! Then show that those five matrices are linearly independent. (Assume a combination gives zero, and check entries to prove each term is zero.) The five permutations are a basis for the subspace of 3 by 3 matrices with row and column sums all equal.

40 If $AS = SA$ for the *shift matrix* S, show that A must have this special form:

$$\text{If} \quad \begin{bmatrix} a & b & c \\ d & e & f \\ g & h & i \end{bmatrix} \begin{bmatrix} 0 & 1 & 0 \\ 0 & 0 & 1 \\ 0 & 0 & 0 \end{bmatrix} = \begin{bmatrix} 0 & 1 & 0 \\ 0 & 0 & 1 \\ 0 & 0 & 0 \end{bmatrix} \begin{bmatrix} a & b & c \\ d & e & f \\ g & h & i \end{bmatrix} \quad \text{then } A = \begin{bmatrix} a & b & c \\ 0 & a & b \\ 0 & 0 & a \end{bmatrix}.$$

"The subspace of matrices that commute with the shift S has dimension _____."

41 Which of the following are bases for $\mathbf{R}^3$?

(a) $(1, 2, 0)$ and $(0, 1, -1)$

(b) $(1, 1, -1)$, $(2, 3, 4)$, $(4, 1, -1)$, $(0, 1, -1)$

(c) $(1, 2, 2)$, $(-1, 2, 1)$, $(0, 8, 0)$

(d) $(1, 2, 2)$, $(-1, 2, 1)$, $(0, 8, 6)$

42 Suppose A is 5 by 4 with rank 4. Show that $A\boldsymbol{x} = \boldsymbol{b}$ has no solution when the 5 by 5 matrix $[\, A \;\; \boldsymbol{b} \,]$ is invertible. Show that $A\boldsymbol{x} = \boldsymbol{b}$ is solvable when $[\, A \;\; \boldsymbol{b} \,]$ is singular.

DIMENSIONS OF THE FOUR SUBSPACES ■ 3.6

The main theorem in this chapter connects ***rank*** and ***dimension***. The ***rank*** of a matrix is the number of pivots. The ***dimension*** of a subspace is the number of vectors in a basis. We count pivots or we count basis vectors. *The rank of A reveals the dimensions of all four fundamental subspaces.* Here are the subspaces, including the new one.

Two subspaces come directly from A, and the other two from A^T:

Four Fundamental Subspaces

1. The ***row space*** is $C(A^T)$, a subspace of $\mathbf{R}^n$.

2. The ***column space*** is $C(A)$, a subspace of $\mathbf{R}^m$.

3. The ***nullspace*** is $N(A)$, a subspace of $\mathbf{R}^n$.

4. The ***left nullspace*** is $N(A^T)$, a subspace of $\mathbf{R}^m$. This is our new space.

In this book the column space and nullspace came first. We know $C(A)$ and $N(A)$ pretty well. Now the other two subspaces come forward. The row space contains all combinations of the rows. *This is the column space of* A^T.

For the left nullspace we solve $A^T\boldsymbol{y} = \mathbf{0}$—that system is n by m. *This is the nullspace of* A^T. The vectors $\boldsymbol{y}$ go on the *left* side of A when the equation is written as $\boldsymbol{y}^TA = \mathbf{0}^T$. The matrices A and A^T are usually different. So are their column spaces and their nullspaces. But those spaces are connected in an absolutely beautiful way.

Part 1 of the Fundamental Theorem finds the dimensions of the four subspaces. One fact stands out: ***The row space and column space have the same dimension*** r (the rank of the matrix). The other important fact involves the two nullspaces: *Their dimensions are* $n - r$ *and* $m - r$*, to make up the full dimensions* n *and* m.

Part 2 of the Fundamental Theorem will describe how the four subspaces fit together (two in $\mathbf{R}^n$ and two in $\mathbf{R}^m$). That completes the "right way" to understand $A\boldsymbol{x} = \boldsymbol{b}$. Stay with it—you are doing real mathematics.

The Four Subspaces for R

Suppose A is reduced to its row echelon form R. For that special form, the four subspaces are easy to identify. We will find a basis for each subspace and check its dimension. Then we watch how the subspaces change (or don't change!) as we look back at A. The main point is that *the four dimensions are the same for* A *and* R.

As a specific 3 by 5 example, look at the four subspaces for the echelon matrix R:

$$\begin{matrix} m = 3 \\ n = 5 \\ r = 2 \end{matrix} \qquad \begin{bmatrix} 1 & 3 & 5 & 0 & 9 \\ 0 & 0 & 0 & 1 & 8 \\ 0 & 0 & 0 & 0 & 0 \end{bmatrix} \qquad \begin{matrix} \textbf{pivot rows } 1 \textbf{ and } 2 \\ \\ \textbf{pivot columns } 1 \textbf{ and } 4 \end{matrix}$$

The rank of this matrix R is $r = 2$ (*two pivots*). Take the subspaces in order:

1. The ***row space*** of R has dimension 2, matching the rank.

Reason: The first two rows are a basis. The row space contains combinations of all three rows, but the third row (the zero row) adds nothing new. So rows 1 and 2 span the row space.

The pivot rows 1 and 2 are also independent. That is obvious for this example, and it is always true. If we look only at the pivot columns, we see the r by r identity matrix. There is no way to combine its rows to give the zero row (except by the combination with all coefficients zero). So the r pivot rows are independent and the dimension is r.

The dimension of the row space is r. The nonzero rows of R form a basis.

2. The ***column space*** of R also has dimension $r = 2$.

Reason: The pivot columns 1 and 4 form a basis. They are independent because they start with the r by r identity matrix. No combination of those pivot columns can give the zero column (except the combination with all coefficients zero). And they also span the column space. Every other (free) column is a combination of the pivot columns. Actually the combinations we need are the three special solutions:

Column 2 is 3 (column 1). The special solution is $(-3, 1, 0, 0, 0)$.

Column 3 is 5 (column 1). The special solution is $(-5, 0, 1, 0, 0,)$.

Column 5 is 9 (column 1) $+8$ (column 4). That solution is $(-9, 0, 0, -8, 1)$.

The pivot columns are independent, and they span, so they are a basis for $\boldsymbol{C}(A)$.
The dimension of the column space is r. The pivot columns form a basis.

3. The ***nullspace*** has dimension $n - r = 5 - 2$. There are $n - r = 3$ free variables. Here x_2, x_3, x_5 are free (no pivots in those columns). They yield the three special solutions to $R\boldsymbol{x} = \boldsymbol{0}$. Set a free variable to 1, and solve for x_1 and x_4:

$$s_2 = \begin{bmatrix} -3 \\ 1 \\ 0 \\ 0 \\ 0 \end{bmatrix} \quad s_3 = \begin{bmatrix} -5 \\ 0 \\ 1 \\ 0 \\ 0 \end{bmatrix} \quad s_5 = \begin{bmatrix} -9 \\ 0 \\ 0 \\ -8 \\ 1 \end{bmatrix} \quad \begin{matrix} R\boldsymbol{x} = \boldsymbol{0} \text{ has the} \\ \text{complete solution} \\ \boldsymbol{x} = x_2 s_2 + x_3 s_3 + x_5 s_5 \end{matrix} .$$

There is a special solution for each free variable. With n variables and r pivot variables, that leaves $n - r$ free variables and special solutions:

The nullspace has dimension $n - r$. ***The special solutions form a basis.***

The special solutions are independent, because they contain the identity matrix in rows 2, 3, 5. All solutions are combinations of special solutions, $\boldsymbol{x} = x_2 s_2 + x_3 s_3 + x_5 s_5$, because this gets x_2, x_3 and x_5 in the correct positions. Then the pivot variables x_1 and x_4 are totally determined by the equations $R\boldsymbol{x} = \boldsymbol{0}$.

4. The ***nullspace*** of R^{T} has dimension $m - r = 3 - 2$.

Reason: The equation $R^{\mathrm{T}}\boldsymbol{y} = \boldsymbol{0}$ looks for combinations of the columns of R^{T} *(the rows of R)* that produce zero. This equation $R^{\mathrm{T}}\boldsymbol{y} = \boldsymbol{0}$ or $\boldsymbol{y}^{\mathrm{T}}R = \boldsymbol{0}^{\mathrm{T}}$ is

$$\begin{array}{r}
y_1\,[1,\ 3,\ 5,\ 0,\ 9] \\
+y_2\,[0,\ 0,\ 0,\ 1,\ 8] \\
+y_3\,[0,\ 0,\ 0,\ 0,\ 0] \\
\hline
[0,\ 0,\ 0,\ 0,\ 0]
\end{array} \tag{1}$$

The solutions y_1, y_2, y_3 are pretty clear. We need $y_1 = 0$ and $y_2 = 0$. The variable y_3 is free (it can be anything). The nullspace of R^T contains all vectors $\boldsymbol{y} = (0, 0, y_3)$. It is the line of all multiples of the basis vector $(0, 0, 1)$.

In all cases R ends with $m - r$ zero rows. Every combination of these $m - r$ rows gives zero. These are the *only* combinations of the rows of R that give zero, because the pivot rows are linearly independent. So we can identify the left nullspace of R, which is the nullspace of R^T:

$R^T\boldsymbol{y} = \boldsymbol{0}$: ***The left nullspace has dimension*** $m - r$.
The solutions are $\boldsymbol{y} = (0, \ldots, 0, y_{r+1}, \ldots, y_m)$.

To produce a zero combination, $\boldsymbol{y}$ must start with r zeros. This leaves dimension $m - r$.

Why is this a "*left* nullspace"? The reason is that $R^T\boldsymbol{y} = \boldsymbol{0}$ can be transposed to $\boldsymbol{y}^T R = \boldsymbol{0}^T$. Now $\boldsymbol{y}^T$ is a row vector to the *left* of R. You see the y's in equation (1) multiplying the rows. This subspace came fourth, and some linear algebra books omit it—but that misses the beauty of the whole subject.

In $\mathbf{R}^n$ ***the row space and nullspace have dimensions*** r and $n - r$ (adding to n).
In $\mathbf{R}^m$ ***the column space and left nullspace have dimensions*** r and $m - r$ (total m).

So far this is proved for echelon matrices R. Figure 3.5 shows the same for A.

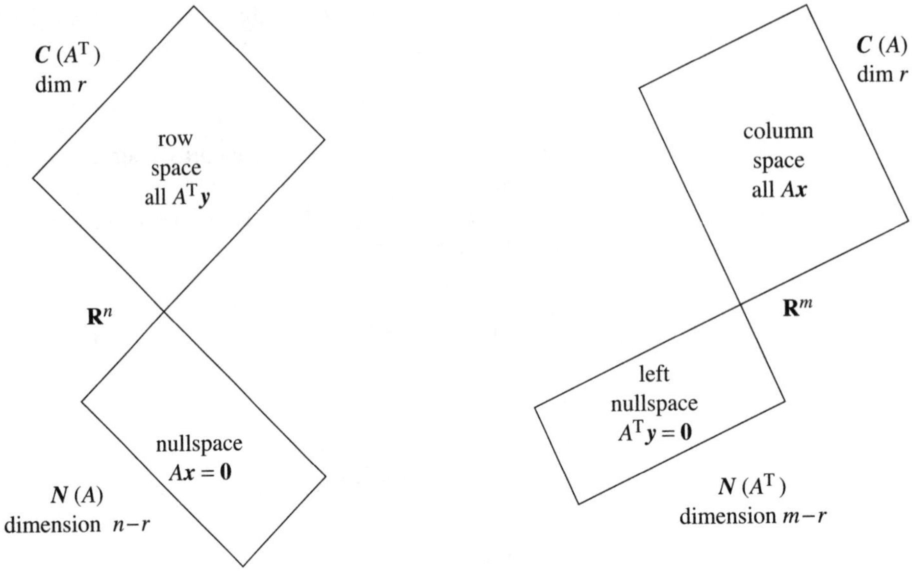

Figure 3.5 The dimensions of the four fundamental subspaces (for R and for A).

The Four Subspaces for A

We have a small job still to do. ***The subspace dimensions for A are the same as for R.*** The job is to explain why. Remember that those matrices are connected by an invertible matrix E (the product of all the elementary matrices that reduce A to R):

$$EA = R \quad \text{and} \quad A = E^{-1}R \tag{2}$$

1 ***A has the same row space as R.*** Same dimension r and same basis.

Reason: Every row of A is a combination of the rows of R. Also every row of R is a combination of the rows of A. In one direction the combinations are given by E^{-1}, in the other direction by E. Elimination changes the rows, but the row *spaces* are identical.

Since A has the same row space as R, we can choose the first r rows of R as a basis. Or we could choose r suitable rows of the original A. They might not always be the *first* r rows of A, because those could be dependent. The good r rows of A are the ones that end up as pivot rows in R.

2 ***The column space of A has dimension r.*** For every matrix this is essential: ***The number of independent columns equals the number of independent rows.***

Wrong reason: "A and R have the same column space." This is false. The columns of R often end in zeros. The columns of A don't often end in zeros. The column spaces are different, but their *dimensions* are the same—equal to r.

Right reason: The same combinations of the columns are zero, for A and R. Say that another way: $A\boldsymbol{x} = \mathbf{0}$ *exactly when* $R\boldsymbol{x} = \mathbf{0}$. So the r independent columns match.

Conclusion The r pivot columns of A are a basis for *its* column space.

3 ***A has the same nullspace as R.*** Same dimension $n - r$ and same basis.

Reason: The elimination steps don't change the solutions. The special solutions are a basis for this nullspace. There are $n - r$ free variables, so the dimension is $n - r$. Notice that $r + (n - r)$ equals n:

(dimension of column space) + (dimension of nullspace) = dimension of $\mathbf{R}^n$.

4 ***The left nullspace of A*** (the nullspace of A^{T}) ***has dimension m − r.***

Reason: A^{T} is just as good a matrix as A. When we know the dimensions for every A, we also know them for A^{T}. Its column space was proved to have dimension r. Since A^{T} is n by m, the "whole space" is now $\mathbf{R}^m$. The counting rule for A was $r + (n - r) = n$. The counting rule for A^{T} is $r + (m - r) = m$. So the nullspace of A^{T} has dimension $m - r$. We now have all details of the main theorem:

***Fundamental Theorem of Linear Algebra*, Part 1**

The column space and row space both have dimension r.
The nullspaces have dimensions n − r and m − r.

By concentrating on *spaces* of vectors, not on individual numbers or vectors, we get these clean rules. You will soon take them for granted—eventually they begin to look obvious. But if you write down an 11 by 17 matrix with 187 nonzero entries, we don't think most people would see why these facts are true:

$$\text{dimension of } \boldsymbol{C}(A) = \text{dimension of } \boldsymbol{C}(A^{\mathrm{T}})$$
$$\text{dimension of } \boldsymbol{C}(A) + \text{dimension of } \boldsymbol{N}(A) = 17.$$

Example 1 $A = [\,1 \;\; 2 \;\; 3\,]$ has $m = 1$ and $n = 3$ and rank $r = 1$.

The row space is a line in $\mathbf{R}^3$. The nullspace is the plane $A\boldsymbol{x} = x_1 + 2x_2 + 3x_3 = 0$. This plane has dimension 2 (which is $3 - 1$). The dimensions add to $1 + 2 = 3$.

The columns of this 1 by 3 matrix are in $\mathbf{R}^1$! The column space is all of $\mathbf{R}^1$. The left nullspace contains only the zero vector. The only solution to $A^{\mathrm{T}}\boldsymbol{y} = \mathbf{0}$ is $\boldsymbol{y} = \mathbf{0}$, the only combination of the row that gives the zero row. Thus $\boldsymbol{N}(A^{\mathrm{T}})$ is $\mathbf{Z}$, the zero space with dimension 0 (which is $m - r$). In $\mathbf{R}^m$ the dimensions add to $1 + 0 = 1$.

Example 2 $A = \begin{bmatrix} 1 & 2 & 3 \\ 1 & 2 & 3 \end{bmatrix}$ has $m = 2$ with $n = 3$ and rank $r = 1$.

The row space is the same line through $(1, 2, 3)$. The nullspace is the same plane $x_1 + 2x_2 + 3x_3 = 0$. Their dimensions still add to $1 + 2 = 3$.

The columns are multiples of the first column $(1, 1)$. But there is more than the zero vector in the left nullspace. The first row minus the second row is the zero row. Therefore $A^{\mathrm{T}}\boldsymbol{y} = \mathbf{0}$ has the solution $\boldsymbol{y} = (1, -1)$. The column space and left nullspace are perpendicular lines in $\mathbf{R}^2$. Their dimensions are 1 and 1, adding to 2:

$$\text{column space} = \text{line through } \begin{bmatrix} 1 \\ 1 \end{bmatrix} \qquad \text{left nullspace} = \text{line through } \begin{bmatrix} 1 \\ -1 \end{bmatrix}.$$

If A has three equal rows, its rank is ______ . What are two of the $\boldsymbol{y}$'s in its left nullspace? ***The*** $\boldsymbol{y}$**'s** ***combine the rows to give the zero row.***

Matrices of Rank One

That last example had rank $r = 1$—and rank one matrices are special. We can describe them all. You will see again that dimension of row space = dimension of column space. When $r = 1$, every row is a multiple of the same row:

$$A = \begin{bmatrix} 1 & 2 & 3 \\ 2 & 4 & 6 \\ -3 & -6 & -9 \\ 0 & 0 & 0 \end{bmatrix} \quad \text{equals} \quad \begin{bmatrix} 1 \\ 2 \\ -3 \\ 0 \end{bmatrix} \quad \text{times} \quad \begin{bmatrix} 1 & 2 & 3 \end{bmatrix}.$$

A column times a row (4 by 1 times 1 by 3) produces a matrix (4 by 3). All rows are multiples of the row $(1, 2, 3)$. All columns are multiples of the column $(1, 2, -3, 0)$. The row space is a line in $\mathbf{R}^n$, and the column space is a line in $\mathbf{R}^m$.

Every rank one matrix has the special form $A = \boldsymbol{u}\,\boldsymbol{v}^{\mathrm{T}}$ ***= column times row.***

The columns are multiples of $\boldsymbol{u}$. The rows are multiples of $\boldsymbol{v}^{\mathrm{T}}$. *The nullspace is the plane perpendicular to* $\boldsymbol{v}$. ($A\boldsymbol{x} = \mathbf{0}$ means that $\boldsymbol{u}(\boldsymbol{v}^{\mathrm{T}}\boldsymbol{x}) = \mathbf{0}$ and then $\boldsymbol{v}^{\mathrm{T}}\boldsymbol{x} = 0$.) It is this perpendicularity of the subspaces that will be Part 2 of the Fundamental Theorem.

■ REVIEW OF THE KEY IDEAS ■

1. The r pivot rows of R are a basis for the row spaces of R and A (same space).
2. The r pivot columns of A (!) are a basis for its column space.
3. The $n-r$ special solutions are a basis for the nullspaces of A and R (same space).
4. The last $m - r$ rows of I are a basis for the left nullspace of R.
5. The last $m - r$ rows of E are a basis for the left nullspace of A.

■ WORKED EXAMPLES ■

3.6 A Find bases and dimensions for all four fundamental subspaces if you know that

$$A = \begin{bmatrix} 1 & 0 & 0 \\ 2 & 1 & 0 \\ 5 & 0 & 1 \end{bmatrix} \begin{bmatrix} 1 & 3 & 0 & 5 \\ 0 & 0 & 1 & 6 \\ 0 & 0 & 0 & 0 \end{bmatrix} = E^{-1}R.$$

By changing only *one* of those numbers, change the dimensions of all four subspaces.

Solution This matrix has pivots in columns 1 and 3. Its rank is $r = 2$.

Row space:	Basis $(1, 3, 0, 5)$ and $(0, 0, 1, 6)$ from R. Dimension 2.
Column space:	Basis $(1, 2, 5)$ and $(0, 1, 0)$ from E^{-1}. Dimension 2.
Nullspace:	Basis $(-3, 1, 0, 0)$ and $(-5, 0, -6, 1)$ from R. Dimension 2.
Nullspace of A^{T}:	Basis $(-5, 0, 1)$ from row 3 of E. Dimension $3 - 2 = 1$.

We need to comment on that left nullspace $N(A^{\mathrm{T}})$. $EA = R$ says that the last row of E combines the three rows of A into the zero row of R. So that last row of E is a basis vector for the left nullspace. If R had *two* zero rows, then the last *two* rows of E would be a basis for the left nullspace (which combines rows of A to give zero rows).

To change all these dimensions we need to change the rank r. The way to do that is to change an entry (any entry) in the last row of R.

3.6 B Suppose you have to put four 1's into a 5 by 6 matrix (all other entries are zero). Describe all the ways to make the dimension of its *row space* as small as possible. Describe all the ways to make the dimension of its *column space* as small as possible. Describe all the ways to make the dimension of its *nullspace* as small as possible. What are those smallest dimensions? What to do if you want the *sum of the dimensions of all four subspaces* as small as possible?

Solution The rank is 1 if the four 1's go into the same row, or into the same column, or into *two rows and two columns* (so $a_{ii} = a_{ij} = a_{ji} = a_{jj} = 1$). Since the column space and row space always have the same dimensions, this answers the first two questions: Dimension 1.

The nullspace has its smallest possible dimension $6 - 4 = 2$ when the rank is $r = 4$. To achieve rank 4, the four 1's must go into four different rows and four different columns. You can't do anything about the sum $r+(n-r)+r+(m-r) = n+m$. It will be $6 + 5 = 11$ no matter how the 1's are placed. The sum is 11 even if there aren't any 1's...

If all the other entries of A are 2's instead of 0's, how do these answers change?

Problem Set 3.6

1 (a) If a 7 by 9 matrix has rank 5, what are the dimensions of the four subspaces? What is the sum of all four dimensions?

(b) If a 3 by 4 matrix has rank 3, what are its column space and left nullspace?

2 Find bases for the four subspaces associated with A and B:

$$A = \begin{bmatrix} 1 & 2 & 4 \\ 2 & 4 & 8 \end{bmatrix} \quad \text{and} \quad B = \begin{bmatrix} 1 & 2 & 4 \\ 2 & 5 & 8 \end{bmatrix}.$$

3 Find a basis for each of the four subspaces associated with

$$A = \begin{bmatrix} 0 & 1 & 2 & 3 & 4 \\ 0 & 1 & 2 & 4 & 6 \\ 0 & 0 & 0 & 1 & 2 \end{bmatrix} = \begin{bmatrix} 1 & 0 & 0 \\ 1 & 1 & 0 \\ 0 & 1 & 1 \end{bmatrix} \begin{bmatrix} 0 & 1 & 2 & 3 & 4 \\ 0 & 0 & 0 & 1 & 2 \\ 0 & 0 & 0 & 0 & 0 \end{bmatrix}.$$

4 Construct a matrix with the required property or explain why this is impossible:

(a) Column space contains $\begin{bmatrix} 1 \\ 1 \\ 0 \end{bmatrix}, \begin{bmatrix} 0 \\ 0 \\ 1 \end{bmatrix}$, row space contains $\begin{bmatrix} 1 \\ 2 \end{bmatrix}, \begin{bmatrix} 2 \\ 5 \end{bmatrix}$.

(b) Column space has basis $\begin{bmatrix} 1 \\ 1 \\ 3 \end{bmatrix}$, nullspace has basis $\begin{bmatrix} 3 \\ 1 \\ 1 \end{bmatrix}$.

(c) Dimension of nullspace $= 1 +$ dimension of left nullspace.

(d) Left nullspace contains $\begin{bmatrix} 1 \\ 3 \end{bmatrix}$, row space contains $\begin{bmatrix} 3 \\ 1 \end{bmatrix}$.

(e) Row space $=$ column space, nullspace $\neq$ left nullspace.

5 If $\mathbf{V}$ is the subspace spanned by $(1, 1, 1)$ and $(2, 1, 0)$, find a matrix A that has $\mathbf{V}$ as its row space and a matrix B that has $\mathbf{V}$ as its nullspace.

6 Without elimination, find dimensions and bases for the four subspaces for

$$A = \begin{bmatrix} 0 & 3 & 3 & 3 \\ 0 & 0 & 0 & 0 \\ 0 & 1 & 0 & 1 \end{bmatrix} \quad \text{and} \quad B = \begin{bmatrix} 1 \\ 4 \\ 5 \end{bmatrix}.$$

7 Suppose the 3 by 3 matrix A is invertible. Write down bases for the four subspaces for A, and also for the 3 by 6 matrix $B = [\,A \;\; A\,]$.

8 What are the dimensions of the four subspaces for A, B, and C, if I is the 3 by 3 identity matrix and 0 is the 3 by 2 zero matrix?

$$A = \begin{bmatrix} I & 0 \end{bmatrix} \quad \text{and} \quad B = \begin{bmatrix} I & I \\ 0^{\mathrm{T}} & 0^{\mathrm{T}} \end{bmatrix} \quad \text{and} \quad C = \begin{bmatrix} 0 \end{bmatrix}.$$

9 Which subspaces are the same for these matrices of different sizes?

(a) $[\,A\,]$ and $\begin{bmatrix} A \\ A \end{bmatrix}$ (b) $\begin{bmatrix} A \\ A \end{bmatrix}$ and $\begin{bmatrix} A & A \\ A & A \end{bmatrix}$.

Prove that all three matrices have the same rank r.

10 If the entries of a 3 by 3 matrix are chosen randomly between 0 and 1, what are the most likely dimensions of the four subspaces? What if the matrix is 3 by 5?

11 (Important) A is an m by n matrix of rank r. Suppose there are right sides $\boldsymbol{b}$ for which $A\boldsymbol{x} = \boldsymbol{b}$ has *no solution.*

(a) What are all inequalities ($<$ or $\leq$) that must be true between m, n, and r?

(b) How do you know that $A^{\mathrm{T}}\boldsymbol{y} = \mathbf{0}$ has solutions other than $\boldsymbol{y} = \mathbf{0}$?

12 Construct a matrix with $(1, 0, 1)$ and $(1, 2, 0)$ as a basis for its row space and its column space. Why can't this be a basis for the row space and nullspace?

13 True or false:

(a) If $m = n$ then the row space of A equals the column space.

(b) The matrices A and $-A$ share the same four subspaces.

(c) If A and B share the same four subspaces then A is a multiple of B.

14 Without computing A, find bases for the four fundamental subspaces:

$$A = \begin{bmatrix} 1 & 0 & 0 \\ 6 & 1 & 0 \\ 9 & 8 & 1 \end{bmatrix} \begin{bmatrix} 1 & 2 & 3 & 4 \\ 0 & 1 & 2 & 3 \\ 0 & 0 & 1 & 2 \end{bmatrix}.$$

15 If you exchange the first two rows of A, which of the four subspaces stay the same? If $\boldsymbol{v} = (1, 2, 3, 4)$ is in the column space of A, write down a vector in the column space of the new matrix.

16 Explain why $\boldsymbol{v} = (1, 0, -1)$ cannot be a row of A and also be in the nullspace.

17 Describe the four subspaces of $\mathbf{R}^3$ associated with

$$A = \begin{bmatrix} 0 & 1 & 0 \\ 0 & 0 & 1 \\ 0 & 0 & 0 \end{bmatrix} \quad \text{and} \quad I + A = \begin{bmatrix} 1 & 1 & 0 \\ 0 & 1 & 1 \\ 0 & 0 & 1 \end{bmatrix}.$$

18 (Left nullspace) Add the extra column $\boldsymbol{b}$ and reduce A to echelon form:

$$\begin{bmatrix} A & \boldsymbol{b} \end{bmatrix} = \begin{bmatrix} 1 & 2 & 3 & b_1 \\ 4 & 5 & 6 & b_2 \\ 7 & 8 & 9 & b_3 \end{bmatrix} \quad \to \quad \begin{bmatrix} 1 & 2 & 3 & b_1 \\ 0 & -3 & -6 & b_2 - 4b_1 \\ 0 & 0 & 0 & b_3 - 2b_2 + b_1 \end{bmatrix}.$$

A combination of the rows of A has produced the zero row. What combination is it? (Look at $b_3 - 2b_2 + b_1$ on the right side.) Which vectors are in the nullspace of A^{T} and which are in the nullspace of A?

19 Following the method of Problem 18, reduce A to echelon form and look at zero rows. The $\boldsymbol{b}$ column tells which combinations you have taken of the rows:

(a) $\begin{bmatrix} 1 & 2 & b_1 \\ 3 & 4 & b_2 \\ 4 & 6 & b_3 \end{bmatrix}$ (b) $\begin{bmatrix} 1 & 2 & b_1 \\ 2 & 3 & b_2 \\ 2 & 4 & b_3 \\ 2 & 5 & b_4 \end{bmatrix}$

>From the $\boldsymbol{b}$ column after elimination, read off $m - r$ basis vectors in the left nullspace of A (combinations of rows that give zero).

20 (a) Describe all solutions to $A\boldsymbol{x} = \boldsymbol{0}$ if

$$A = \begin{bmatrix} 1 & 0 & 0 \\ 2 & 1 & 0 \\ 3 & 4 & 1 \end{bmatrix} \begin{bmatrix} 4 & 2 & 0 & 1 \\ 0 & 0 & 1 & 3 \\ 0 & 0 & 0 & 0 \end{bmatrix}.$$

(b) How many independent solutions are there to $A^{\mathrm{T}}\boldsymbol{y} = \boldsymbol{0}$?

(c) Give a basis for the column space of A.

21 Suppose A is the sum of two matrices of rank one: $A = \boldsymbol{u}\boldsymbol{v}^{\mathrm{T}} + \boldsymbol{w}\boldsymbol{z}^{\mathrm{T}}$.

(a) Which vectors span the column space of A?

(b) Which vectors span the row space of A?

(c) The rank is less than 2 if _____ or if _____.

(d) Compute A and its rank if $\boldsymbol{u} = \boldsymbol{z} = (1, 0, 0)$ and $\boldsymbol{v} = \boldsymbol{w} = (0, 0, 1)$.

22 Construct $A = \boldsymbol{u}\boldsymbol{v}^{\mathrm{T}} + \boldsymbol{w}\boldsymbol{z}^{\mathrm{T}}$ whose column space has basis $(1, 2, 4), (2, 2, 1)$ and whose row space has basis $(1, 0, 0), (0, 1, 1)$.

23 Without multiplying matrices, find bases for the row and column spaces of A:

$$A = \begin{bmatrix} 1 & 2 \\ 4 & 5 \\ 2 & 7 \end{bmatrix} \begin{bmatrix} 3 & 0 & 3 \\ 1 & 1 & 2 \end{bmatrix}.$$

How do you know from these shapes that A is not invertible?

24 $A^{\mathrm{T}}\boldsymbol{y} = \boldsymbol{d}$ is solvable when the right side $\boldsymbol{d}$ is in which of the four subspaces? The solution is unique when the _____ contains only the zero vector.

25 True or false (with a reason or a counterexample):

(a) A and A^{T} have the same number of pivots.

(b) A and A^{T} have the same left nullspace.

(c) If the row space equals the column space then $A^{\mathrm{T}} = A$.

(d) If $A^{\mathrm{T}} = -A$ then the row space of A equals the column space.

26 (*Rank of AB*) If $AB = C$, the rows of C are combinations of the rows of _____. So the rank of C is not greater than the rank of _____. Since $B^{\mathrm{T}}A^{\mathrm{T}} = C^{\mathrm{T}}$, the rank of C is also not greater than the rank of _____.

27 If a, b, c are given with $a \neq 0$, how would you choose d so that $A = \left[\begin{smallmatrix} a & b \\ c & d \end{smallmatrix}\right]$ has rank one? Find a basis for the row space and nullspace.

28 Find the ranks of the 8 by 8 checkerboard matrix B and chess matrix C:

$$B = \begin{bmatrix} 1 & 0 & 1 & 0 & 1 & 0 & 1 & 0 \\ 0 & 1 & 0 & 1 & 0 & 1 & 0 & 1 \\ 1 & 0 & 1 & 0 & 1 & 0 & 1 & 0 \\ \bullet & \bullet & \bullet & \bullet & \bullet & \bullet & \bullet & \bullet \\ 0 & 1 & 0 & 1 & 0 & 1 & 0 & 1 \end{bmatrix} \quad \text{and} \quad C = \begin{bmatrix} r & n & b & q & k & b & n & r \\ p & p & p & p & p & p & p & p \\ & & & \text{four zero rows} & & & & \\ p & p & p & p & p & p & p & p \\ r & n & b & q & k & b & n & r \end{bmatrix}$$

The numbers r, n, b, q, k, p are all different. Find bases for the row space and left nullspace of B and C. Challenge problem: Find a basis for the nullspace of C.

29 Can tic-tac-toe be completed (5 ones and 4 zeros in A) so that rank $(A) = 2$ but neither side passed up a winning move?

4

ORTHOGONALITY

ORTHOGONALITY OF THE FOUR SUBSPACES ■ 4.1

Two vectors are orthogonal when their dot product is zero: $\boldsymbol{v} \cdot \boldsymbol{w} = 0$ or $\boldsymbol{v}^{\mathrm{T}}\boldsymbol{w} = 0$. This chapter moves up a level, from orthogonal vectors to ***orthogonal subspaces***. Orthogonal means the same as perpendicular.

Subspaces entered Chapter 3 with a specific purpose—to throw light on $A\boldsymbol{x} = \boldsymbol{b}$. Right away we needed the column space (for $\boldsymbol{b}$) and the nullspace (for $\boldsymbol{x}$). Then the light turned onto A^{T}, uncovering two more subspaces. Those four fundamental subspaces reveal what a matrix really does.

A matrix multiplies a vector: *A times* $\boldsymbol{x}$. At the first level this is only numbers. At the second level $A\boldsymbol{x}$ is a combination of column vectors. The third level shows subspaces. But I don't think you have seen the whole picture until you study Figure 4.1. It fits the subspaces together, to show the hidden reality of A times $\boldsymbol{x}$. The 90° angles between subspaces are new—and we have to say what they mean.

The row space is perpendicular to the nullspace. Every row of A is perpendicular to every solution of $A\boldsymbol{x} = \boldsymbol{0}$. That gives the 90° angle on the left side of the figure. This perpendicularity of subspaces is Part 2 of the Fundamental Theorem of Linear Algebra.

May we add a word about the left nullspace? It is never reached by $A\boldsymbol{x}$, so it might seem useless. But when $\boldsymbol{b}$ is outside the column space—when we want to solve $A\boldsymbol{x} = \boldsymbol{b}$ and can't do it—then this nullspace of A^{T} comes into its own. It contains the error in the "least-squares" solution. That is the key application of linear algebra in this chapter.

Part 1 of the Fundamental Theorem gave the dimensions of the subspaces. The row and column spaces have the same dimension r (they are drawn the same size). The two nullspaces have the remaining dimensions $n - r$ and $m - r$. Now we will show that ***the row space and nullspace are orthogonal subspaces inside*** $\mathbf{R}^n$.

DEFINITION Two subspaces $\boldsymbol{V}$ and $\boldsymbol{W}$ of a vector space are ***orthogonal*** if every vector $\boldsymbol{v}$ in $\boldsymbol{V}$ is perpendicular to every vector $\boldsymbol{w}$ in $\boldsymbol{W}$:

$$\boldsymbol{v} \cdot \boldsymbol{w} = 0 \quad \textit{or} \quad \boldsymbol{v}^{\mathrm{T}}\boldsymbol{w} = 0 \quad \textit{for all } \boldsymbol{v} \textit{ in } \boldsymbol{V} \textit{ and all } \boldsymbol{w} \textit{ in } \boldsymbol{W}.$$

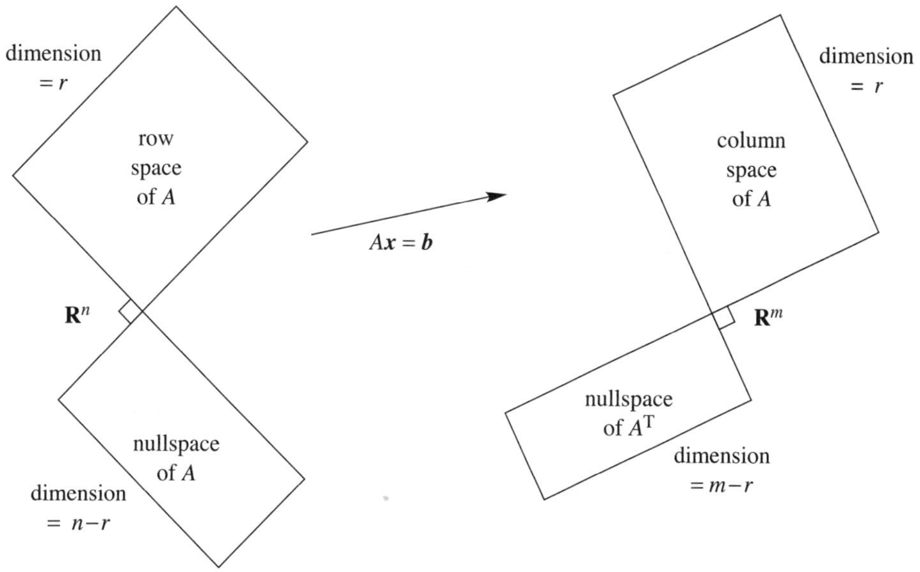

Figure 4.1 Two pairs of orthogonal subspaces. Dimensions add to n and add to m.

Example 1 The floor of your room (extended to infinity) is a subspace $\boldsymbol{V}$. The line where two walls meet is a subspace $\boldsymbol{W}$ (one-dimensional). Those subspaces are orthogonal. Every vector up the meeting line is perpendicular to every vector in the floor. The origin $(0, 0, 0)$ is in the corner. We assume you don't live in a tent.

Example 2 Suppose $\boldsymbol{V}$ is still the floor but $\boldsymbol{W}$ is a wall (a two-dimensional space). The wall and floor look like orthogonal subspaces but they are not! You can find vectors in $\boldsymbol{V}$ and $\boldsymbol{W}$ that are not perpendicular. In fact a vector running along the bottom of the wall is also in the floor. This vector is in both $\boldsymbol{V}$ and $\boldsymbol{W}$—and it is not perpendicular to itself.

When a vector is in two orthogonal subspaces, it *must* be zero. It is perpendicular to itself. It is $\boldsymbol{v}$ and it is $\boldsymbol{w}$, so $\boldsymbol{v}^{\mathrm{T}}\boldsymbol{v} = 0$. This has to be the zero vector.

The crucial examples for linear algebra come from the fundamental subspaces. Zero is the only point where the nullspace meets the row space. The spaces meet at 90°.

4A Every vector $\boldsymbol{x}$ in the nullspace of A is perpendicular to every row of A, because $A\boldsymbol{x} = \boldsymbol{0}$. ***The nullspace and row space are orthogonal subspaces.***

To see why $\boldsymbol{x}$ is perpendicular to the rows, look at $A\boldsymbol{x} = \mathbf{0}$. Each row multiplies $\boldsymbol{x}$:

$$A\boldsymbol{x} = \begin{bmatrix} & \text{row } 1 & \\ & \vdots & \\ & \text{row } m & \end{bmatrix} \begin{bmatrix} \\ \boldsymbol{x} \\ \\ \end{bmatrix} = \begin{bmatrix} 0 \\ \vdots \\ 0 \end{bmatrix}. \tag{1}$$

The first equation says that row 1 is perpendicular to $\boldsymbol{x}$. The last equation says that row m is perpendicular to $\boldsymbol{x}$. *Every row has a zero dot product with* $\boldsymbol{x}$. Then $\boldsymbol{x}$ is perpendicular to every combination of the rows. The whole row space $\boldsymbol{C}(A^{\mathrm{T}})$ is orthogonal to the whole nullspace $\boldsymbol{N}(A)$.

Here is a second proof of that orthogonality for readers who like matrix shorthand. The vectors in the row space are combinations $A^{\mathrm{T}}\boldsymbol{y}$ of the rows. Take the dot product of $A^{\mathrm{T}}\boldsymbol{y}$ with any $\boldsymbol{x}$ in the nullspace. *These vectors are perpendicular*:

$$\boldsymbol{x}^{\mathrm{T}}(A^{\mathrm{T}}\boldsymbol{y}) = (A\boldsymbol{x})^{\mathrm{T}}\boldsymbol{y} = \mathbf{0}^{\mathrm{T}}\boldsymbol{y} = 0. \tag{2}$$

We like the first proof. You can see those rows of A multiplying $\boldsymbol{x}$ to produce zeros in equation (1). The second proof shows why A and A^{T} are both in the Fundamental Theorem. A^{T} goes with $\boldsymbol{y}$ and A goes with $\boldsymbol{x}$. At the end we used $A\boldsymbol{x} = \mathbf{0}$.

Example 3 The rows of A are perpendicular to $\boldsymbol{x} = (1, 1, -1)$ in the nullspace:

$$A\boldsymbol{x} = \begin{bmatrix} 1 & 3 & 4 \\ 5 & 2 & 7 \end{bmatrix} \begin{bmatrix} 1 \\ 1 \\ -1 \end{bmatrix} = \begin{bmatrix} 0 \\ 0 \end{bmatrix} \quad \text{gives the dot products} \quad \begin{matrix} 1 + 3 - 4 = 0 \\ 5 + 2 - 7 = 0 \end{matrix}$$

Now we turn to the other two subspaces. In this example, the column space is all of $\mathbf{R}^2$. The nullspace of A^{T} is only the zero vector. Those two subspaces are also orthogonal.

4B Every vector $\boldsymbol{y}$ in the nullspace of A^{T} is perpendicular to every column of A. ***The left nullspace and the column space are orthogonal in*** $\mathbf{R}^m$.

Apply the original proof to A^{T}. Its nullspace is orthogonal to its row space—and the row space of A^{T} is the column space of A. Q.E.D.

For a visual proof, look at $A^{\mathrm{T}}\boldsymbol{y} = \mathbf{0}$. Each column of A multiplies $\boldsymbol{y}$ to give 0:

$$A^{\mathrm{T}}\boldsymbol{y} = \begin{bmatrix} (\text{column } 1)^{\mathrm{T}} \\ \cdots \\ (\text{column } n)^{\mathrm{T}} \end{bmatrix} \begin{bmatrix} \\ \boldsymbol{y} \\ \\ \end{bmatrix} = \begin{bmatrix} 0 \\ \cdots \\ 0 \end{bmatrix}. \tag{3}$$

The dot product of $\boldsymbol{y}$ with every column of A is zero. Then $\boldsymbol{y}$ in the left nullspace is perpendicular to each column—and to the whole column space.

Orthogonal Complements

Very Important The fundamental subspaces are more than just orthogonal (in pairs). Their dimensions are also right. Two lines could be perpendicular in $\mathbf{R}^3$, but they *could not be* the row space and nullspace of a 3 by 3 matrix. The lines have dimensions 1 and 1, adding to 2. The correct dimensions r and $n-r$ must add to $n=3$. The fundamental subspaces have dimensions 2 and 1, or 3 and 0. The subspaces are not only orthogonal, they are *orthogonal complements.*

DEFINITION The ***orthogonal complement*** of $\boldsymbol{V}$ contains *every* vector that is perpendicular to $\boldsymbol{V}$. This orthogonal subspace is denoted by $\boldsymbol{V}^{\perp}$ (pronounced "$\boldsymbol{V}$ perp").

By this definition, the nullspace is the orthogonal complement of the row space. *Every* $\boldsymbol{x}$ that is perpendicular to the rows satisfies $A\boldsymbol{x}=\mathbf{0}$.

The reverse is also true (automatically). If $\boldsymbol{v}$ is orthogonal to the nullspace, it must be in the row space. Otherwise we could add this $\boldsymbol{v}$ as an extra row of the matrix, without changing its nullspace. The row space would grow, which breaks the law $r+(n-r)=n$. We conclude that $\boldsymbol{N}(A)^{\perp}$ is exactly the row space $\boldsymbol{C}(A^{\mathrm{T}})$.

The left nullspace and column space are not only orthogonal in $\mathbf{R}^m$, they are also orthogonal complements. Their dimensions add to the full dimension m.

***Fundamental Theorem of Linear Algebra*, Part 2**

The nullspace is the orthogonal complement of the row space (in $\mathbf{R}^n$).
The left nullspace is the orthogonal complement of the column space (in $\mathbf{R}^m$).

Part 1 gave the dimensions of the subspaces. Part 2 gives the 90° angles between them. The point of "complements" is that every $\boldsymbol{x}$ can be split into a *row space component* $\boldsymbol{x}_r$ and a *nullspace component* $\boldsymbol{x}_n$. When A multiplies $\boldsymbol{x}=\boldsymbol{x}_r+\boldsymbol{x}_n$, Figure 4.2 shows what happens:

The nullspace component goes to zero: $A\boldsymbol{x}_n=\mathbf{0}$.

The row space component goes to the column space: $A\boldsymbol{x}_r=A\boldsymbol{x}$.

Every vector goes to the column space! Multiplying by A cannot do anything else. But more than that: *Every vector in the column space comes from one and only one vector* $\boldsymbol{x}_r$ *in the row space.* Proof: If $A\boldsymbol{x}_r=A\boldsymbol{x}'_r$, the difference $\boldsymbol{x}_r-\boldsymbol{x}'_r$ is in the nullspace. It is also in the row space, where $\boldsymbol{x}_r$ and $\boldsymbol{x}'_r$ came from. This difference must be the zero vector, because the spaces are perpendicular. Therefore $\boldsymbol{x}_r=\boldsymbol{x}'_r$.

There is an r by r invertible matrix hiding inside A, if we throw away the two nullspaces. From the row space to the column space, A is invertible. The "pseudoinverse" will invert it in Section 7.4.

Example 4 Every diagonal matrix has an r by r invertible submatrix:

$$A=\begin{bmatrix}3&0&0&0&0\\0&5&0&0&0\\0&0&0&0&0\end{bmatrix} \quad \text{contains} \quad \begin{bmatrix}3&0\\0&5\end{bmatrix}.$$

The rank is $r=2$. The other eleven zeros are responsible for the nullspaces.

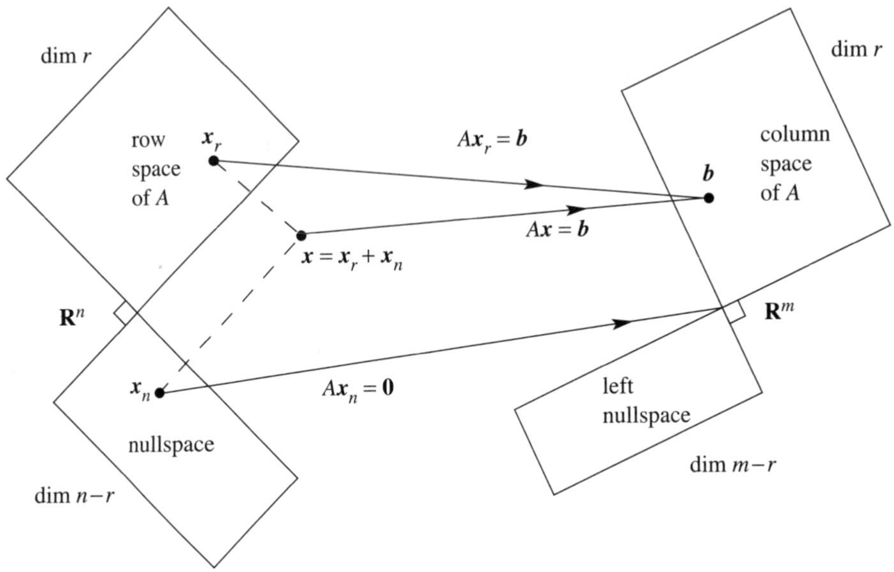

Figure 4.2 The true action of A times $\boldsymbol{x} = \boldsymbol{x}_r + \boldsymbol{x}_n$. Row space vector $\boldsymbol{x}_r$ to column space, nullspace vector $\boldsymbol{x}_n$ to zero.

Section 7.4 will show how every A becomes a diagonal matrix, when we choose the right bases for $\mathbf{R}^n$ and $\mathbf{R}^m$. This ***Singular Value Decomposition*** is a part of the theory that has become extremely important in applications.

Combining Bases from Subspaces

What follows are some valuable facts about bases. They were saved until now—when we are ready to use them. After a week you have a clearer sense of what a basis is (*independent* vectors that *span* the space). Normally we have to check both of these properties. When the count is right, one property implies the other:

4C Any n linearly independent vectors in $\mathbf{R}^n$ must span $\mathbf{R}^n$. They are a basis. Any n vectors that span $\mathbf{R}^n$ must be independent. They are a basis.

Starting with the correct number of vectors, one property of a basis produces the other. This is true in any vector space, but we care most about $\mathbf{R}^n$. When the vectors go into the columns of an n by n *square* matrix A, here are the same two facts:

4D If the n columns of A are independent, they span $\mathbf{R}^n$. So $A\boldsymbol{x} = \boldsymbol{b}$ is solvable. If the n columns span $\mathbf{R}^n$, they are independent. So $A\boldsymbol{x} = \boldsymbol{b}$ has only one solution.

Uniqueness implies existence and existence implies uniqueness. ***Then A is invertible.***

If there are no free variables (uniqueness), there must be n pivots. Then back substitution solves $A\boldsymbol{x} = \boldsymbol{b}$ (existence). Starting in the opposite direction, suppose $A\boldsymbol{x} = \boldsymbol{b}$ can always be solved (existence of solutions). Then elimination produced no zero rows. There are n pivots and no free variables. The nullspace contains only $\boldsymbol{x} = \mathbf{0}$ (uniqueness of solutions).

With a basis for the row space and a basis for the nullspace, we have $r+(n-r) = n$ vectors—the right number. Those n vectors are independent.[2] *Therefore they span* $\mathbf{R}^n$. They are a basis:

Each $\boldsymbol{x}$ in $\mathbf{R}^n$ is the sum $\boldsymbol{x}_r + \boldsymbol{x}_n$ of a row space vector $\boldsymbol{x}_r$ and a nullspace vector $\boldsymbol{x}_n$.

This confirms the splitting in Figure 4.2. It is the key point of orthogonal complements—the dimensions add to n and all vectors are fully accounted for.

Example 5 For $A = \begin{bmatrix} I & I \end{bmatrix} = \begin{bmatrix} 1 & 0 & 1 & 0 \\ 0 & 1 & 0 & 1 \end{bmatrix}$, write any vector $\boldsymbol{x}$ as $\boldsymbol{x}_r + \boldsymbol{x}_n$.

$(1, 0, 1, 0)$ and $(0, 1, 0, 1)$ are a basis for the row space. $(1, 0, -1, 0)$ and $(0, 1, 0, -1)$ are a basis for the nullspace of A. Those four vectors are a basis for $\mathbf{R}^4$. Any $\boldsymbol{x} = (a, b, c, d)$ can be split into $\boldsymbol{x}_r$ in the row space and $\boldsymbol{x}_n$ in the nullspace:

$$\begin{bmatrix} a \\ b \\ c \\ d \end{bmatrix} = \frac{a+c}{2}\begin{bmatrix} 1 \\ 0 \\ 1 \\ 0 \end{bmatrix} + \frac{b+d}{2}\begin{bmatrix} 0 \\ 1 \\ 0 \\ 1 \end{bmatrix} + \frac{a-c}{2}\begin{bmatrix} 1 \\ 0 \\ -1 \\ 0 \end{bmatrix} + \frac{b-d}{2}\begin{bmatrix} 0 \\ 1 \\ 0 \\ -1 \end{bmatrix}.$$

■ REVIEW OF THE KEY IDEAS ■

1. Subspaces $\boldsymbol{V}$ and $\boldsymbol{W}$ are orthogonal if every $\boldsymbol{v}$ in $\boldsymbol{V}$ is orthogonal to every $\boldsymbol{w}$ in $\boldsymbol{W}$.

2. $\boldsymbol{V}$ and $\boldsymbol{W}$ are "orthogonal complements" if $\boldsymbol{W}$ contains **all** vectors perpendicular to $\boldsymbol{V}$ (and vice versa). Inside $\mathbf{R}^n$, the dimensions of $\boldsymbol{V}$ and $\boldsymbol{W}$ add to n.

3. The nullspace $\boldsymbol{N}(A)$ and the row space $\boldsymbol{C}(A^{\mathrm{T}})$ are orthogonal complements, from $A\boldsymbol{x} = \mathbf{0}$. Similarly $\boldsymbol{N}(A^{\mathrm{T}})$ and $\boldsymbol{C}(A)$ are orthogonal complements.

[2] If a combination of the vectors gives $\boldsymbol{x}_r + \boldsymbol{x}_n = \mathbf{0}$, then $\boldsymbol{x}_r = -\boldsymbol{x}_n$ is in both subspaces. It is orthogonal to itself and must be zero. All coefficients of the row space basis and nullspace basis must be zero—which proves independence of the n vectors together.

4. Any n independent vectors in $\mathbf{R}^n$ will span $\mathbf{R}^n$.

5. Every x in $\mathbf{R}^n$ has a nullspace component x_n and a row space component x_r.

WORKED EXAMPLES

4.1 A Suppose S is a six-dimensional subspace of $\boldsymbol{R}^9$. What are the possible dimensions of subspaces orthogonal to S? What are the possible dimensions of the orthogonal complement $S^\perp$ of S? What is the smallest possible size of a matrix A that has row space S? What is the shape of its nullspace matrix N? How could you create a matrix B with extra rows but the same row space? Compare the nullspace matrix for B with the nullspace matrix for A.

Solution If S is six-dimensional in $\boldsymbol{R}^9$, subspaces orthogonal to S can have dimensions $0, 1, 2, 3$. The orthogonal complement $S^\perp$ will be the largest orthogonal subspace, with dimension 3. The smallest matrix A must have 9 columns and 6 rows (its rows are a basis for the 6-dimensional row space S). Its nullspace matrix N will be 9 by 3, since its columns contain a basis for $S^\perp$.

If row 7 of B is a combination of the six rows of A, then B has the same row space as A. It also has the same nullspace matrix N. (The special solutions s_1, s_2, s_3 will be the same. Elimination will change row 7 of B to all zeros.)

4.1 B The equation $x - 4y - 5z = 0$ describes a plane P in $\boldsymbol{R}^3$ (actually a subspace). The plane P is the nullspace $N(A)$ of what 1 by 3 matrix A? Find a basis s_1, s_2 of special solutions of $x - 3y - 4z = 0$ (these would be the columns of the nullspace matrix N). Also find a basis for the line $P^\perp$ that is perpendicular to P. Then split $v = (6, 4, 5)$ into its nullspace component v_n in P and its row space component v_r in $P^\perp$.

Solution The equation $x - 3y - 4z = 0$ is $Ax = \mathbf{0}$ for the 1 by 3 matrix $A = [1 \ \ -3 \ \ -4]$. Columns 2 and 3 are free (no pivots). The special solutions with free variables "1 and 0" are $s_1 = (3, 1, 0)$ and $s_2 = (4, 0, 1)$. The row space of A (which is the line $P^\perp$) certainly has basis $z = (1, -3, -4)$. This is perpendicular to s_1 and s_2 and their plane P.

To split v into $v_n + v_r = (c_1 s_1 + c_2 s_2) + c_3 z$, solve for the numbers c_1, c_2, c_3:

$$\begin{bmatrix} 3 & 4 & 1 \\ 1 & 0 & -3 \\ 0 & 1 & -4 \end{bmatrix} \begin{bmatrix} c_1 \\ c_2 \\ c_3 \end{bmatrix} = \begin{bmatrix} 6 \\ 4 \\ 5 \end{bmatrix} \qquad \begin{array}{l} \text{leads to } \ c_1 = 1,\ c_2 = 1,\ c_3 = -1 \\ v_n = s_1 + s_2 = (7, 1, 1) \ \text{ is in } \ P = N(A) \\ v_r = -z = (-1, 3, 4) \ \text{ is in } \ P^\perp = C(A^T). \end{array}$$

Problem Set 4.1

Questions 1–12 grow out of Figures 4.1 and 4.2.

1 Construct any 2 by 3 matrix of rank one. Copy Figure 4.1 and put one vector in each subspace (two in the nullspace). Which vectors are orthogonal?

2 Redraw Figure 4.2 for a 3 by 2 matrix of rank $r = 2$. Which subspace is $\boldsymbol{Z}$ (zero vector only)? The nullspace part of any vector $\boldsymbol{x}$ in $\mathbf{R}^2$ is $\boldsymbol{x}_n =$ ____.

3 Construct a matrix with the required property or say why that is impossible:

(a) Column space contains $\begin{bmatrix} 1 \\ 2 \\ -3 \end{bmatrix}$ and $\begin{bmatrix} 2 \\ -3 \\ 5 \end{bmatrix}$, nullspace contains $\begin{bmatrix} 1 \\ 1 \\ 1 \end{bmatrix}$

(b) Row space contains $\begin{bmatrix} 1 \\ 2 \\ -3 \end{bmatrix}$ and $\begin{bmatrix} 2 \\ -3 \\ 5 \end{bmatrix}$, nullspace contains $\begin{bmatrix} 1 \\ 1 \\ 1 \end{bmatrix}$

(c) $A\boldsymbol{x} = \begin{bmatrix} 1 \\ 1 \\ 1 \end{bmatrix}$ has a solution and $A^{\mathrm{T}} \begin{bmatrix} 1 \\ 0 \\ 0 \end{bmatrix} = \begin{bmatrix} 0 \\ 0 \\ 0 \end{bmatrix}$

(d) Every row is orthogonal to every column (A is not the zero matrix)

(e) Columns add up to a column of zeros, rows add to a row of 1's.

4 If $AB = 0$ then the columns of B are in the ____ of A. The rows of A are in the ____ of B. Why can't A and B be 3 by 3 matrices of rank 2?

5 (a) If $A\boldsymbol{x} = \boldsymbol{b}$ has a solution and $A^{\mathrm{T}}\boldsymbol{y} = \boldsymbol{0}$, then $\boldsymbol{y}$ is perpendicular to ____.

(b) If $A^{\mathrm{T}}\boldsymbol{y} = \boldsymbol{c}$ has a solution and $A\boldsymbol{x} = \boldsymbol{0}$, then $\boldsymbol{x}$ is perpendicular to ____.

6 This is a system of equations $A\boldsymbol{x} = \boldsymbol{b}$ with *no solution*:

$$\begin{aligned} x + 2y + 2z &= 5 \\ 2x + 2y + 3z &= 5 \\ 3x + 4y + 5z &= 9 \end{aligned}$$

Find numbers y_1, y_2, y_3 to multiply the equations so they add to $0 = 1$. You have found a vector $\boldsymbol{y}$ in which subspace? Its dot product $\boldsymbol{y}^{\mathrm{T}}\boldsymbol{b}$ is 1.

7 Every system with no solution is like the one in Problem 6. There are numbers $y_1, \ldots, y_m$ that multiply the m equations so they add up to $0 = 1$. This is called

Fredholm's Alternative: **Exactly one of these problems has a solution**

$$A\boldsymbol{x} = \boldsymbol{b} \quad \textbf{OR} \quad A^{\mathrm{T}}\boldsymbol{y} = \boldsymbol{0} \text{ with } \boldsymbol{y}^{\mathrm{T}}\boldsymbol{b} = 1.$$

If $\boldsymbol{b}$ is not in the column space of A, it is not orthogonal to the nullspace of A^{T}. Multiply the equations $x_1 - x_2 = 1$ and $x_2 - x_3 = 1$ and $x_1 - x_3 = 1$ by numbers y_1, y_2, y_3 chosen so that the equations add up to $0 = 1$.

8 In Figure 4.2, how do we know that Ax_r is equal to Ax? How do we know that this vector is in the column space? If $A = \begin{bmatrix} 1 & 1 \\ 1 & 1 \end{bmatrix}$ and $x = \begin{bmatrix} 1 \\ 0 \end{bmatrix}$ what is x_r?

9 If Ax is in the nullspace of A^{T} then $Ax = \mathbf{0}$. Reason: Ax is also in the _____ of A and the spaces are _____. *Conclusion*: $A^{\mathrm{T}}A$ *has the same nullspace as* A.

10 Suppose A is a symmetric matrix ($A^{\mathrm{T}} = A$).

(a) Why is its column space perpendicular to its nullspace?

(b) If $Ax = \mathbf{0}$ and $Az = 5z$, which subspaces contain these "eigenvectors" x and z? Symmetric matrices have perpendicular eigenvectors.

11 (Recommended) Draw Figure 4.2 to show each subspace correctly for

$$A = \begin{bmatrix} 1 & 2 \\ 3 & 6 \end{bmatrix} \quad \text{and} \quad B = \begin{bmatrix} 1 & 0 \\ 3 & 0 \end{bmatrix}.$$

12 Find the pieces x_r and x_n and draw Figure 4.2 properly if

$$A = \begin{bmatrix} 1 & -1 \\ 0 & 0 \\ 0 & 0 \end{bmatrix} \quad \text{and} \quad x = \begin{bmatrix} 2 \\ 0 \end{bmatrix}.$$

Questions 13–23 are about orthogonal subspaces.

13 Put bases for the subspaces $\boldsymbol{V}$ and $\boldsymbol{W}$ into the columns of matrices V and W. Explain why the test for orthogonal subspaces can be written $V^{\mathrm{T}}W =$ zero matrix. This matches $v^{\mathrm{T}}w = 0$ for orthogonal vectors.

14 The floor $\boldsymbol{V}$ and the wall $\boldsymbol{W}$ are not orthogonal subspaces, because they share a nonzero vector (along the line where they meet). No planes $\boldsymbol{V}$ and $\boldsymbol{W}$ in $\mathbf{R}^3$ can be orthogonal! Find a vector in the column spaces of both matrices:

$$A = \begin{bmatrix} 1 & 2 \\ 1 & 3 \\ 1 & 2 \end{bmatrix} \quad \text{and} \quad B = \begin{bmatrix} 5 & 4 \\ 6 & 3 \\ 5 & 1 \end{bmatrix}$$

This will be a vector Ax and also $B\widehat{x}$. Think 3 by 4 with the matrix $[\,A \;\; B\,]$.

15 Extend problem 14 to a p-dimensional subspace $\boldsymbol{V}$ and a q-dimensional subspace $\boldsymbol{W}$ of $\mathbf{R}^n$. What inequality on $p+q$ guarantees that $\boldsymbol{V}$ intersects $\boldsymbol{W}$ in a nonzero vector? These subspaces cannot be orthogonal.

16 Prove that every y in $N(A^{\mathrm{T}})$ is perpendicular to every Ax in the column space, using the matrix shorthand of equation (2). Start from $A^{\mathrm{T}}y = \mathbf{0}$.

17 If $\boldsymbol{S}$ is the subspace of $\mathbf{R}^3$ containing only the zero vector, what is $\boldsymbol{S}^{\perp}$? If $\boldsymbol{S}$ is spanned by $(1, 1, 1)$, what is $\boldsymbol{S}^{\perp}$? If $\boldsymbol{S}$ is spanned by $(2, 0, 0)$ and $(0, 0, 3)$, what is $\boldsymbol{S}^{\perp}$?

18 Suppose S only contains two vectors $(1, 5, 1)$ and $(2, 2, 2)$ (not a subspace). Then $S^{\perp}$ is the nullspace of the matrix $A =$ ____. $S^{\perp}$ is a subspace even if S is not.

19 Suppose L is a one-dimensional subspace (a line) in $\mathbf{R}^3$. Its orthogonal complement $L^{\perp}$ is the ____ perpendicular to L. Then $(L^{\perp})^{\perp}$ is a ____ perpendicular to $L^{\perp}$. In fact $(L^{\perp})^{\perp}$ is the same as ____.

20 Suppose V is the whole space $\mathbf{R}^4$. Then $V^{\perp}$ contains only the vector ____. Then $(V^{\perp})^{\perp}$ is ____. So $(V^{\perp})^{\perp}$ is the same as ____.

21 Suppose S is spanned by the vectors $(1, 2, 2, 3)$ and $(1, 3, 3, 2)$. Find two vectors that span $S^{\perp}$. This is the same as solving $Ax = \mathbf{0}$ for which A?

22 If P is the plane of vectors in $\mathbf{R}^4$ satisfying $x_1 + x_2 + x_3 + x_4 = 0$, write a basis for $P^{\perp}$. Construct a matrix that has P as its nullspace.

23 If a subspace S is contained in a subspace V, prove that $S^{\perp}$ contains $V^{\perp}$.

Questions 24–30 are about perpendicular columns and rows.

24 Suppose an n by n matrix is invertible: $AA^{-1} = I$. Then the first column of A^{-1} is orthogonal to the space spanned by which rows of A?

25 Find $A^{\mathrm{T}}A$ if the columns of A are unit vectors, all mutually perpendicular.

26 Construct a 3 by 3 matrix A with no zero entries whose columns are mutually perpendicular. Compute $A^{\mathrm{T}}A$. Why is it a diagonal matrix?

27 The lines $3x + y = b_1$ and $6x + 2y = b_2$ are ____. They are the same line if ____. In that case (b_1, b_2) is perpendicular to the vector ____. The nullspace of the matrix is the line $3x + y =$ ____. One particular vector in that nullspace is ____.

28 Why is each of these statements false?

(a) $(1, 1, 1)$ is perpendicular to $(1, 1, -2)$ so the planes $x + y + z = 0$ and $x + y - 2z = 0$ are orthogonal subspaces.

(b) The subspace spanned by $(1, 1, 0, 0, 0)$ and $(0, 0, 0, 1, 1)$ is the orthogonal complement of the subspace spanned by $(1, -1, 0, 0, 0)$ and $(2, -2, 3, 4, -4)$.

(c) If two subspaces meet only in the zero vector, the subspaces are orthogonal.

29 Find a matrix with $\boldsymbol{v} = (1, 2, 3)$ in the row space and column space. Find another matrix with $\boldsymbol{v}$ in the nullspace and column space. Which pairs of subspaces can $\boldsymbol{v}$ *not* be in?

30 Suppose A is 3 by 4 and B is 4 by 5 and $AB = 0$. Prove $\operatorname{rank}(A) + \operatorname{rank}(B) \le 4$.

31 The command $N = \mathsf{null}(A)$ will produce a basis for the nullspace of A. Then the command $B = \mathsf{null}(N')$ will produce a basis for the ____ of A.

PROJECTIONS ■ 4.2

May we start this section with two questions? (In addition to that one.) The first question aims to show that projections are easy to visualize. The second question is about "projection matrices":

1 What are the projections of $\boldsymbol{b} = (2, 3, 4)$ onto the z axis and the xy plane?

2 What matrices produce those projections onto a line and a plane?

When $\boldsymbol{b}$ is projected onto a line, ***its projection p is the part of b along that line.*** If $\boldsymbol{b}$ is projected onto a plane, $\boldsymbol{p}$ is the part in that plane. ***The projection p is Pb.***

There is a projection matrix P that multiplies $\boldsymbol{b}$ to give $\boldsymbol{p}$. This section finds $\boldsymbol{p}$ and P.

The projection onto the z axis we call $\boldsymbol{p}_1$. The second projection drops straight down to the xy plane. The picture in your mind should be Figure 4.3. Start with $\boldsymbol{b} = (2, 3, 4)$. One projection gives $\boldsymbol{p}_1 = (0, 0, 4)$ and the other gives $\boldsymbol{p}_2 = (2, 3, 0)$. Those are the parts of $\boldsymbol{b}$ along the z axis and in the xy plane.

The projection matrices P_1 and P_2 are 3 by 3. They multiply $\boldsymbol{b}$ with 3 components to produce $\boldsymbol{p}$ with 3 components. Projection onto a line comes from a rank one matrix. Projection onto a plane comes from a rank two matrix:

$$\text{Onto the } z \text{ axis:} \quad P_1 = \begin{bmatrix} 0 & 0 & 0 \\ 0 & 0 & 0 \\ 0 & 0 & 1 \end{bmatrix} \qquad \text{Onto the } xy \text{ plane:} \quad P_2 = \begin{bmatrix} 1 & 0 & 0 \\ 0 & 1 & 0 \\ 0 & 0 & 0 \end{bmatrix}.$$

P_1 picks out the z component of every vector. P_2 picks out the x and y components. To find $\boldsymbol{p}_1$ and $\boldsymbol{p}_2$, multiply $\boldsymbol{b}$ by P_1 and P_2 (small $\boldsymbol{p}$ for the vector, capital P for the matrix that produces it):

$$\boldsymbol{p}_1 = P_1\boldsymbol{b} = \begin{bmatrix} 0 & 0 & 0 \\ 0 & 0 & 0 \\ 0 & 0 & 1 \end{bmatrix} \begin{bmatrix} x \\ y \\ z \end{bmatrix} = \begin{bmatrix} 0 \\ 0 \\ z \end{bmatrix} \qquad \boldsymbol{p}_2 = P_2\boldsymbol{b} = \begin{bmatrix} 1 & 0 & 0 \\ 0 & 1 & 0 \\ 0 & 0 & 0 \end{bmatrix} \begin{bmatrix} x \\ y \\ z \end{bmatrix} = \begin{bmatrix} x \\ y \\ 0 \end{bmatrix}.$$

In this case the projections P_1 and P_2 are perpendicular. The xy plane and the z axis are ***orthogonal subspaces***, like the floor of a room and the line between two walls.

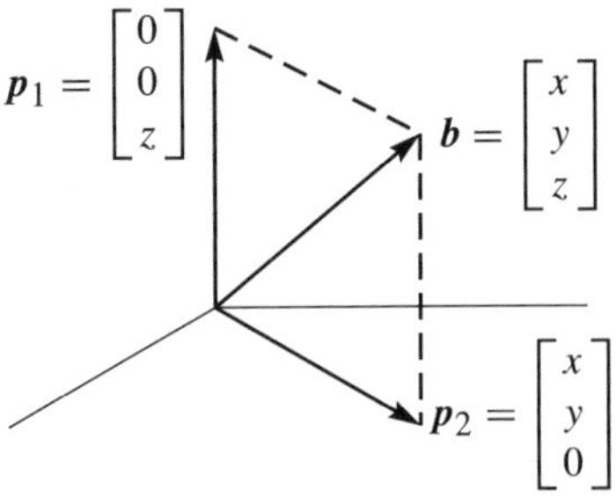

Figure 4.3 The projections of $\boldsymbol{b}$ onto the z axis and the xy plane.

More than that, the line and plane are orthogonal ***complements***. Their dimensions add to $1 + 2 = 3$—every vector $\boldsymbol{b}$ in the whole space is the sum of its parts in the two subspaces. The projections $\boldsymbol{p}_1$ and $\boldsymbol{p}_2$ are exactly those parts:

$$\text{The vectors give } \boldsymbol{p}_1 + \boldsymbol{p}_2 = \boldsymbol{b}. \qquad \text{The matrices give } P_1 + P_2 = I. \tag{1}$$

This is perfect. Our goal is reached—for this example. We have the same goal for any line and any plane and any n-dimensional subspace. The object is to find the part $\boldsymbol{p}$ in each subspace, and the projection matrix P that produces that part $\boldsymbol{p} = P\boldsymbol{b}$. Every subspace of $\mathbf{R}^m$ has its own m by m projection matrix. To compute P, we absolutely need a good description of the subspace that it projects onto.

The best description of a subspace is a basis. We put the basis vectors into the columns of A. ***Now we are projecting onto the column space of*** A! Certainly the z axis is the column space of the 3 by 1 matrix A_1. The xy plane is the column space of A_2. That plane is also the column space of A_3 (a subspace has many bases):

$$A_1 = \begin{bmatrix} 0 \\ 0 \\ 1 \end{bmatrix} \quad \text{and} \quad A_2 = \begin{bmatrix} 1 & 0 \\ 0 & 1 \\ 0 & 0 \end{bmatrix} \quad \text{and} \quad A_3 = \begin{bmatrix} 1 & 2 \\ 2 & 3 \\ 0 & 0 \end{bmatrix}.$$

Our problem is ***to project onto the column space of any*** m ***by*** n ***matrix***. Start with a line (dimension $n = 1$). The matrix A has only one column. Call it $\boldsymbol{a}$.

Projection Onto a Line

We are given a line through the origin, in the direction of $\boldsymbol{a} = (a_1, \ldots, a_m)$. Along that line, we want the point $\boldsymbol{p}$ closest to $\boldsymbol{b} = (b_1, \ldots, b_m)$. The key to projection is orthogonality: ***The line from b to p is perpendicular to the vector a***. This is the dotted line marked $\boldsymbol{e}$ in Figure 4.4—which we now compute by algebra.

The projection $\boldsymbol{p}$ is some multiple of $\boldsymbol{a}$. Call it $\boldsymbol{p} = \widehat{x}\boldsymbol{a} =$ "x hat" times $\boldsymbol{a}$. Our first step is to compute this unknown number $\widehat{x}$. That will give the vector $\boldsymbol{p}$. Then from the formula for $\boldsymbol{p}$, we read off the projection matrix P. These three steps will lead to all projection matrices: *find* $\widehat{x}$, *then find* ***the vector*** $\boldsymbol{p}$, *then find* ***the matrix*** P.

The dotted line $\boldsymbol{b} - \boldsymbol{p}$ is $\boldsymbol{b} - \widehat{x}\boldsymbol{a}$. It is perpendicular to $\boldsymbol{a}$—this will determine $\widehat{x}$. Use the fact that two vectors are perpendicular when their dot product is zero:

$$\boldsymbol{a} \cdot (\boldsymbol{b} - \widehat{x}\boldsymbol{a}) = 0 \quad \text{or} \quad \boldsymbol{a} \cdot \boldsymbol{b} - \widehat{x}\boldsymbol{a} \cdot \boldsymbol{a} = 0 \quad \text{or} \quad \widehat{x} = \frac{\boldsymbol{a} \cdot \boldsymbol{b}}{\boldsymbol{a} \cdot \boldsymbol{a}} = \frac{\boldsymbol{a}^{\mathrm{T}}\boldsymbol{b}}{\boldsymbol{a}^{\mathrm{T}}\boldsymbol{a}}. \tag{2}$$

For vectors the multiplication $\boldsymbol{a}^{\mathrm{T}}\boldsymbol{b}$ is the same as $\boldsymbol{a} \cdot \boldsymbol{b}$. Using the transpose is better, because it applies also to matrices. (We will soon meet $A^{\mathrm{T}}\boldsymbol{b}$.) Our formula for $\widehat{x}$ immediately gives the formula for $\boldsymbol{p}$:

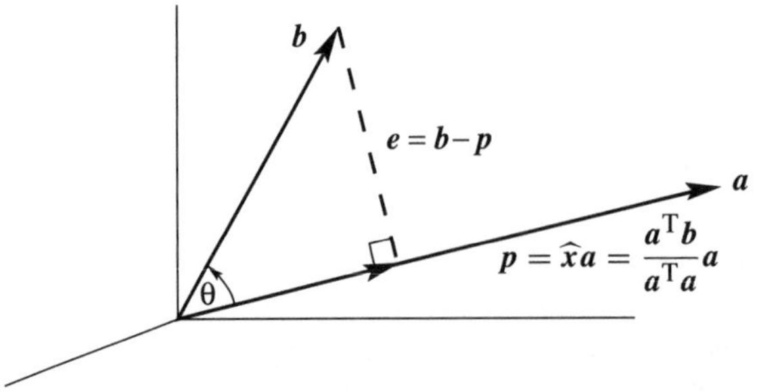

Figure 4.4 The projection $\boldsymbol{p}$, perpendicular to $\boldsymbol{e}$, has length $\|\boldsymbol{b}\|\cos\theta$.

4E The projection of $\boldsymbol{b}$ onto the line through $\boldsymbol{a}$ is the vector $\boldsymbol{p} = \widehat{x}\boldsymbol{a} = \dfrac{\boldsymbol{a}^{\mathrm{T}}\boldsymbol{b}}{\boldsymbol{a}^{\mathrm{T}}\boldsymbol{a}}\boldsymbol{a}$.

Special case 1: If $\boldsymbol{b} = \boldsymbol{a}$ then $\widehat{x} = 1$. The projection of $\boldsymbol{a}$ onto $\boldsymbol{a}$ is itself.

Special case 2: If $\boldsymbol{b}$ is perpendicular to $\boldsymbol{a}$ then $\boldsymbol{a}^{\mathrm{T}}\boldsymbol{b} = 0$. The projection is $\boldsymbol{p} = \boldsymbol{0}$.

Example 1 Project $\boldsymbol{b} = \begin{bmatrix}1\\1\\1\end{bmatrix}$ onto $\boldsymbol{a} = \begin{bmatrix}1\\2\\2\end{bmatrix}$ to find $\boldsymbol{p} = \widehat{x}\boldsymbol{a}$ in Figure 4.4.

Solution The number $\widehat{x}$ is the ratio of $\boldsymbol{a}^{\mathrm{T}}\boldsymbol{b} = 5$ to $\boldsymbol{a}^{\mathrm{T}}\boldsymbol{a} = 9$. So the projection is $\boldsymbol{p} = \frac{5}{9}\boldsymbol{a}$. The error vector between $\boldsymbol{b}$ and $\boldsymbol{p}$ is $\boldsymbol{e} = \boldsymbol{b} - \boldsymbol{p}$. Those vectors $\boldsymbol{p}$ and $\boldsymbol{e}$ will add to $\boldsymbol{b}$:

$$\boldsymbol{p} = \frac{5}{9}\boldsymbol{a} = \left(\frac{5}{9}, \frac{10}{9}, \frac{10}{9}\right) \quad \text{and} \quad \boldsymbol{e} = \boldsymbol{b} - \boldsymbol{p} = \left(\frac{4}{9}, -\frac{1}{9}, -\frac{1}{9}\right).$$

The error $\boldsymbol{e}$ should be perpendicular to $\boldsymbol{a} = (1, 2, 2)$ and it is: $\boldsymbol{e}^{\mathrm{T}}\boldsymbol{a} = \frac{4}{9} - \frac{2}{9} - \frac{2}{9} = 0$.

Look at the right triangle of $\boldsymbol{b}$, $\boldsymbol{p}$, and $\boldsymbol{e}$. The vector $\boldsymbol{b}$ is split into two parts—its component along the line is $\boldsymbol{p}$, its perpendicular part is $\boldsymbol{e}$. Those two sides of a right triangle have length $\|\boldsymbol{b}\|\cos\theta$ and $\|\boldsymbol{b}\|\sin\theta$. Trigonometry matches the dot product:

$$\boldsymbol{p} = \frac{\boldsymbol{a}^{\mathrm{T}}\boldsymbol{b}}{\boldsymbol{a}^{\mathrm{T}}\boldsymbol{a}}\boldsymbol{a} \quad \text{so its length is} \quad \|\boldsymbol{p}\| = \frac{\|\boldsymbol{a}\|\,\|\boldsymbol{b}\|\cos\theta}{\|\boldsymbol{a}\|^2}\|\boldsymbol{a}\| = \|\boldsymbol{b}\|\cos\theta. \tag{3}$$

The dot product is a lot simpler than getting involved with $\cos\theta$ and the length of $\boldsymbol{b}$. The example has square roots in $\cos\theta = 5/3\sqrt{3}$ and $\|\boldsymbol{b}\| = \sqrt{3}$. There are no square roots in the projection $\boldsymbol{p} = \frac{5}{9}\boldsymbol{a}$.

Now comes the ***projection matrix***. In the formula for $\boldsymbol{p}$, what matrix is multiplying $\boldsymbol{b}$? You can see it better if the number $\widehat{x}$ is on the right side of $\boldsymbol{a}$:

$$\boldsymbol{p} = \boldsymbol{a}\widehat{x} = \boldsymbol{a}\frac{\boldsymbol{a}^{\mathrm{T}}\boldsymbol{b}}{\boldsymbol{a}^{\mathrm{T}}\boldsymbol{a}} = P\boldsymbol{b} \quad \textit{when the matrix is} \quad P = \frac{\boldsymbol{a}\boldsymbol{a}^{\mathrm{T}}}{\boldsymbol{a}^{\mathrm{T}}\boldsymbol{a}}.$$

P is a column times a row! The column is $\boldsymbol{a}$, the row is $\boldsymbol{a}^{\mathrm{T}}$. Then divide by the number $\boldsymbol{a}^{\mathrm{T}}\boldsymbol{a}$. The projection matrix P is m by m, but ***its rank is one***. We are projecting onto a one-dimensional subspace, the line through $\boldsymbol{a}$.

Example 2 Find the projection matrix $P = \dfrac{\boldsymbol{a}\boldsymbol{a}^{\mathrm{T}}}{\boldsymbol{a}^{\mathrm{T}}\boldsymbol{a}}$ onto the line through $\boldsymbol{a} = \left[\begin{smallmatrix}1\\2\\2\end{smallmatrix}\right]$.

Solution Multiply column times row and divide by $\boldsymbol{a}^{\mathrm{T}}\boldsymbol{a} = 9$:

$$P = \frac{\boldsymbol{a}\boldsymbol{a}^{\mathrm{T}}}{\boldsymbol{a}^{\mathrm{T}}\boldsymbol{a}} = \frac{1}{9}\begin{bmatrix}1\\2\\2\end{bmatrix}\begin{bmatrix}1 & 2 & 2\end{bmatrix} = \frac{1}{9}\begin{bmatrix}1 & 2 & 2\\2 & 4 & 4\\2 & 4 & 4\end{bmatrix}.$$

This matrix projects *any* vector $\boldsymbol{b}$ onto $\boldsymbol{a}$. Check $\boldsymbol{p} = P\boldsymbol{b}$ for the particular $\boldsymbol{b} = (1, 1, 1)$ in Example 1:

$$\boldsymbol{p} = P\boldsymbol{b} = \frac{1}{9}\begin{bmatrix}1 & 2 & 2\\2 & 4 & 4\\2 & 4 & 4\end{bmatrix}\begin{bmatrix}1\\1\\1\end{bmatrix} = \frac{1}{9}\begin{bmatrix}5\\10\\10\end{bmatrix} \quad \text{which is correct.}$$

If the vector $\boldsymbol{a}$ is doubled, the matrix P stays the same. It still projects onto the same line. If the matrix is squared, P^2 equals P. ***Projecting a second time doesn't change anything***, so $P^2 = P$. The diagonal entries of P add up to $\frac{1}{9}(1+4+4) = 1$.

The matrix $I - P$ should be a projection too. It produces the other side $\boldsymbol{e}$ of the triangle—the perpendicular part of $\boldsymbol{b}$. Note that $(I - P)\boldsymbol{b}$ equals $\boldsymbol{b} - \boldsymbol{p}$ which is $\boldsymbol{e}$. ***When P projects onto one subspace, $I - P$ projects onto the perpendicular subspace.*** Here $I - P$ projects onto the plane perpendicular to $\boldsymbol{a}$.

Now we move beyond projection onto a line. Projecting onto an n-dimensional subspace of $\mathbf{R}^m$ takes more effort. The crucial formulas will be collected in equations (5)–(6)–(7). Basically you need to remember them.

Projection Onto a Subspace

Start with n vectors $\boldsymbol{a}_1, \ldots, \boldsymbol{a}_n$ in $\mathbf{R}^m$. Assume that these $\boldsymbol{a}$'s are linearly independent. ***Problem: Find the combination $\widehat{x}_1\boldsymbol{a}_1 + \cdots + \widehat{x}_n\boldsymbol{a}_n$ that is closest to a given vector $\boldsymbol{b}$.*** We are projecting each $\boldsymbol{b}$ in $\mathbf{R}^m$ onto the subspace spanned by the $\boldsymbol{a}$'s.

With $n = 1$ (only one vector $\boldsymbol{a}_1$) this is projection onto a line. The line is the column space of A, which has just one column. In general the matrix A has n columns $\boldsymbol{a}_1, \ldots, \boldsymbol{a}_n$. Their combinations in $\mathbf{R}^m$ are the vectors $A\boldsymbol{x}$ in the column space. We are looking for the particular combination $\boldsymbol{p} = A\widehat{\boldsymbol{x}}$ (***the projection***) that is closest to $\boldsymbol{b}$. The hat over $\widehat{\boldsymbol{x}}$ indicates the *best* choice, to give the closest vector in the column space. That choice is $\boldsymbol{a}^{\mathrm{T}}\boldsymbol{b}/\boldsymbol{a}^{\mathrm{T}}\boldsymbol{a}$ when $n = 1$. For $n > 1$, the best $\widehat{\boldsymbol{x}}$ is to be found.

We solve this problem for an n-dimensional subspace in three steps: *Find the vector $\widehat{\boldsymbol{x}}$, find the projection $\boldsymbol{p} = A\widehat{\boldsymbol{x}}$, find the matrix P.*

The key is in the geometry! The dotted line in Figure 4.5 goes from $\boldsymbol{b}$ to the nearest point $A\widehat{\boldsymbol{x}}$ in the subspace. ***This error vector $\boldsymbol{b} - A\widehat{\boldsymbol{x}}$ is perpendicular to the subspace.*** The error $\boldsymbol{b} - A\widehat{\boldsymbol{x}}$ makes a right angle with all the vectors $\boldsymbol{a}_1, \ldots, \boldsymbol{a}_n$. That gives the n equations we need to find $\widehat{\boldsymbol{x}}$:

$$\begin{matrix} \boldsymbol{a}_1^{\mathrm{T}}(\boldsymbol{b} - A\widehat{\boldsymbol{x}}) = 0 \\ \vdots \\ \boldsymbol{a}_n^{\mathrm{T}}(\boldsymbol{b} - A\widehat{\boldsymbol{x}}) = 0 \end{matrix} \quad \text{or} \quad \begin{bmatrix} - & \boldsymbol{a}_1^{\mathrm{T}} & - \\ & \vdots & \\ - & \boldsymbol{a}_n^{\mathrm{T}} & - \end{bmatrix} \begin{bmatrix} \\ \boldsymbol{b} - A\widehat{\boldsymbol{x}} \\ \\ \end{bmatrix} = \begin{bmatrix} \\ \boldsymbol{0} \\ \\ \end{bmatrix}. \tag{4}$$

The matrix in those equations is A^{T}. The n equations are exactly $A^{\mathrm{T}}(\boldsymbol{b} - A\widehat{\boldsymbol{x}}) = \boldsymbol{0}$.

Rewrite $A^{\mathrm{T}}(\boldsymbol{b} - A\widehat{\boldsymbol{x}}) = \boldsymbol{0}$ in its famous form $A^{\mathrm{T}}A\widehat{\boldsymbol{x}} = A^{\mathrm{T}}\boldsymbol{b}$. This is the equation for $\widehat{\boldsymbol{x}}$, and the coefficient matrix is $A^{\mathrm{T}}A$. Now we can find $\widehat{\boldsymbol{x}}$ and $\boldsymbol{p}$ and P:

4F The combination $\widehat{x}_1\boldsymbol{a}_1 + \cdots + \widehat{x}_n\boldsymbol{a}_n = A\widehat{\boldsymbol{x}}$ that is closest to $\boldsymbol{b}$ comes from

$$A^{\mathrm{T}}(\boldsymbol{b} - A\widehat{\boldsymbol{x}}) = \boldsymbol{0} \quad \text{or} \quad A^{\mathrm{T}}A\widehat{\boldsymbol{x}} = A^{\mathrm{T}}\boldsymbol{b}. \tag{5}$$

The symmetric matrix $A^{\mathrm{T}}A$ is n by n. It is invertible if the $\boldsymbol{a}$'s are independent. The solution is $\widehat{\boldsymbol{x}} = (A^{\mathrm{T}}A)^{-1}A^{\mathrm{T}}\boldsymbol{b}$. The ***projection*** of $\boldsymbol{b}$ onto the subspace is the vector

$$\boldsymbol{p} = A\widehat{\boldsymbol{x}} = A(A^{\mathrm{T}}A)^{-1}A^{\mathrm{T}}\boldsymbol{b}. \tag{6}$$

This formula shows the n by n ***projection matrix*** that produces $\boldsymbol{p} = P\boldsymbol{b}$:

$$P = A(A^{\mathrm{T}}A)^{-1}A^{\mathrm{T}}. \tag{7}$$

Compare with projection onto a line, when the matrix A has only one column $\boldsymbol{a}$:

$$\widehat{x} = \frac{\boldsymbol{a}^{\mathrm{T}}\boldsymbol{b}}{\boldsymbol{a}^{\mathrm{T}}\boldsymbol{a}} \quad \text{and} \quad \boldsymbol{p} = \boldsymbol{a}\frac{\boldsymbol{a}^{\mathrm{T}}\boldsymbol{b}}{\boldsymbol{a}^{\mathrm{T}}\boldsymbol{a}} \quad \text{and} \quad P = \frac{\boldsymbol{a}\boldsymbol{a}^{\mathrm{T}}}{\boldsymbol{a}^{\mathrm{T}}\boldsymbol{a}}.$$

Those formulas are identical with (5) and (6) and (7)! The number $\boldsymbol{a}^{\mathrm{T}}\boldsymbol{a}$ becomes the matrix $A^{\mathrm{T}}A$. When it is a number, we divide by it. When it is a matrix, we invert it. The new formulas contain $(A^{\mathrm{T}}A)^{-1}$ instead of $1/\boldsymbol{a}^{\mathrm{T}}\boldsymbol{a}$. The linear independence of the columns $\boldsymbol{a}_1, \ldots, \boldsymbol{a}_n$ will guarantee that this inverse matrix exists.

The key step was $A^{\mathrm{T}}(\boldsymbol{b} - A\widehat{\boldsymbol{x}}) = \boldsymbol{0}$. We used geometry ($\boldsymbol{e}$ is perpendicular to all the $\boldsymbol{a}$'s). Linear algebra gives this "normal equation" too, in a very quick way:

1. Our subspace is the column space of A.

2. The error vector $\boldsymbol{b} - A\widehat{\boldsymbol{x}}$ is perpendicular to that column space.

3. Therefore $\boldsymbol{b} - A\widehat{\boldsymbol{x}}$ is in the left nullspace. This means $A^{\mathrm{T}}(\boldsymbol{b} - A\widehat{\boldsymbol{x}}) = \boldsymbol{0}$.

The left nullspace is important in projections. This nullspace of A^{T} contains the error vector $\boldsymbol{e} = \boldsymbol{b} - A\widehat{\boldsymbol{x}}$. The vector $\boldsymbol{b}$ is being split into the projection $\boldsymbol{p}$ and the error $\boldsymbol{e} = \boldsymbol{b} - \boldsymbol{p}$. Figure 4.5 shows the right triangle with sides $\boldsymbol{p}$, $\boldsymbol{e}$, and $\boldsymbol{b}$.

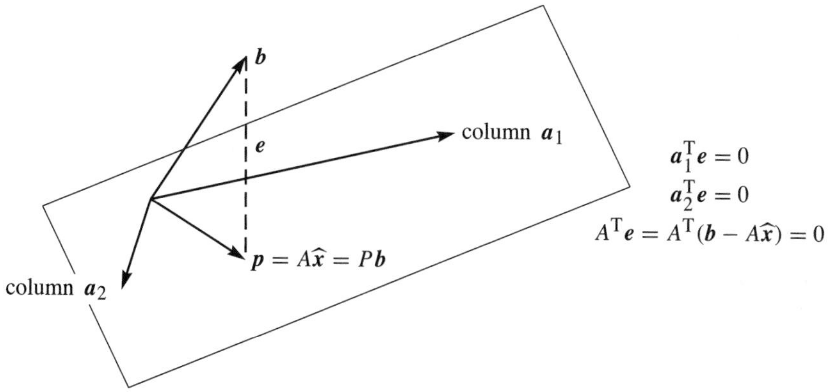

Figure 4.5 The projection $\boldsymbol{p}$ is the nearest point to $\boldsymbol{b}$ in the column space of A. The perpendicular error $\boldsymbol{e}$ must be in the nullspace of A^{T}.

Example 3 If $A = \begin{bmatrix} 1 & 0 \\ 1 & 1 \\ 1 & 2 \end{bmatrix}$ and $\boldsymbol{b} = \begin{bmatrix} 6 \\ 0 \\ 0 \end{bmatrix}$ find $\hat{\boldsymbol{x}}$ and $\boldsymbol{p}$ and P.

Solution Compute the square matrix $A^{\mathrm{T}}A$ and also the vector $A^{\mathrm{T}}\boldsymbol{b}$:

$$A^{\mathrm{T}}A = \begin{bmatrix} 1 & 1 & 1 \\ 0 & 1 & 2 \end{bmatrix} \begin{bmatrix} 1 & 0 \\ 1 & 1 \\ 1 & 2 \end{bmatrix} = \begin{bmatrix} 3 & 3 \\ 3 & 5 \end{bmatrix} \quad \text{and} \quad A^{\mathrm{T}}b = \begin{bmatrix} 1 & 1 & 1 \\ 0 & 1 & 2 \end{bmatrix} \begin{bmatrix} 6 \\ 0 \\ 0 \end{bmatrix} = \begin{bmatrix} 6 \\ 0 \end{bmatrix}.$$

Now solve the normal equation $A^{\mathrm{T}}A\hat{\boldsymbol{x}} = A^{\mathrm{T}}\boldsymbol{b}$ to find $\hat{\boldsymbol{x}}$:

$$\begin{bmatrix} 3 & 3 \\ 3 & 5 \end{bmatrix} \begin{bmatrix} \hat{x}_1 \\ \hat{x}_2 \end{bmatrix} = \begin{bmatrix} 6 \\ 0 \end{bmatrix} \quad \text{gives} \quad \hat{\boldsymbol{x}} = \begin{bmatrix} \hat{x}_1 \\ \hat{x}_2 \end{bmatrix} = \begin{bmatrix} 5 \\ -3 \end{bmatrix}. \tag{8}$$

The combination $\boldsymbol{p} = A\hat{\boldsymbol{x}}$ is the projection of $\boldsymbol{b}$ onto the column space of A:

$$\boldsymbol{p} = 5\begin{bmatrix} 1 \\ 1 \\ 1 \end{bmatrix} - 3\begin{bmatrix} 0 \\ 1 \\ 2 \end{bmatrix} = \begin{bmatrix} 5 \\ 2 \\ -1 \end{bmatrix}. \quad \text{The error is} \quad \boldsymbol{e} = \boldsymbol{b} - \boldsymbol{p} = \begin{bmatrix} 1 \\ -2 \\ 1 \end{bmatrix}. \tag{9}$$

That solves the problem for one particular $\boldsymbol{b}$. To solve it for every $\boldsymbol{b}$, compute the matrix $P = A(A^{\mathrm{T}}A)^{-1}A^{\mathrm{T}}$. The determinant of $A^{\mathrm{T}}A$ is $15 - 9 = 6$; $(A^{\mathrm{T}}A)^{-1}$ is easy. Then multiply A times $(A^{\mathrm{T}}A)^{-1}$ times A^{T} to reach P:

$$(A^{\mathrm{T}}A)^{-1} = \frac{1}{6}\begin{bmatrix} 5 & -3 \\ -3 & 3 \end{bmatrix} \quad \text{and} \quad P = \frac{1}{6}\begin{bmatrix} 5 & 2 & -1 \\ 2 & 2 & 2 \\ -1 & 2 & 5 \end{bmatrix}. \tag{10}$$

Two checks on the calculation. First, the error $e = (1, -2, 1)$ is perpendicular to both columns $(1, 1, 1)$ and $(0, 1, 2)$. Second, the final P times $b = (6, 0, 0)$ correctly gives $p = (5, 2, -1)$. We must also have $P^2 = P$, because a second projection doesn't change the first projection.

Warning The matrix $P = A(A^TA)^{-1}A^T$ is deceptive. You might try to split $(A^TA)^{-1}$ into A^{-1} times $(A^T)^{-1}$. If you make that mistake, and substitute it into P, you will find $P = AA^{-1}(A^T)^{-1}A^T$. Apparently everything cancels. This looks like $P = I$, the identity matrix. We want to say why this is wrong.

The matrix A is rectangular. It has no inverse matrix. We cannot split $(A^TA)^{-1}$ into A^{-1} times $(A^T)^{-1}$ because there is no A^{-1} in the first place.

In our experience, a problem that involves a rectangular matrix almost always leads to A^TA. We cannot split up its inverse, since A^{-1} and $(A^T)^{-1}$ don't exist. What does exist is the inverse of the square matrix A^TA. This fact is so crucial that we state it clearly and give a proof.

4G A^TA is invertible if and only if A has linearly independent columns.

Proof A^TA is a square matrix (n by n). For every matrix A, we will now show that ***A^TA has the same nullspace as A***. When the columns of A are linearly independent, its nullspace contains only the zero vector. Then A^TA, with this same nullspace, is invertible.

Let A be any matrix. If x is in its nullspace, then $Ax = \mathbf{0}$. Multiplying by A^T gives $A^TAx = \mathbf{0}$. So x is also in the nullspace of A^TA.

Now start with the nullspace of A^TA. From $A^TAx = \mathbf{0}$ we must prove that $Ax = \mathbf{0}$. We can't multiply by $(A^T)^{-1}$, which generally doesn't exist. Just multiply by x^T:

$$(x^T)A^TAx = 0 \quad \text{or} \quad (Ax)^T(Ax) = 0 \quad \text{or} \quad \|Ax\|^2 = 0.$$

The vector Ax has length zero. Therefore $Ax = \mathbf{0}$. Every vector x in one nullspace is in the other nullspace. If A has dependent columns, so does A^TA. If A has independent columns, so does A^TA. This is the good case:

When A has independent columns, A^TA is ***square, symmetric,*** and ***invertible.***

To repeat for emphasis: A^TA is (n by m) times (m by n). It is square (n by n). It is symmetric, because its transpose is $(A^TA)^T = A^T(A^T)^T$ which equals A^TA. We just proved that A^TA is invertible—provided A has independent columns. Watch the difference between dependent and independent columns:

$$\underset{A^T}{\begin{bmatrix} 1 & 1 & 0 \\ 2 & 2 & 0 \end{bmatrix}} \underset{\text{dependent}}{\underset{A}{\begin{bmatrix} 1 & 2 \\ 1 & 2 \\ 0 & 0 \end{bmatrix}}} = \underset{\text{singular}}{\underset{A^TA}{\begin{bmatrix} 2 & 4 \\ 4 & 8 \end{bmatrix}}} \qquad \underset{A^T}{\begin{bmatrix} 1 & 1 & 0 \\ 2 & 2 & 1 \end{bmatrix}} \underset{\text{indep.}}{\underset{A}{\begin{bmatrix} 1 & 2 \\ 1 & 2 \\ 0 & 1 \end{bmatrix}}} = \underset{\text{invertible}}{\underset{A^TA}{\begin{bmatrix} 2 & 4 \\ 4 & 9 \end{bmatrix}}}$$

Very brief summary To find the projection $p = \widehat{x}_1 a_1 + \cdots + \widehat{x}_n a_n$, solve $A^T A\widehat{x} = A^T b$. This gives $\widehat{x}$. The projection is $A\widehat{x}$ and the error is $e = b - p = b - A\widehat{x}$. The projection matrix $P = A(A^T A)^{-1} A^T$ gives $p = Pb$.

This matrix satisfies $P^2 = P$. ***The distance from b to the subspace is*** $\|e\|$.

■ REVIEW OF THE KEY IDEAS ■

1. The projection of b onto the line through a is $p = a\widehat{x} = a(a^T b/a^T a)$.
2. The rank one projection matrix $P = aa^T/a^T a$ multiplies b to produce p.
3. Projecting b onto a subspace leaves $e = b - p$ perpendicular to the subspace.
4. When the columns of A are a basis, the equation $A^T A\widehat{x} = A^T b$ leads to $\widehat{x}$ and $p = A\widehat{x}$.
5. The projection matrix $P = A(A^T A)^{-1} A^T$ has $P^T = P$ and $P^2 = P$. Another projection leaves $p = Pb$ unchanged so $P^2 = P$.

■ WORKED EXAMPLES ■

4.2 A Project the vector $b = (3, 4, 4)$ onto the line through $a = (2, 2, 1)$ and then onto the plane that also contains $a^* = (1, 0, 0)$. Check that the first error vector $b - p$ is perpendicular to a, and the second error vector $b - p^*$ is also perpendicular to a^*. Find the 3 by 3 projection matrix P onto that plane. Find a vector e^* whose projection onto the plane of a and a^* is the zero vector.

Solution The projection of $b = (3, 4, 4)$ onto the line through $a = (2, 2, 1)$ is $2a$:

$$p = \frac{b^T a}{a^T a} a = \frac{18}{9}(2, 2, 1) = (4, 4, 2).$$

The error vector $e = b - p = (-1, 0, 2)$ is perpendicular to a. So p is correct.

The plane containing $a = (2, 2, 1)$ and $a^* = (1, 0, 0)$ is the column space of A:

$$A = \begin{bmatrix} 2 & 1 \\ 2 & 0 \\ 1 & 0 \end{bmatrix} \quad A^T A = \begin{bmatrix} 9 & 2 \\ 2 & 1 \end{bmatrix} \quad (A^T A)^{-1} = \frac{1}{5}\begin{bmatrix} 1 & -2 \\ -2 & 9 \end{bmatrix} \quad P = \begin{bmatrix} 1 & 0 & 0 \\ 0 & .8 & .4 \\ 0 & .4 & .2 \end{bmatrix}$$

Then $p^* = Pb = (3, 4.8, 2.4)$ and $e^* = b - p^* = (0, -.8, 1.6)$ is perpendicular to a and a^*. This vector e^* is in the nullspace of P and its projection is zero! Note $P^2 = P$.

4.2 B Suppose your pulse is measured at $x = 70$ beats per minute, then at $x = 80$, then at $x = 120$. Those three equations $Ax = \boldsymbol{b}$ in one unknown have $A^{\mathrm{T}} = [1\ 1\ 1]$ and $\boldsymbol{b} = (70, 80, 120)$. ***The best $\widehat{x}$ is the _____ of 70, 80, 120.*** Use calculus and projection:

1 Minimize $E = (x-70)^2 + (x-80)^2 + (x-120)^2$ by solving $dE/dx = 0$.

2 Project $\boldsymbol{b} = (70, 80, 120)$ onto $\boldsymbol{a} = (1, 1, 1)$ to find $\widehat{x} = \boldsymbol{a}^{\mathrm{T}}\boldsymbol{b}/\boldsymbol{a}^{\mathrm{T}}\boldsymbol{a}$.

In *recursive* least squares, a new fourth measurement 130 changes $\widehat{x}_{\text{old}}$ to $\widehat{x}_{\text{new}}$. Compute $\widehat{x}_{\text{new}}$ and verify the *update formula* $\widehat{x}_{\text{new}} = \widehat{x}_{\text{old}} + \frac{1}{4}(130 - \widehat{x}_{\text{old}})$. Going from 999 to 1000 measurements, $\widehat{x}_{\text{new}} = \widehat{x}_{\text{old}} + \frac{1}{1000}(b_{1000} - \widehat{x}_{\text{old}})$ would only need $\widehat{x}_{\text{old}}$ and the latest value b_{1000}. We don't have to average all 1000 numbers!

Solution The closest horizontal line to the heights 70, 80, 120 is the *average* $\widehat{x} = 90$:

$$\textbf{Calculus}: \quad \frac{dE}{dx} = 2(x-70) + 2(x-80) + 2(x-120) = 0 \text{ gives } \widehat{x} = \frac{70+80+120}{3}$$

$$\textbf{Projection}: \quad \widehat{x} = \frac{\boldsymbol{a}^{\mathrm{T}}\boldsymbol{b}}{\boldsymbol{a}^{\mathrm{T}}\boldsymbol{a}} = \frac{(1,1,1)^{\mathrm{T}}(70,80,120)}{(1,1,1)^{\mathrm{T}}(1,1,1)} = \frac{70+80+120}{3}$$

$A^{\mathrm{T}}A\widehat{x} = A^{\mathrm{T}}\boldsymbol{b}$ is 1 by 1 because A has only one column $(1, 1, 1)$. The new measurement $b_4 = 130$ adds a fourth equation and $\widehat{x}$ is updated to 100, either by averaging b_1, b_2, b_3, b_4 or by recursively using the old average of b_1, b_2, b_3:

$$\widehat{x}_{\text{new}} = \frac{70+80+120+130}{4} = 100 \quad \text{is also} \quad \widehat{x}_{\text{old}} + \frac{1}{4}(b_4 - \widehat{x}_{\text{old}}) = 90 + \frac{1}{4}(40).$$

The update from 999 to 1000 measurements shows the "gain matrix" $\frac{1}{1000}$ in a **Kalman filter** multiplying the prediction error $b_{\text{new}} - \widehat{x}_{\text{old}}$. Notice $\frac{1}{1000} = \frac{1}{999} - \frac{1}{999000}$:

$$\widehat{x}_{\text{new}} = \frac{b_1 + \cdots + b_{1000}}{1000} = \frac{b_1 + \cdots + b_{999}}{999} + \frac{1}{1000}\left(b_{1000} - \frac{b_1 + \cdots + b_{999}}{999}\right).$$

Problem Set 4.2

Questions 1–9 ask for projections onto lines. Also errors $\boldsymbol{e} = \boldsymbol{b} - \boldsymbol{p}$ and matrices P.

1 Project the vector $\boldsymbol{b}$ onto the line through $\boldsymbol{a}$. Check that $\boldsymbol{e}$ is perpendicular to $\boldsymbol{a}$:

$$\text{(a)} \quad \boldsymbol{b} = \begin{bmatrix} 1 \\ 2 \\ 2 \end{bmatrix} \text{ and } \boldsymbol{a} = \begin{bmatrix} 1 \\ 1 \\ 1 \end{bmatrix} \qquad \text{(b)} \quad \boldsymbol{b} = \begin{bmatrix} 1 \\ 3 \\ 1 \end{bmatrix} \text{ and } \boldsymbol{a} = \begin{bmatrix} -1 \\ -3 \\ -1 \end{bmatrix}.$$

2 *Draw* the projection of $\boldsymbol{b}$ onto $\boldsymbol{a}$ and also compute it from $\boldsymbol{p} = \widehat{x}\boldsymbol{a}$:

$$\text{(a)}\quad \boldsymbol{b} = \begin{bmatrix} \cos\theta \\ \sin\theta \end{bmatrix} \quad \text{and} \quad \boldsymbol{a} = \begin{bmatrix} 1 \\ 0 \end{bmatrix} \qquad \text{(b)}\quad \boldsymbol{b} = \begin{bmatrix} 1 \\ 1 \end{bmatrix} \quad \text{and} \quad \boldsymbol{a} = \begin{bmatrix} 1 \\ -1 \end{bmatrix}.$$

3 In Problem 1, find the projection matrix $P = \boldsymbol{a}\boldsymbol{a}^{\mathrm{T}}/\boldsymbol{a}^{\mathrm{T}}\boldsymbol{a}$ onto the line through each vector $\boldsymbol{a}$. Verify in both cases that $P^2 = P$. Multiply $P\boldsymbol{b}$ in each case to compute the projection $\boldsymbol{p}$.

4 Construct the projection matrices P_1 and P_2 onto the lines through the $\boldsymbol{a}$'s in Problem 2. Is it true that $(P_1 + P_2)^2 = P_1 + P_2$? This *would* be true if $P_1 P_2 = 0$.

5 Compute the projection matrices $\boldsymbol{a}\boldsymbol{a}^{\mathrm{T}}/\boldsymbol{a}^{\mathrm{T}}\boldsymbol{a}$ onto the lines through $\boldsymbol{a}_1 = (-1, 2, 2)$ and $\boldsymbol{a}_2 = (2, 2, -1)$. Multiply those projection matrices and explain why their product $P_1 P_2$ is what it is.

6 Project $\boldsymbol{b} = (1, 0, 0)$ onto the lines through $\boldsymbol{a}_1$ and $\boldsymbol{a}_2$ in Problem 5 and also onto $\boldsymbol{a}_3 = (2, -1, 2)$. Add up the three projections $\boldsymbol{p}_1 + \boldsymbol{p}_2 + \boldsymbol{p}_3$.

7 Continuing Problems 5–6, find the projection matrix P_3 onto $\boldsymbol{a}_3 = (2, -1, 2)$. Verify that $P_1 + P_2 + P_3 = I$. The basis $\boldsymbol{a}_1, \boldsymbol{a}_2, \boldsymbol{a}_3$ is orthogonal!

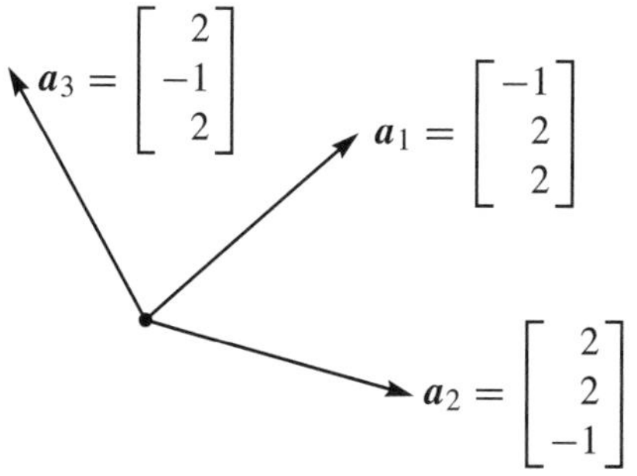

Questions 5–6–7

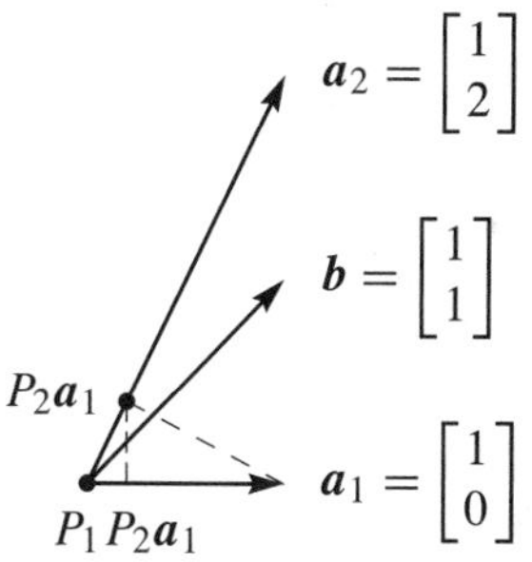

Questions 8–9–10

8 Project the vector $\boldsymbol{b} = (1, 1)$ onto the lines through $\boldsymbol{a}_1 = (1, 0)$ and $\boldsymbol{a}_2 = (1, 2)$. Draw the projections $\boldsymbol{p}_1$ and $\boldsymbol{p}_2$ and add $\boldsymbol{p}_1 + \boldsymbol{p}_2$. The projections do not add to $\boldsymbol{b}$ because the $\boldsymbol{a}$'s are not orthogonal.

9 In Problem 8, the projection of $\boldsymbol{b}$ onto the *plane* of $\boldsymbol{a}_1$ and $\boldsymbol{a}_2$ will equal $\boldsymbol{b}$. Find $P = A(A^{\mathrm{T}}A)^{-1}A^{\mathrm{T}}$ for $A = \begin{bmatrix} \boldsymbol{a}_1 & \boldsymbol{a}_2 \end{bmatrix} = \left[\begin{smallmatrix} 1 & 1 \\ 0 & 2 \end{smallmatrix}\right]$.

10 Project $\boldsymbol{a}_1 = (1, 0)$ onto $\boldsymbol{a}_2 = (1, 2)$. Then project the result back onto $\boldsymbol{a}_1$. Draw these projections and multiply the projection matrices $P_1 P_2$: Is this a projection?

Questions 11–20 ask for projections, and projection matrices, onto subspaces.

11 Project $\boldsymbol{b}$ onto the column space of A by solving $A^{\mathrm{T}}A\widehat{\boldsymbol{x}} = A^{\mathrm{T}}\boldsymbol{b}$ and $\boldsymbol{p} = A\widehat{\boldsymbol{x}}$:

(a) $A = \begin{bmatrix} 1 & 1 \\ 0 & 1 \\ 0 & 0 \end{bmatrix}$ and $b = \begin{bmatrix} 2 \\ 3 \\ 4 \end{bmatrix}$ (b) $A = \begin{bmatrix} 1 & 1 \\ 1 & 1 \\ 0 & 1 \end{bmatrix}$ and $b = \begin{bmatrix} 4 \\ 4 \\ 6 \end{bmatrix}$.

Find $e = b - p$. It should be perpendicular to the columns of A.

12 Compute the projection matrices P_1 and P_2 onto the column spaces in Problem 11. Verify that $P_1 b$ gives the first projection p_1. Also verify $P_2^2 = P_2$.

13 (Quick and Recommended) Suppose A is the 4 by 4 identity matrix with its last column removed. A is 4 by 3. Project $b = (1, 2, 3, 4)$ onto the column space of A. What shape is the projection matrix P and what is P?

14 Suppose b equals 2 times the first column of A. What is the projection of b onto the column space of A? Is $P = I$ for sure in this case? Compute p and P when $b = (0, 2, 4)$ and the columns of A are $(0, 1, 2)$ and $(1, 2, 0)$.

15 If A is doubled, then $P = 2A(4A^TA)^{-1}2A^T$. This is the same as $A(A^TA)^{-1}A^T$. The column space of $2A$ is the same as _____. Is $\widehat{x}$ the same for A and $2A$?

16 What linear combination of $(1, 2, -1)$ and $(1, 0, 1)$ is closest to $b = (2, 1, 1)$?

17 (*Important*) If $P^2 = P$ show that $(I - P)^2 = I - P$. When P projects onto the column space of A, $I - P$ projects onto the _____.

18 (a) If P is the 2 by 2 projection matrix onto the line through $(1, 1)$, then $I - P$ is the projection matrix onto _____.

(b) If P is the 3 by 3 projection matrix onto the line through $(1, 1, 1)$, then $I - P$ is the projection matrix onto _____.

19 To find the projection matrix onto the plane $x - y - 2z = 0$, choose two vectors in that plane and make them the columns of A. The plane should be the column space. Then compute $P = A(A^TA)^{-1}A^T$.

20 To find the projection matrix P onto the same plane $x - y - 2z = 0$, write down a vector e that is perpendicular to that plane. Compute the projection $Q = ee^T/e^Te$ and then $P = I - Q$.

Questions 21–26 show that projection matrices satisfy $P^2 = P$ and $P^T = P$.

21 Multiply the matrix $P = A(A^TA)^{-1}A^T$ by itself. Cancel to prove that $P^2 = P$. Explain why $P(Pb)$ always equals Pb: The vector Pb is in the column space so its projection is _____.

22 Prove that $P = A(A^TA)^{-1}A^T$ is symmetric by computing P^T. Remember that the inverse of a symmetric matrix is symmetric.

23 If A is square and invertible, the warning against splitting $(A^TA)^{-1}$ does not apply. It is true that $AA^{-1}(A^T)^{-1}A^T = I$. *When A is invertible, why is $P = I$?* ***What is the error e?***

24 The nullspace of A^{T} is _____ to the column space $C(A)$. So if $A^{\mathrm{T}}\boldsymbol{b} = \mathbf{0}$, the projection of $\boldsymbol{b}$ onto $C(A)$ should be $\boldsymbol{p} =$ _____. Check that $P = A(A^{\mathrm{T}}A)^{-1}A^{\mathrm{T}}$ gives this answer.

25 The projection matrix P onto an n-dimensional subspace has rank $r = n$. ***Reason:*** The projections $P\boldsymbol{b}$ fill the subspace S. So S is the _____ of P.

26 If an m by m matrix has $A^2 = A$ and its rank is m, prove that $A = I$.

27 The important fact in Theorem **4G** is this: ***If*** $\boldsymbol{A^{\mathrm{T}}Ax = 0}$ ***then*** $\boldsymbol{Ax = 0}$. The vector $A\boldsymbol{x}$ is in the nullspace of _____. $A\boldsymbol{x}$ is always in the column space of _____. To be in both perpendicular spaces, $A\boldsymbol{x}$ must be zero.

28 Use $P^{\mathrm{T}} = P$ and $P^2 = P$ to prove that the length squared of column 2 always equals the diagonal entry p_{22}. This number is $\frac{2}{6} = \frac{4}{36} + \frac{4}{36} + \frac{4}{36}$ for

$$P = \frac{1}{6}\begin{bmatrix} 5 & 2 & -1 \\ 2 & 2 & 2 \\ -1 & 2 & 5 \end{bmatrix}.$$

29 If B has rank m (full row rank, independent rows) show that BB^{T} is invertible.

30 (a) Find the projection matrix P_C onto the column space of A (after looking closely at the matrix!)

$$A = \begin{bmatrix} 3 & 6 & 6 \\ 4 & 8 & 8 \end{bmatrix}$$

(b) Find the 3 by 3 projection matrix P_R onto the row space of A. Multiply $B = P_C A P_R$. Your answer B should be a little surprising—can you explain it?

LEAST SQUARES APPROXIMATIONS ■ 4.3

It often happens that $Ax = b$ has no solution. The usual reason is: *too many equations*. The matrix has more rows than columns. There are more equations than unknowns (m is greater than n). The n columns span a small part of m-dimensional space. Unless all measurements are perfect, b is outside that column space. Elimination reaches an impossible equation and stops. But these are real problems and they need an answer.

To repeat: We cannot always get the error $e = b - Ax$ down to zero. When e is zero, x is an exact solution to $Ax = b$. *When the length of* e *is as small as possible,* $\widehat{x}$ *is a least squares solution*. Our goal in this section is to compute $\widehat{x}$ and use it.

The previous section emphasized p (the projection). This section emphasizes $\widehat{x}$ (the least squares solution). They are connected by $p = A\widehat{x}$. The fundamental equation is still $A^{\mathrm{T}}A\widehat{x} = A^{\mathrm{T}}b$. Here is a short unofficial way to derive it:

When the original $Ax = b$ has no solution, **multiply by A^{T} and solve $A^{\mathrm{T}}A\widehat{x} = A^{\mathrm{T}}b$.**

Example 1 A crucial application of least squares is fitting a straight line to m points. Start with three points: *Find the closest line to the points* $(0, 6)$, $(1, 0)$, *and* $(2, 0)$.

No straight line goes through those points. We are asking for two numbers C and D that satisfy three equations. The line is $b = C + Dt$. Here are the equations at $t = 0, 1, 2$ to match the given values $b = 6, 0, 0$:

The first point is on the line $b = C + Dt$ if	$C + D \cdot 0 = 6$
The second point is on the line $b = C + Dt$ if	$C + D \cdot 1 = 0$
The third point is on the line $b = C + Dt$ if	$C + D \cdot 2 = 0.$

This 3 by 2 system has no solution; $b = (6, 0, 0)$ is not a combination of the columns $(1, 1, 1)$ and $(0, 1, 2)$:

$$A = \begin{bmatrix} 1 & 0 \\ 1 & 1 \\ 1 & 2 \end{bmatrix} \quad x = \begin{bmatrix} C \\ D \end{bmatrix} \quad b = \begin{bmatrix} 6 \\ 0 \\ 0 \end{bmatrix} \quad Ax = b \text{ is } \textit{not} \text{ solvable.}$$

The same numbers were in Example 3 in the last section. We computed $\widehat{x} = (5, -3)$. Those numbers are the best C and D, so $5 - 3t$ is the best line for the 3 points.

In practical problems, there easily could be $m = 100$ points instead of $m = 3$. They don't exactly match any $C + Dt$. Our numbers $6, 0, 0$ exaggerate the error so you can see it clearly.

Minimizing the Error

How do we make the error $e = b - Ax$ as small as possible? This is an important question with a beautiful answer. The best x (called $\widehat{x}$) can be found by geometry or algebra or calculus:

By geometry Every Ax lies in the plane of the columns $(1, 1, 1)$ and $(0, 1, 2)$. In that plane, we look for the point closest to $\boldsymbol{b}$. *The nearest point is the projection* $\boldsymbol{p}$.

The best choice for $A\boldsymbol{x}$ is $\boldsymbol{p}$. The smallest possible error is $\boldsymbol{e} = \boldsymbol{b} - \boldsymbol{p}$. The three points at heights $\boldsymbol{p} = (p_1, p_2, p_3)$ do lie on a line, because $\boldsymbol{p}$ is in the column space. $A\boldsymbol{x} = \boldsymbol{p}$ has the same solution $\widehat{\boldsymbol{x}}$, the best choice for (C, D).

By algebra Every vector $\boldsymbol{b}$ splits into two parts. The part in the column space is $\boldsymbol{p}$. The perpendicular part in the left nullspace is $\boldsymbol{e}$. There is an equation we cannot solve $(A\boldsymbol{x} = \boldsymbol{b})$. There is an equation we do solve (by removing $\boldsymbol{e}$):

$$A\boldsymbol{x} = \boldsymbol{b} = \boldsymbol{p} + \boldsymbol{e} \quad \textit{is impossible;} \qquad A\widehat{\boldsymbol{x}} = \boldsymbol{p} \quad \textit{is solvable.} \tag{1}$$

The solution $A\widehat{\boldsymbol{x}} = \boldsymbol{p}$ makes the error as small as possible, because for any $\boldsymbol{x}$:

$$\|A\boldsymbol{x} - \boldsymbol{b}\|^2 = \|A\boldsymbol{x} - \boldsymbol{p}\|^2 + \|\boldsymbol{e}\|^2. \tag{2}$$

This is the law $c^2 = a^2 + b^2$ for a right triangle. The vector $A\boldsymbol{x} - \boldsymbol{p}$ in the column space is perpendicular to $\boldsymbol{e}$ in the left nullspace. We reduce $A\boldsymbol{x} - \boldsymbol{p}$ to zero by choosing $\boldsymbol{x}$ to be $\widehat{\boldsymbol{x}}$. That leaves the smallest possible errors, namely $\boldsymbol{e} = (e_1, e_2, e_3)$.

Notice what "smallest" means. The *squared length* of $A\boldsymbol{x} - \boldsymbol{b}$ is minimized:

The least squares solution $\widehat{\boldsymbol{x}}$ makes $E = \|A\boldsymbol{x} - \boldsymbol{b}\|^2$ as small as possible.

By calculus Most functions are minimized by calculus! The graph bottoms out and the derivative in every direction is zero. Here the error function to be minimized is a *sum of squares* $e_1^2 + e_2^2 + e_3^2$ (the square of the error in each equation 1, 2, 3):

$$E = \|A\boldsymbol{x} - \boldsymbol{b}\|^2 = (C + D \cdot 0 - 6)^2 + (C + D \cdot 1)^2 + (C + D \cdot 2)^2. \tag{3}$$

The unknowns are C and D. With two unknowns there are *two derivatives*—both zero at the minimum. They are "partial derivatives" because $\partial E/\partial C$ treats D as constant and $\partial E/\partial D$ treats C as constant:

$$\partial E/\partial C = 2(C + D \cdot 0 - 6) \quad + 2(C + D \cdot 1) \quad + 2(C + D \cdot 2) \quad = 0$$

$$\partial E/\partial D = 2(C + D \cdot 0 - 6)(0) + 2(C + D \cdot 1)(1) + 2(C + D \cdot 2)(2) = 0.$$

$\partial E/\partial D$ contains the extra factors $0, 1, 2$ from the chain rule. (The derivative of $(4 + 5x)^2$ is 2 times $4 + 5x$ times an extra 5.) In the C derivative the corresponding factors are $1, 1, 1$, because C is always multiplied by 1. It is no accident that 1, 1, 1 and 0, 1, 2 are the columns of A.

Now cancel 2 from every term above and collect all C's and all D's:

$$\begin{array}{ll}\text{The } C \text{ derivative } \frac{\partial E}{\partial C} \text{ is zero:} & 3C + 3D = 6 \\ \text{The } D \text{ derivative } \frac{\partial E}{\partial D} \text{ is zero:} & 3C + 5D = 0\end{array} \qquad \textbf{This matrix} \begin{bmatrix} 3 & 3 \\ 3 & 5 \end{bmatrix} \textbf{ is } A^{\mathrm{T}}A \quad (4)$$

These equations are identical with $A^{\mathrm{T}}A\widehat{x} = A^{\mathrm{T}}\boldsymbol{b}$. The best C and D are the components of $\widehat{x}$. The equations from calculus are the same as the "normal equations" from linear algebra. These are the key equations of least squares:

The partial derivatives of $\|A\boldsymbol{x} - \boldsymbol{b}\|^2$ *are zero when* $\boldsymbol{A^{\mathrm{T}}A\widehat{x} = A^{\mathrm{T}}b}$.

The solution is $C = 5$ and $D = -3$. Therefore $b = 5 - 3t$ is the best line—it comes closest to the three points. At $t = 0, 1, 2$ this line goes through $\boldsymbol{p} = 5, 2, -1$. It could not go through $\boldsymbol{b} = 6, 0, 0$. The errors are $1, -2, 1$. This is the vector $\boldsymbol{e}$!

Figure 4.6a shows the closest line. It misses by distances $e_1, e_2, e_3 = 1, -2, 1$. *Those are vertical distances.* The least squares line minimizes the total squared error $E = e_1^2 + e_2^2 + e_3^2$.

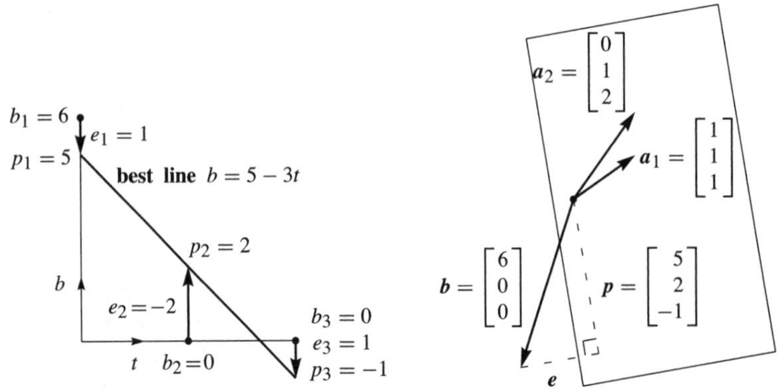

Figure 4.6 Best line and projection: Two pictures, same problem. The line has heights $\boldsymbol{p} = (5, 2, -1)$ with errors $\boldsymbol{e} = (1, -2, 1)$. The equations $A^{\mathrm{T}}A\widehat{x} = A^{\mathrm{T}}\boldsymbol{b}$ give $\widehat{x} = (5, -3)$. The best line is $b = 5 - 3t$ and the projection is $\boldsymbol{p} = 5\boldsymbol{a}_1 - 3\boldsymbol{a}_2$.

Figure 4.6b shows the same problem in 3-dimensional space ($\boldsymbol{bpe}$ space). The vector $\boldsymbol{b}$ is not in the column space of A. That is why we could not solve $A\boldsymbol{x} = \boldsymbol{b}$ and put a line through the three points. The smallest possible error is the perpendicular vector $\boldsymbol{e}$. This is $\boldsymbol{e} = \boldsymbol{b} - A\widehat{x}$, the vector of errors $(1, -2, 1)$ in the three equations—and the distances from the best line. Behind both figures is the fundamental equation $A^{\mathrm{T}}A\widehat{x} = A^{\mathrm{T}}\boldsymbol{b}$.

Notice that the errors $1, -2, 1$ add to zero. The error $\boldsymbol{e} = (e_1, e_2, e_3)$ is perpendicular to the first column $(1, 1, 1)$ in A. The dot product gives $e_1 + e_2 + e_3 = 0$.

The Big Picture

The key figure of this book shows the four subspaces and the true action of a matrix. The vector $\boldsymbol{x}$ on the left side of Figure 4.2 went to $\boldsymbol{b} = A\boldsymbol{x}$ on the right side. In that figure $\boldsymbol{x}$ was split into $\boldsymbol{x}_r + \boldsymbol{x}_n$. There were many solutions to $A\boldsymbol{x} = \boldsymbol{b}$.

In this section the situation is just the opposite. There are *no* solutions to $A\boldsymbol{x} = \boldsymbol{b}$. *Instead of splitting up* $\boldsymbol{x}$ *we are splitting up* $\boldsymbol{b}$. Figure 4.7 shows the big picture for least squares. Instead of $A\boldsymbol{x} = \boldsymbol{b}$ we solve $A\widehat{\boldsymbol{x}} = \boldsymbol{p}$. The error $\boldsymbol{e} = \boldsymbol{b} - \boldsymbol{p}$ is unavoidable.

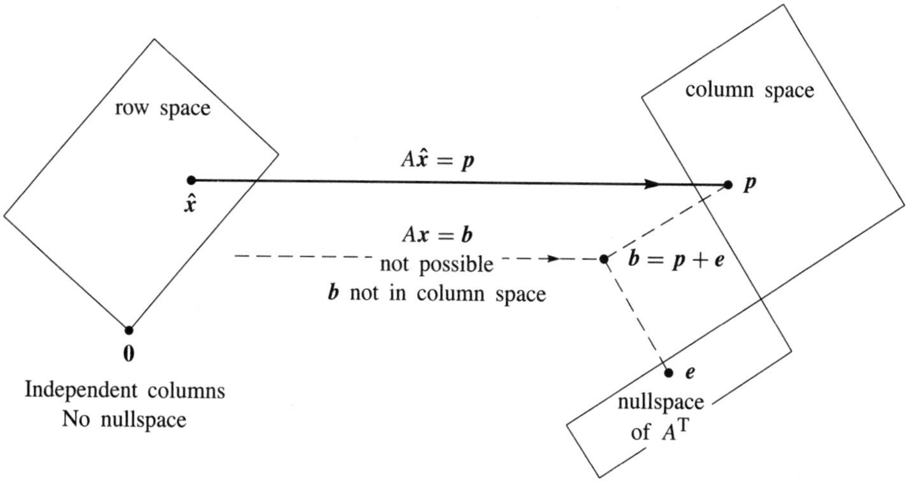

Figure 4.7 The projection $\boldsymbol{p} = A\widehat{\boldsymbol{x}}$ is closest to $\boldsymbol{b}$, so $\widehat{\boldsymbol{x}}$ minimizes $E = \|\boldsymbol{b} - A\boldsymbol{x}\|^2$.

Notice how the nullspace $N(A)$ is very small—just one point. With independent columns, the only solution to $A\boldsymbol{x} = \boldsymbol{0}$ is $\boldsymbol{x} = \boldsymbol{0}$. Then A^TA is invertible. The equation $A^TA\widehat{\boldsymbol{x}} = A^T\boldsymbol{b}$ fully determines the best vector $\widehat{\boldsymbol{x}}$.

Chapter 7 will have the complete picture—all four subspaces included. Every $\boldsymbol{x}$ splits into $\boldsymbol{x}_r + \boldsymbol{x}_n$, and every $\boldsymbol{b}$ splits into $\boldsymbol{p} + \boldsymbol{e}$. The best solution is still $\widehat{\boldsymbol{x}}$ (or $\widehat{\boldsymbol{x}}_r$) in the row space. We can't help $\boldsymbol{e}$ and we don't want $\boldsymbol{x}_n$—this leaves $A\widehat{\boldsymbol{x}} = \boldsymbol{p}$.

Fitting a Straight Line

Fitting a line is the clearest application of least squares. It starts with $m > 2$ points—hopefully near a straight line. At times $t_1, \ldots, t_m$ those points are at heights $b_1, \ldots, b_m$. Figure 4.6a shows the best line $b = C + Dt$, which misses the points by vertical distances $e_1, \ldots, e_m$. No line is perfect, and the least squares line minimizes $E = e_1^2 + \cdots + e_m^2$.

The first example in this section had three points. Now we allow m points (m can be large). The two components of $\widehat{\boldsymbol{x}}$ are still C and D.

A line goes through the m points when we exactly solve $A\boldsymbol{x} = \boldsymbol{b}$. Generally we can't do it. Two unknowns C and D determine a line, so A has only $n = 2$ columns.

To fit the m points, we are trying to solve m equations (and we only want two!):

$$Ax = b \quad \text{is} \quad \begin{aligned} C + Dt_1 &= b_1 \\ C + Dt_2 &= b_2 \\ &\vdots \\ C + Dt_m &= b_m \end{aligned} \qquad \text{with} \qquad A = \begin{bmatrix} 1 & t_1 \\ 1 & t_2 \\ \vdots & \vdots \\ 1 & t_m \end{bmatrix}. \tag{5}$$

The column space is so thin that almost certainly $\boldsymbol{b}$ is outside of it. When $\boldsymbol{b}$ happens to lie in the column space, the points happen to lie on a line. In that case $\boldsymbol{b} = \boldsymbol{p}$. Then $Ax = \boldsymbol{b}$ is solvable and the errors are $\boldsymbol{e} = (0, \ldots, 0)$.

The closest line $C + Dt$ has heights $p_1, \ldots, p_m$ with errors $e_1, \ldots, e_m$.

$A^{\mathrm{T}}A\widehat{x} = A^{\mathrm{T}}b$ will give $\widehat{x} = (C, D)$. The errors are $e_i = b_i - C - Dt_i$.

Fitting points by a straight line is so important that we give the two equations $A^{\mathrm{T}}A\widehat{x} = A^{\mathrm{T}}\boldsymbol{b}$, once and for all. The two columns of A are independent (unless all times t_i are the same). So we turn to least squares and solve $A^{\mathrm{T}}A\widehat{x} = A^{\mathrm{T}}\boldsymbol{b}$. The "dot-product matrix" $A^{\mathrm{T}}A$ is 2 by 2:

$$A^{\mathrm{T}}A = \begin{bmatrix} 1 & \cdots & 1 \\ t_1 & \cdots & t_m \end{bmatrix} \begin{bmatrix} 1 & t_1 \\ \vdots & \vdots \\ 1 & t_m \end{bmatrix} = \begin{bmatrix} m & \sum t_i \\ \sum t_i & \sum t_i^2 \end{bmatrix}. \tag{6}$$

On the right side of the normal equation is the 2 by 1 vector $A^{\mathrm{T}}\boldsymbol{b}$:

$$A^{\mathrm{T}}\boldsymbol{b} = \begin{bmatrix} 1 & \cdots & 1 \\ t_1 & \cdots & t_m \end{bmatrix} \begin{bmatrix} b_1 \\ \vdots \\ b_m \end{bmatrix} = \begin{bmatrix} \sum b_i \\ \sum t_i b_i \end{bmatrix}. \tag{7}$$

In a specific problem, all these numbers are given. A formula for C and D is coming.

4H The line $C + Dt$ which minimizes $e_1^2 + \cdots + e_m^2$ is determined by $A^{\mathrm{T}}A\widehat{x} = A^{\mathrm{T}}\boldsymbol{b}$:

$$\begin{bmatrix} m & \sum t_i \\ \sum t_i & \sum t_i^2 \end{bmatrix} \begin{bmatrix} C \\ D \end{bmatrix} = \begin{bmatrix} \sum b_i \\ \sum t_i b_i \end{bmatrix}. \tag{8}$$

The vertical errors at the m points on the line are the components of $\boldsymbol{e} = \boldsymbol{b} - \boldsymbol{p}$. This error vector (the *residual*) $\boldsymbol{b} - A\widehat{x}$ is perpendicular to the columns of A (geometry). It is in the nullspace of A^{T} (linear algebra). The best $\widehat{x} = (C, D)$ minimizes the total error E, the sum of squares:

$$E(x) = \|Ax - \boldsymbol{b}\|^2 = (C + Dt_1 - b_1)^2 + \cdots + (C + Dt_m - b_m)^2.$$

When calculus sets the derivatives $\partial E/\partial C$ and $\partial E/\partial D$ to zero, it produces 4H.

Other least squares problems have more than two unknowns. Fitting by the best parabola has $n = 3$ coefficients C, D, E (see below). In general we are fitting m data points by n parameters $x_1, \ldots, x_n$. The matrix A has n columns and $n < m$. The derivatives of $\|A\boldsymbol{x} - \boldsymbol{b}\|^2$ give the n equations $A^{\mathrm{T}}A\widehat{\boldsymbol{x}} = A^{\mathrm{T}}\boldsymbol{b}$. **The derivative of a square is linear**—this is why the method of least squares is so popular.

Example 2 A has *orthogonal columns* when the measurement times t_i add to zero. Suppose $b = 1, 2, 4$ at times $t = -2, 0, 2$. Those times add to zero. The dot product with the other column $1, 1, 1$ is zero:

$$\begin{aligned} C + D(-2) &= 1 \\ C + D(0) &= 2 \\ C + D(2) &= 4 \end{aligned} \quad \text{or} \quad A\boldsymbol{x} = \begin{bmatrix} 1 & -2 \\ 1 & 0 \\ 1 & 2 \end{bmatrix} \begin{bmatrix} C \\ D \end{bmatrix} = \begin{bmatrix} 1 \\ 2 \\ 4 \end{bmatrix}.$$

Look at the matrix $A^{\mathrm{T}}A$ in the least squares equation for $\widehat{\boldsymbol{x}}$:

$$A^{\mathrm{T}}A\widehat{\boldsymbol{x}} = A^{\mathrm{T}}\boldsymbol{b} \quad \text{is} \quad \begin{bmatrix} 3 & 0 \\ 0 & 8 \end{bmatrix} \begin{bmatrix} C \\ D \end{bmatrix} = \begin{bmatrix} 7 \\ 6 \end{bmatrix}.$$

Main point: Now $A^{\mathrm{T}}A$ *is diagonal.* We can solve separately for $C = \frac{7}{3}$ and $D = \frac{6}{8}$. The zeros in $A^{\mathrm{T}}A$ are dot products of perpendicular columns in A. The diagonal matrix $A^{\mathrm{T}}A$, with entries $m = 3$ and $t_1^2 + t_2^2 + t_3^2 = 8$, is virtually as good as the identity matrix.

Orthogonal columns are so helpful that it is worth moving the time origin to produce them. To do that, subtract away the average time $\widehat{t} = (t_1 + \cdots + t_m)/m$. The shifted times $T_i = t_i - \widehat{t}$ add to zero. With the columns now orthogonal, $A^{\mathrm{T}}A$ is diagonal. Its entries are m and $T_1^2 + \cdots + T_m^2$. The best C and D have direct formulas:

$$C = \frac{b_1 + \cdots + b_m}{m} \quad \text{and} \quad D = \frac{b_1T_1 + \cdots + b_mT_m}{T_1^2 + \cdots + T_m^2}. \tag{9}$$

The best line is $C + DT$ ***or*** $C + D(t - \widehat{t})$. The time shift that makes $A^{\mathrm{T}}A$ diagonal is an example of the Gram-Schmidt process: *orthogonalize the columns in advance.*

Fitting by a Parabola

If we throw a ball, it would be crazy to fit the path by a straight line. A parabola $b = C + Dt + Et^2$ allows the ball to go up and come down again (b is the height at time t). The actual path is not a perfect parabola, but the whole theory of projectiles starts with that approximation.

When Galileo dropped a stone from the Leaning Tower of Pisa, it accelerated. The distance contains a quadratic term $\frac{1}{2}gt^2$. (Galileo's point was that the stone's mass is not involved.) Without that term we could never send a satellite into the right orbit. But even with a nonlinear function like t^2, the unknowns C, D, E appear linearly. Choosing the best parabola is still a problem in linear algebra.

Problem Fit heights $b_1, \ldots, b_m$ at times $t_1, \ldots, t_m$ by a parabola $b = C + Dt + Et^2$.

With $m > 3$ points, the m equations for an exact fit are generally unsolvable:

$$\begin{aligned} C + Dt_1 + Et_1^2 &= b_1 \\ &\vdots \\ C + Dt_m + Et_m^2 &= b_m \end{aligned} \qquad \text{has the } m \text{ by 3 matrix} \qquad A = \begin{bmatrix} 1 & t_1 & t_1^2 \\ \vdots & \vdots & \vdots \\ 1 & t_m & t_m^2 \end{bmatrix}. \tag{10}$$

Least squares The best parabola chooses $\widehat{x} = (C, D, E)$ to satisfy the three normal equations $A^{\mathrm{T}} A \widehat{x} = A^{\mathrm{T}} b$.

May I ask you to convert this to a problem of projection? The column space of A has dimension ______. The projection of b is $p = A\widehat{x}$, which combines the three columns using the coefficients C, D, E. The error at the first data point is $e_1 = b_1 - C - Dt_1 - Et_1^2$. The total squared error is $e_1^2 +$ ______. If you prefer to minimize by calculus, take the partial derivatives of E with respect to ______, ______, ______. These three derivatives will be zero when $\widehat{x} = (C, D, E)$ solves the 3 by 3 system of equations ______.

Section 8.5 has more least squares applications. The big one is Fourier series—approximating functions instead of vectors. The error to be minimized changes from a sum $e_1^2 + \cdots + e_m^2$ to an integral. We will find the straight line closest to $f(x)$.

Example 3 For a parabola $b = C + Dt + Et^2$ to go through the three heights $b = 6, 0, 0$ when $t = 0, 1, 2$, the equations are

$$\begin{aligned} C + D \cdot 0 + E \cdot 0^2 &= 6 \\ C + D \cdot 1 + E \cdot 1^2 &= 0 \\ C + D \cdot 2 + E \cdot 2^2 &= 0. \end{aligned} \tag{11}$$

This is $Ax = b$. We can solve it exactly. Three data points give three equations and a square matrix. The solution is $x = (C, D, E) = (6, -9, 3)$. The parabola through the three points in Figure 4.8a is $b = 6 - 9t + 3t^2$.

What does this mean for projection? The matrix has three columns, which span the whole space $\mathbf{R}^3$. The projection matrix is the identity matrix! The projection of b is b. The error is zero. We didn't need $A^{\mathrm{T}} A \widehat{x} = A^{\mathrm{T}} b$, because we solved $Ax = b$. Of course we could multiply by A^{T}, but there is no reason to do it.

Figure 4.8a also shows a fourth point b_4 at time t_4. If that falls on the parabola, the new $Ax = b$ (four equations) is still solvable. When the fourth point is not on the parabola, we turn to $A^{\mathrm{T}} A \widehat{x} = A^{\mathrm{T}} b$. Will the least squares parabola stay the same, with all the error at the fourth point? Not likely!

The smallest error vector (e_1, e_2, e_3, e_4) is perpendicular to $(1, 1, 1, 1)$, the first column of A. Least squares balances out the four errors, and they add to zero.

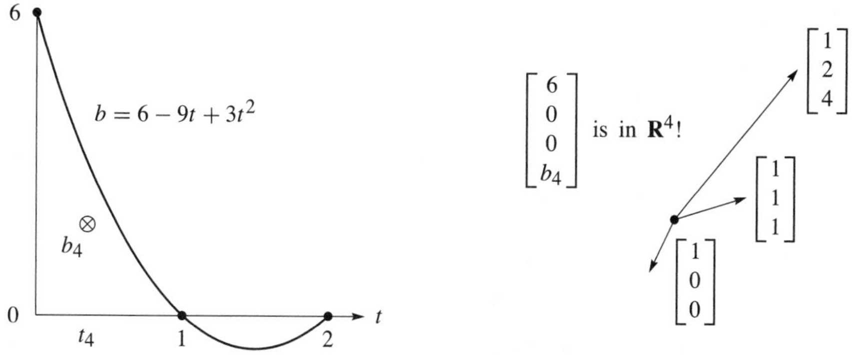

Figure 4.8 From Example 3: An exact fit of the parabola through three points means $\boldsymbol{p} = \boldsymbol{b}$ and $\boldsymbol{e} = \mathbf{0}$. The fourth point will require least squares.

■ REVIEW OF THE KEY IDEAS ■

1. The least squares solution $\widehat{\boldsymbol{x}}$ minimizes $E = \|A\boldsymbol{x} - \boldsymbol{b}\|^2$. This is the sum of squares of the errors in the m equations $(m > n)$.

2. The best $\widehat{\boldsymbol{x}}$ comes from the normal equations $A^{\mathrm{T}}A\widehat{\boldsymbol{x}} = A^{\mathrm{T}}\boldsymbol{b}$.

3. To fit m points by a line $b = C + Dt$, the two normal equations give C and D.

4. The heights of the best line are $\boldsymbol{p} = (p_1, \ldots, p_m)$. The vertical distances to the data points are the errors $\boldsymbol{e} = (e_1, \ldots, e_m)$.

5. If we try to fit m points by a combination of $n < m$ functions, the m equations are generally unsolvable. The n normal equations give the least squares solution.

■ WORKED EXAMPLES ■

4.3 A Start with nine measurements b_1 to b_9, *all zero*, at times $t = 1, \ldots, 9$. The tenth measurement $b_{10} = 40$ is an outlier. Find the best *horizontal line* $y = C$ to fit the ten points $(1, 0), (2, 0), \ldots, (9, 0), (10, 40)$ using three measures for the error E: **(1)** Least squares $e_1^2 + \cdots + e_{10}^2$ **(2)** Least maximum error $|e_{\max}|$ **(3)** Least sum of errors $|e_1| + \cdots + |e_{10}|$.

Then find the least squares straight line $C + Dt$ through those ten points. What happens to C and D if you multiply the b_i by 3 and then add 30 to get $\boldsymbol{b}_{\text{new}} = (30, 30, \ldots, 150)$? What is the best line if you multiply the times $t_i = 1, \ldots, 10$ by 2 and then add 10 to get $t_{\text{new}} = 12, 14, \ldots, 30$?

Solution **(1)** The least squares fit to $0, 0, \ldots, 0, 40$ by a horizontal line is the average $C = \frac{40}{10} = 4$. **(2)** The least maximum error requires $C = 20$, halfway between

0 and 40. **(3)** The least sum requires $C = 0$ (!!). The sum of errors $9|C| + |40 - C|$ would increase if C moves up from zero.

The least sum comes from the *median* measurement (the median of $0, \ldots, 0, 40$ is zero). Changing the best $y = C = b_{\text{median}}$ increases half the errors and decreases half. Many statisticians feel that the least squares solution is too heavily influenced by outliers like $b_{10} = 40$, and they prefer least sum. But the equations become nonlinear.

The least squares straight line $C+Dt$ requires $A^{\mathrm{T}}A$ and $A^{\mathrm{T}}\boldsymbol{b}$ with $t = 1, \ldots, 10$:

$$A^{\mathrm{T}}A = \begin{bmatrix} 10 & \sum t_i \\ \sum t_i & \sum t_i^2 \end{bmatrix} = \begin{bmatrix} 10 & 55 \\ 55 & 385 \end{bmatrix} \qquad A^{\mathrm{T}}\boldsymbol{b} = \begin{bmatrix} \sum b_i \\ \sum t_i b_i \end{bmatrix} = \begin{bmatrix} 40 \\ 400 \end{bmatrix}$$

Those come from equation (9). Then $A^{\mathrm{T}}A\widehat{\boldsymbol{x}} = A^{\mathrm{T}}\boldsymbol{b}$ gives $C = -8$ and $D = 24/11$. Linearity allows us to rescale the measurements $\boldsymbol{b} = (0, 0, \ldots, 40)$. Multiplying $\boldsymbol{b}$ by 3 will multiply C and D by 3. Adding 30 to all b_i will add 30 to C.

Multiplying the times t_i by 2 will *divide* D by 2 (so the line reaches the same heights at the new times). Adding 10 to all times will replace t by $t - 10$. The new line $C+D(\frac{t-10}{2})$ reaches the same heights at $t = 12, 14, \ldots, 30$ (with the same errors) that it previously did at $t = 1, 2, \ldots, 10$. In linear algebra language, these matrices A_{old} and A_{new} have the same column space (why?) so no change in the projection:

$$\begin{bmatrix} 1 & 1 & 1 & 1 & 1 & 1 & 1 & 1 & 1 & 1 \\ 1 & 2 & 3 & 4 & 5 & 6 & 7 & 8 & 9 & 10 \end{bmatrix}^{\mathrm{T}} \quad \begin{bmatrix} 1 & 1 & 1 & 1 & 1 & 1 & 1 & 1 & 1 & 1 \\ 12 & 14 & 16 & 18 & 20 & 22 & 24 & 26 & 28 & 30 \end{bmatrix}^{\mathrm{T}}$$

4.3 B Find the parabola $C + Dt + Et^2$ that comes closest (least squares error) to the values $\boldsymbol{b} = (0, 0, 1, 0, 0)$ at the times $t = -2, -1, 0, 1, 2$. First write down the five equations $A\boldsymbol{x} = \boldsymbol{b}$ in three unknowns $\boldsymbol{x} = (C, D, E)$ for a parabola to go through the five points. No solution because no such parabola exists. Solve $A\boldsymbol{x} = \boldsymbol{b}$ by least squares (using $A^{\mathrm{T}}A\widehat{\boldsymbol{x}} = A^{\mathrm{T}}\boldsymbol{b}$).

I would predict that $D = 0$. Why should the best parabola be symmetric around $t = 0$? In $A^{\mathrm{T}}A\widehat{\boldsymbol{x}} = A^{\mathrm{T}}\boldsymbol{b}$, equation 2 for D should uncouple from equations 1 and 3.

Solution The five equations $A\boldsymbol{x} = \boldsymbol{b}$ and the 3 by 3 matrix $A^{\mathrm{T}}A$ are

$$\begin{aligned} C + D\,(-2) + E\,(-2)^2 &= 0 \\ C + D\,(-1) + E\,(-1)^2 &= 0 \\ C + D\quad(0) + E\quad(0)^2 &= 0 \\ C + D\quad(1) + E\quad(1)^2 &= 0 \\ C + D\quad(2) + E\quad(2)^2 &= 0 \end{aligned} \qquad A = \begin{bmatrix} 1 & -2 & 4 \\ 1 & -1 & 1 \\ 1 & 0 & 0 \\ 1 & 1 & 1 \\ 1 & 2 & 4 \end{bmatrix} \qquad A^{\mathrm{T}}A = \begin{bmatrix} 5 & 0 & 10 \\ 0 & 10 & 0 \\ 10 & 0 & 34 \end{bmatrix}$$

Those zeros in $A^{\mathrm{T}}A$ mean that column 2 of A is orthogonal to columns 1 and 3. We see this directly in A (because the times $-2, -1, 0, 1, 2$ are symmetric). The best C, D, E in the parabola $C + Dt + Et^2$ come from $A^{\mathrm{T}}A\widehat{\boldsymbol{x}} = A^{\mathrm{T}}\boldsymbol{b}$, and equation 2 for D is uncoupled:

$$\begin{bmatrix} 5 & 0 & 10 \\ 0 & 10 & 0 \\ 10 & 0 & 34 \end{bmatrix} \begin{bmatrix} C \\ D \\ E \end{bmatrix} = \begin{bmatrix} 1 \\ 0 \\ 0 \end{bmatrix} \quad \text{leads to} \quad \begin{aligned} C &= \frac{34}{70} = \frac{17}{35} \\ D &= 0 \text{ as predicted} \\ E &= -\frac{10}{70} = -\frac{1}{7} \end{aligned}$$

The symmetry of t's uncoupled equation 2. The symmetry of $\boldsymbol{b} = (0, 0, 1, 0, 0)$ made its right side zero. Symmetric inputs produced a symmetric parabola $\frac{17}{35} - \frac{1}{7}t^2$. Column 3 can be orthogonalized by subtracting its projection $(2, 2, 2, 2, 2)$ onto column 1:

$$A_{\text{new}}\,\widehat{\boldsymbol{x}}_{\text{new}} = \boldsymbol{b} \quad \text{is} \quad \begin{bmatrix} 1 & -2 & 2 \\ 1 & -1 & -1 \\ 1 & 0 & -2 \\ 1 & 1 & -1 \\ 1 & 2 & 2 \end{bmatrix} \begin{bmatrix} C+2E \\ D \\ E \end{bmatrix} = \begin{bmatrix} 0 \\ 0 \\ 1 \\ 0 \\ 0 \end{bmatrix} \qquad \textbf{Notice new third column}$$

$$\left(A_{\text{new}}^{\mathrm{T}} A_{\text{new}}\right) \widehat{\boldsymbol{x}}_{\text{new}} = A_{\text{new}}^{\mathrm{T}}\boldsymbol{b} \quad \text{is} \quad \begin{bmatrix} 5 & 0 & 0 \\ 0 & 10 & 0 \\ 0 & 0 & 14 \end{bmatrix} \begin{bmatrix} C+2E \\ D \\ E \end{bmatrix} = \begin{bmatrix} 1 \\ 0 \\ -2 \end{bmatrix}.$$

All equations are now uncoupled! A_{new} has orthogonal columns. Immediately $14E = -2$ and $E = -\frac{1}{7}$ and $D = 0$. Then $C + 2E = \frac{1}{5}$ gives $C = \frac{1}{5} + \frac{2}{7} = \frac{17}{35}$ as before. $A^{\mathrm{T}}A$ becomes easy when the work of orthogonalization (which is Gram-Schmidt) is done first.

Problem Set 4.3

Problems 1–11 use four data points $\boldsymbol{b} = (0, 8, 8, 20)$ to bring out the key ideas.

1 With $b = 0, 8, 8, 20$ at $t = 0, 1, 3, 4$, set up and solve the normal equations $A^{\mathrm{T}}A\widehat{\boldsymbol{x}} = A^{\mathrm{T}}\boldsymbol{b}$. For the best straight line in Figure 4.9a, find its four heights p_i and four errors e_i. What is the minimum value $E = e_1^2 + e_2^2 + e_3^2 + e_4^2$?

2 (Line $C + Dt$ does go through p's) With $b = 0, 8, 8, 20$ at times $t = 0, 1, 3, 4$, write down the four equations $A\boldsymbol{x} = \boldsymbol{b}$ (unsolvable). Change the measurements to $p = 1, 5, 13, 17$ and find an exact solution to $A\widehat{\boldsymbol{x}} = \boldsymbol{p}$.

3 Check that $\boldsymbol{e} = \boldsymbol{b} - \boldsymbol{p} = (-1, 3, -5, 3)$ is perpendicular to both columns of A. What is the shortest distance $\|\boldsymbol{e}\|$ from $\boldsymbol{b}$ to the column space of A?

4 (By calculus) Write down $E = \|A\boldsymbol{x} - \boldsymbol{b}\|^2$ as a sum of four squares—the last one is $(C + 4D - 20)^2$. Find the derivative equations $\partial E/\partial C = 0$ and $\partial E/\partial D = 0$. Divide by 2 to obtain the normal equations $A^{\mathrm{T}}A\widehat{\boldsymbol{x}} = A^{\mathrm{T}}\boldsymbol{b}$.

5 Find the height C of the best *horizontal line* to fit $\boldsymbol{b} = (0, 8, 8, 20)$. An exact fit would solve the unsolvable equations $C = 0$, $C = 8$, $C = 8$, $C = 20$. Find the 4 by 1 matrix A in these equations and solve $A^{\mathrm{T}}A\widehat{x} = A^{\mathrm{T}}\boldsymbol{b}$. Draw the horizontal line at height $\widehat{x} = C$ and the four errors in $\boldsymbol{e}$.

6 Project $\boldsymbol{b} = (0, 8, 8, 20)$ onto the line through $\boldsymbol{a} = (1, 1, 1, 1)$. Find $\widehat{x} = \boldsymbol{a}^{\mathrm{T}}\boldsymbol{b}/\boldsymbol{a}^{\mathrm{T}}\boldsymbol{a}$ and the projection $\boldsymbol{p} = \widehat{x}\boldsymbol{a}$. Check that $\boldsymbol{e} = \boldsymbol{b} - \boldsymbol{p}$ is perpendicular to $\boldsymbol{a}$, and find the shortest distance $\|\boldsymbol{e}\|$ from $\boldsymbol{b}$ to the line through $\boldsymbol{a}$.

7 Find the closest line $b = Dt$, *through the origin*, to the same four points. An exact fit would solve $D \cdot 0 = 0$, $D \cdot 1 = 8$, $D \cdot 3 = 8$, $D \cdot 4 = 20$. Find the 4 by 1 matrix and solve $A^{\mathrm{T}}A\widehat{x} = A^{\mathrm{T}}\boldsymbol{b}$. Redraw Figure 4.9a showing the best line $b = Dt$ and the e's.

8 Project $\boldsymbol{b} = (0, 8, 8, 20)$ onto the line through $\boldsymbol{a} = (0, 1, 3, 4)$. Find $\widehat{x} = D$ and $\boldsymbol{p} = \widehat{x}\boldsymbol{a}$. The best C in Problems 5–6 and the best D in Problems 7–8 do *not* agree with the best (C, D) in Problems 1–4. That is because $(1, 1, 1, 1)$ and $(0, 1, 3, 4)$ are ______ perpendicular.

9 For the closest parabola $b = C + Dt + Et^2$ to the same four points, write down the unsolvable equations $A\boldsymbol{x} = \boldsymbol{b}$ in three unknowns $\boldsymbol{x} = (C, D, E)$. Set up the three normal equations $A^{\mathrm{T}}A\widehat{x} = A^{\mathrm{T}}\boldsymbol{b}$ (solution not required). In Figure 4.9a you are now fitting a parabola to 4 points—what is happening in Figure 4.9b?

10 For the closest cubic $b = C + Dt + Et^2 + Ft^3$ to the same four points, write down the four equations $A\boldsymbol{x} = \boldsymbol{b}$. Solve them by elimination. In Figure 4.9a this cubic now goes exactly through the points. What are $\boldsymbol{p}$ and $\boldsymbol{e}$?

11 The average of the four times is $\widehat{t} = \frac{1}{4}(0 + 1 + 3 + 4) = 2$. The average of the four b's is $\widehat{b} = \frac{1}{4}(0 + 8 + 8 + 20) = 9$.

(a) Verify that the best line goes through the center point $(\widehat{t}, \widehat{b}) = (2, 9)$.

(b) Explain why $C + D\widehat{t} = \widehat{b}$ comes from the first equation in $A^{\mathrm{T}}A\widehat{x} = A^{\mathrm{T}}\boldsymbol{b}$.

Questions 12–16 introduce basic ideas of statistics—the foundation for least squares.

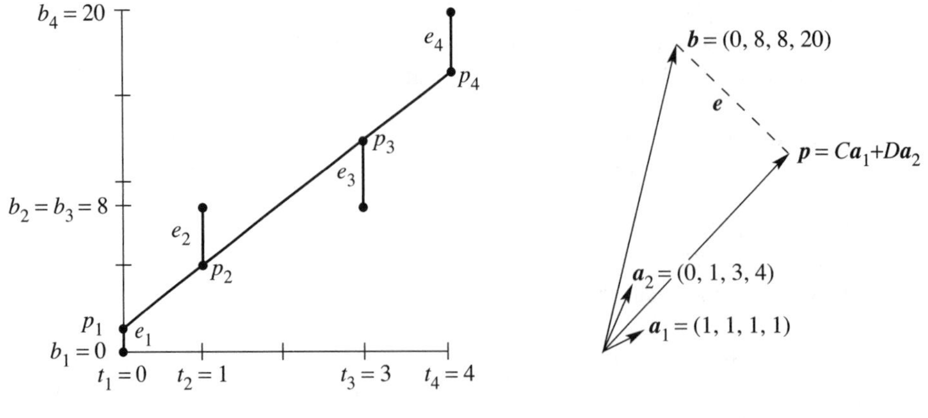

Figure 4.9 **Problems 1–11**: The closest line $C + Dt$ matches $C\boldsymbol{a}_1 + D\boldsymbol{a}_2$ in $\mathbf{R}^4$.

12 (Recommended) This problem projects $\boldsymbol{b} = (b_1, \ldots, b_m)$ onto the line through $\boldsymbol{a} = (1, \ldots, 1)$. We solve m equations $\boldsymbol{a}x = \boldsymbol{b}$ in 1 unknown (by least squares).

(a) Solve $\boldsymbol{a}^{\mathrm{T}}\boldsymbol{a}\widehat{x} = \boldsymbol{a}^{\mathrm{T}}\boldsymbol{b}$ to show that $\widehat{x}$ is the *mean* (the average) of the $\boldsymbol{b}$'s.

(b) Find $\boldsymbol{e} = \boldsymbol{b} - \boldsymbol{a}\widehat{x}$ and the *variance* $\|\boldsymbol{e}\|^2$ and the *standard deviation* $\|\boldsymbol{e}\|$.

(c) The horizontal line $\widehat{\boldsymbol{b}} = 3$ is closest to $\boldsymbol{b} = (1, 2, 6)$. Check that $\boldsymbol{p} = (3, 3, 3)$ is perpendicular to $\boldsymbol{e}$ and find the matrix P.

13 First assumption behind least squares: Each measurement error has ***mean zero***. Multiply the 8 error vectors $\boldsymbol{b} - A\boldsymbol{x} = (\pm 1, \pm 1, \pm 1)$ by $(A^{\mathrm{T}}A)^{-1}A^{\mathrm{T}}$ to show that the 8 vectors $\widehat{\boldsymbol{x}} - \boldsymbol{x}$ also average to zero. The estimate $\widehat{\boldsymbol{x}}$ is *unbiased*.

14 Second assumption behind least squares: The m errors e_i are independent with variance σ^2, so the average of $(\boldsymbol{b} - A\boldsymbol{x})(\boldsymbol{b} - A\boldsymbol{x})^{\mathrm{T}}$ is $\sigma^2 I$. Multiply on the left by $(A^{\mathrm{T}}A)^{-1}A^{\mathrm{T}}$ and on the right by $A(A^{\mathrm{T}}A)^{-1}$ to show that the average of $(\widehat{\boldsymbol{x}} - \boldsymbol{x})(\widehat{\boldsymbol{x}} - \boldsymbol{x})^{\mathrm{T}}$ is $\sigma^2(A^{\mathrm{T}}A)^{-1}$. This is the ***covariance matrix*** for the error in $\widehat{\boldsymbol{x}}$.

15 A doctor takes 4 readings of your heart rate. The best solution to $x = b_1, \ldots, x = b_4$ is the average $\widehat{x}$ of $b_1, \ldots, b_4$. The matrix A is a column of 1's. Problem 14 gives the expected error $(\widehat{x} - x)^2$ as $\sigma^2(A^{\mathrm{T}}A)^{-1} =$ ______. By averaging, the variance drops from σ^2 to $\sigma^2/4$.

16 If you know the average $\widehat{x}_9$ of 9 numbers $b_1, \ldots, b_9$, how can you quickly find the average $\widehat{x}_{10}$ with one more number b_{10}? The idea of *recursive* least squares is to avoid adding 10 numbers. What coefficient correctly gives $\widehat{x}_{10}$?

$$\widehat{x}_{10} = \tfrac{1}{10}b_{10} + \underline{\qquad}\,\widehat{x}_9 = \tfrac{1}{10}(b_1 + \cdots + b_{10}).$$

Questions 17–25 give more practice with $\widehat{x}$ and p and e. Note Question 26.

17 Write down three equations for the line $b = C + Dt$ to go through $b = 7$ at $t = -1$, $b = 7$ at $t = 1$, and $b = 21$ at $t = 2$. Find the least squares solution $\widehat{\boldsymbol{x}} = (C, D)$ and draw the closest line.

18 Find the projection $\boldsymbol{p} = A\widehat{\boldsymbol{x}}$ in Problem 17. This gives the three heights of the closest line. Show that the error vector is $\boldsymbol{e} = (2, -6, 4)$.

19 Suppose the measurements at $t = -1, 1, 2$ are the errors $2, -6, 4$ in Problem 18. Compute $\widehat{\boldsymbol{x}}$ and the closest line to these new measurements. Explain the answer: $\boldsymbol{b} = (2, -6, 4)$ is perpendicular to ______ so the projection is $\boldsymbol{p} = \boldsymbol{0}$.

20 Suppose the measurements at $t = -1, 1, 2$ are $\boldsymbol{b} = (5, 13, 17)$. Compute $\widehat{\boldsymbol{x}}$ and the closest line and $\boldsymbol{e}$. The error is $\boldsymbol{e} = \boldsymbol{0}$ because this $\boldsymbol{b}$ is ______.

21 Which of the four subspaces contains the error vector $\boldsymbol{e}$? Which contains $\boldsymbol{p}$? Which contains $\widehat{\boldsymbol{x}}$? What is the nullspace of A?

22 Find the best line $C + Dt$ to fit $b = 4, 2, -1, 0, 0$ at times $t = -2, -1, 0, 1, 2$.

23 (Distance between lines) The points $P = (x, x, x)$ are on a line through $(1, 1, 1)$ and $Q = (y, 3y, -1)$ are on another line. Choose x and y to minimize the squared distance $\|P - Q\|^2$.

24 Is the error vector $\boldsymbol{e}$ orthogonal to $\boldsymbol{b}$ or $\boldsymbol{p}$ or $\boldsymbol{e}$ or $\widehat{\boldsymbol{x}}$? Show that $\|\boldsymbol{e}\|^2$ equals $\boldsymbol{e}^{\mathrm{T}}\boldsymbol{b}$ which equals $\boldsymbol{b}^{\mathrm{T}}\boldsymbol{b} - \boldsymbol{p}^{\mathrm{T}}\boldsymbol{b}$. This is the smallest total error E.

25 The derivatives of $\|A\boldsymbol{x}\|^2$ with respect to the variables $x_1, \ldots, x_n$ fill the vector $2A^{\mathrm{T}}A\boldsymbol{x}$. The derivatives of $2\boldsymbol{b}^{\mathrm{T}}A\boldsymbol{x}$ fill the vector $2A^{\mathrm{T}}\boldsymbol{b}$. So the derivatives of $\|A\boldsymbol{x} - \boldsymbol{b}\|^2$ are zero when ______.

26 *What condition on* $(t_1, b_1), (t_2, b_2), (t_3, b_3)$ *puts those three points onto a straight line*? A column space answer is: (b_1, b_2, b_3) must be a combination of $(1, 1, 1)$ and (t_1, t_2, t_3). Try to reach a specific equation connecting the t's and b's. I should have thought of this question sooner!

27 Find the *plane* that gives the best fit to the 4 values $\boldsymbol{b} = (0, 1, 3, 4)$ at the corners $(1, 0)$ and $(0, 1)$ and $(-1, 0)$ and $(0, -1)$ of a square. The equations $C + Dx + Ey = b$ at those 4 points are $A\boldsymbol{x} = \boldsymbol{b}$ with 3 unknowns $\boldsymbol{x} = (C, D, E)$. At the center $(0, 0)$ of the square, show that $C + Dx + Ey =$ average of the b's.

ORTHOGONAL BASES AND GRAM-SCHMIDT ■ 4.4

This section has two goals. The first is to see how orthogonality can make calculations simpler. Dot products are zero—so $A^{\mathrm{T}}A$ becomes a diagonal matrix. The second goal is to ***construct orthogonal vectors***. We will pick combinations of the original vectors to produce right angles. Those original vectors are the columns of A, probably *not* orthogonal. The orthogonal vectors will be the columns of a new matrix Q.

You know from Chapter 3 what a basis consists of—independent vectors that span the space. The basis vectors could meet at any angle (except 0° and 180°). But every time we visualize axes, they are perpendicular. *In our imagination, the coordinate axes are practically always orthogonal.* This simplifies the picture and it greatly simplifies the computations.

The vectors $\boldsymbol{q}_1, \ldots, \boldsymbol{q}_n$ are ***orthogonal*** when their dot products $\boldsymbol{q}_i \cdot \boldsymbol{q}_j$ are zero. More exactly $\boldsymbol{q}_i^{\mathrm{T}}\boldsymbol{q}_j = 0$ whenever $i \neq j$. With one more step—just divide each vector by its length—the vectors become ***orthogonal unit vectors***. Their lengths are all 1. Then the basis is called ***orthonormal***.

DEFINITION The vectors $\boldsymbol{q}_1, \ldots, \boldsymbol{q}_n$ are ***orthonormal*** if

$$\boldsymbol{q}_i^{\mathrm{T}}\boldsymbol{q}_j = \begin{cases} 0 & \text{when } i \neq j \quad (\textbf{\textit{orthogonal}} \text{ vectors}) \\ 1 & \text{when } i = j \quad (\textbf{\textit{unit}} \text{ vectors: } \|\boldsymbol{q}_i\| = 1) \end{cases}$$

A matrix with orthonormal columns is assigned the special letter Q.

The matrix Q is easy to work with because $Q^{\mathrm{T}}Q = I$. This repeats in matrix language that the columns $\boldsymbol{q}_1, \ldots, \boldsymbol{q}_n$ are orthonormal. It is equation (1) below, and Q is not required to be square.

When Q is square, $Q^{\mathrm{T}}Q = I$ means that $Q^{\mathrm{T}} = Q^{-1}$: ***transpose = inverse***.

4I ***A matrix Q with orthonormal columns satisfies $Q^{\mathrm{T}}Q = I$:***

$$Q^{\mathrm{T}}Q = \begin{bmatrix} - & \boldsymbol{q}_1^{\mathrm{T}} & - \\ - & \boldsymbol{q}_2^{\mathrm{T}} & - \\ & & \\ - & \boldsymbol{q}_n^{\mathrm{T}} & - \end{bmatrix} \begin{bmatrix} | & | & & | \\ \boldsymbol{q}_1 & \boldsymbol{q}_2 & & \boldsymbol{q}_n \\ | & | & & | \end{bmatrix} = \begin{bmatrix} 1 & 0 & \cdots & 0 \\ 0 & 1 & \cdots & 0 \\ \vdots & \vdots & \ddots & \vdots \\ 0 & 0 & \cdots & 1 \end{bmatrix} = I. \quad (1)$$

When row i of Q^{T} multiplies column j of Q, the dot product is $\boldsymbol{q}_i^{\mathrm{T}}\boldsymbol{q}_j$. Off the diagonal $(i \neq j)$ that dot product is zero by orthogonality. On the diagonal $(i = j)$ the unit vectors give $\boldsymbol{q}_i^{\mathrm{T}}\boldsymbol{q}_i = \|\boldsymbol{q}_i\|^2 = 1$.

If the columns are only orthogonal (not unit vectors), then $Q^{\mathrm{T}}Q$ is a diagonal matrix (not the identity matrix). We wouldn't use the letter Q. But this matrix is almost as good. The important thing is orthogonality—then it is easy to produce unit vectors.

To repeat: $Q^{\mathrm{T}}Q = I$ even when Q is rectangular. In that case Q^{T} is only an inverse from the left. For square matrices we also have $QQ^{\mathrm{T}} = I$, so Q^{T} is the two-sided inverse of Q. The rows of a square Q are orthonormal like the columns. ***The inverse is the transpose***. In this square case we call Q an ***orthogonal matrix***.[2]

Here are three examples of orthogonal matrices—rotation and permutation and reflection. The quickest test is to check $Q^{\mathrm{T}}Q = I$.

Example 1 **(Rotation)** Q rotates every vector in the plane through the angle θ:

$$Q = \begin{bmatrix} \cos\theta & -\sin\theta \\ \sin\theta & \cos\theta \end{bmatrix} \quad \text{and} \quad Q^{\mathrm{T}} = Q^{-1} = \begin{bmatrix} \cos\theta & \sin\theta \\ -\sin\theta & \cos\theta \end{bmatrix}.$$

The columns of Q are orthogonal (take their dot product). They are unit vectors because $\sin^2\theta + \cos^2\theta = 1$. Those columns give an ***orthonormal basis*** for the plane $\mathbf{R}^2$. The standard basis vectors $\boldsymbol{i}$ and $\boldsymbol{j}$ are rotated through θ (see Figure 4.10a).

Q^{-1} rotates vectors back through $-\theta$. It agrees with Q^{T}, because the cosine of $-\theta$ is the cosine of θ, and $\sin(-\theta) = -\sin\theta$. We have $Q^{\mathrm{T}}Q = I$ and $QQ^{\mathrm{T}} = I$.

Example 2 **(Permutation)** These matrices change the order to (y, z, x) and (y, x):

$$\begin{bmatrix} 0 & 1 & 0 \\ 0 & 0 & 1 \\ 1 & 0 & 0 \end{bmatrix} \begin{bmatrix} x \\ y \\ z \end{bmatrix} = \begin{bmatrix} y \\ z \\ x \end{bmatrix} \quad \text{and} \quad \begin{bmatrix} 0 & 1 \\ 1 & 0 \end{bmatrix} \begin{bmatrix} x \\ y \end{bmatrix} = \begin{bmatrix} y \\ x \end{bmatrix}.$$

All columns of these Q's are unit vectors (their lengths are obviously 1). They are also orthogonal (the 1's appear in different places). *The inverse of a permutation matrix is its transpose*. The inverse puts the components back into their original order:

$$\textbf{Inverse = transpose:} \quad \begin{bmatrix} 0 & 0 & 1 \\ 1 & 0 & 0 \\ 0 & 1 & 0 \end{bmatrix} \begin{bmatrix} y \\ z \\ x \end{bmatrix} = \begin{bmatrix} x \\ y \\ z \end{bmatrix} \quad \text{and} \quad \begin{bmatrix} 0 & 1 \\ 1 & 0 \end{bmatrix} \begin{bmatrix} y \\ x \end{bmatrix} = \begin{bmatrix} x \\ y \end{bmatrix}.$$

Every permutation matrix is an orthogonal matrix.

Example 3 **(Reflection)** If $\boldsymbol{u}$ is any unit vector, set $Q = I - 2\boldsymbol{u}\boldsymbol{u}^{\mathrm{T}}$. Notice that $\boldsymbol{u}\boldsymbol{u}^{\mathrm{T}}$ is a matrix while $\boldsymbol{u}^{\mathrm{T}}\boldsymbol{u}$ is the number $\|\boldsymbol{u}\|^2 = 1$. Then Q^{T} and Q^{-1} both equal Q:

$$Q^{\mathrm{T}} = I - 2\boldsymbol{u}\boldsymbol{u}^{\mathrm{T}} = Q \quad \text{and} \quad Q^{\mathrm{T}}Q = I - 4\boldsymbol{u}\boldsymbol{u}^{\mathrm{T}} + 4\boldsymbol{u}\boldsymbol{u}^{\mathrm{T}}\boldsymbol{u}\boldsymbol{u}^{\mathrm{T}} = I. \tag{2}$$

---

[2] "Orthonormal matrix" would have been a better name for Q, but it's not used. Any matrix with orthonormal columns has the letter Q, but we only call it an *orthogonal matrix* when it is square.

Reflection matrices $I - 2uu^T$ are symmetric and also orthogonal. If you square them, you get the identity matrix: $Q^2 = Q^TQ = I$. Reflecting twice through a mirror brings back the original. Notice $u^Tu = 1$ inside $4uu^Tuu^T$ in equation (2).

As examples choose two unit vectors, $u = (1,0)$ and then $u = (1/\sqrt{2}, -1/\sqrt{2})$. Compute $2uu^T$ (column times row) and subtract from I to get Q:

$$Q_1 = I - 2\begin{bmatrix}1\\0\end{bmatrix}\begin{bmatrix}1 & 0\end{bmatrix} = \begin{bmatrix}-1 & 0\\0 & 1\end{bmatrix} \quad \text{and} \quad Q_2 = I - 2\begin{bmatrix}.5 & -.5\\-.5 & .5\end{bmatrix} = \begin{bmatrix}0 & 1\\1 & 0\end{bmatrix}.$$

Q_1 reflects $(1,0)$ across the y axis to $(-1,0)$. Every vector (x,y) goes into its image $(-x,y)$, and the y axis is the mirror:

$$\textbf{Reflection from } Q_1\text{:} \quad \begin{bmatrix}-1 & 0\\0 & 1\end{bmatrix}\begin{bmatrix}x\\y\end{bmatrix} = \begin{bmatrix}-x\\y\end{bmatrix}.$$

Q_2 is reflection across the 45° line. Every (x,y) goes to (y,x)—this was the permutation in Example 2. A vector like $(3,3)$ doesn't move when you exchange 3 and 3—it is on the mirror line. Figure 4.10b shows the 45° mirror.

Rotations preserve the length of a vector. So do reflections. So do permutations. So does multiplication by any orthogonal matrix—***lengths and angles don't change***.

4J ***If Q has orthonormal columns ($Q^TQ = I$), it leaves lengths unchanged:***

$$\|Qx\| = \|x\| \text{ for every vector } x. \tag{3}$$

Q also preserves dot products: $(Qx)^T(Qy) = x^TQ^TQy = x^Ty$. Just use $Q^TQ = I$!

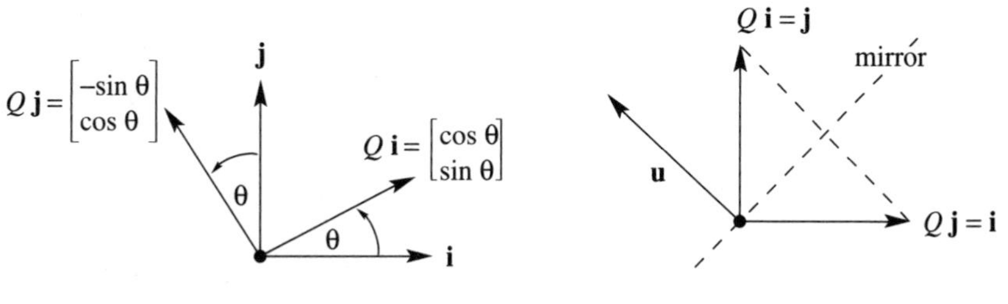

Figure 4.10 Rotation by $Q = \begin{bmatrix}c & -s\\s & c\end{bmatrix}$ and reflection across 45° by $Q = \begin{bmatrix}0 & 1\\1 & 0\end{bmatrix}$.

Proof $\|Qx\|^2$ equals $\|x\|^2$ because $(Qx)^{\mathrm{T}}(Qx) = x^{\mathrm{T}}Q^{\mathrm{T}}Qx = x^{\mathrm{T}}Ix = x^{\mathrm{T}}x$. Orthogonal matrices are excellent for computations—numbers can never grow too large when lengths of vectors are fixed. Good computer codes use Q's as much as possible. That makes them numerically stable.

Projections Using Orthogonal Bases: Q Replaces A

This chapter is about projections onto subspaces. We developed the equations for $\widehat{x}$ and p and P. When the columns of A were a basis for the subspace, all formulas involved $A^{\mathrm{T}}A$. The entries of $A^{\mathrm{T}}A$ are the dot products $a_i^{\mathrm{T}}a_j$.

Suppose the basis vectors are actually orthonormal. The a's become q's. Then $A^{\mathrm{T}}A$ simplifies to $Q^{\mathrm{T}}Q = I$. Look at the improvements in $\widehat{x}$ and p and P. Instead of $Q^{\mathrm{T}}Q$ we print a blank for the identity matrix:

$$\underline{\quad\quad}\,\widehat{x} = Q^{\mathrm{T}}b \quad \text{and} \quad p = Q\widehat{x} \quad \text{and} \quad P = Q\,\underline{\quad\quad}\,Q^{\mathrm{T}}. \tag{4}$$

The least squares solution of $Qx = b$ is $\widehat{x} = Q^{\mathrm{T}}b$. The projection matrix is $P = QQ^{\mathrm{T}}$.

There are no matrices to invert. This is the point of an orthonormal basis. The best $\widehat{x} = Q^{\mathrm{T}}b$ just has dot products of b with the rows of Q^{T}, which are the q's:

$$\widehat{x} = \begin{bmatrix} - & q_1^{\mathrm{T}} & - \\ & \vdots & \\ - & q_n^{\mathrm{T}} & - \end{bmatrix}\begin{bmatrix} \\ b \\ \\ \end{bmatrix} = \begin{bmatrix} q_1^{\mathrm{T}}b \\ \vdots \\ q_n^{\mathrm{T}}b \end{bmatrix} \qquad \text{(dot products)}$$

We have n separate 1-dimensional projections. The "coupling matrix" or "correlation matrix" $A^{\mathrm{T}}A$ is now $Q^{\mathrm{T}}Q = I$. There is no coupling. Here is $p = Q\widehat{x}$:

$$\textbf{\textit{Projection}} \qquad p = \begin{bmatrix} | & & | \\ q_1 & \cdots & q_n \\ | & & | \end{bmatrix}\begin{bmatrix} q_1^{\mathrm{T}}b \\ \vdots \\ q_n^{\mathrm{T}}b \end{bmatrix} = q_1(q_1^{\mathrm{T}}b) + \cdots + q_n(q_n^{\mathrm{T}}b). \tag{5}$$

Important case: When Q is square and $m = n$, the subspace is the whole space. Then $Q^{\mathrm{T}} = Q^{-1}$ and $\widehat{x} = Q^{\mathrm{T}}b$ is the same as $x = Q^{-1}b$. The solution is exact! The projection of b onto the whole space is b itself. In this case $P = QQ^{\mathrm{T}} = I$.

You may think that projection onto the whole space is not worth mentioning. But when $p = b$, our formula assembles b out of its 1-dimensional projections. If $q_1, \ldots, q_n$ is an orthonormal basis for the whole space, so Q is square, then every b is the sum of its components along the q's:

$$b = q_1(q_1^{\mathrm{T}}b) + q_2(q_2^{\mathrm{T}}b) + \cdots + q_n(q_n^{\mathrm{T}}b). \tag{6}$$

That is $QQ^{\mathrm{T}} = I$. It is the foundation of Fourier series and all the great "transforms" of applied mathematics. They break vectors or functions into perpendicular pieces. Then by adding the pieces, the inverse transform puts the function back together.

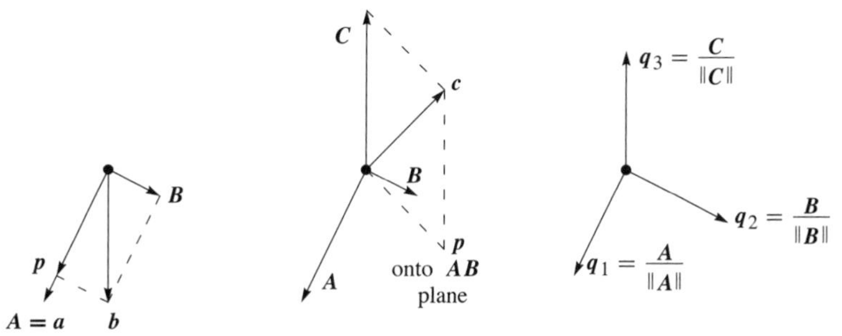

Figure 4.11 First project $\boldsymbol{b}$ onto the line through $\boldsymbol{a}$ and find $\boldsymbol{B}$ as $\boldsymbol{b}-\boldsymbol{p}$. Then project $\boldsymbol{c}$ onto the $\boldsymbol{AB}$ plane and find $\boldsymbol{C}$ as $\boldsymbol{c}-\boldsymbol{p}$. Then divide by $\|\boldsymbol{A}\|$, $\|\boldsymbol{B}\|$, and $\|\boldsymbol{C}\|$.

Example 4 The columns of this matrix Q are orthonormal vectors $\boldsymbol{q}_1, \boldsymbol{q}_2, \boldsymbol{q}_3$:

$$Q = \frac{1}{3}\begin{bmatrix} -1 & 2 & 2 \\ 2 & -1 & 2 \\ 2 & 2 & -1 \end{bmatrix} \quad \text{has first column} \quad \boldsymbol{q}_1 = \begin{bmatrix} -\frac{1}{3} \\ \frac{2}{3} \\ \frac{2}{3} \end{bmatrix}.$$

The separate projections of $\boldsymbol{b} = (0, 0, 1)$ onto $\boldsymbol{q}_1$ and $\boldsymbol{q}_2$ and $\boldsymbol{q}_3$ are

$$\boldsymbol{q}_1(\boldsymbol{q}_1^{\mathrm{T}}\boldsymbol{b}) = \tfrac{2}{3}\boldsymbol{q}_1 \quad \text{and} \quad \boldsymbol{q}_2(\boldsymbol{q}_2^{\mathrm{T}}\boldsymbol{b}) = \tfrac{2}{3}\boldsymbol{q}_2 \quad \text{and} \quad \boldsymbol{q}_3(\boldsymbol{q}_3^{\mathrm{T}}\boldsymbol{b}) = -\tfrac{1}{3}\boldsymbol{q}_3.$$

The sum of the first two is the projection of $\boldsymbol{b}$ onto the *plane* of $\boldsymbol{q}_1$ and $\boldsymbol{q}_2$. The sum of all three is the projection of $\boldsymbol{b}$ onto the *whole space*—which is $\boldsymbol{b}$ itself:

$$\tfrac{2}{3}\boldsymbol{q}_1 + \tfrac{2}{3}\boldsymbol{q}_2 - \tfrac{1}{3}\boldsymbol{q}_3 = \frac{1}{9}\begin{bmatrix} -2+4-2 \\ 4-2-2 \\ 4+4+1 \end{bmatrix} = \begin{bmatrix} 0 \\ 0 \\ 1 \end{bmatrix} = \boldsymbol{b}.$$

The Gram-Schmidt Process

The point of this section is that "orthogonal is good." Projections and least squares always involve $A^{\mathrm{T}}A$. When this matrix becomes $Q^{\mathrm{T}}Q = I$, the inverse is no problem. The one-dimensional projections are uncoupled. The best $\widehat{\boldsymbol{x}}$ is $Q^{\mathrm{T}}\boldsymbol{b}$ (n separate dot products). For this to be true, we had to say "*If* the vectors are orthonormal." ***Now we find a way to create orthonormal vectors.***

Start with three independent vectors $\boldsymbol{a}, \boldsymbol{b}, \boldsymbol{c}$. We intend to construct three orthogonal vectors $\boldsymbol{A}, \boldsymbol{B}, \boldsymbol{C}$. Then (at the end is easiest) we divide $\boldsymbol{A}, \boldsymbol{B}, \boldsymbol{C}$ by their lengths. That produces three orthonormal vectors $\boldsymbol{q}_1 = \boldsymbol{A}/\|\boldsymbol{A}\|$, $\boldsymbol{q}_2 = \boldsymbol{B}/\|\boldsymbol{B}\|$, $\boldsymbol{q}_3 = \boldsymbol{C}/\|\boldsymbol{C}\|$.

Gram-Schmidt Begin by choosing $\boldsymbol{A} = \boldsymbol{a}$. This first direction is accepted. The next direction $\boldsymbol{B}$ must be perpendicular to $\boldsymbol{A}$. ***Start with b and subtract its projection along***

A. This leaves the perpendicular part, which is the orthogonal vector B:

$$\textbf{Gram-Schmidt idea} \qquad B = b - \frac{A^{\mathrm{T}}b}{A^{\mathrm{T}}A}A. \tag{7}$$

A and B are orthogonal in Figure 4.11. Take the dot product with A to verify that $A^{\mathrm{T}}B = A^{\mathrm{T}}b - A^{\mathrm{T}}b = 0$. This vector B is what we have called the error vector e, perpendicular to A. Notice that B in equation (7) is not zero (otherwise a and b would be dependent). The directions A and B are now set.

The third direction starts with c. This is not a combination of A and B (because c is not a combination of a and b). But most likely c is not perpendicular to A and B. So subtract off its components in those two directions to get C:

$$C = c - \frac{A^{\mathrm{T}}c}{A^{\mathrm{T}}A}A - \frac{B^{\mathrm{T}}c}{B^{\mathrm{T}}B}B. \tag{8}$$

This is the one and only idea of the Gram-Schmidt process. ***Subtract from every new vector its projections in the directions already set.*** That idea is repeated at every step.[2] If we also had a fourth vector d, we would subtract its projections onto A, B, C to get D. At the end, divide the orthogonal vectors A, B, C, D by their lengths. The resulting vectors q_1, q_2, q_3, q_4 are orthonormal.

Example 5 Suppose the independent non-orthogonal vectors a, b, c are

$$a = \begin{bmatrix} 1 \\ -1 \\ 0 \end{bmatrix} \quad \text{and} \quad b = \begin{bmatrix} 2 \\ 0 \\ -2 \end{bmatrix} \quad \text{and} \quad c = \begin{bmatrix} 3 \\ -3 \\ 3 \end{bmatrix}.$$

Then $A = a$ has $A^{\mathrm{T}}A = 2$. Subtract from b its projection along $A = (1, -1, 0)$:

$$B = b - \frac{A^{\mathrm{T}}b}{A^{\mathrm{T}}A}A = b - \tfrac{2}{2}A = \begin{bmatrix} 1 \\ 1 \\ -2 \end{bmatrix}.$$

Check: $A^{\mathrm{T}}B = 0$ as required. Now subtract two projections from c to get C:

$$C = c - \frac{A^{\mathrm{T}}c}{A^{\mathrm{T}}A}A - \frac{B^{\mathrm{T}}c}{B^{\mathrm{T}}B}B = c - \tfrac{6}{2}A + \tfrac{6}{6}B = \begin{bmatrix} 1 \\ 1 \\ 1 \end{bmatrix}.$$

Check: $C = (1, 1, 1)$ is perpendicular to A and B. Finally convert A, B, C to unit vectors (length 1, orthonormal). The lengths of A, B, C are $\sqrt{2}$ and $\sqrt{6}$ and $\sqrt{3}$. Divide by those lengths, for an orthonormal basis:

$$q_1 = \frac{1}{\sqrt{2}}\begin{bmatrix} 1 \\ -1 \\ 0 \end{bmatrix} \quad \text{and} \quad q_2 = \frac{1}{\sqrt{6}}\begin{bmatrix} 1 \\ 1 \\ -2 \end{bmatrix} \quad \text{and} \quad q_3 = \frac{1}{\sqrt{3}}\begin{bmatrix} 1 \\ 1 \\ 1 \end{bmatrix}.$$

Usually A, B, C contain fractions. Almost always q_1, q_2, q_3 contain square roots.

[2] I think Gram had the idea. I don't really know where Schmidt came in.

The Factorization $A = QR$

We started with a matrix A, whose columns were $\boldsymbol{a}, \boldsymbol{b}, \boldsymbol{c}$. We ended with a matrix Q, whose columns are $\boldsymbol{q}_1, \boldsymbol{q}_2, \boldsymbol{q}_3$. How are those matrices related? Since the vectors $\boldsymbol{a}, \boldsymbol{b}, \boldsymbol{c}$ are combinations of the $\boldsymbol{q}$'s (and vice versa), there must be a third matrix connecting A to Q. Call it R.

The first step was $\boldsymbol{q}_1 = \boldsymbol{a}/\|\boldsymbol{a}\|$ (other vectors not involved). The second step was equation (7), where $\boldsymbol{b}$ is a combination of $\boldsymbol{A}$ and $\boldsymbol{B}$. At that stage $\boldsymbol{C}$ and $\boldsymbol{q}_3$ were not involved. This non-involvement of later vectors is the key point of Gram-Schmidt:

- The vectors $\boldsymbol{a}$ and $\boldsymbol{A}$ and $\boldsymbol{q}_1$ are all along a single line.
- The vectors $\boldsymbol{a}, \boldsymbol{b}$ and $\boldsymbol{A}, \boldsymbol{B}$ and $\boldsymbol{q}_1, \boldsymbol{q}_2$ are all in the same plane.
- The vectors $\boldsymbol{a}, \boldsymbol{b}, \boldsymbol{c}$ and $\boldsymbol{A}, \boldsymbol{B}, \boldsymbol{C}$ and $\boldsymbol{q}_1, \boldsymbol{q}_2, \boldsymbol{q}_3$ are in one subspace (dimension 3).

At every step $\boldsymbol{a}_1, \ldots, \boldsymbol{a}_k$ are combinations of $\boldsymbol{q}_1, \ldots, \boldsymbol{q}_k$. Later $\boldsymbol{q}$'s are not involved. The connecting matrix R is ***triangular***, and we have $A = QR$:

$$\begin{bmatrix} \boldsymbol{a} & \boldsymbol{b} & \boldsymbol{c} \end{bmatrix} = \begin{bmatrix} \boldsymbol{q}_1 & \boldsymbol{q}_2 & \boldsymbol{q}_3 \end{bmatrix} \begin{bmatrix} \boldsymbol{q}_1^{\mathrm{T}}\boldsymbol{a} & \boldsymbol{q}_1^{\mathrm{T}}\boldsymbol{b} & \boldsymbol{q}_1^{\mathrm{T}}\boldsymbol{c} \\ & \boldsymbol{q}_2^{\mathrm{T}}\boldsymbol{b} & \boldsymbol{q}_2^{\mathrm{T}}\boldsymbol{c} \\ & & \boldsymbol{q}_3^{\mathrm{T}}\boldsymbol{c} \end{bmatrix} \quad \text{or} \quad A = QR. \tag{9}$$

$A = QR$ is Gram-Schmidt in a nutshell. Multiply by Q^{T} to see why $R = Q^{\mathrm{T}}A$.

4K (Gram-Schmidt) From independent vectors $\boldsymbol{a}_1, \ldots, \boldsymbol{a}_n$, Gram-Schmidt constructs orthonormal vectors $\boldsymbol{q}_1, \ldots, \boldsymbol{q}_n$. The matrices with these columns satisfy $A = QR$. Then $R = Q^{\mathrm{T}}A$ is ***triangular*** because later $\boldsymbol{q}$'s are orthogonal to earlier $\boldsymbol{a}$'s.

Here are the $\boldsymbol{a}$'s and $\boldsymbol{q}$'s from the example. The i, j entry of $R = Q^{\mathrm{T}}A$ is row i of Q^{T} times column j of A. This is the dot product of $\boldsymbol{q}_i$ with $\boldsymbol{a}_j$:

$$A = \begin{bmatrix} 1 & 2 & 3 \\ -1 & 0 & -3 \\ 0 & -2 & 3 \end{bmatrix} = \begin{bmatrix} 1/\sqrt{2} & 1/\sqrt{6} & 1/\sqrt{3} \\ -1/\sqrt{2} & 1/\sqrt{6} & 1/\sqrt{3} \\ 0 & -2/\sqrt{6} & 1/\sqrt{3} \end{bmatrix} \begin{bmatrix} \sqrt{2} & \sqrt{2} & \sqrt{18} \\ 0 & \sqrt{6} & -\sqrt{6} \\ 0 & 0 & \sqrt{3} \end{bmatrix} = QR.$$

The lengths of $\boldsymbol{A}, \boldsymbol{B}, \boldsymbol{C}$ are the numbers $\sqrt{2}, \sqrt{6}, \sqrt{3}$ on the diagonal of R. Because of the square roots, QR looks less beautiful than LU. Both factorizations are absolutely central to calculations in linear algebra.

Any m by n matrix A with independent columns can be factored into QR. The m by n matrix Q has orthonormal columns, and the square matrix R is upper triangular with positive diagonal. We must not forget why this is useful for least squares: $\boldsymbol{A}^{\mathrm{T}}\boldsymbol{A}$ ***equals*** $\boldsymbol{R}^{\mathrm{T}}\boldsymbol{Q}^{\mathrm{T}}\boldsymbol{Q}\boldsymbol{R} = \boldsymbol{R}^{\mathrm{T}}\boldsymbol{R}$. The least squares equation $A^{\mathrm{T}}A\widehat{x} = A^{\mathrm{T}}\boldsymbol{b}$ simplifies to

$$R^{\mathrm{T}}R\widehat{x} = R^{\mathrm{T}}Q^{\mathrm{T}}\boldsymbol{b} \qquad \text{or} \qquad R\widehat{x} = Q^{\mathrm{T}}\boldsymbol{b}. \tag{10}$$

Instead of solving $Ax = b$, which is impossible, we solve $R\widehat{x} = Q^{\mathrm{T}}b$ by back substitution—which is very fast. The real cost is the mn^2 multiplications in the Gram-Schmidt process, which are needed to construct the orthogonal Q and the triangular R.

Here is an informal code. It executes equations (11) and (12), for $k = 1$ then $k = 2$ and eventually $k = n$. Equation (11) normalizes to unit vectors: For $k = 1, \ldots, n$

$$r_{kk} = \left(\sum_{i=1}^{m} a_{ik}^2\right)^{1/2} \quad \text{and} \quad q_{ik} = \frac{a_{ik}}{r_{kk}} \quad \text{for} \quad i = 1, \ldots, m. \tag{11}$$

Equation (12) subtracts from $\boldsymbol{a}_j$ its projection onto $\boldsymbol{q}_k$: For $j = k+1, \ldots, n$

$$r_{kj} = \sum_{i=1}^{m} q_{ik} a_{ij} \quad \text{and} \quad a_{ij} = a_{ij} - q_{ik} r_{kj} \quad \text{for} \quad i = 1, \ldots, m. \tag{12}$$

Starting from $\boldsymbol{a}, \boldsymbol{b}, \boldsymbol{c} = \boldsymbol{a}_1, \boldsymbol{a}_2, \boldsymbol{a}_3$ this code will construct $\boldsymbol{q}_1$, $\boldsymbol{B}$, $\boldsymbol{q}_2$, $\boldsymbol{C}$, $\boldsymbol{q}_3$:

1 $\quad \boldsymbol{q}_1 = \boldsymbol{a}_1/\|\boldsymbol{a}_1\|$ in (11)

2 $\quad \boldsymbol{B} = \boldsymbol{a}_2 - (\boldsymbol{q}_1^{\mathrm{T}}\boldsymbol{a}_2)\boldsymbol{q}_1$ and $\boldsymbol{C}^* = \boldsymbol{a}_3 - (\boldsymbol{q}_1^{\mathrm{T}}\boldsymbol{a}_3)\boldsymbol{q}_1$ in (12)

3 $\quad \boldsymbol{q}_2 = \boldsymbol{B}/\|\boldsymbol{B}\|$ in (11)

4 $\quad \boldsymbol{C} = \boldsymbol{C}^* - (\boldsymbol{q}_2^{\mathrm{T}}\boldsymbol{C}^*)\boldsymbol{q}_2$ in (12)

5 $\quad \boldsymbol{q}_3 = \boldsymbol{C}/\|\boldsymbol{C}\|$ in (11)

Equation (12) subtracts off projections as soon as the new vector $\boldsymbol{q}_k$ is found. This change to "subtract one projection at a time" is called ***modified Gram-Schmidt***. It is numerically more stable than equation (8) which subtracts all projections at once.

■ REVIEW OF THE KEY IDEAS ■

1. If the orthonormal vectors $\boldsymbol{q}_1, \ldots, \boldsymbol{q}_n$ are the columns of Q, then $\boldsymbol{q}_i^{\mathrm{T}}\boldsymbol{q}_j = 0$ and $\boldsymbol{q}_i^{\mathrm{T}}\boldsymbol{q}_i = 1$ translate into $Q^{\mathrm{T}}Q = I$.

2. If Q is square (an ***orthogonal matrix***) then $Q^{\mathrm{T}} = Q^{-1}$.

3. The length of $Q\boldsymbol{x}$ equals the length of $\boldsymbol{x}$: $\|Q\boldsymbol{x}\| = \|\boldsymbol{x}\|$.

4. The projection onto the column space spanned by the $\boldsymbol{q}$'s is $P = QQ^{\mathrm{T}}$.

5. If Q is square then $P = I$ and every $\boldsymbol{b} = \boldsymbol{q}_1(\boldsymbol{q}_1^{\mathrm{T}}\boldsymbol{b}) + \cdots + \boldsymbol{q}_n(\boldsymbol{q}_n^{\mathrm{T}}\boldsymbol{b})$.

6. Gram-Schmidt produces orthonormal vectors $\boldsymbol{q}_1, \boldsymbol{q}_2, \boldsymbol{q}_3$ from independent $\boldsymbol{a}, \boldsymbol{b}, \boldsymbol{c}$. In matrix form this is the factorization $A = QR = $ (orthogonal Q)(triangular R).

■ WORKED EXAMPLES ■

4.4 A Add two more columns with all entries 1 or -1, so the columns of this 4 by 4 "Hadamard matrix" are orthogonal. How do you turn H into an *orthogonal matrix* Q?

$$H = \begin{bmatrix} 1 & 1 & x & x \\ 1 & 1 & x & x \\ 1 & -1 & x & x \\ 1 & -1 & x & x \end{bmatrix} \quad \text{and} \quad Q = \begin{bmatrix} & & & \\ & & & \\ & & & \\ & & & \end{bmatrix}$$

Why can't a 5 by 5 matrix have orthogonal columns of 1's and -1's? Actually the next possible size is 8 by 8, constructed from four blocks:

The block matrix $H_8 = \begin{bmatrix} H & H \\ H & -H \end{bmatrix}$ is a Hadamard matrix with orthogonal columns. What is the product $H_8^{\mathrm{T}} H_8$?

The projection of $\boldsymbol{b} = (6, 0, 0, 2)$ onto the first column of H is $\boldsymbol{p}_1 = (2, 2, 2, 2)$ and the projection onto the second column is $\boldsymbol{p}_2 = (1, 1, -1, -1)$. What is the projection $\boldsymbol{p}_{1,2}$ of $\boldsymbol{b}$ onto the 2-dimensional space spanned by the first two columns, and why?

Solution Columns 3 and 4 of this H could be multiplied by -1 or exchanged:

$$H = \begin{bmatrix} 1 & 1 & 1 & 1 \\ 1 & 1 & -1 & -1 \\ 1 & -1 & 1 & -1 \\ 1 & -1 & -1 & 1 \end{bmatrix} \text{ has orthogonal columns.} \quad Q = \frac{H}{2} \text{ has orthonormal columns.}$$

Dividing by 2 gives unit vectors in Q. Orthogonality for 5 by 5 is impossible because the dot product of columns would have five 1's and/or -1's and could not add to zero. The 8 by 8 matrix H_8 does have orthogonal columns (of length $\sqrt{8}$). Then Q_8 will be $H_8/\sqrt{8}$:

$$H_8^{\mathrm{T}} H_8 = \begin{bmatrix} H^{\mathrm{T}} & H^{\mathrm{T}} \\ H^{\mathrm{T}} & -H^{\mathrm{T}} \end{bmatrix} \begin{bmatrix} H & H \\ H & -H \end{bmatrix} = \begin{bmatrix} 2H^{\mathrm{T}}H & 0 \\ 0 & 2H^{\mathrm{T}}H \end{bmatrix} = \begin{bmatrix} 8I & 0 \\ 0 & 8I \end{bmatrix}.$$

When columns are orthogonal, we can project $(6, 0, 0, 2)$ onto $(1, 1, 1, 1)$ and $(1, 1, -1, -1)$ and add:

$$\text{Projection } \boldsymbol{p}_{1,2} = \boldsymbol{p}_1 + \boldsymbol{p}_2 = (2, 2, 2, 2) + (1, 1, -1, -1) = (3, 3, 1, 1).$$

This is the value of orthogonal columns. A quick proof of $\boldsymbol{p}_{1,2} = \boldsymbol{p}_1 + \boldsymbol{p}_2$ is to check that columns 1 and 2 (call them $\boldsymbol{a}_1$ and $\boldsymbol{a}_2$) are perpendicular to the error $\boldsymbol{e} = \boldsymbol{b} - \boldsymbol{p}_1 - \boldsymbol{p}_2$:

$$\boldsymbol{e} = \boldsymbol{b} - \frac{\boldsymbol{a}_1^{\mathrm{T}}\boldsymbol{b}}{\boldsymbol{a}_1^{\mathrm{T}}\boldsymbol{a}_1}\boldsymbol{a}_1 - \frac{\boldsymbol{a}_2^{\mathrm{T}}\boldsymbol{b}}{\boldsymbol{a}_2^{\mathrm{T}}\boldsymbol{a}_2}\boldsymbol{a}_2 \quad \text{and} \quad \boldsymbol{a}_1^{\mathrm{T}}\boldsymbol{e} = \boldsymbol{a}_1^{\mathrm{T}}\boldsymbol{b} - \frac{\boldsymbol{a}_1^{\mathrm{T}}\boldsymbol{b}}{\boldsymbol{a}_1^{\mathrm{T}}\boldsymbol{a}_1}\boldsymbol{a}_1^{\mathrm{T}}\boldsymbol{a}_1 = 0 \quad \text{and also} \quad \boldsymbol{a}_2^{\mathrm{T}}\boldsymbol{e} = 0.$$

So $\boldsymbol{p}_1 + \boldsymbol{p}_2$ is in the space of $\boldsymbol{a}_1$ and $\boldsymbol{a}_2$, and *its error* $\boldsymbol{e}$ *is perpendicular to that space.*

The Gram-Schmidt process on those orthogonal columns $\boldsymbol{a}_1$ and $\boldsymbol{a}_2$ would be happy with their directions. It would only divide by their lengths. *But if* $\boldsymbol{a}_1$ *and* $\boldsymbol{a}_2$ *are not orthogonal, the projection* $\boldsymbol{p}_{1,2}$ *is not generally* $\boldsymbol{p}_1+\boldsymbol{p}_2$. For example, if $\boldsymbol{b}=\boldsymbol{a}_1$ then $\boldsymbol{p}_1=\boldsymbol{b}$ and $\boldsymbol{p}_{1,2}=\boldsymbol{b}$ but $\boldsymbol{p}_2\neq\mathbf{0}$.

Problem Set 4.4

Problems 1–12 are about orthogonal vectors and orthogonal matrices.

1 Are these pairs of vectors orthonormal or only orthogonal or only independent?

$$\text{(a)}\quad \begin{bmatrix}1\\0\end{bmatrix} \text{ and } \begin{bmatrix}-1\\1\end{bmatrix} \qquad \text{(b)}\quad \begin{bmatrix}.6\\.8\end{bmatrix} \text{ and } \begin{bmatrix}.4\\-.3\end{bmatrix} \qquad \text{(c)}\quad \begin{bmatrix}\cos\theta\\ \sin\theta\end{bmatrix} \text{ and } \begin{bmatrix}-\sin\theta\\ \cos\theta\end{bmatrix}.$$

Change the second vector when necessary to produce orthonormal vectors.

2 The vectors $(2,2,-1)$ and $(-1,2,2)$ are orthogonal. Divide them by their lengths to find orthonormal vectors $\boldsymbol{q}_1$ and $\boldsymbol{q}_2$. Put those into the columns of Q and multiply $Q^{\mathrm{T}}Q$ and QQ^{T}.

3 (a) If A has three orthogonal columns each of length 4, what is $A^{\mathrm{T}}A$?

(b) If A has three orthogonal columns of lengths $1, 2, 3$, what is $A^{\mathrm{T}}A$?

4 Give an example of each of the following:

(a) A matrix Q that has orthonormal columns but $QQ^{\mathrm{T}}\neq I$.

(b) Two orthogonal vectors that are not linearly independent.

(c) An orthonormal basis for $\mathbf{R}^4$, where every component is $\frac{1}{2}$ or $-\frac{1}{2}$.

5 Find two orthogonal vectors in the plane $x+y+2z=0$. Make them orthonormal.

6 If Q_1 and Q_2 are orthogonal matrices, show that their product Q_1Q_2 is also an orthogonal matrix. (Use $Q^{\mathrm{T}}Q=I$.)

7 If Q has orthonormal columns, what is the least squares solution $\widehat{\boldsymbol{x}}$ to $Q\boldsymbol{x}=\boldsymbol{b}$?

8 If $\boldsymbol{q}_1$ and $\boldsymbol{q}_2$ are orthonormal vectors in $\mathbf{R}^5$, what combination ____ $\boldsymbol{q}_1+$ ____ $\boldsymbol{q}_2$ is closest to a given vector $\boldsymbol{b}$?

9 (a) Compute $P=QQ^{\mathrm{T}}$ when $\boldsymbol{q}_1=(.8,.6,0)$ and $\boldsymbol{q}_2=(-.6,.8,0)$. Verify that $P^2=P$.

(b) Prove that always $(QQ^{\mathrm{T}})(QQ^{\mathrm{T}})=QQ^{\mathrm{T}}$ by using $Q^{\mathrm{T}}Q=I$. Then $P=QQ^{\mathrm{T}}$ is the projection matrix onto the column space of Q.

10 Orthonormal vectors are automatically linearly independent. Two proofs:

(a) Vector proof: When $c_1\boldsymbol{q}_1 + c_2\boldsymbol{q}_2 + c_3\boldsymbol{q}_3 = \mathbf{0}$, what dot product leads to $c_1 = 0$? Similarly $c_2 = 0$ and $c_3 = 0$. Thus the $\boldsymbol{q}$'s are independent.

(b) Matrix proof: Show that $Q\boldsymbol{x} = \mathbf{0}$ leads to $\boldsymbol{x} = \mathbf{0}$. Since Q may be rectangular, you can use Q^{T} but not Q^{-1}.

11 (a) Find orthonormal vectors $\boldsymbol{q}_1$ and $\boldsymbol{q}_2$ in the plane of $\boldsymbol{a} = (1, 3, 4, 5, 7)$ and $\boldsymbol{b} = (-6, 6, 8, 0, 8)$.

(b) Which vector in this plane is closest to $(1, 0, 0, 0, 0)$?

12 If $\boldsymbol{a}_1, \boldsymbol{a}_2, \boldsymbol{a}_3$ is a basis for $\mathbf{R}^3$, any vector $\boldsymbol{b}$ can be written as

$$\boldsymbol{b} = x_1\boldsymbol{a}_1 + x_2\boldsymbol{a}_2 + x_3\boldsymbol{a}_3 \qquad \text{or} \qquad \begin{bmatrix} & & \\ \boldsymbol{a}_1 & \boldsymbol{a}_2 & \boldsymbol{a}_3 \\ & & \end{bmatrix}\begin{bmatrix} x_1 \\ x_2 \\ x_3 \end{bmatrix} = \boldsymbol{b}.$$

(a) Suppose the $\boldsymbol{a}$'s are orthonormal. Show that $x_1 = \boldsymbol{a}_1^{\mathrm{T}}\boldsymbol{b}$.

(b) Suppose the $\boldsymbol{a}$'s are orthogonal. Show that $x_1 = \boldsymbol{a}_1^{\mathrm{T}}\boldsymbol{b}/\boldsymbol{a}_1^{\mathrm{T}}\boldsymbol{a}_1$.

(c) If the $\boldsymbol{a}$'s are independent, x_1 is the first component of _____ times $\boldsymbol{b}$.

Problems 13–25 are about the Gram-Schmidt process and $A = QR$.

13 What multiple of $\boldsymbol{a} = \left[\begin{smallmatrix} 1 \\ 1 \end{smallmatrix}\right]$ should be subtracted from $\boldsymbol{b} = \left[\begin{smallmatrix} 4 \\ 0 \end{smallmatrix}\right]$ to make the result $\boldsymbol{B}$ orthogonal to $\boldsymbol{a}$? Sketch a figure to show $\boldsymbol{a}$, $\boldsymbol{b}$, and $\boldsymbol{B}$.

14 Complete the Gram-Schmidt process in Problem 13 by computing $\boldsymbol{q}_1 = \boldsymbol{a}/\|\boldsymbol{a}\|$ and $\boldsymbol{q}_2 = \boldsymbol{B}/\|\boldsymbol{B}\|$ and factoring into QR:

$$\begin{bmatrix} 1 & 4 \\ 1 & 0 \end{bmatrix} = \begin{bmatrix} \boldsymbol{q}_1 & \boldsymbol{q}_2 \end{bmatrix}\begin{bmatrix} \|\boldsymbol{a}\| & ? \\ 0 & \|\boldsymbol{B}\| \end{bmatrix}.$$

15 (a) Find orthonormal vectors $\boldsymbol{q}_1, \boldsymbol{q}_2, \boldsymbol{q}_3$ such that $\boldsymbol{q}_1, \boldsymbol{q}_2$ span the column space of

$$A = \begin{bmatrix} 1 & 1 \\ 2 & -1 \\ -2 & 4 \end{bmatrix}.$$

(b) Which of the four fundamental subspaces contains $\boldsymbol{q}_3$?

(c) Solve $A\boldsymbol{x} = (1, 2, 7)$ by least squares.

16 What multiple of $\boldsymbol{a} = (4, 5, 2, 2)$ is closest to $\boldsymbol{b} = (1, 2, 0, 0)$? Find orthonormal vectors $\boldsymbol{q}_1$ and $\boldsymbol{q}_2$ in the plane of $\boldsymbol{a}$ and $\boldsymbol{b}$.

17 Find the projection of $\boldsymbol{b}$ onto the line through $\boldsymbol{a}$:

$$\boldsymbol{a} = \begin{bmatrix} 1 \\ 1 \\ 1 \end{bmatrix} \quad \text{and} \quad \boldsymbol{b} = \begin{bmatrix} 1 \\ 3 \\ 5 \end{bmatrix} \quad \text{and} \quad \boldsymbol{p} = ? \quad \text{and} \quad \boldsymbol{e} = \boldsymbol{b} - \boldsymbol{p} = ?$$

Compute the orthonormal vectors $\boldsymbol{q}_1 = \boldsymbol{a}/\|\boldsymbol{a}\|$ and $\boldsymbol{q}_2 = \boldsymbol{e}/\|\boldsymbol{e}\|$.

18 (Recommended) Find orthogonal vectors $\boldsymbol{A}, \boldsymbol{B}, \boldsymbol{C}$ by Gram-Schmidt from $\boldsymbol{a}, \boldsymbol{b}, \boldsymbol{c}$:

$$\boldsymbol{a} = (1, -1, 0, 0) \qquad \boldsymbol{b} = (0, 1, -1, 0) \qquad \boldsymbol{c} = (0, 0, 1, -1).$$

$\boldsymbol{A}, \boldsymbol{B}, \boldsymbol{C}$ and $\boldsymbol{a}, \boldsymbol{b}, \boldsymbol{c}$ are bases for the vectors perpendicular to $\boldsymbol{d} = (1, 1, 1, 1)$.

19 If $A = QR$ then $A^{\mathrm{T}}A = R^{\mathrm{T}}R =$ _____ triangular times _____ triangular. *Gram-Schmidt on A corresponds to elimination on $A^{\mathrm{T}}A$.* Compare the pivots for $A^{\mathrm{T}}A$ with $\|\boldsymbol{a}\|^2 = 3$ and $\|\boldsymbol{e}\|^2 = 8$ in Problem 17:

$$A = \begin{bmatrix} 1 & 1 \\ 1 & 3 \\ 1 & 5 \end{bmatrix} \quad \text{and} \quad A^{\mathrm{T}}A = \begin{bmatrix} 3 & 9 \\ 9 & 35 \end{bmatrix}.$$

20 True or false (give an example in either case):

(a) Q^{-1} is an orthogonal matrix when Q is an orthogonal matrix.

(b) If Q (3 by 2) has orthonormal columns then $\|Q\boldsymbol{x}\|$ always equals $\|\boldsymbol{x}\|$.

21 Find an orthonormal basis for the column space of A:

$$A = \begin{bmatrix} 1 & -2 \\ 1 & 0 \\ 1 & 1 \\ 1 & 3 \end{bmatrix} \quad \text{and} \quad \boldsymbol{b} = \begin{bmatrix} -4 \\ -3 \\ 3 \\ 0 \end{bmatrix}.$$

Then compute the projection of $\boldsymbol{b}$ onto that column space.

22 Find orthogonal vectors $\boldsymbol{A}, \boldsymbol{B}, \boldsymbol{C}$ by Gram-Schmidt from

$$\boldsymbol{a} = \begin{bmatrix} 1 \\ 1 \\ 2 \end{bmatrix} \quad \text{and} \quad \boldsymbol{b} = \begin{bmatrix} 1 \\ -1 \\ 0 \end{bmatrix} \quad \text{and} \quad \boldsymbol{c} = \begin{bmatrix} 1 \\ 0 \\ 4 \end{bmatrix}.$$

23 Find $\boldsymbol{q}_1, \boldsymbol{q}_2, \boldsymbol{q}_3$ (orthonormal) as combinations of $\boldsymbol{a}, \boldsymbol{b}, \boldsymbol{c}$ (independent columns). Then write A as QR:

$$A = \begin{bmatrix} 1 & 2 & 4 \\ 0 & 0 & 5 \\ 0 & 3 & 6 \end{bmatrix}.$$

24 (a) Find a basis for the subspace $\boldsymbol{S}$ in $\mathbf{R}^4$ spanned by all solutions of

$$x_1 + x_2 + x_3 - x_4 = 0.$$

(b) Find a basis for the orthogonal complement $\boldsymbol{S}^{\perp}$.

(c) Find $\boldsymbol{b}_1$ in $\boldsymbol{S}$ and $\boldsymbol{b}_2$ in $\boldsymbol{S}^{\perp}$ so that $\boldsymbol{b}_1 + \boldsymbol{b}_2 = \boldsymbol{b} = (1, 1, 1, 1)$.

25 If $ad - bc > 0$, the entries in $A = QR$ are

$$\begin{bmatrix} a & b \\ c & d \end{bmatrix} = \frac{\begin{bmatrix} a & -c \\ c & a \end{bmatrix}}{\sqrt{a^2+c^2}} \frac{\begin{bmatrix} a^2+c^2 & ab+cd \\ 0 & ad-bc \end{bmatrix}}{\sqrt{a^2+c^2}}.$$

Write $A = QR$ when $a, b, c, d = 2, 1, 1, 1$ and also $1, 1, 1, 1$. Which entry of R becomes zero when the columns are dependent and Gram-Schmidt breaks down?

Problems 26–29 use the QR code in equations (11–12). It executes Gram-Schmidt.

26 Show why $\boldsymbol{C}$ (found via $\boldsymbol{C}^*$ in the steps after (12)) is equal to $\boldsymbol{C}$ in equation (8).

27 Equation (8) subtracts from $\boldsymbol{c}$ its components along $\boldsymbol{A}$ and $\boldsymbol{B}$. Why not subtract the components along $\boldsymbol{a}$ and along $\boldsymbol{b}$?

28 Write a working code and apply it to $\boldsymbol{a} = (2, 2, -1)$, $\boldsymbol{b} = (0, -3, 3)$, $\boldsymbol{c} = (1, 0, 0)$. What are the $\boldsymbol{q}$'s?

29 Where are the mn^2 multiplications in equations (11) and (12)?

Problems 30–35 involve orthogonal matrices that are special.

30 The first four ***wavelets*** are in the columns of this wavelet matrix W:

$$W = \frac{1}{2}\begin{bmatrix} 1 & 1 & \sqrt{2} & 0 \\ 1 & 1 & -\sqrt{2} & 0 \\ 1 & -1 & 0 & \sqrt{2} \\ 1 & -1 & 0 & -\sqrt{2} \end{bmatrix}.$$

What is special about the columns? Find the inverse wavelet transform W^{-1}.

31 (a) Choose c so that Q is an orthogonal matrix:

$$Q = c\begin{bmatrix} 1 & -1 & -1 & -1 \\ -1 & 1 & -1 & -1 \\ -1 & -1 & 1 & -1 \\ -1 & -1 & -1 & 1 \end{bmatrix}.$$

(b) Change the first row and column to all 1's and fill in another orthogonal Q.

32 Project $\boldsymbol{b} = (1, 1, 1, 1)$ onto the first column in Problem 31(a). Then project $\boldsymbol{b}$ onto the plane of the first two columns.

33 If $\boldsymbol{u}$ is a unit vector, then $Q = I - 2\boldsymbol{u}\boldsymbol{u}^{\mathrm{T}}$ is a Householder reflection matrix (Example 3). Find Q_1 from $\boldsymbol{u} = (0, 1)$ and Q_2 from $\boldsymbol{u} = (0, \sqrt{2}/2, \sqrt{2}/2)$. Draw the reflections when Q_1 and Q_2 multiply (x, y) and (x, y, z).

34 $Q = I - 2\boldsymbol{u}\boldsymbol{u}^{\mathrm{T}}$ is a reflection matrix when $\boldsymbol{u}^{\mathrm{T}}\boldsymbol{u} = 1$.

(a) Show that $Q\boldsymbol{u} = -\boldsymbol{u}$. The mirror is perpendicular to $\boldsymbol{u}$.

(b) Find $Q\boldsymbol{v}$ when $\boldsymbol{u}^{\mathrm{T}}\boldsymbol{v} = 0$. The mirror contains $\boldsymbol{v}$. It reflects to itself.

35 (MATLAB) Factor $[\,Q, R\,] = \mathbf{qr}(A)$ for $A = \mathbf{eye}(4) - \mathbf{diag}([\,1 \;\; 1 \;\; 1\,], -1)$. Can you renormalize the orthogonal columns of Q to get nice integer components?

36 Find all matrices that are both orthogonal and lower triangular.

5

DETERMINANTS

THE PROPERTIES OF DETERMINANTS ■ 5.1

The determinant of a square matrix is a single number. That number contains an amazing amount of information about the matrix. It tells immediately whether the matrix is invertible. ***The determinant is zero when the matrix has no inverse***. When A is invertible, the determinant of A^{-1} is $1/(\det A)$. If $\det A = 2$ then $\det A^{-1} = \frac{1}{2}$. In fact the determinant leads to a formula for every entry in A^{-1}.

This is one use for determinants—to find formulas for inverse matrices and pivots and solutions $A^{-1}\boldsymbol{b}$. For a matrix of numbers, we seldom use those formulas. (Or rather, we use elimination as the quickest way to the answer.) For a matrix with entries a, b, c, d, its determinant shows how A^{-1} changes as A changes:

$$A = \begin{bmatrix} a & b \\ c & d \end{bmatrix} \quad \text{has inverse} \quad A^{-1} = \frac{1}{ad - bc} \begin{bmatrix} d & -b \\ -c & a \end{bmatrix}. \tag{1}$$

Multiply those matrices to get I. The determinant of A is $ad - bc$. When $\det A = 0$, we are asked to divide by zero and we can't—then A has no inverse. (The rows are parallel when $a/c = b/d$. This gives $ad = bc$ and a zero determinant.) Dependent rows lead to $\det A = 0$.

The determinant is also connected to the pivots. For a 2 by 2 matrix the pivots are a and $d - (c/a)b$. ***The product of the pivots is the determinant***:

$$a\left(d - \frac{c}{a}b\right) = ad - bc \quad \text{which is} \quad \det A.$$

After a row exchange the pivots are c and $b - (a/c)d$. Those pivots multiply to give $bc - ad$. The row exchange reversed the sign of the determinant.

Looking ahead The determinant of an n by n matrix can be found in three ways:

1 Multiply the n pivots (times 1 or -1). This is the pivot formula.
2 Add up $n!$ terms (times 1 or -1). This is the "big" formula.
3 Combine n smaller determinants (times 1 or -1). This is the cofactor formula.

You see that plus or minus signs—the decisions between 1 and -1—play a big part in determinants. That comes from the following rule for n by n matrices:

The determinant changes sign when two rows* (or two columns) *are exchanged.

The identity matrix has determinant $+1$. Exchange two rows and $\det P = -1$. Exchange two more rows and the new permutation has $\det P = +1$. Half of all permutations are *even* ($\det P = 1$) and half are *odd* ($\det P = -1$). Starting from I, half of the P's involve an even number of exchanges and half require an odd number. In the 2 by 2 case, ad has a plus sign and bc has minus—coming from the row exchange:

$$\det \begin{bmatrix} 1 & 0 \\ 0 & 1 \end{bmatrix} = 1 \quad \text{and} \quad \det \begin{bmatrix} 0 & 1 \\ 1 & 0 \end{bmatrix} = -1.$$

The other essential rule is linearity—but a warning comes first. Linearity does not mean that $\det(A + B) = \det A + \det B$. ***This is absolutely false***. That kind of linearity is not even true when $A = I$ and $B = I$. The false rule would say that $\det 2I = 1 + 1 = 2$. The true rule is $\det 2I = 2^n$. Determinants are multiplied by 2^n (not just by 2) when matrices are multiplied by 2.

We don't intend to define the determinant by its formulas. It is better to start with its properties—*sign reversal and linearity*. The properties are simple (Section 5.1). They prepare for the formulas (Section 5.2). Then come the applications, including these three:

(1) Determinants give A^{-1} and $A^{-1}\boldsymbol{b}$ (this formula is called *Cramer's Rule*).

(2) When the edges of a box are the rows of A, the *volume* is $|\det A|$.

(3) The numbers λ for which $A - \lambda I$ is singular and $\det(A - \lambda I) = 0$ are the *eigenvalues* of A. This is the most important application and it fills Chapter 6.

The Properties of the Determinant

There are three basic properties (rules 1, 2, 3). By using those rules we can compute the determinant of any square matrix A. ***This number is written in two ways,* det *A* *and* |*A*|.** Notice: Brackets for the matrix, straight bars for its determinant. When A is a 2 by 2 matrix, the three properties lead to the answer we expect:

$$\text{The determinant of} \quad \begin{bmatrix} a & b \\ c & d \end{bmatrix} \quad \text{is} \quad \begin{vmatrix} a & b \\ c & d \end{vmatrix} = ad - bc.$$

We will check each rule against this 2 by 2 formula, but do not forget: The rules apply to any n by n matrix. When we prove that properties 4–10 follow from 1–3, the proof must apply to all square matrices.

Property 1 (the easiest rule) matches the determinant of I with the volume of a unit cube.

1 ***The determinant of the n by n identity matrix is 1.***

$$\begin{vmatrix} 1 & 0 \\ 0 & 1 \end{vmatrix} = 1 \qquad \text{and} \qquad \begin{vmatrix} 1 & & \\ & \ddots & \\ & & 1 \end{vmatrix} = 1.$$

2 ***The determinant changes sign when two rows are exchanged*** (sign reversal):

$$\text{Check:} \quad \begin{vmatrix} c & d \\ a & b \end{vmatrix} = -\begin{vmatrix} a & b \\ c & d \end{vmatrix} \quad \text{(both sides equal } bc - ad\text{).}$$

Because of this rule, we can find $\det P$ for any permutation matrix. Just exchange rows of I until you reach P. Then $\det P = +1$ for an ***even*** number of row exchanges and $\det P = -1$ for an ***odd*** number.

The third rule has to make the big jump to the determinants of all matrices.

3 ***The determinant is a linear function of each row separately*** (all other rows stay fixed). If the first row is multiplied by t, the determinant is multiplied by t. If first rows are added, determinants are added. This rule only applies when the other rows do not change! Notice how c and d stay the same:

multiply row 1 by any number t: $$\begin{vmatrix} ta & tb \\ c & d \end{vmatrix} = t\begin{vmatrix} a & b \\ c & d \end{vmatrix}$$

add row 1 of A to row 1 of A': $$\begin{vmatrix} a+a' & b+b' \\ c & d \end{vmatrix} = \begin{vmatrix} a & b \\ c & d \end{vmatrix} + \begin{vmatrix} a' & b' \\ c & d \end{vmatrix}.$$

In the first case, both sides are $tad - tbc$. Then t factors out. In the second case, both sides are $ad + a'd - bc - b'c$. These rules still apply when A is n by n, and the last $n - 1$ rows don't change. May we emphasize rule 3 with numbers:

$$\begin{vmatrix} 5 & 0 & 0 \\ 0 & 1 & 0 \\ 0 & 0 & 1 \end{vmatrix} = 5\begin{vmatrix} 1 & 0 & 0 \\ 0 & 1 & 0 \\ 0 & 0 & 1 \end{vmatrix} \quad \text{and} \quad \begin{vmatrix} 1 & 2 & 3 \\ 0 & 1 & 0 \\ 0 & 0 & 1 \end{vmatrix} = \begin{vmatrix} 1 & 0 & 0 \\ 0 & 1 & 0 \\ 0 & 0 & 1 \end{vmatrix} + \begin{vmatrix} 0 & 2 & 3 \\ 0 & 1 & 0 \\ 0 & 0 & 1 \end{vmatrix}.$$

By itself, rule 3 does not say what any of those determinants are. But with rule 1, the first determinant is 5 (and the second is 1).

Combining multiplication and addition, we get any linear combination in the first row: t(row 1 of A) $+\, t'$(row 1 of A'). With this combined row, the determinant is t times $\det A$ plus t' times $\det A'$. The other rows must stay the same.

This rule does not mean that $\det 2I = 2 \det I$. To obtain $2I$ we have to multiply *both* rows by 2, and the factor 2 comes out both times:

$$\begin{vmatrix} 2 & 0 \\ 0 & 2 \end{vmatrix} = 2^2 = 4 \qquad \text{and} \qquad \begin{vmatrix} t & 0 \\ 0 & t \end{vmatrix} = t^2.$$

This is just like area and volume. Expand a rectangle by 2 and its area increases by 4. Expand an n-dimensional box by t and its volume increases by t^n. The connection is no accident—we will see how *determinants equal volumes*.

Pay special attention to rules 1–3. They completely determine the number $\det A$—but for a big matrix that fact is not obvious. We could stop here to find a formula for n by n determinants. It would be a little complicated—we prefer to go gradually. Instead we write down other properties which follow directly from the first three. These extra rules make determinants much easier to work with.

4 ***If two rows of A are equal, then*** $\det A = 0$.

$$\text{Check 2 by 2:} \quad \begin{vmatrix} a & b \\ a & b \end{vmatrix} = 0.$$

Rule 4 follows from rule 2. (Remember we must use the rules and not the 2 by 2 formula.) *Exchange the two equal rows.* The determinant D is supposed to change sign. But also D has to stay the same, because the matrix is not changed. The only number with $-D = D$ is $D = 0$—this must be the determinant. (Note: In Boolean algebra the reasoning fails, because $-1 = 1$. Then D is defined by rules 1, 3, 4.)

A matrix with two equal rows has no inverse. Rule 4 makes $\det A = 0$. But matrices can be singular and determinants can be zero without having equal rows! Rule 5 will be the key. We can do row operations without changing $\det A$.

5 ***Subtracting a multiple of one row from another row leaves*** $\det A$ ***unchanged.***

$$\begin{vmatrix} a & b \\ c - \ell a & d - \ell b \end{vmatrix} = \begin{vmatrix} a & b \\ c & d \end{vmatrix}.$$

Linearity splits the left side into the right side plus another term $-\ell \begin{vmatrix} a & b \\ a & b \end{vmatrix}$. This extra term is zero by rule 4. Therefore rule 5 is correct. Note how the second row changes while the first row stays the same—as required by rule 3.

Conclusion *The determinant is not changed by the usual elimination steps from A to U. Thus* $\det A$ *equals* $\det U$. If we can find determinants of triangular matrices U, we can find determinants of all matrices A. Every row exchange reverses the sign, so always $\det A = \pm \det U$. Rule 5 has narrowed the problem to triangular matrices.

6 ***A matrix with a row of zeros has*** $\det A = 0$.

$$\begin{vmatrix} 0 & 0 \\ c & d \end{vmatrix} = 0 \quad \text{and} \quad \begin{vmatrix} a & b \\ 0 & 0 \end{vmatrix} = 0.$$

For an easy proof, add some other row to the zero row. The determinant is not changed (rule 5). But the matrix now has two equal rows. So $\det A = 0$ by rule 4.

7 ***If A is triangular then*** $\det A = a_{11}a_{22}\cdots a_{nn}$ ***= product of diagonal entries.***

$$\begin{vmatrix} a & b \\ 0 & d \end{vmatrix} = ad \quad \text{and also} \quad \begin{vmatrix} a & 0 \\ c & d \end{vmatrix} = ad.$$

Suppose all diagonal entries of A are nonzero. Eliminate the off-diagonal entries by the usual steps. (If A is lower triangular, subtract multiples of each row from lower rows. If A is upper triangular, subtract from higher rows.) By rule 5 the determinant is not changed—and now the matrix is diagonal:

$$\text{We must still prove that} \quad \det \begin{bmatrix} a_{11} & & & 0 \\ & a_{22} & & \\ & & \ddots & \\ 0 & & & a_{nn} \end{bmatrix} = a_{11}a_{22}\cdots a_{nn}.$$

For this we apply rules 1 and 3. Factor a_{11} from the first row. Then factor a_{22} from the second row. Eventually factor a_{nn} from the last row. The determinant is a_{11} times a_{22} times $\cdots$ times a_{nn} times $\det I$. Then rule 1 (used at last!) is $\det I = 1$.

What if a diagonal entry a_{ii} is zero? Then the triangular A is singular. Elimination produces a *zero row*. By rule 5 the determinant is unchanged, and by rule 6 a zero row means $\det A = 0$. Thus rule 7 is proved—triangular matrices have easy determinants.

8 ***If A is singular then*** $\det A = 0$. ***If A is invertible then*** $\det A \neq 0$.

$$\begin{bmatrix} a & b \\ c & d \end{bmatrix} \quad \text{is singular if and only if} \quad ad - bc = 0.$$

Proof Elimination goes from A to U. If A is singular then U has a zero row. The rules give $\det A = \det U = 0$. If A is invertible then U has the pivots along its diagonal. The product of nonzero pivots (using rule 7) gives a nonzero determinant:

$$\det A = \pm \det U = \pm \text{ (product of the pivots).} \tag{2}$$

The pivots of a 2 by 2 matrix (if $a \neq 0$) are a and $d - (bc/a)$:

$$\textbf{The determinant is} \quad \begin{vmatrix} a & b \\ c & d \end{vmatrix} = \begin{vmatrix} a & b \\ 0 & d - (bc/a) \end{vmatrix} = ad - bc.$$

This is the first formula for the determinant. MATLAB would use it to find $\det A$ from the pivots. The plus or minus sign depends on whether the number of row exchanges is even or odd. In other words, $+1$ or -1 is the determinant of the permutation matrix P that exchanges rows. With no row exchanges, the number zero is even and $P = I$ and $\det A = \det U =$ *product of pivots*. Always $\det L = 1$, because L is triangular with 1's on the diagonal. What we have is this:

$$\text{If} \quad PA = LU \quad \text{then} \quad \det P \; \det A = \det L \; \det U. \tag{3}$$

Again, $\det P = \pm 1$ and $\det A = \pm \det U$. Equation (3) is our first case of rule 9.

9 ***The determinant of AB is*** $\det A$ ***times*** $\det B$: $|AB| = |A|\,|B|$.

$$\begin{vmatrix} a & b \\ c & d \end{vmatrix} \begin{vmatrix} p & q \\ r & s \end{vmatrix} = \begin{vmatrix} ap + br & aq + bs \\ cp + dr & cq + ds \end{vmatrix}.$$

When the matrix B is A^{-1}, *this rule says that the determinant of* A^{-1} *is* $1/\det A$:

$$AA^{-1} = I \quad \text{so} \quad (\det A)(\det A^{-1}) = \det I = 1.$$

This product rule is the most intricate so far. We could check the 2 by 2 case by algebra:

$$|A|\,|B| = (ad - bc)(ps - qr) = (ap + br)(cq + ds) - (aq + bs)(cp + dr) = |AB|.$$

For the n by n case, here is a snappy proof that $|AB| = |A|\,|B|$. When $|B|$ is not zero, consider the ratio $D(A) = |AB|/|B|$. If this ratio has properties 1,2,3—which we now check—it has to be the determinant $|A|$.

Property 1 (*Determinant of* I) If $A = I$ then the ratio becomes $|B|/|B| = 1$.

Property 2 (*Sign reversal*) When two rows of A are exchanged, so are the same two rows of AB. Therefore $|AB|$ changes sign and so does the ratio $|AB|/|B|$.

Property 3 (*Linearity*) When row 1 of A is multiplied by t, so is row 1 of AB. This multiplies $|AB|$ by t and multiplies the ratio by t—as desired.

If row 1 of A is added to row 1 of A', then row 1 of AB is added to row 1 of $A'B$. By rule 3, the determinants add. After dividing by $|B|$, the ratios add.

Conclusion This ratio $|AB|/|B|$ has the same three properties that define $|A|$. Therefore it equals $|A|$. This proves the product rule $|AB| = |A|\,|B|$. The case $|B| = 0$ is separate and easy, because AB is singular when B is singular. The rule $|AB| = |A|\,|B|$ becomes $0 = 0$.

10 ***The transpose*** A^{T} ***has the same determinant as*** A.

$$\text{Check:} \quad \begin{vmatrix} a & b \\ c & d \end{vmatrix} = \begin{vmatrix} a & c \\ b & d \end{vmatrix} \quad \text{since both sides equal} \quad ad - bc.$$

The equation $|A^{\mathrm{T}}| = |A|$ becomes $0 = 0$ when A is singular (we know that A^{T} is also singular). Otherwise A has the usual factorization $PA = LU$. Transposing both sides gives $A^{\mathrm{T}}P^{\mathrm{T}} = U^{\mathrm{T}}L^{\mathrm{T}}$. The proof of $|A| = |A^{\mathrm{T}}|$ comes by using rule 9 for products:

$$\text{Compare} \quad \det P \det A = \det L \det U \quad \text{with} \quad \det A^{\mathrm{T}} \det P^{\mathrm{T}} = \det U^{\mathrm{T}} \det L^{\mathrm{T}}.$$

First, $\det L = \det L^{\mathrm{T}} = 1$ (both have 1's on the diagonal). Second, $\det U = \det U^{\mathrm{T}}$ (transposing leaves the main diagonal unchanged, and triangular determinants only involve that diagonal). Third, $\det P = \det P^{\mathrm{T}}$ (permutations have $P^{\mathrm{T}} = P^{-1}$, so $|P|\,|P^{\mathrm{T}}| = 1$ by rule 9; thus $|P|$ and $|P^{\mathrm{T}}|$ both equal 1 or both equal -1). Fourth and finally, the comparison proves that $\det A$ equals $\det A^{\mathrm{T}}$.

Important comment Rule 10 practically doubles our list of properties. Every rule for the rows can apply also to the columns (just by transposing, since $|A| = |A^{\mathrm{T}}|$). The determinant changes sign when two columns are exchanged. *A zero column or two* *equal columns will make the determinant zero.* If a column is multiplied by t, so is the determinant. The determinant is a linear function of each column separately.

It is time to stop. The list of properties is long enough. Next we find and use an explicit formula for the determinant.

■ REVIEW OF THE KEY IDEAS ■

1. The determinant is defined by $\det I = 1$, sign reversal, and linearity in each row.

2. After elimination $\det A$ is $\pm$ (product of the pivots).

3. The determinant is zero exactly when A is not invertible.

4. Two remarkable properties are $\det AB = (\det A)(\det B)$ and $\det A^{\mathrm{T}} = \det A$.

■ WORKED EXAMPLES ■

5.1 A Apply these operations to A and find the determinants of M_1, M_2, M_3, M_4:

In M_1, each a_{ij} is multiplied by $(-1)^{i+j}$. This gives the sign pattern shown below.

In M_2, rows $1, 2, 3$ of A are *subtracted* from rows $2, 3, 1$.

In M_3, rows $1, 2, 3$ of A are *added* to rows $2, 3, 1$.

The i, j entry of M_4 is (row i of A) $\cdot$ (row j of A).

How are the determinants of M_1, M_2, M_3, M_4 related to the determinant of A?

$$\begin{bmatrix} a_{11} & -a_{12} & a_{13} \\ -a_{21} & a_{22} & -a_{23} \\ a_{31} & -a_{32} & a_{33} \end{bmatrix} \quad \begin{bmatrix} \text{row } 1 - \text{row } 3 \\ \text{row } 2 - \text{row } 1 \\ \text{row } 3 - \text{row } 2 \end{bmatrix} \quad \begin{bmatrix} \text{row } 1 + \text{row } 3 \\ \text{row } 2 + \text{row } 1 \\ \text{row } 3 + \text{row } 2 \end{bmatrix} \quad \begin{bmatrix} \text{row } 1 \cdot \text{row } 1 & \cdot & \cdot \\ \text{row } 2 \cdot \text{row } 1 & \cdot & \cdot \\ \text{row } 3 \cdot \text{row } 1 & \cdot & \cdot \end{bmatrix}$$

Solution The four determinants are $\det A$, 0, $2 \det A$, and $(\det A)^2$. Here are reasons:

$$M_1 = \begin{bmatrix} 1 & & \\ & -1 & \\ & & 1 \end{bmatrix} \begin{bmatrix} a_{11} & a_{12} & a_{13} \\ a_{21} & a_{22} & a_{23} \\ a_{31} & a_{32} & a_{33} \end{bmatrix} \begin{bmatrix} 1 & & \\ & -1 & \\ & & 1 \end{bmatrix} \quad \text{so } \det M_1 = (-1)(\det A)(-1).$$

The matrix M_2 is singular because its rows add to the zero row. Then $\det M_2 = 0$.

The matrix M_3 can be split into *eight matrices* by Rule 3 (linearity in each row):

$$\det M_3 = \begin{vmatrix} \text{row } 1 \\ \text{row } 2 \\ \text{row } 3 \end{vmatrix} + \begin{vmatrix} \text{row } 3 \\ \text{row } 2 \\ \text{row } 3 \end{vmatrix} + \begin{vmatrix} \text{row } 1 \\ \text{row } 1 \\ \text{row } 3 \end{vmatrix} + \cdots + \begin{vmatrix} \text{row } 3 \\ \text{row } 1 \\ \text{row } 2 \end{vmatrix}.$$

All but the first and last have repeated rows and zero determinant. The first is A and the last has *two* row exchanges. So $\det M_3 = \det A + \det A$. (Try $A = I$.)

The matrix M_4 is exactly AA^{T}. Its determinant is $(\det A)(\det A^{\mathrm{T}}) = (\det A)^2$.

5.1 B Find the determinant of A by subtracting row 1 from row 2, then column 3 from column 2, then row or column exchanges to make the matrix lower triangular:

$$A = \begin{bmatrix} 1 & 1 & 1 \\ a & 1 & 1 \\ 0 & b & 1 \end{bmatrix} \quad \text{is singular for which } a \text{ and } b?$$

Solution Subtract row 1 from row 2, then column 3 from column 2. *Two* exchanges make the matrix triangular. Then $\det A = (a-1)(b-1)$.

$$A \longrightarrow \begin{bmatrix} 1 & 0 & 1 \\ a-1 & 0 & 0 \\ 0 & b-1 & 1 \end{bmatrix} \xrightarrow[\text{columns } 2 \leftrightarrow 3]{\text{rows } 1 \leftrightarrow 2} \begin{bmatrix} a-1 & 0 & 0 \\ 1 & 1 & 0 \\ 0 & 1 & b-1 \end{bmatrix}.$$

Note that $a = 1$ gives equal rows in A and $b = 1$ gives equal columns. So not surprising that $(a-1)$ and $(b-1)$ are factors of $\det A$.

Problem Set 5.1

Questions 1–12 are about the rules for determinants.

1 If a 4 by 4 matrix has $\det A = \frac{1}{2}$, find $\det(2A)$ and $\det(-A)$ and $\det(A^2)$ and $\det(A^{-1})$.

2 If a 3 by 3 matrix has $\det A = -1$, find $\det(\frac{1}{2}A)$ and $\det(-A)$ and $\det(A^2)$ and $\det(A^{-1})$.

3 True or false, with a reason if true or a counterexample if false:

(a) The determinant of $I + A$ is $1 + \det A$.

(b) The determinant of ABC is $|A|\,|B|\,|C|$.

(c) The determinant of $4A$ is $4|A|$.

(d) The determinant of $AB - BA$ is zero. (Try an example.)

4 Which row exchanges show that these "reverse identity matrices" J_3 and J_4 have $|J_3| = -1$ but $|J_4| = +1$?

$$\det \begin{bmatrix} 0 & 0 & 1 \\ 0 & 1 & 0 \\ 1 & 0 & 0 \end{bmatrix} = -1 \quad \text{but} \quad \det \begin{bmatrix} 0 & 0 & 0 & 1 \\ 0 & 0 & 1 & 0 \\ 0 & 1 & 0 & 0 \\ 1 & 0 & 0 & 0 \end{bmatrix} = +1.$$

5 For $n = 5, 6, 7$, count the row exchanges to permute the reverse identity J_n to the identity matrix I_n. Propose a rule for every size n and predict whether J_{101} has determinant $+1$ or -1.

6 Show how Rule 6 (determinant $= 0$ if a row is all zero) comes from Rule 3.

7 Find the determinants of rotations and reflections:

$$Q = \begin{bmatrix} \cos\theta & -\sin\theta \\ \sin\theta & \cos\theta \end{bmatrix} \quad \text{and} \quad Q = \begin{bmatrix} 1 - 2\cos^2\theta & -2\cos\theta\sin\theta \\ -2\cos\theta\sin\theta & 1 - 2\sin^2\theta \end{bmatrix}.$$

8 Prove that every orthogonal matrix ($Q^{\mathrm{T}}Q = I$) has determinant 1 or -1.

(a) Use the product rule $|AB| = |A|\,|B|$ and the transpose rule $|Q| = |Q^{\mathrm{T}}|$.

(b) Use only the product rule. If $|\det Q| > 1$ then $\det Q^n = (\det Q)^n$ blows up. How do you know this can't happen to Q^n?

9 Do these matrices have determinant 0, 1, 2, or 3?

$$A = \begin{bmatrix} 0 & 0 & 1 \\ 1 & 0 & 0 \\ 0 & 1 & 0 \end{bmatrix} \quad B = \begin{bmatrix} 0 & 1 & 1 \\ 1 & 0 & 1 \\ 1 & 1 & 0 \end{bmatrix} \quad C = \begin{bmatrix} 1 & 1 & 1 \\ 1 & 1 & 1 \\ 1 & 1 & 1 \end{bmatrix}.$$

10 If the entries in every row of A add to zero, solve $Ax = \mathbf{0}$ to prove $\det A = 0$. If those entries add to one, show that $\det(A - I) = 0$. Does this mean $\det A = 1$?

11 Suppose that $CD = -DC$ and find the flaw in this reasoning: Taking determinants gives $|C|\,|D| = -|D|\,|C|$. Therefore $|C| = 0$ or $|D| = 0$. One or both of the matrices must be singular. (That is not true.)

12 The inverse of a 2 by 2 matrix seems to have determinant $= 1$:

$$\det A^{-1} = \det \frac{1}{ad - bc}\begin{bmatrix} d & -b \\ -c & a \end{bmatrix} = \frac{ad - bc}{ad - bc} = 1.$$

What is wrong with this calculation? What is the correct $\det A^{-1}$?

Questions 13–27 use the rules to compute specific determinants.

13 Reduce A to U and find $\det A =$ product of the pivots:

$$A = \begin{bmatrix} 1 & 1 & 1 \\ 1 & 2 & 2 \\ 1 & 2 & 3 \end{bmatrix} \qquad A = \begin{bmatrix} 1 & 2 & 3 \\ 2 & 2 & 3 \\ 3 & 3 & 3 \end{bmatrix}.$$

14 By applying row operations to produce an upper triangular U, compute

$$\det \begin{bmatrix} 1 & 2 & 3 & 0 \\ 2 & 6 & 6 & 1 \\ -1 & 0 & 0 & 3 \\ 0 & 2 & 0 & 7 \end{bmatrix} \quad \text{and} \quad \det \begin{bmatrix} 2 & -1 & 0 & 0 \\ -1 & 2 & -1 & 0 \\ 0 & -1 & 2 & -1 \\ 0 & 0 & -1 & 2 \end{bmatrix}.$$

15 Use row operations to simplify and compute these determinants:

$$\det \begin{bmatrix} 101 & 201 & 301 \\ 102 & 202 & 302 \\ 103 & 203 & 303 \end{bmatrix} \quad \text{and} \quad \det \begin{bmatrix} 1 & t & t^2 \\ t & 1 & t \\ t^2 & t & 1 \end{bmatrix}.$$

16 Find the determinants of a rank one matrix and a skew-symmetric matrix:

$$A = \begin{bmatrix} 1 \\ 2 \\ 3 \end{bmatrix} \begin{bmatrix} 1 & -4 & 5 \end{bmatrix} \quad \text{and} \quad K = \begin{bmatrix} 0 & 1 & 3 \\ -1 & 0 & 4 \\ -3 & -4 & 0 \end{bmatrix}.$$

17 A skew-symmetric matrix has $K^{\mathrm{T}} = -K$. Insert a, b, c for $1, 3, 4$ in Question 16 and show that $|K| = 0$. Write down a 4 by 4 example with $|K| = 1$.

18 Use row operations to show that the 3 by 3 "Vandermonde determinant" is

$$\det \begin{bmatrix} 1 & a & a^2 \\ 1 & b & b^2 \\ 1 & c & c^2 \end{bmatrix} = (b-a)(c-a)(c-b).$$

19 Find the determinants of U and U^{-1} and U^2:

$$U = \begin{bmatrix} 1 & 4 & 6 \\ 0 & 2 & 5 \\ 0 & 0 & 3 \end{bmatrix} \quad \text{and} \quad U = \begin{bmatrix} a & b \\ 0 & d \end{bmatrix}.$$

20 Suppose you do two row operations at once, going from

$$\begin{bmatrix} a & b \\ c & d \end{bmatrix} \quad \text{to} \quad \begin{bmatrix} a - Lc & b - Ld \\ c - la & d - lb \end{bmatrix}.$$

Find the second determinant. Does it equal $ad - bc$?

21 *Row exchange*: Add row 1 of A to row 2, then subtract row 2 from row 1. Then add row 1 to row 2 and multiply row 1 by -1 to reach B. Which rules show

$$\det B = \begin{vmatrix} c & d \\ a & b \end{vmatrix} \quad \text{equals} \quad -\det A = -\begin{vmatrix} a & b \\ c & d \end{vmatrix}?$$

Those rules could replace Rule 2 in the definition of the determinant.

22 From $ad - bc$, find the determinants of A and A^{-1} and $A - \lambda I$:

$$A = \begin{bmatrix} 2 & 1 \\ 1 & 2 \end{bmatrix} \quad \text{and} \quad A^{-1} = \frac{1}{3}\begin{bmatrix} 2 & -1 \\ -1 & 2 \end{bmatrix} \quad \text{and} \quad A - \lambda I = \begin{bmatrix} 2-\lambda & 1 \\ 1 & 2-\lambda \end{bmatrix}.$$

Which two numbers λ lead to $\det(A - \lambda I) = 0$? Write down the matrix $A - \lambda I$ for each of those numbers λ—it should not be invertible.

23 From $A = \left[\begin{smallmatrix} 4 & 1 \\ 2 & 3 \end{smallmatrix}\right]$ find A^2 and A^{-1} and $A - \lambda I$ and their determinants. Which two numbers λ lead to $|A - \lambda I| = 0$?

24 Elimination reduces A to U. Then $A = LU$:

$$A = \begin{bmatrix} 3 & 3 & 4 \\ 6 & 8 & 7 \\ -3 & 5 & -9 \end{bmatrix} = \begin{bmatrix} 1 & 0 & 0 \\ 2 & 1 & 0 \\ -1 & 4 & 1 \end{bmatrix} \begin{bmatrix} 3 & 3 & 4 \\ 0 & 2 & -1 \\ 0 & 0 & -1 \end{bmatrix} = LU.$$

Find the determinants of L, U, A, $U^{-1}L^{-1}$, and $U^{-1}L^{-1}A$.

25 If the i, j entry of A is i times j, show that $\det A = 0$. (Exception when $A = [\,1\,]$.)

26 If the i, j entry of A is $i + j$, show that $\det A = 0$. (Exception when $n = 1$ or 2.)

27 Compute the determinants of these matrices by row operations:

$$A = \begin{bmatrix} 0 & a & 0 \\ 0 & 0 & b \\ c & 0 & 0 \end{bmatrix} \quad \text{and} \quad B = \begin{bmatrix} 0 & a & 0 & 0 \\ 0 & 0 & b & 0 \\ 0 & 0 & 0 & c \\ d & 0 & 0 & 0 \end{bmatrix} \quad \text{and} \quad C = \begin{bmatrix} a & a & a \\ a & b & b \\ a & b & c \end{bmatrix}.$$

28 True or false (give a reason if true or a 2 by 2 example if false):

(a) If A is not invertible then AB is not invertible.

(b) The determinant of A is always the product of its pivots.

(c) The determinant of $A - B$ equals $\det A - \det B$.

(d) AB and BA have the same determinant.

29 What is wrong with this proof that projection matrices have $\det P = 1$?

$$P = A(A^{\mathrm{T}}A)^{-1}A^{\mathrm{T}} \quad \text{so} \quad |P| = |A|\frac{1}{|A^{\mathrm{T}}||A|}|A^{\mathrm{T}}| = 1.$$

30 (Calculus question) Show that the partial derivatives of $\ln(\det A)$ give A^{-1}!

$$f(a,b,c,d) = \ln(ad - bc) \qquad \text{leads to} \qquad \begin{bmatrix} \partial f/\partial a & \partial f/\partial c \\ \partial f/\partial b & \partial f/\partial d \end{bmatrix} = A^{-1}.$$

31 (MATLAB) The Hilbert matrix **hilb**(n) has i, j entry equal to $1/(i+j-1)$. Print the determinants of **hilb**(1), **hilb**$(2), \ldots,$ **hilb**(10). Hilbert matrices are hard to work with! What are the pivots?

32 (MATLAB) What is a typical determinant (experimentally) of **rand**(n) and **randn**(n) for $n = 50, 100, 200, 400$? (And what does "Inf" mean in MATLAB?)

33 (MATLAB) Find the largest determinant of a 6 by 6 matrix of 1's and -1's.

34 If you know that $\det A = 6$, what is the determinant of B?

$$\det A = \begin{vmatrix} \text{row } 1 \\ \text{row } 2 \\ \text{row } 3 \end{vmatrix} = 6 \qquad \det B = \begin{vmatrix} \text{row } 3 + \text{row } 2 + \text{row } 1 \\ \text{row } 2 + \text{row } 1 \\ \text{row } 1 \end{vmatrix} = ?$$

PERMUTATIONS AND COFACTORS ■ 5.2

A computer finds the determinant from the pivots. This section explains two other ways to do it. There is a "big formula" using all $n!$ permutations. There is a "cofactor formula" using determinants of size $n-1$. The best example is my favorite 4 by 4 matrix:

$$A = \begin{bmatrix} 2 & -1 & 0 & 0 \\ -1 & 2 & -1 & 0 \\ 0 & -1 & 2 & -1 \\ 0 & 0 & -1 & 2 \end{bmatrix} \quad \text{has} \quad \det A = 5.$$

We can find this determinant in all three ways: ***pivots, big formula, cofactors.***

1. The product of the pivots is $2 \cdot \frac{3}{2} \cdot \frac{4}{3} \cdot \frac{5}{4}$. Cancellation produces 5.

2. The "big formula" in equation (8) has $4! = 24$ terms. Only five terms are nonzero:

$$\det A = 16 - 4 - 4 - 4 + 1 = 5.$$

The 16 comes from $2 \cdot 2 \cdot 2 \cdot 2$ on the diagonal of A. Where do -4 and $+1$ come from? When you can find those five terms, you have understood formula (8).

3. The numbers $2, -1, 0, 0$ in the first row multiply their cofactors 4, 3, 2, 1 from the other rows. That gives $2 \cdot 4 - 1 \cdot 3 = 5$. Those cofactors are 3 by 3 determinants. They use the rows and columns that are *not* used by the entry in the first row. ***Every term in a determinant uses each row and column once!***

The Pivot Formula

Elimination leaves the pivots $d_1, \ldots, d_n$ on the diagonal of the upper triangular U. If no row exchanges are involved, ***multiply those pivots*** to find the determinant:

$$\det A = (\det L)(\det U) = (1)(d_1 d_2 \cdots d_n). \tag{1}$$

This formula for $\det A$ appeared in the previous section, with the further possibility of row exchanges. The permutation matrix in $PA = LU$ has determinant -1 or $+1$. This factor $\det P = \pm 1$ enters the determinant of A:

$$(\det P)(\det A) = (\det L)(\det U) \quad \text{gives} \quad \det A = \pm(d_1 d_2 \cdots d_n). \tag{2}$$

When A has fewer than n pivots, $\det A = 0$ by Rule 8. The matrix is singular.

Example 1 A row exchange produces pivots 4, 2, 1 and that important minus sign:

$$A = \begin{bmatrix} 0 & 0 & 1 \\ 0 & 2 & 3 \\ 4 & 5 & 6 \end{bmatrix} \qquad PA = \begin{bmatrix} 4 & 5 & 6 \\ 0 & 2 & 3 \\ 0 & 0 & 1 \end{bmatrix} \qquad \det A = -(4)(2)(1) = -8.$$

The odd number of row exchanges (namely one exchange) means that $\det P = -1$.

The next example has no row exchanges. It may be the first matrix we factored into LU (when it was 3 by 3). What is remarkable is that we can go directly to n by n. Pivots give the determinant. We will also see how determinants give the pivots.

Example 2 The first pivots of this tridiagonal matrix A are $2, \frac{3}{2}, \frac{4}{3}$. The next are $\frac{5}{4}$ and $\frac{6}{5}$ and eventually $\frac{n+1}{n}$. Factoring this n by n matrix reveals its determinant:

$$\begin{bmatrix} 2 & -1 & & & \\ -1 & 2 & -1 & & \\ & -1 & 2 & \cdot & \\ & & \cdot & \cdot & -1 \\ & & & -1 & 2 \end{bmatrix} = \begin{bmatrix} 1 & & & & \\ -\frac{1}{2} & 1 & & & \\ & -\frac{2}{3} & 1 & & \\ & & \cdot & \cdot & \\ & & & -\frac{n-1}{n} & 1 \end{bmatrix} \begin{bmatrix} \mathbf{2} & -1 & & & \\ & \mathbf{\frac{3}{2}} & -1 & & \\ & & \mathbf{\frac{4}{3}} & -1 & \\ & & & \cdot & \cdot \\ & & & & \mathbf{\frac{n+1}{n}} \end{bmatrix}$$

The pivots are on the diagonal of U (the last matrix). When 2 and $\frac{3}{2}$ and $\frac{4}{3}$ and $\frac{5}{4}$ are multiplied, the fractions cancel. The determinant of the 4 by 4 matrix is 5. The 3 by 3 determinant is 4. *The n by n determinant is $n+1$*:

$$\det A = (\mathbf{2})\left(\mathbf{\frac{3}{2}}\right)\left(\mathbf{\frac{4}{3}}\right)\cdots\left(\mathbf{\frac{n+1}{n}}\right) = n+1.$$

Important point: The first pivots depend only on the *upper left corner* of the original matrix A. This is a rule for all matrices without row exchanges.

> The first k pivots come from the k by k matrix A_k in the top left corner of A. ***The determinant of that corner submatrix*** A_k ***is*** $d_1 d_2 \cdots d_k$.

The 1 by 1 matrix A_1 contains the very first pivot d_1. This is $\det A_1$. The 2 by 2 matrix in the corner has $\det A_2 = d_1 d_2$. Eventually the n by n determinant uses the product of all n pivots to give $\det A_n$ which is $\det A$.

Elimination deals with the corner matrix A_k while starting on the whole matrix. We assume no row exchanges—then $A = LU$ and $A_k = L_k U_k$. Dividing one determinant by the previous determinant ($\det A_k$ divided by $\det A_{k-1}$) cancels everything but the latest pivot d_k. ***This gives a ratio of determinants formula for the pivots***:

Pivots from determinants **The kth pivot is** $d_k = \dfrac{d_1 d_2 \cdots d_k}{d_1 d_2 \cdots d_{k-1}} = \dfrac{\det A_k}{\det A_{k-1}}$. (3)

In the $-1, 2, -1$ matrices this ratio correctly gives the pivots $\frac{2}{1}, \frac{3}{2}, \frac{4}{3}, \ldots, \frac{n+1}{n}$. The Hilbert matrices in Problem 5.1.31 also build from the upper left corner.

We don't need row exchanges when all these corner submatrices have $\det A_k \neq 0$.

The Big Formula for Determinants

Pivots are good for computing. They concentrate a lot of information—enough to find the determinant. But it is hard to connect them to the original a_{ij}. That part will be clearer if we go back to rules 1-2-3, linearity and sign reversal and $\det I = 1$. We want to derive a single explicit formula for the determinant, directly from the entries a_{ij}.

The formula has n! terms. Its size grows fast because $n! = 1, 2, 6, 24, 120, \ldots$. For $n = 11$ there are forty million terms. For $n = 2$, the two terms are ad and bc. Half the terms have minus signs (as in $-bc$). The other half have plus signs (as in ad). For $n = 3$ there are $3! = (3)(2)(1)$ terms. Here are those six terms:

$$\text{3 by 3 determinant} \quad \begin{vmatrix} a_{11} & a_{12} & a_{13} \\ a_{21} & a_{22} & a_{23} \\ a_{31} & a_{32} & a_{33} \end{vmatrix} = \begin{matrix} +a_{11}a_{22}a_{33} + a_{12}a_{23}a_{31} + a_{13}a_{21}a_{32} \\ -a_{11}a_{23}a_{32} - a_{12}a_{21}a_{33} - a_{13}a_{22}a_{31}. \end{matrix} \tag{4}$$

Notice the pattern. Each product like $a_{11}a_{23}a_{32}$ has ***one entry from each row***. It also has ***one entry from each column***. The column order 1, 3, 2 means that this particular term comes with a minus sign. The column order 3, 1, 2 in $a_{13}a_{21}a_{32}$ has a plus sign. It will be "permutations" that tell us the sign.

The next step ($n = 4$) brings $4! = 24$ terms. There are 24 ways to choose one entry from each row and column. Down the main diagonal, $a_{11}a_{22}a_{33}a_{44}$ with column order $1, 2, 3, 4$ always has a plus sign. That is the "identity permutation".

To derive the big formula I start with $n = 2$. The goal is to reach $ad - bc$ in a systematic way. Break each row into two simpler rows:

$$\begin{bmatrix} a & b \end{bmatrix} = \begin{bmatrix} a & 0 \end{bmatrix} + \begin{bmatrix} 0 & b \end{bmatrix} \quad \text{and} \quad \begin{bmatrix} c & d \end{bmatrix} = \begin{bmatrix} c & 0 \end{bmatrix} + \begin{bmatrix} 0 & d \end{bmatrix}.$$

Now apply linearity, first in row 1 (with row 2 fixed) and then in row 2 (with row 1 fixed):

$$\begin{aligned} \begin{vmatrix} a & b \\ c & d \end{vmatrix} &= \begin{vmatrix} a & 0 \\ c & d \end{vmatrix} + \begin{vmatrix} 0 & b \\ c & d \end{vmatrix} \\ &= \begin{vmatrix} a & 0 \\ c & 0 \end{vmatrix} + \begin{vmatrix} a & 0 \\ 0 & d \end{vmatrix} + \begin{vmatrix} 0 & b \\ c & 0 \end{vmatrix} + \begin{vmatrix} 0 & b \\ 0 & d \end{vmatrix}. \end{aligned} \tag{5}$$

The last line has $2^2 = 4$ determinants. The first and fourth are zero because their rows are dependent—one row is a multiple of the other row. We are left with $2! = 2$ determinants to compute:

$$\begin{vmatrix} a & 0 \\ 0 & d \end{vmatrix} + \begin{vmatrix} 0 & b \\ c & 0 \end{vmatrix} = ad \begin{vmatrix} 1 & 0 \\ 0 & 1 \end{vmatrix} + bc \begin{vmatrix} 0 & 1 \\ 1 & 0 \end{vmatrix} = ad - bc.$$

The splitting led to permutation matrices. Their determinants give a plus or minus sign. The 1's are multiplied by numbers that come from A. The permutation tells the column sequence, in this case $(1, 2)$ or $(2, 1)$.

Now try $n = 3$. Each row splits into 3 simpler rows like $\begin{bmatrix} a_{11} & 0 & 0 \end{bmatrix}$. Using linearity in each row, $\det A$ splits into $3^3 = 27$ simple determinants. If a column choice is repeated—for example if we also choose $\begin{bmatrix} a_{21} & 0 & 0 \end{bmatrix}$—then the simple determinant is zero. We pay attention only when ***the nonzero terms come from different columns***.

$$\begin{vmatrix} a_{11} & a_{12} & a_{13} \\ a_{21} & a_{22} & a_{23} \\ a_{31} & a_{32} & a_{33} \end{vmatrix} = \begin{vmatrix} a_{11} & & \\ & a_{22} & \\ & & a_{33} \end{vmatrix} + \begin{vmatrix} & a_{12} & \\ & & a_{23} \\ a_{31} & & \end{vmatrix} + \begin{vmatrix} & & a_{13} \\ a_{21} & & \\ & a_{32} & \end{vmatrix}$$
$$+ \begin{vmatrix} a_{11} & & \\ & & a_{23} \\ & a_{32} & \end{vmatrix} + \begin{vmatrix} & a_{12} & \\ a_{21} & & \\ & & a_{33} \end{vmatrix} + \begin{vmatrix} & & a_{13} \\ & a_{22} & \\ a_{31} & & \end{vmatrix}.$$

There are $3! = 6$ ***ways to order the columns, so six determinants.*** The six permutations of $(1, 2, 3)$ include the identity permutation $(1, 2, 3)$ from $P = I$:

$$\textbf{Column numbers} = (1,2,3), (2,3,1), (3,1,2), (1,3,2), (2,1,3), (3,2,1). \tag{6}$$

The last three are *odd permutations* (one exchange). The first three are *even permutations* (0 or 2 exchanges). When the column sequence is (α, β, ω), we have chosen the entries $a_{1\alpha}a_{2\beta}a_{3\omega}$—and the column sequence comes with a plus or minus sign. The determinant of A is now split into six simple terms. Factor out the a_{ij}:

$$\det A = a_{11}a_{22}a_{33}\begin{vmatrix} 1 & & \\ & 1 & \\ & & 1 \end{vmatrix} + a_{12}a_{23}a_{31}\begin{vmatrix} & 1 & \\ & & 1 \\ 1 & & \end{vmatrix} + a_{13}a_{21}a_{32}\begin{vmatrix} & & 1 \\ 1 & & \\ & 1 & \end{vmatrix}$$
$$+ a_{11}a_{23}a_{32}\begin{vmatrix} 1 & & \\ & & 1 \\ & 1 & \end{vmatrix} + a_{12}a_{21}a_{33}\begin{vmatrix} & 1 & \\ 1 & & \\ & & 1 \end{vmatrix} + a_{13}a_{22}a_{31}\begin{vmatrix} & & 1 \\ & 1 & \\ 1 & & \end{vmatrix}. \tag{7}$$

The first three (even) permutations have $\det P = +1$, the last three (odd) permutations have $\det P = -1$. We have proved the 3 by 3 formula in a systematic way.

Now you can see the n by n formula. There are $n!$ orderings of the columns. The columns $(1, 2, \ldots, n)$ go in each possible order $(\alpha, \beta, \ldots, \omega)$. Taking $a_{1\alpha}$ from row 1 and $a_{2\beta}$ from row 2 and eventually $a_{n\omega}$ from row n, the determinant contains the product $a_{1\alpha}a_{2\beta}\cdots a_{n\omega}$ times $+1$ or -1. Half the column orderings have sign -1.

The complete determinant of A is the sum of these $n!$ simple determinants, times 1 or -1. The simple determinants $a_{1\alpha}a_{2\beta}\cdots a_{n\omega}$ choose ***one entry from every row and column***:

$$\det A = \text{sum over all } \boldsymbol{n!} \text{ column permutations } P = (\alpha, \beta, \ldots, \omega)$$
$$= \sum (\det P) a_{1\alpha}a_{2\beta}\cdots a_{n\omega} = \textbf{BIG FORMULA}. \tag{8}$$

The 2 by 2 case is $+a_{11}a_{22} - a_{12}a_{21}$ (which is $ad - bc$). Here P is $(1, 2)$ or $(2, 1)$.

The 3 by 3 case has three products "down to the right" (see Problem 30) and three products "down to the left". Warning: Many people believe they should follow this pattern in the 4 by 4 case. They only take 8 products—but we need 24.

Example 3 (Determinant of U) When U is upper triangular, only one of the $n!$ products can be nonzero. This one term comes from the diagonal: $\det U = +u_{11}u_{22}\cdots u_{nn}$. All other column orderings pick at least one entry below the diagonal, where U has zeros. As soon as we pick a number like $u_{21} = 0$ from below the diagonal, that term in equation (8) is sure to be zero.

Of course $\det I = 1$. The only nonzero term is $+(1)(1)\cdots(1)$ from the diagonal.

Example 4 Suppose Z is the identity matrix except for column 3. Then

$$\text{determinant of } Z = \begin{vmatrix} 1 & 0 & a & 0 \\ 0 & 1 & b & 0 \\ 0 & 0 & c & 0 \\ 0 & 0 & d & 1 \end{vmatrix} = c. \tag{9}$$

The term $(1)(1)(c)(1)$ comes from the main diagonal with a plus sign. There are 23 other products (choosing one factor from each row and column) but they are all zero. Reason: If we pick a, b, or d from column 3, that column is used up. Then the only available choice from row 3 is zero.

Here is a different reason for the same answer. If $c = 0$, then Z has a row of zeros and $\det Z = c = 0$ is correct. If c is not zero, *use elimination.* Subtract multiples of row 3 from the other rows, to knock out a, b, d. That leaves a diagonal matrix and $\det Z = c$.

This example will soon be used for "Cramer's Rule". If we move a, b, c, d into the first column of Z, the determinant is $\det Z = a$. (*Why*?) Changing one column of I leaves Z with an easy determinant, coming from its main diagonal only.

Example 5 Suppose A has 1's just above and below the main diagonal. Here $n = 4$:

$$A_4 = \begin{bmatrix} 0 & \mathbf{1} & 0 & 0 \\ \mathbf{1} & 0 & \mathbf{1} & 0 \\ 0 & \mathbf{1} & 0 & \mathbf{1} \\ 0 & 0 & \mathbf{1} & 0 \end{bmatrix} \quad \text{and} \quad P_4 = \begin{bmatrix} 0 & \mathbf{1} & 0 & 0 \\ \mathbf{1} & 0 & 0 & 0 \\ 0 & 0 & 0 & \mathbf{1} \\ 0 & 0 & \mathbf{1} & 0 \end{bmatrix} \quad \text{have determinant 1.}$$

The only nonzero choice in the first row is column 2. The only nonzero choice in row 4 is column 3. Then rows 2 and 3 must choose columns 1 and 4. In other words P_4 is the only permutation that picks out nonzeros in A_4. The determinant of P_4 is $+1$ (two exchanges to reach $2, 1, 4, 3$). Therefore $\det A_4 = +1$.

Determinant by Cofactors

Formula (8) is a direct definition of the determinant. It gives you everything at once—but you have to digest it. Somehow this sum of $n!$ terms must satisfy rules 1-2-3 (then all the other properties follow). The easiest is $\det I = 1$, already checked. The rule of linearity becomes clear, if you separate out the factor a_{11} or a_{12} or $a_{1\alpha}$ that comes from the first row. With $n = 3$ we separate the determinant into

$$\det A = a_{11}\,(a_{22}a_{33} - a_{23}a_{32}) + a_{12}\,(a_{23}a_{31} - a_{21}a_{33}) + a_{13}\,(a_{21}a_{32} - a_{22}a_{31}). \tag{10}$$

Those three quantities in parentheses are called ***"cofactors"***. They are 2 by 2 determinants, coming from matrices in rows 2 and 3. The first row contributes the factors a_{11}, a_{12}, a_{13}. The lower rows contribute the cofactors C_{11}, C_{12}, C_{13}. Certainly the determinant $a_{11}C_{11} + a_{12}C_{12} + a_{13}C_{13}$ depends linearly on a_{11}, a_{12}, a_{13}—this is rule 3.

The cofactor of a_{11} is $C_{11} = a_{22}a_{33} - a_{23}a_{32}$. You can see it in this splitting:

$$\begin{vmatrix} a_{11} & a_{12} & a_{13} \\ a_{21} & a_{22} & a_{23} \\ a_{31} & a_{32} & a_{33} \end{vmatrix} = \begin{vmatrix} a_{11} & & \\ & a_{22} & a_{23} \\ & a_{32} & a_{33} \end{vmatrix} + \begin{vmatrix} & a_{12} & \\ a_{21} & & a_{23} \\ a_{31} & & a_{33} \end{vmatrix} + \begin{vmatrix} & & a_{13} \\ a_{21} & a_{22} & \\ a_{31} & a_{32} & \end{vmatrix}.$$

We are still choosing ***one entry from each row and column***. Since a_{11} uses up row 1 and column 1, that leaves a 2 by 2 determinant as its cofactor.

As always, we have to watch signs. The 2 by 2 determinant that goes with a_{12} looks like $a_{21}a_{33} - a_{23}a_{31}$. But in the cofactor C_{12}, *its sign is reversed.* Then $a_{12}C_{12}$ is the correct 3 by 3 determinant. The sign pattern for cofactors along the first row is plus-minus-plus-minus. ***You cross out row 1 and column j to get a submatrix M_{1j} of size $n-1$.*** Multiply its determinant by $(-1)^{1+j}$ to get the cofactor:

$$\text{The cofactors along row 1 are } C_{1j} = (-1)^{1+j} \det M_{1j}.$$

$$\textbf{The cofactor expansion is } \det A = a_{11}C_{11} + a_{12}C_{12} + \cdots + a_{1n}C_{1n}. \qquad (11)$$

In the big formula (8), the terms that multiply a_{11} combine to give $\det M_{11}$. The sign is $(-1)^{1+1}$, meaning *plus*. Equation (11) is another form of equation (8) and also equation (10), with factors from row 1 multiplying cofactors from the other rows.

Note Whatever is possible for row 1 is possible for row i. The entries a_{ij} in that row also have cofactors C_{ij}. Those are determinants of order $n-1$, multiplied by $(-1)^{i+j}$. Since a_{ij} accounts for row i and column j, ***the submatrix M_{ij} throws out row i and column j***. The display shows a_{43} and M_{43} (with row 4 and column 3 crossed out). The sign $(-1)^{4+3}$ multiplies the determinant of M_{43} to give C_{43}. The sign matrix shows the $\pm$ pattern:

$$A = \begin{bmatrix} \bullet & \bullet & & \bullet \\ \bullet & \bullet & & \bullet \\ \bullet & \bullet & & \bullet \\ & & a_{43} & \end{bmatrix} \qquad \text{signs } (-1)^{i+j} = \begin{bmatrix} + & - & + & - \\ - & + & - & + \\ + & - & + & - \\ - & + & - & + \end{bmatrix}.$$

5A The determinant is the dot product of any row i of A with its cofactors:

$$\textbf{COFACTOR FORMULA} \qquad \det A = a_{i1}C_{i1} + a_{i2}C_{i2} + \cdots + a_{in}C_{in}. \qquad (12)$$

Each cofactor C_{ij} (order $n-1$, without row i and column j) includes its correct sign:

$$C_{ij} = (-1)^{i+j} \det M_{ij}.$$

A determinant of order n is a combination of determinants of order $n-1$. A recursive person would keep going. Each subdeterminant breaks into determinants of order $n-2$. *We could define all determinants via equation* (12). This rule goes from order n to $n-1$ to $n-2$ and eventually to order 1. Define the 1 by 1 determinant $|a|$ to be the number a. Then the cofactor method is complete.

We preferred to construct $\det A$ from its properties (linearity, sign reversal, and $\det I = 1$). The big formula (8) and the cofactor formulas (10)–(12) follow from those properties. One last formula comes from the rule that $\det A = \det A^{\mathrm{T}}$. We can expand in cofactors, *down a column* instead of across a row. Down column j the entries are a_{1j} to a_{nj}. The cofactors are C_{1j} to C_{nj}. The determinant is the dot product:

$$\textbf{Cofactors down column } j: \qquad \det A = a_{1j}C_{1j} + a_{2j}C_{2j} + \cdots + a_{nj}C_{nj}. \tag{13}$$

Cofactors are most useful when the matrices have many zeros—as in the next examples.

Example 6 The -1, 2, -1 matrix has only two nonzeros in its first row. So only two cofactors C_{11} and C_{12} are involved in the determinant. I will highlight C_{12}:

$$\begin{vmatrix} 2 & -1 & & \\ \mathbf{-1} & 2 & \mathbf{-1} & \\ & -1 & \mathbf{2} & \mathbf{-1} \\ & & \mathbf{-1} & \mathbf{2} \end{vmatrix} = 2\begin{vmatrix} 2 & -1 & \\ -1 & 2 & -1 \\ & -1 & 2 \end{vmatrix} - (-1)\begin{vmatrix} \mathbf{-1} & \mathbf{-1} & \\ & \mathbf{2} & \mathbf{-1} \\ & \mathbf{-1} & \mathbf{2} \end{vmatrix}. \tag{14}$$

You see 2 times C_{11} first on the right, from crossing out row 1 and column 1. This cofactor has exactly the same $-1, 2, -1$ pattern as the original A—but one size smaller.

To compute the boldface C_{12}, *use cofactors down its first column.* The only nonzero is at the top. That contributes another -1 (so we are back to minus). Its cofactor is the $-1, 2, -1$ determinant which is 2 by 2, *two sizes smaller* than the original A.

Summary Equation (14) gives the 4 by 4 determinant D_4 from $2D_3$ minus D_2. ***Each*** D_n ***(the*** $-1, 2, -1$ ***determinant of order*** n***) comes from*** D_{n-1} ***and*** D_{n-2}:

$$D_4 = 2D_3 - D_2 \qquad \text{and generally} \qquad D_n = 2D_{n-1} - D_{n-2}. \tag{15}$$

Direct calculation gives $D_2 = 3$ and $D_3 = 4$. Therefore $D_4 = 2(4) - 3 = 5$. These determinants 3, 4, 5 fit the formula $\boldsymbol{D_n = n+1}$. That "special tridiagonal answer" also came from the product of pivots in Example 2.

The idea behind cofactors is to reduce the order one step at a time. The determinants $D_n = n+1$ obey the recursion formula $n+1 = 2n - (n-1)$. As they must.

Example 7 This is the same matrix, except the first entry (upper left) is now 1:

$$B_4 = \begin{bmatrix} 1 & -1 & & \\ -1 & 2 & -1 & \\ & -1 & 2 & -1 \\ & & -1 & 2 \end{bmatrix}.$$

All pivots of this matrix turn out to be 1. So its determinant is 1. How does that come from cofactors? Expanding on row 1, the cofactors all agree with Example 6. Just change $a_{11} = 2$ to $b_{11} = 1$:

$$\det B_4 = D_3 - D_2 \qquad \text{instead of} \qquad \det A_4 = 2D_3 - D_2.$$

The determinant of B_4 is $4 - 3 = 1$. The determinant of every B_n is $n - (n-1) = 1$. Problem 13 asks you to use cofactors of the *last* row. You still find $\det B_n = 1$.

■ REVIEW OF THE KEY IDEAS ■

1. With no row exchanges, $\det A$ = (*product of the pivots*). In the upper left corner, $\det A_k$ = (*product of the first k pivots*).

2. Every term in the big formula (8) uses each row and column once. Half of the $n!$ terms have plus signs (when $\det P = +1$) and half have minus signs.

3. The cofactor C_{ij} is $(-1)^{i+j}$ times the smaller determinant that omits row i and column j (because a_{ij} uses that row and column).

4. The determinant is the dot product of any row of A with its row of cofactors. When a row of A has a lot of zeros, we only need a few cofactors.

■ WORKED EXAMPLES ■

5.2 A A *Hessenberg matrix* is a triangular matrix with one extra diagonal. Use cofactors of row 1 to show that the 4 by 4 determinant satisfies Fibonacci's rule $|H_4| = |H_3| + |H_2|$. The same rule will continue for all sizes, $|H_n| = |H_{n-1}| + |H_{n-2}|$. Which Fibonacci number is $|H_n|$?

$$H_2 = \begin{bmatrix} 2 & 1 \\ 1 & 2 \end{bmatrix} \qquad H_3 = \begin{bmatrix} 2 & 1 & \\ 1 & 2 & 1 \\ 1 & 1 & 2 \end{bmatrix} \qquad H_4 = \begin{bmatrix} 2 & 1 & & \\ \mathbf{1} & 2 & \mathbf{1} & \\ \mathbf{1} & 1 & \mathbf{2} & \mathbf{1} \\ \mathbf{1} & 1 & \mathbf{1} & \mathbf{2} \end{bmatrix}$$

Solution The cofactor C_{11} for H_4 is the determinant $|H_3|$. We also need C_{12} (in boldface):

$$C_{12} = -\begin{vmatrix} \mathbf{1} & \mathbf{1} & \mathbf{0} \\ \mathbf{1} & \mathbf{2} & \mathbf{1} \\ \mathbf{1} & \mathbf{1} & \mathbf{2} \end{vmatrix} = -\begin{vmatrix} 2 & 1 & 0 \\ 1 & 2 & 1 \\ 1 & 1 & 2 \end{vmatrix} + \begin{vmatrix} 1 & 0 & 0 \\ 1 & 2 & 1 \\ 1 & 1 & 2 \end{vmatrix}$$

Rows 2 and 3 stayed the same and we used linearity in row 1. The two determinants on the right are $-|H_3|$ and $+|H_2|$. Then the 4 by 4 determinant is

$$|H_4| = 2C_{11} + 1C_{12} = 2|H_3| - |H_3| + |H_2| = |H_3| + |H_2|.$$

The actual numbers are $|H_2| = 3$ and $|H_3| = 5$ (and of course $|H_1| = 2$). Since $|H_n|$ follows Fibonacci's rule $|H_{n-1}| + |H_{n-2}|$, it must be $|H_n| = F_{n+2}$.

5.2 B These questions use the $\pm$ signs (even and odd P's) in the big formula for $\det A$:

1. If A is the 10 by 10 all-ones matrix, how does the big formula give $\det A = 0$?

2. If you multiply all $n!$ permutations together into a single P, is it odd or even?

3. If you multiply each a_{ij} by the fraction $\frac{i}{j}$, why is $\det A$ unchanged?

Solution In Question **1**, with all $a_{ij} = 1$, all the products in the big formula (8) will be 1. Half of them come with a plus sign, and half with minus. So they cancel to leave $\det A = 0$. (Of course the all-ones matrix is singular.)

In Question **2**, multiplying $\begin{bmatrix} 1 & 0 \\ 0 & 1 \end{bmatrix}\begin{bmatrix} 1 & 0 \\ 0 & 1 \end{bmatrix}$ gives an odd permutation. Also for 3 by 3, the three odd permutations multiply (in any order) to give *odd*. But for $n > 3$ the product of all permutations will be *even*. There are $n!/2$ odd permutations and that is an even number as soon as it includes the factor 4.

In Question **3**, each a_{ij} is multiplied by i/j. So each product $a_{1\alpha}a_{2\beta}\cdots a_{n\omega}$ in the big formula is multiplied by all the row numbers $i = 1, 2, \ldots, n$ and divided by all the column numbers $j = 1, 2, \ldots, n$. (The columns come in some permuted order!) Then each product is unchanged and $\det A$ stays the same.

Another approach to Question **3**: We are multiplying the matrix A by the diagonal matrix $D = \mathbf{diag}(1:n)$ when row i is multiplied by i. And we are postmultiplying by D^{-1} when column j is divided by j. The determinant of DAD^{-1} is $\det A$ by the product rule.

Problem Set 5.2

Problems 1–10 use the big formula with $n!$ terms: $|A| = \sum \pm a_{1\alpha}a_{2\beta}\cdots a_{n\omega}$.

1 Compute the determinants of A, B, C from six terms. Are their rows independent?

$$A = \begin{bmatrix} 1 & 2 & 3 \\ 3 & 1 & 2 \\ 3 & 2 & 1 \end{bmatrix} \quad B = \begin{bmatrix} 1 & 2 & 3 \\ 4 & 4 & 4 \\ 5 & 6 & 7 \end{bmatrix} \quad C = \begin{bmatrix} 1 & 1 & 1 \\ 1 & 1 & 0 \\ 1 & 0 & 0 \end{bmatrix}.$$

2 Compute the determinants of A, B, C. Are their columns independent?

$$A = \begin{bmatrix} 1 & 1 & 0 \\ 1 & 0 & 1 \\ 0 & 1 & 1 \end{bmatrix} \quad B = \begin{bmatrix} 1 & 2 & 3 \\ 4 & 5 & 6 \\ 7 & 8 & 9 \end{bmatrix} \quad C = \begin{bmatrix} A & 0 \\ 0 & B \end{bmatrix}.$$

3 Show that $\det A = 0$, regardless of the five nonzeros marked by x's:

$$A = \begin{bmatrix} x & x & x \\ 0 & 0 & x \\ 0 & 0 & x \end{bmatrix}. \qquad \text{(What is the rank of } A\text{?)}$$

4 This problem shows in two ways that $\det A = 0$ (the x's are any numbers):

$$A = \begin{bmatrix} x & x & x & x & x \\ x & x & x & x & x \\ 0 & 0 & 0 & x & x \\ 0 & 0 & 0 & x & x \\ 0 & 0 & 0 & x & x \end{bmatrix}.$$

(a) How do you know that the rows are linearly dependent?

(b) Explain why all 120 terms are zero in the big formula for $\det A$.

5 Find two ways to choose nonzeros from four different rows and columns:

$$A = \begin{bmatrix} 1 & 0 & 0 & 1 \\ 0 & 1 & 1 & 1 \\ 1 & 1 & 0 & 1 \\ 1 & 0 & 0 & 1 \end{bmatrix} \qquad B = \begin{bmatrix} 1 & 0 & 0 & 2 \\ 0 & 3 & 4 & 5 \\ 5 & 4 & 0 & 3 \\ 2 & 0 & 0 & 1 \end{bmatrix} \qquad (B \text{ has the same zeros as } A).$$

Is $\det A$ equal to $1 + 1$ or $1 - 1$ or $-1 - 1$? What is $\det B$?

6 Place the smallest number of zeros in a 4 by 4 matrix that will guarantee $\det A = 0$. Place as many zeros as possible while still allowing $\det A \neq 0$.

7 (a) If $a_{11} = a_{22} = a_{33} = 0$, how many of the six terms in $\det A$ will be zero?

(b) If $a_{11} = a_{22} = a_{33} = a_{44} = 0$, how many of the 24 products $a_{1j}a_{2k}a_{3l}a_{4m}$ are sure to be zero?

8 How many 5 by 5 permutation matrices have $\det P = +1$? Those are even permutations. Find one that needs four exchanges to reach the identity matrix.

9 If $\det A$ is not zero, at least one of the $n!$ terms in formula (8) is not zero. Deduce that some ordering of the rows of A leaves no zeros on the diagonal. (Don't use P from elimination; that PA can have zeros on the diagonal.)

10 Show that 4 is the largest determinant for a 3 by 3 matrix of 1's and -1's.

11 How many permutations of $(1, 2, 3, 4)$ are even and what are they? Extra credit: What are all the possible 4 by 4 determinants of $I + P_{\text{even}}$?

Problems 12–24 use cofactors $C_{ij} = (-1)^{i+j} \det M_{ij}$. Remove row i and column j.

12 Find all cofactors and put them into a cofactor matrix C. Find $\det B$ by cofactors:

$$A = \begin{bmatrix} 2 & 1 \\ 3 & 6 \end{bmatrix} \qquad B = \begin{bmatrix} 1 & 2 & 3 \\ 4 & 5 & 6 \\ 7 & 0 & 0 \end{bmatrix}.$$

13 Find the cofactor matrix C and multiply A times C^{T}. Compare AC^{T} with A^{-1}:

$$A = \begin{bmatrix} 2 & -1 & 0 \\ -1 & 2 & -1 \\ 0 & -1 & 2 \end{bmatrix} \qquad A^{-1} = \frac{1}{4}\begin{bmatrix} 3 & 2 & 1 \\ 2 & 4 & 2 \\ 1 & 2 & 3 \end{bmatrix}.$$

14 The matrix B_n is the $-1, 2, -1$ matrix A_n except that $b_{11} = 1$ instead of $a_{11} = 2$. Using cofactors of the *last* row of B_4 show that $|B_4| = 2|B_3| - |B_2|$ and find $|B_4|$:

$$B_4 = \begin{bmatrix} 1 & -1 & & \\ -1 & 2 & -1 & \\ & -1 & 2 & -1 \\ & & -1 & 2 \end{bmatrix} \qquad B_3 = \begin{bmatrix} 1 & -1 & \\ -1 & 2 & -1 \\ & -1 & 2 \end{bmatrix}.$$

The recursion $|B_n| = 2|B_{n-1}| - |B_{n-2}|$ is satisfied when every $|B_n| = 1$. This recursion is the same as for the A's. The difference is in the starting values $1, 1, 1$ for $n = 1, 2, 3$.

15 The n by n determinant C_n has 1's above and below the main diagonal:

$$C_1 = |0| \qquad C_2 = \begin{vmatrix} 0 & 1 \\ 1 & 0 \end{vmatrix} \qquad C_3 = \begin{vmatrix} 0 & 1 & 0 \\ 1 & 0 & 1 \\ 0 & 1 & 0 \end{vmatrix} \qquad C_4 = \begin{vmatrix} 0 & 1 & 0 & 0 \\ 1 & 0 & 1 & 0 \\ 0 & 1 & 0 & 1 \\ 0 & 0 & 1 & 0 \end{vmatrix}.$$

(a) What are these determinants C_1, C_2, C_3, C_4?

(b) By cofactors find the relation between C_n and C_{n-1} and C_{n-2}. Find C_{10}.

16 The matrices in Problem 15 have 1's just above and below the main diagonal. Going down the matrix, which order of columns (if any) gives all 1's? Explain why that permutation is *even* for $n = 4, 8, 12, \ldots$ and *odd* for $n = 2, 6, 10, \ldots$. Then

$$C_n = 0 \text{ (odd } n) \qquad C_n = 1 \ (n = 4, 8, \cdots) \qquad C_n = -1 \ (n = 2, 6, \cdots).$$

17 The tridiagonal 1, 1, 1 matrix of order n has determinant E_n:

$$E_1 = |1| \quad E_2 = \begin{vmatrix} 1 & 1 \\ 1 & 1 \end{vmatrix} \quad E_3 = \begin{vmatrix} 1 & 1 & 0 \\ 1 & 1 & 1 \\ 0 & 1 & 1 \end{vmatrix} \quad E_4 = \begin{vmatrix} 1 & 1 & 0 & 0 \\ 1 & 1 & 1 & 0 \\ 0 & 1 & 1 & 1 \\ 0 & 0 & 1 & 1 \end{vmatrix}.$$

(a) By cofactors show that $E_n = E_{n-1} - E_{n-2}$.

(b) Starting from $E_1 = 1$ and $E_2 = 0$ find $E_3, E_4, \ldots, E_8$.

(c) By noticing how these numbers eventually repeat, find E_{100}.

18 F_n is the determinant of the 1, 1, −1 tridiagonal matrix of order n:

$$F_2 = \begin{vmatrix} 1 & -1 \\ 1 & 1 \end{vmatrix} = 2 \quad F_3 = \begin{vmatrix} 1 & -1 & 0 \\ 1 & 1 & -1 \\ 0 & 1 & 1 \end{vmatrix} = 3 \quad F_4 = \begin{vmatrix} 1 & -1 & & \\ 1 & 1 & -1 & \\ & 1 & 1 & -1 \\ & & 1 & 1 \end{vmatrix} \neq 4.$$

Expand in cofactors to show that $F_n = F_{n-1} + F_{n-2}$. These determinants are *Fibonacci numbers* $1, 2, 3, 5, 8, 13, \ldots$. The sequence usually starts 1, 1, 2, 3 (with two 1's) so our F_n is the usual F_{n+1}.

19 Go back to B_n in Problem 14. It is the same as A_n except for $b_{11} = 1$. So use linearity in the first row, where $[1 \ \ -1 \ \ 0]$ equals $[2 \ \ -1 \ \ 0]$ minus $[1 \ \ 0 \ \ 0]$:

$$|B_n| = \begin{vmatrix} 1 & -1 & 0 \\ -1 & & \\ & A_{n-1} & \\ 0 & & \end{vmatrix} = \begin{vmatrix} 2 & -1 & 0 \\ -1 & & \\ & A_{n-1} & \\ 0 & & \end{vmatrix} - \begin{vmatrix} 1 & 0 & 0 \\ -1 & & \\ & A_{n-1} & \\ 0 & & \end{vmatrix}.$$

Linearity gives $|B_n| = |A_n| - |A_{n-1}| =$ _____.

20 Explain why the 4 by 4 Vandermonde determinant contains x^3 but not x^4 or x^5:

$$V_4 = \det \begin{bmatrix} 1 & a & a^2 & a^3 \\ 1 & b & b^2 & b^3 \\ 1 & c & c^2 & c^3 \\ 1 & x & x^2 & x^3 \end{bmatrix}.$$

The determinant is zero at $x =$ _____, _____, and _____. The cofactor of x^3 is $V_3 = (b-a)(c-a)(c-b)$. Then $V_4 = (b-a)(c-a)(c-b)(x-a)(x-b)(x-c)$.

21 Find G_2 and G_3 and then by row operations G_4. Can you predict G_n?

$$G_2 = \begin{vmatrix} 0 & 1 \\ 1 & 0 \end{vmatrix} \quad G_3 = \begin{vmatrix} 0 & 1 & 1 \\ 1 & 0 & 1 \\ 1 & 1 & 0 \end{vmatrix} \quad G_4 = \begin{vmatrix} 0 & 1 & 1 & 1 \\ 1 & 0 & 1 & 1 \\ 1 & 1 & 0 & 1 \\ 1 & 1 & 1 & 0 \end{vmatrix}.$$

22 Compute the determinants S_1, S_2, S_3 of these $1, 3, 1$ tridiagonal matrices:

$$S_1 = \begin{vmatrix} 3 \end{vmatrix} \qquad S_2 = \begin{vmatrix} 3 & 1 \\ 1 & 3 \end{vmatrix} \qquad S_3 = \begin{vmatrix} 3 & 1 & 0 \\ 1 & 3 & 1 \\ 0 & 1 & 3 \end{vmatrix}$$

Make a Fibonacci guess for S_4 and verify that you are right.

23 Cofactors of the $1, 3, 1$ matrices in Problem 22 give a recursion $S_n = 3S_{n-1} - S_{n-2}$.

Challenge: *Show that S_n is the Fibonacci number F_{2n+2} by proving $F_{2n+2} = 3F_{2n} - F_{2n-2}$.* Keep using the Fibonacci's rule $F_k = F_{k-1} + F_{k-2}$.

24 Change 3 to 2 in the upper left corner of the matrices in Problem 22. Why does that subtract S_{n-1} from the determinant S_n? Show that the determinants become the Fibonacci numbers $2, 5, 13$ (always F_{2n+1}).

Problems 25–28 are about block matrices and block determinants.

25 With 2 by 2 blocks in 4 by 4 matrices, you cannot always use block determinants:

$$\begin{vmatrix} A & B \\ 0 & D \end{vmatrix} = |A|\,|D| \qquad \text{but} \qquad \begin{vmatrix} A & B \\ C & D \end{vmatrix} \neq |A|\,|D| - |C|\,|B|.$$

(a) Why is the first statement true? Somehow B doesn't enter.

(b) Show by example that equality fails (as shown) when C enters.

(c) Show by example that the answer $\det(AD - CB)$ is also wrong.

26 With block multiplication, $A = LU$ has $A_k = L_kU_k$ in the top left corner:

$$A = \begin{bmatrix} A_k & * \\ * & * \end{bmatrix} = \begin{bmatrix} L_k & 0 \\ * & * \end{bmatrix} \begin{bmatrix} U_k & * \\ 0 & * \end{bmatrix}.$$

(a) Suppose the first three pivots of A are $2, 3, -1$. What are the determinants of L_1, L_2, L_3 (with diagonal 1's) and U_1, U_2, U_3 and A_1, A_2, A_3?

(b) If A_1, A_2, A_3 have determinants $5, 6, 7$ find the three pivots from equation (3).

27 Block elimination subtracts CA^{-1} times the first row $[\,A \;\; B\,]$ from the second row $[C \;\; D]$. This leaves the *Schur complement* $D - CA^{-1}B$ in the corner:

$$\begin{bmatrix} I & 0 \\ -CA^{-1} & I \end{bmatrix} \begin{bmatrix} A & B \\ C & D \end{bmatrix} = \begin{bmatrix} A & B \\ 0 & D - CA^{-1}B \end{bmatrix}.$$

Take determinants of these block matrices to prove correct rules for square blocks:

$$\begin{vmatrix} A & B \\ C & D \end{vmatrix} = \underset{\text{if } A^{-1} \text{ exists}}{|A|\,|D - CA^{-1}B|} = \underset{\text{if } AC = CA.}{|AD - CB|}$$

28 If A is m by n and B is n by m, block multiplication gives $\det M = \det AB$:

$$M = \begin{bmatrix} 0 & A \\ -B & I \end{bmatrix} = \begin{bmatrix} AB & A \\ 0 & I \end{bmatrix} \begin{bmatrix} I & 0 \\ -B & I \end{bmatrix}.$$

If A is a single row and B is a single column what is $\det M$? If A is a column and B is a row what is $\det M$? Do a 3 by 3 example of each.

29 (A calculus question based on the cofactor expansion) Show that the derivative of $\det A$ with respect to a_{11} is the cofactor C_{11}. The other entries are fixed—we are only changing a_{11}.

30 A 3 by 3 determinant has three products "down to the right" and three "down to the left" with minus signs. Compute the six terms in the figure to find D. Then explain without determinants why this matrix is or is not invertible:

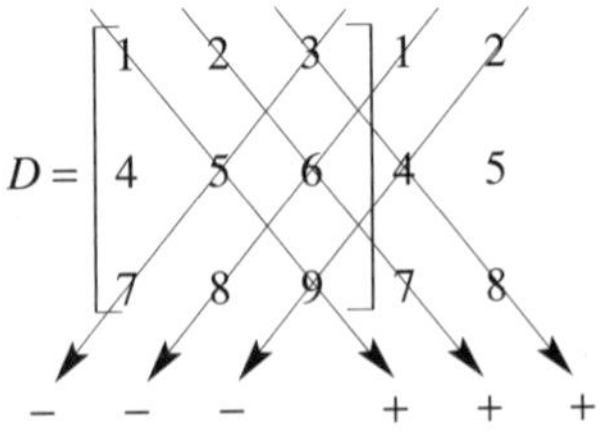

31 For E_4 in Problem 17, five of the $4! = 24$ terms in the big formula (8) are nonzero. Find those five terms to show that $E_4 = -1$.

32 For the 4 by 4 tridiagonal matrix (entries -1, 2, -1) find the five terms in the big formula that give $\det A = 16 - 4 - 4 - 4 + 1$.

33 Find the determinant of this cyclic P by cofactors of row 1 and then the "big formula". How many exchanges reorder 4, 1, 2, 3 into 1, 2, 3, 4? Is $|P^2| = 1$ or -1?

$$P = \begin{bmatrix} 0 & 0 & 0 & 1 \\ 1 & 0 & 0 & 0 \\ 0 & 1 & 0 & 0 \\ 0 & 0 & 1 & 0 \end{bmatrix} \qquad P^2 = \begin{bmatrix} 0 & 0 & 1 & 0 \\ 0 & 0 & 0 & 1 \\ 1 & 0 & 0 & 0 \\ 0 & 1 & 0 & 0 \end{bmatrix} = \begin{bmatrix} 0 & I \\ I & 0 \end{bmatrix}.$$

34 The $-1, 2, -1$ matrix is 2*eye(n)−diag(ones($n-1$, 1), 1)−diag(ones($n-1$, 1), −1). Change $A(1,1)$ to 1 so $\det A = 1$. Predict the entries of A^{-1} based on $n = 3$ and test the prediction for $n = 4$.

35 (MATLAB) The $-1, 2, -1$ matrices have determinant $n+1$. Compute $(n+1)A^{-1}$ for $n = 3$ and 4, and verify your guess for $n = 5$. (Inverses of tridiagonal matrices have the rank one form $\boldsymbol{u}\boldsymbol{v}^{\mathrm{T}}$ above the diagonal.)

36 The symmetric Pascal matrices have determinant 1. If I subtract 1 from the n, n entry, why does the determinant become zero? (Use rule 3 or cofactors in row n.)

$$\det \begin{bmatrix} 1 & 1 & 1 & 1 \\ 1 & 2 & 3 & 4 \\ 1 & 3 & 6 & 10 \\ 1 & 4 & 10 & 20 \end{bmatrix} = 1 \text{ (known)} \qquad \det \begin{bmatrix} 1 & 1 & 1 & 1 \\ 1 & 2 & 3 & 4 \\ 1 & 3 & 6 & 10 \\ 1 & 4 & 10 & \mathbf{19} \end{bmatrix} = \mathbf{0} \text{ (to explain).}$$

CRAMER'S RULE, INVERSES, AND VOLUMES ■ 5.3

This section applies determinants to solve $Ax = b$ and also to invert A. In the entries of A^{-1}, you will see $\det A$ in every denominator—we divide by it. (If $\det A = 0$ then we can't divide and A^{-1} doesn't exist.) Each number in A^{-1} is a determinant divided by another determinant. So is every component of $\boldsymbol{x} = A^{-1}\boldsymbol{b}$.

Cramer's Rule solves $A\boldsymbol{x} = \boldsymbol{b}$. A neat idea gives the first component x_1. Replacing the first column of I by $\boldsymbol{x}$ gives a matrix with determinant x_1. When you multiply by A, the first column becomes $A\boldsymbol{x}$ which is $\boldsymbol{b}$. The other columns are copied from A:

$$\begin{bmatrix} & & \\ & A & \\ & & \end{bmatrix} \begin{bmatrix} x_1 & 0 & 0 \\ x_2 & 1 & 0 \\ x_3 & 0 & 1 \end{bmatrix} = \begin{bmatrix} b_1 & a_{12} & a_{13} \\ b_2 & a_{22} & a_{23} \\ b_3 & a_{32} & a_{33} \end{bmatrix} = B_1. \qquad (1)$$

We multiplied a column at a time. *Now take determinants.* The product rule is:

$$(\det A)(x_1) = \det B_1 \qquad \text{or} \qquad x_1 = \frac{\det B_1}{\det A}. \qquad (2)$$

This is the first component of $\boldsymbol{x}$ in Cramer's Rule! Changing a column of A gives B_1.

To find x_2, put the vector $\boldsymbol{x}$ into the *second* column of the identity matrix:

$$\begin{bmatrix} \boldsymbol{a}_1 & \boldsymbol{a}_2 & \boldsymbol{a}_3 \end{bmatrix} \begin{bmatrix} 1 & x_1 & 0 \\ 0 & x_2 & 0 \\ 0 & x_3 & 1 \end{bmatrix} = \begin{bmatrix} \boldsymbol{a}_1 & \boldsymbol{b} & \boldsymbol{a}_3 \end{bmatrix} = B_2. \qquad (3)$$

Take determinants to find $(\det A)(x_2) = \det B_2$. This gives x_2 in Cramer's Rule:

5B (CRAMER's RULE) If $\det A$ is not zero, $A\boldsymbol{x} = \boldsymbol{b}$ has the unique solution

$$x_1 = \frac{\det B_1}{\det A} \qquad x_2 = \frac{\det B_2}{\det A} \qquad \cdots \qquad x_n = \frac{\det B_n}{\det A}$$

The matrix B_j has the jth column of A replaced by the vector $\boldsymbol{b}$.

A MATLAB program for Cramer's rule only needs one line to find B_j and x_j:

$$x(j) = \det([A(:,1:j-1)\ \boldsymbol{b}\ A(:,j+1:n)])/\det(A)$$

To solve an n by n system, Cramer's Rule evaluates $n+1$ determinants (of A and the n different B's). When each one is the sum of $n!$ terms—applying the "big formula" with all permutations—this makes a total of $(n+1)!$ terms. *It would be crazy to solve equations that way.* But we do finally have an explicit formula for the solution $\boldsymbol{x}$.

Example 1 Use Cramer's Rule (it needs four determinants) to solve

$$\begin{aligned} x_1+x_2+x_3 &= 1 \\ -2x_1+x_2 \quad &= 0 \\ -4x_1 \quad +x_3 &= 0 \end{aligned} \quad \text{with} \quad \det A = \begin{vmatrix} 1 & 1 & 1 \\ -2 & 1 & 0 \\ -4 & 0 & 1 \end{vmatrix} = 7.$$

The right side $(1,0,0)$ goes into columns 1, 2, 3 to produce the matrices B_1, B_2, B_3:

$$|B_1| = \begin{vmatrix} 1 & 1 & 1 \\ 0 & 1 & 0 \\ 0 & 0 & 1 \end{vmatrix} = 1 \qquad |B_2| = \begin{vmatrix} 1 & 1 & 1 \\ -2 & 0 & 0 \\ -4 & 0 & 1 \end{vmatrix} = 2 \qquad |B_3| = \begin{vmatrix} 1 & 1 & 1 \\ -2 & 1 & 0 \\ -4 & 0 & 0 \end{vmatrix} = 4.$$

Cramer's Rule takes ratios to find the components of $\boldsymbol{x}$. Always divide by $\det A$:

$$x_1 = \frac{|B_1|}{|A|} = \frac{1}{7} \qquad x_2 = \frac{|B_2|}{|A|} = \frac{2}{7} \qquad x_3 = \frac{|B_3|}{|A|} = \frac{4}{7}.$$

I always substitute x_1, x_2, x_3 back into the equations, to check the calculations.

A Formula for A^{-1}

In Example 1, the right side $\boldsymbol{b}$ was the first column of I. *The solution* $(\frac{1}{7}, \frac{2}{7}, \frac{4}{7})$ must be the first column of A^{-1}. Then the first column of $AA^{-1} = I$ is correct.

Important point: When $\boldsymbol{b} = (1,0,0)$ replaces a column of A, the determinant is 1 times a *cofactor*. Look back to see how the determinants $|B_j|$ are 2 by 2 cofactors of A:

$|B_1| = 1$ is the cofactor $C_{11} = \begin{vmatrix} 1 & 0 \\ 0 & 1 \end{vmatrix}$ Cross out row 1, column 1

$|B_2| = 2$ is the cofactor $C_{12} = -\begin{vmatrix} -2 & 0 \\ -4 & 1 \end{vmatrix}$ Cross out row 1, column 2

$|B_3| = 4$ is the cofactor $C_{13} = \begin{vmatrix} -2 & 1 \\ -4 & 0 \end{vmatrix}$ Cross out row 1, column 3

Main point: ***The numerators in*** $\boldsymbol{A^{-1}}$ ***are cofactors. They are divided by*** **det** $\boldsymbol{A}$.

For the second column of A^{-1}, change $\boldsymbol{b}$ to $(0,1,0)$. *The determinants of* B_1, B_2, B_3 *are cofactors* (in bold) *from crossing out row* 2. The cofactor signs are $(-)(+)(-)$:

$$\begin{vmatrix} 0 & \mathbf{1} & \mathbf{1} \\ 1 & 1 & 0 \\ 0 & \mathbf{0} & \mathbf{1} \end{vmatrix} = -1 \qquad \begin{vmatrix} \mathbf{1} & 0 & \mathbf{1} \\ -2 & 1 & 0 \\ \mathbf{-4} & 0 & \mathbf{1} \end{vmatrix} = 5 \qquad \begin{vmatrix} \mathbf{1} & \mathbf{1} & 0 \\ -2 & 1 & 1 \\ \mathbf{-4} & \mathbf{0} & 0 \end{vmatrix} = -4.$$

Divide $-1, 5, -4$ ***by*** $|A| = 7$ ***to get the second column of*** A^{-1}.

For the third column of A^{-1}, the right side is $\boldsymbol{b} = (0, 0, 1)$. The determinants of the three B's become cofactors of the *third row*. Those are $-1, -2, 3$. We always divide by $|A| = 7$. Now we have all columns of A^{-1}:

$$\textbf{Inverse matrix}\quad A^{-1} = \frac{1}{7}\begin{bmatrix} 1 & -1 & -1 \\ 2 & 5 & -2 \\ 4 & -4 & 3 \end{bmatrix} = \frac{\text{cofactors of } A}{\text{determinant of } A}.$$

Summary In solving $AA^{-1} = I$, the columns of I lead to the columns of A^{-1}. Then Cramer's Rule using $\boldsymbol{b}$ = columns of I gives the short formula (4) for A^{-1}. We will include a separate direct proof of this formula below.

5C ***The*** i, j ***entry of*** A^{-1} ***is the cofactor*** C_{ji} (not C_{ij}) ***divided by*** $\det A$:

$$\textbf{FORMULA FOR } A^{-1} \qquad (A^{-1})_{ij} = \frac{C_{ji}}{\det A} \quad \text{and} \quad A^{-1} = \frac{C^{\mathrm{T}}}{\det A}. \tag{4}$$

The cofactors C_{ij} go into the *"cofactor matrix"* C. ***Its transpose leads to*** A^{-1}.

To compute the i, j entry of A^{-1}, cross out row j and column i of A. Multiply the determinant by $(-1)^{i+j}$ to get the cofactor, and divide by $\det A$.

Example 2 The matrix $A = \begin{bmatrix} a & b \\ c & d \end{bmatrix}$ has cofactor matrix $C = \begin{bmatrix} d & -c \\ -b & a \end{bmatrix}$. Look at A times the transpose of C:

$$AC^{\mathrm{T}} = \begin{bmatrix} a & b \\ c & d \end{bmatrix}\begin{bmatrix} d & -b \\ -c & a \end{bmatrix} = \begin{bmatrix} ad-bc & 0 \\ 0 & ad-bc \end{bmatrix}. \tag{5}$$

The matrix on the right is $\det A$ times I. So divide by $\det A$. Then A times $C^{\mathrm{T}}/\det A$ is I, which reveals A^{-1}:

$$A^{-1} \text{ is } \frac{C^{\mathrm{T}}}{\det A} \quad \text{which is} \quad \frac{1}{ad-bc}\begin{bmatrix} d & -b \\ -c & a \end{bmatrix}. \tag{6}$$

This 2 by 2 example uses letters. The 3 by 3 example used numbers. Inverting a 4 by 4 matrix would need sixteen cofactors (each one is a 3 by 3 determinant). Elimination is faster—but now we know an explicit formula for A^{-1}.

Direct proof of the formula $A^{-1} = C^{\mathrm{T}}/\det A$ The idea is to multiply A times C^{T}:

$$\begin{bmatrix} a_{11} & a_{12} & a_{13} \\ a_{21} & a_{22} & a_{23} \\ a_{31} & a_{32} & a_{33} \end{bmatrix}\begin{bmatrix} C_{11} & C_{21} & C_{31} \\ C_{12} & C_{22} & C_{32} \\ C_{13} & C_{23} & C_{33} \end{bmatrix} = \begin{bmatrix} \det A & 0 & 0 \\ 0 & \det A & 0 \\ 0 & 0 & \det A \end{bmatrix}. \tag{7}$$

Row 1 of A times column 1 of the cofactors yields the first $\det A$ on the right:

$$a_{11}C_{11} + a_{12}C_{12} + a_{13}C_{13} = \det A \quad \text{by the cofactor rule.}$$

Similarly row 2 of A times column 2 of C^{T} yields $\det A$. The entries a_{2j} are multiplying cofactors C_{2j} as they should, to give the determinant.

How to explain the zeros off the main diagonal in equation (7)? Rows of A are multiplying cofactors from *different* rows. Row 2 of A times column 1 of C^{T} gives zero, but why?

$$a_{21}C_{11} + a_{22}C_{12} + a_{23}C_{13} = 0. \tag{8}$$

Answer: This is the cofactor rule for a new matrix, when the second row of A is copied into its first row. The new matrix A^* has two equal rows, so $\det A^* = 0$ in equation (8). Notice that A^* has the same cofactors C_{11}, C_{12}, C_{13} as A—because all rows agree after the first row. Thus the remarkable matrix multiplication (7) is correct:

$$AC^{\mathrm{T}} = (\det A)I \qquad \text{or} \qquad A^{-1} = \frac{C^{\mathrm{T}}}{\det A}.$$

Example 3 A triangular matrix of 1's has determinant 1. Then A^{-1} contains cofactors:

$$A = \begin{bmatrix} 1 & 0 & 0 & 0 \\ 1 & 1 & 0 & 0 \\ 1 & 1 & 1 & 0 \\ 1 & 1 & 1 & 1 \end{bmatrix} \quad \text{has inverse} \quad A^{-1} = \frac{C^{\mathrm{T}}}{1} = \begin{bmatrix} 1 & 0 & 0 & 0 \\ -1 & 1 & 0 & 0 \\ 0 & -1 & 1 & 0 \\ 0 & 0 & -1 & 1 \end{bmatrix}.$$

Cross out row 1 and column 1 of A to see the 3 by 3 cofactor $C_{11} = 1$. Now cross out row 1 and column 2 for C_{12}. The 3 by 3 submatrix is still triangular with determinant 1. But the cofactor C_{12} is -1 because of the sign $(-1)^{1+2}$. This number -1 goes into the (2, 1) entry of A^{-1}—don't forget to transpose C!

The inverse of a triangular matrix is triangular. Cofactors give a reason why.

Example 4 If all cofactors are nonzero, is A sure to be invertible? *No way.*

Example 5 Here is part of a direct computation of A^{-1} (see Problem 14):

$$A = \begin{bmatrix} 0 & 1 & 3 \\ 1 & 0 & 1 \\ 2 & 1 & 0 \end{bmatrix} \quad \text{and} \quad \begin{matrix} |A| = 5 \\ C_{12} = -(-2) \\ C_{22} = -6 \end{matrix} \quad \text{and} \quad A^{-1} = \frac{1}{5}\begin{bmatrix} x & x & x \\ 2 & -6 & x \\ x & x & x \end{bmatrix}.$$

Area of a Triangle

Everybody knows the area of a rectangle—base times height. The area of a triangle is *half* the base times the height. But here is a question that those formulas don't answer. ***If we know the corners (x_1, y_1) and (x_2, y_2) and (x_3, y_3) of a triangle, what is the area?***

Using the corners to find the base and height is not a good way. Determinants are much better. There are square roots in the base and height, but they cancel out in the good formula. ***The area of a triangle is half of a 3 by 3 determinant.*** If one corner is at the origin, say $(x_3, y_3) = (0, 0)$, the determinant is only 2 by 2.

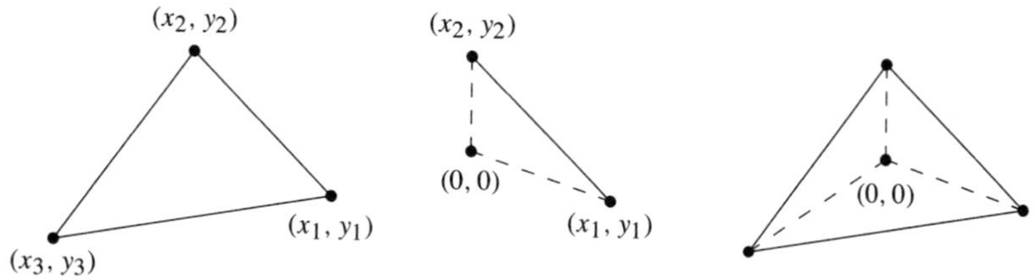

Figure 5.1 General triangle; special triangle from $(0,0)$; general from three specials.

The triangle with corners (x_1, y_1) and (x_2, y_2) and (x_3, y_3) has area $= \frac{1}{2}$ (determinant):

Area of triangle $\frac{1}{2}\begin{vmatrix} x_1 & y_1 & 1 \\ x_2 & y_2 & 1 \\ x_3 & y_3 & 1 \end{vmatrix}$ $\qquad$ $\text{Area} = \frac{1}{2}\begin{vmatrix} x_1 & y_1 \\ x_2 & y_2 \end{vmatrix}$ when $(x_3, y_3) = (0, 0)$.

When you set $x_3 = y_3 = 0$ in the 3 by 3 determinant, you get the 2 by 2 determinant. These formulas have no square roots—they are reasonable to memorize. The 3 by 3 determinant breaks into a sum of three 2 by 2's, just as the third triangle in Figure 5.1 breaks into three triangles from $(0,0)$:

$$\text{Area} = \frac{1}{2}\begin{vmatrix} x_1 & y_1 & 1 \\ x_2 & y_2 & 1 \\ x_3 & y_3 & 1 \end{vmatrix} = \begin{array}{l} +\frac{1}{2}(x_1y_2 - x_2y_1) \\ +\frac{1}{2}(x_2y_3 - x_3y_2) \\ +\frac{1}{2}(x_3y_1 - x_1y_3). \end{array} \tag{9}$$

This shows the area of the general triangle as the sum of three special areas. If $(0,0)$ is outside the triangle, two of the special areas can be negative—but the sum is still correct. The real problem is to explain the special area $\frac{1}{2}(x_1y_2 - x_2y_1)$.

Why is this the area of a triangle? We can remove the factor $\frac{1}{2}$ and change to a parallelogram (twice as big, because the parallelogram contains two equal triangles). We now prove that the parallelogram area is the determinant $x_1y_2 - x_2y_1$. This area in Figure 5.2 is 11, and therefore the triangle has area $\frac{11}{2}$.

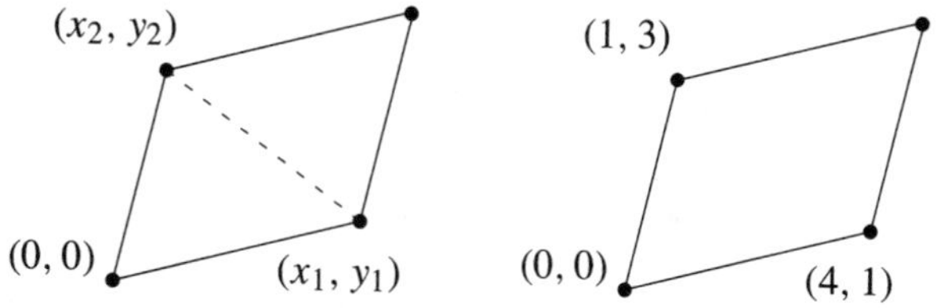

Parallelogram

$$\text{Area} = \begin{vmatrix} 4 & 1 \\ 1 & 3 \end{vmatrix} = 11$$

Triangle: Area $= \frac{11}{2}$

Figure 5.2 A triangle is half of a parallelogram. Area is half of a determinant.

Proof that a parallelogram starting from* $(0,0)$ *has area* $=2$ *by* 2 *determinant.

There are many proofs but this one fits with the book. We show that the area has the same properties 1-2-3 as the determinant. Then area = determinant! Remember that those three rules defined the determinant and led to all its other properties.

1 When $A = I$, the parallelogram becomes the unit square. Its area is $\det I = 1$.

2 When rows are exchanged, the determinant reverses sign. The absolute value (positive area) stays the same—it is the same parallelogram.

3 If row 1 is multiplied by t, Figure 5.3a shows that the area is also multiplied by t. Suppose a new row (x_1', y_1') is added to (x_1, y_1) (keeping row 2 fixed). Figure 5.3b shows that the solid parallelogram areas add to the dotted parallelogram area (because the two triangles completed by dotted lines are the same).

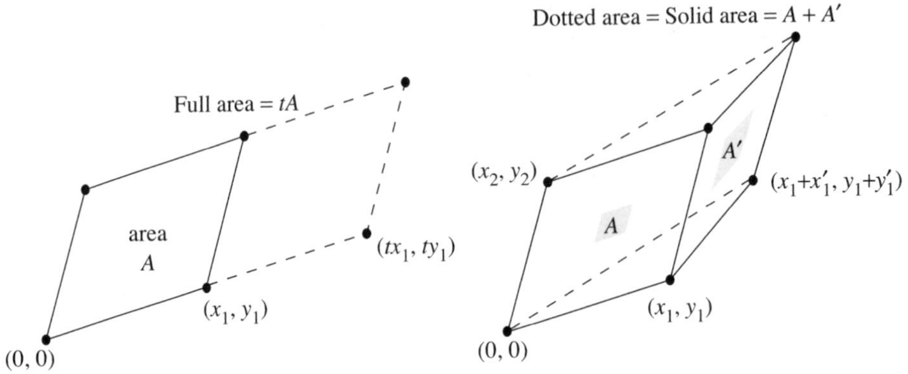

Figure 5.3 Areas obey the rule of linearity (keeping the side (x_2, y_2) constant).

That is an exotic proof, when we could use plane geometry. But the proof has a major attraction—it applies in n dimensions. The n edges going out from the origin are given by the rows of an n by n matrix. This is like the triangle with two edges going out from $(0, 0)$. The box is completed by more edges, just as the parallelogram was completed from a triangle. Figure 5.4 shows a three-dimensional box—whose edges are not at right angles.

***The volume of the box in Figure 5.4 equals the absolute value of* det *A*.** Our proof checks again that rules 1–3 for determinants are also obeyed by volumes. When an edge is stretched by a factor t, the volume is multiplied by t. When edge 1 is added to edge $1'$, the new box has edge $1 + 1'$. Its volume is the sum of the two original volumes. This is Figure 5.3b lifted into three dimensions or n dimensions. I would draw the boxes but this paper is only two-dimensional.

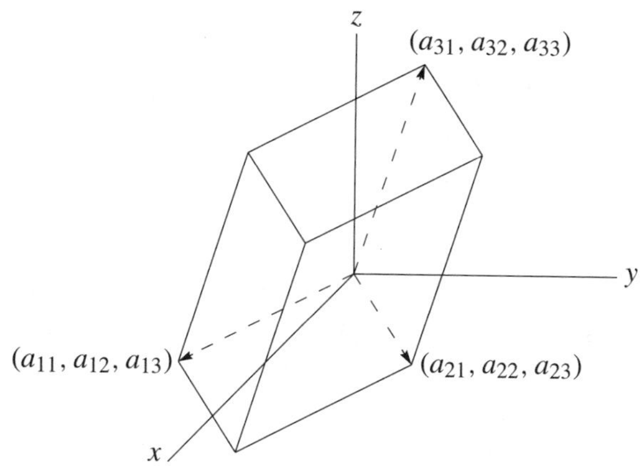

Figure 5.4 Three-dimensional box formed from the three rows of A.

The unit cube has volume $= 1$, which is $\det I$. This leaves only rule 2 to be checked. Row exchanges or edge exchanges leave the same box and the same absolute volume. The determinant changes sign, to indicate whether the edges are a *right-handed triple* ($\det A > 0$) or a *left-handed triple* ($\det A < 0$). The box volume follows the rules for determinants, so volume of the box (or parallelepiped) $=$ absolute value of the determinant.

Example 6 Suppose a rectangular box (90° angles) has side lengths r, s, and t. Its volume is r times s times t. The diagonal matrix with entries r, s, and t produces those three sides. Then $\det A$ also equals $r\,s\,t$.

Example 7 In calculus, the box is infinitesimally small! To integrate over a circle, we might change x and y to r and θ. Those are polar coordinates: $x = r\cos\theta$ and $y = r\sin\theta$. The area of a "polar box" is a determinant J times $dr\,d\theta$:

$$J = \begin{vmatrix} \partial x/\partial r & \partial x/\partial \theta \\ \partial y/\partial r & \partial y/\partial \theta \end{vmatrix} = \begin{vmatrix} \cos\theta & -r\sin\theta \\ \sin\theta & r\cos\theta \end{vmatrix} = r.$$

This determinant is the r in the small area $dA = r\,dr\,d\theta$. The stretching factor J goes into double integrals just as dx/du goes into an ordinary integral $\int dx = \int (dx/du)\,du$. For triple integrals the Jacobian matrix J with nine derivatives will be 3 by 3.

The Cross Product

This is an extra (and optional) application, special for three dimensions. Start with vectors $\boldsymbol{u} = (u_1, u_2, u_3)$ and $\boldsymbol{v} = (v_1, v_2, v_3)$. These pages are about their *cross product*. Unlike the dot product, which is a number, the cross product is a vector—also in three dimensions. It is written $\boldsymbol{u} \times \boldsymbol{v}$ and pronounced "$\boldsymbol{u}$ cross $\boldsymbol{v}$." We will quickly give the components of this vector, and also the properties that make it useful in geometry and physics.

This time we bite the bullet, and write down the formula before the properties.

DEFINITION The ***cross product*** of $\boldsymbol{u} = (u_1, u_2, u_3)$ and $\boldsymbol{v} = (v_1, v_2, v_3)$ is the vector

$$\boldsymbol{u} \times \boldsymbol{v} = \begin{vmatrix} \boldsymbol{i} & \boldsymbol{j} & \boldsymbol{k} \\ u_1 & u_2 & u_3 \\ v_1 & v_2 & v_3 \end{vmatrix} = (u_2v_3 - u_3v_2)\boldsymbol{i} + (u_3v_1 - u_1v_3)\boldsymbol{j} + (u_1v_2 - u_2v_1)\boldsymbol{k}. \tag{10}$$

This vector is perpendicular to u and v. The cross product $\boldsymbol{v} \times \boldsymbol{u}$ is $-(\boldsymbol{u} \times \boldsymbol{v})$.

Comment The 3 by 3 determinant is the easiest way to remember $\boldsymbol{u} \times \boldsymbol{v}$. It is not especially legal, because the first row contains vectors $\boldsymbol{i}, \boldsymbol{j}, \boldsymbol{k}$ and the other rows contain numbers. In the determinant, the vector $\boldsymbol{i} = (1, 0, 0)$ multiplies u_2v_3 and $-u_3v_2$. The result is $(u_2v_3 - u_3v_2, 0, 0)$, which displays the first component of the cross product.

Notice the cyclic pattern of the subscripts: 2 and 3 give component 1, then 3 and 1 give component 2, then 1 and 2 give component 3. This completes the definition of $\boldsymbol{u} \times \boldsymbol{v}$. Now we list the properties of the cross product:

Property 1 $\boldsymbol{v} \times \boldsymbol{u}$ reverses rows 2 and 3 in the determinant so it equals $-(\boldsymbol{u} \times \boldsymbol{v})$.

Property 2 The cross product $\boldsymbol{u} \times \boldsymbol{v}$ is perpendicular to $\boldsymbol{u}$ (and also to $\boldsymbol{v}$). The direct proof is to watch terms cancel. Perpendicularity is a zero dot product:

$$\boldsymbol{u} \cdot (\boldsymbol{u} \times \boldsymbol{v}) = u_1(u_2v_3 - u_3v_2) + u_2(u_3v_1 - u_1v_3) + u_3(u_1v_2 - u_2v_1) = 0. \tag{11}$$

The determinant now has rows $\boldsymbol{u}, \boldsymbol{u}$ and $\boldsymbol{v}$ so it is zero.

Property 3 The cross product of any vector with itself (two equal rows) is $\boldsymbol{u} \times \boldsymbol{u} = \boldsymbol{0}$.

When $\boldsymbol{u}$ and $\boldsymbol{v}$ are parallel, the cross product is zero. When $\boldsymbol{u}$ and $\boldsymbol{v}$ are perpendicular, the dot product is zero. One involves $\sin\theta$ and the other involves $\cos\theta$:

$$\|\boldsymbol{u} \times \boldsymbol{v}\| = \|\boldsymbol{u}\|\,\|\boldsymbol{v}\|\,|\sin\theta| \quad \text{and} \quad |\boldsymbol{u} \cdot \boldsymbol{v}| = \|\boldsymbol{u}\|\,\|\boldsymbol{v}\|\,|\cos\theta|. \tag{12}$$

Example 8 Since $\boldsymbol{u} = (3, 2, 0)$ and $\boldsymbol{v} = (1, 4, 0)$ are in the xy plane, $\boldsymbol{u} \times \boldsymbol{v}$ goes up the z axis:

$$\boldsymbol{u} \times \boldsymbol{v} = \begin{vmatrix} \boldsymbol{i} & \boldsymbol{j} & \boldsymbol{k} \\ 3 & 2 & 0 \\ 1 & 4 & 0 \end{vmatrix} = 10\boldsymbol{k}. \quad \text{The cross product is } \boldsymbol{u} \times \boldsymbol{v} = (0, 0, 10).$$

The length of u × v equals the area of the parallelogram with sides u and v. This will be important: In this example the area is 10.

Example 9 The cross product of $\boldsymbol{u} = (1, 1, 1)$ and $\boldsymbol{v} = (1, 1, 2)$ is $(1, -1, 0)$:

$$\begin{vmatrix} \boldsymbol{i} & \boldsymbol{j} & \boldsymbol{k} \\ 1 & 1 & 1 \\ 1 & 1 & 2 \end{vmatrix} = \boldsymbol{i}\begin{vmatrix} 1 & 1 \\ 1 & 2 \end{vmatrix} - \boldsymbol{j}\begin{vmatrix} 1 & 1 \\ 1 & 2 \end{vmatrix} + \boldsymbol{k}\begin{vmatrix} 1 & 1 \\ 1 & 1 \end{vmatrix} = \boldsymbol{i} - \boldsymbol{j}.$$

This vector $(1, -1, 0)$ is perpendicular to $(1, 1, 1)$ and $(1, 1, 2)$ as predicted. Area $= \sqrt{2}$.

Example 10 The cross product of $(1, 0, 0)$ and $(0, 1, 0)$ obeys the *right hand rule*. It goes up not down:

$$\begin{vmatrix} \boldsymbol{i} & \boldsymbol{j} & \boldsymbol{k} \\ 1 & 0 & 0 \\ 0 & 1 & 0 \end{vmatrix} = \boldsymbol{k}$$

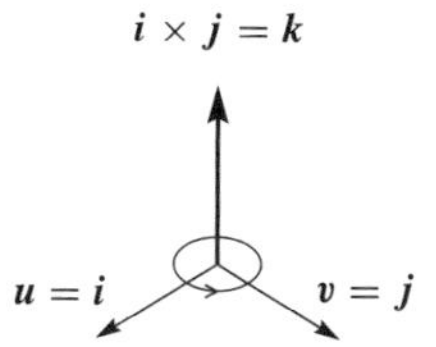

Rule $\boldsymbol{u} \times \boldsymbol{v}$ points along your right thumb when the fingers curl from $\boldsymbol{u}$ to $\boldsymbol{v}$.

Thus $\boldsymbol{i} \times \boldsymbol{j} = \boldsymbol{k}$. The right hand rule also gives $\boldsymbol{j} \times \boldsymbol{k} = \boldsymbol{i}$ and $\boldsymbol{k} \times \boldsymbol{i} = \boldsymbol{j}$. Note the cyclic order. In the opposite order (anti-cyclic) the thumb is reversed and the cross product goes the other way: $\boldsymbol{k} \times \boldsymbol{j} = -\boldsymbol{i}$ and $\boldsymbol{i} \times \boldsymbol{k} = -\boldsymbol{j}$ and $\boldsymbol{j} \times \boldsymbol{i} = -\boldsymbol{k}$. You see the three plus signs and three minus signs from a 3 by 3 determinant.

The definition of $\boldsymbol{u} \times \boldsymbol{v}$ can be based on vectors instead of their components:

DEFINITION The ***cross product*** is a vector with length $\|\boldsymbol{u}\|\,\|\boldsymbol{v}\|\,|\sin\theta|$. Its direction is perpendicular to $\boldsymbol{u}$ and $\boldsymbol{v}$. It points "up" or "down" by the right hand rule.

This definition appeals to physicists, who hate to choose axes and coordinates. They see (u_1, u_2, u_3) as the position of a mass and (F_x, F_y, F_z) as a force acting on it. If $\boldsymbol{F}$ is parallel to $\boldsymbol{u}$, then $\boldsymbol{u} \times \boldsymbol{F} = \boldsymbol{0}$—there is no turning. The mass is pushed out or pulled in. The cross product $\boldsymbol{u} \times \boldsymbol{F}$ is the turning force or *torque*. It points along the turning axis (perpendicular to $\boldsymbol{u}$ and $\boldsymbol{F}$). Its length $\|\boldsymbol{u}\|\,\|\boldsymbol{F}\|\sin\theta$ measures the "moment" that produces turning.

Triple Product = Determinant = Volume

Since $\boldsymbol{u} \times \boldsymbol{v}$ is a vector, we can take its dot product with a third vector $\boldsymbol{w}$. That produces the ***triple product*** $(\boldsymbol{u} \times \boldsymbol{v}) \cdot \boldsymbol{w}$. It is called a "scalar" triple product, because it is a number. In fact it is a determinant:

$$(\boldsymbol{u} \times \boldsymbol{v}) \cdot \boldsymbol{w} = \begin{vmatrix} w_1 & w_2 & w_3 \\ u_1 & u_2 & u_3 \\ v_1 & v_2 & v_3 \end{vmatrix} = \begin{vmatrix} u_1 & u_2 & u_3 \\ v_1 & v_2 & v_3 \\ w_1 & w_2 & w_3 \end{vmatrix}. \tag{13}$$

We can put $\boldsymbol{w}$ in the top or bottom row. The two determinants are the same because ______ row exchanges go from one to the other. Notice when this determinant is zero:

$(\boldsymbol{u} \times \boldsymbol{v}) \cdot \boldsymbol{w} = 0$ exactly when the vectors $\boldsymbol{u}, \boldsymbol{v}, \boldsymbol{w}$ lie in the *same plane*.

First reason $\boldsymbol{u} \times \boldsymbol{v}$ is perpendicular to that plane so its dot product with $\boldsymbol{w}$ is zero.

Second reason Three vectors in a plane are dependent. The matrix is singular ($\det = 0$).

Third reason Zero volume when the $\boldsymbol{u}, \boldsymbol{v}, \boldsymbol{w}$ box is squashed onto a plane.

It is remarkable that $(\boldsymbol{u} \times \boldsymbol{v}) \cdot \boldsymbol{w}$ equals the volume of the box with sides $\boldsymbol{u}, \boldsymbol{v}, \boldsymbol{w}$. This 3 by 3 determinant carries tremendous information. Like $ad - bc$ for a 2 by 2 matrix, it separates invertible from singular. Chapter 6 will be looking for singular.

■ REVIEW OF THE KEY IDEAS ■

1. Cramer's Rule solves $A\boldsymbol{x} = \boldsymbol{b}$ by ratios like $x_1 = |B_1|/|A| = |\boldsymbol{b}\, \boldsymbol{a}_2 \cdots \boldsymbol{a}_n|/|A|$.
2. When C is the cofactor matrix for A, the inverse is $A^{-1} = C^{\mathrm{T}}/\det A$.
3. The volume of a box is $|\det A|$, when the box edges are the rows of A.
4. Area and volume are needed to change variables in double and triple integrals.
5. In $\boldsymbol{R}^3$, the cross product $\boldsymbol{u} \times \boldsymbol{v}$ is perpendicular to $\boldsymbol{u}$ and $\boldsymbol{v}$.

■ WORKED EXAMPLES ■

5.3 A Use Cramer's Rule with ratios $\det B_j/\det A$ to solve $A\boldsymbol{x} = \boldsymbol{b}$. Also find the inverse matrix $A^{-1} = C^{\mathrm{T}}/\det A$. Why is the solution $\boldsymbol{x}$ in the first part the same as column 3 of A^{-1}? Which cofactors are involved in computing that column $\boldsymbol{x}$?

$$A\boldsymbol{x} = \boldsymbol{b} \quad \text{is} \quad \begin{bmatrix} 2 & 6 & 2 \\ 1 & 4 & 2 \\ 5 & 9 & 0 \end{bmatrix} \begin{bmatrix} x \\ y \\ z \end{bmatrix} = \begin{bmatrix} 0 \\ 0 \\ 1 \end{bmatrix}.$$

Find the volumes of the boxes whose edges are *columns* of A and then rows of A^{-1}.

Solution The determinants of the B_j (with right side $\boldsymbol{b}$ placed in column j) are

$$|B_1| = \begin{vmatrix} 0 & 6 & 2 \\ 0 & 4 & 2 \\ 1 & 9 & 0 \end{vmatrix} = 4 \qquad |B_2| = \begin{vmatrix} 2 & 0 & 2 \\ 1 & 0 & 2 \\ 5 & 1 & 0 \end{vmatrix} = -2 \qquad |B_3| = \begin{vmatrix} 2 & 6 & 0 \\ 1 & 4 & 0 \\ 5 & 9 & 1 \end{vmatrix} = 2.$$

Those are cofactors C_{31}, C_{32}, C_{33} of row 3. Their dot product with row 3 is $\det A$:

$$\det A = a_{31}C_{31} + a_{32}C_{32} + a_{33}C_{33} = (5, 9, 0) \cdot (4, -2, 2) = 2.$$

The three ratios $\det B_j/\det A$ give the three components of $\boldsymbol{x} = (2, -1, 1)$. This $\boldsymbol{x}$ is the third column of A^{-1} because $\boldsymbol{b} = (0, 0, 1)$ is the third column of I. The cofactors along the other rows of A, divided by $\det A = 2$, give the other columns of A^{-1}:

$$A^{-1} = \frac{C^{\mathrm{T}}}{\det A} = \frac{1}{2}\begin{bmatrix} -18 & 18 & 4 \\ 10 & -10 & -2 \\ -11 & 12 & 2 \end{bmatrix}. \quad \text{Multiply to check} \quad AA^{-1} = I$$

The box from the columns of A has volume $= \det A = 2$ (the same as the box from the rows, since $|A^{\mathrm{T}}| = |A|$). The box from rows of A^{-1} has volume $|A^{-1}| = 1/|A| = \frac{1}{2}$.

5.3 B If A is singular, the equation $AC^{\mathrm{T}} = (\det A)I$ becomes $AC^{\mathrm{T}} =$ zero matrix. *Then each column of C^{T} is in the nullspace of A.* Those columns contain cofactors along rows of A. So the cofactors quickly find the nullspace of a 3 by 3 matrix—my apologies that this comes so late!

Solve $A\boldsymbol{x} = \mathbf{0}$ by $\boldsymbol{x} =$ cofactors along a row, for these singular matrices of rank 2:

$$A = \begin{bmatrix} 1 & 4 & 7 \\ 2 & 3 & 9 \\ 2 & 2 & 8 \end{bmatrix} \qquad A = \begin{bmatrix} 1 & 1 & 2 \\ 1 & 1 & 1 \\ 1 & 1 & 1 \end{bmatrix}$$

Any nonzero column of C^{T} will give the desired solution to $A\boldsymbol{x} = \mathbf{0}$ (with rank 2, A has at least one nonzero cofactor). If A has rank 1 we get $\boldsymbol{x} = \mathbf{0}$ and the idea will not work.

Solution The first matrix has these cofactors along its top row (note the minus sign):

$$\begin{vmatrix} 3 & 9 \\ 2 & 8 \end{vmatrix} = 6 \qquad -\begin{vmatrix} 2 & 9 \\ 2 & 8 \end{vmatrix} = 2 \qquad \begin{vmatrix} 2 & 3 \\ 2 & 2 \end{vmatrix} = -2$$

Then $\boldsymbol{x} = (6, 2, -2)$ solves $A\boldsymbol{x} = \mathbf{0}$. The cofactors along the second row are $(-18, -6, 6)$ which is just $-3\boldsymbol{x}$. This is also in the one-dimensional nullspace of A.

The second matrix has *zero cofactors* along its first row. The nullvector $\boldsymbol{x} = (0, 0, 0)$ is not interesting. The cofactors of row 2 give $\boldsymbol{x} = (1, -1, 0)$ which solves $A\boldsymbol{x} = \mathbf{0}$.

Every n by n matrix of rank $n - 1$ has at least one nonzero cofactor by Problem 3.3.9. But for rank $n - 2$, all cofactors are zero. In that case cofactors only find $\boldsymbol{x} = \mathbf{0}$.

Problem Set 5.3

Problems 1–5 are about Cramer's Rule for $\boldsymbol{x} = A^{-1}\boldsymbol{b}$.

1 Solve these linear equations by Cramer's Rule $x_j = \det B_j / \det A$:

(a) $\begin{aligned} 2x_1 + 5x_2 &= 1 \\ x_1 + 4x_2 &= 2 \end{aligned}$ (b) $\begin{aligned} 2x_1 + x_2 \phantom{{}+ x_3} &= 1 \\ x_1 + 2x_2 + x_3 &= 0 \\ x_2 + 2x_3 &= 0. \end{aligned}$

2 Use Cramer's Rule to solve for y (only). Call the 3 by 3 determinant D:

(a) $\begin{aligned} ax + by &= 1 \\ cx + dy &= 0 \end{aligned}$ (b) $\begin{aligned} ax + by + cz &= 1 \\ dx + ey + fz &= 0 \\ gx + hy + iz &= 0. \end{aligned}$

3 Cramer's Rule breaks down when $\det A = 0$. Example (a) has no solution while (b) has infinitely many. What are the ratios $x_j = \det B_j / \det A$ in these two cases?

(a) $\begin{aligned} 2x_1 + 3x_2 &= 1 \\ 4x_1 + 6x_2 &= 1 \end{aligned}$ (parallel lines) (b) $\begin{aligned} 2x_1 + 3x_2 &= 1 \\ 4x_1 + 6x_2 &= 2 \end{aligned}$ (same line)

4 *Quick proof of Cramer's rule.* The determinant is a linear function of column 1. It is zero if two columns are equal. When $\boldsymbol{b} = A\boldsymbol{x} = x_1\boldsymbol{a}_1 + x_2\boldsymbol{a}_2 + x_3\boldsymbol{a}_3$ goes into the first column of A, the determinant of this matrix B_1 is

$$|\boldsymbol{b} \quad \boldsymbol{a}_2 \quad \boldsymbol{a}_3| = |x_1\boldsymbol{a}_1 + x_2\boldsymbol{a}_2 + x_3\boldsymbol{a}_3 \quad \boldsymbol{a}_2 \quad \boldsymbol{a}_3| = x_1|\boldsymbol{a}_1 \quad \boldsymbol{a}_2 \quad \boldsymbol{a}_3| = x_1 \det A.$$

(a) What formula for x_1 comes from left side = right side?

(b) What steps lead to the middle equation?

5 If the right side $\boldsymbol{b}$ is the first column of A, solve the 3 by 3 system $A\boldsymbol{x} = \boldsymbol{b}$. How does each determinant in Cramer's Rule lead to this solution $\boldsymbol{x}$?

Problems 6–16 are about $A^{-1} = C^{\mathrm{T}} / \det A$. Remember to transpose C.

6 Find A^{-1} from the cofactor formula $C^{\mathrm{T}} / \det A$. Use symmetry in part (b).

(a) $A = \begin{bmatrix} 1 & 2 & 0 \\ 0 & 3 & 0 \\ 0 & 4 & 1 \end{bmatrix}$ (b) $A = \begin{bmatrix} 2 & -1 & 0 \\ -1 & 2 & -1 \\ 0 & -1 & 2 \end{bmatrix}$.

7 If all the cofactors are zero, how do you know that A has no inverse? If none of the cofactors are zero, is A sure to be invertible?

8 Find the cofactors of A and multiply AC^{T} to find $\det A$:

$$A = \begin{bmatrix} 1 & 1 & 4 \\ 1 & 2 & 2 \\ 1 & 2 & 5 \end{bmatrix} \quad \text{and} \quad C = \begin{bmatrix} 6 & -3 & 0 \\ \cdot & \cdot & \cdot \\ \cdot & \cdot & \cdot \end{bmatrix} \quad \text{and } AC^{\mathrm{T}} = \underline{\qquad}.$$

If you change that 4 to 100, why is $\det A$ unchanged?

9 Suppose $\det A = 1$ and you know all the cofactors. How can you find A?

10 From the formula $AC^{\mathrm{T}} = (\det A)I$ show that $\det C = (\det A)^{n-1}$.

11 (for professors only) If you know all 16 cofactors of a 4 by 4 invertible matrix A, how would you find A?

12 If all entries of A are integers, and $\det A = 1$ or -1, prove that all entries of A^{-1} are integers. Give a 2 by 2 example.

13 If all entries of A and A^{-1} are integers, prove that $\det A = 1$ or -1. Hint: What is $\det A$ times $\det A^{-1}$?

14 Complete the calculation of A^{-1} by cofactors in Example 5.

15 L is lower triangular and S is symmetric. Assume they are invertible:

$$L = \begin{bmatrix} a & 0 & 0 \\ b & c & 0 \\ d & e & f \end{bmatrix} \qquad S = \begin{bmatrix} a & b & d \\ b & c & e \\ d & e & f \end{bmatrix}.$$

(a) Which three cofactors of L are zero? Then L^{-1} is lower triangular.

(b) Which three pairs of cofactors of S are equal? Then S^{-1} is symmetric.

16 For $n = 5$ the matrix C contains _____ cofactors and each 4 by 4 cofactor contains _____ terms and each term needs _____ multiplications. Compare with $5^3 = 125$ for the Gauss-Jordan computation of A^{-1} in Section 2.4.

Problems 17–27 are about area and volume by determinants.

17 (a) Find the area of the parallelogram with edges $\boldsymbol{v} = (3, 2)$ and $\boldsymbol{w} = (1, 4)$.

(b) Find the area of the triangle with sides $\boldsymbol{v}$, $\boldsymbol{w}$, and $\boldsymbol{v} + \boldsymbol{w}$. Draw it.

(c) Find the area of the triangle with sides $\boldsymbol{v}$, $\boldsymbol{w}$, and $\boldsymbol{w} - \boldsymbol{v}$. Draw it.

18 A box has edges from $(0, 0, 0)$ to $(3, 1, 1)$ and $(1, 3, 1)$ and $(1, 1, 3)$. Find its volume and also find the area of each parallelogram face using $\|\boldsymbol{u} \times \boldsymbol{v}\|$.

19 (a) The corners of a triangle are $(2, 1)$ and $(3, 4)$ and $(0, 5)$. What is the area?

(b) Add a corner at $(-1, 0)$ to make a lopsided region (four sides). What is the area?

20 The parallelogram with sides $(2, 1)$ and $(2, 3)$ has the same area as the parallelogram with sides $(2, 2)$ and $(1, 3)$. Find those areas from 2 by 2 determinants and say why they must be equal. (I can't see why from a picture. Please write to me if you do.)

21 The Hadamard matrix H has orthogonal rows. The box is a hypercube!

$$\text{What is} \quad |H| = \begin{vmatrix} 1 & 1 & 1 & 1 \\ 1 & 1 & -1 & -1 \\ 1 & -1 & -1 & 1 \\ 1 & -1 & 1 & -1 \end{vmatrix} = \text{volume of a hypercube in } \mathbf{R}^4?$$

22 If the columns of a 4 by 4 matrix have lengths L_1, L_2, L_3, L_4, what is the largest possible value for the determinant (based on volume)? If all entries are 1 or -1, what are those lengths and the maximum determinant?

23 Show by a picture how a rectangle with area x_1y_2 minus a rectangle with area x_2y_1 produces the same area as our parallelogram.

24 When the edge vectors $\boldsymbol{a}, \boldsymbol{b}, \boldsymbol{c}$ are perpendicular, the volume of the box is $\|\boldsymbol{a}\|$ times $\|\boldsymbol{b}\|$ times $\|\boldsymbol{c}\|$. The matrix $A^{\mathrm{T}}A$ is ______. Find $\det A^{\mathrm{T}}A$ and $\det A$.

25 The box with edges $\boldsymbol{i}$ and $\boldsymbol{j}$ and $\boldsymbol{w} = 2\boldsymbol{i} + 3\boldsymbol{j} + 4\boldsymbol{k}$ has height ______. What is the volume? What is the matrix with this determinant? What is $\boldsymbol{i} \times \boldsymbol{j}$ and what is its dot product with $\boldsymbol{w}$?

26 An n-dimensional cube has how many corners? How many edges? How many $(n-1)$-dimensional faces? The cube whose edges are the rows of $2I$ has volume ______. A hypercube computer has parallel processors at the corners with connections along the edges.

27 The triangle with corners $(0,0)$, $(1,0)$, $(0,1)$ has area $\frac{1}{2}$. The pyramid with four corners $(0,0,0)$, $(1,0,0)$, $(0,1,0)$, $(0,0,1)$ has volume ______. What is the volume of a pyramid in $\mathbf{R}^4$ with five corners at $(0,0,0,0)$ and the rows of I?

Problems 28–31 are about areas dA and volumes dV in calculus.

28 Polar coordinates satisfy $x = r\cos\theta$ and $y = r\sin\theta$. Polar area $J\,dr\,d\theta$ includes J:

$$J = \begin{vmatrix} \partial x/\partial r & \partial x/\partial \theta \\ \partial y/\partial r & \partial y/\partial \theta \end{vmatrix} = \begin{vmatrix} \cos\theta & -r\sin\theta \\ \sin\theta & r\cos\theta \end{vmatrix}.$$

The two columns are orthogonal. Their lengths are ______. Thus $J =$ ______.

29 Spherical coordinates ρ, ϕ, θ satisfy $x = \rho\sin\phi\cos\theta$ and $y = \rho\sin\phi\sin\theta$ and $z = \rho\cos\phi$. Find the 3 by 3 matrix of partial derivatives: $\partial x/\partial\rho$, $\partial x/\partial\phi$, $\partial x/\partial\theta$ in row 1. Simplify its determinant to $J = \rho^2\sin\phi$. Then dV in a sphere is $\rho^2\sin\phi\,d\rho\,d\phi d\theta$.

30 The matrix that connects r, θ to x, y is in Problem 27. Invert that 2 by 2 matrix:

$$J^{-1} = \begin{vmatrix} \partial r/\partial x & \partial r/\partial y \\ \partial \theta/\partial x & \partial \theta/\partial y \end{vmatrix} = \begin{vmatrix} \cos\theta & ? \\ ? & ? \end{vmatrix} = ?$$

It is surprising that $\partial r/\partial x = \partial x/\partial r$ (***Calculus***, Gilbert Strang, p. 501). Multiplying the matrices in 28 and 30 gives the chain rule $\frac{\partial x}{\partial x} = \frac{\partial x}{\partial r}\frac{\partial r}{\partial x} + \frac{\partial x}{\partial \theta}\frac{\partial \theta}{\partial x} = 1$.

31 The triangle with corners $(0,0)$, $(6,0)$, and $(1,4)$ has area ______. When you rotate it by $\theta = 60^\circ$ the area is ______. The determinant of the rotation matrix is

$$J = \begin{vmatrix} \cos\theta & -\sin\theta \\ \sin\theta & \cos\theta \end{vmatrix} = \begin{vmatrix} \frac{1}{2} & ? \\ ? & ? \end{vmatrix} = ?$$

Problems 32–39 are about the triple product $(\boldsymbol{u} \times \boldsymbol{v}) \cdot \boldsymbol{w}$ in three dimensions.

32 A box has base area $\|\boldsymbol{u} \times \boldsymbol{v}\|$. Its perpendicular height is $\|\boldsymbol{w}\| \cos\theta$. Base area times height = volume = $\|\boldsymbol{u} \times \boldsymbol{v}\| \, \|\boldsymbol{w}\| \cos\theta$ which is $(\boldsymbol{u} \times \boldsymbol{v}) \cdot \boldsymbol{w}$. Compute base area, height, and volume for $\boldsymbol{u} = (2, 4, 0)$, $\boldsymbol{v} = (-1, 3, 0)$, $\boldsymbol{w} = (1, 2, 2)$.

33 The volume of the same box is given more directly by a 3 by 3 determinant. Evaluate that determinant.

34 Expand the 3 by 3 determinant in equation (13) in cofactors of its row u_1, u_2, u_3. This expansion is the dot product of $\boldsymbol{u}$ with the vector ____.

35 Which of the triple products $(\boldsymbol{u} \times \boldsymbol{w}) \cdot \boldsymbol{v}$ and $(\boldsymbol{w} \times \boldsymbol{u}) \cdot \boldsymbol{v}$ and $(\boldsymbol{v} \times \boldsymbol{w}) \cdot \boldsymbol{u}$ are the same as $(\boldsymbol{u} \times \boldsymbol{v}) \cdot \boldsymbol{w}$? Which orders of the rows $\boldsymbol{u}$, $\boldsymbol{v}$, $\boldsymbol{w}$ give the correct determinant?

36 Let $P = (1, 0, -1)$ and $Q = (1, 1, 1)$ and $R = (2, 2, 1)$. Choose S so that $PQRS$ is a parallelogram and compute its area. Choose T, U, V so that $OPQRSTUV$ is a tilted box and compute its volume.

37 Suppose (x, y, z) and $(1, 1, 0)$ and $(1, 2, 1)$ lie on a plane through the origin. What determinant is zero? What equation does this give for the plane?

38 Suppose (x, y, z) is a linear combination of $(2, 3, 1)$ and $(1, 2, 3)$. What determinant is zero? What equation does this give for the plane of all combinations?

39 (a) Explain from volumes why $\det 2A = 2^n \det A$ for n by n matrices.

(b) For what size matrix is the false statement $\det A + \det A = \det(A + A)$ true?

6

EIGENVALUES AND EIGENVECTORS

INTRODUCTION TO EIGENVALUES ■ 6.1

Linear equations $Ax = b$ come from steady state problems. Eigenvalues have their greatest importance in *dynamic problems*. The solution of $du/dt = Au$ is changing with time—growing or decaying or oscillating. We can't find it by elimination. This chapter enters a new part of linear algebra. All matrices in this chapter are square.

A good model comes from the powers $A, A^2, A^3, \ldots$ of a matrix. Suppose you need the hundredth power A^{100}. The starting matrix A becomes unrecognizable after a few steps:

$$\underset{A}{\begin{bmatrix} .8 & .3 \\ .2 & .7 \end{bmatrix}} \quad \underset{A^2}{\begin{bmatrix} .70 & .45 \\ .30 & .55 \end{bmatrix}} \quad \underset{A^3}{\begin{bmatrix} .650 & .525 \\ .350 & .475 \end{bmatrix}} \quad \cdots \quad \underset{A^{100}}{\begin{bmatrix} .6000 & .6000 \\ .4000 & .4000 \end{bmatrix}}$$

A^{100} was found by using the *eigenvalues* of A, not by multiplying 100 matrices. Those eigenvalues are a new way to see into the heart of a matrix.

To explain eigenvalues, we first explain eigenvectors. Almost all vectors change direction, when they are multiplied by A. ***Certain exceptional vectors x are in the same direction as Ax. Those are the "eigenvectors"***. Multiply an eigenvector by A, and the vector Ax is a number λ times the original x.

The basic equation is $Ax = \lambda x$. The number λ is the ***"eigenvalue"***. It tells whether the special vector x is stretched or shrunk or reversed or left unchanged—when it is multiplied by A. We may find $\lambda = 2$ or $\frac{1}{2}$ or -1 or 1. The eigenvalue λ could be zero! Then $Ax = 0x$ means that this eigenvector x is in the nullspace.

If A is the identity matrix, every vector has $Ax = x$. All vectors are eigenvectors. The eigenvalue (the number lambda) is $\lambda = 1$. This is unusual to say the least. Most 2 by 2 matrices have *two* eigenvector directions and *two* eigenvalues. This section teaches how to compute the x's and λ's. It can come early in the course because we only need the determinant of a 2 by 2 matrix.

For the matrix A in our model above, here are eigenvectors x_1 and x_2. Multiplying those vectors by A gives x_1 and $\frac{1}{2}x_2$. The eigenvalues are $\lambda_1 = 1$ and $\lambda_2 = \frac{1}{2}$:

$$x_1 = \begin{bmatrix} .6 \\ .4 \end{bmatrix} \quad \text{and} \quad Ax_1 = \begin{bmatrix} .8 & .3 \\ .2 & .7 \end{bmatrix}\begin{bmatrix} .6 \\ .4 \end{bmatrix} = x_1 \quad (Ax = x \text{ means that } \lambda_1 = 1)$$

$$x_2 = \begin{bmatrix} 1 \\ -1 \end{bmatrix} \quad \text{and} \quad Ax_2 = \begin{bmatrix} .8 & .3 \\ .2 & .7 \end{bmatrix}\begin{bmatrix} 1 \\ -1 \end{bmatrix} = \begin{bmatrix} .5 \\ -.5 \end{bmatrix} \quad (\text{this is } \tfrac{1}{2}x_2 \text{ so } \lambda_2 = \tfrac{1}{2}).$$

If we again multiply x_1 by A, we still get x_1. Every power of A will give $A^n x_1 = x_1$. Multiplying x_2 by A gave $\frac{1}{2}x_2$, and if we multiply again we get $(\frac{1}{2})^2 x_2$. ***When A is squared, the eigenvectors x_1 and x_2 stay the same.*** The λ's are now 1^2 and $(\frac{1}{2})^2$. ***The eigenvalues are squared!*** This pattern keeps going, because the eigenvectors stay in their own directions (Figure 6.1) and never get mixed. The eigenvectors of A^{100} are the same x_1 and x_2. The eigenvalues of A^{100} are $1^{100} = 1$ and $(\frac{1}{2})^{100} =$ very small number.

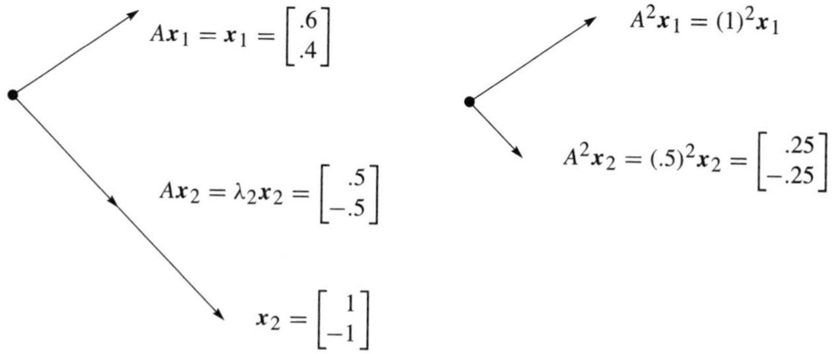

Figure 6.1 The eigenvectors keep their directions. A^2 has eigenvalues 1^2 and $(.5)^2$.

Other vectors do change direction. But all other vectors are combinations of the two eigenvectors. The first column of A is the combination $x_1 + (.2)x_2$:

$$\begin{bmatrix} .8 \\ .2 \end{bmatrix} \quad \text{is} \quad x_1 + (.2)x_2 = \begin{bmatrix} .6 \\ .4 \end{bmatrix} + \begin{bmatrix} .2 \\ -.2 \end{bmatrix}. \tag{1}$$

Multiplying by A gives the first column of A^2. Do it separately for x_1 and $(.2)x_2$. Of course $Ax_1 = x_1$. And A multiplies x_2 by its eigenvalue $\frac{1}{2}$:

$$A\begin{bmatrix} .8 \\ .2 \end{bmatrix} = \begin{bmatrix} .7 \\ .3 \end{bmatrix} \quad \text{is} \quad x_1 + \frac{1}{2}(.2)x_2 = \begin{bmatrix} .6 \\ .4 \end{bmatrix} + \begin{bmatrix} .1 \\ -.1 \end{bmatrix}.$$

Each eigenvector is multiplied by its eigenvalue, when we multiply by A. We didn't need these eigenvectors to find A^2. But it is the good way to do 99 multiplications.

At every step $\boldsymbol{x}_1$ is unchanged and $\boldsymbol{x}_2$ is multiplied by $(\frac{1}{2})$, so we have $(\frac{1}{2})^{99}$:

$$A^{99}\begin{bmatrix}.8\\.2\end{bmatrix} \quad \text{is really} \quad \boldsymbol{x}_1 + (.2)(\frac{1}{2})^{99}\boldsymbol{x}_2 = \begin{bmatrix}.6\\.4\end{bmatrix} + \begin{bmatrix}\text{very}\\\text{small}\\\text{vector}\end{bmatrix}.$$

This is the first column of A^{100}. The number we originally wrote as .6000 was not exact. We left out $(.2)(\frac{1}{2})^{99}$ which wouldn't show up for 30 decimal places.

The eigenvector $\boldsymbol{x}_1$ is a "steady state" that doesn't change (because $\lambda_1 = 1$). The eigenvector $\boldsymbol{x}_2$ is a "decaying mode" that virtually disappears (because $\lambda_2 = .5$). The higher the power of A, the closer its columns approach the steady state.

We mention that this particular A is a ***Markov matrix***. Its entries are positive and every column adds to 1. Those facts guarantee that the largest eigenvalue is $\lambda = 1$ (as we found). Its eigenvector $\boldsymbol{x}_1 = (.6, .4)$ is the *steady state*—which all columns of A^k will approach. Section 8.3 shows how Markov matrices appear in applications.

For projections we can spot the steady state ($\lambda = 1$) and the nullspace ($\lambda = 0$).

Example 1 **The projection matrix $P = \begin{bmatrix}.5 & .5\\ .5 & .5\end{bmatrix}$ has eigenvalues 1 and 0.**

Its eigenvectors are $\boldsymbol{x}_1 = (1, 1)$ and $\boldsymbol{x}_2 = (1, -1)$. For those vectors, $P\boldsymbol{x}_1 = \boldsymbol{x}_1$ (steady state) and $P\boldsymbol{x}_2 = \boldsymbol{0}$ (nullspace). This example illustrates three things that we mention now:

1. Each column of P adds to 1, so $\lambda = 1$ is an eigenvalue.

2. P is singular, so $\lambda = 0$ is an eigenvalue.

3. P is symmetric, so its eigenvectors $(1, 1)$ and $(1, -1)$ are perpendicular.

The only possible eigenvalues of a projection matrix are 0 and 1. The eigenvectors for $\lambda = 0$ (which means $P\boldsymbol{x} = 0\boldsymbol{x}$) fill up the nullspace. The eigenvectors for $\lambda = 1$ (which means $P\boldsymbol{x} = \boldsymbol{x}$) fill up the column space. The nullspace is projected to zero. The column space projects onto itself.

An in-between vector like $\boldsymbol{v} = (3, 1)$ partly disappears and partly stays:

$$\boldsymbol{v} = \begin{bmatrix}1\\-1\end{bmatrix} + \begin{bmatrix}2\\2\end{bmatrix} \quad \text{projects onto} \quad P\boldsymbol{v} = \begin{bmatrix}0\\0\end{bmatrix} + \begin{bmatrix}2\\2\end{bmatrix}.$$

The projection keeps the column space part of $\boldsymbol{v}$ and destroys the nullspace part. To emphasize: *Special properties of a matrix lead to special eigenvalues and eigenvectors.* That is a major theme of this chapter (it is captured in a table at the very end).

Projections have $\lambda = 0$ and 1. Permutations have all $|\lambda| = 1$. The next matrix R (a reflection and at the same time a permutation) is also special.

Example 2 **The reflection matrix $R = \left[\begin{smallmatrix} 0 & 1 \\ 1 & 0 \end{smallmatrix}\right]$ has eigenvalues 1 and −1.**
The eigenvector $(1, 1)$ is unchanged by R. The second eigenvector is $(1, -1)$—its signs are reversed by R. A matrix with no negative entries can still have a negative eigenvalue! The eigenvectors for R are the same as for P, because $R = 2P - I$:

$$R = 2P - I \quad \text{or} \quad \begin{bmatrix} 0 & 1 \\ 1 & 0 \end{bmatrix} = 2\begin{bmatrix} .5 & .5 \\ .5 & .5 \end{bmatrix} - \begin{bmatrix} 1 & 0 \\ 0 & 1 \end{bmatrix}. \tag{2}$$

Here is the point. If $P\boldsymbol{x} = \lambda\boldsymbol{x}$ then $2P\boldsymbol{x} = 2\lambda\boldsymbol{x}$. The eigenvalues are doubled when the matrix is doubled. Now subtract $I\boldsymbol{x} = \boldsymbol{x}$. The result is $(2P - I)\boldsymbol{x} = (2\lambda - 1)\boldsymbol{x}$. ***When a matrix is shifted by I, each λ is shifted by*** 1. No change in eigenvectors.

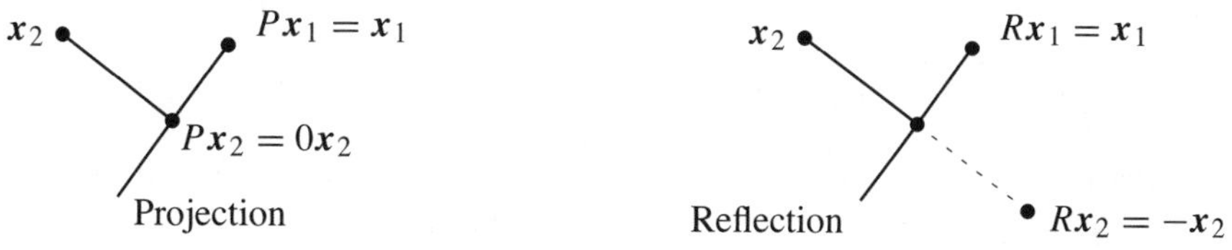

Figure 6.2 Projections have eigenvalues 1 and 0. Reflections have $\lambda = 1$ and -1. A typical $\boldsymbol{x}$ changes direction, but not the eigenvectors $\boldsymbol{x}_1$ and $\boldsymbol{x}_2$.

The eigenvalues are related exactly as the matrices are related:

$$R = 2P - I \qquad \text{so the eigenvalues of } R \text{ are} \qquad 2(1) - 1 = 1 \text{ and } 2(0) - 1 = -1.$$

The eigenvalues of R^2 are λ^2. In this case $R^2 = I$. Check $(1)^2 = 1$ and $(-1)^2 = 1$.

The Equation for the Eigenvalues

In small examples we found λ's and $\boldsymbol{x}$'s by trial and error. Now we use determinants and linear algebra. *This is the key calculation in the chapter*—***to solve*** $A\boldsymbol{x} = \lambda\boldsymbol{x}$.

First move $\lambda\boldsymbol{x}$ to the left side. Write the equation $A\boldsymbol{x} = \lambda\boldsymbol{x}$ as $(A - \lambda I)\boldsymbol{x} = \boldsymbol{0}$. The matrix $A - \lambda I$ times the eigenvector $\boldsymbol{x}$ is the zero vector. ***The eigenvectors make up the nullspace of*** $A - \lambda I$! When we know an eigenvalue λ, we find an eigenvector by solving $(A - \lambda I)\boldsymbol{x} = \boldsymbol{0}$.

Eigenvalues first. If $(A - \lambda I)\boldsymbol{x} = \boldsymbol{0}$ has a nonzero solution, $A - \lambda I$ is not invertible. ***The determinant of $A - \lambda I$ must be zero.*** This is how to recognize an eigenvalue λ:

6A The number λ is an eigenvalue of A if and only if $A - \lambda I$ is singular:

$$\det(A - \lambda I) = 0. \tag{3}$$

This "characteristic equation" involves only λ, not $\boldsymbol{x}$. When A is n by n, $\det(A-\lambda I)=0$ is an equation of degree n. Then A has n eigenvalues and each λ leads to $\boldsymbol{x}$:

$$\textbf{For each } \lambda \textbf{ solve} \quad (A-\lambda I)\boldsymbol{x}=\mathbf{0} \quad \text{or} \quad A\boldsymbol{x}=\lambda\boldsymbol{x} \quad \textbf{to find an eigenvector } \boldsymbol{x}. \tag{4}$$

Example 3 $A=\begin{bmatrix}1&2\\2&4\end{bmatrix}$ is already singular (zero determinant). Find its λ's and $\boldsymbol{x}$'s.

When A is singular, $\lambda=0$ is one of the eigenvalues. The equation $A\boldsymbol{x}=0\boldsymbol{x}$ has solutions. They are the eigenvectors for $\lambda=0$. But here is the way to find *all* λ's and $\boldsymbol{x}$'s! Always subtract λI from A:

$$\textit{Subtract } \lambda \textit{ from the diagonal to find} \quad A-\lambda I=\begin{bmatrix}1-\lambda & 2\\ 2 & 4-\lambda\end{bmatrix}.$$

Take the determinant "ad − bc" of this 2 ***by*** 2 ***matrix.*** From $1-\lambda$ times $4-\lambda$, the "ad" part is $\lambda^2-5\lambda+4$. The "bc" part, not containing λ, is 2 times 2.

$$\det\begin{bmatrix}1-\lambda & 2\\ 2 & 4-\lambda\end{bmatrix}=(1-\lambda)(4-\lambda)-(2)(2)=\lambda^2-5\lambda. \tag{5}$$

Set this determinant $\boldsymbol{\lambda^2-5\lambda}$ ***to zero.*** One solution is $\lambda=0$ (as expected, since A is singular). Factoring into λ times $\lambda-5$, the other root is $\lambda=5$:

$$\det(A-\lambda I)=\lambda^2-5\lambda=0 \quad \text{yields the eigenvalues} \quad \lambda_1=0 \quad \text{and} \quad \lambda_2=5.$$

Now find the eigenvectors. Solve $(A-\lambda I)\boldsymbol{x}=\mathbf{0}$ separately for $\lambda_1=0$ and $\lambda_2=5$:

$$(A-0I)\boldsymbol{x}=\begin{bmatrix}1&2\\2&4\end{bmatrix}\begin{bmatrix}y\\z\end{bmatrix}=\begin{bmatrix}0\\0\end{bmatrix} \text{ yields an eigenvector } \begin{bmatrix}y\\z\end{bmatrix}=\begin{bmatrix}2\\-1\end{bmatrix} \text{ for } \lambda_1=0$$

$$(A-5I)\boldsymbol{x}=\begin{bmatrix}-4&2\\2&-1\end{bmatrix}\begin{bmatrix}y\\z\end{bmatrix}=\begin{bmatrix}0\\0\end{bmatrix} \text{ yields an eigenvector } \begin{bmatrix}y\\z\end{bmatrix}=\begin{bmatrix}1\\2\end{bmatrix} \text{ for } \lambda_2=5.$$

The matrices $A-0I$ and $A-5I$ are singular (because 0 and 5 are eigenvalues). The eigenvectors $(2,-1)$ and $(1,2)$ are in the nullspaces: $(A-\lambda I)\boldsymbol{x}=\mathbf{0}$ is $A\boldsymbol{x}=\lambda\boldsymbol{x}$.

We need to emphasize: *There is nothing exceptional about* $\lambda=0$. Like every other number, zero might be an eigenvalue and it might not. If A is singular, it is. The eigenvectors fill the nullspace: $A\boldsymbol{x}=0\boldsymbol{x}=\mathbf{0}$. If A is invertible, zero is not an eigenvalue. We shift A by a multiple of I to *make it singular*. In the example, the shifted matrix $A-5I$ was singular and 5 was the other eigenvalue.

Summary To solve the eigenvalue problem for an n by n matrix, follow these steps:

1. ***Compute the determinant of*** $A-\lambda I$. With λ subtracted along the diagonal, this determinant starts with λ^n or $-\lambda^n$. It is a polynomial in λ of degree n.

2. ***Find the roots of this polynomial***, by solving $\det(A-\lambda I)=0$. The n roots are the n eigenvalues of A. They make $A-\lambda I$ singular.

3. For each eigenvalue λ, ***solve*** $(A-\lambda I)\boldsymbol{x}=\mathbf{0}$ ***to find an eigenvector*** $\boldsymbol{x}$.

A note on quick computations, when A is 2 by 2. The determinant of $A - \lambda I$ is a quadratic (starting with λ^2). >From factoring or the quadratic formula, we find its two roots (the eigenvalues). Then the eigenvectors come immediately from $A - \lambda I$. This matrix is singular, so both rows are multiples of a vector (a, b). *The eigenvector is any multiple of* $(b, -a)$. The example had $\lambda = 0$ and $\lambda = 5$:

$$\lambda = 0: \text{ rows of } A - 0I \text{ in the direction } (1, 2); \text{ eigenvector in the direction } (2, -1)$$

$$\lambda = 5: \text{ rows of } A - 5I \text{ in the direction } (-4, 2); \text{ eigenvector in the direction } (2, 4).$$

Previously we wrote that last eigenvector as $(1, 2)$. Both $(1, 2)$ and $(2, 4)$ are correct. There is a whole *line of eigenvectors*—any nonzero multiple of $\boldsymbol{x}$ is as good as $\boldsymbol{x}$. MATLAB 's **eig**(A) divides by the length, to make the eigenvector into a unit vector.

We end with a warning. Some 2 by 2 matrices have only *one* line of eigenvectors. This can only happen when two eigenvalues are equal. (On the other hand $A = I$ has equal eigenvalues and plenty of eigenvectors.) Similarly some n by n matrices don't have n independent eigenvectors. Without n eigenvectors, we don't have a basis. We can't write every $\boldsymbol{v}$ as a combination of eigenvectors. In the language of the next section, we can't diagonalize a matrix without n independent eigenvectors.

Good News, Bad News

Bad news first: If you add a row of A to another row, or exchange rows, the eigenvalues usually change. *Elimination does not preserve the* λ's. The triangular U has *its* eigenvalues sitting along the diagonal—they are the pivots. But they are not the eigenvalues of A! Eigenvalues are changed when row 1 is added to row 2:

$$U = \begin{bmatrix} 1 & 1 \\ 0 & 0 \end{bmatrix} \text{ has } \lambda = 0 \text{ and } \lambda = 1; \quad A = \begin{bmatrix} 1 & 1 \\ 1 & 1 \end{bmatrix} \text{ has } \lambda = 0 \text{ and } \lambda = 2.$$

Good news second: The *product* λ_1 *times* λ_2 *and the sum* $\lambda_1 + \lambda_2$ *can be found quickly from the matrix.* For this A, the product is 0 times 2. That agrees with the determinant (which is 0). The sum of eigenvalues is $0 + 2$. That agrees with the sum down the main diagonal (which is $1 + 1$). These quick checks always work:

6B ***The product of the n eigenvalues equals the determinant of A.***

6C ***The sum of the n eigenvalues equals the sum of the n diagonal entries of A.*** This sum along the main diagonal is called the ***trace*** of A:

$$\lambda_1 + \lambda_2 + \cdots + \lambda_n = \textbf{\textit{trace}} = a_{11} + a_{22} + \cdots + a_{nn}. \tag{6}$$

Those checks are very useful. They are proved in Problems 16–17 and again in the next section. They don't remove the pain of computing λ's. But when the computation is wrong, they generally tell us so. To compute correct λ's, go back to $\det(A - \lambda I) = 0$.

The determinant test makes the *product* of the λ's equal to the *product* of the pivots (assuming no row exchanges). But the sum of the λ's is not the sum of the pivots—as the example showed. The individual λ's have almost nothing to do with the individual pivots. In this new part of linear algebra, the key equation is really *nonlinear*: λ multiplies $\boldsymbol{x}$.

Imaginary Eigenvalues

One more bit of news (not too terrible). The eigenvalues might not be real numbers.

Example 4 ***The 90° rotation*** $Q = \left[\begin{smallmatrix} 0 & 1 \\ -1 & 0 \end{smallmatrix}\right]$ ***has no real eigenvectors or eigenvalues.*** *No vector $Q\boldsymbol{x}$ stays in the same direction as $\boldsymbol{x}$* (except the zero vector which is useless). There cannot be an eigenvector, unless we go to ***imaginary numbers***. Which we do.

To see how i can help, look at Q^2 which is $-I$. If Q is rotation through 90°, then Q^2 is rotation through 180°. Its eigenvalues are -1 and -1. (Certainly $-I\boldsymbol{x} = -1\boldsymbol{x}$.) Squaring Q is supposed to square its eigenvalues λ, so we must have $\lambda^2 = -1$. *The eigenvalues of the* 90° *rotation matrix Q are $+i$ and $-i$*, because $i^2 = -1$.

Those λ's come as usual from $\det(Q - \lambda I) = 0$. This equation gives $\lambda^2 + 1 = 0$. Its roots are $\lambda_1 = i$ and $\lambda_2 = -i$. They add to zero (which is the trace of Q). The product is $(i)(-i) = 1$ (which is the determinant).

We meet the imaginary number i also in the eigenvectors of Q:

$$\begin{bmatrix} 0 & 1 \\ -1 & 0 \end{bmatrix} \begin{bmatrix} 1 \\ i \end{bmatrix} = i \begin{bmatrix} 1 \\ i \end{bmatrix} \quad \text{and} \quad \begin{bmatrix} 0 & 1 \\ -1 & 0 \end{bmatrix} \begin{bmatrix} i \\ 1 \end{bmatrix} = -i \begin{bmatrix} i \\ 1 \end{bmatrix}.$$

Somehow these complex vectors $\boldsymbol{x}_1 = (1, i)$ and $\boldsymbol{x}_2 = (i, 1)$ keep their direction as they are rotated. Don't ask me how. This example makes the all-important point that real matrices can easily have complex eigenvalues. The particular eigenvalues i and $-i$ also illustrate two special properties of Q:

1. Q is an orthogonal matrix so the absolute value of each λ is $|\lambda| = 1$.

2. Q is a skew-symmetric matrix so each λ is pure imaginary.

A symmetric matrix ($A^{\mathrm{T}} = A$) can be compared to a real number. A skew-symmetric matrix ($A^{\mathrm{T}} = -A$) can be compared to an imaginary number. An orthogonal matrix ($A^{\mathrm{T}}A = I$) can be compared to a complex number with $|\lambda| = 1$. For the eigenvalues those are more than analogies—they are theorems to be proved in Section 6.4. The eigenvectors for all these special matrices are perpendicular. Somehow $(i, 1)$ and $(1, i)$ are perpendicular (in Chapter 10).

Eigshow

There is a MATLAB demo (just type **eigshow**), displaying the eigenvalue problem for a 2 by 2 matrix. It starts with the unit vector $x = (1, 0)$. *The mouse makes this vector move around the unit circle.* At the same time the screen shows Ax, in color and also moving. Possibly Ax is ahead of x. Possibly Ax is behind x. *Sometimes Ax is parallel to x.* At that parallel moment, $Ax = \lambda x$ (twice in the second figure).

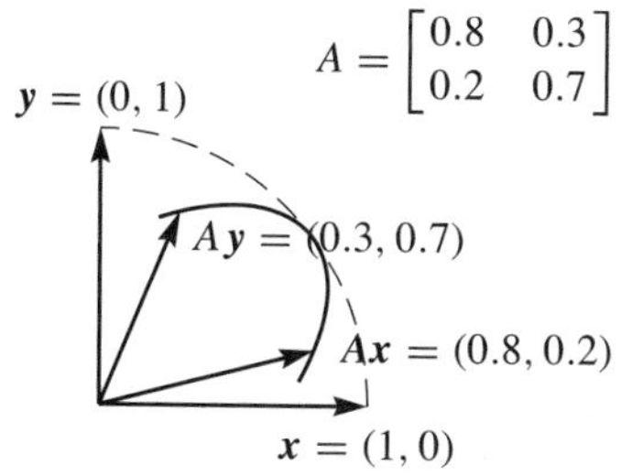

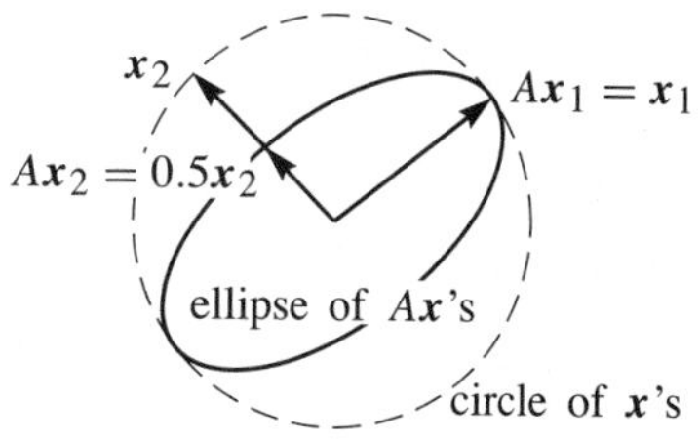

The eigenvalue λ is the length of Ax, when the unit eigenvector x is parallel. The built-in choices for A illustrate three possibilities:

1. There are *no real eigenvectors. Ax stays behind or ahead of x.* This means the eigenvalues and eigenvectors are complex, as they are for the rotation Q.

2. There is only *one* line of eigenvectors (unusual). The moving directions Ax and x meet but don't cross. This happens for the last 2 by 2 matrix below.

3. There are eigenvectors in *two* independent directions. This is typical! Ax crosses x at the first eigenvector x_1, and it crosses back at the second eigenvector x_2.

Suppose A is singular (rank one). Its column space is a line. The vector Ax has to stay on that line while x circles around. One eigenvector x is along the line. Another eigenvector appears when $Ax_2 = \mathbf{0}$. Zero is an eigenvalue of a singular matrix.

You can mentally follow x and Ax for these six matrices. How many eigenvectors and where? When does Ax go clockwise, instead of counterclockwise with x?

$$A = \begin{bmatrix} 2 & 0 \\ 0 & 1 \end{bmatrix} \quad \begin{bmatrix} 2 & 0 \\ 0 & -1 \end{bmatrix} \quad \begin{bmatrix} 0 & 1 \\ 1 & 0 \end{bmatrix} \quad \begin{bmatrix} 0 & 1 \\ -1 & 0 \end{bmatrix} \quad \begin{bmatrix} 1 & 1 \\ 1 & 1 \end{bmatrix} \begin{bmatrix} 1 & 1 \\ 0 & 1 \end{bmatrix}$$

■ REVIEW OF THE KEY IDEAS ■

1. $Ax = \lambda x$ says that x keeps the same direction when multiplied by A.
2. $Ax = \lambda x$ also says that $\det(A - \lambda I) = 0$. This determines n eigenvalues.
3. The eigenvalues of A^2 and A^{-1} are λ^2 and λ^{-1}, with the same eigenvectors.
4. The sum and product of the λ's equal the trace (sum of a_{ii}) and determinant.
5. Special matrices like projections P and rotations Q have special eigenvalues !

■ WORKED EXAMPLES ■

6.1 A Find the eigenvalues and eigenvectors of A and A^2 and A^{-1} and $A + 4I$:

$$A = \begin{bmatrix} 2 & -1 \\ -1 & 2 \end{bmatrix} \quad \text{and} \quad A^2 = \begin{bmatrix} 5 & -4 \\ -4 & 5 \end{bmatrix}.$$

Check the trace $\lambda_1 + \lambda_2$ and the determinant $\lambda_1 \lambda_2$ for A and also A^2.

Solution The eigenvalues of A come from $\det(A - \lambda I) = 0$:

$$\det(A - \lambda I) = \begin{vmatrix} 2-\lambda & -1 \\ -1 & 2-\lambda \end{vmatrix} = \lambda^2 - 4\lambda + 3 = 0.$$

This factors into $(\lambda - 1)(\lambda - 3) = 0$ so the eigenvalues of A are $\lambda_1 = 1$ and $\lambda_2 = 3$. For the trace, the sum $2 + 2$ agrees with $1 + 3$. The determinant 3 agrees with the product $\lambda_1 \lambda_2 = 3$. The eigenvectors come separately by solving $(A - \lambda I)x = \mathbf{0}$ which is $Ax = \lambda x$:

$$\boldsymbol{\lambda = 1:} \quad (A - I)x = \begin{bmatrix} 1 & -1 \\ -1 & 1 \end{bmatrix} \begin{bmatrix} x \\ y \end{bmatrix} = \begin{bmatrix} 0 \\ 0 \end{bmatrix} \text{ gives the eigenvector } x_1 = \begin{bmatrix} 1 \\ 1 \end{bmatrix}$$

$$\boldsymbol{\lambda = 3:} \quad (A - 3I)x = \begin{bmatrix} -1 & -1 \\ -1 & -1 \end{bmatrix} \begin{bmatrix} x \\ y \end{bmatrix} = \begin{bmatrix} 0 \\ 0 \end{bmatrix} \text{ gives the eigenvector } x_2 = \begin{bmatrix} 1 \\ -1 \end{bmatrix}$$

A^2 and A^{-1} and $A + 4I$ keep the *same eigenvectors*. Their eigenvalues are λ^2, λ^{-1}, $\lambda + 4$:

$$A^2 \text{ has } 1^2 = 1 \text{ and } 3^2 = 9 \quad A^{-1} \text{ has } \frac{1}{1} \text{ and } \frac{1}{3} \quad A + 4I \text{ has } 1 + 4 = 5 \text{ and } 3 + 4 = 7$$

The trace of A^2 is $5 + 5 = 1 + 9 = 10$. The determinant is $25 - 16 = 9$.

Notes for later sections: A has *orthogonal eigenvectors* (Section 6.4 on symmetric matrices). A can be *diagonalized* (Section 6.2). A is *similar* to any 2 by 2 matrix with eigenvalues 1 and 3 (Section 6.6). A is a *positive definite matrix* (Section 6.5) since $A = A^{\mathrm{T}}$ and the λ's are positive.

6.1 B For which real numbers c does this matrix A have (**a**) two real eigenvalues and eigenvectors (**b**) a repeated eigenvalue with only one eigenvector (**c**) two complex eigenvalues and eigenvectors?

$$A = \begin{bmatrix} 2 & -c \\ -1 & 2 \end{bmatrix} \qquad A^{\mathrm{T}}A = \begin{bmatrix} 5 & -2c-2 \\ -2c-2 & 4+c^2 \end{bmatrix}.$$

What is the determinant of $A^{\mathrm{T}}A$ by the product rule? What is its trace? How do you know that $A^{\mathrm{T}}A$ doesn't have a negative eigenvalue?

Solution The determinant of A is $4-c$. The determinant of $A-\lambda I$ is

$$\det\begin{bmatrix} 2-\lambda & -c \\ -1 & 2-\lambda \end{bmatrix} = \lambda^2 - 4\lambda + (4-c) = 0.$$

The formula for the roots of a quadratic is

$$\lambda = \frac{-b \pm \sqrt{b^2-4ac}}{2} = \frac{4 \pm \sqrt{16-16+4c}}{2} = 2 \pm \sqrt{c}.$$

Check the trace (it is 4) and the determinant $(2+\sqrt{c})(2-\sqrt{c}) = 4-c$. The eigenvalues are real and different for $c > 0$. There are two independent eigenvectors $(\sqrt{c}, 1)$ and $(-\sqrt{c}, 1)$. Both roots become $\lambda = 2$ for $c = 0$, and there is only one independent eigenvector $(0, 1)$. Both eigenvalues are complex for $c < 0$ and the eigenvectors $(\sqrt{c}, 1)$ and $(-\sqrt{c}, 1)$ become complex.

The determinant of $A^{\mathrm{T}}A$ is $\det(A^{\mathrm{T}})\det(A) = (4-c)^2$. The trace of $A^{\mathrm{T}}A$ is $5+4+c^2$. If one eigenvalue is negative, the other must be positive to produce this trace $\lambda_1+\lambda_2 = 9+c^2$. But then negative times positive would give a negative determinant. In fact every $A^{\mathrm{T}}A$ has real nonnegative eigenvalues (Section 6.5).

Problem Set 6.1

1 The example at the start of the chapter has

$$A = \begin{bmatrix} .8 & .3 \\ .2 & .7 \end{bmatrix} \quad \text{and} \quad A^2 = \begin{bmatrix} .70 & .45 \\ .30 & .55 \end{bmatrix} \quad \text{and} \quad A^\infty = \begin{bmatrix} .6 & .6 \\ .4 & .4 \end{bmatrix}.$$

The matrix A^2 is halfway between A and A^∞. Explain why $A^2 = \frac{1}{2}(A + A^\infty)$ from the eigenvalues and eigenvectors of these three matrices.

(a) Show from A how a row exchange can produce different eigenvalues.

(b) Why is a zero eigenvalue *not* changed by the steps of elimination?

2 Find the eigenvalues and the eigenvectors of these two matrices:

$$A = \begin{bmatrix} 1 & 4 \\ 2 & 3 \end{bmatrix} \quad \text{and} \quad A+I = \begin{bmatrix} 2 & 4 \\ 2 & 4 \end{bmatrix}.$$

$A+I$ has the _____ eigenvectors as A. Its eigenvalues are _____ by 1.

3 Compute the eigenvalues and eigenvectors of A and A^{-1}:

$$A = \begin{bmatrix} 0 & 2 \\ 2 & 3 \end{bmatrix} \quad \text{and} \quad A^{-1} = \begin{bmatrix} -3/4 & 1/2 \\ 1/2 & 0 \end{bmatrix}.$$

A^{-1} has the ______ eigenvectors as A. When A has eigenvalues λ_1 and λ_2, its inverse has eigenvalues ______.

4 Compute the eigenvalues and eigenvectors of A and A^2:

$$A = \begin{bmatrix} -1 & 3 \\ 2 & 0 \end{bmatrix} \quad \text{and} \quad A^2 = \begin{bmatrix} 7 & -3 \\ -2 & 6 \end{bmatrix}.$$

A^2 has the same ______ as A. When A has eigenvalues λ_1 and λ_2, A^2 has eigenvalues ______.

5 Find the eigenvalues of A and B and $A + B$:

$$A = \begin{bmatrix} 1 & 0 \\ 1 & 1 \end{bmatrix} \quad \text{and} \quad B = \begin{bmatrix} 1 & 1 \\ 0 & 1 \end{bmatrix} \quad \text{and} \quad A + B = \begin{bmatrix} 2 & 1 \\ 1 & 2 \end{bmatrix}.$$

Eigenvalues of $A+B$ (are equal to)(are not equal to) eigenvalues of A plus eigenvalues of B.

6 Find the eigenvalues of A and B and AB and BA:

$$A = \begin{bmatrix} 1 & 0 \\ 1 & 1 \end{bmatrix} \quad \text{and} \quad B = \begin{bmatrix} 1 & 1 \\ 0 & 1 \end{bmatrix} \quad \text{and} \quad AB = \begin{bmatrix} 1 & 1 \\ 1 & 2 \end{bmatrix} \quad \text{and} \quad BA = \begin{bmatrix} 2 & 1 \\ 1 & 1 \end{bmatrix}.$$

Eigenvalues of AB (are equal to)(are not equal to) eigenvalues of A times eigenvalues of B. Eigenvalues of AB (are equal to)(are not equal to) eigenvalues of BA.

7 Elimination produces $A = LU$. The eigenvalues of U are on its diagonal; they are the ______. The eigenvalues of L are on its diagonal; they are all ______. The eigenvalues of A are not the same as ______.

8 (a) If you know $\boldsymbol{x}$ is an eigenvector, the way to find λ is to ______.

(b) If you know λ is an eigenvalue, the way to find $\boldsymbol{x}$ is to ______.

9 What do you do to $A\boldsymbol{x} = \lambda\boldsymbol{x}$, in order to prove (a), (b), and (c)?

(a) λ^2 is an eigenvalue of A^2, as in Problem 4.

(b) λ^{-1} is an eigenvalue of A^{-1}, as in Problem 3.

(c) $\lambda + 1$ is an eigenvalue of $A + I$, as in Problem 2.

10 Find the eigenvalues and eigenvectors for both of these Markov matrices A and A^∞. Explain why A^{100} is close to A^∞:

$$A = \begin{bmatrix} .6 & .2 \\ .4 & .8 \end{bmatrix} \quad \text{and} \quad A^\infty = \begin{bmatrix} 1/3 & 1/3 \\ 2/3 & 2/3 \end{bmatrix}.$$

11 Here is a strange fact about 2 by 2 matrices with eigenvalues $\lambda_1 \neq \lambda_2$: The columns of $A - \lambda_1 I$ are multiples of the eigenvector $\boldsymbol{x}_2$. Any idea why this should be?

12 Find the eigenvalues and eigenvectors for the projection matrices P and P^{100}:

$$P = \begin{bmatrix} .2 & .4 & 0 \\ .4 & .8 & 0 \\ 0 & 0 & 1 \end{bmatrix}.$$

If two eigenvectors share the same λ, so do all their linear combinations. Find an eigenvector of P with no zero components.

13 From the unit vector $\boldsymbol{u} = \left(\frac{1}{6}, \frac{1}{6}, \frac{3}{6}, \frac{5}{6}\right)$ construct the rank one projection matrix $P = \boldsymbol{u}\boldsymbol{u}^{\mathrm{T}}$.

(a) Show that $P\boldsymbol{u} = \boldsymbol{u}$. Then $\boldsymbol{u}$ is an eigenvector with $\lambda = 1$.

(b) If $\boldsymbol{v}$ is perpendicular to $\boldsymbol{u}$ show that $P\boldsymbol{v} = \boldsymbol{0}$. Then $\lambda = 0$.

(c) Find three independent eigenvectors of P all with eigenvalue $\lambda = 0$.

14 Solve $\det(Q - \lambda I) = 0$ by the quadratic formula to reach $\lambda = \cos\theta \pm i \sin\theta$:

$$Q = \begin{bmatrix} \cos\theta & -\sin\theta \\ \sin\theta & \cos\theta \end{bmatrix} \quad \text{rotates the } xy \text{ plane by the angle } \theta.$$

Find the eigenvectors of Q by solving $(Q - \lambda I)\boldsymbol{x} = \boldsymbol{0}$. Use $i^2 = -1$.

15 Every permutation matrix leaves $\boldsymbol{x} = (1, 1, \ldots, 1)$ unchanged. Then $\lambda = 1$. Find two more λ's for these permutations:

$$P = \begin{bmatrix} 0 & 1 & 0 \\ 0 & 0 & 1 \\ 1 & 0 & 0 \end{bmatrix} \quad \text{and} \quad P = \begin{bmatrix} 0 & 0 & 1 \\ 0 & 1 & 0 \\ 1 & 0 & 0 \end{bmatrix}.$$

16 Prove that the determinant of A equals the product $\lambda_1 \lambda_2 \cdots \lambda_n$. Start with the polynomial $\det(A - \lambda I)$ separated into its n factors. Then set $\lambda =$ _____:

$$\det(A - \lambda I) = (\lambda_1 - \lambda)(\lambda_2 - \lambda) \cdots (\lambda_n - \lambda) \quad \text{so} \quad \det A = \underline{\qquad}.$$

17 The sum of the diagonal entries (the *trace*) equals the sum of the eigenvalues:

$$A = \begin{bmatrix} a & b \\ c & d \end{bmatrix} \quad \text{has} \quad \det(A - \lambda I) = \lambda^2 - (a + d)\lambda + ad - bc = 0.$$

If A has $\lambda_1 = 3$ and $\lambda_2 = 4$ then $\det(A - \lambda I) =$ _____. The quadratic formula gives the eigenvalues $\lambda = (a + d + \sqrt{})/2$ and $\lambda =$ _____. Their sum is _____.

18 If A has $\lambda_1 = 4$ and $\lambda_2 = 5$ then $\det(A - \lambda I) = (\lambda - 4)(\lambda - 5) = \lambda^2 - 9\lambda + 20$. Find three matrices that have trace $a + d = 9$ and determinant 20 and $\lambda = 4, 5$.

19 A 3 by 3 matrix B is known to have eigenvalues $0, 1, 2$. This information is enough to find three of these:

(a) the rank of B

(b) the determinant of $B^{\mathrm{T}}B$

(c) the eigenvalues of $B^{\mathrm{T}}B$

(d) the eigenvalues of $(B + I)^{-1}$.

20 Choose the second row of $A = \begin{bmatrix} 0 & 1 \\ * & * \end{bmatrix}$ so that A has eigenvalues 4 and 7.

21 Choose a, b, c, so that $\det(A - \lambda I) = 9\lambda - \lambda^3$. Then the eigenvalues are $-3, 0, 3$:

$$A = \begin{bmatrix} 0 & 1 & 0 \\ 0 & 0 & 1 \\ a & b & c \end{bmatrix}.$$

22 ***The eigenvalues of A equal the eigenvalues of*** A^{T}. This is because $\det(A - \lambda I)$ equals $\det(A^{\mathrm{T}} - \lambda I)$. That is true because ______. Show by an example that the eigenvectors of A and A^{T} are *not* the same.

23 Construct any 3 by 3 Markov matrix M: positive entries down each column add to 1. If $\boldsymbol{e} = (1, 1, 1)$ verify that $M^{\mathrm{T}}\boldsymbol{e} = \boldsymbol{e}$. By Problem 22, $\lambda = 1$ is also an eigenvalue of M. Challenge: A 3 by 3 singular Markov matrix with trace $\frac{1}{2}$ has eigenvalues $\lambda =$ ______.

24 Find three 2 by 2 matrices that have $\lambda_1 = \lambda_2 = 0$. The trace is zero and the determinant is zero. The matrix A might not be 0 but check that $A^2 = 0$.

25 This matrix is singular with rank one. Find three λ's and three eigenvectors:

$$A = \begin{bmatrix} 1 \\ 2 \\ 1 \end{bmatrix} \begin{bmatrix} 2 & 1 & 2 \end{bmatrix} = \begin{bmatrix} 2 & 1 & 2 \\ 4 & 2 & 4 \\ 2 & 1 & 2 \end{bmatrix}.$$

26 Suppose A and B have the same eigenvalues $\lambda_1, \ldots, \lambda_n$ with the same independent eigenvectors $\boldsymbol{x}_1, \ldots, \boldsymbol{x}_n$. Then $A = B$. *Reason*: Any vector $\boldsymbol{x}$ is a combination $c_1\boldsymbol{x}_1 + \cdots + c_n\boldsymbol{x}_n$. What is $A\boldsymbol{x}$? What is $B\boldsymbol{x}$?

27 The block B has eigenvalues $1, 2$ and C has eigenvalues $3, 4$ and D has eigenvalues $5, 7$. Find the eigenvalues of the 4 by 4 matrix A:

$$A = \begin{bmatrix} B & C \\ 0 & D \end{bmatrix} = \begin{bmatrix} 0 & 1 & 3 & 0 \\ -2 & 3 & 0 & 4 \\ 0 & 0 & 6 & 1 \\ 0 & 0 & 1 & 6 \end{bmatrix}.$$

28 Find the rank and the four eigenvalues of

$$A = \begin{bmatrix} 1 & 1 & 1 & 1 \\ 1 & 1 & 1 & 1 \\ 1 & 1 & 1 & 1 \\ 1 & 1 & 1 & 1 \end{bmatrix} \quad \text{and} \quad C = \begin{bmatrix} 1 & 0 & 1 & 0 \\ 0 & 1 & 0 & 1 \\ 1 & 0 & 1 & 0 \\ 0 & 1 & 0 & 1 \end{bmatrix}.$$

29 Subtract I from the previous A. Find the λ's and then the determinant:

$$B = A - I = \begin{bmatrix} 0 & 1 & 1 & 1 \\ 1 & 0 & 1 & 1 \\ 1 & 1 & 0 & 1 \\ 1 & 1 & 1 & 0 \end{bmatrix}.$$

When A (all ones) is 5 by 5, the eigenvalues of A and $B = A - I$ are ____ and ____.

30 (Review) Find the eigenvalues of A, B, and C:

$$A = \begin{bmatrix} 1 & 2 & 3 \\ 0 & 4 & 5 \\ 0 & 0 & 6 \end{bmatrix} \quad \text{and} \quad B = \begin{bmatrix} 0 & 0 & 1 \\ 0 & 2 & 0 \\ 3 & 0 & 0 \end{bmatrix} \quad \text{and} \quad C = \begin{bmatrix} 2 & 2 & 2 \\ 2 & 2 & 2 \\ 2 & 2 & 2 \end{bmatrix}.$$

31 When $a + b = c + d$ show that $(1, 1)$ is an eigenvector and find both eigenvalues of

$$A = \begin{bmatrix} a & b \\ c & d \end{bmatrix}.$$

32 When P exchanges rows 1 and 2 *and* columns 1 and 2, the eigenvalues don't change. Find eigenvectors of A and PAP for $\lambda = 11$:

$$A = \begin{bmatrix} 1 & 2 & 1 \\ 3 & 6 & 3 \\ 4 & 8 & 4 \end{bmatrix} \quad \text{and} \quad PAP = \begin{bmatrix} 6 & 3 & 3 \\ 2 & 1 & 1 \\ 8 & 4 & 4 \end{bmatrix}.$$

33 Suppose A has eigenvalues $0, 3, 5$ with independent eigenvectors $\boldsymbol{u}, \boldsymbol{v}, \boldsymbol{w}$.

(a) Give a basis for the nullspace and a basis for the column space.

(b) Find a particular solution to $A\boldsymbol{x} = \boldsymbol{v} + \boldsymbol{w}$. Find all solutions.

(c) Show that $A\boldsymbol{x} = \boldsymbol{u}$ has no solution. (If it did then ____ would be in the column space.)

34 *Is there a real 2 by 2 matrix* (other than I) *with* $A^3 = I$? Its eigenvalues must satisfy $\lambda^3 = I$. They can be $e^{2\pi i/3}$ and $e^{-2\pi i/3}$. What trace and determinant would this give? Construct A.

35 Find the eigenvalues of this permutation matrix P. Which vectors are not changed by the permutation? They are eigenvectors for $\lambda = 1$. Can you find two more eigenvectors?

$$P = \begin{bmatrix} 0 & 1 & 0 \\ 0 & 0 & 1 \\ 1 & 0 & 0 \end{bmatrix}.$$

36 There are six 3 by 3 permutation matrices P. What numbers can be the *determinants* of P? What numbers can be *pivots*? What numbers can be the *trace* of P? What *four numbers* can be eigenvalues of P?

DIAGONALIZING A MATRIX ■ 6.2

When $\boldsymbol{x}$ is an eigenvector, multiplication by A is just multiplication by a single number: $A\boldsymbol{x} = \lambda\boldsymbol{x}$. All the difficulties of matrices are swept away. Instead of an interconnected system, we can follow the eigenvectors separately. It is like having a *diagonal matrix*, with no off-diagonal interconnections. The 100th power of a diagonal matrix is easy.

The point of this section is very direct. ***The matrix A turns into a diagonal matrix Λ when we use the eigenvectors properly***. This is the matrix form of our key idea. We start right off with that one essential computation.

6D Diagonalization Suppose the n by n matrix A has n linearly independent eigenvectors $\boldsymbol{x}_1, \ldots, \boldsymbol{x}_n$. Put them into the columns of an ***eigenvector matrix*** S. Then $S^{-1}AS$ is the ***eigenvalue matrix*** Λ:

$$S^{-1}AS = \Lambda = \begin{bmatrix} \lambda_1 & & \\ & \ddots & \\ & & \lambda_n \end{bmatrix}. \tag{1}$$

The matrix A is "diagonalized." We use capital lambda for the eigenvalue matrix, because of the small λ's (the eigenvalues) on its diagonal.

Proof Multiply A times its eigenvectors, which are the columns of S. The first column of AS is $A\boldsymbol{x}_1$. That is $\lambda_1\boldsymbol{x}_1$. Each column of S is multiplied by its eigenvalue:

$$AS = A\begin{bmatrix} \boldsymbol{x}_1 & \cdots & \boldsymbol{x}_n \end{bmatrix} = \begin{bmatrix} \lambda_1\boldsymbol{x}_1 & \cdots & \lambda_n\boldsymbol{x}_n \end{bmatrix}.$$

The trick is to split this matrix AS into S times Λ:

$$\begin{bmatrix} \lambda_1 x_1 & \cdots & \lambda_n x_n \end{bmatrix} = \begin{bmatrix} x_1 & \cdots & x_n \end{bmatrix} \begin{bmatrix} \lambda_1 & & \\ & \ddots & \\ & & \lambda_n \end{bmatrix} = S\Lambda.$$

Keep those matrices in the right order! Then λ_1 multiplies the first column x_1, as shown. The diagonalization is complete, and we can write $AS = S\Lambda$ in two good ways:

$$AS = S\Lambda \quad \text{is} \quad S^{-1}AS = \Lambda \quad \text{or} \quad A = S\Lambda S^{-1}. \tag{2}$$

The matrix S has an inverse, because its columns (the eigenvectors of A) were assumed to be linearly independent. *Without n independent eigenvectors, we can't diagonalize.*

The matrices A and Λ have the same eigenvalues $\lambda_1, \ldots, \lambda_n$. The eigenvectors are different. The job of the original eigenvectors was to diagonalize A—those eigenvectors of A went into S. The new eigenvectors, for the diagonal matrix Λ, are just the columns of I. By diagonalizing A and reaching Λ, we can solve differential equations or difference equations or even $Ax = b$.

Example 1 The projection matrix $P = \begin{bmatrix} .5 & .5 \\ .5 & .5 \end{bmatrix}$ has $\lambda = 1$ and 0. Put the eigenvectors $(1, 1)$ and $(-1, 1)$ into S. Then $S^{-1}PS$ is the eigenvalue matrix Λ:

$$\underset{S^{-1}}{\begin{bmatrix} .5 & .5 \\ -.5 & .5 \end{bmatrix}} \underset{P}{\begin{bmatrix} .5 & .5 \\ .5 & .5 \end{bmatrix}} \underset{S}{\begin{bmatrix} 1 & -1 \\ 1 & 1 \end{bmatrix}} = \underset{\Lambda}{\begin{bmatrix} 1 & 0 \\ 0 & 0 \end{bmatrix}}$$

The original projection satisfied $P^2 = P$. The new projection satisfies $\Lambda^2 = \Lambda$. The column space has swung around from $(1, 1)$ to $(1, 0)$. The nullspace has swung around from $(-1, 1)$ to $(0, 1)$. Diagonalization lines up the eigenvectors with the xy axes.

Here are four small remarks about diagonalization, before the applications.

Remark 1 Suppose the numbers $\lambda_1, \ldots, \lambda_n$ are all different. Then it is automatic that the eigenvectors $x_1, \ldots, x_n$ are independent. See **6E** below. Therefore *any matrix that has no repeated eigenvalues can be diagonalized.*

Remark 2 The eigenvector matrix S is not unique. ***We can multiply eigenvectors by any nonzero constants.*** Suppose we multiply the columns of S by 5 and -1. Divide the rows of S^{-1} by 5 and -1 to find the new inverse:

$$S_{\text{new}}^{-1} P S_{\text{new}} = \begin{bmatrix} .1 & .1 \\ .5 & -.5 \end{bmatrix} \begin{bmatrix} .5 & .5 \\ .5 & .5 \end{bmatrix} \begin{bmatrix} 5 & 1 \\ 5 & -1 \end{bmatrix} = \begin{bmatrix} 1 & 0 \\ 0 & 0 \end{bmatrix} = \text{same } \Lambda.$$

The extreme case is $A = I$, when every vector is an eigenvector. Any invertible matrix S can be the eigenvector matrix. Then $S^{-1}IS = I$ (which is Λ).

Remark 3 To diagonalize A we *must* use an eigenvector matrix. From $S^{-1}AS = \Lambda$ we know that $AS = S\Lambda$. Suppose the first column of S is x. Then the first columns of AS and $S\Lambda$ are Ax and $\lambda_1 x$. For those to be equal, x must be an eigenvector.

The eigenvectors in S come in the same order as the eigenvalues in Λ. To reverse the order in S and Λ, put $(-1, 1)$ before $(1, 1)$:

$$\begin{bmatrix} -.5 & .5 \\ .5 & .5 \end{bmatrix}\begin{bmatrix} .5 & .5 \\ .5 & .5 \end{bmatrix}\begin{bmatrix} -1 & 1 \\ 1 & 1 \end{bmatrix} = \begin{bmatrix} 0 & 0 \\ 0 & 1 \end{bmatrix} = \Lambda \text{ in the new order } 0, 1.$$

Remark 4 (repeated warning for repeated eigenvalues) Some matrices have too few eigenvectors. *Those matrices are not diagonalizable.* Here are two examples:

$$A = \begin{bmatrix} 1 & -1 \\ 1 & -1 \end{bmatrix} \quad \text{and} \quad A = \begin{bmatrix} 0 & 1 \\ 0 & 0 \end{bmatrix}.$$

Their eigenvalues happen to be 0 and 0. Nothing is special about $\lambda = 0$—it is the repetition of λ that counts. All eigenvectors of the second matrix are multiples of $(1, 0)$!

$$Ax = 0x \quad \text{means} \quad \begin{bmatrix} 0 & 1 \\ 0 & 0 \end{bmatrix}\begin{bmatrix} \\ x \\ \\ \end{bmatrix} = \begin{bmatrix} 0 \\ 0 \end{bmatrix} \quad \text{and} \quad x = c\begin{bmatrix} 1 \\ 0 \end{bmatrix}.$$

There is no second eigenvector, so the unusual matrix A cannot be diagonalized. This matrix is the best example to test any statement about eigenvectors. In many true-false questions, this matrix leads to *false*.

Remember that there is no connection between invertibility and diagonalizability:

- ***Invertibility*** is concerned with the ***eigenvalues*** (zero or not).
- ***Diagonalizability*** is concerned with the ***eigenvectors*** (too few or enough).

Each eigenvalue has at least one eigenvector! If $(A - \lambda I)x = \mathbf{0}$ leads you to $x = \mathbf{0}$, then λ is *not* an eigenvalue. Look for a mistake in solving $\det(A - \lambda I) = 0$. **The eigenvectors for n different λ's are independent and A is diagonalizable.**

6E (Independent x from different λ) Eigenvectors $x_1, \ldots, x_j$ that correspond to distinct (all different) eigenvalues are linearly independent. An n by n matrix that has n different eigenvalues (no repeated λ's) must be diagonalizable.

Proof Suppose $c_1x_1 + c_2x_2 = \mathbf{0}$. Multiply by A to find $c_1\lambda_1x_1 + c_2\lambda_2x_2 = \mathbf{0}$. Multiply by λ_2 to find $c_1\lambda_2x_1 + c_2\lambda_2x_2 = \mathbf{0}$. Now subtract one from the other:

$$\text{Subtraction leaves} \quad (\lambda_1 - \lambda_2)c_1x_1 = \mathbf{0}. \quad \text{Therefore } c_1 = 0.$$

Since the λ's are different and $x_1 \neq \mathbf{0}$, we are forced to the conclusion that $c_1 = 0$. Similarly $c_2 = 0$. No other combination gives $c_1x_1 + c_2x_2 = \mathbf{0}$, so the eigenvectors x_1 and x_2 must be independent.

This proof extends directly to j eigenvectors. Suppose $c_1\boldsymbol{x}_1 + \cdots + c_j\boldsymbol{x}_j = \boldsymbol{0}$. Multiply by A, multiply by λ_j, and subtract. This removes $\boldsymbol{x}_j$. Now multiply by A and by λ_{j-1} and subtract. This removes $\boldsymbol{x}_{j-1}$. Eventually only $\boldsymbol{x}_1$ is left:

$$(\lambda_1 - \lambda_2)\cdots(\lambda_1 - \lambda_j)c_1\boldsymbol{x}_1 = \boldsymbol{0} \quad \text{which forces} \quad c_1 = 0. \tag{3}$$

Similarly every $c_i = 0$. When the λ's are all different, the eigenvectors are independent.

With n different eigenvalues, the full set of eigenvectors goes into the columns of the eigenvector matrix S. Then A is diagonalized.

Example 2 The Markov matrix $A = \begin{bmatrix} .8 & .3 \\ .2 & .7 \end{bmatrix}$ in the last section had $\lambda_1 = 1$ and $\lambda_2 = .5$. Here is $A = S\Lambda S^{-1}$ with those eigenvalues in Λ:

$$\begin{bmatrix} .8 & .3 \\ .2 & .7 \end{bmatrix} = \begin{bmatrix} .6 & 1 \\ .4 & -1 \end{bmatrix} \begin{bmatrix} 1 & 0 \\ 0 & .5 \end{bmatrix} \begin{bmatrix} 1 & 1 \\ .4 & -.6 \end{bmatrix} = S\Lambda S^{-1}.$$

The eigenvectors $(.6, .4)$ and $(1, -1)$ are in the columns of S. They are also the eigenvectors of A^2, because $A^2\boldsymbol{x} = A\lambda\boldsymbol{x} = \lambda^2\boldsymbol{x}$. Then A^2 has the same S, and ***the eigenvalue matrix of A^2 is Λ^2***:

$$A^2 = S\Lambda S^{-1}S\Lambda S^{-1} = S\Lambda^2 S^{-1}.$$

Just keep going, and you see why the high powers A^k approach a "steady state":

$$A^k = S\Lambda^k S^{-1} = \begin{bmatrix} .6 & 1 \\ .4 & -1 \end{bmatrix} \begin{bmatrix} 1^k & 0 \\ 0 & (.5)^k \end{bmatrix} \begin{bmatrix} 1 & 1 \\ .4 & -.6 \end{bmatrix}.$$

As k gets larger, $(.5)^k$ gets smaller. In the limit it disappears completely. That limit is

$$A^\infty = \begin{bmatrix} .6 & 1 \\ .4 & -1 \end{bmatrix} \begin{bmatrix} 1 & 0 \\ 0 & 0 \end{bmatrix} \begin{bmatrix} 1 & 1 \\ .4 & -.6 \end{bmatrix} = \begin{bmatrix} .6 & .6 \\ .4 & .4 \end{bmatrix}.$$

The limit has the eigenvector $\boldsymbol{x}_1$ in both columns. We saw this A^∞ on the very first page of the chapter. Now we see it more quickly from powers like $A^{100} = S\Lambda^{100}S^{-1}$.

Eigenvalues of AB and $A + B$

The first guess about the eigenvalues of AB is not true. An eigenvalue λ of A times an eigenvalue β of B usually does *not* give an eigenvalue of AB. It is very tempting to think it should. *Here is a false proof*:

$$AB\boldsymbol{x} = A\beta\boldsymbol{x} = \beta A\boldsymbol{x} = \beta\lambda\boldsymbol{x}. \tag{4}$$

It seems that β times λ is an eigenvalue. When x is an eigenvector for A and B, this proof is correct. ***The mistake is to expect that A and B automatically share the same eigenvector x.*** Usually they don't. Eigenvectors of A are not generally eigenvectors of B. A and B can have all eigenvalues $\lambda = 0$ and $\beta = 0$ while 1 is an eigenvalue of AB:

$$A = \begin{bmatrix} 0 & 1 \\ 0 & 0 \end{bmatrix} \quad \text{and} \quad B = \begin{bmatrix} 0 & 0 \\ 1 & 0 \end{bmatrix}; \quad \text{then} \quad AB = \begin{bmatrix} 1 & 0 \\ 0 & 0 \end{bmatrix} \quad \text{and} \quad A + B = \begin{bmatrix} 0 & 1 \\ 1 & 0 \end{bmatrix}.$$

For the same reason, the eigenvalues of $A+B$ are generally not $\lambda+\beta$. Here $\lambda+\beta=0$ while $A+B$ has eigenvalues 1 and -1. (At least they add to zero.)

The false proof suggests what is true. Suppose x really is an eigenvector for both A and B. Then we do have $ABx = \lambda\beta x$. Sometimes all n eigenvectors are shared, and we *can* multiply eigenvalues. The test $AB = BA$ for shared eigenvectors is important in quantum mechanics—time out to mention this application of linear algebra:

6F ***Commuting matrices share eigenvectors*** Suppose A and B can be diagonalized. They share the same eigenvector matrix S if and only if $AB = BA$.

Heisenberg's uncertainty principle In quantum mechanics, the position matrix P and the momentum matrix Q do not commute. In fact $QP - PQ = I$ (these are infinite matrices). Then we cannot have $Px = \mathbf{0}$ at the same time as $Q\mathbf{x} = \mathbf{0}$ (unless $x = \mathbf{0}$). If we knew the position exactly, we could not also know the momentum exactly. Problem 32 derives Heisenberg's uncertainty principle from the Schwarz inequality.

Fibonacci Numbers

We present a famous example, where eigenvalues tell how fast the Fibonacci numbers grow. ***Every new Fibonacci number is the sum of the two previous F's***:

The sequence $0, 1, 1, 2, 3, 5, 8, 13, \ldots$ ***comes from*** $F_{k+2} = F_{k+1} + F_k$.

These numbers turn up in a fantastic variety of applications. Plants and trees grow in a spiral pattern, and a pear tree has 8 growths for every 3 turns. For a willow those numbers can be 13 and 5. The champion is a sunflower of Daniel O'Connell, which had 233 seeds in 144 loops. Those are the Fibonacci numbers F_{13} and F_{12}. Our problem is more basic.

Problem: Find the Fibonacci number F_{100} The slow way is to apply the rule $F_{k+2} = F_{k+1} + F_k$ one step at a time. By adding $F_6 = 8$ to $F_7 = 13$ we reach $F_8 = 21$. Eventually we come to F_{100}. Linear algebra gives a better way.

The key is to begin with a matrix equation $\mathbf{u}_{k+1} = A\boldsymbol{u}_k$. That is a one-step rule for vectors, while Fibonacci gave a two-step rule for scalars. We match them by putting two Fibonacci numbers into a vector:

$$\text{Let } \boldsymbol{u}_k = \begin{bmatrix} F_{k+1} \\ F_k \end{bmatrix}. \text{ The rule } \begin{matrix} F_{k+2} = F_{k+1} + F_k \\ F_{k+1} = F_{k+1} \end{matrix} \text{ becomes } \boldsymbol{u}_{k+1} = \begin{bmatrix} 1 & 1 \\ 1 & 0 \end{bmatrix} \boldsymbol{u}_k. \tag{5}$$

Every step multiplies by $A = \begin{bmatrix} 1 & 1 \\ 1 & 0 \end{bmatrix}$. After 100 steps we reach $\boldsymbol{u}_{100} = A^{100}\boldsymbol{u}_0$:

$$\boldsymbol{u}_0 = \begin{bmatrix} 1 \\ 0 \end{bmatrix}, \quad \boldsymbol{u}_1 = \begin{bmatrix} 1 \\ 1 \end{bmatrix}, \quad \boldsymbol{u}_2 = \begin{bmatrix} 2 \\ 1 \end{bmatrix}, \quad \boldsymbol{u}_3 = \begin{bmatrix} 3 \\ 2 \end{bmatrix}, \quad \ldots, \quad \boldsymbol{u}_{100} = \begin{bmatrix} F_{101} \\ F_{100} \end{bmatrix}.$$

This problem is just right for eigenvalues. Subtract λ from the diagonal of A:

$$A - \lambda I = \begin{bmatrix} 1-\lambda & 1 \\ 1 & -\lambda \end{bmatrix} \quad \text{leads to} \quad \det(A - \lambda I) = \lambda^2 - \lambda - 1.$$

The equation $\lambda^2 - \lambda - 1 = 0$ is solved by the quadratic formula $\left(-b \pm \sqrt{b^2 - 4ac}\right)/2a$:

$$\lambda_1 = \frac{1+\sqrt{5}}{2} \approx 1.618 \quad \text{and} \quad \lambda_2 = \frac{1-\sqrt{5}}{2} \approx -.618.$$

These eigenvalues λ_1 and λ_2 lead to eigenvectors $\boldsymbol{x}_1$ and $\boldsymbol{x}_2$. This completes step 1:

$$\begin{bmatrix} 1-\lambda_1 & 1 \\ 1 & -\lambda_1 \end{bmatrix} \begin{bmatrix} \boldsymbol{x}_1 \end{bmatrix} = \begin{bmatrix} 0 \\ 0 \end{bmatrix} \quad \text{when} \quad \boldsymbol{x}_1 = \begin{bmatrix} \lambda_1 \\ 1 \end{bmatrix}$$

$$\begin{bmatrix} 1-\lambda_2 & 1 \\ 1 & -\lambda_2 \end{bmatrix} \begin{bmatrix} \boldsymbol{x}_2 \end{bmatrix} = \begin{bmatrix} 0 \\ 0 \end{bmatrix} \quad \text{when} \quad \boldsymbol{x}_2 = \begin{bmatrix} \lambda_2 \\ 1 \end{bmatrix}.$$

Step 2 finds the combination of those eigenvectors that gives $\boldsymbol{u}_0 = (1, 0)$:

$$\begin{bmatrix} 1 \\ 0 \end{bmatrix} = \frac{1}{\lambda_1 - \lambda_2}\left(\begin{bmatrix} \lambda_1 \\ 1 \end{bmatrix} - \begin{bmatrix} \lambda_2 \\ 1 \end{bmatrix} \right) \quad \text{or} \quad \boldsymbol{u}_0 = \frac{\boldsymbol{x}_1 - \boldsymbol{x}_2}{\lambda_1 - \lambda_2}. \tag{6}$$

Step 3 multiplies $\boldsymbol{u}_0$ by A^{100} to find $\boldsymbol{u}_{100}$. The eigenvectors $\boldsymbol{x}_1$ and $\boldsymbol{x}_2$ stay separate! They are multiplied by $(\lambda_1)^{100}$ and $(\lambda_2)^{100}$:

$$\boldsymbol{u}_{100} = \frac{(\lambda_1)^{100}\boldsymbol{x}_1 - (\lambda_2)^{100}\boldsymbol{x}_2}{\lambda_1 - \lambda_2}. \tag{7}$$

We want F_{100} = second component of $\boldsymbol{u}_{100}$. The second components of $\boldsymbol{x}_1$ and $\boldsymbol{x}_2$ are 1. Substitute the numbers λ_1 and λ_2 into equation (7), to find $\lambda_1 - \lambda_2 = \sqrt{5}$ and F_{100}:

$$F_{100} = \frac{1}{\sqrt{5}}\left[\left(\frac{1+\sqrt{5}}{2}\right)^{100} - \left(\frac{1-\sqrt{5}}{2}\right)^{100} \right] \approx 3.54 \cdot 10^{20}. \tag{8}$$

Is this a whole number? *Yes.* The fractions and square roots must disappear, because Fibonacci's rule $F_{k+2} = F_{k+1} + F_k$ stays with integers. The second term in (8) is less than $\frac{1}{2}$, so it must move the first term to the nearest whole number:

$$k\text{th Fibonacci number} \quad = \quad \text{nearest integer to} \quad \frac{1}{\sqrt{5}}\left(\frac{1+\sqrt{5}}{2}\right)^k. \tag{9}$$

The ratio of F_6 to F_5 is $8/5 = 1.6$. The ratio F_{101}/F_{100} must be very close to $(1+\sqrt{5})/2$. The Greeks called this number the *"golden mean"*. For some reason a rectangle with sides 1.618 and 1 looks especially graceful.

Matrix Powers A^k

Fibonacci's example is a typical difference equation $\boldsymbol{u}_{k+1} = A\mathbf{u}_k$. ***Each step multiplies by A.*** The solution is $\boldsymbol{u}_k = A^k\boldsymbol{u}_0$. We want to make clear how diagonalizing the matrix gives a quick way to compute A^k.

The eigenvector matrix S produces $A = S\Lambda S^{-1}$. This is a factorization of the matrix, like $A = LU$ or $A = QR$. The new factorization is perfectly suited to computing powers, because ***every time*** $\boldsymbol{S^{-1}}$ ***multiplies S we get I***:

$$\begin{aligned} A^2 &= S\Lambda S^{-1}S\Lambda S^{-1} = S\Lambda^2 S^{-1} \\ A^k &= (S\Lambda S^{-1})\cdots(S\Lambda S^{-1}) = S\Lambda^k S^{-1}. \end{aligned}$$

The eigenvector matrix for A^k is still S, and the eigenvalue matrix is Λ^k. We knew that. The eigenvectors don't change, and the eigenvalues are taken to the kth power. When A is diagonalized, $A^k\boldsymbol{u}_0$ is easy. Here are steps 1,2,3 (taken for Fibonacci):

1. Find the eigenvalues of A and look for n independent eigenvectors.

2. Write $\boldsymbol{u}_0$ as a combination $c_1\boldsymbol{x}_1 + \cdots + c_n\boldsymbol{x}_n$ of the eigenvectors.

3. Multiply each eigenvector $\boldsymbol{x}_i$ by $(\lambda_i)^k$. Then

$$\boldsymbol{u}_k = A^k\boldsymbol{u}_0 = c_1(\lambda_1)^k\boldsymbol{x}_1 + \cdots + c_n(\lambda_n)^k\boldsymbol{x}_n. \tag{10}$$

In matrix language A^k is $(S\Lambda S^{-1})^k$ which is S times Λ^k times S^{-1}. In vector language, the eigenvectors in S lead to the c's:

$$\boldsymbol{u}_0 = c_1\boldsymbol{x}_1 + \cdots + c_n\boldsymbol{x}_n = \begin{bmatrix} \boldsymbol{x}_1 & \cdots & \boldsymbol{x}_n \end{bmatrix}\begin{bmatrix} c_1 \\ \vdots \\ c_n \end{bmatrix}. \qquad \text{This says that} \quad \boldsymbol{u}_0 = S\mathbf{c}.$$

The coefficients in Step 2 are $\mathbf{c} = S^{-1}\boldsymbol{u}_0$. Then Step 3 multiplies by Λ^k. The combination $\boldsymbol{u}_k = \sum c_i(\lambda_i)^k\boldsymbol{x}_i$ in (10) is the product of S and Λ^k and $\boldsymbol{c}$:

$$A^k\boldsymbol{u}_0 = S\Lambda^k S^{-1}\boldsymbol{u}_0 = S\Lambda^k\mathbf{c} = \begin{bmatrix} \boldsymbol{x}_1 & \cdots & \boldsymbol{x}_n \end{bmatrix}\begin{bmatrix} (\lambda_1)^k & & \\ & \ddots & \\ & & (\lambda_n)^k \end{bmatrix}\begin{bmatrix} c_1 \\ \vdots \\ c_n \end{bmatrix}. \tag{11}$$

This result is exactly $\boldsymbol{u}_k = c_1(\lambda_1)^k\boldsymbol{x}_1 + \cdots + c_n(\lambda_n)^k\boldsymbol{x}_n$. It solves $\boldsymbol{u}_{k+1} = A\boldsymbol{u}_k$.

Example 3 Compute $A^k = S\Lambda^k S^{-1}$ when S and Λ and S^{-1} contain whole numbers:

$$A = \begin{bmatrix} 1 & 1 \\ 0 & 2 \end{bmatrix} \quad \text{has} \quad \lambda_1 = 1 \quad \text{and} \quad \boldsymbol{x}_1 = \begin{bmatrix} 1 \\ 0 \end{bmatrix}, \quad \lambda_2 = 2 \quad \text{and} \quad \boldsymbol{x}_2 = \begin{bmatrix} 1 \\ 1 \end{bmatrix}.$$

A is triangular, with 1 and 2 on the diagonal. A^k is also triangular, with 1 and 2^k on the diagonal. Those numbers stay separate in Λ^k. They are combined in A^k:

$$A^k = S\Lambda^k S^{-1} = \begin{bmatrix} 1 & 1 \\ 0 & 1 \end{bmatrix} \begin{bmatrix} 1^k & \\ & 2^k \end{bmatrix} \begin{bmatrix} 1 & -1 \\ 0 & 1 \end{bmatrix} = \begin{bmatrix} 1 & 2^k - 1 \\ 0 & 2^k \end{bmatrix}.$$

With $k = 1$ we get A. With $k = 0$ we get I. With $k = -1$ we get A^{-1}.

Note The zeroth power of every nonsingular matrix is $A^0 = I$. The product $S\Lambda^0 S^{-1}$ becomes SIS^{-1} which is I. Every λ to the zeroth power is 1. But the rule breaks down when $\lambda = 0$. Then 0^0 is not determined. We don't know A^0 when A is singular.

Nondiagonalizable Matrices (Optional)

Suppose λ is an eigenvalue of A. We discover that fact in two ways:

1. Eigenvectors (geometric) There are nonzero solutions to $A\boldsymbol{x} = \lambda\boldsymbol{x}$.
2. Eigenvalues (algebraic) The determinant of $A - \lambda I$ is zero.

The number λ may be a simple eigenvalue or a multiple eigenvalue, and we want to know its ***multiplicity***. Most eigenvalues have multiplicity $M = 1$ (simple eigenvalues). Then there is a single line of eigenvectors, and $\det(A - \lambda I)$ does not have a double factor. For exceptional matrices, an eigenvalue can be ***repeated***. Then there are two different ways to count its multiplicity:

1. (Geometric Multiplicity = GM) Count the independent eigenvectors for λ. This is the dimension of the nullspace of $A - \lambda I$.
2. (Algebraic Multiplicity = AM) Count the repetitions of λ among the eigenvalues. Look at the n roots of $\det(A - \lambda I) = 0$.

The following matrix A is the standard example of trouble. Its eigenvalue $\lambda = 0$ is repeated. It is a double eigenvalue (AM $= 2$) with only one eigenvector (GM $= 1$). The geometric multiplicity can be below the algebraic multiplicity—it is never larger:

$$A = \begin{bmatrix} 0 & 1 \\ 0 & 0 \end{bmatrix} \quad \text{has} \quad \det(A - \lambda I) = \begin{vmatrix} -\lambda & 1 \\ 0 & -\lambda \end{vmatrix} = \lambda^2.$$

There "should" be two eigenvectors, because $\lambda^2 = 0$ has a double root. The double factor λ^2 makes AM $= 2$. But there is only one eigenvector $\boldsymbol{x} = (1, 0)$. ***This shortage of eigenvectors when* GM $<$ AM *means that A is not diagonalizable.***

The vector called "repeats" in the Teaching Code **eigval** gives the algebraic multiplicity AM for each eigenvalue. When repeats = [1 1. . . 1] we know that the n eigenvalues are all different. A is certainly diagonalizable in that case. The sum of all components in "repeats" is always n, because the nth degree equation $\det(A-\lambda I)=0$ always has n roots (counting repetitions).

The diagonal matrix **D** in the Teaching Code **eigvec** gives the geometric multiplicity GM for each eigenvalue. This counts the independent eigenvectors. The total number of independent eigenvectors might be less than n. The n by n matrix A is diagonalizable if and only if this total number is n.

We have to emphasize: There is nothing special about $\lambda = 0$. It makes for easy computations, but these three matrices also have the same shortage of eigenvectors. Their repeated eigenvalue is $\lambda = 5$. Traces are 10, determinants are 25:

$$A = \begin{bmatrix} 5 & 1 \\ 0 & 5 \end{bmatrix} \quad \text{and} \quad A = \begin{bmatrix} 6 & -1 \\ 1 & 4 \end{bmatrix} \quad \text{and} \quad A = \begin{bmatrix} 7 & 2 \\ -2 & 3 \end{bmatrix}.$$

Those all have $\det(A-\lambda I) = (\lambda-5)^2$. The algebraic multiplicity is AM = 2. But $A - 5I$ has rank $r = 1$. The geometric multiplicity is GM = 1. There is only one eigenvector for $\lambda = 5$, and these matrices are not diagonalizable.

■ REVIEW OF THE KEY IDEAS ■

1. If A has n independent eigenvectors (they go into the columns of S), then $S^{-1}AS$ is diagonal: $S^{-1}AS = \Lambda$ and $A = S\Lambda S^{-1}$.

2. The powers of A are $A^k = S\Lambda^k S^{-1}$. The eigenvectors in S are unchanged.

3. The eigenvalues of A^k are $(\lambda_1)^k, \ldots, (\lambda_n)^k$. The eigenvalues of A^{-1} are $1/\lambda_i$.

4. The solution to $\boldsymbol{u}_{k+1} = A\boldsymbol{u}_k$ starting from $\boldsymbol{u}_0$ is $\boldsymbol{u}_k = A^k\boldsymbol{u}_0 = S\Lambda^k S^{-1}\boldsymbol{u}_0$:

$$\boldsymbol{u}_k = c_1(\lambda_1)^k\boldsymbol{x}_1 + \cdots + c_n(\lambda_n)^k\boldsymbol{x}_n \quad \text{provided} \quad \boldsymbol{u}_0 = c_1\boldsymbol{x}_1 + \cdots + c_n\boldsymbol{x}_n.$$

5. A is diagonalizable if every eigenvalue has enough eigenvectors (GM=AM).

■ WORKED EXAMPLES ■

6.2 A The Lucas numbers are like the Fibonacci numbers except they start with $L_1 = 1$ and $L_2 = 3$. Following the rule $L_{k+2} = L_{k+1} + L_k$, the next Lucas numbers are $4, 7, 11, 18$. Show that the Lucas number L_{100} is $\lambda_1^{100} + \lambda_2^{100}$.

Solution $u_{k+1} = \begin{bmatrix} 1 & 1 \\ 1 & 0 \end{bmatrix} u_k$ is the same as for Fibonacci, because $L_{k+2} = L_{k+1} + L_k$ is the same rule (with different starting values). We can copy equation (5):

$$\text{Let } u_k = \begin{bmatrix} L_{k+1} \\ L_k \end{bmatrix}. \text{ The rule } \begin{matrix} L_{k+2} = L_{k+1} + L_k \\ L_{k+1} = L_{k+1} \end{matrix} \text{ becomes } u_{k+1} = \begin{bmatrix} 1 & 1 \\ 1 & 0 \end{bmatrix} u_k.$$

The eigenvalues and eigenvectors of $A = \begin{bmatrix} 1 & 1 \\ 1 & 0 \end{bmatrix}$ still come from $\lambda^2 = \lambda + 1$:

$$\lambda_1 = \frac{1+\sqrt{5}}{2} \text{ and } x_1 = \begin{bmatrix} \lambda_1 \\ 1 \end{bmatrix} \qquad \lambda_2 = \frac{1-\sqrt{5}}{2} \text{ and } x_2 = \begin{bmatrix} \lambda_2 \\ 1 \end{bmatrix}.$$

Now solve $c_1 x_1 + c_2 x_2 = u_1 = (3, 1)$. The coefficients are $c_1 = \lambda_1$ and $c_2 = \lambda_2$! Check:

$$\lambda_1 x_1 + \lambda_2 x_2 = \begin{bmatrix} \lambda_1 & \lambda_2 \\ 1 & 1 \end{bmatrix} \begin{bmatrix} \lambda_1 \\ \lambda_2 \end{bmatrix} = \begin{bmatrix} \lambda_1^2 + \lambda_2^2 \\ \lambda_1 + \lambda_2 \end{bmatrix} = \begin{bmatrix} \text{trace of } A^2 \\ \text{trace of } A \end{bmatrix} = \begin{bmatrix} 3 \\ 1 \end{bmatrix} = u_1.$$

The solution $u_{100} = A^{99} u_1$ tells us the Lucas numbers (L_{101}, L_{100}). The second components of x_1 and x_2 are 1, so the second component of u_{100} is

$$L_{100} = c_1 \lambda_1^{99} + c_2 \lambda_2^{99} = \lambda_1^{100} + \lambda_2^{100}.$$

Every $L_k = \lambda_1^k + \lambda_2^k$ is a whole number (why)? Since λ_2 is very small, L_k must be close to λ_1^k. Lucas starts faster than Fibonacci, and ends up larger by a factor near $\sqrt{5}$.

6.2 B Find all eigenvector matrices S that diagonalize A (rank 1) to give $S^{-1}AS = \Lambda$:

$$A = \begin{bmatrix} 1 & 1 & 1 \\ 1 & 1 & 1 \\ 1 & 1 & 1 \end{bmatrix} = \begin{bmatrix} 1 \\ 1 \\ 1 \end{bmatrix} \begin{bmatrix} 1 & 1 & 1 \end{bmatrix}$$

What is A^n? Which matrices B commute with A (so that $AB = BA$)?

Solution Since A has rank 1, its nullspace is a two-dimensional plane. Any vector with $x + y + z = 0$ (components adding to zero) solves $Ax = \mathbf{0}$. So $\lambda = 0$ is an eigenvalue with multiplicity 2. There are two independent eigenvectors ($GM = 2$). The other eigenvalue must be $\lambda = 3$ because the trace of A is $1 + 1 + 1 = 3$. Check these λ's:

$$\det(A - \lambda I) = \begin{vmatrix} 1-\lambda & 1 & 1 \\ 1 & 1-\lambda & 1 \\ 1 & 1 & 1-\lambda \end{vmatrix} = (1-\lambda)^3 + 2 - 3(1-\lambda) = -\lambda^3 + 3\lambda^2.$$

Then $\lambda^2(3-\lambda)=0$ and the eigenvalues are $\lambda_1=0, \lambda_2=0, \lambda_3=3$. The eigenvectors for $\lambda=3$ are multiples of $\boldsymbol{x}_3=(1,1,1)$. The eigenvectors for $\lambda_1=\lambda_2=0$ are *any two independent vectors in the plane* $x+y+z=0$. These are the columns of all possible eigenvector matrices S:

$$S=\begin{bmatrix} x & X & c \\ y & Y & c \\ -x-y & -X-Y & c \end{bmatrix} \quad \text{and} \quad S^{-1}AS=\Lambda=\begin{bmatrix} 0 & 0 & 0 \\ 0 & 0 & 0 \\ 0 & 0 & 3 \end{bmatrix}$$

where $c \neq 0$ and $xY \neq yX$. The powers A^n come quickly by multiplication:

$$A^2=\begin{bmatrix} 3 & 3 & 3 \\ 3 & 3 & 3 \\ 3 & 3 & 3 \end{bmatrix}=3A \quad \text{and} \quad A^n=3^{n-1}A.$$

To find matrices B that commute with A, look at AB and BA. The 1's in A produce the column sums C_1, C_2, C_3 and the row sums R_1, R_2, R_3 of B:

$$AB=\text{column sums}=\begin{bmatrix} C_1 & C_2 & C_3 \\ C_1 & C_2 & C_3 \\ C_1 & C_2 & C_3 \end{bmatrix} \qquad BA=\text{row sums}=\begin{bmatrix} R_1 & R_1 & R_1 \\ R_2 & R_2 & R_2 \\ R_3 & R_3 & R_3 \end{bmatrix}$$

If $AB=BA$, *all six column and row sums of B must be the same*. One possible B is A itself, since $AA=AA$. B is any linear combination of permutation matrices!

This is a 5-dimensional space (Problem 3.5.39) of matrices that commute with A. All B's share the eigenvector $(1,1,1)$. Their other eigenvectors are in the plane $x+y+z=0$. Three degrees of freedom in the λ's and two in the unit eigenvectors.

Problem Set 6.2

Questions 1–8 are about the eigenvalue and eigenvector matrices.

1 Factor these two matrices into $A=S\Lambda S^{-1}$:

$$A=\begin{bmatrix} 1 & 2 \\ 0 & 3 \end{bmatrix} \quad \text{and} \quad A=\begin{bmatrix} 1 & 1 \\ 2 & 2 \end{bmatrix}.$$

2 If $A=S\Lambda S^{-1}$ then $A^3=(\quad)(\quad)(\quad)$ and $A^{-1}=(\quad)(\quad)(\quad)$.

3 If A has $\lambda_1=2$ with eigenvector $\boldsymbol{x}_1=\begin{bmatrix} 1 \\ 0 \end{bmatrix}$ and $\lambda_2=5$ with $\boldsymbol{x}_2=\begin{bmatrix} 1 \\ 1 \end{bmatrix}$, use $S\Lambda S^{-1}$ to find A. No other matrix has the same λ's and $\boldsymbol{x}$'s.

4 Suppose $A=S\Lambda S^{-1}$. What is the eigenvalue matrix for $A+2I$? What is the eigenvector matrix? Check that $A+2I=(\quad)(\quad)(\quad)^{-1}$.

5 True or false: If the columns of S (eigenvectors of A) are linearly independent, then

(a) A is invertible (b) A is diagonalizable

(c) S is invertible (d) S is diagonalizable.

6 If the eigenvectors of A are the columns of I, then A is a _____ matrix. If the eigenvector matrix S is triangular, then S^{-1} is triangular. Prove that A is also triangular.

7 Describe all matrices S that diagonalize this matrix A:

$$A = \begin{bmatrix} 4 & 0 \\ 1 & 2 \end{bmatrix}.$$

Then describe all matrices that diagonalize A^{-1}.

8 Write down the most general matrix that has eigenvectors $\begin{bmatrix} 1 \\ 1 \end{bmatrix}$ and $\begin{bmatrix} 1 \\ -1 \end{bmatrix}$.

Questions 9–14 are about Fibonacci and Gibonacci numbers.

9 For the Fibonacci matrix $A = \begin{bmatrix} 1 & 1 \\ 1 & 0 \end{bmatrix}$, compute A^2 and A^3 and A^4. Then use the text and a calculator to find F_{20}.

10 Suppose each number G_{k+2} is the *average* of the two previous numbers G_{k+1} and G_k. Then $G_{k+2} = \frac{1}{2}(G_{k+1} + G_k)$:

$$\begin{aligned} G_{k+2} &= \tfrac{1}{2}G_{k+1} + \tfrac{1}{2}G_k \\ G_{k+1} &= G_{k+1} \end{aligned} \quad \text{is} \quad \begin{bmatrix} G_{k+2} \\ G_{k+1} \end{bmatrix} = \begin{bmatrix} & A & \end{bmatrix} \begin{bmatrix} G_{k+1} \\ G_k \end{bmatrix}.$$

(a) Find the eigenvalues and eigenvectors of A.

(b) Find the limit as $n \to \infty$ of the matrices $A^n = S\Lambda^n S^{-1}$.

(c) If $G_0 = 0$ and $G_1 = 1$ show that the Gibonacci numbers approach $\frac{2}{3}$.

11 Diagonalize the Fibonacci matrix by completing S^{-1}:

$$\begin{bmatrix} 1 & 1 \\ 1 & 0 \end{bmatrix} = \begin{bmatrix} \lambda_1 & \lambda_2 \\ 1 & 1 \end{bmatrix} \begin{bmatrix} \lambda_1 & 0 \\ 0 & \lambda_2 \end{bmatrix} \begin{bmatrix} & \\ & \end{bmatrix}.$$

Do the multiplication $S\Lambda^k S^{-1}\begin{bmatrix} 1 \\ 0 \end{bmatrix}$ to find its second component. This is the kth Fibonacci number $F_k = (\lambda_1^k - \lambda_2^k)/(\lambda_1 - \lambda_2)$.

12 The numbers λ_1^k and λ_2^k satisfy the Fibonacci rule $F_{k+2} = F_{k+1} + F_k$:

$$\lambda_1^{k+2} = \lambda_1^{k+1} + \lambda_1^k \quad \text{and} \quad \lambda_2^{k+2} = \lambda_2^{k+1} + \lambda_2^k.$$

Prove this by using the original equation for the λ's. Then any combination of λ_1^k and λ_2^k satisfies the rule. The combination $F_k = (\lambda_1^k - \lambda_2^k)/(\lambda_1 - \lambda_2)$ gives the right start $F_0 = 0$ and $F_1 = 1$.

13 Lucas started with $L_0 = 2$ and $L_1 = 1$. The rule $L_{k+2} = L_{k+1} + L_k$ is the same, so Fibonacci's matrix A is the same. Add its eigenvectors $\boldsymbol{x}_1 + \boldsymbol{x}_2$:

$$\begin{bmatrix} \lambda_1 \\ 1 \end{bmatrix} + \begin{bmatrix} \lambda_2 \\ 1 \end{bmatrix} = \begin{bmatrix} \frac{1}{2}(1+\sqrt{5}) \\ 1 \end{bmatrix} + \begin{bmatrix} \frac{1}{2}(1-\sqrt{5}) \\ 1 \end{bmatrix} = \begin{bmatrix} 1 \\ 2 \end{bmatrix} = \begin{bmatrix} L_1 \\ L_0 \end{bmatrix}.$$

After 10 steps the second component of $A^{10}(\boldsymbol{x}_1 + \boldsymbol{x}_2)$ is $\lambda_1^{10} + \lambda_2^{10}$. Compute that Lucas number L_{10} by $L_{k+2} = L_{k+1} + L_k$, and compute approximately by λ_1^{10}.

14 Prove that every third Fibonacci number in $0, 1, 1, 2, 3, \ldots$ is even.

Questions 15–18 are about diagonalizability.

15 True or false: If the eigenvalues of A are $2, 2, 5$ then the matrix is certainly

(a) invertible (b) diagonalizable (c) not diagonalizable.

16 True or false: If the only eigenvectors of A are multiples of $(1, 4)$ then A has

(a) no inverse (b) a repeated eigenvalue (c) no diagonalization $S\Lambda S^{-1}$.

17 Complete these matrices so that $\det A = 25$. Then check that $\lambda = 5$ is repeated—the determinant of $A - \lambda I$ is $(\lambda - 5)^2$. Find an eigenvector with $A\boldsymbol{x} = 5\boldsymbol{x}$. These matrices will not be diagonalizable because there is no second line of eigenvectors.

$$A = \begin{bmatrix} 8 & \\ & 2 \end{bmatrix} \quad \text{and} \quad A = \begin{bmatrix} 9 & 4 \\ & 1 \end{bmatrix} \quad \text{and} \quad A = \begin{bmatrix} 10 & 5 \\ -5 & \end{bmatrix}$$

18 The matrix $A = \left[\begin{smallmatrix} 3 & 1 \\ 0 & 3 \end{smallmatrix}\right]$ is not diagonalizable because the rank of $A - 3I$ is ______. Change one entry to make A diagonalizable. Which entries could you change?

Questions 19–23 are about powers of matrices.

19 $A^k = S\Lambda^k S^{-1}$ approaches the zero matrix as $k \to \infty$ if and only if every λ has absolute value less than ______. Which of these matrices has $A^k \to 0$?

$$A = \begin{bmatrix} .6 & .4 \\ .4 & .6 \end{bmatrix} \quad \text{and} \quad B = \begin{bmatrix} .6 & .9 \\ .1 & .6 \end{bmatrix}.$$

20 (Recommended) Find Λ and S to diagonalize A in Problem 19. What is the limit of Λ^k as $k \to \infty$? What is the limit of $S\Lambda^k S^{-1}$? In the columns of this limiting matrix you see the ______.

21 Find Λ and S to diagonalize B in Problem 19. What is $B^{10}\boldsymbol{u}_0$ for these $\boldsymbol{u}_0$?

$$\boldsymbol{u}_0 = \begin{bmatrix} 3 \\ 1 \end{bmatrix} \quad \text{and} \quad \boldsymbol{u}_0 = \begin{bmatrix} 3 \\ -1 \end{bmatrix} \quad \text{and} \quad \boldsymbol{u}_0 = \begin{bmatrix} 6 \\ 0 \end{bmatrix}.$$

22 Diagonalize A and compute $S\Lambda^k S^{-1}$ to prove this formula for A^k:

$$A = \begin{bmatrix} 2 & 1 \\ 1 & 2 \end{bmatrix} \quad \text{has} \quad A^k = \frac{1}{2}\begin{bmatrix} 3^k + 1 & 3^k - 1 \\ 3^k - 1 & 3^k + 1 \end{bmatrix}.$$

23 Diagonalize B and compute $S\Lambda^k S^{-1}$ to prove this formula for B^k:

$$B = \begin{bmatrix} 3 & 1 \\ 0 & 2 \end{bmatrix} \quad \text{has} \quad B^k = \begin{bmatrix} 3^k & 3^k - 2^k \\ 0 & 2^k \end{bmatrix}.$$

Questions 24–29 are new applications of $A = S\Lambda S^{-1}$.

24 Suppose that $A = S\Lambda S^{-1}$. Take determinants to prove that $\det A = \lambda_1\lambda_2\cdots\lambda_n =$ product of λ's. This quick proof only works when A is ____.

25 Show that trace $AB =$ trace BA, by adding the diagonal entries of AB and BA:

$$A = \begin{bmatrix} a & b \\ c & d \end{bmatrix} \quad \text{and} \quad B = \begin{bmatrix} q & r \\ s & t \end{bmatrix}.$$

Choose A as S and B as ΛS^{-1}. Then $S\Lambda S^{-1}$ has the same trace as $\Lambda S^{-1}S$. The trace of A equals the trace of Λ which is ____.

26 $AB - BA = I$ is impossible since the left side has trace $=$ ____. But find an elimination matrix so that $A = E$ and $B = E^{\mathrm{T}}$ give

$$AB - BA = \begin{bmatrix} -1 & 0 \\ 0 & 1 \end{bmatrix} \quad \text{which has trace zero.}$$

27 If $A = S\Lambda S^{-1}$, diagonalize the block matrix $B = \begin{bmatrix} \mathbf{A} & \mathbf{0} \\ \mathbf{0} & \mathbf{2A} \end{bmatrix}$. Find its eigenvalue and eigenvector matrices.

28 Consider all 4 by 4 matrices A that are diagonalized by the same fixed eigenvector matrix S. Show that the A's form a subspace (cA and $A_1 + A_2$ have this same S). What is this subspace when $S = I$? What is its dimension?

29 Suppose $A^2 = A$. On the left side A multiplies each column of A. Which of our four subspaces contains eigenvectors with $\lambda = 1$? Which subspace contains eigenvectors with $\lambda = 0$? From the dimensions of those subspaces, A has a full set of independent eigenvectors and can be diagonalized.

30 (Recommended) Suppose $A\boldsymbol{x} = \lambda\boldsymbol{x}$. If $\lambda = 0$ then $\boldsymbol{x}$ is in the nullspace. If $\lambda \neq 0$ then $\boldsymbol{x}$ is in the column space. Those spaces have dimensions $(n - r) + r = n$. So why doesn't every square matrix have n linearly independent eigenvectors?

31 The eigenvalues of A are 1 and 9, the eigenvalues of B are -1 and 9:

$$A = \begin{bmatrix} 5 & 4 \\ 4 & 5 \end{bmatrix} \quad \text{and} \quad B = \begin{bmatrix} 4 & 5 \\ 5 & 4 \end{bmatrix}.$$

Find a matrix square root of A from $R = S\sqrt{\Lambda}\,S^{-1}$. Why is there no real matrix square root of B?

32 (**Heisenberg's Uncertainty Principle**) $AB - BA = I$ can happen for infinite matrices with $A = A^{\mathrm{T}}$ and $B = -B^{\mathrm{T}}$. Then

$$\boldsymbol{x}^{\mathrm{T}}\boldsymbol{x} = \boldsymbol{x}^{\mathrm{T}}AB\boldsymbol{x} - \boldsymbol{x}^{\mathrm{T}}BA\boldsymbol{x} \leq 2\|A\boldsymbol{x}\|\,\|B\boldsymbol{x}\|.$$

Explain that last step by using the Schwarz inequality. Then the inequality says that $\|A\boldsymbol{x}\|/\|\boldsymbol{x}\|$ times $\|B\boldsymbol{x}\|/\|\boldsymbol{x}\|$ is at least $\frac{1}{2}$. It is impossible to get the position error and momentum error both very small.

33 If A and B have the same λ's with the same independent eigenvectors, their factorizations into _____ are the same. So $A = B$.

34 Suppose the same S diagonalizes both A and B, so that $A = S\Lambda_1 S^{-1}$ and $B = S\Lambda_2 S^{-1}$. Prove that $AB = BA$.

35 Substitute $A = S\Lambda S^{-1}$ into the product $(A - \lambda_1 I)(A - \lambda_2 I)\cdots(A - \lambda_n I)$ and explain why this produces the zero matrix. We are substituting the matrix A for the number λ in the polynomial $p(\lambda) = \det(A - \lambda I)$. The ***Cayley-Hamilton Theorem*** says that this product is always $p(A) =$ *zero matrix*, even if A is not diagonalizable.

36 Test the Cayley-Hamilton Theorem on Fibonacci's matrix $A = \begin{bmatrix} 1 & 1 \\ 1 & 0 \end{bmatrix}$. The theorem predicts that $A^2 - A - I = 0$, since the polynomial $\det(A - \lambda I)$ is $\lambda^2 - \lambda - 1$.

37 If $A = \begin{bmatrix} a & b \\ 0 & d \end{bmatrix}$ then $\det(A - \lambda I)$ is $(\lambda - a)(\lambda - d)$. Check the Cayley-Hamilton statement that $(A - aI)(A - dI) =$ *zero matrix*.

38 (a) When do the eigenvectors for $\lambda = 0$ span the nullspace $\boldsymbol{N}(A)$?

(b) When do all the eigenvectors for $\lambda \neq 0$ span the column space $\boldsymbol{C}(A)$?

39 Find the eigenvalues and eigenvectors and the kth power of A. Worked Example **2.4 C** described A as the "adjacency matrix" for this 3-node graph. The i, j entry of A^k counts the k-step paths from i to j—what is the 2, 2 entry of A^4 and which 4-step paths along edges of the graph begin and end at node 2?

$$A = \begin{bmatrix} 1 & 1 & 1 \\ 1 & 0 & 0 \\ 1 & 0 & 0 \end{bmatrix}$$

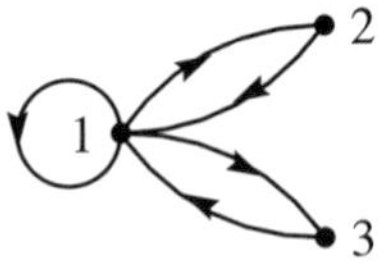

40 If $A = \left[\begin{smallmatrix}1&0\\0&2\end{smallmatrix}\right]$ and $AB = BA$, show that $B = \left[\begin{smallmatrix}a&b\\c&d\end{smallmatrix}\right]$ is also a diagonal matrix. B has the same eigen_____ as A but different eigen_____ . These diagonal matrices B form a two-dimensional subspace of matrix space. $AB - BA = 0$ gives four equations for the unknowns $\boldsymbol{a}, \boldsymbol{b}, \boldsymbol{c}, \boldsymbol{d}$—find the rank of the 4 by 4 matrix.

41 If A is 5 by 5, then $AB - BA =$ zero matrix gives 25 equations for the 25 entries in B. How do you know that the 25 by 25 matrix is singular (and there is always a nonzero solution B)?

42 The powers A^k approach zero if all $|\lambda_i| < 1$ and they blow up if any $|\lambda_i| > 1$. Peter Lax gives these striking examples in his book *Linear Algebra*:

$$A = \begin{bmatrix} 3 & 2 \\ 1 & 4 \end{bmatrix} \quad B = \begin{bmatrix} 3 & 2 \\ -5 & -3 \end{bmatrix} \quad C = \begin{bmatrix} 5 & 7 \\ -3 & -4 \end{bmatrix} \quad D = \begin{bmatrix} 5 & 6.9 \\ -3 & -4 \end{bmatrix}$$

$$\|A^{1024}\| > 10^{700} \quad B^{1024} = I \quad C^{1024} = -C \quad \|D^{1024}\| < 10^{-78}$$

Find the eigenvalues $\lambda = e^{i\theta}$ of B and C to show $B^4 = I$ and $C^3 = -I$.

APPLICATIONS TO DIFFERENTIAL EQUATIONS ■ 6.3

Eigenvalues and eigenvectors and $A = S\Lambda S^{-1}$ are perfect for matrix powers A^k. They are also perfect for differential equations. This section is mostly linear algebra, but to read it you need one fact from calculus: *The derivative of* $e^{\lambda t}$ *is* $\lambda e^{\lambda t}$. It helps to know what e is, but I am not even sure that is essential. The whole point of the section is this: **To convert differential equations into linear algebra.**

The ordinary scalar equation $du/dt = u$ is solved by $u = e^t$. The equation $du/dt = 4u$ is solved by $u = e^{4t}$. Constant coefficient equations have exponential solutions!

$$\frac{du}{dt} = \lambda u \quad \text{has the solutions} \quad u(t) = Ce^{\lambda t}. \tag{1}$$

The number C turns up on both sides of $du/dt = \lambda u$. At $t = 0$ the solution $Ce^{\lambda t}$ reduces to C (because $e^0 = 1$). By choosing $C = u(0)$, ***the solution that starts from*** $u(0)$ ***at*** $t = 0$ ***is*** $u(0)e^{\lambda t}$.

We just solved a 1 by 1 problem. Linear algebra moves to n by n. The unknown is a vector $\boldsymbol{u}$ (now boldface). It starts from the initial vector $\boldsymbol{u}(0)$, which is given. The n equations contain a square matrix A:

Problem

$$\text{Solve} \quad \frac{d\boldsymbol{u}}{dt} = A\boldsymbol{u} \quad \text{starting from the vector} \quad \boldsymbol{u}(0) \quad \text{at} \quad t = 0. \tag{2}$$

This system of differential equations is *linear*. If $\boldsymbol{u}(t)$ and $\boldsymbol{v}(t)$ are solutions, so is $C\boldsymbol{u}(t) + D\boldsymbol{v}(t)$. We will need n constants like C and D to match the n components of $\boldsymbol{u}(0)$. Our first job is to find n "pure exponential solutions" to the equation $d\boldsymbol{u}/dt = A\boldsymbol{u}$.

Notice that A is a *constant* matrix. In other linear equations, A changes as t changes. In nonlinear equations, A changes as $\boldsymbol{u}$ changes. We don't have either of these difficulties. Equation (2) is "linear with constant coefficients". Those and only those are the differential equations that we will convert directly to linear algebra. The main point will be: ***Solve linear constant coefficient equations by exponentials*** $e^{\lambda t}\boldsymbol{x}$.

Solution of $d\boldsymbol{u}/dt = A\boldsymbol{u}$

Our pure exponential solution will be $e^{\lambda t}$ times a fixed vector $\boldsymbol{x}$. You may guess that λ is an eigenvalue of A, and $\boldsymbol{x}$ is the eigenvector. Substitute $\boldsymbol{u}(t) = e^{\lambda t}\boldsymbol{x}$ into the equation $d\boldsymbol{u}/dt = A\boldsymbol{u}$ to prove you are right (the factor $e^{\lambda t}$ will cancel):

$$A\boldsymbol{u} = Ae^{\lambda t}\boldsymbol{x} \quad \text{agrees with} \quad \frac{d\boldsymbol{u}}{dt} = \lambda e^{\lambda t}\boldsymbol{x} \quad \text{provided} \quad A\boldsymbol{x} = \lambda\boldsymbol{x}. \tag{3}$$

All components of this special solution $u = e^{\lambda t}x$ share the same $e^{\lambda t}$. The solution grows when $\lambda > 0$. It decays when $\lambda < 0$. In general λ can be a complex number. Then the real part of λ decides growth or decay, while the imaginary part gives oscillation like a sine wave.

Example 1 Solve $\frac{du}{dt} = Au = \begin{bmatrix} 0 & 1 \\ 1 & 0 \end{bmatrix} u$ starting from $u(0) = \begin{bmatrix} 4 \\ 2 \end{bmatrix}$.

This is a vector equation for u. It contains two scalar equations for the components y and z. They are "coupled together" because the matrix is not diagonal:

$$\textbf{Equation } \frac{du}{dt} = Au \qquad \frac{d}{dt}\begin{bmatrix} y \\ z \end{bmatrix} = \begin{bmatrix} 0 & 1 \\ 1 & 0 \end{bmatrix}\begin{bmatrix} y \\ z \end{bmatrix} \text{ means that } \frac{dy}{dt} = z \text{ and } \frac{dz}{dt} = y.$$

The idea of eigenvectors is to combine those equations in a way that gets back to 1 by 1 problems. The combinations $y + z$ and $y - z$ will do it:

$$\frac{d}{dt}(y+z) = z + y \qquad \text{and} \qquad \frac{d}{dt}(y-z) = -(y-z).$$

The combination $y + z$ grows like e^t, because it has $\lambda = 1$. The combination $y - z$ decays like e^{-t}, because it has $\lambda = -1$. Here is the point: We don't have to juggle the original equations $du/dt = Au$, looking for these special combinations. The eigenvectors and eigenvalues do it for us.

This matrix A has eigenvalues 1 and -1. Here are two eigenvectors:

$$\begin{bmatrix} 0 & 1 \\ 1 & 0 \end{bmatrix}\begin{bmatrix} 1 \\ 1 \end{bmatrix} = \begin{bmatrix} 1 \\ 1 \end{bmatrix} = x_1 \qquad \text{and} \qquad \begin{bmatrix} 0 & 1 \\ 1 & 0 \end{bmatrix}\begin{bmatrix} 1 \\ -1 \end{bmatrix} = \begin{bmatrix} -1 \\ 1 \end{bmatrix} = -x_2.$$

The pure exponential solutions u_1 and u_2 take the form $e^{\lambda t}x$ with $\lambda = 1$ and -1:

$$u_1(t) = e^{\lambda_1 t}x_1 = e^t\begin{bmatrix} 1 \\ 1 \end{bmatrix} \qquad \text{and} \qquad u_2(t) = e^{\lambda_2 t}x_2 = e^{-t}\begin{bmatrix} 1 \\ -1 \end{bmatrix}. \tag{4}$$

Notice: These u's are eigenvectors. They satisfy $Au_1 = u_1$ and $Au_2 = -u_2$, just like x_1 and x_2. The factors e^t and e^{-t} change with time. Those factors give $du_1/dt = u_1 = Au_1$ and $du_2/dt = -u_2 = Au_2$. We have two solutions to $du/dt = Au$. To find all other solutions, multiply those special solutions by any C and D and add:

$$\textbf{\textit{General solution}} \quad u(t) = Ce^t\begin{bmatrix} 1 \\ 1 \end{bmatrix} + De^{-t}\begin{bmatrix} 1 \\ -1 \end{bmatrix} = \begin{bmatrix} Ce^t + De^{-t} \\ Ce^t - De^{-t} \end{bmatrix}. \tag{5}$$

With these constants C and D, we can match any starting vector $u(0)$. Set $t = 0$ and $e^0 = 1$. The problem asked for $u(0) = (4, 2)$:

$$C\begin{bmatrix} 1 \\ 1 \end{bmatrix} + D\begin{bmatrix} 1 \\ -1 \end{bmatrix} = \begin{bmatrix} 4 \\ 2 \end{bmatrix} \quad \text{yields} \quad C = 3 \quad \text{and} \quad D = 1.$$

With $C = 3$ and $D = 1$ in the solution (5), the initial value problem is solved.

We summarize the steps. The same three steps that solved $u_{k+1} = Au_k$ now solve $du/dt = Au$. The powers A^k led to λ^k. The differential equation leads to $e^{\lambda t}$:

1. Find the eigenvalues λ_i and n independent eigenvectors $\boldsymbol{x}_i$ of A.

2. Write $\boldsymbol{u}(0)$ as a combination $c_1\boldsymbol{x}_1 + \cdots + c_n\boldsymbol{x}_n$ of the eigenvectors.

3. Multiply each eigenvector $\boldsymbol{x}_i$ by $e^{\lambda_i t}$. Then $\boldsymbol{u}(t)$ is the combination

$$\boldsymbol{u}(t) = c_1 e^{\lambda_1 t}\boldsymbol{x}_1 + \cdots + c_n e^{\lambda_n t}\boldsymbol{x}_n. \tag{6}$$

Example 2 Solve $d\boldsymbol{u}/dt = A\boldsymbol{u}$ knowing the eigenvalues $\lambda = 1, 2, 3$ of A:

$$\frac{d\boldsymbol{u}}{dt} = \begin{bmatrix} 1 & 1 & 1 \\ 0 & 2 & 1 \\ 0 & 0 & 3 \end{bmatrix} \boldsymbol{u} \quad \text{starting from} \quad \boldsymbol{u}(0) = \begin{bmatrix} 6 \\ 5 \\ 4 \end{bmatrix}.$$

Step 1 The eigenvectors are $\boldsymbol{x}_1 = (1, 0, 0)$ and $\boldsymbol{x}_2 = (1, 1, 0)$ and $\boldsymbol{x}_3 = (1, 1, 1)$.

Step 2 The vector $\boldsymbol{u}(0) = (6, 5, 4)$ is $\boldsymbol{x}_1 + \boldsymbol{x}_2 + 4\boldsymbol{x}_3$. Thus $(c_1, c_2, c_3) = (1, 1, 4)$.

Step 3 The pure exponential solutions are $e^t\boldsymbol{x}_1$ and $e^{2t}\boldsymbol{x}_2$ and $e^{3t}\boldsymbol{x}_3$.

Solution: The combination that starts from $\boldsymbol{u}(0)$ is $\boldsymbol{u}(t) = e^t\boldsymbol{x}_1 + e^{2t}\boldsymbol{x}_2 + 4e^{3t}\boldsymbol{x}_3$. The coefficients 1, 1, 4 came from solving the linear equation $c_1\boldsymbol{x}_1 + c_2\boldsymbol{x}_2 + c_3\boldsymbol{x}_3 = \boldsymbol{u}(0)$:

$$\begin{bmatrix} & & \\ \boldsymbol{x}_1 & \boldsymbol{x}_2 & \boldsymbol{x}_3 \\ & & \end{bmatrix} \begin{bmatrix} c_1 \\ c_2 \\ c_3 \end{bmatrix} = \begin{bmatrix} 4 \\ 5 \\ 6 \end{bmatrix} \quad \text{which is} \quad S\boldsymbol{c} = \boldsymbol{u}(0). \tag{7}$$

You now have the basic idea—how to solve $d\boldsymbol{u}/dt = A\boldsymbol{u}$. The rest of this section goes further. We solve equations that contain *second* derivatives, because they arise so often in applications. We also decide whether $\boldsymbol{u}(t)$ approaches zero or blows up or just oscillates. At the end comes the *matrix exponential* e^{At}. Then $e^{At}\boldsymbol{u}(0)$ solves the equation $d\boldsymbol{u}/dt = A\boldsymbol{u}$ in the same way that $A^k\boldsymbol{u}_0$ solves the equation $\boldsymbol{u}_{k+1} = A\boldsymbol{u}_k$.

All these steps use the λ's and $\boldsymbol{x}$'s. With extra time, this section makes a strong connection to the whole topic of differential equations. It solves the constant coefficient problems that turn into linear algebra. Use this section to clarify these simplest but most important differential equations—whose solution is completely based on $e^{\lambda t}$.

Second Order Equations

The most important equation in mechanics is $my'' + by' + ky = 0$. The first term is the mass m times the acceleration $a = y''$. This term ma balances the force F (*Newton's Law!*). The force includes the damping $-by'$ and the elastic restoring force $-ky$, proportional to distance moved. This is a second-order equation because it contains the second derivative $y'' = d^2y/dt^2$. It is still linear with constant coefficients m, b, k.

In a differential equations course, the method of solution is to substitute $y = e^{\lambda t}$. Each derivative brings down a factor λ. We want $y = e^{\lambda t}$ to solve the equation:

$$m\frac{d^2y}{dt^2} + b\frac{dy}{dt} + ky = (m\lambda^2 + b\lambda + k)e^{\lambda t} = 0. \tag{8}$$

Everything depends on $m\lambda^2 + b\lambda + k = 0$. This equation for λ has two roots λ_1 and λ_2. Then the equation for y has two pure solutions $y_1 = e^{\lambda_1 t}$ and $y_2 = e^{\lambda_2 t}$. Their combinations $c_1y_1 + c_2y_2$ give the complete solution.

In a linear algebra course we expect matrices and eigenvalues. Therefore we turn the scalar equation (with y'') into a vector equation (first derivative only!). Suppose $m = 1$. The unknown vector $\boldsymbol{u}$ has components y and y'. The equation is $d\boldsymbol{u}/dt = A\boldsymbol{u}$:

$$\begin{aligned}\frac{dy}{dt} &= y' \\ \frac{dy'}{dt} &= -ky - by'\end{aligned} \quad \text{converts to} \quad \frac{d}{dt}\begin{bmatrix} y \\ y' \end{bmatrix} = \begin{bmatrix} 0 & 1 \\ -k & -b \end{bmatrix}\begin{bmatrix} y \\ y' \end{bmatrix}. \tag{9}$$

The first equation $dy/dt = y'$ is trivial (but true). The second equation connects y'' to y' and y. Together the equations connect $\boldsymbol{u}'$ to $\boldsymbol{u}$. So we solve by eigenvalues:

$$A - \lambda I = \begin{bmatrix} -\lambda & 1 \\ -k & -b-\lambda \end{bmatrix} \quad \text{has determinant} \quad \lambda^2 + b\lambda + k = 0.$$

The equation for the λ*'s is the same*! It is still $\lambda^2 + b\lambda + k = 0$, since $m = 1$. The roots λ_1 and λ_2 are now *eigenvalues of* A. The eigenvectors and the complete solution are

$$\boldsymbol{x}_1 = \begin{bmatrix} 1 \\ \lambda_1 \end{bmatrix} \quad \boldsymbol{x}_2 = \begin{bmatrix} 1 \\ \lambda_2 \end{bmatrix} \quad \boldsymbol{u}(t) = c_1e^{\lambda_1 t}\begin{bmatrix} 1 \\ \lambda_1 \end{bmatrix} + c_2e^{\lambda_2 t}\begin{bmatrix} 1 \\ \lambda_2 \end{bmatrix}.$$

In the first component of $\boldsymbol{u}(t)$, you see $y = c_1e^{\lambda_1 t} + c_2e^{\lambda_2 t}$—the same solution as before.

It can't be anything else. In the second component of $\boldsymbol{u}(t)$ you see dy/dt. The vector problem is completely consistent with the scalar problem.

Note 1 Real engineering and real physics deal with systems (not just a single mass at one point). The unknown $\boldsymbol{y}$ is a vector. The coefficient of $\boldsymbol{y}''$ is a *mass matrix* M, not a number m. The coefficient of $\boldsymbol{y}$ is a *stiffness matrix* K, not a number k. The coefficient of $\boldsymbol{y}'$ is a damping matrix which might be zero.

The equation $M\boldsymbol{y}'' + K\boldsymbol{y} = \boldsymbol{f}$ is a major part of computational mechanics. It is controlled by the eigenvalues of $M^{-1}K$ in $K\boldsymbol{x} = \lambda M\boldsymbol{x}$.

Note 2 In linear algebra the serious danger is a shortage of eigenvectors. Our eigenvectors $(1, \lambda_1)$ and $(1, \lambda_2)$ are the same if $\lambda_1 = \lambda_2$. Then we can't diagonalize A. In this case we don't yet have a complete solution to $d\boldsymbol{u}/dt = A\boldsymbol{u}$.

In differential equations the danger is also a repeated λ. After $y = e^{\lambda t}$, a second solution has to be found. It turns out to be $y = te^{\lambda t}$.

This "impure" solution (with the extra t) will soon appear also in e^{At}.

Example 3 Solve $y'' + 4y' + 3y = 0$ by substituting $e^{\lambda t}$ and also by linear algebra.

Solution Substituting $y = e^{\lambda t}$ yields $(\lambda^2 + 4\lambda + 3)e^{\lambda t} = 0$. The quadratic factors into $\lambda^2+4\lambda+3 = (\lambda+1)(\lambda+3) = 0$. Therefore $\lambda_1 = -1$ and $\lambda_2 = -3$. The pure solutions are $y_1 = e^{-t}$ and $y_2 = e^{-3t}$. The complete solution $c_1y_1 + c_2y_2$ approaches zero.

To use linear algebra we set $\boldsymbol{u} = (y, y')$. This leads to a vector equation $\boldsymbol{u}' = A\boldsymbol{u}$:

$$\begin{aligned} dy/dt &= y' \\ dy'/dt &= -3y - 4y' \end{aligned} \quad \text{converts to} \quad \frac{d\boldsymbol{u}}{dt} = \begin{bmatrix} 0 & 1 \\ -3 & -4 \end{bmatrix} \boldsymbol{u}.$$

This A is called a "companion matrix" and we find its eigenvalues:

$$|A - \lambda I| = \begin{vmatrix} -\lambda & 1 \\ -3 & -4-\lambda \end{vmatrix} = \lambda^2 + 4\lambda + 3 = 0.$$

The λ's are still -1 and -3 and the solution is the same. With constant coefficients and pure exponentials, calculus goes back to algebra.

Stability of 2 by 2 Matrices

For the solution of $d\boldsymbol{u}/dt = A\boldsymbol{u}$, there is a fundamental question. *Does the solution approach $\boldsymbol{u} = \boldsymbol{0}$ as $t \to \infty$?* Is the problem *stable*? Example 3 was certainly stable, because both pure solutions e^{-t} and e^{-3t} approach zero. Stability depends on the eigenvalues -1 and -3, and the eigenvalues depend on A.

The complete solution $\boldsymbol{u}(t)$ is built from pure solutions $e^{\lambda t}\boldsymbol{x}$. If the eigenvalue λ is real, we know exactly when $e^{\lambda t}$ approaches zero: *The number λ must be negative.* If the eigenvalue is a complex number $\lambda = r + is$, *the real part r must be negative.* When $e^{\lambda t}$ splits into $e^{rt}e^{ist}$, the factor e^{ist} has absolute value fixed at 1:

$$e^{ist} = \cos st + i \sin st \quad \text{has} \quad |e^{ist}|^2 = \cos^2 st + \sin^2 st = 1.$$

The factor e^{rt} controls growth ($r > 0$ is instability) or decay ($r < 0$ is stability).

The question is: ***Which matrices have negative eigenvalues?*** More accurately, when are the ***real parts*** of the λ's all negative? 2 by 2 matrices allow a clear answer.

6G Stability The matrix $A = \begin{bmatrix} a & b \\ c & d \end{bmatrix}$ is ***stable*** and $\boldsymbol{u}(t) \to \boldsymbol{0}$ when the eigenvalues have negative real parts. The matrix A must pass two tests:

The trace $T = a + d$ must be negative.
The determinant $D = ad - bc$ must be positive.

Reason If the λ's are real and negative, their sum is negative. This is the trace T. Their product is positive. This is the determinant D. The argument also goes in the reverse direction. If $D = \lambda_1\lambda_2$ is positive, then λ_1 and λ_2 have the same sign. If $T = \lambda_1 + \lambda_2$ is negative, that sign will be negative. We can test T and D.

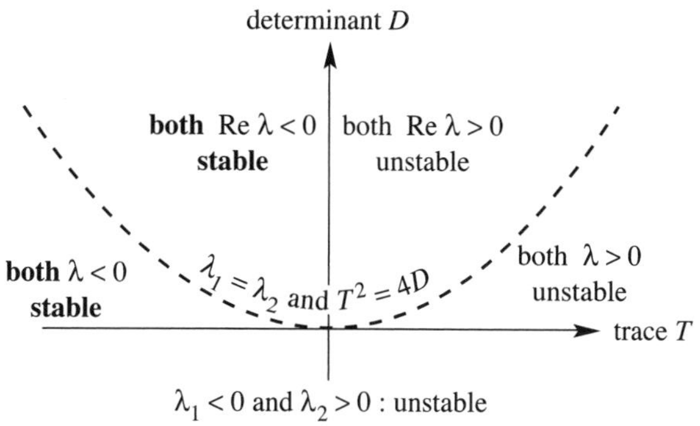

Figure 6.3 A 2 by 2 matrix is stable ($\boldsymbol{u}(t) \to \mathbf{0}$) when $T < 0$ and $D > 0$.

If the λ's are complex numbers, they must have the form $r + is$ and $r - is$. Otherwise T and D will not be real. The determinant D is automatically positive, since $(r + is)(r - is) = r^2 + s^2$. The trace T is $r + is + r - is = 2r$. So a negative trace means that the real part r is negative and the matrix is stable. Q.E.D.

Figure 6.3 shows the parabola $T^2 = 4D$ which separates real from complex eigenvalues. Solving $\lambda^2 - T\lambda + D = 0$ leads to $\sqrt{T^2 - 4D}$. This is real below the parabola and imaginary above it. The stable region is the *upper left quarter* of the figure—where the trace T is negative and the determinant D is positive.

Example 4 Which of these matrices is stable?

$$A_1 = \begin{bmatrix} 0 & -1 \\ -2 & -3 \end{bmatrix} \qquad A_2 = \begin{bmatrix} 4 & 5 \\ -6 & -7 \end{bmatrix} \qquad A_3 = \begin{bmatrix} -8 & 8 \\ 8 & -8 \end{bmatrix}.$$

The Exponential of a Matrix

We return briefly to write the solution $\boldsymbol{u}(t)$ in a new form $e^{At}\boldsymbol{u}(0)$. This gives a perfect parallel with $A^k\boldsymbol{u}_0$ in the previous section. First we have to say what e^{At} means.

The matrix e^{At} has a matrix in the exponent. To define e^{At}, copy e^x. The direct definition of e^x is by the infinite series $1 + x + \frac{1}{2}x^2 + \frac{1}{6}x^3 + \cdots$. When you substitute the matrix At for x, this series defines the matrix e^{At}:

Matrix exponential $$e^{At} = I + At + \tfrac{1}{2}(At)^2 + \tfrac{1}{6}(At)^3 + \cdots. \tag{10}$$

Its t derivative is $$A + A^2t + \tfrac{1}{2}A^3t^2 + \cdots = Ae^{At}.$$

The number that divides $(At)^n$ is "n factorial." This is $n! = (1)(2)\cdots(n-1)(n)$. The factorials after $1, 2, 6$ are $4! = 24$ and $5! = 120$. They grow quickly. The series always converges and its derivative is always Ae^{At}. Therefore $e^{At}\boldsymbol{u}(0)$ solves the differential equation with one quick formula—*even if there is a shortage of eigenvectors.*

This chapter emphasizes how to find $\boldsymbol{u}(t) = e^{At}\boldsymbol{u}(0)$ by diagonalization. Assume A does have n eigenvectors, so it is diagonalizable. Substitute $A = S\Lambda S^{-1}$ into the series for e^{At}. Whenever $S\Lambda S^{-1}S\Lambda S^{-1}$ appears, cancel $S^{-1}S$ in the middle:

$$\begin{aligned} e^{At} &= I + S\Lambda S^{-1}t + \tfrac{1}{2}(S\Lambda S^{-1}t)(S\Lambda S^{-1}t) + \cdots \\ &= S\,[I + \Lambda t + \tfrac{1}{2}(\Lambda t)^2 + \cdots]\,S^{-1} \\ &= Se^{\Lambda t}S^{-1}. \end{aligned} \tag{11}$$

That equation says: e^{At} *equals* $Se^{\Lambda t}S^{-1}$. To compute e^{At}, compute the λ's as usual. Then Λ is a diagonal matrix and so is $e^{\Lambda t}$—the numbers $e^{\lambda_i t}$ are on its diagonal. Multiply $Se^{\Lambda t}S^{-1}\boldsymbol{u}(0)$ to recognize the new solution $\boldsymbol{u}(t) = e^{At}\boldsymbol{u}(0)$. It is the old solution in terms of eigenvalues in Λ and eigenvectors in S:

$$e^{At}\boldsymbol{u}(0) = Se^{\Lambda t}S^{-1}\boldsymbol{u}(0) = \begin{bmatrix} & & \\ \boldsymbol{x}_1 & \cdots & \boldsymbol{x}_n \\ & & \end{bmatrix} \begin{bmatrix} e^{\lambda_1 t} & & \\ & \ddots & \\ & & e^{\lambda_n t} \end{bmatrix} \begin{bmatrix} c_1 \\ \vdots \\ c_n \end{bmatrix}. \tag{12}$$

A combination $c_1\boldsymbol{x}_1 + \cdots + c_n\boldsymbol{x}_n$ is $S\boldsymbol{c}$. This matches the starting value when $\boldsymbol{u}(0) = S\boldsymbol{c}$. The column $\boldsymbol{c} = S^{-1}\boldsymbol{u}(0)$ at the end of equation (12) brings back the best form

$$\boldsymbol{u}(t) = \begin{bmatrix} & & \\ \boldsymbol{x}_1 & \cdots & \boldsymbol{x}_n \\ & & \end{bmatrix} \begin{bmatrix} c_1e^{\lambda_1 t} \\ \vdots \\ c_ne^{\lambda_n t} \end{bmatrix} = c_1e^{\lambda_1 t}\boldsymbol{x}_1 + \cdots + c_ne^{\lambda_n t}\boldsymbol{x}_n. \tag{13}$$

This $e^{At}\boldsymbol{u}(0)$ is the same answer that came from our three steps:

1. Find the λ's and $\boldsymbol{x}$'s, eigenvalues and eigenvectors.

2. Write $\boldsymbol{u}(0) = c_1\boldsymbol{x}_1 + \cdots + c_n\boldsymbol{x}_n$. Here we need n eigenvectors.

3. Multiply each $\boldsymbol{x}_i$ by $e^{\lambda_i t}$. The solution is a combination of pure solutions:

$$\boldsymbol{u}(t) = c_1e^{\lambda_1 t}\boldsymbol{x}_1 + \cdots + c_ne^{\lambda_n t}\boldsymbol{x}_n. \tag{14}$$

Example 5 Use the series to find e^{At} for $A = \left[\begin{smallmatrix} 0 & 1 \\ -1 & 0 \end{smallmatrix}\right]$. Notice that $A^4 = I$:

$$A = \begin{bmatrix} & 1 \\ -1 & \end{bmatrix} \quad A^2 = \begin{bmatrix} -1 & \\ & -1 \end{bmatrix} \quad A^3 = \begin{bmatrix} & -1 \\ 1 & \end{bmatrix} \quad A^4 = \begin{bmatrix} 1 & \\ & 1 \end{bmatrix}.$$

A^5, A^6, A^7, A^8 will repeat these four matrices. The top right corner has $1, 0, -1, 0$ repeating over and over. The infinite series for e^{At} contains $t/1!, 0, -t^3/3!, 0$. In other words $t - \frac{1}{6}t^3$ starts that top right corner, and $1 - \frac{1}{2}t^2$ is in the top left:

$$I + At + \tfrac{1}{2}(At)^2 + \tfrac{1}{6}(At)^3 + \cdots = \begin{bmatrix} 1 - \frac{1}{2}t^2 + \cdots & t - \frac{1}{6}t^3 + \cdots \\ -t + \frac{1}{6}t^3 - \cdots & 1 - \frac{1}{2}t^2 + \cdots \end{bmatrix}.$$

On the left side is the series for e^{At}. The top row of the matrix shows the series for $\cos t$ and $\sin t$. We have found e^{At} directly:

$$e^{At} = \begin{bmatrix} \cos t & \sin t \\ -\sin t & \cos t \end{bmatrix}. \tag{15}$$

At $t = 0$ this gives $e^0 = I$. Most important, the derivative of e^{At} is Ae^{At}:

$$\frac{d}{dt}\begin{bmatrix} \cos t & \sin t \\ -\sin t & \cos t \end{bmatrix} = \begin{bmatrix} -\sin t & \cos t \\ -\cos t & -\sin t \end{bmatrix} = \begin{bmatrix} 0 & 1 \\ -1 & 0 \end{bmatrix}\begin{bmatrix} \cos t & \sin t \\ -\sin t & \cos t \end{bmatrix}.$$

A is a skew-symmetric matrix ($A^{\mathrm{T}} = -A$). Its exponential e^{At} is an orthogonal matrix. The eigenvalues of A are i and $-i$. The eigenvalues of e^{At} are e^{it} and e^{-it}. This illustrates two general rules:

1 ***The eigenvalues of e^{At} are $e^{\lambda t}$.***

2 ***When A is skew-symmetric, e^{At} is orthogonal.***

Example 6 Solve $\dfrac{d\boldsymbol{u}}{dt} = A\boldsymbol{u} = \begin{bmatrix} 1 & 1 \\ 0 & 2 \end{bmatrix}\boldsymbol{u}$ starting from $\boldsymbol{u}(0) = \begin{bmatrix} 2 \\ 1 \end{bmatrix}$ at $t = 0$.

Solution The eigenvalues 1 and 2 are on the diagonal of A (since A is triangular). The eigenvectors are $\boldsymbol{x}_1 = (1, 0)$ and $\boldsymbol{x}_2 = (1, 1)$:

$$\begin{bmatrix} 1 & 1 \\ 0 & 2 \end{bmatrix}\begin{bmatrix} 1 \\ 0 \end{bmatrix} = \begin{bmatrix} 1 \\ 0 \end{bmatrix} \quad \text{and} \quad \begin{bmatrix} 1 & 1 \\ 0 & 2 \end{bmatrix}\begin{bmatrix} 1 \\ 1 \end{bmatrix} = 2\begin{bmatrix} 1 \\ 1 \end{bmatrix}.$$

Step 2 writes $\boldsymbol{u}(0)$ as a combination $\boldsymbol{x}_1 + \boldsymbol{x}_2$ of these eigenvectors. This is $S\boldsymbol{c} = \boldsymbol{u}(0)$. In this case $c_1 = c_2 = 1$. Then $\boldsymbol{u}(t)$ is the same combination of pure exponential solutions:

$$\boldsymbol{u}(t) = e^t\begin{bmatrix} 1 \\ 0 \end{bmatrix} + e^{2t}\begin{bmatrix} 1 \\ 1 \end{bmatrix}.$$

That is the clearest solution $e^{\lambda_1 t}\boldsymbol{x}_1 + e^{\lambda_2 t}\boldsymbol{x}_2$. In matrix form, the eigenvectors go into S:

$$\boldsymbol{u}(t) = Se^{\Lambda t}S^{-1}\boldsymbol{u}(0) \text{ is } \begin{bmatrix} 1 & 1 \\ 0 & 1 \end{bmatrix}\begin{bmatrix} e^t & \\ & e^{2t} \end{bmatrix}\begin{bmatrix} 1 & 1 \\ 0 & 1 \end{bmatrix}^{-1}\boldsymbol{u}(0) = \begin{bmatrix} e^t & e^{2t} + e^t \\ 0 & e^{2t} \end{bmatrix}\boldsymbol{u}(0).$$

That last matrix is e^{At}. It's not bad to see what a matrix exponential looks like (this is a particularly nice one). The situation is the same as for $A\boldsymbol{x} = \boldsymbol{b}$ and inverses. We don't really need A^{-1} to find $\boldsymbol{x}$, and we don't need e^{At} to solve $d\boldsymbol{u}/dt = A\boldsymbol{u}$. But as quick formulas for the answers, $A^{-1}\boldsymbol{b}$ and $e^{At}\boldsymbol{u}(0)$ are unbeatable.

■ REVIEW OF THE KEY IDEAS ■

1. The equation $\boldsymbol{u}' = A\boldsymbol{u}$ is linear with constant coefficients, starting from $\boldsymbol{u}(0)$.

2. Its solution is usually a combination of exponentials, involving each λ and $\boldsymbol{x}$:

$$\boldsymbol{u}(t) = c_1 e^{\lambda_1 t}\boldsymbol{x}_1 + \cdots + c_n e^{\lambda_n t}\boldsymbol{x}_n.$$

3. The constants $c_1, \ldots, c_n$ are determined by $\boldsymbol{u}(0) = c_1\boldsymbol{x}_1 + \cdots + c_n\boldsymbol{x}_n$.

4. The solution approaches zero (stability) if Real part $(\lambda) < 0$ for every λ.

5. The solution is always $\boldsymbol{u}(t) = e^{At}\boldsymbol{u}(0)$, with the matrix exponential e^{At}.

6. Equations involving y'' reduce to $\boldsymbol{u}' = A\boldsymbol{u}$ by combining y' and y into $\boldsymbol{u} = (y', y)$.

■ WORKED EXAMPLES ■

6.3 A Find the eigenvalues and eigenvectors of A and write $\boldsymbol{u}(0) = (2, 0, 2)$ as a combination $c_1\boldsymbol{x}_1 + c_2\boldsymbol{x}_2 + c_3\boldsymbol{x}_3$ of the eigenvectors. Then solve both equations:

$$\frac{d\boldsymbol{u}}{dt} = A\boldsymbol{u} = \begin{bmatrix} -2 & 1 & 0 \\ 1 & -2 & 1 \\ 0 & 1 & -2 \end{bmatrix}\boldsymbol{u} \quad \text{and also} \quad \frac{d^2\boldsymbol{u}}{dt^2} = A\boldsymbol{u} \quad \text{with} \quad \frac{d\boldsymbol{u}}{dt}(0) = \boldsymbol{0}.$$

The $1, -2, 1$ diagonals make A into a *second difference matrix* (like a second derivative). So the first equation $\boldsymbol{u}' = A\boldsymbol{u}$ is like the *heat equation* $\partial u/\partial t = \partial^2 u/\partial x^2$. Its solution $u(t)$ will decay as the heat diffuses out. The second equation $\boldsymbol{u}'' = A\boldsymbol{u}$ is like the *wave equation* $\partial^2 u/\partial t^2 = \partial^2 u/\partial x^2$. Its solution will oscillate like a string on a violin.

Solution The eigenvalues and eigenvectors come from $\det(A - \lambda I) = 0$:

$$\det(A - \lambda I) = \begin{vmatrix} -2-\lambda & 1 & 0 \\ 1 & -2-\lambda & 1 \\ 0 & 1 & -2-\lambda \end{vmatrix} = (-2-\lambda)^3 - 2(-2-\lambda) = 0.$$

One eigenvalue is $\lambda = -2$, when $-2 - \lambda$ is zero. Factor out $-2 - \lambda$ to leave $(-2 - \lambda)^2 - 2 = 0$ or $\lambda^2 + 4\lambda + 2 = 0$. The other eigenvalues (also negative) are $\lambda = -2 \pm \sqrt{2}$.

The eigenvectors are found separately:

$$\lambda = -2: \qquad (A+2I)\boldsymbol{x} = \begin{bmatrix} 0 & 1 & 0 \\ 1 & 0 & 1 \\ 0 & 1 & 0 \end{bmatrix} \begin{bmatrix} x \\ y \\ z \end{bmatrix} = \begin{bmatrix} 0 \\ 0 \\ 0 \end{bmatrix} \quad \text{for } \boldsymbol{x}_1 = \begin{bmatrix} \sqrt{2} \\ 0 \\ -\sqrt{2} \end{bmatrix}$$

$$\lambda = -2-\sqrt{2}: \quad (A-\lambda I)\boldsymbol{x} = \begin{bmatrix} \sqrt{2} & 1 & 0 \\ 1 & \sqrt{2} & 1 \\ 0 & 1 & \sqrt{2} \end{bmatrix} \begin{bmatrix} x \\ y \\ z \end{bmatrix} = \begin{bmatrix} 0 \\ 0 \\ 0 \end{bmatrix} \quad \text{for } \boldsymbol{x}_2 = \begin{bmatrix} 1 \\ -\sqrt{2} \\ 1 \end{bmatrix}$$

$$\lambda = -2+\sqrt{2}: \quad (A-\lambda I)\boldsymbol{x} = \begin{bmatrix} -\sqrt{2} & 1 & 0 \\ 1 & -\sqrt{2} & 1 \\ 0 & 1 & -\sqrt{2} \end{bmatrix} \begin{bmatrix} x \\ y \\ z \end{bmatrix} = \begin{bmatrix} 0 \\ 0 \\ 0 \end{bmatrix} \quad \text{for } \boldsymbol{x}_3 = \begin{bmatrix} 1 \\ \sqrt{2} \\ 1 \end{bmatrix}$$

All those eigenvectors have length 2, so $\frac{1}{2}\boldsymbol{x}_1, \frac{1}{2}\boldsymbol{x}_2, \frac{1}{2}\boldsymbol{x}_3$ are unit vectors. These eigenvectors are *orthogonal* (proved in Section 6.4 for every real symmetric matrix A). Expand $\boldsymbol{u}(0)$ as a combination $c_1\boldsymbol{x}_1 + c_2\boldsymbol{x}_2 + c_3\boldsymbol{x}_3$ (then $c_1 = 0$ and $c_2 = c_3 = 1$):

$$\begin{bmatrix} \boldsymbol{x}_1 & \boldsymbol{x}_2 & \boldsymbol{x}_3 \end{bmatrix} \begin{bmatrix} c_1 \\ c_2 \\ c_3 \end{bmatrix} = \boldsymbol{u}(0) \quad \text{is} \quad \begin{bmatrix} \sqrt{2} & 1 & 1 \\ 0 & -\sqrt{2} & \sqrt{2} \\ -\sqrt{2} & 1 & 1 \end{bmatrix} \begin{bmatrix} 0 \\ 1 \\ 1 \end{bmatrix} = \begin{bmatrix} 2 \\ 0 \\ 2 \end{bmatrix}.$$

From $\boldsymbol{u}(0) = \boldsymbol{x}_2 + \boldsymbol{x}_3$ the solution decays to $\boldsymbol{u}(t) = e^{\lambda_2 t}\boldsymbol{x}_2 + e^{\lambda_3 t}\boldsymbol{x}_3$. Since all λ's are negative, $\boldsymbol{u}(t)$ approaches zero (*stability*). The least negative $\lambda = -2+\sqrt{2}$ gives the decay rate. This is like Problem 6.3.5 except people are in three rooms (change people to temperature for the heat equation). The rate $\boldsymbol{u}'$ of movement between rooms is the population difference or temperature difference. The total going *into* the first room is $u_2 - 2u_1$ as required by $A\boldsymbol{u}$. Eventually $\boldsymbol{u}(t) \to \boldsymbol{0}$ and the rooms empty out.

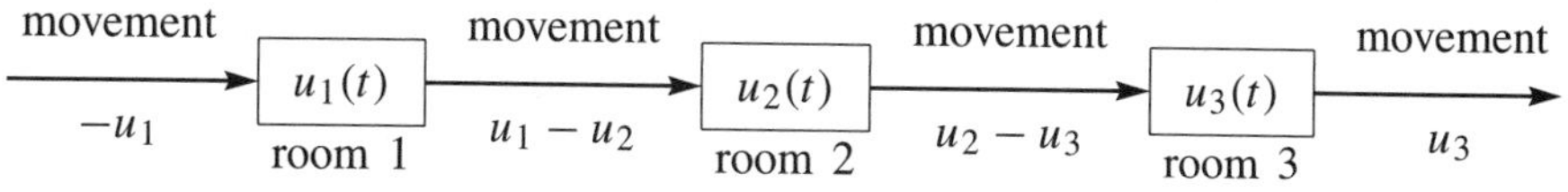

Turn now to the "wave equation" $d^2\boldsymbol{u}/dt^2 = A\boldsymbol{u}$ (not developed in the text). The same eigenvectors lead to oscillations $e^{i\omega t}\boldsymbol{x}$ and $e^{-i\omega t}\boldsymbol{x}$ with frequencies from $\omega^2 = -\lambda$:

$$\frac{d^2}{dt^2}(e^{i\omega t}\boldsymbol{x}) = A(e^{i\omega t}\boldsymbol{x}) \qquad \text{becomes} \qquad (i\omega)^2 e^{i\omega t}\boldsymbol{x} = \lambda e^{i\omega t}\boldsymbol{x} \quad \text{and} \quad \omega^2 = -\lambda.$$

There are two square roots of $-\lambda$, so we have $e^{i\omega t}\boldsymbol{x}$ and $e^{-i\omega t}\boldsymbol{x}$. With three eigenvectors this makes *six* solutions. A combination will match the six components of $\boldsymbol{u}(0)$ and $\boldsymbol{u}'(0) =$ velocity. Since $\boldsymbol{u}' = \boldsymbol{0}$ in this problem, $e^{i\omega t}\boldsymbol{x}$ combines with $e^{-i\omega t}\boldsymbol{x}$ into $2\cos\omega t\,\boldsymbol{x}$. Our particular $\boldsymbol{u}(0)$ is again $\boldsymbol{x}_2 + \boldsymbol{x}_3$, and the solution oscillates:

$$\boldsymbol{u}(t) = 2(\cos\omega_2 t)\boldsymbol{x}_2 + 2(\cos\omega_3 t)\boldsymbol{x}_3 \quad \text{with} \quad (\omega_2)^2 = 2+\sqrt{2} \quad \text{and} \quad (\omega_3)^2 = 2-\sqrt{2}.$$

Each λ is negative, so $\omega^2 = -\lambda$ gives two real frequencies. A symmetric matrix like A with negative eigenvalues is a *negative definite matrix*. (Section 6.5 takes a more positive viewpoint, for *positive definite matrices*.) Matrices like A and $-A$ are the key to all the engineering applications in Section 8.1.

6.3 B Solve the four equations $da/dt = 0, db/dt = a, dc/dt = 2b, dz/dt = 3c$ in that order starting from $\boldsymbol{u}(0) = (a(0), b(0), c(0), z(0))$. Note the matrix in $\boldsymbol{u}' = A\boldsymbol{u}$ and solve the same equations by the matrix exponential in $\boldsymbol{u}(t) = e^{At}\boldsymbol{u}(0)$:

$$\frac{d}{dt}\begin{bmatrix} a \\ b \\ c \\ z \end{bmatrix} = \begin{bmatrix} 0 & 0 & 0 & 0 \\ 1 & 0 & 0 & 0 \\ 0 & 2 & 0 & 0 \\ 0 & 0 & 3 & 0 \end{bmatrix}\begin{bmatrix} a \\ b \\ c \\ z \end{bmatrix} \quad \text{is} \quad \frac{d\boldsymbol{u}}{dt} = A\boldsymbol{u}.$$

First find A^2, A^3, A^4 and then $e^{At} = I + At + \frac{1}{2}(At)^2 + \frac{1}{6}(At)^3$. Why does the series stop there? Check that e^A is Pascal's triangular matrix at $t = 1$, and the derivative of e^{At} at $t = 0$ is A. Finally, verify $(e^A)(e^A) = (e^{2A})$. Why is this true for any A?

Solution Integrate $da/dt = 0$, then $db/dt = a$, then $dc/dt = 2b$ and $dz/dt = 3c$:

$$\begin{aligned} a(t) &= a(0) \\ b(t) &= ta(0) + b(0) \\ c(t) &= t^2a(0) + 2tb(0) + c(0) \\ z(t) &= t^3a(0) + 3t^2b(0) + 3tc(0) + z(0) \end{aligned} \qquad \text{which must match } e^{At}\boldsymbol{u}(0).$$

The powers of A are all zero after A^3. So the series for e^{At} stops after four terms:

$$A = \begin{bmatrix} 0 & 0 & 0 & 0 \\ 1 & 0 & 0 & 0 \\ 0 & 2 & 0 & 0 \\ 0 & 0 & 3 & 0 \end{bmatrix} \quad A^2 = \begin{bmatrix} 0 & 0 & 0 & 0 \\ 0 & 0 & 0 & 0 \\ 2 & 0 & 0 & 0 \\ 0 & 6 & 0 & 0 \end{bmatrix} \quad A^3 = \begin{bmatrix} 0 & 0 & 0 & 0 \\ 0 & 0 & 0 & 0 \\ 0 & 0 & 0 & 0 \\ 6 & 0 & 0 & 0 \end{bmatrix}$$

The diagonals move down at each step and disappear for A^4. (There must be a diagonal-diagonal rule to go with the row-column and column-row rules for multiplying matrices.) The matrix exponential is the same one that multiplied $(a(0), b(0), c(0), z(0))$ above:

$$e^{At} = I + At + \frac{(At)^2}{2} + \frac{(At)^3}{6} = \begin{bmatrix} 1 & & & \\ t & 1 & & \\ t^2 & 2t & 1 & \\ t^3 & 3t^2 & 3t & 1 \end{bmatrix}$$

At $t = 1$, e^A is *Pascal's triangular matrix* P_L. The derivative of e^{At} at $t = 0$ is A:

$$\lim_{t\to 0}\frac{e^{At} - I}{t} = \lim_{t\to 0}\begin{bmatrix} 0 & 0 & 0 & 0 \\ 1 & 0 & 0 & 0 \\ t & 2 & 0 & 0 \\ t^2 & 3t & 3 & 0 \end{bmatrix} = \begin{bmatrix} 0 & 0 & 0 & 0 \\ 1 & 0 & 0 & 0 \\ 0 & 2 & 0 & 0 \\ 0 & 0 & 3 & 0 \end{bmatrix} = A.$$

A is the ***matrix logarithm*** of Pascal's e^A! The inverse of Pascal is e^{-A} with two negative diagonals. The square of e^A is always e^{2A} (and also $e^{As}e^{At} = e^{A(s+t)}$) for many reasons:

Solving with e^A from $t=0$ to 1 and then 1 to 2 agrees with e^{2A} from 0 to 2. The squared series $(I+A+\frac{A^2}{2}+\cdots)^2$ agrees with $I+2A+\frac{(2A)^2}{2}+\cdots=e^{2A}$. If A can be diagonalized (this A can't!) then $(Se^{\Lambda}S^{-1})(Se^{\Lambda}S^{-1})=Se^{2\Lambda}S^{-1}$.

Problem Set 6.3

1 Find λ's and $\boldsymbol{x}$'s so that $\boldsymbol{u}=e^{\lambda t}\boldsymbol{x}$ solves

$$\frac{d\boldsymbol{u}}{dt}=\begin{bmatrix}4 & 3\\ 0 & 1\end{bmatrix}\boldsymbol{u}.$$

What combination $\boldsymbol{u}=c_1e^{\lambda_1 t}\boldsymbol{x}_1+c_2e^{\lambda_2 t}\boldsymbol{x}_2$ starts from $\boldsymbol{u}(0)=(5,-2)$?

2 Solve Problem 1 for $\boldsymbol{u}=(y,z)$ by back substitution:

$$\text{First solve } \frac{dz}{dt}=z \text{ starting from } z(0)=-2.$$

$$\text{Then solve } \frac{dy}{dt}=4y+3z \text{ starting from } y(0)=5.$$

The solution for y will be a combination of e^{4t} and e^t.

3 Find A to change the scalar equation $y''=5y'+4y$ into a vector equation for $\boldsymbol{u}=(y,y')$:

$$\frac{d\boldsymbol{u}}{dt}=\begin{bmatrix}y'\\ y''\end{bmatrix}=\begin{bmatrix} & \\ & \end{bmatrix}\begin{bmatrix}y\\ y'\end{bmatrix}=A\boldsymbol{u}.$$

What are the eigenvalues of A? Find them also by substituting $y=e^{\lambda t}$ into $y''=5y'+4y$.

4 The rabbit and wolf populations show fast growth of rabbits (from $6r$) but loss to wolves (from $-2w$):

$$\frac{dr}{dt}=6r-2w \quad\text{and}\quad \frac{dw}{dt}=2r+w.$$

Find the eigenvalues and eigenvectors. If $r(0)=w(0)=30$ what are the populations at time t? After a long time, is the ratio of rabbits to wolves 1 to 2 or is it 2 to 1?

5 A door is opened between rooms that hold $v(0)=30$ people and $w(0)=10$ people. The movement between rooms is proportional to the difference $v-w$:

$$\frac{dv}{dt}=w-v \quad\text{and}\quad \frac{dw}{dt}=v-w.$$

Show that the total $v+w$ is constant (40 people). Find the matrix in $d\boldsymbol{u}/dt=A\boldsymbol{u}$ and its eigenvalues and eigenvectors. What are v and w at $t=1$?

6 Reverse the diffusion of people in Problem 5 to $d\boldsymbol{u}/dt = -A\boldsymbol{u}$:

$$\frac{dv}{dt} = v - w \quad \text{and} \quad \frac{dw}{dt} = w - v.$$

The total $v + w$ still remains constant. How are the λ's changed now that A is changed to $-A$? But show that $v(t)$ grows to infinity from $v(0) = 30$.

7 The solution to $y'' = 0$ is a straight line $y = C + Dt$. Convert to a matrix equation:

$$\frac{d}{dt}\begin{bmatrix} y \\ y' \end{bmatrix} = \begin{bmatrix} 0 & 1 \\ 0 & 0 \end{bmatrix}\begin{bmatrix} y \\ y' \end{bmatrix} \text{ has the solution } \begin{bmatrix} y \\ y' \end{bmatrix} = e^{At}\begin{bmatrix} y(0) \\ y'(0) \end{bmatrix}.$$

This matrix A cannot be diagonalized. Find A^2 and compute $e^{At} = I + At + \frac{1}{2}A^2t^2 + \cdots$. Multiply your e^{At} times $(y(0), y'(0))$ to check the straight line $y(t) = y(0) + y'(0)t$.

8 Substitute $y = e^{\lambda t}$ into $y'' = 6y' - 9y$ to show that $\lambda = 3$ is a repeated root. This is trouble; we need a second solution after e^{3t}. The matrix equation is

$$\frac{d}{dt}\begin{bmatrix} y \\ y' \end{bmatrix} = \begin{bmatrix} 0 & 1 \\ -9 & 6 \end{bmatrix}\begin{bmatrix} y \\ y' \end{bmatrix}.$$

Show that this matrix has $\lambda = 3, 3$ and only one line of eigenvectors. *Trouble here too.* Show that the second solution is $y = te^{3t}$.

9 Figure out how to write $my'' + by' + ky = 0$ as a vector equation $M\boldsymbol{u}' = A\boldsymbol{u}$.

10 The matrix in this question is skew-symmetric ($A^{\mathrm{T}} = -A$):

$$\frac{d\boldsymbol{u}}{dt} = \begin{bmatrix} 0 & c & -b \\ -c & 0 & a \\ b & -a & 0 \end{bmatrix}\boldsymbol{u} \quad \text{or} \quad \begin{aligned} u_1' &= cu_2 - bu_3 \\ u_2' &= au_3 - cu_1 \\ u_3' &= bu_1 - au_2. \end{aligned}$$

(a) The derivative of $\|\boldsymbol{u}(t)\|^2 = u_1^2 + u_2^2 + u_3^2$ is $2u_1u_1' + 2u_2u_2' + 2u_3u_3'$. Substitute u_1', u_2', u_3' to get *zero*. Then $\|\boldsymbol{u}(t)\|^2$ stays equal to $\|\boldsymbol{u}(0)\|^2$.

(b) *When A is skew-symmetric, $Q = e^{At}$ is orthogonal.* Prove $Q^{\mathrm{T}} = e^{-At}$ from the series for $Q = e^{At}$. Then $Q^{\mathrm{T}}Q = I$.

11 (a) Write $(1, 0)$ as a combination $c_1\boldsymbol{x}_1 + c_2\boldsymbol{x}_2$ of these two eigenvectors of A:

$$\begin{bmatrix} 0 & 1 \\ -1 & 0 \end{bmatrix}\begin{bmatrix} 1 \\ i \end{bmatrix} = i\begin{bmatrix} 1 \\ i \end{bmatrix} \qquad \begin{bmatrix} 0 & 1 \\ -1 & 0 \end{bmatrix}\begin{bmatrix} 1 \\ -i \end{bmatrix} = -i\begin{bmatrix} 1 \\ -i \end{bmatrix}.$$

(b) The solution to $d\boldsymbol{u}/dt = A\boldsymbol{u}$ starting from $(1, 0)$ is $c_1e^{it}\boldsymbol{x}_1 + c_2e^{-it}\boldsymbol{x}_2$. Substitute $e^{it} = \cos t + i \sin t$ and $e^{-it} = \cos t - i \sin t$ to find $\boldsymbol{u}(t)$.

12 (a) Write down two familiar functions that solve the equation $d^2y/dt^2 = -y$. Which one starts with $y(0) = 1$ and $y'(0) = 0$?

(b) This second-order equation $y'' = -y$ produces a vector equation $\boldsymbol{u}' = A\boldsymbol{u}$:

$$\boldsymbol{u} = \begin{bmatrix} y \\ y' \end{bmatrix} \qquad \frac{d\boldsymbol{u}}{dt} = \begin{bmatrix} y' \\ y'' \end{bmatrix} = \begin{bmatrix} 0 & 1 \\ -1 & 0 \end{bmatrix} \begin{bmatrix} y \\ y' \end{bmatrix} = A\boldsymbol{u}.$$

Put $y(t)$ from part (a) into $\boldsymbol{u}(t) = (y, y')$. This solves Problem 11 again.

13 A particular solution to $d\boldsymbol{u}/dt = A\boldsymbol{u} - \boldsymbol{b}$ is $\boldsymbol{u}_p = A^{-1}\boldsymbol{b}$, if A is invertible. The solutions to $d\boldsymbol{u}/dt = A\boldsymbol{u}$ give $\boldsymbol{u}_n$. Find the complete solution $\boldsymbol{u}_p + \boldsymbol{u}_n$ to

(a) $\dfrac{du}{dt} = 2u - 8$ (b) $\dfrac{d\boldsymbol{u}}{dt} = \begin{bmatrix} 2 & 0 \\ 0 & 3 \end{bmatrix} \boldsymbol{u} - \begin{bmatrix} 8 \\ 6 \end{bmatrix}.$

14 If c is not an eigenvalue of A, substitute $\boldsymbol{u} = e^{ct}\boldsymbol{v}$ and find $\boldsymbol{v}$ to solve $d\boldsymbol{u}/dt = A\boldsymbol{u} - e^{ct}\boldsymbol{b}$. This $\boldsymbol{u} = e^{ct}\boldsymbol{v}$ is a particular solution. How does it break down when c is an eigenvalue?

15 Find a matrix A to illustrate each of the unstable regions in Figure 6.4:

(a) $\lambda_1 < 0$ and $\lambda_2 > 0$

(b) $\lambda_1 > 0$ and $\lambda_2 > 0$

(c) Complex λ's with real part $a > 0$.

Questions 16–25 are about the matrix exponential e^{At}.

16 Write five terms of the infinite series for e^{At}. Take the t derivative of each term. Show that you have four terms of Ae^{At}. Conclusion: $e^{At}\boldsymbol{u}_0$ solves $\boldsymbol{u}' = A\boldsymbol{u}$.

17 The matrix $B = \left[\begin{smallmatrix} 0 & -1 \\ 0 & 0 \end{smallmatrix}\right]$ has $B^2 = 0$. Find e^{Bt} from a (short) infinite series. Check that the derivative of e^{Bt} is Be^{Bt}.

18 Starting from $\boldsymbol{u}(0)$ the solution at time T is $e^{AT}\boldsymbol{u}(0)$. Go an additional time t to reach $e^{At}\left(e^{AT}\boldsymbol{u}(0)\right)$. This solution at time $t + T$ can also be written as ____. Conclusion: e^{At} times e^{AT} equals ____.

19 Write $A = \left[\begin{smallmatrix} 1 & 1 \\ 0 & 0 \end{smallmatrix}\right]$ in the form $S\Lambda S^{-1}$. Find e^{At} from $Se^{\Lambda t}S^{-1}$.

20 If $A^2 = A$ show that the infinite series produces $e^{At} = I + (e^t - 1)A$. For $A = \left[\begin{smallmatrix} 1 & 1 \\ 0 & 0 \end{smallmatrix}\right]$ in Problem 19 this gives $e^{At} =$ ____.

21 Generally $e^A e^B$ is different from $e^B e^A$. They are both different from e^{A+B}. Check this using Problems 19–20 and 17:

$$A = \begin{bmatrix} 1 & 1 \\ 0 & 0 \end{bmatrix} \qquad B = \begin{bmatrix} 0 & -1 \\ 0 & 0 \end{bmatrix} \qquad A + B = \begin{bmatrix} 1 & 0 \\ 0 & 0 \end{bmatrix}.$$

22 Write $A = \begin{bmatrix} 1 & 1 \\ 0 & 3 \end{bmatrix}$ as $S\Lambda S^{-1}$. Multiply $Se^{\Lambda t}S^{-1}$ to find the matrix exponential e^{At}. Check e^{At} when $t = 0$.

23 Put $A = \begin{bmatrix} 1 & 3 \\ 0 & 0 \end{bmatrix}$ into the infinite series to find e^{At}. First compute A^2!

$$e^{At} = \begin{bmatrix} 1 & 0 \\ 0 & 1 \end{bmatrix} + \begin{bmatrix} t & 3t \\ 0 & 0 \end{bmatrix} + \frac{1}{2}\begin{bmatrix} & \\ & \end{bmatrix} + \cdots = \begin{bmatrix} e^t & \\ 0 & \end{bmatrix}.$$

24 Give two reasons why the matrix exponential e^{At} is never singular:

(a) Write down its inverse.

(b) Write down its eigenvalues. If $A\boldsymbol{x} = \lambda\boldsymbol{x}$ then $e^{At}\boldsymbol{x} =$ _____ $\boldsymbol{x}$.

25 Find a solution $x(t)$, $y(t)$ that gets large as $t \to \infty$. To avoid this instability a scientist exchanged the two equations:

$$\begin{aligned} dx/dt &= 0x - 4y \\ dy/dt &= -2x + 2y \end{aligned} \quad \text{becomes} \quad \begin{aligned} dy/dt &= -2x + 2y \\ dx/dt &= 0x - 4y. \end{aligned}$$

Now the matrix $\begin{bmatrix} -2 & 2 \\ 0 & -4 \end{bmatrix}$ is stable. It has negative eigenvalues. Comment on this.

SYMMETRIC MATRICES ■ 6.4

For projection onto a line, the eigenvalues are 1 and 0. Eigenvectors are *on* the line (where $P\boldsymbol{x} = \boldsymbol{x}$) and *perpendicular* to the line (where $P\boldsymbol{x} = \boldsymbol{0}$). Now we open up to all other ***symmetric matrices***. It is no exaggeration to say that these are the most important matrices the world will ever see—in the theory of linear algebra and also in the applications. We come immediately to the key questions about symmetry. Not only the questions, but also the answers.

What is special about $A\boldsymbol{x} = \lambda\boldsymbol{x}$ ***when A is symmetric***? We are looking for special properties of the eigenvalues λ and the eigenvectors $\boldsymbol{x}$ when $A = A^{\mathrm{T}}$.

The diagonalization $A = S\Lambda S^{-1}$ will reflect the symmetry of A. We get some hint by transposing to $A^{\mathrm{T}} = (S^{-1})^{\mathrm{T}}\Lambda S^{\mathrm{T}}$. Those are the same since $A = A^{\mathrm{T}}$. Possibly S^{-1} in the first form equals S^{T} in the second form. Then $S^{\mathrm{T}}S = I$. That makes each eigenvector in S orthogonal to the other eigenvectors. The key facts get first place in the Table at the end of this chapter, and here they are:

1. A symmetric matrix has only ***real eigenvalues***.
2. The ***eigenvectors*** can be chosen ***orthonormal***.

Those orthonormal eigenvectors go into the columns of S. There are n of them (independent because they are orthonormal). Every symmetric matrix can be diagonalized.

Its eigenvector matrix S becomes an orthogonal matrix Q. Orthogonal matrices have $Q^{-1} = Q^{\mathrm{T}}$—what we suspected about S is true. To remember it we write $S = Q$, when we choose orthonormal eigenvectors.

Why do we use the word "choose"? Because the eigenvectors do not *have* to be unit vectors. Their lengths are at our disposal. We will choose unit vectors—eigenvectors of length one, which are orthonormal and not just orthogonal. Then $A = S\Lambda S^{-1}$ is in its special and particular form $Q\Lambda Q^{\mathrm{T}}$ for symmetric matrices:

6H (Spectral Theorem) Every symmetric matrix has the factorization $A = Q\Lambda Q^{\mathrm{T}}$ with real eigenvalues in Λ and orthonormal eigenvectors in Q:

$$A = Q\Lambda Q^{-1} = Q\Lambda Q^{\mathrm{T}} \quad \text{with} \quad Q^{-1} = Q^{\mathrm{T}}.$$

It is easy to see that $Q\Lambda Q^{\mathrm{T}}$ is symmetric. Take its transpose. You get $(Q^{\mathrm{T}})^{\mathrm{T}}\Lambda^{\mathrm{T}}Q^{\mathrm{T}}$, which is $Q\Lambda Q^{\mathrm{T}}$ again. The harder part is to prove that every symmetric matrix has real λ's and orthonormal $\boldsymbol{x}$'s. This is the *"spectral theorem"* in mathematics and the *"principal axis theorem"* in geometry and physics. We have to prove it! No choice. I will approach the proof in three steps:

1. By an example (which only proves that $A = Q\Lambda Q^{\mathrm{T}}$ might be true)

2. By calculating the 2 by 2 case (which convinces most fair-minded people)

3. By a proof when no eigenvalues are repeated (leaving only real diehards).

The diehards are worried about repeated eigenvalues. Are there still n orthonormal eigenvectors? *Yes, there are*. They go into the columns of S (which becomes Q). The last page before the problems outlines this fourth and final step.

We now take steps 1 and 2. In a sense they are optional. The 2 by 2 case is mostly for fun, since it is included in the final n by n case.

Example 1 Find the λ's and $\boldsymbol{x}$'s when $A = \begin{bmatrix} 1 & 2 \\ 2 & 4 \end{bmatrix}$ and $A - \lambda I = \begin{bmatrix} 1-\lambda & 2 \\ 2 & 4-\lambda \end{bmatrix}$.

Solution The equation $\det(A - \lambda I) = 0$ is $\lambda^2 - 5\lambda = 0$. The eigenvalues are 0 and 5 (*both real*). We can see them directly: $\lambda = 0$ is an eigenvalue because A is singular, and $\lambda = 5$ is the other eigenvalue so that $0 + 5$ agrees with $1 + 4$. This is the *trace* down the diagonal of A.

Two eigenvectors are $(2, -1)$ and $(1, 2)$—orthogonal but not yet orthonormal. The eigenvector for $\lambda = 0$ is in the *nullspace* of A. The eigenvector for $\lambda = 5$ is in the *column space*. We ask ourselves, why are the nullspace and column space perpendicular? The Fundamental Theorem says that the nullspace is perpendicular to the ***row space***—not the column space. But our matrix is *symmetric*! Its row and column spaces are the same. Its eigenvectors $(2, -1)$ and $(1, 2)$ must be (and are) perpendicular.

These eigenvectors have length $\sqrt{5}$. Divide them by $\sqrt{5}$ to get unit vectors. Put those into the columns of S (which is Q). Then $Q^{-1}AQ$ is Λ and $Q^{-1} = Q^{\mathrm{T}}$:

$$Q^{-1}AQ = \frac{\begin{bmatrix} 2 & -1 \\ 1 & 2 \end{bmatrix}}{\sqrt{5}} \begin{bmatrix} 1 & 2 \\ 2 & 4 \end{bmatrix} \frac{\begin{bmatrix} 2 & 1 \\ -1 & 2 \end{bmatrix}}{\sqrt{5}} = \begin{bmatrix} 0 & 0 \\ 0 & 5 \end{bmatrix} = \Lambda.$$

Now comes the calculation for *any* 2 by 2 symmetric matrix $[a \;\; b\,; \;\; b \;\; c]$. First, ***real eigenvalues***. Second, ***perpendicular eigenvectors***. The λ's come from

$$\det \begin{bmatrix} a-\lambda & b \\ b & c-\lambda \end{bmatrix} = \lambda^2 - (a+c)\lambda + (ac - b^2) = 0. \tag{1}$$

The test for real roots of $A\lambda^2 + B\lambda + C = 0$ is based on $B^2 - 4AC$. *This must not be negative*, or its square root in the quadratic formula would be imaginary. Our equation has $A = 1$ and $B = -(a+c)$ and $C = ac - b^2$. Look at $B^2 - 4AC$:

Real eigenvalues: $(a+c)^2 - 4(ac - b^2)$ must not be negative.

Rewrite that as $a^2 + 2ac + c^2 - 4ac + 4b^2$. Rewrite again as $(a-c)^2 + 4b^2$. Those squares are not negative! So the roots λ_1 and λ_2 (the eigenvalues) are certainly real.

Perpendicular eigenvectors: Compute $\boldsymbol{x}_1$ and $\boldsymbol{x}_2$ and their dot product:

$$(A - \lambda_1 I)\boldsymbol{x}_1 = \begin{bmatrix} a-\lambda_1 & b \\ b & c-\lambda_1 \end{bmatrix} \begin{bmatrix} \boldsymbol{x}_1 \end{bmatrix} = \mathbf{0} \quad \text{so} \quad \boldsymbol{x}_1 = \begin{bmatrix} b \\ \lambda_1 - a \end{bmatrix} \quad \begin{matrix} \text{from} \\ \text{first} \\ \text{row} \end{matrix}$$

$$(A - \lambda_2 I)\boldsymbol{x}_2 = \begin{bmatrix} a-\lambda_2 & b \\ b & c-\lambda_2 \end{bmatrix} \begin{bmatrix} \boldsymbol{x}_2 \end{bmatrix} = \mathbf{0} \quad \text{so} \quad \boldsymbol{x}_2 = \begin{bmatrix} \lambda_2 - c \\ b \end{bmatrix} \quad \begin{matrix} \text{from} \\ \text{second} \\ \text{row} \end{matrix}$$

The dot product of $\boldsymbol{x}_1$ and $\boldsymbol{x}_2$ proves that these eigenvectors are perpendicular:

$$\boldsymbol{x}_1 \cdot \boldsymbol{x}_2 = b(\lambda_2 - c) + (\lambda_1 - a)b = b(\lambda_1 + \lambda_2 - a - c) = 0. \tag{2}$$

This is zero because $\lambda_1 + \lambda_2$ equals the trace $a + c$. Thus $\boldsymbol{x}_1 \cdot \boldsymbol{x}_2 = 0$. Eagle eyes might notice the special case $a = c, b = 0$ when $\boldsymbol{x}_1 = \boldsymbol{x}_2 = \mathbf{0}$. This case has repeated eigenvalues, as in $A = I$. It still has perpendicular eigenvectors $(1,0)$ and $(0,1)$.

Now comes the general n by n case, with real λ's and perpendicular eigenvectors.

6I Real Eigenvalues The eigenvalues of a real symmetric matrix are real.

Proof Suppose that $A\boldsymbol{x} = \lambda\boldsymbol{x}$. Until we know otherwise, λ might be a complex number $a + ib$ (a and b real). *Its complex conjugate is* $\bar{\lambda} = a - ib$. Similarly the components of $\boldsymbol{x}$ may be complex numbers, and switching the signs of their imaginary parts gives $\bar{x}$. The good thing is that $\bar{\lambda}$ times $\bar{x}$ is always the conjugate of λ times x. So take conjugates of $Ax = \lambda x$, remembering that A is real:

$$Ax = \lambda x \quad \text{leads to} \quad A\bar{x} = \bar{\lambda}\bar{x}. \qquad \text{Transpose to} \quad \bar{x}^{\mathrm{T}}A = \bar{x}^{\mathrm{T}}\bar{\lambda}. \tag{3}$$

Now take the dot product of the first equation with $\bar{x}$ and the last equation with x:

$$\bar{x}^{\mathrm{T}}Ax = \bar{x}^{\mathrm{T}}\lambda x \qquad \text{and also} \qquad \bar{x}^{\mathrm{T}}Ax = \bar{x}^{\mathrm{T}}\bar{\lambda}x. \tag{4}$$

The left sides are the same so the right sides are equal. One equation has λ, the other has $\bar{\lambda}$. They multiply $\bar{x}^{\mathrm{T}}x$ which is not zero—it is the squared length of the eigenvector. *Therefore λ must equal $\bar{\lambda}$*, and $a + ib$ equals $a - ib$. The imaginary part is $b = 0$. Q.E.D.

The eigenvectors come from solving the real equation $(A - \lambda I)x = 0$. So the x's are also real. The important fact is that they are perpendicular.

6J Orthogonal Eigenvectors Eigenvectors of a real symmetric matrix (when they correspond to different λ's) are always perpendicular.

A has real eigenvalues and n real orthogonal eigenvectors if and only if $A = A^{\mathrm{T}}$.

Proof Suppose $Ax = \lambda_1 x$ and $Ay = \lambda_2 y$ and $A = A^{\mathrm{T}}$. Take dot products of the first equation with y and the second with x:

$$(\lambda_1 x)^{\mathrm{T}}y = (Ax)^{\mathrm{T}}y = x^{\mathrm{T}}A^{\mathrm{T}}y = x^{\mathrm{T}}Ay = x^{\mathrm{T}}\lambda_2 y. \tag{5}$$

The left side is $x^{\mathrm{T}}\lambda_1 y$, the right side is $x^{\mathrm{T}}\lambda_2 y$. Since $\lambda_1 \neq \lambda_2$, this proves that $x^{\mathrm{T}}y = 0$. The eigenvector x (for λ_1) is perpendicular to the eigenvector y (for λ_2).

Example 2 Find the λ's and x's for this symmetric matrix with trace zero:

$$A = \begin{bmatrix} -3 & 4 \\ 4 & 3 \end{bmatrix} \quad \text{has} \quad \det(A - \lambda I) = \begin{vmatrix} -3-\lambda & 4 \\ 4 & 3-\lambda \end{vmatrix} = \lambda^2 - 25.$$

The roots of $\lambda^2 - 25 = 0$ are $\lambda_1 = 5$ and $\lambda_2 = -5$ (both real). The eigenvectors $x_1 = (1, 2)$ and $x_2 = (-2, 1)$ are perpendicular. To make them into unit vectors, divide by their lengths $\sqrt{5}$. The new x_1 and x_2 are the columns of Q, and Q^{-1} equals Q^{T}:

$$A = Q\Lambda Q^T = \frac{\begin{bmatrix} 1 & -2 \\ 2 & 1 \end{bmatrix}}{\sqrt{5}} \begin{bmatrix} 5 & 0 \\ 0 & -5 \end{bmatrix} \frac{\begin{bmatrix} 1 & 2 \\ -2 & 1 \end{bmatrix}}{\sqrt{5}}.$$

This example shows the main goal of this section—***to diagonalize symmetric matrices A by orthogonal eigenvector matrices $S = Q$***:

6H (repeated) Every symmetric matrix A has a complete set of orthogonal eigenvectors:

$$A = S\Lambda S^{-1} \quad \text{becomes} \quad A = Q\Lambda Q^{\mathrm{T}}.$$

If $A = A^{\mathrm{T}}$ has a double eigenvalue λ, there are two independent eigenvectors. We use Gram-Schmidt to make them orthogonal. The Teaching Code **eigvec** does this for each eigenspace of A, whatever its dimension. The eigenvectors go into the columns of Q.

One more step. Every 2 by 2 symmetric matrix looks like

$$A = Q\Lambda Q^{\mathrm{T}} = \begin{bmatrix} \boldsymbol{x}_1 & \boldsymbol{x}_2 \end{bmatrix} \begin{bmatrix} \lambda_1 & \\ & \lambda_2 \end{bmatrix} \begin{bmatrix} \boldsymbol{x}_1^{\mathrm{T}} \\ \boldsymbol{x}_2^{\mathrm{T}} \end{bmatrix}. \tag{6}$$

The columns $\boldsymbol{x}_1$ and $\boldsymbol{x}_2$ times the rows $\lambda_1\boldsymbol{x}_1^{\mathrm{T}}$ and $\lambda_2\boldsymbol{x}_2^{\mathrm{T}}$ produce A:

$$A = \lambda_1\boldsymbol{x}_1\boldsymbol{x}_1^{\mathrm{T}} + \lambda_2\boldsymbol{x}_2\boldsymbol{x}_2^{\mathrm{T}}. \tag{7}$$

This is the great factorization $Q\Lambda Q^{\mathrm{T}}$, written in terms of λ's and $\boldsymbol{x}$'s. When the symmetric matrix is n by n, there are n columns in Q multiplying n rows in Q^{T}. The n pieces are $\lambda_i\boldsymbol{x}_i\boldsymbol{x}_i^{\mathrm{T}}$. Those are matrices! Equation (7) for our example is

$$A = \begin{bmatrix} -3 & 4 \\ 4 & 3 \end{bmatrix} = 5\begin{bmatrix} 1/5 & 2/5 \\ 2/5 & 4/5 \end{bmatrix} - 5\begin{bmatrix} 4/5 & -2/5 \\ -2/5 & 1/5 \end{bmatrix}. \tag{8}$$

On the right, each $\boldsymbol{x}_i\boldsymbol{x}_i^{\mathrm{T}}$ is a ***projection matrix***. It is like $\boldsymbol{u}\boldsymbol{u}^{\mathrm{T}}$ in Chapter 4. The spectral theorem for symmetric matrices says that A is a combination of projection matrices:

$$A = \lambda_1 P_1 + \cdots + \lambda_n P_n \qquad \lambda_i = \text{eigenvalue}, \quad P_i = \text{projection onto eigenspace}.$$

Complex Eigenvalues of Real Matrices

Equation (3) went from $A\boldsymbol{x} = \lambda\boldsymbol{x}$ to $A\overline{\boldsymbol{x}} = \overline{\lambda}\overline{\boldsymbol{x}}$. In the end, λ and $\boldsymbol{x}$ were real. Those two equations were the same. But a *non*symmetric matrix can easily produce λ and $\boldsymbol{x}$ that are complex. In this case, $A\overline{\boldsymbol{x}} = \overline{\lambda}\overline{\boldsymbol{x}}$ is different from $A\boldsymbol{x} = \lambda\boldsymbol{x}$. It gives us a new eigenvalue (which is $\overline{\lambda}$) and a new eigenvector (which is $\overline{\boldsymbol{x}}$):

For real matrices, complex λ's and x's come in "conjugate pairs."

$$\textit{If} \quad A\boldsymbol{x} = \lambda\boldsymbol{x} \quad \textit{then} \quad A\overline{\boldsymbol{x}} = \overline{\lambda}\overline{\boldsymbol{x}}.$$

Example 3 $A = \begin{bmatrix} \cos\theta & -\sin\theta \\ \sin\theta & \cos\theta \end{bmatrix}$ has $\lambda_1 = \cos\theta + i\sin\theta$ and $\lambda_2 = \cos\theta - i\sin\theta$.

Those eigenvalues are conjugate to each other. They are λ and $\overline{\lambda}$, because the imaginary part $\sin\theta$ switches sign. The eigenvectors must be $\boldsymbol{x}$ and $\overline{\boldsymbol{x}}$, because A is real:

$$
\begin{aligned}
A\boldsymbol{x} &= \begin{bmatrix} \cos\theta & -\sin\theta \\ \sin\theta & \cos\theta \end{bmatrix}\begin{bmatrix} 1 \\ -i \end{bmatrix} = (\cos\theta + i\sin\theta)\begin{bmatrix} 1 \\ -i \end{bmatrix} \\
A\overline{\boldsymbol{x}} &= \begin{bmatrix} \cos\theta & -\sin\theta \\ \sin\theta & \cos\theta \end{bmatrix}\begin{bmatrix} 1 \\ i \end{bmatrix} = (\cos\theta - i\sin\theta)\begin{bmatrix} 1 \\ i \end{bmatrix}.
\end{aligned}
\tag{9}
$$

One is $A\boldsymbol{x} = \lambda\boldsymbol{x}$, the other is $A\overline{\boldsymbol{x}} = \overline{\lambda}\overline{\boldsymbol{x}}$. The eigenvectors are $(1, -i)$ and $(1, i)$. For any real matrix the λ's and also the $\boldsymbol{x}$'s are complex conjugates.

For this rotation matrix the absolute value is $|\lambda| = 1$, because $\cos^2\theta + \sin^2\theta = 1$. ***This fact*** $|\lambda| = 1$ ***holds for the eigenvalues of every orthogonal matrix.***

We apologize that a touch of complex numbers slipped in. They are unavoidable even when the matrix is real. Chapter 10 goes beyond complex numbers λ and complex vectors $\boldsymbol{x}$ to complex matrices A. Then you have the whole picture.

We end with two optional discussions.

Eigenvalues versus Pivots

The eigenvalues of A are very different from the pivots. For eigenvalues, we solve $\det(A - \lambda I) = 0$. For pivots, we use elimination. The only connection so far is this:

product of pivots = determinant = product of eigenvalues.

We are assuming a full set of pivots $d_1, \cdots, d_n$. There are n real eigenvalues $\lambda_1, \cdots, \lambda_n$. The d's and λ's are not the same, but they come from the same matrix. This paragraph is about a hidden relation for symmetric matrices: ***The pivots and the eigenvalues have the same signs.***

> **6K** ***If** A **is symmetric the number of positive*** (negative) ***eigenvalues equals the number of positive*** (negative) ***pivots.***

Example 4 This symmetric matrix A has one positive eigenvalue and one positive pivot:

$$A = \begin{bmatrix} 1 & 3 \\ 3 & 1 \end{bmatrix} \quad \begin{matrix} \text{has pivots 1 and } -8 \\ \text{eigenvalues 4 and } -2. \end{matrix}$$

The signs of the pivots match the signs of the eigenvalues, one plus and one minus. This could be false when the matrix is not symmetric:

$$B = \begin{bmatrix} 1 & 6 \\ -1 & -4 \end{bmatrix} \quad \begin{matrix} \text{has pivots 1 and 2} \\ \text{eigenvalues } -1 \text{ and } -2. \end{matrix}$$

The pivots of B are positive, the eigenvalues are negative. The diagonal has both signs! The diagonal entries are a third set of numbers and we say nothing about them.

Here is a proof that the pivots and eigenvalues have matching signs, when $A = A^{\mathrm{T}}$. You see it best when the pivots are divided out of the rows of U, and $A = LDL^{\mathrm{T}}$. The diagonal pivot matrix D goes between triangular matrices L and L^{T}:

$$\begin{bmatrix} 1 & 3 \\ 3 & 1 \end{bmatrix} = \begin{bmatrix} 1 & 0 \\ 3 & 1 \end{bmatrix} \begin{bmatrix} 1 & \\ & -8 \end{bmatrix} \begin{bmatrix} 1 & 3 \\ 0 & 1 \end{bmatrix} \quad \textbf{This is } A = LDL^{\mathrm{T}}\textbf{. It is symmetric.}$$

The special event is the appearance of L^{T}. This only happens for symmetric matrices, because LDL^{T} is always symmetric. (Take its transpose to get LDL^{T} again.)

Watch the eigenvalues when L and L^{T} move toward the identity matrix. At the start, the eigenvalues of LDL^{T} are 4 and -2. At the end, the eigenvalues of IDI^{T} are 1 and -8 (the pivots!). The eigenvalues are changing, as the "3" in L moves to zero. But to change *sign*, an eigenvalue would have to cross zero. The matrix would at that moment be singular. Our changing matrix always has pivots 1 and -8, so it is never singular. The signs cannot change, as the λ's move to the d's.

We repeat the proof for any $A = LDL^{\mathrm{T}}$. Move L toward I, by moving the off-diagonal entries to zero. The pivots are not changing and not zero. The eigenvalues λ of LDL^{T} change to the eigenvalues d of IDI^{T}. Since these eigenvalues cannot cross zero as they move into the pivots, their signs cannot change. Q.E.D.

This connects the two halves of applied linear algebra—pivots and eigenvalues.

All Symmetric Matrices are Diagonalizable

When no eigenvalues of A are repeated, the eigenvectors are sure to be independent. Then A can be diagonalized. But a repeated eigenvalue can produce a shortage of eigenvectors. This *sometimes* happens for nonsymmetric matrices. It *never* happens for symmetric matrices. ***There are always enough eigenvectors to diagonalize*** $A = A^{\mathrm{T}}$.

Here are three matrices, all with $\lambda = -1$ and 1 and 1 (a repeated eigenvalue):

$$A = \begin{bmatrix} 0 & 1 & 0 \\ 1 & 0 & 0 \\ 0 & 0 & 1 \end{bmatrix} \quad B = \begin{bmatrix} -1 & 0 & 1 \\ 0 & 1 & 0 \\ 0 & 0 & 1 \end{bmatrix} \quad C = \begin{bmatrix} -1 & 0 & 0 \\ 0 & 1 & 1 \\ 0 & 0 & 1 \end{bmatrix}.$$

A is symmetric. We guarantee that it can be diagonalized. The nonsymmetric B can also be diagonalized. The nonsymmetric C has only two eigenvectors, not three. It cannot be diagonalized.

One way to deal with repeated eigenvalues is to separate them a little. Change the lower right corner of A, B, C from 1 to d. The eigenvalues are -1 and 1 and d. The three eigenvectors are independent. But when d reaches 1, *two eigenvectors of C collapse into one*. Its eigenvector matrix S loses invertibility:

$$\textbf{Eigenvectors of } C: \begin{bmatrix} 1 & 0 & 0 \\ 0 & 1 & 1 \\ 0 & 0 & d-1 \end{bmatrix} \text{ approaches } \begin{bmatrix} 1 & 0 & 0 \\ 0 & 1 & 1 \\ 0 & 0 & 0 \end{bmatrix} \quad \textbf{Only two eigenvectors!}$$

This cannot happen when $A = A^{\mathrm{T}}$. Reason: *The eigenvectors stay perpendicular.* They cannot collapse as $d \to 1$. In our example the eigenvectors don't even change:

$$\begin{bmatrix} 0 & 1 & 0 \\ 1 & 0 & 0 \\ 0 & 0 & d \end{bmatrix} \quad \text{has orthogonal eigenvectors = columns of} \quad S = \begin{bmatrix} 1 & 1 & 0 \\ -1 & 1 & 0 \\ 0 & 0 & 1 \end{bmatrix}.$$

Final note The eigenvectors of a skew-symmetric matrix ($A^{\mathrm{T}} = -A$) are perpendicular. The eigenvectors of an orthogonal matrix ($Q^{\mathrm{T}} = Q^{-1}$) are also perpendicular. The best matrices have perpendicular eigenvectors! They are all diagonalizable. I stop there.

The reason for stopping is that the eigenvectors may contain complex numbers. We need Chapter 10 to say what "perpendicular" means. When $\boldsymbol{x}$ and $\boldsymbol{y}$ are complex vectors, the test is no longer $\boldsymbol{x}^{\mathrm{T}}\boldsymbol{y} = 0$. It will change to $\overline{\boldsymbol{x}}^{\mathrm{T}}\boldsymbol{y} = 0$. So we can't prove anything now—but we can reveal the answer. ***A real matrix has perpendicular eigenvectors if and only if*** $\boldsymbol{A^{\mathrm{T}}A = AA^{\mathrm{T}}}$. Symmetric and skew-symmetric and orthogonal matrices are included among these "normal" matrices. They may be called *normal* but they are special. The very best are symmetric.

■ REVIEW OF THE KEY IDEAS ■

1. A symmetric matrix has *real eigenvalues* and *perpendicular eigenvectors*.
2. Diagonalization becomes $A = Q\Lambda Q^{\mathrm{T}}$ with an orthogonal matrix Q.
3. All symmetric matrices are diagonalizable, even with repeated eigenvalues.
4. The signs of the eigenvalues match the signs of the pivots, when $A = A^{\mathrm{T}}$.

■ WORKED EXAMPLES ■

6.4 A Find the eigenvalues of A_3 and B_4, and check the orthogonality of their first two eigenvectors. Graph these eigenvectors to see discrete sines and cosines:

$$A_3 = \begin{bmatrix} 2 & -1 & 0 \\ -1 & 2 & -1 \\ 0 & -1 & 2 \end{bmatrix} \qquad B_4 = \begin{bmatrix} 1 & -1 & & \\ -1 & 2 & -1 & \\ & -1 & 2 & -1 \\ & & -1 & 1 \end{bmatrix}$$

The $-1, 2, -1$ pattern in both matrices is a "second difference". Section 8.1 will explain how this is like a second derivative. Then $A\boldsymbol{x} = \lambda\boldsymbol{x}$ and $B\boldsymbol{x} = \lambda\boldsymbol{x}$ are like $d^2x/dt^2 = \lambda x$. This has eigenvectors $x = \sin kt$ and $x = \cos kt$ that are the bases for Fourier series. The matrices lead to "discrete sines" and "discrete cosines" that are the bases for the *Discrete Fourier Transform*. This DFT is absolutely central to all areas of digital signal processing. The favorite choice for JPEG in image processing has been B_8.

Solution The eigenvalues of A_3 are $\lambda = 2 - \sqrt{2}$ and 2 and $2 + \sqrt{2}$. Their sum is 6 (the trace of A_3) and their product is 4 (the determinant). The eigenvector matrix S gives the "Discrete Sine Transform" and the graph shows how the components of the first two eigenvectors fall onto sine curves. Please draw the third eigenvector onto a third sine curve!

$$S = \begin{bmatrix} 1 & \sqrt{2} & 1 \\ \sqrt{2} & 0 & -\sqrt{2} \\ 1 & -\sqrt{2} & 1 \end{bmatrix}$$

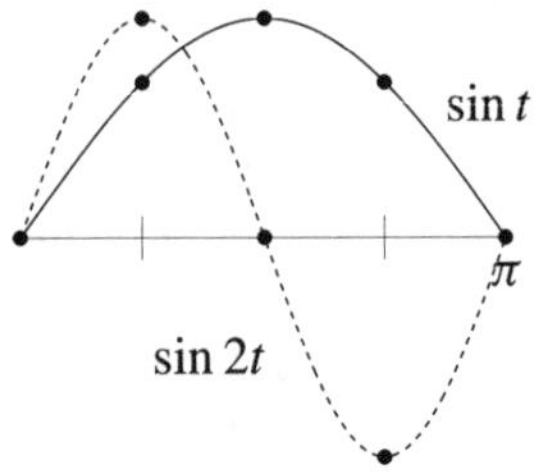

The eigenvalues of B_4 are $\lambda = 2 - \sqrt{2}$ and 2 and $2 + \sqrt{2}$ and 0 (the same as for A_3, plus the zero eigenvalue). The trace is still 6, but the determinant is now zero. The eigenvector matrix C gives the 4-point "Discrete Cosine Transform" and the graph shows how the first two eigenvectors fall onto cosine curves. (Please plot the third eigenvector!) These eigenvectors match cosines at the *halfway points* $\frac{\pi}{8}, \frac{3\pi}{8}, \frac{5\pi}{8}, \frac{7\pi}{8}$.

$$C = \begin{bmatrix} 1 & 1 & 1 & 1 \\ 1 & \sqrt{2}-1 & -1 & 1-\sqrt{2} \\ 1 & 1-\sqrt{2} & -1 & \sqrt{2}-1 \\ 1 & -1 & 1 & -1 \end{bmatrix}$$

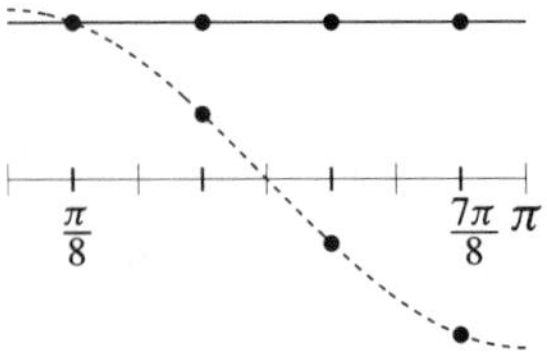

Both S and C have orthogonal columns (these are the eigenvectors of the symmetric A_3 and B_4). When we multiply an input signal by S or C, we split that signal into pure frequencies—like separating a musical chord into pure notes. This Discrete Fourier Transform is the most useful and insightful transform in all of signal processing. We are seeing the sines and cosines (DST and DCT) that go into the DFT. Of course these beautiful patterns continue for larger matrices. Here is a MATLAB code to create B_8 and its eigenvector matrix C_8 and plot the first four eigenvectors onto cosine curves:

$n = 8$; $e =$ ones$(n-1, 1)$; $B = 2*$ eye$(n)-$diag$(e, -1)-$diag$(e, 1)$; $B(1, 1) = 1$; $B(n, n) = 1$; $[C, \Lambda] =$ eig(B); plot$(C(\,:\,, 1{:}4), '-o')$

Problem Set 6.4

1 Write A as $M + N$, symmetric matrix plus skew-symmetric matrix:

$$A = \begin{bmatrix} 1 & 2 & 4 \\ 4 & 3 & 0 \\ 8 & 6 & 5 \end{bmatrix} = M + N \qquad (M^T = M, N^T = -N).$$

For any square matrix, $M = \frac{A+A^T}{2}$ and $N =$ ____ add up to A.

2 If C is symmetric prove that $A^{\mathrm{T}}CA$ is also symmetric. (Transpose it.) When A is 6 by 3, what are the shapes of C and $A^{\mathrm{T}}CA$?

3 Find the eigenvalues and the unit eigenvectors of

$$A = \begin{bmatrix} 1 & 1 & 1 \\ 1 & 0 & 0 \\ 1 & 0 & 0 \end{bmatrix}.$$

4 Find an orthogonal matrix Q that diagonalizes $A = \begin{bmatrix} -2 & 6 \\ 6 & 7 \end{bmatrix}$.

5 Find an orthogonal matrix Q that diagonalizes this symmetric matrix:

$$A = \begin{bmatrix} 1 & 0 & 2 \\ 0 & -1 & -2 \\ 2 & -2 & 0 \end{bmatrix}.$$

6 Find *all* orthogonal matrices that diagonalize $A = \begin{bmatrix} 9 & 12 \\ 12 & 16 \end{bmatrix}$.

7 (a) Find a symmetric matrix $\begin{bmatrix} 1 & b \\ b & 1 \end{bmatrix}$ that has a negative eigenvalue.

(b) How do you know it must have a negative pivot?

(c) How do you know it can't have two negative eigenvalues?

8 If $A^3 = 0$ then the eigenvalues of A must be _____. Give an example that has $A \neq 0$. But if A is symmetric, diagonalize it to prove that A must be zero.

9 If $\lambda = a + ib$ is an eigenvalue of a real matrix A, then its conjugate $\bar{\lambda} = a - ib$ is also an eigenvalue. (If $Ax = \lambda x$ then also $A\bar{x} = \bar{\lambda}\bar{x}$.) Prove that every real 3 by 3 matrix has a real eigenvalue.

10 Here is a quick "proof" that the eigenvalues of all real matrices are real:

$$Ax = \lambda x \quad \text{gives} \quad x^{\mathrm{T}}Ax = \lambda x^{\mathrm{T}}x \quad \text{so} \quad \lambda = \frac{x^{\mathrm{T}}Ax}{x^{\mathrm{T}}x} \quad \text{is real.}$$

Find the flaw in this reasoning—a hidden assumption that is not justified.

11 Write A and B in the form $\lambda_1 x_1 x_1^{\mathrm{T}} + \lambda_2 x_2 x_2^{\mathrm{T}}$ of the spectral theorem $Q\Lambda Q^T$:

$$A = \begin{bmatrix} 3 & 1 \\ 1 & 3 \end{bmatrix} \qquad B = \begin{bmatrix} 9 & 12 \\ 12 & 16 \end{bmatrix} \qquad (\text{keep } \|x_1\| = \|x_2\| = 1).$$

12 Every 2 by 2 symmetric matrix is $\lambda_1 x_1 x_1^{\mathrm{T}} + \lambda_2 x_2 x_2^{\mathrm{T}} = \lambda_1 P_1 + \lambda_2 P_2$. Explain $P_1 + P_2 = x_1 x_1^{\mathrm{T}} + x_2 x_2^{\mathrm{T}} = I$ from columns times rows, $P_1 P_2 = x_1 x_1^{\mathrm{T}} + x_2 x_2^{\mathrm{T}} = 0$ and from row times column.

13 What are the eigenvalues of $A = \begin{bmatrix} 0 & b \\ -b & 0 \end{bmatrix}$? Create a 3 by 3 skew-symmetric matrix ($A^{\mathrm{T}} = -A$) and verify that its eigenvalues are all imaginary.

14 This matrix M is skew symmetric and also ______. Then its eigenvalues are all pure imaginary and they have $|\lambda| = 1$. ($\|M\boldsymbol{x}\| = \|\boldsymbol{x}\|$ for every $\boldsymbol{x}$ so $\|\lambda\boldsymbol{x}\| = \|\boldsymbol{x}\|$ for eigenvectors.) Find all four eigenvalues of

$$M = \frac{1}{\sqrt{3}} \begin{bmatrix} 0 & 1 & 1 & 1 \\ -1 & 0 & -1 & 1 \\ -1 & 1 & 0 & -1 \\ -1 & -1 & 1 & 0 \end{bmatrix}.$$

15 Show that A (symmetric but complex) does not have two independent eigenvectors:

$$A = \begin{bmatrix} i & 1 \\ 1 & -i \end{bmatrix} \text{ is not diagonalizable; } \det(A - \lambda I) = \lambda^2.$$

$A^{\mathrm{T}} = A$ is not such a special property for complex matrices. The good property is $\overline{A}^{\mathrm{T}} = A$ (Section 10.2). Then all λ's are real and eigenvectors are orthogonal.

16 Even if A is rectangular, the block matrix $B = \begin{bmatrix} 0 & A \\ A^{\mathrm{T}} & 0 \end{bmatrix}$ is symmetric:

$$B\boldsymbol{x} = \lambda\boldsymbol{x} \quad \text{is} \quad \begin{bmatrix} 0 & A \\ A^{\mathrm{T}} & 0 \end{bmatrix} \begin{bmatrix} \boldsymbol{y} \\ \boldsymbol{z} \end{bmatrix} = \lambda \begin{bmatrix} \boldsymbol{y} \\ \boldsymbol{z} \end{bmatrix} \quad \text{which is} \quad \begin{matrix} A\boldsymbol{z} = \lambda\boldsymbol{y} \\ A^{\mathrm{T}}\boldsymbol{y} = \lambda\boldsymbol{z}. \end{matrix}$$

(a) Show that $-\lambda$ is also an eigenvalue, with the eigenvector $(\boldsymbol{y}, -\boldsymbol{z})$.

(b) Show that $A^{\mathrm{T}}A\boldsymbol{z} = \lambda^2\boldsymbol{z}$, so that λ^2 is an eigenvalue of $A^{\mathrm{T}}A$.

(c) If $A = I$ (2 by 2) find all four eigenvalues and eigenvectors of B.

17 If $A = \begin{bmatrix} 1 \\ 1 \end{bmatrix}$ in Problem 16, find all three eigenvalues and eigenvectors of B.

18 ***Another proof that eigenvectors are perpendicular when $A = A^{\mathrm{T}}$.*** Suppose $A\boldsymbol{x} = \lambda\boldsymbol{x}$ and $A\boldsymbol{y} = 0\boldsymbol{y}$ and $\lambda \neq 0$. Then $\boldsymbol{y}$ is in the nullspace and $\boldsymbol{x}$ is in the column space. They are perpendicular because ______. Go carefully—why are these subspaces orthogonal? If the second eigenvalue is a nonzero number β, apply this argument to $A - \beta I$. The eigenvalue moves to zero and the eigenvectors stay the same—so they are perpendicular.

19 Find the eigenvector matrix S for this matrix B. Show that it doesn't collapse at $d = 1$, even though $\lambda = 1$ is repeated. Are the eigenvectors perpendicular?

$$B = \begin{bmatrix} -1 & 0 & 1 \\ 0 & 1 & 0 \\ 0 & 0 & d \end{bmatrix} \quad \text{has} \quad \lambda = -1, 1, d.$$

20 From the trace and the determinant find the eigenvalues of

$$A = \begin{bmatrix} -3 & 4 \\ 4 & 3 \end{bmatrix}.$$

Compare the signs of the λ's with the signs of the pivots.

21 True or false. Give a reason or a counterexample.

(a) A matrix with real eigenvalues and eigenvectors is symmetric.

(b) A matrix with real eigenvalues and orthogonal eigenvectors is symmetric.

(c) The inverse of a symmetric matrix is symmetric.

(d) The eigenvector matrix S of a symmetric matrix is symmetric.

22 A *normal matrix* has $A^{\mathrm{T}}A = AA^{\mathrm{T}}$; it has orthogonal eigenvectors. Why is every skew-symmetric matrix normal? Why is every orthogonal matrix normal? When is $\left[\begin{smallmatrix} a & 1 \\ -1 & d \end{smallmatrix}\right]$ normal?

23 (A paradox for instructors) If $AA^{\mathrm{T}} = A^{\mathrm{T}}A$ then A and A^{T} share the same eigenvectors (true). A and A^{T} always share the same eigenvalues. Find the flaw in this conclusion: They must have the same S and Λ. Therefore A equals A^{T}.

24 (Recommended) Which of these classes of matrices do A and B belong to: Invertible, orthogonal, projection, permutation, diagonalizable, Markov?

$$A = \begin{bmatrix} 0 & 0 & 1 \\ 0 & 1 & 0 \\ 1 & 0 & 0 \end{bmatrix} \qquad B = \frac{1}{3}\begin{bmatrix} 1 & 1 & 1 \\ 1 & 1 & 1 \\ 1 & 1 & 1 \end{bmatrix}.$$

Which of these factorizations are possible for A and B: LU, QR, $S\Lambda S^{-1}$, $Q\Lambda Q^{\mathrm{T}}$?

25 What number b in $\left[\begin{smallmatrix} 2 & b \\ 1 & 0 \end{smallmatrix}\right]$ makes $A = Q\Lambda Q^{\mathrm{T}}$ possible? What number makes $A = S\Lambda S^{-1}$ impossible? What number makes A^{-1} impossible?

26 Find all 2 by 2 matrices that are orthogonal and also symmetric. Which two numbers can be eigenvalues?

27 This A is nearly symmetric. But its eigenvectors are far from orthogonal:

$$A = \begin{bmatrix} 1 & 10^{-15} \\ 0 & 1+10^{-15} \end{bmatrix} \quad \text{has eigenvectors} \quad \begin{bmatrix} 1 \\ 0 \end{bmatrix} \quad \text{and} \quad [?]$$

What is the angle between the eigenvectors?

28 (MATLAB) Take two symmetric matrices with different eigenvectors, say $A = \left[\begin{smallmatrix} 1 & 0 \\ 0 & 2 \end{smallmatrix}\right]$ and $B = \left[\begin{smallmatrix} 8 & 1 \\ 1 & 0 \end{smallmatrix}\right]$. Graph the eigenvalues $\lambda_1(A+tB)$ and $\lambda_2(A+tB)$ for $-8 < t < 8$. Peter Lax says on page 113 of *Linear Algebra* that λ_1 and λ_2 appear to be on a collision course at certain values of t. "Yet at the last minute they turn aside." How close do they come?

29 My file **scarymatlab** shows what can happen when roundoff destroys symmetry:

$$A = [\,1\,1\,1\,1\,1;1:5\,]'; \; B = A' * A; \; P = A * \text{inv}(B) * A'; \; [\,Q, \;\; E\,] = \text{eig}(P);$$

B is exactly symmetric and P should be, but isn't. Multiplying $Q' * Q$ will show two eigenvectors of P with dot product .9999 instead of 0.

POSITIVE DEFINITE MATRICES ■ 6.5

This section concentrates on ***symmetric matrices that have positive eigenvalues***. If symmetry makes a matrix important, this extra property (*all* $\lambda > 0$) makes it special. When we say special, we don't mean rare. Symmetric matrices with positive eigenvalues enter all kinds of applications of linear algebra. They are called ***positive definite***.

The first problem is to recognize these matrices. You may say, just find the eigenvalues and test $\lambda > 0$. That is exactly what we want to avoid. Calculating eigenvalues is work. When the λ's are needed, we can compute them. But if we just want to know that they are positive, there are faster ways. Here are the two goals of this section:

- To find *quick tests* on a symmetric matrix that guarantee *positive eigenvalues.*
- To explain two applications of positive definiteness.

The matrix A is symmetric so the λ's are automatically real.

Start with 2 by 2. When does $A = \left[\begin{smallmatrix} a & b \\ b & c \end{smallmatrix}\right]$ ***have*** $\lambda_1 > 0$ ***and*** $\lambda_2 > 0$?

6L ***The eigenvalues of A are positive if and only if*** $a > 0$ ***and*** $ac - b^2 > 0$.

$A = \left[\begin{smallmatrix} 4 & 5 \\ 5 & 7 \end{smallmatrix}\right]$ has $a = 4$ and $ac - b^2 = 28 - 25 = 3$. So A has positive eigenvalues. The test is failed by $\left[\begin{smallmatrix} 4 & 5 \\ 5 & 6 \end{smallmatrix}\right]$ and also failed by $\left[\begin{smallmatrix} -1 & 0 \\ 0 & -7 \end{smallmatrix}\right]$. One failure is because the determinant is $24 - 25 < 0$. The other failure is because $a = -1$. The determinant of $+7$ is not enough to pass, because the test has *two parts*: the 1 by 1 determinant a and the 2 by 2 determinant.

Proof of the 2 by 2 tests: If $\lambda_1 > 0$ and $\lambda_2 > 0$, then their product $\lambda_1\lambda_2$ and sum $\lambda_1 + \lambda_2$ are positive. Their product is the determinant so $ac - b^2 > 0$. Their sum is the trace so $a + c > 0$. Then a and c are both positive (if one were not positive then $ac - b^2 > 0$ would have failed).

Now start with $a > 0$ and $ac - b^2 > 0$. Together they force $c > 0$. Since $\lambda_1\lambda_2$ is the positive determinant, the λ's have the same sign. Since the trace is $a + c > 0$, *that sign must be* $+$.

This test on a and $ac - b^2$ uses determinants. The next test requires *positive pivots*.

6M ***The eigenvalues of $A = A^{\mathrm{T}}$ are positive if and only if the pivots are positive:***

$$a > 0 \qquad \text{and} \qquad \frac{ac - b^2}{a} > 0.$$

The point is to recognize that ratio of positive numbers as the *second pivot* of A:

$$\begin{bmatrix} a & b \\ b & c \end{bmatrix} \xrightarrow[\text{The multiplier is } b/a]{\text{The first pivot is } a} \begin{bmatrix} a & b \\ 0 & c - \frac{b}{a}b \end{bmatrix} \qquad \textbf{\textit{The second pivot is}} \quad c - \frac{b^2}{a} = \frac{ac - b^2}{a}$$

This connects two big parts of linear algebra. ***Positive eigenvalues mean positive pivots and vice versa.*** We gave a proof for all symmetric matrices in the last section (**Theorem 6K**). So the pivots give a quick test for $\lambda > 0$. They are a lot faster to compute than the eigenvalues. It is very satisfying to see pivots and determinants and eigenvalues come together in this course.

Example 1 This matrix has $a = 1$ (positive). But $ac - b^2 = 3 - 2^2$ is negative:

$$\begin{bmatrix} 1 & 2 \\ 2 & 3 \end{bmatrix} \quad \text{has a negative determinant and pivot. So a } \textbf{negative eigenvalue}.$$

The pivots are 1 and -1. The eigenvalues also multiply to give -1. One eigenvalue is negative (we don't want its formula, which has a square root, just its sign).

Here is a different way to look at symmetric matrices with positive eigenvalues. From $A\boldsymbol{x} = \lambda\boldsymbol{x}$, multiply by $\boldsymbol{x}^{\mathrm{T}}$ to get $\boldsymbol{x}^{\mathrm{T}}A\boldsymbol{x} = \lambda\boldsymbol{x}^{\mathrm{T}}\boldsymbol{x}$. The right side is a positive λ times a positive $\boldsymbol{x}^{\mathrm{T}}\boldsymbol{x} = \|\boldsymbol{x}\|^2$. So $\boldsymbol{x}^{\mathrm{T}}A\boldsymbol{x}$ is positive for any eigenvector.

The new idea is that ***this number $\boldsymbol{x}^{\mathrm{T}}A\boldsymbol{x}$ is positive for all nonzero vectors $\boldsymbol{x}$***, not just the eigenvectors. Matrices with this property $\boldsymbol{x}^{\mathrm{T}}A\boldsymbol{x} > 0$ are ***positive definite matrices***. We will prove that exactly these matrices have positive eigenvalues and pivots.

Definition The matrix A is ***positive definite*** if $\boldsymbol{x}^{\mathrm{T}}A\boldsymbol{x} > 0$ for every nonzero vector:

$$\boldsymbol{x}^{\mathrm{T}}A\boldsymbol{x} = \begin{bmatrix} x & y \end{bmatrix} \begin{bmatrix} a & b \\ b & c \end{bmatrix} \begin{bmatrix} x \\ y \end{bmatrix} = ax^2 + 2bxy + cy^2 > 0.$$

$\boldsymbol{x}^{\mathrm{T}}A\boldsymbol{x}$ is a number (1 by 1 matrix). The four entries a, b, b, c give the four parts of $\boldsymbol{x}^{\mathrm{T}}A\boldsymbol{x}$. From a and c come the pure squares ax^2 and cy^2. From b and b off the diagonal come the cross terms bxy and byx (the same). Adding those four parts gives $\boldsymbol{x}^{\mathrm{T}}A\boldsymbol{x}$:

$$f(x, y) = \boldsymbol{x}^{\mathrm{T}}A\boldsymbol{x} = ax^2 + 2bxy + cy^2 \quad \text{is "second degree."}$$

The rest of this book has been linear (mostly $A\boldsymbol{x}$). Now the degree has gone from 1 to 2. The *second* derivatives of $ax^2+2bxy+cy^2$ are constant. Those second derivatives are $2a$, $2b$, $2b$, $2c$. They go into the ***second derivative matrix*** $2A$:

$$\frac{\partial f}{\partial x} = 2ax + 2by \qquad \text{and} \qquad \begin{bmatrix} \dfrac{\partial^2 f}{\partial x^2} & \dfrac{\partial^2 f}{\partial y \partial x} \\ \dfrac{\partial^2 f}{\partial x \partial y} & \dfrac{\partial^2 f}{\partial y^2} \end{bmatrix} = \begin{bmatrix} 2a & 2b \\ 2b & 2c \end{bmatrix} = 2A.$$
$$\frac{\partial f}{\partial y} = 2bx + 2cy$$

This is the 2 by 2 version of what everybody knows for 1 by 1. There the function is ax^2, its slope is $2ax$, and its second derivative is $2a$. Now the function is $\boldsymbol{x}^{\mathrm{T}}A\boldsymbol{x}$, its first derivatives are in the vector $2A\boldsymbol{x}$, and its second derivatives are in the matrix $2A$. Third derivatives are all zero.

First Application: Test for a Minimum

Where does calculus use second derivatives? When f'' is positive, the curve bends up from its tangent line. The point $x = 0$ is a minimum point of $y = x^2$. It is a maximum point of $y = -x^2$. To decide minimum versus maximum for a one-variable function $f(x)$, calculus looks at its second derivative.

For a two-variable function $f(x, y)$, the *matrix of second derivatives* holds the key. One number is not enough to decide minimum versus maximum (versus saddle point). ***The function*** $f = \boldsymbol{x}^{\mathrm{T}}A\boldsymbol{x}$ ***has a minimum at*** $x = y = 0$ ***if and only if*** A ***is positive definite***. The statement "A is a positive definite matrix" is the 2 by 2 version of "a is a positive number".

Example 2 This matrix A is positive definite. We test by pivots or determinants:

$$A = \begin{bmatrix} 1 & 2 \\ 2 & 7 \end{bmatrix} \quad \text{has positive pivots and determinants (1 and 3).}$$

More directly, $\boldsymbol{x}^{\mathrm{T}}A\boldsymbol{x} = x^2 + 4xy + 7y^2$ is positive because it is a **sum of squares**:

$$\text{Rewrite} \quad x^2 + 4xy + 7y^2 \quad \text{as} \quad (x + 2y)^2 + 3y^2.$$

The pivots 1 and 3 multiply those squares. This is no accident! By the algebra of "completing the square," this always happens. So when the pivots are positive, the quadratic function $f(x, y) = \boldsymbol{x}^{\mathrm{T}}A\boldsymbol{x}$ is guaranteed to be positive: a sum of squares.

Comparing our examples $\left[\begin{smallmatrix} 1 & 2 \\ 2 & 3 \end{smallmatrix}\right]$ and $\left[\begin{smallmatrix} 1 & 2 \\ 2 & 7 \end{smallmatrix}\right]$, the only difference is that change from 3 to 7. The borderline is $a_{22} = 4$. Above 4, the matrix is positive definite. At $a_{22} = 4$, the borderline matrix is only ***semidefinite***. Then (> 0) changes to (≥ 0):

$$\begin{bmatrix} 1 & 2 \\ 2 & 4 \end{bmatrix} \quad \text{is singular. It has eigenvalues 5 and 0.}$$

This matrix has $a > 0$ but $ac - b^2 = 0$. Not quite positive definite!

We will summarize this section so far. We have four ways to recognize a positive definite matrix. Right now it is only 2 by 2.

6N When a 2 by 2 symmetric matrix has one of these four properties, it has them all:

1. Both of the eigenvalues are positive.

2. The 1 by 1 and 2 by 2 determinants are positive: $a > 0$ and $ac - b^2 > 0$.

3. The pivots are positive: $a > 0$ and $(ac - b^2)/a > 0$.

4. The function $\boldsymbol{x}^{\mathrm{T}} A \boldsymbol{x} = ax^2 + 2bxy + cy^2$ is positive except at $(0, 0)$.

When A has one (therefore all) of these four properties, it is a ***positive definite matrix***.

Note We deal only with symmetric matrices. The cross derivative $\partial^2 f/\partial x \partial y$ always equals $\partial^2 f/\partial y \partial x$. For $f(x, y, z)$ the nine second derivatives fill a symmetric 3 by 3 matrix. It is positive definite when the three pivots (and the three eigenvalues, and the three determinants) are positive. **When the first derivatives $\partial f/\partial x$ and $\partial f/\partial y$ are zero and the second derivative matrix is positive definite, we have found a local minimum.**

Example 3 Is $f(x, y) = x^2 + 8xy + 3y^2$ everywhere positive—except at $(0, 0)$?

Solution The second derivatives are $f_{xx} = 2$ and $f_{xy} = f_{yx} = 8$ and $f_{yy} = 6$, all positive. But the test is not positive derivatives. We look for positive *definiteness*. The answer is *no*, this function is not always positive. By trial and error we locate a point $x = 1$, $y = -1$ where $f(1, -1) = 1 - 8 + 3 = -4$. Better to do linear algebra, and apply the exact tests to the matrix that produced $f(x, y)$:

$$x^2 + 8xy + 3y^2 = \begin{bmatrix} x & y \end{bmatrix} \begin{bmatrix} 1 & 4 \\ 4 & 3 \end{bmatrix} \begin{bmatrix} x \\ y \end{bmatrix}.$$

The matrix has $ac - b^2 = 3 - 16$. The pivots are 1 and -13. The eigenvalues are _____ (we don't need them). The matrix is not positive definite.

Note how $8xy$ comes from $a_{12} = 4$ above the diagonal and $a_{21} = 4$ symmetrically below. That matrix multiplication in $\boldsymbol{x}^{\mathrm{T}} A \boldsymbol{x}$ makes the function appear.

Main point The sign of b is not the essential thing. The cross derivative $\partial^2 f/\partial x \partial y$ can be positive or negative—the test uses b^2. The *size* of b, compared to a and c, decides whether A is positive definite and the function $f(x, y)$ has a minimum.

Example 4 For which numbers c is $x^2 + 8xy + cy^2$ always positive (or zero)?

Solution The matrix is $A = \begin{bmatrix} 1 & 4 \\ 4 & c \end{bmatrix}$. Again $a = 1$ passes the first test. The second test has $ac - b^2 = c - 16$. For a positive definite matrix we need $c > 16$.

The "semidefinite" borderline is $c = 16$. At that point $\begin{bmatrix} 1 & 4 \\ 4 & 16 \end{bmatrix}$ has $\lambda = 17$ and 0, determinants 1 and 0, pivots 1 and ______. The function $x^2 + 8xy + 16y^2$ is $(x+4y)^2$. Its graph does not go below zero, but it stays at zero all along the line $x + 4y = 0$. This is close to positive definite, but each test just misses: $\boldsymbol{x}^{\mathrm{T}} A \boldsymbol{x}$ ***equals* zero for the vector** $\boldsymbol{x} = (4, -1)$. So A is only semidefinite.

Example 5 When A is positive definite, write $f(x, y)$ as a sum of two squares.

Solution This is called "completing the square." The part $ax^2 + 2bxy$ is correct in the first square $a\left(x + \frac{b}{a}y\right)^2$. But that ends with a final $a\left(\frac{b}{a}y\right)^2$. To stay even, this added amount b^2y^2/a has to be subtracted off from cy^2 at the end:

Completing the Square
$$ax^2 + 2bxy + cy^2 = a\left(x + \frac{b}{a}y\right)^2 + \left(\frac{ac - b^2}{a}\right)y^2. \tag{1}$$

After that gentle touch of algebra, the situation is clearer. The two squares (never negative) are multiplied by numbers that could be positive or negative. *Those numbers a and $(ac - b^2)/a$ are the pivots!* So positive pivots give a sum of squares and a positive definite matrix. Think back to the pivots and multipliers in LDL^{T}:

$$A = \begin{bmatrix} a & b \\ b & c \end{bmatrix} = \begin{bmatrix} 1 & 0 \\ b/a & 1 \end{bmatrix} \begin{bmatrix} a & \\ & (ac - b^2)/a \end{bmatrix} \begin{bmatrix} 1 & b/a \\ 0 & 1 \end{bmatrix}. \tag{2}$$

To complete the square, we started with a and b. *Elimination does exactly the same.* It starts with the first column. Inside $\left(x + \frac{b}{a}y\right)^2$ are the numbers 1 and $\frac{b}{a}$ from L.

Every positive definite symmetric matrix factors into $A = LDL^{\mathrm{T}}$ with positive pivots in D. The "Cholesky factorization" is $A = (L\sqrt{D})(L\sqrt{D})^{\mathrm{T}}$.

Important to compare $A = LDL^{\mathrm{T}}$ with $A = Q\Lambda Q^{\mathrm{T}}$. One is based on pivots (in D). The other is based on eigenvalues (in Λ). *Please* do not think that pivots equal eigenvalues. Their signs are the same, but the numbers are entirely different.

Positive Definite Matrices: n by n

For a 2 by 2 matrix, the "positive definite test" uses *eigenvalues* or *determinants* or *pivots*. All those numbers must be positive. We hope and expect that the same three tests carry over to n by n symmetric matrices. They do.

6O When a symmetric matrix has one of these four properties, it has them all:

1. All ***n eigenvalues*** are positive.

2. All ***n upper left determinants*** are positive.

3. All ***n pivots*** are positive.

4. $\boldsymbol{x}^{\mathrm{T}} A \boldsymbol{x}$ is positive except at $\boldsymbol{x} = \mathbf{0}$. The matrix A is ***positive definite.***

The "upper left determinants" are 1 by 1, 2 by 2, . . ., n by n. The last one is the determinant of the complete matrix A. This remarkable theorem ties together the whole linear algebra course—at least for symmetric matrices. We believe that two examples are more helpful than a proof (we nearly have a proof already). Then we give two applications.

Example 6 Test these matrices A and A^* for positive definiteness:

$$A = \begin{bmatrix} 2 & -1 & 0 \\ -1 & 2 & -1 \\ 0 & -1 & 2 \end{bmatrix} \quad \text{and} \quad A^* = \begin{bmatrix} 2 & -1 & b \\ -1 & 2 & -1 \\ b & -1 & 2 \end{bmatrix}.$$

Solution This A is an old friend (or enemy). Its pivots are 2 and $\frac{3}{2}$ and $\frac{4}{3}$, all positive. Its upper left determinants are 2 and 3 and 4, all positive. Its eigenvalues are $2-\sqrt{2}$ and 2 and $2+\sqrt{2}$, all positive. That completes tests **1**, **2**, and **3**.

We can write $\boldsymbol{x}^{\mathrm{T}} A \boldsymbol{x}$ as a sum of *three* squares (since $n = 3$). The pivots 2, $\frac{3}{2}$, $\frac{4}{3}$ appear outside the squares. The multipliers $-\frac{1}{2}$ and $-\frac{2}{3}$ in L are inside the squares:

$$\begin{aligned} \boldsymbol{x}^{\mathrm{T}} A \boldsymbol{x} &= 2\left(x_1^2 - x_1x_2 + x_2^2 - x_2x_3 + x_3^2\right) \\ &= 2\left(x_1 - \tfrac{1}{2}x_2\right)^2 + \tfrac{3}{2}\left(x_2 - \tfrac{2}{3}x_3\right)^2 + \tfrac{4}{3}(x_3)^2 > 0. \qquad \textbf{This is positive.} \end{aligned}$$

Go to the second matrix A^*. *The determinant test is easiest.* The 1 by 1 determinant is 2, the 2 by 2 determinant is 3. The 3 by 3 determinant comes from the whole A^*:

$$\det A^* = 4 + 2b - 2b^2 = (1+b)(4-2b) \quad \text{must be positive.}$$

At $b = -1$ and $b = 2$ we get $\det A^* = 0$. In those cases A^* is positive *semi*definite (no inverse, zero eigenvalue, $\boldsymbol{x}^{\mathrm{T}} A^* \boldsymbol{x} \geq 0$). *Between $b = -1$ and $b = 2$ the matrix is positive definite.* The corner entry $b = 0$ in the first matrix A was safely between.

Second Application: The Ellipse $ax^2 + 2bxy + cy^2 = 1$

Think of a tilted ellipse centered at $(0, 0)$, as in Figure 6.4a. Turn it to line up with the coordinate axes. That is Figure 6.4b. These two pictures show the geometry behind the factorization $A = Q\Lambda Q^{-1} = Q\Lambda Q^{\mathrm{T}}$:

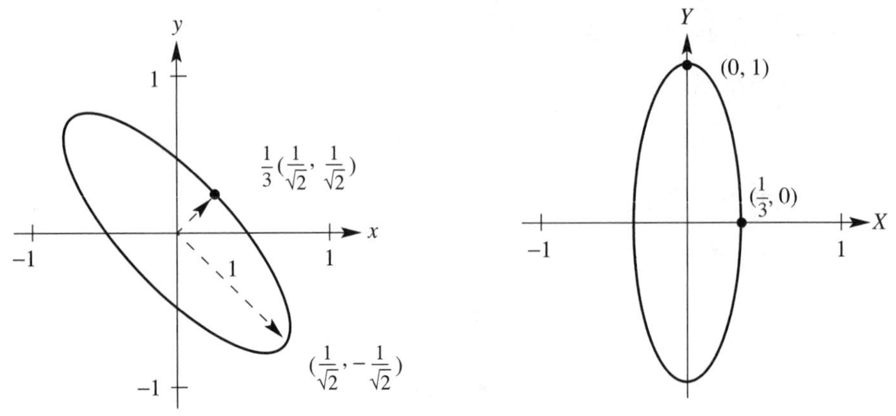

Figure 6.4 The tilted ellipse $5x^2 + 8xy + 5y^2 = 1$. Lined up it is $9X^2 + Y^2 = 1$.

1. The tilted ellipse is associated with A. Its equation is $\boldsymbol{x}^{\mathrm{T}}A\boldsymbol{x} = 1$.
2. The lined-up ellipse is associated with Λ. Its equation is $\boldsymbol{X}^{\mathrm{T}}\Lambda\boldsymbol{X} = 1$.
3. The rotation matrix from $\boldsymbol{x}$ to $\boldsymbol{X}$ that lines up the ellipse is Q.

Example 7 Find the axes of this tilted ellipse $5x^2 + 8xy + 5y^2 = 1$.

Solution Start with the positive definite matrix that matches this equation:

$$\text{The function is } \begin{bmatrix} x & y \end{bmatrix}\begin{bmatrix} 5 & 4 \\ 4 & 5 \end{bmatrix}\begin{bmatrix} x \\ y \end{bmatrix} = 1. \qquad \text{The matrix is } A = \begin{bmatrix} 5 & 4 \\ 4 & 5 \end{bmatrix}.$$

The eigenvalues of A are $\lambda_1 = 9$ and $\lambda_2 = 1$. The eigenvectors are $\begin{bmatrix} 1 \\ 1 \end{bmatrix}$ and $\begin{bmatrix} 1 \\ -1 \end{bmatrix}$. To make them unit vectors, divide by $\sqrt{2}$. Then $A = Q\Lambda Q^{\mathrm{T}}$ is

$$\begin{bmatrix} 5 & 4 \\ 4 & 5 \end{bmatrix} = \frac{1}{\sqrt{2}}\begin{bmatrix} 1 & 1 \\ 1 & -1 \end{bmatrix}\begin{bmatrix} 9 & 0 \\ 0 & 1 \end{bmatrix}\frac{1}{\sqrt{2}}\begin{bmatrix} 1 & 1 \\ 1 & -1 \end{bmatrix}.$$

Now multiply by $\begin{bmatrix} x & y \end{bmatrix}$ on the left and $\begin{bmatrix} x \\ y \end{bmatrix}$ on the right to get back the function $\boldsymbol{x}^{\mathrm{T}}A\boldsymbol{x}$:

$$5x^2 + 8xy + 5y^2 = 9\left(\frac{x+y}{\sqrt{2}}\right)^2 + 1\left(\frac{x-y}{\sqrt{2}}\right)^2. \qquad (3)$$

The function is again a sum of two squares. But this is different from completing the square. The coefficients are not the pivots 5 and 9/5 from D, they are the eigenvalues 9 and 1 from Λ. Inside *these* squares are the eigenvectors $(1, 1)/\sqrt{2}$ and $(1, -1)/\sqrt{2}$.

The axes of the tilted ellipse point along the eigenvectors. This explains why $A = Q\Lambda Q^{\mathrm{T}}$ is called the "principal axis theorem"—it displays the axes. Not only the axis directions (from the eigenvectors) but also the axis lengths (from the eigenvalues). To see it all, use capital letters for the new coordinates that line up the ellipse:

$$\frac{x+y}{\sqrt{2}} = X \qquad \text{and} \qquad \frac{x-y}{\sqrt{2}} = Y.$$

The ellipse becomes $9X^2 + Y^2 = 1$. The largest value of X^2 is $1/9$. The point at the end of the shorter axis has $X = 1/3$ and $Y = 0$. Notice: The *bigger* eigenvalue λ_1 gives the *shorter* axis, of half-length $1/\sqrt{\lambda_1} = 1/3$. The point at the end of the major axis has $X = 0$ and $Y = 1$. The smaller eigenvalue $\lambda_2 = 1$ gives the greater length $1/\sqrt{\lambda_2} = 1$.

In the xy system, the axes are along the eigenvectors of A. In the XY system, the axes are along the eigenvectors of Λ—the coordinate axes. Everything comes from the diagonalization $A = Q\Lambda Q^{\mathrm{T}}$.

6P Suppose $A = Q\Lambda Q^{\mathrm{T}}$ is positive definite. The graph of $\boldsymbol{x}^{\mathrm{T}} A\boldsymbol{x} = 1$ is an ellipse:

$$\begin{bmatrix} x & y \end{bmatrix} Q\Lambda Q^{\mathrm{T}} \begin{bmatrix} x \\ y \end{bmatrix} = \begin{bmatrix} X & Y \end{bmatrix} \Lambda \begin{bmatrix} X \\ Y \end{bmatrix} = \lambda_1 X^2 + \lambda_2 Y^2 = 1.$$

The half-lengths of the axes are $1/\sqrt{\lambda_1}$ and $1/\sqrt{\lambda_2}$.

For an ellipse, A must be positive definite. $A = I$ gives the circle $x^2 + y^2 = 1$. If an eigenvalue is negative (exchange 4's and 5's in A), we don't have an ellipse. The sum of squares becomes a *difference of squares*: $9X^2 - Y^2 = 1$. This is a *hyperbola*. For a negative definite matrix like $A = -I$, the graph of $-x^2 - y^2 = 1$ has no points at all.

■ REVIEW OF THE KEY IDEAS ■

1. Positive definite matrices have positive eigenvalues and positive pivots.

2. A quick test is given by the upper left determinants: $a > 0$ and $ac - b^2 > 0$.

3. The quadratic function $f = \boldsymbol{x}^{\mathrm{T}} A\boldsymbol{x}$ then has a minimum at $\boldsymbol{x} = \boldsymbol{0}$:

$$\boldsymbol{x}^{\mathrm{T}} A\boldsymbol{x} = ax^2 + 2bxy + cy^2 \text{ is positive except at } (x, y) = (0, 0).$$

4. The ellipse $\boldsymbol{x}^{\mathrm{T}} A\boldsymbol{x} = 1$ has its axes along the eigenvectors of A.

5. *Coming*: $A^{\mathrm{T}} A$ is automatically positive definite if A has independent columns.

■ WORKED EXAMPLES ■

6.5 A The great factorizations of a symmetric matrix are $A = LDL^{\mathrm{T}}$ from pivots and multipliers, and $A = Q\Lambda Q^{\mathrm{T}}$ from eigenvalues and eigenvectors. Show that $\boldsymbol{x}^{\mathrm{T}}A\boldsymbol{x} > 0$ for all nonzero $\boldsymbol{x}$ exactly when the pivots and eigenvalues are positive. Try these n by n tests on pascal(6) and ones(6) and hilb(6) and other matrices in MATLAB's gallery.

Solution To prove $\boldsymbol{x}^{\mathrm{T}}A\boldsymbol{x} > 0$ put parentheses into $\boldsymbol{x}^{\mathrm{T}}LDL^{\mathrm{T}}\boldsymbol{x}$ and $\boldsymbol{x}^{\mathrm{T}}Q\Lambda Q^{\mathrm{T}}\boldsymbol{x}$:

$$\boldsymbol{x}^{\mathrm{T}}A\boldsymbol{x} = (L^{\mathrm{T}}\boldsymbol{x})^{\mathrm{T}}D(L^{\mathrm{T}}\boldsymbol{x}) \qquad \text{and} \qquad \boldsymbol{x}^{\mathrm{T}}A\boldsymbol{x} = (Q^{\mathrm{T}}\boldsymbol{x})^{\mathrm{T}}\Lambda(Q^{\mathrm{T}}\boldsymbol{x}).$$

If $\boldsymbol{x}$ is nonzero, then $\boldsymbol{y} = L^{\mathrm{T}}\boldsymbol{x}$ and $\boldsymbol{z} = Q^{\mathrm{T}}\boldsymbol{x}$ are nonzero (those matrices are invertible). So $\boldsymbol{x}^{\mathrm{T}}A\boldsymbol{x} = \boldsymbol{y}^{\mathrm{T}}D\boldsymbol{y} = \boldsymbol{z}^{\mathrm{T}}\Lambda\boldsymbol{z}$ becomes a sum of squares and A is positive definite:

$$\begin{aligned} \boldsymbol{x}^{\mathrm{T}}A\boldsymbol{x} &= \boldsymbol{y}^{\mathrm{T}}D\boldsymbol{y} &&= d_1y_1^2 + \cdots + d_ny_n^2 &&> 0 \\ \boldsymbol{x}^{\mathrm{T}}A\boldsymbol{x} &= \boldsymbol{z}^{\mathrm{T}}\Lambda\boldsymbol{z} &&= \lambda_1z_1^2 + \cdots + \lambda_nz_n^2 &&> 0 \end{aligned}$$

Honesty makes me add one little comment to this fast and beautiful proof. A zero in the pivot position would force a row exchange and a permutation matrix P. So the factorization might be $PAP^{\mathrm{T}} = LDL^{\mathrm{T}}$ (we exchange columns with P^{T} to maintain symmetry). Now the fast proof applies to $A = (P^{-1}L)D(P^{-1}L)^{\mathrm{T}}$ with no problem.

MATLAB has a gallery of unusual matrices (type help gallery) and here are four:

pascal(6) is positive definite because all its pivots are 1 (Worked Example **2.6 A**).

ones(6) is positive *semidefinite* because its eigenvalues are $0, 0, 0, 0, 0, 6$.

hilb(6) is positive definite even though eig(hilb(6)) shows two eigenvalues very near zero. In fact $\boldsymbol{x}^{\mathrm{T}}$ hilb(6) $\boldsymbol{x} = \int_0^1 (x_1 + x_2s + \cdots + x_6s^5)^2\, ds > 0$.

rand(6)+rand(6)′ can be positive definite or not (experiments give only 1 in 10000):

$n = 20000$; $p = 0$; for $k = 1:n$, $A =$ rand(6); $p = p +$ all(eig($A + A'$) > 0); end, $p\,/\,n$

6.5 B Find conditions on the blocks $A = A^{\mathrm{T}}$ and $C = C^{\mathrm{T}}$ and B of this matrix M:

When is the symmetric block matrix $M = \begin{bmatrix} A & B \\ B^{\mathrm{T}} & C \end{bmatrix}$ **positive definite?**

Solution Test M for positive pivots, starting in the upper left corner. The first pivots of M are the pivots of A! **First condition** The block A must be positive definite.

Multiply the first row of M by $B^{\mathrm{T}}A^{-1}$ and subtract from the second row to get a block of zeros. The *Schur complement* $S = C - B^{\mathrm{T}}A^{-1}B$ appears in the corner:

$$\begin{bmatrix} I & 0 \\ -B^{\mathrm{T}}A^{-1} & I \end{bmatrix}\begin{bmatrix} A & B \\ B^{\mathrm{T}} & C \end{bmatrix} = \begin{bmatrix} A & B \\ 0 & C - B^{\mathrm{T}}A^{-1}B \end{bmatrix} = \begin{bmatrix} A & B \\ 0 & S \end{bmatrix}$$

The last pivots of M are pivots of S! **Second condition** S must be positive definite.

The two conditions are exactly like $a > 0$ and $c > b^2/a$, except they apply to blocks.

Problem Set 6.5

Problems 1–13 are about tests for positive definiteness.

1 Which of A_1, A_2, A_3, A_4 has two positive eigenvalues? Use the test, don't compute the λ's. Find an $\boldsymbol{x}$ so that $\boldsymbol{x}^{\mathrm{T}}A_1\boldsymbol{x} < 0$.

$$A_1 = \begin{bmatrix} 5 & 6 \\ 6 & 7 \end{bmatrix} \quad A_2 = \begin{bmatrix} -1 & -2 \\ -2 & -5 \end{bmatrix} \quad A_3 = \begin{bmatrix} 1 & 10 \\ 10 & 100 \end{bmatrix} \quad A_4 = \begin{bmatrix} 1 & 10 \\ 10 & 101 \end{bmatrix}.$$

2 For which numbers b and c are these matrices positive definite?

$$A = \begin{bmatrix} 1 & b \\ b & 9 \end{bmatrix} \quad \text{and} \quad A = \begin{bmatrix} 2 & 4 \\ 4 & c \end{bmatrix}.$$

With the pivots in D and multiplier in L, factor each A into LDL^{T}.

3 What is the quadratic $f = ax^2+2bxy+cy^2$ for each of these matrices? Complete the square to write f as a sum of one or two squares $d_1(\quad)^2 + d_2(\quad)^2$.

$$A = \begin{bmatrix} 1 & 2 \\ 2 & 9 \end{bmatrix} \quad \text{and} \quad A = \begin{bmatrix} 1 & 3 \\ 3 & 9 \end{bmatrix}.$$

4 Show that $f(x, y) = x^2 + 4xy + 3y^2$ does not have a minimum at $(0, 0)$ even though it has positive coefficients. Write f as a *difference* of squares and find a point (x, y) where f is negative.

5 The function $f(x, y) = 2xy$ certainly has a saddle point and not a minimum at $(0, 0)$. What symmetric matrix A produces this f? What are its eigenvalues?

6 (*Important*) If A has independent columns then $A^{\mathrm{T}}A$ is square and symmetric and invertible (Section 4.2). ***Rewrite*** $\boldsymbol{x}^{\mathrm{T}}\boldsymbol{A}^{\mathrm{T}}\boldsymbol{A}\boldsymbol{x}$ ***to show why it is positive except when*** $\boldsymbol{x} = \boldsymbol{0}$**.** Then $A^{\mathrm{T}}A$ is more than invertible, it is positive definite.

7 Test to see if $A^{\mathrm{T}}A$ is positive definite in each case:

$$A = \begin{bmatrix} 1 & 2 \\ 0 & 3 \end{bmatrix} \quad \text{and} \quad A = \begin{bmatrix} 1 & 1 \\ 1 & 2 \\ 2 & 1 \end{bmatrix} \quad \text{and} \quad A = \begin{bmatrix} 1 & 1 & 2 \\ 1 & 2 & 1 \end{bmatrix}.$$

8 The function $f(x, y) = 3(x+2y)^2 + 4y^2$ is positive except at $(0, 0)$. What is the matrix in $f = [\,x \;\; y\,]A[\,x \;\; y\,]^{\mathrm{T}}$? Check that the pivots of A are 3 and 4.

9 Find the 3 by 3 matrix A and its pivots, rank, eigenvalues, and determinant:

$$\begin{bmatrix} x_1 & x_2 & x_3 \end{bmatrix} \begin{bmatrix} & & \\ & A & \\ & & \end{bmatrix} \begin{bmatrix} x_1 \\ x_2 \\ x_3 \end{bmatrix} = 4(x_1 - x_2 + 2x_3)^2.$$

10 Which 3 by 3 symmetric matrices A produce these functions $f = \boldsymbol{x}^{\mathrm{T}}A\boldsymbol{x}$? Why is the first matrix positive definite but not the second one?

(a) $f = 2(x_1^2 + x_2^2 + x_3^2 - x_1x_2 - x_2x_3)$

(b) $f = 2(x_1^2 + x_2^2 + x_3^2 - x_1x_2 - x_1x_3 - x_2x_3)$.

11 Compute the three upper left determinants to establish positive definiteness. Verify that their ratios give the second and third pivots.

$$A = \begin{bmatrix} 2 & 2 & 0 \\ 2 & 5 & 3 \\ 0 & 3 & 8 \end{bmatrix}.$$

12 For what numbers c and d are A and B positive definite? Test the 3 determinants:

$$A = \begin{bmatrix} c & 1 & 1 \\ 1 & c & 1 \\ 1 & 1 & c \end{bmatrix} \quad \text{and} \quad B = \begin{bmatrix} 1 & 2 & 3 \\ 2 & d & 4 \\ 3 & 4 & 5 \end{bmatrix}.$$

13 Find a matrix with $a > 0$ and $c > 0$ and $a+c > 2b$ that has a negative eigenvalue.

Problems 14–20 are about applications of the tests.

14 *If A is positive definite then A^{-1} is positive definite.* Best proof: The eigenvalues of A^{-1} are positive because _____. *Second proof* (only for 2 by 2):

$$\text{The entries of } A^{-1} = \frac{1}{ac - b^2}\begin{bmatrix} c & -b \\ -b & a \end{bmatrix} \quad \text{pass the determinant tests} \quad \_\_\_\_\_.$$

15 *If A and B are positive definite, then $A+B$ is positive definite.* Pivots and eigenvalues are not convenient for $A + B$. Much better to prove $\boldsymbol{x}^{\mathrm{T}}(A + B)\boldsymbol{x} > 0$.

16 A positive definite matrix cannot have a zero (or even worse, a negative number) on its diagonal. Show that this matrix fails to have $\boldsymbol{x}^{\mathrm{T}}A\boldsymbol{x} > 0$:

$$\begin{bmatrix} x_1 & x_2 & x_3 \end{bmatrix}\begin{bmatrix} 4 & 1 & 1 \\ 1 & 0 & 2 \\ 1 & 2 & 5 \end{bmatrix}\begin{bmatrix} x_1 \\ x_2 \\ x_3 \end{bmatrix} \quad \text{is not positive when} \quad (x_1, x_2, x_3) = (\quad,\quad,\quad).$$

17 A diagonal entry a_{jj} of a symmetric matrix cannot be smaller than all the λ's. If it were, then $A - a_{jj}I$ would have _____ eigenvalues and would be positive definite. But $A - a_{jj}I$ has a _____ on the main diagonal.

18 If $A\boldsymbol{x} = \lambda\boldsymbol{x}$ then $\boldsymbol{x}^{\mathrm{T}}A\boldsymbol{x} =$ _____. If A is positive definite, prove that $\lambda > 0$.

19 Reverse Problem 18 to show that *if all* $\lambda > 0$ *then* $\boldsymbol{x}^{\mathrm{T}}A\boldsymbol{x} > 0$. We must do this for *every* nonzero $\boldsymbol{x}$, not just the eigenvectors. So write $\boldsymbol{x}$ as a combination of the eigenvectors and explain why all "cross terms" are $\boldsymbol{x}_i^{\mathrm{T}}\boldsymbol{x}_j = 0$:

$$\boldsymbol{x}^{\mathrm{T}}A\boldsymbol{x} = (c_1\boldsymbol{x}_1 + \cdots + c_n\boldsymbol{x}_n)^{\mathrm{T}}(c_1\lambda_1\boldsymbol{x}_1 + \cdots + c_n\lambda_n x_n) = c_1^2\lambda_1\boldsymbol{x}_1^{\mathrm{T}}\boldsymbol{x}_1 + \cdots + c_n^2\lambda_n\boldsymbol{x}_n^{\mathrm{T}}\boldsymbol{x}_n > 0.$$

20 Give a quick reason why each of these statements is true:

(a) Every positive definite matrix is invertible.

(b) The only positive definite projection matrix is $P = I$.

(c) A diagonal matrix with positive diagonal entries is positive definite.

(d) A symmetric matrix with a positive determinant might not be positive definite!

Problems 21–24 use the eigenvalues; Problems 25–27 are based on pivots.

21 For which s and t do A and B have all $\lambda > 0$ (therefore positive definite)?

$$A = \begin{bmatrix} s & -4 & -4 \\ -4 & s & -4 \\ -4 & -4 & s \end{bmatrix} \quad \text{and} \quad B = \begin{bmatrix} t & 3 & 0 \\ 3 & t & 4 \\ 0 & 4 & t \end{bmatrix}.$$

22 From $A = Q\Lambda Q^{\mathrm{T}}$ compute the positive definite symmetric square root $Q\Lambda^{1/2}Q^{\mathrm{T}}$ of each matrix. Check that this square root gives $R^2 = A$:

$$A = \begin{bmatrix} 5 & 4 \\ 4 & 5 \end{bmatrix} \quad \text{and} \quad A = \begin{bmatrix} 10 & 6 \\ 6 & 10 \end{bmatrix}.$$

23 You may have seen the equation for an ellipse as $\left(\frac{x}{a}\right)^2 + \left(\frac{y}{b}\right)^2 = 1$. What are a and b when the equation is written as $\lambda_1 x^2 + \lambda_2 y^2 = 1$? The ellipse $9x^2 + 16y^2 = 1$ has axes with half-lengths $a =$ _____ and $b =$ _____.

24 Draw the tilted ellipse $x^2 + xy + y^2 = 1$ and find the half-lengths of its axes from the eigenvalues of the corresponding A.

25 With positive pivots in D, the factorization $A = LDL^{\mathrm{T}}$ becomes $L\sqrt{D}\sqrt{D}L^{\mathrm{T}}$. (Square roots of the pivots give $D = \sqrt{D}\sqrt{D}$.) Then $C = L\sqrt{D}$ yields the ***Cholesky factorization*** $A = CC^{\mathrm{T}}$ which is "symmetrized LU":

$$\text{From} \quad C = \begin{bmatrix} 3 & 0 \\ 1 & 2 \end{bmatrix} \quad \text{find } A. \qquad \text{From} \quad A = \begin{bmatrix} 4 & 8 \\ 8 & 25 \end{bmatrix} \quad \text{find } C.$$

26 In the Cholesky factorization $A = CC^{\mathrm{T}}$, with $C = L\sqrt{D}$, the square roots of the pivots are on the diagonal of C. Find C (lower triangular) for

$$A = \begin{bmatrix} 9 & 0 & 0 \\ 0 & 1 & 2 \\ 0 & 2 & 8 \end{bmatrix} \quad \text{and} \quad A = \begin{bmatrix} 1 & 1 & 1 \\ 1 & 2 & 2 \\ 1 & 2 & 7 \end{bmatrix}.$$

27 The symmetric factorization $A = LDL^{\mathrm{T}}$ means that $\boldsymbol{x}^{\mathrm{T}}A\boldsymbol{x} = \boldsymbol{x}^{\mathrm{T}}LDL^{\mathrm{T}}\boldsymbol{x}$:

$$\begin{bmatrix} x & y \end{bmatrix} \begin{bmatrix} a & b \\ b & c \end{bmatrix} \begin{bmatrix} x \\ y \end{bmatrix} = \begin{bmatrix} x & y \end{bmatrix} \begin{bmatrix} 1 & 0 \\ b/a & 1 \end{bmatrix} \begin{bmatrix} a & 0 \\ 0 & (ac - b^2)/a \end{bmatrix} \begin{bmatrix} 1 & b/a \\ 0 & 1 \end{bmatrix} \begin{bmatrix} x \\ y \end{bmatrix}.$$

The left side is $ax^2 + 2bxy + cy^2$. The right side is $a\left(x + \frac{b}{a}y\right)^2 +$ _____ y^2. The second pivot completes the square! Test with $a = 2$, $b = 4$, $c = 10$.

28 Without multiplying $A = \begin{bmatrix} \cos\theta & -\sin\theta \\ \sin\theta & \cos\theta \end{bmatrix} \begin{bmatrix} 2 & 0 \\ 0 & 5 \end{bmatrix} \begin{bmatrix} \cos\theta & \sin\theta \\ -\sin\theta & \cos\theta \end{bmatrix}$, find

(a) the determinant of A (b) the eigenvalues of A
(c) the eigenvectors of A (d) a reason why A is symmetric positive definite.

29 For $f_1(x, y) = \frac{1}{4}x^4 + x^2y + y^2$ and $f_2(x, y) = x^3 + xy - x$ find the second derivative matrices A_1 and A_2:

$$A = \begin{bmatrix} \partial^2 f/\partial x^2 & \partial^2 f/\partial x \partial y \\ \partial^2 f/\partial y \partial x & \partial^2 f/\partial y^2 \end{bmatrix}.$$

A_1 is positive definite so f_1 is concave up (= convex). Find the minimum point of f_1 and the saddle point of f_2 (look where first derivatives are zero).

30 The graph of $z = x^2 + y^2$ is a bowl opening upward. *The graph* of $z = x^2 - y^2$ *is a saddle*. The graph of $z = -x^2 - y^2$ is a bowl opening downward. What is a test on a, b, c for $z = ax^2 + 2bxy + cy^2$ to have a saddle at $(0, 0)$?

31 Which values of c give a bowl and which give a saddle point for the graph of $z = 4x^2 + 12xy + cy^2$? Describe this graph at the borderline value of c.

32 A *group* of nonsingular matrices includes AB and A^{-1} if it includes A and B. "Products and inverses stay in the group." Which of these sets are groups (updating Problem 2.7.37)? Positive definite symmetric matrices A, orthogonal matrices Q, all exponentials e^{tA} of a fixed matrix A, matrices P with positive eigenvalues, matrices D with determinant 1. Invent a "subgroup" of one of these groups (not the identity I by itself—this is the smallest group).

SIMILAR MATRICES ■ 6.6

The key step in this chapter was to diagonalize a matrix. That was done by S—the eigenvector matrix. The diagonal matrix $S^{-1}AS$ is Λ—the eigenvalue matrix. But diagonalization was not possible for every A. Some matrices resisted and we had to leave them alone. They had too few eigenvectors to produce S. In this new section, S remains the best choice when we can find it, ***but we allow any invertible matrix M.***

Starting from A we go to $M^{-1}AM$. This new matrix may happen to be diagonal—more likely not. It still shares important properties of A. No matter which M we choose, ***the eigenvalues stay the same***. The matrices A and $M^{-1}AM$ are called "similar". A typical matrix A is similar to a whole family of other matrices because there are so many choices of M.

DEFINITION Let M be any invertible matrix. Then $B = M^{-1}AM$ is ***similar*** to A.

If $B = M^{-1}AM$ then immediately $A = MBM^{-1}$. That means: If B is similar to A then A is similar to B. The matrix in this reverse direction is M^{-1}—just as good as M.

A diagonalizable matrix is similar to Λ. In that special case M is S. We have $A = S\Lambda S^{-1}$ and $\Lambda = S^{-1}AS$. They certainly have the same eigenvalues! This section is opening up to other similar matrices $B = M^{-1}AM$.

The combination $M^{-1}AM$ appears when we change variables in a differential equation. Start with an equation for $\boldsymbol{u}$ and set $\boldsymbol{u} = M\boldsymbol{v}$:

$$\frac{d\boldsymbol{u}}{dt} = A\boldsymbol{u} \quad \text{becomes} \quad M\frac{d\boldsymbol{v}}{dt} = AM\boldsymbol{v} \quad \text{which is} \quad \frac{d\boldsymbol{v}}{dt} = M^{-1}AM\boldsymbol{v}.$$

The original coefficient matrix was A, the new one at the right is $M^{-1}AM$. Changing variables leads to a similar matrix. When $M = S$ the new system is diagonal—the maximum in simplicity. But other choices of M also make the new system easier to solve. Since we can always go back to $\boldsymbol{u}$, similar matrices have to give the same growth or decay. More precisely, *the eigenvalues of A and B are the same.*

6Q (No change in λ's) Similar matrices A and $M^{-1}AM$ have the same eigenvalues. If $\boldsymbol{x}$ is an eigenvector of A then $M^{-1}\boldsymbol{x}$ is an eigenvector of $B = M^{-1}AM$.

The proof is quick, since $B = M^{-1}AM$ gives $A = MBM^{-1}$. Suppose $A\boldsymbol{x} = \lambda\boldsymbol{x}$:

$$MBM^{-1}\boldsymbol{x} = \lambda\boldsymbol{x} \quad \text{means that} \quad BM^{-1}\boldsymbol{x} = \lambda M^{-1}\boldsymbol{x}.$$

The eigenvalue of B is the same λ. The eigenvector is now $M^{-1}\boldsymbol{x}$.

The following example finds three matrices that are similar to one projection matrix.

Example 1

The projection $A = \begin{bmatrix} .5 & .5 \\ .5 & .5 \end{bmatrix}$ is similar to $\Lambda = S^{-1}AS = \begin{bmatrix} 1 & 0 \\ 0 & 0 \end{bmatrix}$

Now choose $M = \begin{bmatrix} 1 & 0 \\ 1 & 2 \end{bmatrix}$: the similar matrix $M^{-1}AM$ is $\begin{bmatrix} 1 & 1 \\ 0 & 0 \end{bmatrix}$.

Also choose $M = \begin{bmatrix} 0 & -1 \\ 1 & 0 \end{bmatrix}$: the similar matrix $M^{-1}AM$ is $\begin{bmatrix} .5 & -.5 \\ -.5 & .5 \end{bmatrix}$.

These matrices $M^{-1}AM$ all have the same eigenvalues 1 and 0. ***Every 2 by 2 matrix with those eigenvalues is similar to A.*** The eigenvectors change with M.

The eigenvalues in that example are 1 and 0, *not repeated*. This makes life easy. Repeated eigenvalues are harder. The next example has eigenvalues 0 and 0. The zero matrix shares those eigenvalues, but it is in a family by itself: $M^{-1}0M = 0$.

The following matrix A is similar to every nonzero matrix with eigenvalues 0 and 0.

Example 2

$$A = \begin{bmatrix} 0 & 1 \\ 0 & 0 \end{bmatrix} \text{ is similar to every matrix } B = \begin{bmatrix} cd & d^2 \\ -c^2 & -cd \end{bmatrix} \text{ except } B = 0.$$

These matrices B all have zero determinant (like A). They all have rank one (like A). Their trace is $cd - cd = 0$. Their eigenvalues are 0 and 0 (like A). I chose any $M = \left[\begin{smallmatrix} a & b \\ c & d \end{smallmatrix}\right]$ with $ad - bc = 1$, and $B = M^{-1}AM$.

These matrices B can't be diagonalized. In fact A is as close to diagonal as possible. It is the "***Jordan form***" for the family of matrices B. This is the outstanding member (my class says "Godfather") of the family. The Jordan form $J = A$ is as near as we can come to diagonalizing these matrices, when there is only one eigenvector.

Chapter 7 will explain another approach to similar matrices. Instead of changing variables by $\boldsymbol{u} = M\boldsymbol{v}$, we "***change the basis***". In this approach, similar matrices will represent the same transformation of n-dimensional space. When we choose a basis for $\mathbf{R}^n$, we get a matrix. The standard basis vectors ($M = I$) lead to $I^{-1}AI$ which is A. Other bases lead to similar matrices $B = M^{-1}AM$.

In this "similarity transformation" from A to B, some things change and some don't. Here is a table to show connections between similar matrices A and B:

Not changed	**Changed**
Eigenvalues	Eigenvectors
Trace and determinant	Nullspace
Rank	Column space
Number of independent eigenvectors	Row space
	Left nullspace
Jordan form	Singular values

The eigenvalues don't change for similar matrices; the eigenvectors do. The trace is the sum of the λ's (unchanged). The determinant is the product of the same λ's.[1] The nullspace consists of the eigenvectors for $\lambda = 0$ (if any), so it can change. Its dimension $n - r$ does not change! The *number* of eigenvectors stays the same for each λ, while the vectors themselves are multiplied by M^{-1}.

The *singular values* depend on $A^{\mathrm{T}}A$, which definitely changes. They come in the next section. The table suggests good exercises in linear algebra. But the last entry in the unchanged column—the *Jordan form*—is more than an exercise. We lead up to it with one more example of similar matrices.

Example 3 This Jordan matrix J has triple eigenvalue $5, 5, 5$. Its only eigenvectors are multiples of $(1, 0, 0)$! Algebraic multiplicity 3, geometric multiplicity 1:

$$\text{If} \quad J = \begin{bmatrix} 5 & 1 & 0 \\ 0 & 5 & 1 \\ 0 & 0 & 5 \end{bmatrix} \quad \text{then} \quad J - 5I = \begin{bmatrix} 0 & 1 & 0 \\ 0 & 0 & 1 \\ 0 & 0 & 0 \end{bmatrix} \quad \text{has rank 2.}$$

Every similar matrix $B = M^{-1}JM$ has the same triple eigenvalue $5, 5, 5$. Also $B - 5I$ must have the same rank 2. Its nullspace has dimension $3 - 2 = 1$. So each similar matrix B also has only one independent eigenvector.

The transpose matrix J^{T} has the same eigenvalues $5, 5, 5$, and $J^{\mathrm{T}} - 5I$ has the same rank 2. ***Jordan's theory says that J^{T} is similar to J.*** The matrix that produces the similarity happens to be the *reverse identity* M:

$$J^{\mathrm{T}} = M^{-1}JM \quad \text{is} \quad \begin{bmatrix} 5 & 0 & 0 \\ 1 & 5 & 0 \\ 0 & 1 & 5 \end{bmatrix} = \begin{bmatrix} & & 1 \\ & 1 & \\ 1 & & \end{bmatrix} \begin{bmatrix} 5 & 1 & 0 \\ 0 & 5 & 1 \\ 0 & 0 & 5 \end{bmatrix} \begin{bmatrix} & & 1 \\ & 1 & \\ 1 & & \end{bmatrix}.$$

All blank entries are zero. An eigenvector of J^{T} is $M^{-1}(1, 0, 0) = (0, 0, 1)$. There is one line of eigenvectors $(x_1, 0, 0)$ for J and another line $(0, 0, x_3)$ for J^{T}.

The key fact is that this matrix J is similar to *every* matrix A with eigenvalues $5, 5, 5$ and *one line* of eigenvectors. There is an M with $M^{-1}AM = J$.

Example 4 Since J is as close to diagonal as we can get, the equation $d\boldsymbol{u}/dt = J\boldsymbol{u}$ cannot be simplified by changing variables. We must solve it as it stands:

$$\frac{d\boldsymbol{u}}{dt} = J\boldsymbol{u} = \begin{bmatrix} 5 & 1 & 0 \\ 0 & 5 & 1 \\ 0 & 0 & 5 \end{bmatrix} \begin{bmatrix} x \\ y \\ z \end{bmatrix} \quad \text{is} \quad \begin{aligned} dx/dt &= 5x + y \\ dy/dt &= 5y + z \\ dz/dt &= 5z. \end{aligned}$$

[1] The determinant is unchanged because $\det B = (\det M^{-1})(\det A)(\det M) = \det A$.

The system is triangular. We think naturally of back substitution. Solve the last equation and work upwards. Main point: *All solutions contain* e^{5t}:

$$\begin{aligned}
\frac{dz}{dt} &= 5z & &\text{yields} \quad z = z(0)e^{5t} \\
\frac{dy}{dt} &= 5y + z & &\text{yields} \quad y = \big(y(0) + tz(0)\big)e^{5t} \\
\frac{dx}{dt} &= 5x + y & &\text{yields} \quad x = \big(x(0) + ty(0) + \tfrac{1}{2}t^2 z(0)\big)e^{5t}.
\end{aligned}$$

The two missing eigenvectors are responsible for the te^{5t} and t^2e^{5t} terms in y and z. The factors t and t^2 enter because $\lambda = 5$ is a triple eigenvalue with one eigenvector.

The Jordan Form

For every A, we want to choose M so that $M^{-1}AM$ is as ***nearly diagonal as possible***. When A has a full set of n eigenvectors, they go into the columns of M. Then $M = S$. The matrix $S^{-1}AS$ is diagonal, period. This matrix is the Jordan form of A—when A can be diagonalized. In the general case, eigenvectors are missing and Λ can't be reached.

Suppose A has s independent eigenvectors. Then it is similar to a matrix with s blocks. Each block is like J in Example 3. *The eigenvalue is on the diagonal and the diagonal above it contains* 1's. This block accounts for one eigenvector of A. When there are n eigenvectors and n blocks, they are all 1 by 1. In that case J is Λ.

6R (Jordan form) If A has s independent eigenvectors, it is similar to a matrix J that has s Jordan blocks on its diagonal: There is a matrix M such that

$$M^{-1}AM = \begin{bmatrix} J_1 & & \\ & \ddots & \\ & & J_s \end{bmatrix} = J. \tag{1}$$

Each block in J has one eigenvalue λ_i, one eigenvector, and 1's above the diagonal:

$$J_i = \begin{bmatrix} \lambda_i & 1 & & \\ & \cdot & \cdot & \\ & & \cdot & 1 \\ & & & \lambda_i \end{bmatrix}. \tag{2}$$

A is similar to B if they share the same Jordan form J—not otherwise.

This is the big theorem about matrix similarity. In every family of similar matrices, we are picking one outstanding member called J. It is nearly diagonal (or if possible completely diagonal). For that J, we can solve $d\boldsymbol{u}/dt = J\boldsymbol{u}$ as in Example 4.

We can take powers J^k as in Problems 9–10. Every other matrix in the family has the form $A = MJM^{-1}$. The connection through M solves $d\boldsymbol{u}/dt = A\boldsymbol{u}$.

The point you must see is that $MJM^{-1}MJM^{-1} = MJ^2M^{-1}$. That cancellation of $M^{-1}M$ in the middle has been used through this chapter (when M was S). We found A^{100} from $S\Lambda^{100}S^{-1}$—by diagonalizing the matrix. Now we can't quite diagonalize A. So we use $MJ^{100}M^{-1}$ instead.

Jordan's Theorem **6R** is proved in my textbook *Linear Algebra and Its Applications*, published by Brooks-Cole. Please refer to that book (or more advanced books) for the proof. The reasoning is rather intricate and in actual computations the Jordan form is not at all popular—its calculation is not stable. A slight change in A will separate the repeated eigenvalues and remove the off-diagonal 1's—switching to a diagonal Λ. Proved or not, you have caught the central idea of similarity—to make A as simple as possible while preserving its essential properties.

■ REVIEW OF THE KEY IDEAS ■

1. B is similar to A if $B = M^{-1}AM$, for some invertible matrix M.

2. Similar matrices have the same eigenvalues. Eigenvectors are multiplied by M^{-1}.

3. If A has n independent eigenvectors then A is similar to Λ (take $M = S$).

4. Every matrix is similar to a Jordan matrix J (which has Λ as its diagonal part). J has a block for each eigenvector, and 1's for missing eigenvectors.

■ WORKED EXAMPLES ■

6.6 A The 4 by 4 triangular Pascal matrix P_L and its inverse (alternating diagonals) are

$$P_L = \begin{bmatrix} 1 & 0 & 0 & 0 \\ 1 & 1 & 0 & 0 \\ 1 & 2 & 1 & 0 \\ 1 & 3 & 3 & 1 \end{bmatrix} \quad \text{and} \quad P_L^{-1} = \begin{bmatrix} 1 & 0 & 0 & 0 \\ -1 & 1 & 0 & 0 \\ 1 & -2 & 1 & 0 \\ -1 & 3 & -3 & 1 \end{bmatrix}.$$

Check that P_L and P_L^{-1} have the same eigenvalues. Find a diagonal matrix D with alternating signs that gives $P_L^{-1} = D^{-1}P_LD$, so P_L is similar to P_L^{-1}. Show that P_LD with columns of alternating signs is its own inverse. P_LD is pascal(4, 1) in MATLAB.

Since P_L and P_L^{-1} are similar they have the same Jordan form J. Find J by checking the number of independent eigenvectors of P_L with $\lambda = 1$.

Solution The triangular matrices P_L and P_L^{-1} both have $\lambda = 1, 1, 1, 1$ on their main diagonals. Choose D with alternating 1 and -1 on its diagonal. *D equals D^{-1}*:

$$D^{-1}P_LD = \begin{bmatrix} -1 & & & \\ & 1 & & \\ & & -1 & \\ & & & 1 \end{bmatrix}\begin{bmatrix} 1 & 0 & 0 & 0 \\ 1 & 1 & 0 & 0 \\ 1 & 2 & 1 & 0 \\ 1 & 3 & 3 & 1 \end{bmatrix}\begin{bmatrix} -1 & & & \\ & 1 & & \\ & & -1 & \\ & & & 1 \end{bmatrix} = P_L^{-1}.$$

Check: Changing signs in rows 1 and 3 of P_L, and columns 1 and 3, produces the four negative entries in P_L^{-1}. We are multiplying row i by $(-1)^i$ and column j by $(-1)^j$, which gives the alternating diagonals. Then $P_LD = \text{pascal}(n, 1)$ has *columns with alternating signs* and equals its own inverse!

$$(P_LD)(P_LD) = P_LD^{-1}P_LD = P_LP_L^{-1} = I.$$

P_L has only one line of eigenvectors $\boldsymbol{x} = (0, 0, 0, x_4)$, with $\lambda = 1$. The rank of $P_L - I$ is certainly 3. So its Jordan form J has only *one block* (also with $\lambda = 1$):

$$P_L \text{ and also } P_L^{-1} \text{ are somehow similar to Jordan's } J = \begin{bmatrix} 1 & 1 & 0 & 0 \\ 0 & 1 & 1 & 0 \\ 0 & 0 & 1 & 1 \\ 0 & 0 & 0 & 1 \end{bmatrix}$$

6.6 B If A is similar to A^{-1}, explain why its eigenvalues come in reciprocal pairs $\lambda = a$ and $\lambda = 1/a$. The 3 by 3 Pascal matrix P_S has paired eigenvalues $4+\sqrt{15}, 4-\sqrt{15}, 1$. Use $P_L^{-1} = D^{-1}P_LD$ and the symmetric factorization $P_S = P_LP_L^{\mathrm{T}}$ in Worked Example **2.6 A** to prove that *P_S is similar to P_S^{-1}*.

Solution When A has nonzero eigenvalues $\lambda_1, \ldots, \lambda_n$, its inverse has eigenvalues $\lambda_1^{-1}, \ldots, \lambda_n^{-1}$. Reason: Multiply $A\boldsymbol{x} = \lambda\boldsymbol{x}$ by A^{-1} and λ^{-1} to get $A^{-1}\boldsymbol{x} = \lambda^{-1}\boldsymbol{x}$.

If A and A^{-1} are similar they have the same set of eigenvalues. So an even number of λ's must pair off in the form a and $1/a$. The product $(4+\sqrt{15})(4-\sqrt{15}) = 16 - 15 = 1$ shows that $4+\sqrt{15}, 4-\sqrt{15}, 1$ do pair off properly.

The symmetric Pascal matrices have paired eigenvalues because P_S is similar to P_S^{-1}. To prove this similarity, using $D = D^{-1} = D^{\mathrm{T}}$, start from $P_S = P_LP_L^{\mathrm{T}}$:

$$P_S^{-1} = (P_L^{\mathrm{T}})^{-1}(P_L^{-1}) = (D^{-1}P_LD)^{\mathrm{T}}(D^{-1}P_LD) = D^{-1}P_L^{\mathrm{T}}P_LD = (P_LD)^{-1}(P_LP_L^{\mathrm{T}})(P_LD).$$

This is $P_S^{-1} = M^{-1}P_SM$ (similar matrices!) for the matrix $M = P_LD$.

The eigenvalues of larger matrices P_S don't have nice formulas. But eig(pascal (n)) will confirm that those eigenvalues come in reciprocal pairs a and $1/a$. The Jordan form of P_S is the diagonal Λ, because symmetric matrices always have a complete set of eigenvectors.

Problem Set 6.6

1 If $B = M^{-1}AM$ and also $C = N^{-1}BN$, what matrix T gives $C = T^{-1}AT$? Conclusion: If B is similar to A and C is similar to B, then ____.

2 If $C = F^{-1}AF$ and also $C = G^{-1}BG$, what matrix M gives $B = M^{-1}AM$? Conclusion: If C is similar to A and also to B then ____.

3 Show that A and B are similar by finding M so that $B = M^{-1}AM$:

$$A = \begin{bmatrix} 1 & 0 \\ 1 & 0 \end{bmatrix} \quad \text{and} \quad B = \begin{bmatrix} 0 & 1 \\ 0 & 1 \end{bmatrix}$$

$$A = \begin{bmatrix} 1 & 1 \\ 1 & 1 \end{bmatrix} \quad \text{and} \quad B = \begin{bmatrix} 1 & -1 \\ -1 & 1 \end{bmatrix}$$

$$A = \begin{bmatrix} 1 & 2 \\ 3 & 4 \end{bmatrix} \quad \text{and} \quad B = \begin{bmatrix} 4 & 3 \\ 2 & 1 \end{bmatrix}.$$

4 If a 2 by 2 matrix A has eigenvalues 0 and 1, why is it similar to $\Lambda = \left[\begin{smallmatrix} 1 & 0 \\ 0 & 0 \end{smallmatrix}\right]$? Deduce from Problem 2 that all 2 by 2 matrices with those eigenvalues are similar.

5 Which of these six matrices are similar? Check their eigenvalues.

$$\begin{bmatrix} 1 & 0 \\ 0 & 1 \end{bmatrix} \quad \begin{bmatrix} 0 & 1 \\ 1 & 0 \end{bmatrix} \quad \begin{bmatrix} 1 & 1 \\ 0 & 0 \end{bmatrix} \quad \begin{bmatrix} 0 & 0 \\ 1 & 1 \end{bmatrix} \quad \begin{bmatrix} 1 & 0 \\ 1 & 0 \end{bmatrix} \quad \begin{bmatrix} 0 & 1 \\ 0 & 1 \end{bmatrix}.$$

6 There are sixteen 2 by 2 matrices whose entries are 0's and 1's. Similar matrices go into the same family. How many families? How many matrices (total 16) in each family?

7 (a) If $\boldsymbol{x}$ is in the nullspace of A show that $M^{-1}\boldsymbol{x}$ is in the nullspace of $M^{-1}AM$.

(b) The nullspaces of A and $M^{-1}AM$ have the same (vectors)(basis)(dimension).

8 If A and B have the exactly the same eigenvalues and eigenvectors, does $A = B$? With n independent eigenvectors we do have $A = B$. Find $A \neq B$ when both have eigenvalues $0, 0$ but only one line of eigenvectors $(x_1, 0)$.

9 By direct multiplication find A^2 and A^3 and A^5 when

$$A = \begin{bmatrix} 1 & 1 \\ 0 & 1 \end{bmatrix}.$$

Guess the form of A^k. Set $k = 0$ to find A^0 and $k = -1$ to find A^{-1}.

Questions 10–14 are about the Jordan form.

10 By direct multiplication, find J^2 and J^3 when

$$J = \begin{bmatrix} c & 1 \\ 0 & c \end{bmatrix}.$$

Guess the form of J^k. Set $k = 0$ to find J^0. Set $k = -1$ to find J^{-1}.

11 The text solved $d\boldsymbol{u}/dt = J\boldsymbol{u}$ for a 3 by 3 Jordan block J. Add a fourth equation $dw/dt = 5w + x$. Follow the pattern of solutions for z, y, x to find w.

12 These Jordan matrices have eigenvalues $0, 0, 0, 0$. They have two eigenvectors (one from each block). But the block sizes don't match and they are *not similar*:

$$J = \left[\begin{array}{cc|cc} 0 & 1 & 0 & 0 \\ 0 & 0 & 0 & 0 \\ \hline 0 & 0 & 0 & 1 \\ 0 & 0 & 0 & 0 \end{array}\right] \quad \text{and} \quad K = \left[\begin{array}{ccc|c} 0 & 1 & 0 & 0 \\ 0 & 0 & 1 & 0 \\ 0 & 0 & 0 & 0 \\ \hline 0 & 0 & 0 & 0 \end{array}\right].$$

For any matrix M, compare JM with MK. If they are equal show that M is not invertible. Then $M^{-1}JM = K$ is impossible.

13 Prove that A^{T} *is always similar to* A (we knew the λ's are the same):

1. For one Jordan block J_i: Find M_i so that $M_i^{-1} J_i M_i = J_i^{\mathrm{T}}$ (see Example 3).
2. For any J with blocks J_i: Build M_0 from blocks so that $M_0^{-1} J M_0 = J^{\mathrm{T}}$.
3. For any $A = MJM^{-1}$: Show that A^{T} is similar to J^{T} and so to J and to A.

14 Find two more matrices similar to J in Example 3.

15 Prove that $\det(A - \lambda I) = \det(M^{-1}AM - \lambda I)$. (You could write $I = M^{-1}M$ and factor out $\det M^{-1}$ and $\det M$.) This says that A and $M^{-1}AM$ have the same characteristic polynomial. So their roots are the same eigenvalues.

16 Which pairs are similar? Choose a, b, c, d to prove that the other pairs aren't:

$$\begin{bmatrix} a & b \\ c & d \end{bmatrix} \quad \begin{bmatrix} b & a \\ d & c \end{bmatrix} \quad \begin{bmatrix} c & d \\ a & b \end{bmatrix} \quad \begin{bmatrix} d & c \\ b & a \end{bmatrix}.$$

17 True or false, with a good reason:

(a) An invertible matrix can't be similar to a singular matrix.

(b) A symmetric matrix can't be similar to a nonsymmetric matrix.

(c) A can't be similar to $-A$ unless $A = 0$.

(d) A can't be similar to $A + I$.

18 If B is invertible prove that AB *has the same eigenvalues as* BA.

19 If A is 6 by 4 and B is 4 by 6, AB and BA have different sizes. But still

$$\begin{bmatrix} I & -A \\ 0 & I \end{bmatrix}\begin{bmatrix} AB & 0 \\ B & 0 \end{bmatrix}\begin{bmatrix} I & A \\ 0 & I \end{bmatrix} = \begin{bmatrix} 0 & 0 \\ B & BA \end{bmatrix} = G.$$

(a) What sizes are the blocks of G? They are the same in each matrix.

(b) This equation is $M^{-1}FM = G$, so F and G have the same 10 eigenvalues. F has the eigenvalues of AB plus 4 zeros; G has the eigenvalues of BA plus 6 zeros. ***AB has the same eigenvalues as BA plus _____ zeros.***

20 Why are these statements all true?

(a) If A is similar to B then A^2 is similar to B^2.

(b) A^2 and B^2 can be similar when A and B are not similar (try $\lambda = 0, 0$).

(c) $\left[\begin{smallmatrix} 3 & 0 \\ 0 & 4 \end{smallmatrix}\right]$ is similar to $\left[\begin{smallmatrix} 3 & 1 \\ 0 & 4 \end{smallmatrix}\right]$.

(d) $\left[\begin{smallmatrix} 3 & 0 \\ 0 & 3 \end{smallmatrix}\right]$ is not similar to $\left[\begin{smallmatrix} 3 & 1 \\ 0 & 3 \end{smallmatrix}\right]$.

(e) If we exchange rows 1 and 2 of A, and then exchange columns 1 and 2, **the eigenvalues stay the same**.

21 If J is the 5 by 5 Jordan block with $\lambda = 0$, find J^2 and count its eigenvectors and find its Jordan form (two blocks).

SINGULAR VALUE DECOMPOSITION (SVD) ■ 6.7

The Singular Value Decomposition is a highlight of linear algebra. A is any m by n matrix, square or rectangular. We will diagonalize it, but not by $S^{-1}AS$. Its row space is r-dimensional (inside $\mathbf{R}^n$). Its column space is also r-dimensional (inside $\mathbf{R}^m$). We are going to choose special ***orthonormal bases*** $\boldsymbol{v}_1, \ldots, \boldsymbol{v}_r$ for the row space and $\boldsymbol{u}_1, \ldots, \boldsymbol{u}_r$ for the column space. For those bases, we want each $A\boldsymbol{v}_i$ to be in the direction of $\boldsymbol{u}_i$. In matrix form, ***these equations*** $A\boldsymbol{v}_i = \sigma_i \boldsymbol{u}_i$ ***become*** $AV = U\Sigma$ ***or*** $A = U\Sigma V^{\mathrm{T}}$. This is the SVD.

Image Compression

Unusually, I am going to stop the theory and describe applications. This is the century of data, and often that data is stored in a matrix. *A digital image is really a matrix of pixel values.* Each little picture element or "pixel" has a gray scale number between black and white (it has three numbers for a color picture). The picture might have $512 = 2^9$ pixels in each row and $256 = 2^8$ pixels down each column. We have a 256 by 512 pixel matrix with 2^{17} entries! To store one picture, the computer has no problem. But if you go in for a CT scan or Magnetic Resonance, you produce an image at every cross section—a ton of data. If the pictures are frames in a movie, 30 frames a second means 108,000 images per hour. Compression is especially needed for high definition digital TV, or the equipment could not keep up in real time.

What is compression? We want to replace those 2^{17} matrix entries by a smaller number, *without losing picture quality*. A simple way would be to use larger pixels—replace groups of four pixels by their average value. This is 4 : 1 compression. But if we carry it further, like 16 : 1, our image becomes "blocky". We want to replace the mn entries by a smaller number, in a way that the human visual system won't notice.

Compression is a billion dollar problem and everyone has ideas. Later in this book I will describe Fourier transforms (used in **jpeg**) and wavelets (now in **JPEG2000**). Here we try an SVD approach: *Replace the* 256 *by* 512 *pixel matrix by* ***a matrix of rank one****: a column times a row.* If this is successful, the storage requirement for a column and row becomes $256 + 512$ (plus instead of times!). The compression ratio $(256)(512)/(256 + 512)$ is better than 170 : 1. This is more than we hope for. We may actually use five matrices of rank one (so a matrix approximation of rank 5). The compression is still 34 : 1 and the crucial question is the picture quality.

Where does the SVD come in? *The best rank one approximation to A is the matrix $\sigma_1 \boldsymbol{u}_1 \boldsymbol{v}_1^{\mathrm{T}}$.* It uses the largest singular value σ_1 and its left and right singular vectors $\boldsymbol{u}_1$ and $\boldsymbol{v}_1$. The best rank 5 approximation includes $\sigma_2 \boldsymbol{u}_2 \boldsymbol{v}_2^{\mathrm{T}} + \cdots + \sigma_5 \boldsymbol{u}_5 \boldsymbol{v}_5^{\mathrm{T}}$. If we can compute those $\boldsymbol{u}$'s and $\boldsymbol{v}$'s quickly (a big "if" since you will see them as eigenvectors for $A^{\mathrm{T}}A$ and AA^{T}) then this SVD algorithm is competitive.

I will mention a different matrix, one that a library needs to compress. The rows correspond to key words. The columns correspond to titles in the library. The entry in this *word-title matrix* is $a_{ij} = 1$ if word i is in title j (otherwise $a_{ij} = 0$). We might normalize the columns to be unit vectors, so that long titles don't get an advantage.

Instead of the title, we might use a table of contents or an abstract that better captures the content. (Other books might share the title "*Introduction to Linear Algebra*". If you are searching for the SVD, you want the right book.) Instead of $a_{ij} = 1$, the entries of A can include the *frequency* of the search words in each document.

Once the indexing matrix is created, the search is a linear algebra problem. If we use 100,000 words from an English dictionary and 2,000,000,000 web pages as documents, it is a long search. We need a shortcut. This matrix has to be compressed. I will now explain the SVD approach, which gives an optimal low rank approximation to A. (It works better for library matrices than for natural images.) There is an ever-present tradeoff in the cost to compute the $\boldsymbol{u}$'s and $\boldsymbol{v}$'s, and I hope you will invent a better way.

The Bases and the SVD

Start with a 2 by 2 matrix. Let its rank be $r = 2$, so this matrix A is invertible. Its row space is the plane $\mathbf{R}^2$. We want $\boldsymbol{v}_1$ and $\boldsymbol{v}_2$ to be perpendicular unit vectors, an orthonormal basis. ***We also want $A\boldsymbol{v}_1$ and $A\boldsymbol{v}_2$ to be perpendicular.*** (This is the tricky part. It is what makes the bases special.) Then the unit vectors $\boldsymbol{u}_1 = A\boldsymbol{v}_1/\|A\boldsymbol{v}_1\|$ and $\boldsymbol{u}_2 = A\boldsymbol{v}_2/\|A\boldsymbol{v}_2\|$ will be orthonormal. As a specific example, we work with the unsymmetric matrix

$$A = \begin{bmatrix} 2 & 2 \\ -1 & 1 \end{bmatrix}. \tag{1}$$

First point Why not choose one orthogonal basis in Q, instead of two in U and V? *Because no orthogonal matrix Q will make $Q^{-1}AQ$ diagonal.* We need $U^{-1}AV$.

Second point Why not choose the eigenvectors of A as the basis? *Because that basis is not orthonormal.* A is not symmetric and we need *two different* orthogonal matrices.

We are aiming for orthonormal bases that diagonalize A. The two bases will be different— one basis cannot do it. When the inputs are $\boldsymbol{v}_1$ and $\boldsymbol{v}_2$, the outputs are $A\boldsymbol{v}_1$ and $A\boldsymbol{v}_2$. We want those to line up with $\boldsymbol{u}_1$ and $\boldsymbol{u}_2$. ***The basis vectors have to give $A\boldsymbol{v}_1 = \sigma_1\boldsymbol{u}_1$ and also $A\boldsymbol{v}_2 = \sigma_2\boldsymbol{u}_2$.*** The "singular values" σ_1 and σ_2 are the lengths $\|A\boldsymbol{v}_1\|$ and $\|A\boldsymbol{v}_2\|$. With $\boldsymbol{v}_1$ and $\boldsymbol{v}_2$ as columns of V you see what we are asking for:

$$A\begin{bmatrix} \boldsymbol{v}_1 & \boldsymbol{v}_2 \end{bmatrix} = \begin{bmatrix} \sigma_1\boldsymbol{u}_1 & \sigma_2\boldsymbol{u}_2 \end{bmatrix} = \begin{bmatrix} \boldsymbol{u}_1 & \boldsymbol{u}_2 \end{bmatrix}\begin{bmatrix} \sigma_1 & \\ & \sigma_2 \end{bmatrix}. \tag{2}$$

In matrix notation that is $AV = U\Sigma$, or $U^{-1}AV = \Sigma$, or $U^{\mathrm{T}}AV = \Sigma$. The diagonal matrix Σ is like Λ (capital sigma versus capital lambda). Σ contains the *singular values* σ_1, σ_2, which are different from the eigenvalues λ_1, λ_2 in Λ.

The difference comes from U and V. When they both equal S, we have $S^{-1}AS = \Lambda$. The matrix is diagonalized. But the eigenvectors in S are not generally orthonormal. The new requirement is that *U and V must be orthogonal matrices.*

Orthonormal basis $$V^{\mathrm{T}}V = \begin{bmatrix} - & v_1^{\mathrm{T}} & - \\ - & v_2^{\mathrm{T}} & - \end{bmatrix}\begin{bmatrix} v_1 & v_2 \end{bmatrix} = \begin{bmatrix} 1 & 0 \\ 0 & 1 \end{bmatrix}. \quad (3)$$

Thus $V^{\mathrm{T}}V = I$ which means $V^{\mathrm{T}} = V^{-1}$. Similarly $U^{\mathrm{T}}U = I$ and $U^{\mathrm{T}} = U^{-1}$.

6R The ***Singular Value Decomposition*** **(SVD)** has orthogonal matrices U and V:

$$AV = U\Sigma \quad \text{and then} \quad A = U\Sigma V^{-1} = U\Sigma V^{\mathrm{T}}. \quad (4)$$

This is the new factorization of A: ***orthogonal*** times ***diagonal*** times ***orthogonal.***

There is a neat way to remove U and see V by itself: *Multiply A^{T} times A.*

$$A^{\mathrm{T}}A = (U\Sigma V^{\mathrm{T}})^{\mathrm{T}}(U\Sigma V^{\mathrm{T}}) = V\Sigma^{\mathrm{T}}\Sigma V^{\mathrm{T}}. \quad (5)$$

$U^{\mathrm{T}}U$ disappears because it equals I. Then Σ^{T} is next to Σ. Multiplying those diagonal matrices gives σ_1^2 and σ_2^2. That leaves an ordinary diagonalization of the crucial symmetric matrix $A^{\mathrm{T}}A$, whose eigenvalues are σ_1^2 and σ_2^2:

$$A^{\mathrm{T}}A = V\begin{bmatrix} \sigma_1^2 & 0 \\ 0 & \sigma_2^2 \end{bmatrix}V^{\mathrm{T}}. \quad (6)$$

This is exactly like $A = Q\Lambda Q^{\mathrm{T}}$. But the symmetric matrix is not A itself. Now the symmetric matrix is $A^{\mathrm{T}}A$! *And the columns of V are the eigenvectors of $A^{\mathrm{T}}A$.*

This tells us how to find V. We are ready to complete the example.

Example 1 Find the singular value decomposition of $A = \left[\begin{smallmatrix} 2 & 2 \\ -1 & 1 \end{smallmatrix}\right]$.

Solution Compute $A^{\mathrm{T}}A$ and its eigenvectors. Then make them unit vectors:

$$A^{\mathrm{T}}A = \begin{bmatrix} 5 & 3 \\ 3 & 5 \end{bmatrix} \quad \text{has unit eigenvectors} \quad v_1 = \begin{bmatrix} 1/\sqrt{2} \\ 1/\sqrt{2} \end{bmatrix} \quad \text{and} \quad v_2 = \begin{bmatrix} -1/\sqrt{2} \\ 1/\sqrt{2} \end{bmatrix}.$$

The eigenvalues of $A^{\mathrm{T}}A$ are 8 and 2. The v's are perpendicular, because eigenvectors of every symmetric matrix are perpendicular—and $A^{\mathrm{T}}A$ is automatically symmetric.

What about u_1 and u_2? They are quick to find, because Av_1 is going to be in the direction of u_1 and Av_2 is in the direction of u_2:

$$Av_1 = \begin{bmatrix} 2 & 2 \\ -1 & 1 \end{bmatrix}\begin{bmatrix} 1/\sqrt{2} \\ 1/\sqrt{2} \end{bmatrix} = \begin{bmatrix} 2\sqrt{2} \\ 0 \end{bmatrix}. \quad \text{The unit vector is} \quad u_1 = \begin{bmatrix} 1 \\ 0 \end{bmatrix}.$$

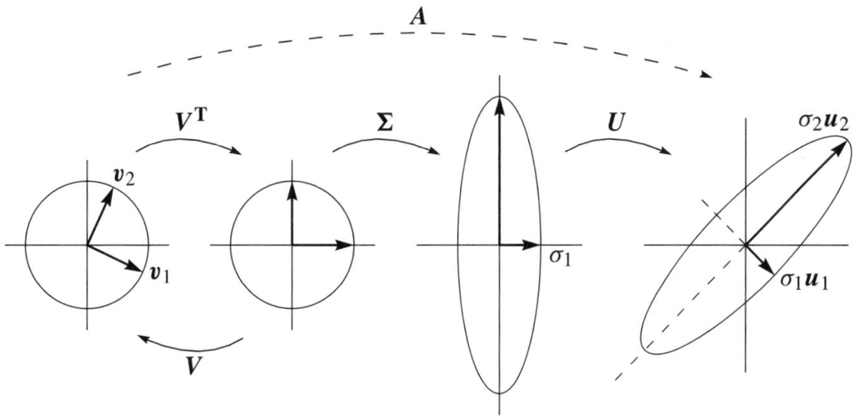

Figure 6.5 U and V are rotations and reflections. Σ is a stretching matrix.

Clearly $A\boldsymbol{v}_1$ is the same as $2\sqrt{2}\boldsymbol{u}_1$. The first singular value is $\sigma_1 = 2\sqrt{2}$. Then $\sigma_1^2 = 8$, which is the eigenvalue of $A^{\mathrm{T}}A$. We have $A\boldsymbol{v}_1 = \sigma_1\boldsymbol{u}_1$ exactly as required. Similarly

$$A\boldsymbol{v}_2 = \begin{bmatrix} 2 & 2 \\ -1 & 1 \end{bmatrix}\begin{bmatrix} -1/\sqrt{2} \\ 1/\sqrt{2} \end{bmatrix} = \begin{bmatrix} 0 \\ \sqrt{2} \end{bmatrix}. \quad \text{The unit vector is} \quad \boldsymbol{u}_2 = \begin{bmatrix} 0 \\ 1 \end{bmatrix}.$$

Now $A\boldsymbol{v}_2$ is $\sqrt{2}\boldsymbol{u}_2$. The second singular value is $\sigma_2 = \sqrt{2}$. And σ_2^2 agrees with the other eigenvalue 2 of $A^{\mathrm{T}}A$. We have completed the SVD:

$$A = U\Sigma V^{\mathrm{T}} \quad \text{is} \quad \begin{bmatrix} 2 & 2 \\ -1 & 1 \end{bmatrix} = \begin{bmatrix} 1 & 0 \\ 0 & 1 \end{bmatrix}\begin{bmatrix} 2\sqrt{2} & \\ & \sqrt{2} \end{bmatrix}\begin{bmatrix} 1/\sqrt{2} & 1/\sqrt{2} \\ -1/\sqrt{2} & 1/\sqrt{2} \end{bmatrix}. \tag{7}$$

This matrix, and every invertible 2 by 2 matrix, ***transforms the unit circle to an ellipse***. You can see that in the figure, which was created by Cliff Long and Tom Hern.

One final point about that example. We found the $\boldsymbol{u}$'s from the $\boldsymbol{v}$'s. Could we find the $\boldsymbol{u}$'s directly? *Yes*, by multiplying AA^{T} instead of $A^{\mathrm{T}}A$:

$$AA^{\mathrm{T}} = (U\Sigma V^{\mathrm{T}})(V\Sigma^{\mathrm{T}}U^{\mathrm{T}}) = U\Sigma\Sigma^{\mathrm{T}}U^{\mathrm{T}}. \tag{8}$$

This time it is $V^{\mathrm{T}}V = I$ that disappears. Multiplying $\Sigma\Sigma^{\mathrm{T}}$ gives σ_1^2 and σ_2^2 as before. *The columns of U are the eigenvectors of AA^{T}*:

$$AA^{\mathrm{T}} = \begin{bmatrix} 2 & 2 \\ -1 & 1 \end{bmatrix}\begin{bmatrix} 2 & -1 \\ 2 & 1 \end{bmatrix} = \begin{bmatrix} 8 & 0 \\ 0 & 2 \end{bmatrix}.$$

This matrix happens to be diagonal. Its eigenvectors are $(1, 0)$ and $(0, 1)$. This agrees with $\boldsymbol{u}_1$ and $\boldsymbol{u}_2$ found earlier. Why should we take the first eigenvector to be $(1, 0)$ instead of $(0, 1)$? Because we have to follow the order of the eigenvalues. Notice that AA^{T} has the same eigenvalues (8 and 2) as $A^{\mathrm{T}}A$. The singular values are $\sqrt{8}$ and $\sqrt{2}$.

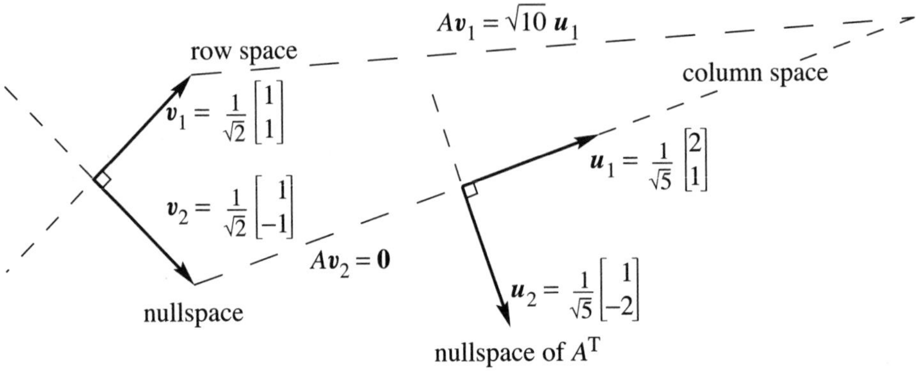

Figure 6.6 The SVD chooses orthonormal bases for 4 subspaces so that $A\boldsymbol{v}_i = \sigma_i \boldsymbol{u}_i$.

Example 2 Find the SVD of the singular matrix $A = \begin{bmatrix}2&2\\1&1\end{bmatrix}$. The rank is $r = 1$. The row space has only one basis vector $\boldsymbol{v}_1$. The column space has only one basis vector $\boldsymbol{u}_1$. We can see those vectors $(1, 1)$ and $(2, 1)$ in A, and make them into unit vectors:

$$\textbf{Row space} \quad \boldsymbol{v}_1 = \frac{1}{\sqrt{2}}\begin{bmatrix}1\\1\end{bmatrix} \qquad \textbf{Column space} \quad \boldsymbol{u}_1 = \frac{1}{\sqrt{5}}\begin{bmatrix}2\\1\end{bmatrix}$$

Then $A\boldsymbol{v}_1$ must equal $\sigma_1\boldsymbol{u}_1$. It does, with $\sigma_1 = \sqrt{10}$. This A is $\sigma_1\boldsymbol{u}_1\boldsymbol{v}_1^{\mathrm{T}}$ with rank 1.

The SVD could stop after the row basis and column basis (it usually doesn't). It is customary for U and V to be square. The matrices need a second column. The vector $\boldsymbol{v}_2$ must be orthogonal to $\boldsymbol{v}_1$, and $\boldsymbol{u}_2$ must be orthogonal to $\boldsymbol{u}_1$:

$$\boldsymbol{v}_2 = \frac{1}{\sqrt{2}}\begin{bmatrix}1\\-1\end{bmatrix} \quad \text{and} \quad \boldsymbol{u}_2 = \frac{1}{\sqrt{5}}\begin{bmatrix}1\\-2\end{bmatrix}.$$

The vector $\boldsymbol{v}_2$ is in the nullspace. It is perpendicular to $\boldsymbol{v}_1$ in the row space. Multiply by A to get $A\boldsymbol{v}_2 = \mathbf{0}$. We could say that the second singular value is $\sigma_2 = 0$, but singular values are like pivots—only the r nonzeros are counted.

All three matrices U, Σ, V are 2 by 2 in the complete SVD:

$$\begin{bmatrix}2&2\\1&1\end{bmatrix} = U\Sigma V^{\mathrm{T}} = \frac{1}{\sqrt{5}}\begin{bmatrix}2&1\\1&-2\end{bmatrix}\begin{bmatrix}\sqrt{10}&0\\0&0\end{bmatrix}\frac{1}{\sqrt{2}}\begin{bmatrix}1&1\\1&-1\end{bmatrix}. \tag{9}$$

6S ***The matrices U and V contain orthonormal bases for all four subspaces:***

first	r	columns of V :	*row space* of A
last	$n-r$	columns of V :	*nullspace* of A
first	r	columns of U :	*column space* of A
last	$m-r$	columns of U :	*nullspace* of A^{T}.

The first columns $\boldsymbol{v}_1, \ldots, \boldsymbol{v}_r$ and $\boldsymbol{u}_1, \ldots, \boldsymbol{u}_r$ are eigenvectors of $A^{\mathrm{T}}A$ and AA^{T}. Then $A\boldsymbol{v}_i$ falls in the direction of $\boldsymbol{u}_i$, and we now explain why. The last $\boldsymbol{v}$'s and $\boldsymbol{u}$'s (in the nullspaces) are easier. As long as those are orthonormal, the SVD will be correct.

Proof of SVD: Start from $A^{\mathrm{T}}A\boldsymbol{v}_i = \sigma_i^2\boldsymbol{v}_i$, which gives the $\boldsymbol{v}$'s and σ's. To prove that $A\boldsymbol{v}_i = \sigma_i\boldsymbol{u}_i$, the key steps are to multiply by $\boldsymbol{v}_i^{\mathrm{T}}$ and by A:

$$\boldsymbol{v}_i^{\mathrm{T}}A^{\mathrm{T}}A\boldsymbol{v}_i = \sigma_i^2\boldsymbol{v}_i^{\mathrm{T}}\boldsymbol{v}_i \quad \text{gives} \quad \|A\boldsymbol{v}_i\|^2 = \sigma_i^2 \quad \text{so that} \quad \|A\boldsymbol{v}_i\| = \sigma_i \tag{10}$$

$$AA^{\mathrm{T}}A\boldsymbol{v}_i = \sigma_i^2 A\boldsymbol{v}_i \quad \text{gives} \quad \boldsymbol{u}_i = A\boldsymbol{v}_i/\sigma_i \quad \text{as a unit eigenvector of} \quad AA^{\mathrm{T}}. \tag{11}$$

Equation (10) used the small trick of placing parentheses in $(\boldsymbol{v}_i^{\mathrm{T}}A^{\mathrm{T}})(A\boldsymbol{v}_i)$. This is a vector $A\boldsymbol{v}_i$ times its transpose, giving $\|A\boldsymbol{v}_i\|^2$. Equation (11) placed the parentheses in $(AA^{\mathrm{T}})(A\boldsymbol{v}_i)$. This shows that $A\boldsymbol{v}_i$ is an eigenvector of AA^{T}. We divide by its length σ_i to get the unit vector $\boldsymbol{u}_i = A\boldsymbol{v}_i/\sigma_i$. This is the equation $A\boldsymbol{v}_i = \sigma_i\boldsymbol{u}_i$ that we want! It says that A is diagonalized by these outstanding bases.

I will give you my opinion directly. The SVD is the climax of this linear algebra course. I think of it as the final step in the Fundamental Theorem. First come the *dimensions* of the four subspaces. Then their *orthogonality*. Then the *orthonormal bases which diagonalize A*. It is all in the formula $A = U\Sigma V^{\mathrm{T}}$. More applications are coming—they are certainly important—but you have made it to the top.

Eigshow (Part 2)

Section 6.1 described the MATLAB demo called **eigshow**. The first option is *eig*, when $\boldsymbol{x}$ moves in a circle and $A\boldsymbol{x}$ follows on an ellipse. The second option is *svd*, when two vectors $\boldsymbol{x}$ and $\boldsymbol{y}$ stay perpendicular as they travel around a circle. Then $A\boldsymbol{x}$ and $A\boldsymbol{y}$ move too (not usually perpendicular). There are four vectors on the screen.

The SVD is seen graphically when $A\boldsymbol{x}$ is perpendicular to $A\boldsymbol{y}$. Their directions at that moment give an orthonormal basis $\boldsymbol{u}_1, \boldsymbol{u}_2$. Their lengths give the singular values σ_1, σ_2. The vectors $\boldsymbol{x}$ and $\boldsymbol{y}$ at that same moment are the orthonormal basis $\boldsymbol{v}_1, \boldsymbol{v}_2$.

The Java demo on **web.mit.edu/18.06/www** shows $A\boldsymbol{v}_1 = \sigma_1\boldsymbol{u}_1$ and $A\boldsymbol{v}_2 = \sigma_2\boldsymbol{u}_2$. In matrix language that is $AV = U\Sigma$. This is the SVD.

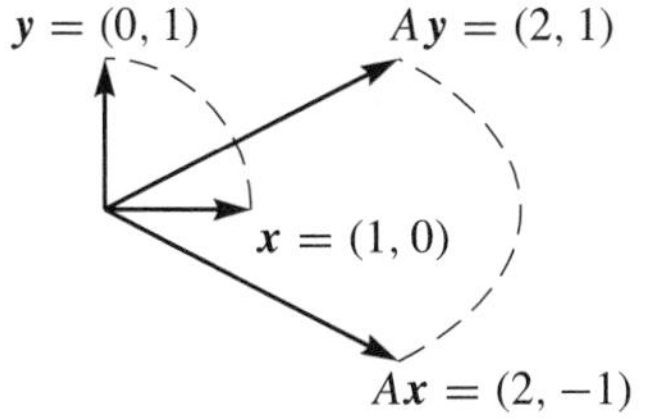

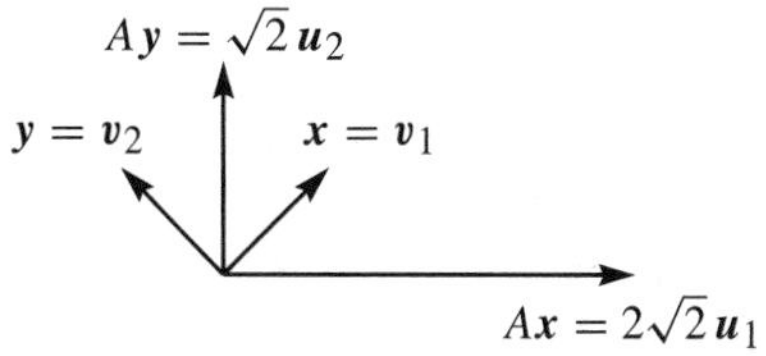

Searching the Web

I will end with an application of the SVD to web search engines. When you type a search word, you get a list of related web sites in order of importance. (Regrettably, typing "SVD" produced 13 non-mathematical SVD's before the real one. "Cofactors" was even worse but "cofactor" had one good entry. "Four subspaces" did much better.) The HITS algorithm that we describe is one way to produce that ranked list. It begins with about 200 sites found from an index of key words, and after that we look only at links between pages. HITS is link-based not content-based.

Start with the 200 sites and all sites that link to them and all sites they link to. That is our list, to be put in order. Importance can be measured in two ways:

1. The site is an ***authority***: *links come from many sites.* Especially from hubs.

2. The site is a ***hub***: *it links to many sites in the list.* Especially to authorities.

We want numbers $\boldsymbol{x}_1, \ldots, \boldsymbol{x}_N$ to rank the authorities and $\boldsymbol{y}_1, \ldots, \boldsymbol{y}_N$ to rank the hubs. Start with a simple count: $\boldsymbol{x}_i^{\mathbf{0}}$ and $\boldsymbol{y}_i^{\mathbf{0}}$ count the links into and out of site i.

Here is the point: *A good authority has links from important sites* (like hubs). Links from universities count more heavily than links from friends. *A good hub is linked to important sites* (like authorities). A link to amazon.com means more than a link to wellesleycambridge.com. The rankings $\boldsymbol{x}^{\mathbf{0}}$ and $\boldsymbol{y}^{\mathbf{0}}$ from counting links are updated to $\boldsymbol{x}^{\mathbf{1}}$ and $\boldsymbol{y}^{\mathbf{1}}$ by taking account of *good* links (measuring their quality by $\boldsymbol{x}^{\mathbf{0}}$ and $\boldsymbol{y}^{\mathbf{0}}$):

$$\textbf{Authority values} \quad x_i^1 = \sum_{j \text{ links to } i} y_j^0 \qquad\qquad \textbf{Hub values} \quad y_i^1 = \sum_{i \text{ links to } j} x_j^0 \tag{12}$$

In matrix language those are $\boldsymbol{x}^{\mathbf{1}} = A^{\mathrm{T}}\boldsymbol{y}^{\mathbf{0}}$ and $\boldsymbol{y}^{\mathbf{1}} = A\boldsymbol{x}^{\mathbf{0}}$. The matrix A contains 1's and 0's, with $a_{ij} = 1$ when i links to j. In the language of graphs, A is an "adjacency matrix" for the World Wide Web. It is pretty large.

The algorithm doesn't stop there. The new $\boldsymbol{x}^{\mathbf{1}}$ and $\boldsymbol{y}^{\mathbf{1}}$ give better rankings, but not the best. Take another step like (12) to $\boldsymbol{x}^{\mathbf{2}}$ and $\boldsymbol{y}^{\mathbf{2}}$. *Notice how $A^{\mathrm{T}}A$ and AA^{T} appear*:

$$\boldsymbol{x}^2 = A^{\mathrm{T}}\boldsymbol{y}^1 = A^{\mathrm{T}}A\boldsymbol{x}^0 \qquad \text{and} \qquad \boldsymbol{y}^2 = A^{\mathrm{T}}\boldsymbol{x}^1 = AA^{\mathrm{T}}\boldsymbol{y}^0. \tag{13}$$

In two steps we are multiplying $\boldsymbol{x}^{\mathbf{0}}$ by $A^{\mathrm{T}}A$ and $\boldsymbol{y}^{\mathbf{0}}$ by AA^{T}. In twenty steps we are multiplying by $(A^{\mathrm{T}}A)^{10}$ and $(AA^{\mathrm{T}})^{10}$. When we take these powers, the largest eigenvalue σ_1^2 begins to dominate. And the vectors $\boldsymbol{x}$ and $\boldsymbol{y}$ gradually line up with the leading eigenvectors $\boldsymbol{v}_1$ and $\boldsymbol{u}_1$ of $A^{\mathrm{T}}A$ and AA^{T}. We are computing the top terms in the SVD iteratively, by the *power method* that is further discussed in Section 9.3. It is wonderful that linear algebra helps to understand the Web.

Google actually creates rankings by a random walk that follows web links. The more often this random walk goes to a site, the higher the ranking. The frequency of visits gives the leading eigenvector ($\lambda = 1$) of the normalized adjacency matrix for the

Web. ***That matrix has 2.7 billion rows and columns, from 2.7 billion web sites.*** This is the largest eigenvalue problem ever solved.

Some details are on the Web, but many important techniques are secrets of *Google*: `www.mathworks.com/company/newsletter/clevescorner/oct02_cleve.shtml` Probably *Google* starts with last month's eigenvector as a first approximation, and runs the random walk very fast. To get a high ranking, you want a lot of links from important sites. The HITS algorithm is described in the 1999 *Scientific American* (June 16). But I don't think the SVD is mentioned there. . .

■ REVIEW OF THE KEY IDEAS ■

1. The SVD factors A into $U\Sigma V^{\mathrm{T}}$, with r singular values $\sigma_1 \geq \ldots \geq \sigma_r > 0$.

2. The numbers $\sigma_1^2, \ldots, \sigma_r^2$ are the nonzero eigenvalues of AA^{T} and $A^{\mathrm{T}}A$.

3. The orthonormal columns of U and V are eigenvectors of AA^{T} and $A^{\mathrm{T}}A$.

4. Those columns are orthonormal bases for the four fundamental subspaces of A.

5. Those bases diagonalize the matrix: $A\boldsymbol{v}_i = \sigma_i \boldsymbol{u}_i$ for $i \leq r$. This is $AV = U\Sigma$.

■ WORKED EXAMPLES ■

6.7 A Identify by name these decompositions $A = \boldsymbol{c}_1\boldsymbol{r}_1 + \cdots + \boldsymbol{c}_n\boldsymbol{r}_n$ of an n by n matrix into n rank one matrices (column $\boldsymbol{c}$ times row $\boldsymbol{r}$):

1. *Orthogonal* columns $\boldsymbol{c}_1, \ldots, \boldsymbol{c}_n$ and *orthogonal* rows $\boldsymbol{r}_1, \ldots, \boldsymbol{r}_n$
2. *Orthogonal* columns $\boldsymbol{c}_1, \ldots, \boldsymbol{c}_n$ and *triangular* rows $\boldsymbol{r}_1, \ldots, \boldsymbol{r}_n$
3. *Triangular* columns $\boldsymbol{c}_1, \ldots, \boldsymbol{c}_n$ and *triangular* rows $\boldsymbol{r}_1, \ldots, \boldsymbol{r}_n$

Triangular means that $\boldsymbol{c}_i$ and $\boldsymbol{r}_i$ have zeros before component i. The matrix C with columns $\boldsymbol{c}_i$ is lower triangular, the matrix R with rows $\boldsymbol{r}_i$ is upper triangular. Where do the rank and the pivots and singular values come into this picture?

Solution These three splittings $A = CR$ are basic to linear algebra, pure or applied:

1. Singular Value Decomposition $A = U\Sigma V^{\mathrm{T}}$ (*orthogonal* U, *orthogonal* ΣV^{T})

2. Gram-Schmidt Orthogonalization $A = QR$ (*orthogonal* Q, *triangular* R)

3. Gaussian Elimination $A = LU$ (*triangular* L, *triangular* U)

When A (possibly rectangular) has rank r, we need only r rank one matrices (not n).

With orthonormal rows in V^{T}, the σ's in Σ come in: $A = \sigma_1 \boldsymbol{c}_1 \boldsymbol{r}_1 + \cdots + \sigma_n \boldsymbol{c}_n \boldsymbol{r}_n$. With diagonal 1's in L and U, the pivots d_i come in: $A = LDU = d_1 \boldsymbol{c}_1 \boldsymbol{r}_1 + \cdots + d_n \boldsymbol{c}_n \boldsymbol{r}_n$. With the diagonal of R placed in H, QR becomes $QHR = h_1 \boldsymbol{c}_1 \boldsymbol{r}_1 + \cdots + h_n \boldsymbol{c}_n \boldsymbol{r}_n$. These numbers h_i have no standard name and I propose "heights". Each h_i tells the height of column i above the base from the first $i-1$ columns. The volume of the full n-dimensional box comes from $A = U\Sigma V^{\mathrm{T}} = LDU = QHR$:

$$|\,\det A\,| = |\textit{ product of } \sigma\text{'s }| = |\textit{ product of } d\text{'s }| = |\textit{ product of } h\text{'s }|.$$

Problem Set 6.7

Problems 1–3 compute the SVD of a square singular matrix A.

1 Compute $A^{\mathrm{T}}A$ and its eigenvalues $\sigma_1^2, 0$ and unit eigenvectors $\boldsymbol{v}_1, \boldsymbol{v}_2$:

$$A = \begin{bmatrix} 1 & 4 \\ 2 & 8 \end{bmatrix}.$$

2 (a) Compute AA^{T} and its eigenvalues $\sigma_1^2, 0$ and unit eigenvectors $\boldsymbol{u}_1, \boldsymbol{u}_2$.

(b) Verify from Problem 1 that $A\boldsymbol{v}_1 = \sigma_1 \boldsymbol{u}_1$. Find all entries in the SVD:

$$\begin{bmatrix} 1 & 4 \\ 2 & 8 \end{bmatrix} = \begin{bmatrix} \boldsymbol{u}_1 & \boldsymbol{u}_2 \end{bmatrix} \begin{bmatrix} \sigma_1 & \\ & 0 \end{bmatrix} \begin{bmatrix} \boldsymbol{v}_1 & \boldsymbol{v}_2 \end{bmatrix}^{\mathrm{T}}.$$

3 Write down orthonormal bases for the four fundamental subspaces of this A.

Problems 4–7 ask for the SVD of matrices of rank 2.

4 (a) Find the eigenvalues and unit eigenvectors of $A^{\mathrm{T}}A$ and AA^{T} for the Fibonacci matrix

$$A = \begin{bmatrix} 1 & 1 \\ 1 & 0 \end{bmatrix}$$

(b) Construct the singular value decomposition of A.

5 Show that the vectors in Problem 4 satisfy $A\boldsymbol{v}_1 = \sigma_1 \boldsymbol{u}_1$ and $A\boldsymbol{v}_2 = \sigma_2 \boldsymbol{u}_2$.

6 Use the SVD part of the MATLAB demo **eigshow** to find the same vectors $\boldsymbol{v}_1$ and $\boldsymbol{v}_2$ graphically.

7 Compute $A^{\mathrm{T}}A$ and AA^{T} and their eigenvalues and unit eigenvectors for

$$A = \begin{bmatrix} 1 & 1 & 0 \\ 0 & 1 & 1 \end{bmatrix}.$$

Multiply the three matrices $U\Sigma V^{\mathrm{T}}$ to recover A.

Problems 8–15 bring out the underlying ideas of the SVD.

8 Suppose $\boldsymbol{u}_1, \ldots, \boldsymbol{u}_n$ and $\boldsymbol{v}_1, \ldots, \boldsymbol{v}_n$ are orthonormal bases for $\mathbf{R}^n$. Construct the matrix A that transforms each $\boldsymbol{v}_j$ into $\boldsymbol{u}_j$ to give $A\boldsymbol{v}_1 = \boldsymbol{u}_1, \ldots, A\boldsymbol{v}_n = \boldsymbol{u}_n$.

9 Construct the matrix with rank one that has $Av = 12u$ for $v = \frac{1}{2}(1, 1, 1, 1)$ and $u = \frac{1}{3}(2, 2, 1)$.Its only singular value is $\sigma_1 =$ ____.

10 Suppose A has orthogonal columns $w_1, w_2, \ldots, w_n$ of lengths $\sigma_1, \sigma_2, \ldots, \sigma_n$. What are U, Σ, and V in the SVD?

11 Explain how the SVD expresses the matrix A as the sum of r rank one matrices:

$$A = \sigma_1 u_1 v_1^{\mathrm{T}} + \cdots + \sigma_r u_r v_r^{\mathrm{T}}.$$

12 Suppose A is a 2 by 2 symmetric matrix with unit eigenvectors u_1 and u_2. If its eigenvalues are $\lambda_1 = 3$ and $\lambda_2 = -2$, what are the matrices U, Σ, V^{T} in its SVD?

13 If $A = QR$ with an orthonormal matrix Q, then the SVD of A is almost the same as the SVD of R. Which of the three matrices in the SVD is changed because of Q?

14 Suppose A is invertible (with $\sigma_1 > \sigma_2 > 0$). Change A by as small a matrix as possible to produce a singular matrix A_0. Hint: U and V do not change:

$$A = \begin{bmatrix} u_1 & u_2 \end{bmatrix} \begin{bmatrix} \sigma_1 & \\ & \sigma_2 \end{bmatrix} \begin{bmatrix} v_1 & v_2 \end{bmatrix}^{\mathrm{T}}$$

15 (a) If A changes to $4A$, what is the change in the SVD?

(b) What is the SVD for A^{T} and for A^{-1}?

16 Why doesn't the SVD for $A + I$ just use $\Sigma + I$?

17 (MATLAB) Run a random walk starting from web site $x(1) = 1$ and record the visits to each site. From the site $x(k-1) = 1, 2, 3$, or 4 the code chooses $x(k)$ with probabilities given by column $x(k-1)$ of A. At the end p gives the fraction of time at each site from a histogram (and $Ap \approx p$—please check this steady state eigenvector):

$$A = [0\ \ .1\ \ .2\ \ .7;\ \ .05\ \ 0\ \ .15\ \ .8;\ \ .15\ \ .25\ \ 0\ \ .6;\ \ .1\ \ .3\ \ .6\ \ 0]' =$$

Markov matrix

```
n = 1000;  x = zeros(1, n);  x(1) = 1;
for k = 2 : n  x(k) = min(find(rand<cumsum(A(:, x(k − 1)))));  end
p = hist(x, 1 : 4)/n
```

How are the properties of a matrix reflected in its eigenvalues and eigenvectors? This question is fundamental throughout Chapter 6. A table that organizes the key facts may be helpful. For each class of matrices, here are the special properties of the eigenvalues λ_i and eigenvectors x_i.

Symmetric: $A^{\mathrm{T}} = A$	real λ's	orthogonal $\boldsymbol{x}_i^{\mathrm{T}}\boldsymbol{x}_j = 0$
Orthogonal: $Q^{\mathrm{T}} = Q^{-1}$	all $\lvert\lambda\rvert = 1$	orthogonal $\overline{\boldsymbol{x}}_i^{\mathrm{T}}\boldsymbol{x}_j = 0$
Skew-symmetric: $A^{\mathrm{T}} = -A$	imaginary λ's	orthogonal $\overline{\boldsymbol{x}}_i^{\mathrm{T}}\boldsymbol{x}_j = 0$
Complex Hermitian: $\overline{A}^{\mathrm{T}} = A$	real λ's	orthogonal $\overline{\boldsymbol{x}}_i^{\mathrm{T}}\boldsymbol{x}_j = 0$
Positive Definite: $\boldsymbol{x}^{\mathrm{T}}A\boldsymbol{x} > 0$	all $\lambda > 0$	orthogonal
Markov: $m_{ij} > 0, \sum_{i=1}^{n} m_{ij} = 1$	$\lambda_{\max} = 1$	steady state $\boldsymbol{x} > 0$
Similar: $B = M^{-1}AM$	$\lambda(B) = \lambda(A)$	$\boldsymbol{x}(B) = M^{-1}\boldsymbol{x}(A)$
Projection: $P = P^2 = P^{\mathrm{T}}$	$\lambda = 1;\ 0$	column space; nullspace
Reflection: $I - 2\boldsymbol{u}\boldsymbol{u}^{\mathrm{T}}$	$\lambda = -1;\ 1, .., 1$	$\boldsymbol{u};\ \boldsymbol{u}^{\perp}$
Rank One: $\boldsymbol{u}\boldsymbol{v}^{\mathrm{T}}$	$\lambda = \boldsymbol{v}^{\mathrm{T}}\boldsymbol{u};\ 0, .., 0$	$\boldsymbol{u};\ \boldsymbol{v}^{\perp}$
Inverse: A^{-1}	$1/\lambda(A)$	eigenvectors of A
Shift: $A + cI$	$\lambda(A) + c$	eigenvectors of A
Stable Powers: $A^n \to 0$	all $\lvert\lambda\rvert < 1$	
Stable Exponential: $e^{At} \to 0$	all $Re\ \lambda < 0$	
Cyclic Permutation: $P(1, .., n) = (2, .., n, 1)$	$\lambda_k = e^{2\pi ik/n}$	$\boldsymbol{x}_k = (1, \lambda_k, \ldots, \lambda_k^{n-1})$
Tridiagonal: $-1, 2, -1$ on diagonals	$\lambda_k = 2 - 2\cos\dfrac{k\pi}{n+1}$	$\boldsymbol{x}_k = \left(\sin\dfrac{k\pi}{n+1}, \sin\dfrac{2k\pi}{n+1}, \ldots\right)$
Diagonalizable: $S\Lambda S^{-1}$	diagonal of Λ	columns of S are independent
Symmetric: $Q\Lambda Q^{\mathrm{T}}$	diagonal of Λ (real)	columns of Q are orthonormal
Jordan: $J = M^{-1}AM$	diagonal of J	each block gives $\boldsymbol{x} = (0, .., 1, .., 0)$
Every Matrix: $A = U\Sigma V^{\mathrm{T}}$	$\operatorname{rank}(A) = \operatorname{rank}(\Sigma)$	eigenvectors of $A^{\mathrm{T}}A$, AA^{T} in V, U

7

LINEAR TRANSFORMATIONS

THE IDEA OF A LINEAR TRANSFORMATION ■ 7.1

When a matrix A multiplies a vector $\boldsymbol{v}$, it "transforms" $\boldsymbol{v}$ into another vector $A\boldsymbol{v}$ ***In goes*** $\boldsymbol{v}$, ***out comes*** $A\boldsymbol{v}$. This transformation follows the same idea as a function. In goes a number x, out comes $f(x)$. For one vector $\boldsymbol{v}$ or one number x, we multiply by the matrix or we evaluate the function. The deeper goal is to see all $\boldsymbol{v}$'s at once. We are transforming the whole space when we multiply every $\boldsymbol{v}$ by A.

Start again with a matrix A. It transforms $\boldsymbol{v}$ to $A\boldsymbol{v}$. It transforms $\boldsymbol{w}$ to $A\boldsymbol{w}$. Then we *know* what happens to $\boldsymbol{u} = \boldsymbol{v} + \boldsymbol{w}$. There is no doubt about $A\boldsymbol{u}$, it has to equal $A\boldsymbol{v} + A\boldsymbol{w}$. Matrix multiplication $T(\boldsymbol{v}) = A\boldsymbol{v}$ gives a ***linear transformation***:

DEFINITION A transformation T assigns an output $T(\boldsymbol{v})$ to each input vector $\boldsymbol{v}$. The transformation is ***linear*** if it meets these requirements for all $\boldsymbol{v}$ and $\boldsymbol{w}$:

(a) $T(\boldsymbol{v} + \boldsymbol{w}) = T(\boldsymbol{v}) + T(\boldsymbol{w})$ (b) $T(c\boldsymbol{v}) = cT(\boldsymbol{v})$ for all c.

If the input is $\boldsymbol{v} = \boldsymbol{0}$, the output must be $T(\boldsymbol{v}) = \boldsymbol{0}$. We combine (a) and (b) into one:

Linearity: $T(c\boldsymbol{v} + d\boldsymbol{w})$ ***must equal*** $cT(\boldsymbol{v}) + dT(\boldsymbol{w})$.

Again I test matrix multiplication: $A(c\boldsymbol{v} + d\boldsymbol{w}) = cA\boldsymbol{v} + dA\boldsymbol{w}$ is *true*.

A linear transformation is highly restricted. Suppose T adds $\boldsymbol{u}_0$ to every vector. Then $T(\boldsymbol{v}) = \boldsymbol{v} + \boldsymbol{u}_0$ and $T(\boldsymbol{w}) = \boldsymbol{w} + \boldsymbol{u}_0$. This isn't good, or at least *it isn't linear*. Applying T to $\boldsymbol{v} + \boldsymbol{w}$ produces $\boldsymbol{v} + \boldsymbol{w} + \boldsymbol{u}_0$. That is not the same as $T(\boldsymbol{v}) + T(\boldsymbol{w})$:

$$\boldsymbol{v} + \boldsymbol{w} + \boldsymbol{u}_0 \quad \text{is different from} \quad T(\boldsymbol{v}) + T(\boldsymbol{w}) = \boldsymbol{v} + \boldsymbol{u}_0 + \boldsymbol{w} + \boldsymbol{u}_0.$$

The exception is when $\boldsymbol{u}_0 = \boldsymbol{0}$. The transformation reduces to $T(\boldsymbol{v}) = \boldsymbol{v}$. This is the ***identity transformation*** (nothing moves, as in multiplication by I). That is certainly linear. In this case the input space $\mathbf{V}$ is the same as the output space $\mathbf{W}$.

The linear-plus-shift transformation $T(\boldsymbol{v}) = A\boldsymbol{v} + \boldsymbol{u}_0$ is called *"affine."* Straight lines stay straight although T is not linear. Computer graphics works with affine transformations. The shift to computer graphics is in Section 8.6.

Example 1 Choose a fixed vector $\boldsymbol{a} = (1, 3, 4)$, and let $T(\boldsymbol{v})$ be the dot product $\boldsymbol{a} \cdot \boldsymbol{v}$:

The input is $\boldsymbol{v} = (v_1, v_2, v_3)$. The output is $T(\boldsymbol{v}) = \boldsymbol{a} \cdot \boldsymbol{v} = v_1 + 3v_2 + 4v_3$.

This is linear. The inputs $\boldsymbol{v}$ come from three-dimensional space, so $\mathbf{V} = \mathbf{R}^3$. The outputs are just numbers, so the output space is $\mathbf{W} = \mathbf{R}^1$. We are multiplying by the row matrix $A = [\,1\ \ 3\ \ 4\,]$. Then $T(\boldsymbol{v}) = A\boldsymbol{v}$.

You will get good at recognizing which transformations are linear. If the output involves squares or products or lengths, v_1^2 or $v_1 v_2$ or $\|\boldsymbol{v}\|$, then T is not linear.

Example 2 The length $T(\boldsymbol{v}) = \|\boldsymbol{v}\|$ is not linear. Requirement (a) for linearity would be $\|\boldsymbol{v} + \boldsymbol{w}\| = \|\boldsymbol{v}\| + \|\boldsymbol{w}\|$. Requirement (b) would be $\|c\boldsymbol{v}\| = c\|\boldsymbol{v}\|$. Both are false!

Not (a): The sides of a triangle satisfy an *inequality* $\|\boldsymbol{v} + \boldsymbol{w}\| \leq \|\boldsymbol{v}\| + \|\boldsymbol{w}\|$.

Not (b): The length $\|-\boldsymbol{v}\|$ is not $-\|\boldsymbol{v}\|$. For negative c, we fail.

Example 3 (Important) T is the transformation that *rotates every vector by* $30°$. The domain is the xy plane (where the input vector $\boldsymbol{v}$ is). The range is also the xy plane (where the rotated vector $T(\boldsymbol{v})$ is). We described T without mentioning a matrix: just rotate the plane by $30°$.

Is rotation linear? *Yes it is.* We can rotate two vectors and add the results. The sum of rotations $T(\boldsymbol{v}) + T(\boldsymbol{w})$ is the same as the rotation $T(\boldsymbol{v} + \boldsymbol{w})$ of the sum. The whole plane is turning together, in this linear transformation.

Note Transformations have a language of their own. Where there is no matrix, we can't talk about a column space. But the idea can be rescued and used. The column space consisted of all outputs $A\boldsymbol{v}$. The nullspace consisted of all inputs for which $A\boldsymbol{v} = \mathbf{0}$. Translate those into "range" and "kernel":

Range of T = set of all outputs $T(\boldsymbol{v})$: corresponds to column space

Kernel of T = set of all inputs for which $T(\boldsymbol{v}) = \mathbf{0}$: corresponds to nullspace.

The range is in the output space $\mathbf{W}$. The kernel is in the input space $\mathbf{V}$. When T is multiplication by a matrix, $T(\boldsymbol{v}) = A\boldsymbol{v}$, you can translate to column space and nullspace.

For an m by n matrix, the nullspace is a subspace of $\mathbf{V} = \mathbf{R}^n$. The column space is a subspace of ______. The range might or might not be the whole output space $\mathbf{W}$.

Examples of Transformations (mostly linear)

Example 4 Project every 3-dimensional vector down onto the xy plane. The range is that plane, which contains every $T(\boldsymbol{v})$. The kernel is the z axis (which projects down to zero). This projection is linear.

Example 5 Project every 3-dimensional vector onto the horizontal plane $z = 1$. The vector $\boldsymbol{v} = (x, y, z)$ is transformed to $T(\boldsymbol{v}) = (x, y, 1)$. This transformation is not linear. Why not? It doesn't even transform $\boldsymbol{v} = \mathbf{0}$ into $T(\boldsymbol{v}) = \mathbf{0}$.

Multiply every 3-dimensional vector by a 3 by 3 matrix A. This is definitely a linear transformation!

$$T(\boldsymbol{v}+\boldsymbol{w}) = A(\boldsymbol{v}+\boldsymbol{w}) \quad \text{which does equal} \quad A\boldsymbol{v}+A\boldsymbol{w} = T(\boldsymbol{v})+T(\boldsymbol{w}).$$

Example 6 Suppose A is an *invertible matrix*. The kernel of T is the zero vector; the range $\mathbf{W}$ equals the domain $\mathbf{V}$. Another linear transformation is multiplication by A^{-1}. This is the ***inverse transformation*** T^{-1}, which brings every vector $T(\boldsymbol{v})$ back to $\boldsymbol{v}$:

$$T^{-1}(T(\boldsymbol{v})) = \boldsymbol{v} \quad \text{matches the matrix multiplication} \quad A^{-1}(A\boldsymbol{v}) = \boldsymbol{v}.$$

We are reaching an unavoidable question. ***Are all linear transformations produced by matrices?*** Each m by n matrix does produce a linear transformation from $\mathbf{V} = \mathbf{R}^n$ to $\mathbf{W} = \mathbf{R}^m$. Our question is the converse. When a linear T is described as a "rotation" or "projection" or ". . .", is there always a matrix hiding behind T?

The answer is *yes*. This is an approach to linear algebra that doesn't start with matrices. The next section shows that we still end up with matrices.

Linear Transformations of the Plane

It is more interesting to *see* a transformation than to define it. When a 2 by 2 matrix A multiplies all vectors in $\mathbf{R}^2$, we can watch how it acts. Start with a "house" that has eleven endpoints. Those eleven vectors $\boldsymbol{v}$ are transformed into eleven vectors $A\boldsymbol{v}$. Straight lines between $\boldsymbol{v}$'s become straight lines between the transformed vectors $A\boldsymbol{v}$. (The transformation from house to house is linear!) Applying A to a standard house produces a new house—possibly stretched or rotated or otherwise unlivable.

This part of the book is visual, not theoretical. We will show six houses and the matrices that produce them. The columns of H are the eleven circled points of the first house. (H is 2 by 12, so **plot2d** will connect the 11th circle to the first.) The 11 points in the house matrix H are multiplied by A to produce the other houses. The houses on the cover of the book were produced this way (before Christine Curtis turned them into a quilt for Professor Curtis). H is in the Teaching Code **house**.

$$H = \begin{bmatrix} -6 & -6 & -7 & 0 & 7 & 6 & 6 & -3 & -3 & 0 & 0 & -6 \\ -7 & 2 & 1 & 8 & 1 & 2 & -7 & -7 & -2 & -2 & -7 & -7 \end{bmatrix}.$$

■ REVIEW OF THE KEY IDEAS ■

1. A transformation T takes each $\boldsymbol{v}$ in the input space to $T(\boldsymbol{v})$ in the output space.

2. Linearity requires that $T(c_1\boldsymbol{v}_1+\cdots+c_n\boldsymbol{v}_n) = c_1\,T(\boldsymbol{v}_1)+\cdots+c_n\,T(\boldsymbol{v}_n)$.

3. The transformation $T(\boldsymbol{v}) = A\boldsymbol{v}+\boldsymbol{v}_0$ is linear only if $\boldsymbol{v}_0 = \mathbf{0}$!

4. The quilt on the book cover shows $T(\text{house}) = AH$ for nine matrices A.

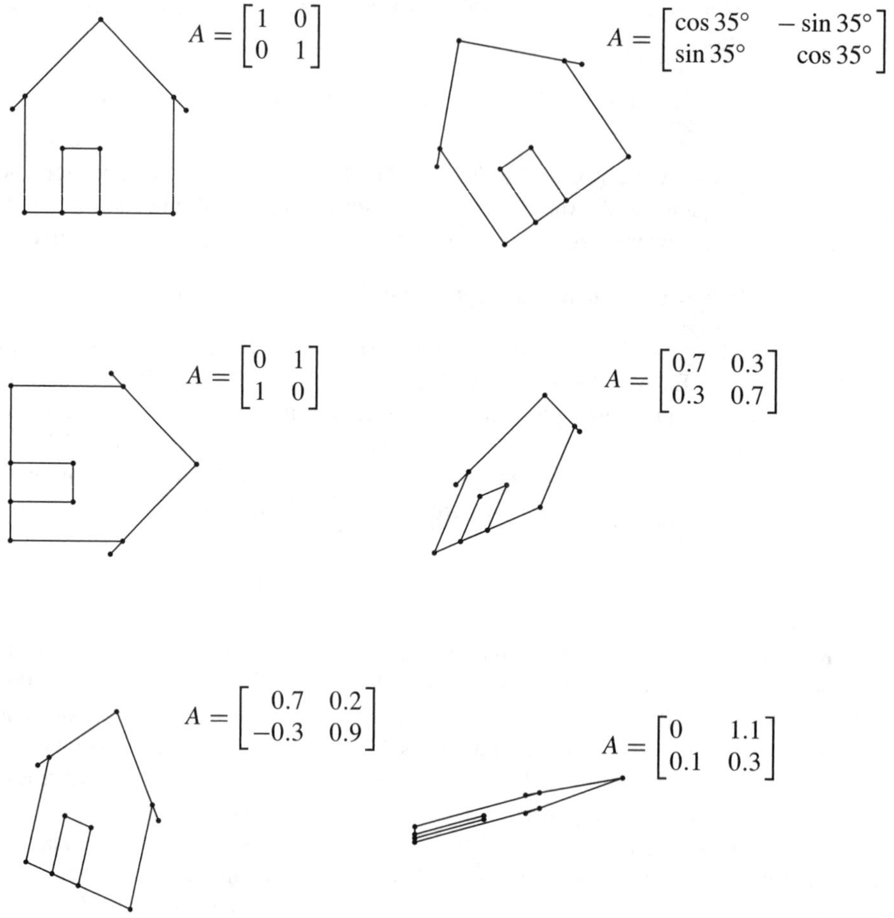

Figure 7.1 Linear transformations of a house drawn by **plot2d**$(A * H)$.

■ WORKED EXAMPLES ■

7.1 A The matrix $\left[\begin{smallmatrix} 1 & 0 \\ 3 & 1 \end{smallmatrix}\right]$ gives a ***shearing transformation*** $T(x, y) = (x, 3x + y)$. Draw the xy plane and show what happens to $(1, 0)$ and $(2, 0)$ on the x axis. What happens to the points on the vertical lines $x = 0$ and $x = a$? If the inputs fill the unit square $0 \leq x \leq 1$, $0 \leq y \leq 1$, draw the outputs (the transformed square).

Solution The points $(1, 0)$ and $(2, 0)$ on the x axis transform by T to $(1, 3)$ and $(2, 6)$. The horizontal x axis transforms to the straight line with slope 3 (going through $(0, 0)$ of course). The points on the y axis are *not moved* because $T(0, y) = (0, y)$. The y axis is the line of eigenvectors of T with $\lambda = 1$.

The vertical line $x = a$ is moved up by $3a$, since $3a$ is added to the y component. This is the "shearing". Vertical lines slide higher and higher as you go from left to right.

The unit square has one side on the y axis (unchanged). The opposite side from $(1, 0)$ to $(1, 1)$ moves upward, to go from $(1, 3)$ to $(1, 4)$. The transformed square has a lower side from $(0, 0)$ to $(1, 3)$ and a parallel upper side from $(0, 1)$ to $(1, 4)$. It is a parallelogram. Multiplication by any A transforms squares to parallelograms!

7.1 B A **nonlinear transformation** T is invertible if every $\boldsymbol{b}$ in the output space comes from exactly one $\boldsymbol{x}$ in the input space: $T(\boldsymbol{x}) = \boldsymbol{b}$ always has exactly one solution. Which of these transformations (on real numbers x) is invertible and what is T^{-1}? None are linear, not even T_3. When you solve $T(\boldsymbol{x}) = \boldsymbol{b}$, you are inverting T:

$$T_1(x) = x^2 \quad T_2(x) = x^3 \quad T_3(x) = x + 9 \quad T_4(x) = e^x \quad T_5(x) = \frac{1}{x} \text{ for nonzero } x\text{'s}$$

Solution T_1 is not invertible because $x^2 = 1$ has *two* solutions (and $x^2 = -1$ has *no* solution). T_4 is not invertible because $e^x = -1$ has no solution. (If the output space changes to *positive* b's then the inverse of $e^x = b$ is $x = \ln b$.) Notice that $T_5^2 =$ identity. But $T_3^2(x) = x + 18$. What are $T_2^2(x)$ and T_4^2?

T_2, T_3, T_5 are invertible. The solutions to $x^3 = b$ and $x + 9 = b$ and $\frac{1}{x} = b$ are unique:

$$x = T_2^{-1}(b) = b^{1/3} \qquad x = T_3^{-1}(b) = b - 9 \qquad x = T_5^{-1}(b) = \frac{1}{b}$$

Problem Set 7.1

1 A linear transformation must leave the zero vector fixed: $T(\mathbf{0}) = \mathbf{0}$. Prove this from $T(\boldsymbol{v} + \boldsymbol{w}) = T(\boldsymbol{v}) + T(\boldsymbol{w})$ by choosing $\boldsymbol{w} =$ _____. Prove it also from requirement (b) by choosing $c =$ _____.

2 Requirement (b) gives $T(c\boldsymbol{v}) = cT(\boldsymbol{v})$ and also $T(d\boldsymbol{w}) = dT(\boldsymbol{w})$. Then by addition, requirement (a) gives $T(\quad) = (\quad)$. What is $T(c\boldsymbol{v} + d\boldsymbol{w} + e\boldsymbol{u})$?

3 Which of these transformations is not linear? The input is $\boldsymbol{v} = (v_1, v_2)$:

(a) $T(\boldsymbol{v}) = (v_2, v_1)$ (b) $T(\boldsymbol{v}) = (v_1, v_1)$ (c) $T(\boldsymbol{v}) = (0, v_1)$

(d) $T(\boldsymbol{v}) = (0, 1)$.

4 If S and T are linear transformations, is $S(T(\boldsymbol{v}))$ linear or quadratic?

(a) (Special case) If $S(\boldsymbol{v}) = \boldsymbol{v}$ and $T(\boldsymbol{v}) = \boldsymbol{v}$, then $S(T(\boldsymbol{v})) = \boldsymbol{v}$ or $\boldsymbol{v}^2$?

(b) (General case) $S(\boldsymbol{w}_1+\boldsymbol{w}_2) = S(\boldsymbol{w}_1)+S(\boldsymbol{w}_2)$ and $T(\boldsymbol{v}_1+\boldsymbol{v}_2) = T(\boldsymbol{v}_1)+T(\boldsymbol{v}_2)$ combine into

$$S(T(\boldsymbol{v}_1 + \boldsymbol{v}_2)) = S(\underline{\qquad}) = \underline{\qquad} + \underline{\qquad}.$$

5 Suppose $T(\boldsymbol{v}) = \boldsymbol{v}$ except that $T(0, v_2) = (0, 0)$. Show that this transformation satisfies $T(c\boldsymbol{v}) = cT(\boldsymbol{v})$ but not $T(\boldsymbol{v} + \boldsymbol{w}) = T(\boldsymbol{v}) + T(\boldsymbol{w})$.

6 Which of these transformations satisfy $T(\boldsymbol{v}+\boldsymbol{w}) = T(\boldsymbol{v})+T(\boldsymbol{w})$ and which satisfy $T(c\boldsymbol{v}) = cT(\boldsymbol{v})$?

(a) $T(\boldsymbol{v}) = \boldsymbol{v}/\|\boldsymbol{v}\|$ (b) $T(\boldsymbol{v}) = v_1+v_2+v_3$ (c) $T(\boldsymbol{v}) = (v_1, 2v_2, 3v_3)$

(d) $T(\boldsymbol{v}) =$ largest component of $\boldsymbol{v}$.

7 For these transformations of $\mathbf{V} = \mathbf{R}^2$ to $\mathbf{W} = \mathbf{R}^2$, find $T(T(\boldsymbol{v}))$. Is this transformation T^2 linear?

(a) $T(\boldsymbol{v}) = -\boldsymbol{v}$ (b) $T(\boldsymbol{v}) = \boldsymbol{v} + (1, 1)$

(c) $T(\boldsymbol{v}) = 90°$ rotation $= (-v_2, v_1)$

(d) $T(\boldsymbol{v}) =$ projection $= \left(\frac{v_1+v_2}{2}, \frac{v_1+v_2}{2}\right)$.

8 Find the range and kernel (like the column space and nullspace) of T:

(a) $T(v_1, v_2) = (v_2, v_1)$ (b) $T(v_1, v_2, v_3) = (v_1, v_2)$

(c) $T(v_1, v_2) = (0, 0)$ (d) $T(v_1, v_2) = (v_1, v_1)$.

9 The "cyclic" transformation T is defined by $T(v_1, v_2, v_3) = (v_2, v_3, v_1)$. What is $T(T(\boldsymbol{v}))$? What is $T^3(\boldsymbol{v})$? What is $T^{100}(\boldsymbol{v})$? Apply T three times and 100 times to $\boldsymbol{v}$.

10 A linear transformation from $\mathbf{V}$ to $\mathbf{W}$ has an *inverse* from $\mathbf{W}$ to $\mathbf{V}$ when the range is all of $\mathbf{W}$ and the kernel contains only $\boldsymbol{v} = \mathbf{0}$. Why are these transformations not invertible?

(a) $T(v_1, v_2) = (v_2, v_2)$ $\quad \mathbf{W} = \mathbf{R}^2$

(b) $T(v_1, v_2) = (v_1, v_2, v_1 + v_2)$ $\quad \mathbf{W} = \mathbf{R}^3$

(c) $T(v_1, v_2) = v_1$ $\quad \mathbf{W} = \mathbf{R}^1$

11 If $T(\boldsymbol{v}) = A\boldsymbol{v}$ and A is m by n, then T is "multiplication by A."

(a) What are the input and output spaces $\mathbf{V}$ and $\mathbf{W}$?

(b) Why is range of T = column space of A?

(c) Why is kernel of T = nullspace of A?

12 Suppose a linear T transforms $(1,1)$ to $(2,2)$ and $(2,0)$ to $(0,0)$. Find $T(\boldsymbol{v})$ when

(a) $\boldsymbol{v}=(2,2)$ (b) $\boldsymbol{v}=(3,1)$ (c) $\boldsymbol{v}=(-1,1)$ (d) $\boldsymbol{v}=(a,b)$.

Problems 13–20 may be harder. The input space V contains all 2 by 2 matrices M.

13 M is any 2 by 2 matrix and $A=\begin{bmatrix}1&2\\3&4\end{bmatrix}$. The transformation T is defined by $T(M)=AM$. What rules of matrix multiplication show that T is linear?

14 Suppose $A=\begin{bmatrix}1&2\\3&5\end{bmatrix}$. Show that the range of T is the whole matrix space $\mathbf{V}$ and the kernel is the zero matrix:

(1) If $AM=0$ prove that M must be the zero matrix.

(2) Find a solution to $AM=B$ for any 2 by 2 matrix B.

15 Suppose $A=\begin{bmatrix}1&2\\3&6\end{bmatrix}$. Show that the identity matrix I is not in the range of T. Find a nonzero matrix M such that $T(M)=AM$ is zero.

16 Suppose T transposes every matrix M. Try to find a matrix A which gives $AM=M^{\mathrm{T}}$ for every M. Show that no matrix A will do it. *To professors:* Is this a linear transformation that doesn't come from a matrix?

17 The transformation T that transposes every matrix is definitely linear. Which of these extra properties are true?

(a) $T^2=$ identity transformation.

(b) The kernel of T is the zero matrix.

(c) Every matrix is in the range of T.

(d) $T(M)=-M$ is impossible.

18 Suppose $T(M)=\begin{bmatrix}1&0\\0&0\end{bmatrix}\begin{bmatrix}\text{M}\end{bmatrix}\begin{bmatrix}0&0\\0&1\end{bmatrix}$. Find a matrix with $T(M)\neq 0$. Describe all matrices with $T(M)=0$ (the kernel of T) and all output matrices $T(M)$ (the range of T).

19 If $A\neq 0$ and $B\neq 0$ then there is a matrix M such that $AMB\neq 0$. Show by example that $M=I$ might fail. For your example find an M that succeeds.

20 If A and B are invertible and $T(M)=AMB$, find $T^{-1}(M)$ in the form $(\quad)M(\quad)$.

Questions 21–27 are about house transformations AH. The output is T (house).

21 How can you tell from the picture of T (house) that A is

(a) a diagonal matrix?

(b) a rank-one matrix?

(c) a lower triangular matrix?

22 Draw a picture of T (house) for these matrices:

$$D = \begin{bmatrix} 2 & 0 \\ 0 & 1 \end{bmatrix} \quad \text{and} \quad A = \begin{bmatrix} .7 & .7 \\ .3 & .3 \end{bmatrix} \quad \text{and} \quad U = \begin{bmatrix} 1 & 1 \\ 0 & 1 \end{bmatrix}.$$

23 What are the conditions on $A = \left[\begin{smallmatrix} \mathbf{a} & \mathbf{b} \\ \mathbf{c} & \mathbf{d} \end{smallmatrix}\right]$ to ensure that T (house) will

(a) sit straight up?

(b) expand the house by 3 in all directions?

(c) rotate the house with no change in its shape?

24 What are the conditions on $\det A = ad - bc$ to ensure that T (house) will

(a) be squashed onto a line?

(b) keep its endpoints in clockwise order (not reflected)?

(c) have the same area as the original house?

If one side of the house stays in place, how do you know that $A = I$?

25 Describe T (house) when $T(\boldsymbol{v}) = -\boldsymbol{v} + (1, 0)$. This T is "affine."

26 Change the house matrix H to add a chimney.

27 This MATLAB program creates a vector of 50 angles called theta, and then draws the unit circle and T (circle) = ellipse. You can change A.

```
A = [2 1;1 2]
theta = [0:2 * pi/50:2 * pi];
circle = [cos(theta); sin(theta)];
ellipse = A * circle;
axis([−4 4 −4 4]); axis('square')
plot(circle(1,:), circle(2,:), ellipse(1,:), ellipse(2,:))
```

28 Add two eyes and a smile to the circle in Problem 27. (If one eye is dark and the other is light, you can tell when the face is reflected across the y axis.) Multiply by matrices A to get new faces.

29 The standard house is drawn by **plot2d**(H). Circles from o and lines from –:

```
x = H(1, :)'; y = H(2, :)';
axis([−1010−1010]), axis('square')
plot(x, y,'o', x, y,'−');
```

Test **plot2d**(A′ ∗ H) and **plot2d**(A′ ∗ A ∗ H) with the matrices in Figure 7.1.

30 Without a computer describe the houses $A * H$ for these matrices A:

$$\begin{bmatrix} 1 & 0 \\ 0 & .1 \end{bmatrix} \quad \text{and} \quad \begin{bmatrix} .5 & .5 \\ .5 & .5 \end{bmatrix} \quad \text{and} \quad \begin{bmatrix} .5 & .5 \\ -.5 & .5 \end{bmatrix} \quad \text{and} \quad \begin{bmatrix} 1 & 1 \\ 1 & 0 \end{bmatrix}.$$

31 What matrices give the houses on the front cover? The second is $A = I$.

THE MATRIX OF A LINEAR TRANSFORMATION ■ 7.2

The next pages assign a matrix to every linear transformation. For ordinary column vectors, the input $\boldsymbol{v}$ is in $\mathbf{V} = \mathbf{R}^n$ and the output $T(\boldsymbol{v})$ is in $\mathbf{W} = \mathbf{R}^m$. The matrix for this transformation T will be m by n.

The standard basis vectors for $\mathbf{R}^n$ and $\mathbf{R}^m$ lead to a standard matrix for T. Then $T(\boldsymbol{v}) = A\boldsymbol{v}$ in the normal way. But these spaces also have other bases, so the same T is represented by other matrices. A main theme of linear algebra is to choose the bases that give the best matrix.

When $\mathbf{V}$ and $\mathbf{W}$ are not $\mathbf{R}^n$ and $\mathbf{R}^m$, they still have bases. Each choice of basis leads to a matrix for T. When the input basis is different from the output basis, the matrix for $T(\boldsymbol{v}) = \boldsymbol{v}$ will not be the identity I. It will be the "change of basis matrix."

Key idea of this section

When we know $T(\boldsymbol{v}_1), \ldots, T(\boldsymbol{v}_n)$ for the basis vectors $\boldsymbol{v}_1, \ldots, \boldsymbol{v}_n$, linearity produces $T(\boldsymbol{v})$ for every other vector $\boldsymbol{v}$.

Reason Every input $\boldsymbol{v}$ is a unique combination $c_1\boldsymbol{v}_1 + \cdots + c_n\boldsymbol{v}_n$ of the basis vectors. Since T is a linear transformation (here is the moment for linearity), the output $T(\boldsymbol{v})$ must be the same combination of the known outputs $T(\boldsymbol{v}_1), \ldots, T(\boldsymbol{v}_n)$:

$$\begin{array}{c}\textit{Suppose } \boldsymbol{v} = c_1\boldsymbol{v}_1 + \cdots + c_n\boldsymbol{v}_n. \\ \textit{Then linearity requires } T(\boldsymbol{v}) = c_1 T(\boldsymbol{v}_1) + \cdots + c_n T(\boldsymbol{v}_n).\end{array} \tag{1}$$

The rule of linearity extends from $c\boldsymbol{v} + d\boldsymbol{w}$ to all combinations $c_1\boldsymbol{v}_1 + \cdots + c_n\boldsymbol{v}_n$. Our first example gives the outputs $T(\boldsymbol{v})$ for the standard basis vectors $(1, 0)$ and $(0, 1)$.

Example 1 Suppose T transforms $\boldsymbol{v}_1 = (1, 0)$ to $T(\boldsymbol{v}_1) = (2, 3, 4)$. Suppose the second basis vector $\boldsymbol{v}_2 = (0, 1)$ goes to $T(\boldsymbol{v}_2) = (5, 5, 5)$. If T is linear from $\mathbf{R}^2$ to $\mathbf{R}^3$ then its "standard matrix" is 3 by 2. Those outputs go into its columns:

$$A = \begin{bmatrix} 2 & 5 \\ 3 & 5 \\ 4 & 5 \end{bmatrix}. \qquad T(\boldsymbol{v}_1 + \boldsymbol{v}_2) = T(\boldsymbol{v}_1) + T(\boldsymbol{v}_2) \quad \text{is} \quad \begin{bmatrix} 2 & 5 \\ 3 & 5 \\ 4 & 5 \end{bmatrix} \begin{bmatrix} 1 \\ 1 \end{bmatrix} = \begin{bmatrix} 7 \\ 8 \\ 9 \end{bmatrix}.$$

Example 2 The derivatives of the functions $1, x, x^2, x^3$ are $0, 1, 2x, 3x^2$. Those are four facts about the transformation T that "***takes the derivative.***" The inputs and outputs are functions! Now add the crucial fact that T is linear:

$$T(\boldsymbol{v}) = \frac{d\boldsymbol{v}}{dx} \qquad \text{obeys the linearity rule} \qquad \frac{d}{dx}(c\boldsymbol{v} + d\boldsymbol{w}) = c\frac{d\boldsymbol{v}}{dx} + d\frac{d\boldsymbol{w}}{dx}.$$

It is exactly this linearity that you use to find all other derivatives. From the derivative of each separate power $1, x, x^2, x^3$ (those are the basis vectors $\boldsymbol{v}_1, \boldsymbol{v}_2, \boldsymbol{v}_3, \boldsymbol{v}_4$) you find the derivative of any polynomial like $4 + x + x^2 + x^3$:

$$\frac{d}{dx}(4 + x + x^2 + x^3) = 1 + 2x + 3x^2 \qquad \text{(because of linearity!)}$$

This example applies T (the derivative d/dx) to the input $4\boldsymbol{v}_1 + \boldsymbol{v}_2 + \boldsymbol{v}_3 + \boldsymbol{v}_4$. Here the input space $\mathbf{V}$ contains all combinations of $1, x, x^2, x^3$. I call them vectors, you might call them functions. Those four vectors are a basis for the space $\mathbf{V}$ of cubic polynomials (degree ≤ 3).

For the nullspace of A, we solved $A\boldsymbol{v} = \mathbf{0}$. For the kernel of the derivative T, we solve $d\boldsymbol{v}/dx = \mathbf{0}$. The solution is $\boldsymbol{v}$ = constant. The nullspace of T is one-dimensional, containing all constant functions like $\boldsymbol{v}_1 = 1$ (the first basis function).

To find the range (or column space), look at all outputs from $T(\boldsymbol{v}) = d\boldsymbol{v}/dx$. The inputs are cubic polynomials $a + bx + cx^2 + dx^3$, so the outputs are *quadratic polynomials* (degree ≤ 2). For the output space $\mathbf{W}$ we have a choice. If $\mathbf{W}$ = cubics, then the range of T (the quadratics) is a subspace. If $\mathbf{W}$ = quadratics, then the range is all of $\mathbf{W}$.

That second choice emphasizes the difference between the domain or input space ($\mathbf{V}$ = cubics) and the image or output space ($\mathbf{W}$ = quadratics). $\mathbf{V}$ has dimension $n = 4$ and $\mathbf{W}$ has dimension $m = 3$. The matrix for T in equation (2) will be 3 by 4.

The range of T is a three-dimensional subspace. The matrix will have rank $r = 3$. The kernel is one-dimensional. The sum $3+1 = 4$ is the dimension of the input space. This was $r + (n - r) = n$ in the Fundamental Theorem of Linear Algebra. Always (***dimension of range***) + (***dimension of kernel***) = ***dimension of*** $\mathbf{V}$.

Example 3 The ***integral*** is the inverse of the derivative. That is the Fundamental Theorem of Calculus. We see it now in linear algebra. The transformation T^{-1} that "takes the integral from 0 to x" is linear! Apply T^{-1} to $1, x, x^2$, which are $\boldsymbol{w}_1, \boldsymbol{w}_2, \boldsymbol{w}_3$:

$$\int_0^x 1\,dx = x, \quad \int_0^x x\,dx = \tfrac{1}{2}x^2, \quad \int_0^x x^2\,dx = \tfrac{1}{3}x^3.$$

By linearity, the integral of $\boldsymbol{w} = B + Cx + Dx^2$ is $T^{-1}(\boldsymbol{w}) = Bx + \frac{1}{2}Cx^2 + \frac{1}{3}Dx^3$. The integral of a quadratic is a cubic. The input space of T^{-1} is the quadratics, the output space is the cubics. ***Integration takes*** $\mathbf{W}$ ***back to*** $\mathbf{V}$. Its matrix will be 4 by 3.

Range of T^{-1} The outputs $Bx + \frac{1}{2}Cx^2 + \frac{1}{3}Dx^3$ are cubics with no constant term.

Kernel of T^{-1} The output is zero only if $B = C = D = 0$. The nullspace is $\mathbf{Z}$, the zero vector. Now $3 + 0 = 3$ is the dimension of the input space $\mathbf{W}$ for T^{-1}.

Matrices for the Derivative and Integral

We will show how the matrices A and A^{-1} copy the derivative T and the integral T^{-1}. This is an excellent example from calculus. Then comes the general rule—how to represent any linear transformation T by a matrix A.

The derivative transforms the space **V** of cubics to the space **W** of quadratics. The basis for **V** is $1, x, x^2, x^3$. The basis for **W** is $1, x, x^2$. *The matrix that "takes the derivative" is* 3 *by* 4*:*

$$A = \begin{bmatrix} 0 & 1 & 0 & 0 \\ 0 & 0 & 2 & 0 \\ 0 & 0 & 0 & 3 \end{bmatrix} = \textbf{matrix form of derivative } T. \tag{2}$$

Why is A the correct matrix? Because ***multiplying by*** A ***agrees with transforming by*** T. The derivative of $v = a + bx + cx^2 + dx^3$ is $T(v) = b + 2cx + 3dx^2$. The same b and $2c$ and $3d$ appear when we multiply by the matrix:

$$\begin{bmatrix} 0 & 1 & 0 & 0 \\ 0 & 0 & 2 & 0 \\ 0 & 0 & 0 & 3 \end{bmatrix} \begin{bmatrix} a \\ b \\ c \\ d \end{bmatrix} = \begin{bmatrix} b \\ 2c \\ 3d \end{bmatrix}. \tag{3}$$

Look also at T^{-1}. The integration matrix is 4 by 3. Watch how the following matrix starts with $w = B + Cx + Dx^2$ and produces its integral $Bx + \frac{1}{2}Cx^2 + \frac{1}{3}Dx^3$:

$$\textbf{\textit{Integration:}} \begin{bmatrix} 0 & 0 & 0 \\ 1 & 0 & 0 \\ 0 & \frac{1}{2} & 0 \\ 0 & 0 & \frac{1}{3} \end{bmatrix} \begin{bmatrix} B \\ C \\ D \end{bmatrix} = \begin{bmatrix} 0 \\ B \\ \frac{1}{2}C \\ \frac{1}{3}D \end{bmatrix}. \tag{4}$$

I want to call that matrix A^{-1}, and I will. But you realize that rectangular matrices don't have inverses. At least they don't have two-sided inverses. This rectangular A has a *one-sided inverse*. The integral is a one-sided inverse of the derivative!

$$AA^{-1} = \begin{bmatrix} 1 & 0 & 0 \\ 0 & 1 & 0 \\ 0 & 0 & 1 \end{bmatrix} \quad \text{but} \quad A^{-1}A = \begin{bmatrix} 0 & 0 & 0 & 0 \\ 0 & 1 & 0 & 0 \\ 0 & 0 & 1 & 0 \\ 0 & 0 & 0 & 1 \end{bmatrix}.$$

If you integrate a function and then differentiate, you get back to the start. So $AA^{-1} = I$. But if you differentiate before integrating, the constant term is lost. ***The integral of the derivative of*** 1 ***is zero:***

$$T^{-1}T(1) = \text{integral of zero function } = 0.$$

This matches $A^{-1}A$, whose first column is all zero. The derivative T has a kernel (the constant functions). Its matrix A has a nullspace. Main point again: Av copies $T(v)$.

Construction of the Matrix

Now we construct a matrix for any linear transformation. Suppose T transforms the space $\mathbf{V}$ (n-dimensional) to the space $\mathbf{W}$ (m-dimensional). We choose a basis $\boldsymbol{v}_1, \ldots, \boldsymbol{v}_n$ for $\mathbf{V}$ and a basis $\boldsymbol{w}_1, \ldots, \boldsymbol{w}_m$ for $\mathbf{W}$. The matrix A will be m by n. To find its first column, apply T to the first basis vector $\boldsymbol{v}_1$:

$T(v_1)$ is a combination $a_{11}w_1 + \cdots + a_{m1}w_m$ ***of the output basis for W.***

These numbers $a_{11}, \ldots, a_{m1}$ go into the first column of A. Transforming $\boldsymbol{v}_1$ to $T(\boldsymbol{v}_1)$ matches multiplying $(1, 0, \ldots, 0)$ by A. It yields that first column of the matrix. When T is the derivative and the first basis vector is 1, its derivative is $T(\boldsymbol{v}_1) = \mathbf{0}$. So for the derivative, the first column of A was all zero.

For the integral, the first basis function is again 1 and its integral is x. This is 1 times the second basis function. So the first column of A^{-1} was $(0, 1, 0, 0)$.

7A Each linear transformation T from $\mathbf{V}$ to $\mathbf{W}$ is represented by a matrix A (after the bases are chosen for $\mathbf{V}$ and $\mathbf{W}$). The jth column of A is found by applying T to the jth basis vector $\boldsymbol{v}_j$:

$$T(\boldsymbol{v}_j) = \text{combination of basis vectors of } \mathbf{W} = a_{1j}\boldsymbol{w}_1 + \cdots + a_{mj}\boldsymbol{w}_m. \qquad (5)$$

These numbers $a_{1j}, \ldots, a_{mj}$ go into column j of A. *The matrix is constructed to get the basis vectors right.* ***Then linearity gets all other vectors right.*** Every $\boldsymbol{v}$ is a combination $c_1\boldsymbol{v}_1 + \cdots + c_n\boldsymbol{v}_n$, and $T(\boldsymbol{v})$ is a combination of the $\boldsymbol{w}$'s. When A multiplies the coefficient vector $\boldsymbol{c} = (c_1, \ldots, c_n)$ in the $\boldsymbol{v}$ combination, $A\boldsymbol{c}$ produces the coefficients in the $T(\boldsymbol{v})$ combination. This is because matrix multiplication (combining columns) is linear like T.

A tells what T does. Every linear transformation can be converted to a matrix. This matrix depends on the bases.

Example 4 ***If the bases change, T is the same but the matrix A is different.***

Suppose we reorder the basis to $x, x^2, x^3, 1$ for the cubics in $\mathbf{V}$. Keep the original basis $1, x, x^2$ for the quadratics in $\mathbf{W}$. Now apply T to the first basis vector $\boldsymbol{v}_1$. The derivative of x is 1. This is the first basis vector of $\mathbf{W}$. So the first column of A looks different:

$$A_{\text{new}} = \begin{bmatrix} 1 & 0 & 0 & 0 \\ 0 & 2 & 0 & 0 \\ 0 & 0 & 3 & 0 \end{bmatrix} = \begin{array}{l} \text{matrix for the derivative } T \\ \text{when the bases change to} \\ x, x^2, x^3, 1 \text{ and } 1, x, x^2. \end{array}$$

When we reorder the basis of $\mathbf{V}$, we reorder the columns of A. The input basis vector $\boldsymbol{v}_j$ is responsible for column j. The output basis vector $\boldsymbol{w}_i$ is responsible for row i. Soon the changes in the bases will be more than permutations.

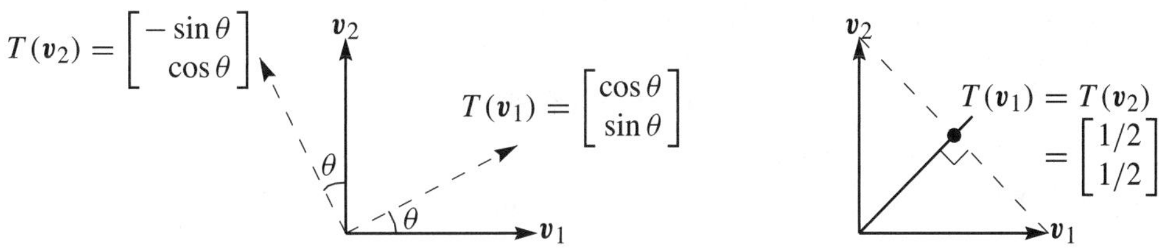

Figure 7.2 Rotation by θ and projection onto the 45° line.

Products AB Match Transformations TS

The examples of derivative and integral made three points. First, linear transformations T are everywhere—in calculus and differential equations and linear algebra. Second, spaces other than $\mathbf{R}^n$ are important—we had functions, cubics, and quadratics. Third, T still boils down to a matrix A. Now we make sure that we can find this matrix.

The next examples have $\mathbf{V} = \mathbf{W}$. We choose the same basis for both spaces. Then we can compare the matrices A^2 and AB with the transformations T^2 and TS.

Example 5 T rotates every plane vector by the same angle θ. Here $\mathbf{V} = \mathbf{W} = \mathbf{R}^2$. Find the rotation matrix A. The answer depends on the basis!

Solution The standard basis is $v_1 = (1, 0)$ and $v_2 = (0, 1)$. To find A, apply T to those basis vectors. In Figure 7.2a, they are rotated by θ. *The first vector* $(1, 0)$ ***swings around to*** $(\cos\theta, \sin\theta)$. This equals $\cos\theta$ times $(1, 0)$ plus $\sin\theta$ times $(0, 1)$. Therefore those numbers $\cos\theta$ and $\sin\theta$ go into the *first column* of A:

$$\begin{bmatrix} \cos\theta & \\ \sin\theta & \end{bmatrix} \text{ shows column 1} \qquad A = \begin{bmatrix} \cos\theta & -\sin\theta \\ \sin\theta & \cos\theta \end{bmatrix} \text{ shows both columns.}$$

For the second column, transform the second vector $(0, 1)$. The figure shows it rotated to $(-\sin\theta, \cos\theta)$. *Those numbers go into the second column.* Multiplying A times $(0, 1)$ produces that column, so A agrees with T.

Example 6 (*Projection*) Suppose T projects every plane vector onto the 45° line. Find its matrix for two different choices of the basis. We will find two matrices.

Solution Start with a specially chosen basis, not drawn in Figure 7.2. The basis vector v_1 is along the 45° line. *It projects to itself.* From $T(v_1) = v_1$, the first column of A contains 1 and 0. The second basis vector v_2 is along the perpendicular line (135°). *This basis vector projects to zero.* So the second column of A contains 0 and 0:

$$\textbf{Projection } A = \begin{bmatrix} 1 & 0 \\ 0 & 0 \end{bmatrix} \quad \text{when } \mathbf{V} \text{ and } \mathbf{W} \text{ have the 45° and 135° basis.}$$

With the basis in the opposite order (135° then 45°), the matrix is _____.

Now take the standard basis $(1,0)$ and $(0,1)$. Figure 7.2b shows how $(1,0)$ projects to $(\frac{1}{2},\frac{1}{2})$. That gives the first column of A. The other basis vector $(0,1)$ also projects to $(\frac{1}{2},\frac{1}{2})$. So the standard matrix for this projection is A:

$$\textbf{Projection } A = \begin{bmatrix} \frac{1}{2} & \frac{1}{2} \\ \frac{1}{2} & \frac{1}{2} \end{bmatrix} \quad \text{for the } \textit{same} \quad T \text{ and the standard basis.}$$

Both A's are projection matrices. If you square A it doesn't change. Projecting twice is the same as projecting once: $T^2 = T$ so $A^2 = A$. Notice what is hidden in that statement: ***The matrix for*** $\boldsymbol{T^2}$ ***is*** $\boldsymbol{A^2}$.

We have come to something important—the real reason for the way matrices are multiplied. *At last we discover why*! Two transformations S and T are represented by two matrices B and A. When we apply T to the output from S, we get the "composition" TS. When we apply A after B, we get the matrix product AB. Matrix multiplication gives the correct matrix AB to represent TS.

The transformation S is from a space $\mathbf{U}$ to $\mathbf{V}$. Its matrix B uses a basis $\boldsymbol{u}_1, \ldots, \boldsymbol{u}_p$ for $\mathbf{U}$ and a basis $\boldsymbol{v}_1, \ldots, \boldsymbol{v}_n$ for $\mathbf{V}$. The matrix is n by p. The transformation T is from $\mathbf{V}$ to $\mathbf{W}$ as before. *Its matrix A must use the same basis $\boldsymbol{v}_1, \ldots, \boldsymbol{v}_n$ for* $\mathbf{V}$—this is the output space for S and the input space for T. Then AB matches TS:

7B Multiplication The linear transformation TS starts with any vector $\boldsymbol{u}$ in $\mathbf{U}$, goes to $S(\boldsymbol{u})$ in $\mathbf{V}$ and then to $T(S(\boldsymbol{u}))$ in $\mathbf{W}$. The matrix AB starts with any $\boldsymbol{x}$ in $\mathbf{R}^p$, goes to $B\boldsymbol{x}$ in $\mathbf{R}^n$ and then to $AB\boldsymbol{x}$ in $\mathbf{R}^m$. The matrix AB correctly represents TS:

$$TS: \quad \mathbf{U} \to \mathbf{V} \to \mathbf{W} \qquad AB: \quad (m \text{ by } n)(n \text{ by } p) = (m \text{ by } p).$$

The input is $\boldsymbol{u} = x_1\boldsymbol{u}_1 + \cdots + x_p\boldsymbol{u}_p$. The output $T(S(\boldsymbol{u}))$ matches the output $AB\boldsymbol{x}$. Product of transformations matches product of matrices. The most important cases are when the spaces $\mathbf{U}, \mathbf{V}, \mathbf{W}$ are the same and their bases are the same. With $m = n = p$ we have square matrices.

Example 7 S rotates the plane by θ and T also rotates by θ. Then TS rotates by 2θ. This transformation T^2 corresponds to the rotation matrix A^2 through 2θ:

$$T = S \qquad A = B \qquad A^2 = \text{rotation by } 2\theta = \begin{bmatrix} \cos 2\theta & -\sin 2\theta \\ \sin 2\theta & \cos 2\theta \end{bmatrix}.$$

By matching (transformation)^2 with (matrix)^2, we pick up the formulas for $\cos 2\theta$ and $\sin 2\theta$. Multiply A times A:

$$\begin{bmatrix} \cos\theta & -\sin\theta \\ \sin\theta & \cos\theta \end{bmatrix} \begin{bmatrix} \cos\theta & -\sin\theta \\ \sin\theta & \cos\theta \end{bmatrix} = \begin{bmatrix} \cos^2\theta - \sin^2\theta & -2\sin\theta\cos\theta \\ 2\sin\theta\cos\theta & \cos^2\theta - \sin^2\theta \end{bmatrix}.$$

Comparing with the display above, $\cos 2\theta = \cos^2\theta - \sin^2\theta$ and $\sin 2\theta = 2\sin\theta\cos\theta$. Trigonometry comes from linear algebra.

Example 8 S rotates by θ and T rotates by $-\theta$. Then $TS = I$ and $AB = I$.

In this case $T(S(\boldsymbol{u}))$ is $\boldsymbol{u}$. We rotate forward and back. For the matrices to match, $AB\boldsymbol{x}$ must be $\boldsymbol{x}$. The two matrices are inverses. Check this by putting $\cos(-\theta) = \cos\theta$ and $\sin(-\theta) = -\sin\theta$ into A:

$$AB = \begin{bmatrix} \cos\theta & \sin\theta \\ -\sin\theta & \cos\theta \end{bmatrix} \begin{bmatrix} \cos\theta & -\sin\theta \\ \sin\theta & \cos\theta \end{bmatrix} = \begin{bmatrix} \cos^2\theta + \sin^2\theta & 0 \\ 0 & \cos^2\theta + \sin^2\theta \end{bmatrix}.$$

By the famous identity for $\cos^2\theta + \sin^2\theta$, this is I.

Earlier T took the derivative and S took the integral. Then TS is the identity but not ST. Therefore AB is the identity matrix but not BA:

$$AB = \begin{bmatrix} 0 & 1 & 0 & 0 \\ 0 & 0 & 2 & 0 \\ 0 & 0 & 0 & 3 \end{bmatrix} \begin{bmatrix} 0 & 0 & 0 \\ 1 & 0 & 0 \\ 0 & \frac{1}{2} & 0 \\ 0 & 0 & \frac{1}{3} \end{bmatrix} = I \quad \text{but} \quad BA = \begin{bmatrix} 0 & 0 & 0 & 0 \\ 0 & 1 & 0 & 0 \\ 0 & 0 & 1 & 0 \\ 0 & 0 & 0 & 1 \end{bmatrix}.$$

The Identity Transformation and Change of Basis

We need the matrix for the special and boring transformation $T(\boldsymbol{v}) = \boldsymbol{v}$. This identity transformation does nothing to $\boldsymbol{v}$. The matrix also does nothing, *provided* the output basis is the same as the input basis. The output $T(\boldsymbol{v}_1)$ is $\boldsymbol{v}_1$. When the bases are the same, this is $\boldsymbol{w}_1$. So the first column of A is $(1, 0, \ldots, 0)$.

When each output $T(\boldsymbol{v}_j) = \boldsymbol{v}_j$ is the same as $\boldsymbol{w}_j$, the matrix is just I.

This seems reasonable: The identity transformation is represented by the identity matrix. But suppose the bases are different. Then $T(\boldsymbol{v}_1) = \boldsymbol{v}_1$ is a combination of the $\boldsymbol{w}$'s. That combination $m_{11}\boldsymbol{w}_1 + \cdots + m_{n1}\boldsymbol{w}_n$ tells us the first column of the matrix M. We will use M (instead of A) for a matrix that represents the identity transformation.

When the outputs $T(\boldsymbol{v}_j) = \boldsymbol{v}_j$ are combinations $\sum_{i=1}^{n} m_{ij}\boldsymbol{w}_i$, the "change of basis matrix" is M.

The basis is changing but the vectors themselves are not changing: $T(\boldsymbol{v}) = \boldsymbol{v}$. When the input has one basis and the output has another basis, the matrix is not I.

Example 9 The input basis is $\boldsymbol{v}_1 = (3, 7)$ and $\boldsymbol{v}_2 = (2, 5)$. The output basis is $\boldsymbol{w}_1 = (1, 0)$ and $\boldsymbol{w}_2 = (0, 1)$. Then the matrix M is easy to compute:

$$\text{The matrix for} \quad T(\boldsymbol{v}) = \boldsymbol{v} \quad \text{is} \quad M = \begin{bmatrix} 3 & 2 \\ 7 & 5 \end{bmatrix}.$$

Reason The first input is $\boldsymbol{v}_1 = (3, 7)$. The output is also $(3, 7)$ but we express it as $3\boldsymbol{w}_1 + 7\boldsymbol{w}_2$. Then the first column of M contains 3 and 7.

This seems too simple to be important. It becomes trickier when the change of basis goes the other way. We get the inverse of the previous matrix M:

Example 10 The input basis is $v_1 = (1,0)$ and $v_2 = (0,1)$. The outputs are just $T(v) = v$. But the output *basis* is $w_1 = (3,7)$ and $w_2 = (2,5)$.

$$\text{The matrix for } T(v) = v \quad \text{is} \quad \begin{bmatrix} 3 & 2 \\ 7 & 5 \end{bmatrix}^{-1} = \begin{bmatrix} 5 & -2 \\ -7 & 3 \end{bmatrix}.$$

Reason The first input is $v_1 = (1,0)$. The output is also v_1 but we express it as $5w_1 - 7w_2$. Check that $5(3,7) - 7(2,5)$ does produce $(1,0)$. We are combining the columns of the previous M to get the columns of I. The matrix to do that is M^{-1}:

$$\begin{bmatrix} w_1 & w_2 \end{bmatrix} \begin{bmatrix} 5 & -2 \\ -7 & 3 \end{bmatrix} = \begin{bmatrix} v_1 & v_2 \end{bmatrix} \quad \text{is} \quad MM^{-1} = I.$$

A mathematician would say that MM^{-1} corresponds to the product of two identity transformations. We start and end with the same basis $(1,0)$ and $(0,1)$. Matrix multiplication must give I. So the two change of basis matrices are inverses.

Warning One mistake about M is very easy to make. Example 9 changes from the basis of v's to the standard columns of I. But matrix multiplication goes the other way. When you multiply M times the columns of I, you get the v's. It seems backward but it is really OK.

One thing is sure. Multiplying A times $(1,0,\ldots,0)$ gives column 1 of the matrix. The novelty of this section is that $(1,0,\ldots,0)$ stands for the first vector v_1, *written in the basis of* v's. Then column 1 of the matrix is that same vector v_1, *written in the standard basis*. This is when we keep $T = I$ and change the basis for $\mathbf{V}$.

In the rest of the book we keep the standard basis and T is multiplication by A.

■ REVIEW OF THE KEY IDEAS ■

1. If we know $T(v_1), \ldots, T(v_n)$ for a basis, linearity determines all other $T(v)$.

2. $\left\{ \begin{array}{ll} \text{Linear transformation } T & \text{Matrix } A \ (m \text{ by } n) \\ \text{Input basis } v_1, \ldots, v_n & \text{represents } T \\ \text{Output basis } w_1, \ldots, w_m & \text{in these bases} \end{array} \right\}$

3. The derivative and integral matrices are one-sided inverses: $d(\text{constant})/dx = 0$:

   $$\text{(Derivative) (Integral)} = I = \text{Fundamental Theorem of Calculus !}$$

4. The change of basis matrix M represents $T(v) = v$. Its columns are the coefficients of the output basis expressed in the input basis: $w_j = m_{1j}v_1 + \cdots + m_{nj}v_n$.

WORKED EXAMPLES

7.2 A Using the standard basis, find the 4 by 4 matrix P that represents a *cyclic permutation* from $\boldsymbol{x} = (x_1, x_2, x_3, x_4)$ to $T(\boldsymbol{x}) = (x_4, x_1, x_2, x_3)$. Find the matrix for T^2. What is the triple shift $T^3(\boldsymbol{x})$ and why is $T^3 = T^{-1}$? Find two real independent eigenvectors of P. What are all the eigenvalues of P?

Solution The first vector $(1, 0, 0, 0)$ in the standard basis transforms to $(0, 1, 0, 0)$ which is the second basis vector. So the first column of P is $(0, 1, 0, 0)$. The other three columns come from transforming the other three standard basis vectors:

$$P = \begin{bmatrix} 0 & 0 & 0 & 1 \\ 1 & 0 & 0 & 0 \\ 0 & 1 & 0 & 0 \\ 0 & 0 & 1 & 0 \end{bmatrix} \qquad \text{Then } P \begin{bmatrix} x_1 \\ x_2 \\ x_3 \\ x_4 \end{bmatrix} = \begin{bmatrix} x_4 \\ x_1 \\ x_2 \\ x_3 \end{bmatrix} \text{ copies } T.$$

Since we used the standard basis, T *is ordinary multiplication by* P. The matrix for T^2 is a "double cyclic shift" P^2 and it produces (x_3, x_4, x_1, x_2).

The triple shift T^3 will transform $\boldsymbol{x} = (x_1, x_2, x_3, x_4)$ to $T^3(\boldsymbol{x}) = (x_2, x_3, x_4, x_1)$. If we apply T once more we are back to the original $\boldsymbol{x}$, so $T^4 =$ identity transformation. For matrices this is $P^4 = I$. This means that $T^3 T =$ identity and $T^3 = T^{-1}$.

Two real eigenvectors of P are $(1, 1, 1, 1)$ with eigenvalue $\lambda = 1$ and $(1, -1, 1, -1)$ with eigenvalue $\lambda = -1$. The shift leaves $(1, 1, 1, 1)$ unchanged and it reverses signs in $(1, -1, 1, -1)$. The other two eigenvalues are $\lambda_3 = i$ and $\lambda_4 = -i$. The determinant of P is $\lambda_1 \lambda_2 \lambda_3 \lambda_4 = -1$ as in Problem 5.2 which used cofactors of the first row.

Notice that the eigenvalues $1, -1, i, -i$ add to zero (the trace of P). They are the four roots of 1, since $\det(P - \lambda I) = \lambda^4 - 1$. They are equally spaced around the unit circle in the complex plane. I think P must be a 90° rotation times a reflection in $\mathbf{R}^4$.

7.2 B The space of 2 by 2 matrices has these four "vectors" as a basis:

$$\boldsymbol{u}_1 = \begin{bmatrix} 1 & 0 \\ 0 & 0 \end{bmatrix} \quad \boldsymbol{u}_2 = \begin{bmatrix} 0 & 1 \\ 0 & 0 \end{bmatrix} \quad \boldsymbol{u}_3 = \begin{bmatrix} 0 & 0 \\ 1 & 0 \end{bmatrix} \quad \boldsymbol{u}_4 = \begin{bmatrix} 0 & 0 \\ 0 & 1 \end{bmatrix}.$$

T is the linear transformation that *transposes* every 2 by 2 matrix. What is the matrix A that represents T in this basis (output basis = input basis)? What is the inverse matrix A^{-1}? What is the transformation T^{-1} that inverts the transpose operation?

Also, T_2 multiplies each matrix by $M = \left[\begin{smallmatrix} \mathbf{a} & \mathbf{b} \\ \mathbf{c} & \mathbf{d} \end{smallmatrix}\right]$. What 4 by 4 matrix A_2 represents T_2?

Solution Transposing those four "basis matrices" permutes them to $\boldsymbol{u}_1, \boldsymbol{u}_3, \boldsymbol{u}_2, \boldsymbol{u}_4$!

$$\begin{aligned} T(\boldsymbol{u}_1) &= \boldsymbol{u}_1 \\ T(\boldsymbol{u}_2) &= \boldsymbol{u}_3 \\ T(\boldsymbol{u}_3) &= \boldsymbol{u}_2 \\ T(\boldsymbol{u}_4) &= \boldsymbol{u}_4 \end{aligned} \quad \text{gives the four columns of } A = \begin{bmatrix} 1 & 0 & 0 & 0 \\ 0 & 0 & 1 & 0 \\ 0 & 1 & 0 & 0 \\ 0 & 0 & 0 & 1 \end{bmatrix}.$$

The inverse matrix A^{-1} is the same as A. The inverse transformation T^{-1} is the same as T. If we transpose and transpose again, the final output equals the original input.

To find the matrix A_2, multiply the basis matrices $\boldsymbol{u}_1, \boldsymbol{u}_2, \boldsymbol{u}_3, \boldsymbol{u}_4$ by M:

$$\begin{bmatrix} a & b \\ c & d \end{bmatrix}\begin{bmatrix} 1 & 0 \\ 0 & 0 \end{bmatrix} = \begin{bmatrix} a & 0 \\ c & 0 \end{bmatrix} = a\boldsymbol{u}_1 + c\boldsymbol{u}_3$$
$$\begin{bmatrix} a & b \\ c & d \end{bmatrix}\begin{bmatrix} 0 & 1 \\ 0 & 0 \end{bmatrix} = \begin{bmatrix} 0 & a \\ 0 & c \end{bmatrix} = a\boldsymbol{u}_2 + c\boldsymbol{u}_4$$
$$\begin{bmatrix} a & b \\ c & d \end{bmatrix}\begin{bmatrix} 0 & 0 \\ 1 & 0 \end{bmatrix} = \begin{bmatrix} b & 0 \\ d & 0 \end{bmatrix} = b\boldsymbol{u}_1 + d\boldsymbol{u}_3$$
$$\begin{bmatrix} a & b \\ c & d \end{bmatrix}\begin{bmatrix} 0 & 0 \\ 0 & 1 \end{bmatrix} = \begin{bmatrix} 0 & b \\ 0 & d \end{bmatrix} = b\boldsymbol{u}_2 + d\boldsymbol{u}_4$$

gives the columns of $A = \begin{bmatrix} a & 0 & b & 0 \\ 0 & a & 0 & b \\ c & 0 & d & 0 \\ 0 & c & 0 & d \end{bmatrix}$.

This A is the "Kronecker product" or "tensor product" of M with I, written $M \otimes I$.

Problem Set 7.2

Questions 1–4 extend the first derivative example to higher derivatives.

1 The transformation S takes the ***second derivative***. Keep $1, x, x^2, x^3$ as the basis $\boldsymbol{v}_1, \boldsymbol{v}_2, \boldsymbol{v}_3, \boldsymbol{v}_4$ and also as $\boldsymbol{w}_1, \boldsymbol{w}_2, \boldsymbol{w}_3, \boldsymbol{w}_4$. Write $S\boldsymbol{v}_1, S\boldsymbol{v}_2, S\boldsymbol{v}_3, S\boldsymbol{v}_4$ in terms of the $\boldsymbol{w}$'s. Find the 4 by 4 matrix B for S.

2 What functions have $\boldsymbol{v}'' = \boldsymbol{0}$? They are in the kernel of the second derivative S. What vectors are in the nullspace of its matrix B in Problem 1?

3 B is not the square of the 4 by 3 first derivative matrix

$$A = \begin{bmatrix} 0 & 1 & 0 & 0 \\ 0 & 0 & 2 & 0 \\ 0 & 0 & 0 & 3 \end{bmatrix}.$$

Add a zero row to A, so that output space = input space. Then compare A^2 with B. Conclusion: For $B = A^2$ we also want output basis = _____ basis. Then $m = n$.

4 (a) The product TS produces the *third* derivative. Add zero rows to make 4 by 4 matrices, then compute AB.

(b) The matrix B^2 corresponds to S^2 = *fourth* derivative. Why is this entirely zero?

Questions 5–10 are about a particular T and its matrix A.

5 With bases $\boldsymbol{v}_1, \boldsymbol{v}_2, \boldsymbol{v}_3$ and $\boldsymbol{w}_1, \boldsymbol{w}_2, \boldsymbol{w}_3$, suppose $T(\boldsymbol{v}_1) = \boldsymbol{w}_2$ and $T(\boldsymbol{v}_2) = T(\boldsymbol{v}_3) = \boldsymbol{w}_1 + \boldsymbol{w}_3$. T is a linear transformation. Find the matrix A.

6 (a) What is the output from T in Question 5 when the input is $\boldsymbol{v}_1 + \boldsymbol{v}_2 + \boldsymbol{v}_3$?

(b) Multiply A times the vector $(1, 1, 1)$.

7 Since $T(\boldsymbol{v}_2) = T(\boldsymbol{v}_3)$, the solutions to $T(\boldsymbol{v}) = \mathbf{0}$ are $\boldsymbol{v} =$ _____. What vectors are in the nullspace of A? Find all solutions to $T(\boldsymbol{v}) = \boldsymbol{w}_2$.

8 Find a vector that is not in the column space of A. Find a combination of $\boldsymbol{w}$'s that is not in the range of T.

9 You don't have enough information to determine T^2. Why not? Why is its matrix not necessarily A^2?

10 Find the rank of A. This is not the dimension of the output space $\mathbf{W}$. It is the dimension of the _____ of T.

Questions 11–14 are about invertible linear transformations.

11 Suppose $T(\boldsymbol{v}_1) = \boldsymbol{w}_1 + \boldsymbol{w}_2 + \boldsymbol{w}_3$ and $T(\boldsymbol{v}_2) = \boldsymbol{w}_2 + \boldsymbol{w}_3$ and $T(\boldsymbol{v}_3) = \boldsymbol{w}_3$. Find the matrix for T using these basis vectors. What input vector $\boldsymbol{v}$ gives $T(\boldsymbol{v}) = \boldsymbol{w}_1$?

12 Invert the matrix A in Problem 11. Also invert the transformation T—what are $T^{-1}(\boldsymbol{w}_1)$ and $T^{-1}(\boldsymbol{w}_2)$ and $T^{-1}(\boldsymbol{w}_3)$? Find all $\boldsymbol{v}$'s that solve $T(\boldsymbol{v}) = \mathbf{0}$.

13 Which of these are true and why is the other one ridiculous?

(a) $T^{-1}T = I$ (b) $T^{-1}(T(\boldsymbol{v}_1)) = \boldsymbol{v}_1$ (c) $T^{-1}(T(\boldsymbol{w}_1)) = \boldsymbol{w}_1$.

14 Suppose the spaces $\mathbf{V}$ and $\mathbf{W}$ have the same basis $\boldsymbol{v}_1, \boldsymbol{v}_2$.

(a) Describe a transformation T (not I) that is its own inverse.

(b) Describe a transformation T (not I) that equals T^2.

(c) Why can't the same T be used for both (a) and (b)?

Questions 15–20 are about changing the basis.

15 (a) What matrix transforms $(1, 0)$ into $(2, 5)$ and transforms $(0, 1)$ to $(1, 3)$?

(b) What matrix transforms $(2, 5)$ to $(1, 0)$ and $(1, 3)$ to $(0, 1)$?

(c) Why does no matrix transform $(2, 6)$ to $(1, 0)$ and $(1, 3)$ to $(0, 1)$?

16 (a) What matrix M transforms $(1, 0)$ and $(0, 1)$ to (r, t) and (s, u)?

(b) What matrix N transforms (a, c) and (b, d) to $(1, 0)$ and $(0, 1)$?

(c) What condition on a, b, c, d will make part (b) impossible?

17 (a) How do M and N in Problem 16 yield the matrix that transforms (a, c) to (r, t) and (b, d) to (s, u)?

(b) What matrix transforms $(2, 5)$ to $(1, 1)$ and $(1, 3)$ to $(0, 2)$?

18 If you keep the same basis vectors but put them in a different order, the change of basis matrix M is a _____ matrix. If you keep the basis vectors in order but change their lengths, M is a _____ matrix.

19 The matrix that rotates the axis vectors $(1, 0)$ and $(0, 1)$ through an angle θ is Q. What are the coordinates (a, b) of the original $(1, 0)$ using the new (rotated) axes? This can be tricky. Draw a figure or solve this equation for a and b:

$$Q = \begin{bmatrix} \cos\theta & -\sin\theta \\ \sin\theta & \cos\theta \end{bmatrix} \qquad \begin{bmatrix} 1 \\ 0 \end{bmatrix} = a \begin{bmatrix} \cos\theta \\ \sin\theta \end{bmatrix} + b \begin{bmatrix} -\sin\theta \\ \cos\theta \end{bmatrix}.$$

20 The matrix that transforms $(1, 0)$ and $(0, 1)$ to $(1, 4)$ and $(1, 5)$ is $M =$ ____. The combination $a(1, 4) + b(1, 5)$ that equals $(1, 0)$ has $(a, b) = (\ ,\ )$. How are those new coordinates of $(1, 0)$ related to M or M^{-1}?

Questions 21–24 are about the space of quadratic polynomials $A + Bx + Cx^2$.

21 The parabola $\boldsymbol{w}_1 = \frac{1}{2}(x^2 + x)$ equals one at $x = 1$ and zero at $x = 0$ and $x = -1$. Find the parabolas $\boldsymbol{w}_2$, $\boldsymbol{w}_3$, and $\boldsymbol{y}(x)$:

(a) $\boldsymbol{w}_2$ equals one at $x = 0$ and zero at $x = 1$ and $x = -1$.

(b) $\boldsymbol{w}_3$ equals one at $x = -1$ and zero at $x = 0$ and $x = 1$.

(c) $y(x)$ equals 4 at $x = 1$ and 5 at $x = 0$ and 6 at $x = -1$. Use $\boldsymbol{w}_1, \boldsymbol{w}_2, \boldsymbol{w}_3$.

22 One basis for second-degree polynomials is $\boldsymbol{v}_1 = 1$ and $\boldsymbol{v}_2 = x$ and $\boldsymbol{v}_3 = x^2$. Another basis is $\boldsymbol{w}_1, \boldsymbol{w}_2, \boldsymbol{w}_3$ from Problem 21. Find two change of basis matrices, from the $\boldsymbol{w}$'s to the $\boldsymbol{v}$'s and from the $\boldsymbol{v}$'s to the $\boldsymbol{w}$'s.

23 What are the three equations for A, B, C if the parabola $Y = A + Bx + Cx^2$ equals 4 at $x = a$ and 5 at $x = b$ and 6 at $x = c$? Find the determinant of the 3 by 3 matrix. For which numbers a, b, c will it be impossible to find this parabola Y?

24 Under what condition on the numbers $m_1, m_2, \ldots, m_9$ do these three parabolas give a basis for the space of all parabolas?

$$\boldsymbol{v}_1(x) = m_1 + m_2 x + m_3 x^2 \text{ and } \boldsymbol{v}_2(x) = m_4 + m_5 x + m_6 x^2 \text{ and}$$
$$\boldsymbol{v}_3(x) = m_7 + m_8 x + m_9 x^2.$$

25 The Gram-Schmidt process changes a basis $\boldsymbol{a}_1, \boldsymbol{a}_2, \boldsymbol{a}_3$ to an orthonormal basis $\boldsymbol{q}_1, \boldsymbol{q}_2, \boldsymbol{q}_3$. These are columns in $A = QR$. Show that R is the change of basis matrix from the $\boldsymbol{a}$'s to the $\boldsymbol{q}$'s ($\boldsymbol{a}_2$ is what combination of $\boldsymbol{q}$'s when $A = QR$?).

26 Elimination changes the rows of A to the rows of U with $A = LU$. Row 2 of A is what combination of the rows of U? Writing $A^{\mathrm{T}} = U^{\mathrm{T}} L^{\mathrm{T}}$ to work with columns, the change of basis matrix is $M = L^{\mathrm{T}}$. (We have bases provided the matrices are ____.)

27 Suppose $\boldsymbol{v}_1, \boldsymbol{v}_2, \boldsymbol{v}_3$ are eigenvectors for T. This means $T(\boldsymbol{v}_i) = \lambda_i \boldsymbol{v}_i$ for $i = 1, 2, 3$. What is the matrix for T when the input and output bases are the $\boldsymbol{v}$'s?

28 Every invertible linear transformation can have I as its matrix. Choose any input basis $\boldsymbol{v}_1, \ldots, \boldsymbol{v}_n$ and for output basis choose $\boldsymbol{w}_i$ to be $T(\boldsymbol{v}_i)$. Why must T be invertible?

Questions 29–32 review some basic linear transformations.

29 Using $\boldsymbol{v}_1 = \boldsymbol{w}_1$ and $\boldsymbol{v}_2 = \boldsymbol{w}_2$ find the standard matrix for these T's:

(a) $T(\boldsymbol{v}_1) = \mathbf{0}$ and $T(\boldsymbol{v}_2) = 3\boldsymbol{v}_1$ (b) $T(\boldsymbol{v}_1) = \boldsymbol{v}_1$ and $T(\boldsymbol{v}_1 + \boldsymbol{v}_2) = \boldsymbol{v}_1$.

30 Suppose T is reflection across the x axis and S is reflection across the y axis. The domain $\mathbf{V}$ is the xy plane. If $\boldsymbol{v} = (x, y)$ what is $S(T(\boldsymbol{v}))$? Find a simpler description of the product ST.

31 Suppose T is reflection across the 45° line, and S is reflection across the y axis. If $\boldsymbol{v} = (2, 1)$ then $T(\boldsymbol{v}) = (1, 2)$. Find $S(T(\boldsymbol{v}))$ and $T(S(\boldsymbol{v}))$. This shows that generally $ST \neq TS$.

32 Show that the product ST of two reflections is a rotation. Multiply these reflection matrices to find the rotation angle:

$$\begin{bmatrix} \cos 2\theta & \sin 2\theta \\ \sin 2\theta & -\cos 2\theta \end{bmatrix} \begin{bmatrix} \cos 2\alpha & \sin 2\alpha \\ \sin 2\alpha & -\cos 2\alpha \end{bmatrix}.$$

33 True or false: If we know $T(\boldsymbol{v})$ for n different nonzero vectors in $\mathbf{R}^n$, then we know $T(\boldsymbol{v})$ for every vector in $\mathbf{R}^n$.

CHANGE OF BASIS ■ 7.3

This section returns to one of the fundamental ideas of linear algebra—*a basis for* $\mathbf{R}^n$. We don't intend to change that idea, but we do intend to change the basis. It often happens (and we will give examples) that one basis is especially suitable for a specific problem. By changing to that basis, the vectors and the matrices reveal the information we want. The whole idea of a *transform* (this book explains the Fourier transform and wavelet transform) is exactly a change of basis.

Remember what it means for the vectors $\boldsymbol{w}_1, \ldots, \boldsymbol{w}_n$ to be a basis for $\mathbf{R}^n$:

1. The $\boldsymbol{w}'s$ are linearly independent.

2. The $n \times n$ matrix W with these columns is invertible.

3. Every vector $\boldsymbol{v}$ in $\mathbf{R}^n$ can be written in exactly one way as a combination of the $\boldsymbol{w}$'s:

$$\boldsymbol{v} = c_1\boldsymbol{w}_1 + c_2\boldsymbol{w}_2 + \cdots + c_n\boldsymbol{w}_n. \tag{1}$$

Here is the key point: Those coefficients $c_1, \ldots, c_n$ completely describe the vector $\boldsymbol{v}$, ***after we have decided on the basis***. Originally, a column vector $\boldsymbol{v}$ just has the components $v_1, \ldots, v_n$. In the new basis of $\boldsymbol{w}$'s, the same vector is described by the different set of numbers $c_1, \ldots c_n$. It takes n numbers to describe each vector and it also requires a choice of basis. The n numbers are the coordinates of $\boldsymbol{v}$ in that basis:

$$\boldsymbol{v} = \begin{bmatrix} v_1 \\ \vdots \\ v_n \end{bmatrix}_{\text{standard basis}} \quad \text{and also} \quad \boldsymbol{v} = \begin{bmatrix} c_1 \\ \vdots \\ c_n \end{bmatrix}_{\text{basis of } \boldsymbol{w}\text{'s}} \tag{2}$$

A basis is a set of axes for $\mathbf{R}^n$. The coordinates $c_1, \ldots, c_n$ tell how far to go along each axis. The axes are at right angles when the $\boldsymbol{w}$'s are orthogonal.

Small point: What is the "*standard basis*"? Those basis vectors are simply the columns of the n by n identity matrix I. These columns $\boldsymbol{e}_1, \ldots, \boldsymbol{e}_n$ are the "default basis." When I write down the vector $\boldsymbol{v} = (2, 4, 5)$ in $\mathbf{R}^3$, I am intending and you are expecting the standard basis (the usual xyz axes, where the coordinates are 2, 4, 5):

$$\boldsymbol{v} = 2\boldsymbol{e}_1 + 4\boldsymbol{e}_2 + 5\boldsymbol{e}_3 = 2\begin{bmatrix} 1 \\ 0 \\ 0 \end{bmatrix} + 4\begin{bmatrix} 0 \\ 1 \\ 0 \end{bmatrix} + 5\begin{bmatrix} 0 \\ 0 \\ 1 \end{bmatrix} = \begin{bmatrix} 2 \\ 4 \\ 5 \end{bmatrix}.$$

The new question is: *What are the coordinates c_1, c_2, c_3 in the new basis $\boldsymbol{w}_1, \boldsymbol{w}_2, \boldsymbol{w}_3$?* As usual we put the basis vectors into the columns of a matrix. This is the *basis matrix* W. Then the fundamental equation $\boldsymbol{v} = c_1\boldsymbol{w}_1 + \ldots + c_n\boldsymbol{w}_n$ has the matrix form $\boldsymbol{v} = W\boldsymbol{c}$. From this we immediately know $\boldsymbol{c} = W^{-1}\boldsymbol{v}$.

7C The coordinates $\boldsymbol{c} = (c_1, \ldots, c_n)$ of $\boldsymbol{v}$ in the basis $\boldsymbol{w}_1, \ldots, \boldsymbol{w}_n$ are given by $\boldsymbol{c} = W^{-1}\boldsymbol{v}$. ***The change of basis matrix W^{-1} is the inverse of the basis matrix W.***

The standard basis has $W = I$. The coordinates in that default basis $\boldsymbol{e}_1, \ldots, \boldsymbol{e}_n$ are the usual components $v_1, \ldots, v_n$. Our first new example is the wavelet basis for $\mathbf{R}^4$.

Example 1 (**Wavelet basis**) Wavelets are little waves. They have different lengths and they are localized at different places. The first basis vector is not actually a wavelet, it is the very useful flat vector of all ones. The others are "Haar wavelets":

$$\boldsymbol{w}_1 = \begin{bmatrix} 1 \\ 1 \\ 1 \\ 1 \end{bmatrix} \quad \boldsymbol{w}_2 = \begin{bmatrix} 1 \\ 1 \\ -1 \\ -1 \end{bmatrix} \quad \boldsymbol{w}_3 = \begin{bmatrix} 1 \\ -1 \\ 0 \\ 0 \end{bmatrix} \quad \boldsymbol{w}_4 = \begin{bmatrix} 0 \\ 0 \\ 1 \\ -1 \end{bmatrix}. \tag{3}$$

Those vectors are *orthogonal*, which is good. You see how $\boldsymbol{w}_3$ is localized in the first half and $\boldsymbol{w}_4$ is localized in the second half. Their coefficients c_3 and c_4 tell us about details in the first half and last half of $\boldsymbol{v}$. The ultimate in localization is the standard basis.

Why do want to change the basis? I think of v_1, v_2, v_3, v_4 as the intensities of a signal. It could be an audio signal, like music on a CD. It could be a medical signal, like an electrocardiogram. Of course $n = 4$ is very short, and $n = 10,000$ is more realistic. We may need to *compress* that long signal, by keeping only the largest 5% of the coefficients. This is 20 : 1 compression and (to give only one of its applications) it makes modern video conferencing possible.

If we keep only 5% of the *standard* basis coefficients, we lose 95% of the signal. In image processing, most of the image disappears. In audio, 95% of the tape goes blank. But if we choose a better basis of $\boldsymbol{w}$'s, 5% of the basis vectors can come very close to the original signal. In image processing and audio coding, you can't see or hear the difference. We don't need the other 95%!

One good basis vector is a flat $(1, 1, 1, 1)$. That part alone can represent the constant background of our image. A short wave like $(0, 0, 1, -1)$ or in higher dimensions $(0, 0, 0, 0, 0, 0, 1, -1)$ represents a detail at the end of the signal.

The three steps of ***transform*** and ***compression*** and ***inverse transform*** are

input $\boldsymbol{v}$	$\rightarrow$	coefficients $\boldsymbol{c}$	$\rightarrow$	compressed $\widehat{\boldsymbol{c}}$	$\rightarrow$	compressed $\widehat{\boldsymbol{v}}$
	[***lossless***]		[***lossy***]		[***reconstruct***]	

In linear algebra, where everything is perfect, we omit the compression step. The output $\widehat{\boldsymbol{v}}$ is exactly the same as the input $\boldsymbol{v}$. The transform gives $\boldsymbol{c} = W^{-1}\boldsymbol{v}$ and the reconstruction brings back $\boldsymbol{v} = W\boldsymbol{c}$. In true signal processing, where nothing is perfect but everything is fast, the transform (lossless) and the compression (which only loses unnecessary information) are absolutely the keys to success. Then $\widehat{\boldsymbol{v}} = W\widehat{\boldsymbol{c}}$.

I will show those steps for a typical vector like $\boldsymbol{v} = (6, 4, 5, 1)$. Its wavelet coefficients are 4, 1, 1, 2. This means that $\boldsymbol{v}$ can be reconstructed from $\boldsymbol{c} = (4, 1, 1, 2)$ using $\boldsymbol{w}_1, \boldsymbol{w}_2, \boldsymbol{w}_3, \boldsymbol{w}_4$. In matrix form the reconstruction is $\boldsymbol{v} = W\boldsymbol{c}$:

$$\begin{bmatrix} 6 \\ 4 \\ 5 \\ 1 \end{bmatrix} = 4\begin{bmatrix} 1 \\ 1 \\ 1 \\ 1 \end{bmatrix} + \begin{bmatrix} 1 \\ 1 \\ -1 \\ -1 \end{bmatrix} + \begin{bmatrix} 1 \\ -1 \\ 0 \\ 0 \end{bmatrix} + 2\begin{bmatrix} 0 \\ 0 \\ 1 \\ -1 \end{bmatrix} = \begin{bmatrix} 1 & 1 & 1 & 0 \\ 1 & 1 & -1 & 0 \\ 1 & -1 & 0 & 1 \\ 1 & -1 & 0 & -1 \end{bmatrix}\begin{bmatrix} 4 \\ 1 \\ 1 \\ 2 \end{bmatrix}. \tag{4}$$

Those coefficients $\boldsymbol{c} = (4, 1, 1, 2)$ are $W^{-1}\boldsymbol{v}$. Inverting this basis matrix W is easy because the $\boldsymbol{w}$'s in its columns are orthogonal. But they are not unit vectors. So the inverse is the transpose divided by the lengths squared, $W^{-1} = (W^{\mathrm{T}}W)^{-1}W^{\mathrm{T}}$:

$$W^{-1} = \begin{bmatrix} \frac{1}{4} & & & \\ & \frac{1}{4} & & \\ & & \frac{1}{2} & \\ & & & \frac{1}{2} \end{bmatrix}\begin{bmatrix} 1 & 1 & 1 & 1 \\ 1 & 1 & -1 & -1 \\ 1 & -1 & 0 & 0 \\ 0 & 0 & 1 & -1 \end{bmatrix}.$$

From the 1's in the first row of $\boldsymbol{c} = W^{-1}\boldsymbol{v}$, notice that c_1 is the average of v_1, v_2, v_3, v_4:

$$c_1 = \frac{6+4+5+1}{4} = 4.$$

Example 2 (**Same wavelet basis by recursion**) I can't resist showing you a faster way to find the c's. The special point of the wavelet basis is that you can pick off the details in c_3 and c_4, before the coarse details in c_2 and the overall average in c_1. A picture will explain this "multiscale" method, which is in Chapter 1 of my textbook with Nguyen on *Wavelets and Filter Banks*:

Split $\boldsymbol{v} = (6, 4, 5, 1)$ ***into averages and waves at small scale and then large scale:***

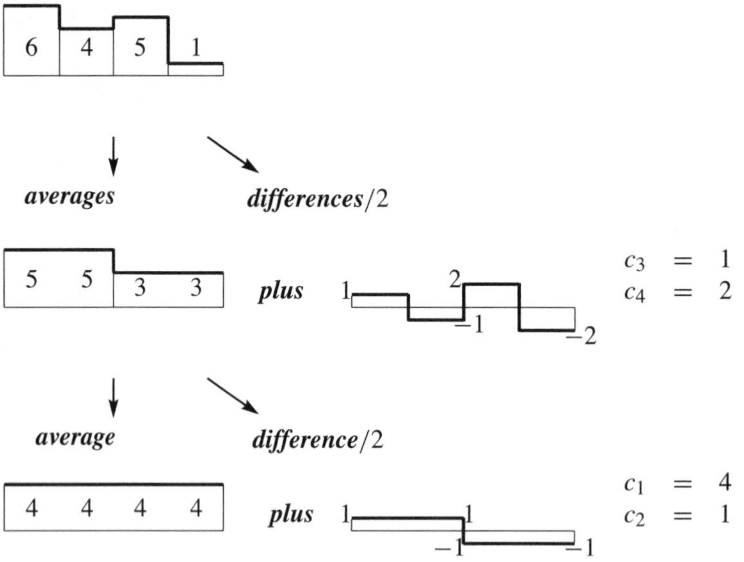

Example 3 (**Fourier basis**) The first thing an electrical engineer does with a signal is to take its Fourier transform. This is a discrete signal (a vector $\boldsymbol{v}$) and we are speaking about its ***Discrete Fourier Transform***. The DFT involves complex numbers. But if we choose $n = 4$, the matrices are small and the only complex numbers are i and i^3.

Notice that $i^3 = -i$ because $i^2 = -1$. A true electrical engineer would write j instead of i. The four basis vectors are in the columns of the Fourier matrix F:

$$F = \begin{bmatrix} 1 & 1 & 1 & 1 \\ 1 & i & i^2 & i^3 \\ 1 & i^2 & i^4 & i^6 \\ 1 & i^3 & i^6 & i^9 \end{bmatrix}.$$

The first column is the useful flat basis vector $(1, 1, 1, 1)$. It represents the average signal or the direct current (the DC term). It is a wave at zero frequency. The third column is $(1, -1, 1, -1)$, which alternates at the highest frequency. *The Fourier transform decomposes the signal into waves at equally spaced frequencies.*

The Fourier matrix F is absolutely the most important complex matrix in mathematics and science and engineering. The last section of this book explains the ***Fast Fourier Transform***: it is a factorization of F into matrices with many zeros. The FFT has revolutionized entire industries, by speeding up the Fourier transform. The beautiful thing is that F^{-1} looks like F, with i changed to $-i$:

$$F^{-1} = \frac{1}{4}\begin{bmatrix} 1 & 1 & 1 & 1 \\ 1 & (-i) & (-i)^2 & (-i)^3 \\ 1 & (-i)^2 & (-i)^4 & (-i)^6 \\ 1 & (-i)^3 & (-i)^6 & (-i)^9 \end{bmatrix} = \frac{1}{4}\overline{F}.$$

The MATLAB command $\boldsymbol{c} = \text{fft}(\boldsymbol{v})$ produces the Fourier coefficients $c_1, \ldots, c_n$ of the vector $\boldsymbol{v}$. It multiplies $\boldsymbol{v}$ by F^{-1} (fast).

The Dual Basis

The columns of W contain the basis vectors $\boldsymbol{w}_1, \ldots, \boldsymbol{w}_n$. To find the coefficients $c_1, \ldots, c_n$ of a vector in this basis, we use the matrix W^{-1}. This subsection just introduces a notation and a new word for the rows of W^{-1}. The vectors in those rows (call them $\boldsymbol{u}_1^{\mathrm{T}}, \ldots, \boldsymbol{u}_n^{\mathrm{T}}$) are the *dual basis*.

The properties of the dual basis reflect $W^{-1}W = I$ and also $WW^{-1} = I$. The product $W^{-1}W$ takes rows of W^{-1} times columns of W, in other words dot products of the $\boldsymbol{u}$'s with the $\boldsymbol{w}$'s. The two bases are "biorthogonal" because we get 1's and 0's:

$$W^{-1}W = \begin{bmatrix} \boldsymbol{u}_1^{\mathrm{T}} \\ \vdots \\ \boldsymbol{u}_n^{\mathrm{T}} \end{bmatrix}\begin{bmatrix} \boldsymbol{w}_1 & \cdots & \boldsymbol{w}_n \end{bmatrix} = I \quad \text{so} \quad \boldsymbol{u}_i^{\mathrm{T}}\boldsymbol{w}_j = \begin{cases} 1 & \text{if } i = j \\ 0 & \text{if } i \neq j \end{cases}$$

For an orthonormal basis, the $\boldsymbol{u}$'s are the same as the $\boldsymbol{w}$'s. We have been calling them $\boldsymbol{q}$'s. The basis of $\boldsymbol{q}$'s is biorthogonal to itself! The rows in W^{-1} are the same as the columns in W. In other words $W^{-1} = W^{\mathrm{T}}$. That is the specially important case of an *orthogonal matrix.*

Other bases are not orthonormal. The axes don't have to be perpendicular. The basis matrix W can be invertible without having orthogonal columns.

When the inverse matrices are in the opposite order $WW^{-1} = I$, we learn something new. The columns are $\boldsymbol{w}_j$, the rows are $\boldsymbol{u}_i^{\mathrm{T}}$, and each product is a rank one matrix. *Multiply columns times rows*:

$$WW^{-1} = \begin{bmatrix} & & \\ \boldsymbol{w}_1 & \cdots & \boldsymbol{w}_n \\ & & \end{bmatrix} \begin{bmatrix} \boldsymbol{u}_1^{\mathrm{T}} \\ \vdots \\ \boldsymbol{u}_n^{\mathrm{T}} \end{bmatrix} = \boldsymbol{w}_1\boldsymbol{u}_1^{\mathrm{T}} + \cdots + \boldsymbol{w}_n\boldsymbol{u}_n^{\mathrm{T}} = I.$$

WW^{-1} is the order that we constantly use to change the basis. The coefficients are in $\boldsymbol{c} = W^{-1}\boldsymbol{v}$. So W^{-1} is the first (with the $\boldsymbol{u}_i^{\mathrm{T}}$ in its rows). Then we reconstruct $\boldsymbol{v}$ from $W\boldsymbol{c}$. Use the $\boldsymbol{u}$'s and $\boldsymbol{w}$'s to state the basic facts that $\boldsymbol{c} = W^{-1}\boldsymbol{v}$ and $\boldsymbol{v} = W\boldsymbol{c} = WW^{-1}\boldsymbol{v}$:

The coefficients are $c_i = \boldsymbol{u}_i^{\mathrm{T}}\boldsymbol{v}$ and the vector is $v = \sum_1^n \boldsymbol{w}_i(\boldsymbol{u}_i^{\mathrm{T}}\boldsymbol{v})$. (5)

The analysis step takes dot products of $\boldsymbol{v}$ with the dual basis to find the c's. The synthesis step adds up the pieces $c_i\boldsymbol{w}_i$ to reconstruct the vector $\boldsymbol{v}$.

■ REVIEW OF THE KEY IDEAS ■

1. The new basis vectors $\boldsymbol{w}_j$ are the columns of an invertible matrix W.

2. The coefficients of $\boldsymbol{v}$ in this new basis are $\boldsymbol{c} = W^{-1}\boldsymbol{v}$ (the analysis step).

3. The vector $\boldsymbol{v}$ is reconstructed as $W\boldsymbol{c} = c_1\boldsymbol{w}_1 + \cdots + c_n\boldsymbol{w}_n$ (the synthesis step).

4. Compression would simplify $\boldsymbol{c}$ to $\widehat{\boldsymbol{c}}$ and we reconstruct $\widehat{\boldsymbol{v}} = \widehat{c}_1\boldsymbol{w}_1 + \cdots + \widehat{c}_n\boldsymbol{w}_n$.

5. The rows of W^{-1} are the dual basis vectors $\boldsymbol{u}_i$ and $c_i = \boldsymbol{u}_i^{\mathrm{T}}\boldsymbol{v}$. Then $\boldsymbol{u}_i^{\mathrm{T}}\boldsymbol{w}_j = \delta_{ij}$.

■ WORKED EXAMPLES ■

7.3 A Match $a_0 + a_1x + a_2x^2$ with $b_0 + b_1(x+1) + b_2(x+1)^2$, to find the 3 by 3 matrix M_1 that connects these coefficients by $\boldsymbol{a} = M_1\boldsymbol{b}$. M_1 will be familiar to Pascal!

The matrix to reverse that change must be M_1^{-1}, and $\boldsymbol{b} = M_1^{-1}\boldsymbol{a}$. This shifts the center of the series back, so $a_0 + a_1(x-1) + a_2(x-1)^2$ equals $b_0 + b_1x + b_2x^2$. Match those quadratics to find M_{-1}, the inverse of Pascal. Also find M_t from $a_0 + a_1x + a_2x^2 = b_0 + b_1(x+t) + b_2(x+t)^2$. Verify that $M_sM_t = M_{s+t}$.

Solution Match $a_0 + a_1x + a_2x$ with $b_0 + b_1(x+1) + b_2(x+1)^2$ to find M_1:

$$\begin{array}{lll} \text{Constant term} & a_0 = & b_0 + b_1 + \ b_2 \\ \text{Coefficient of } x & a_1 = & \qquad b_1 + 2b_2 \\ \text{Coefficient of } x^2 & a_2 = & \qquad\qquad b_2 \end{array} \qquad \begin{bmatrix} a_0 \\ a_1 \\ a_2 \end{bmatrix} = \begin{bmatrix} 1 & 1 & 1 \\ & 1 & 2 \\ & & 1 \end{bmatrix} \begin{bmatrix} b_0 \\ b_1 \\ b_2 \end{bmatrix}$$

By writing $(x+1)^2 = 1 + 2x + x^2$ we see $1, 2, 1$ in this change of basis matrix.

The matrix M_1 is Pascal's upper triangular P_U. Its inverse M_1^{-1} comes by matching $a_0 + a_1(x-1) + a_2(x-1)^2$ with $b_0 + b_1x + b_2x^2$. The constant terms are the same if $a_0 - a_1 + a_2 = b_0$. This gives alternating signs in $M_1^{-1} = M_{-1}$.

$$\textbf{Inverse of } \mathbf{M_1} = M_{-1} = \begin{bmatrix} 1 & -1 & 1 \\ & 1 & -2 \\ & & 1 \end{bmatrix} \qquad \textbf{Shift by t} \quad M_t = \begin{bmatrix} 1 & t & t^2 \\ & 1 & 2t \\ & & 1 \end{bmatrix}.$$

$M_sM_t = M_{s+t}$ and $M_1M_{-1} = M_0 = I$. Pascal fans might wonder if his symmetric matrix P_S also appears in a change of basis. It does, when the new basis has negative powers $(x+1)^{-k}$ (more about this on the course website **web.mit.edu/18.06/www**).

Problem Set 7.3

1 Express the vectors $\boldsymbol{e} = (1,0,0,0)$ and $\boldsymbol{v} = (1,-1,1,-1)$ in the wavelet basis, as in equation (4). The coefficients c_1, c_2, c_3, c_4 solve $W\boldsymbol{c} = \boldsymbol{e}$ and $W\boldsymbol{c} = \boldsymbol{v}$.

2 Follow Example 2 to represent $\boldsymbol{v} = (7,5,3,1)$ in the wavelet basis. Start with

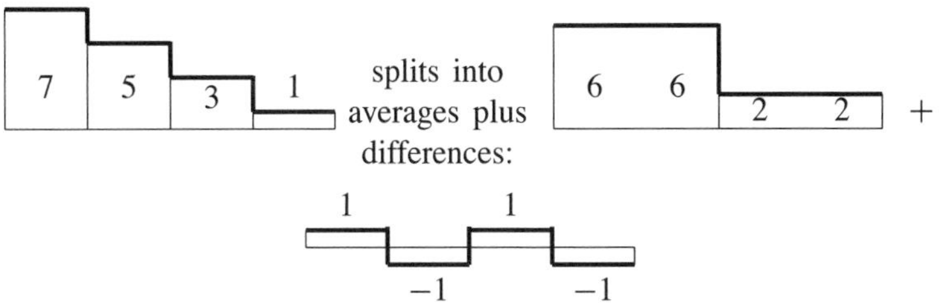

The last step writes 6, 6, 2, 2 as an overall average plus a difference, using $1, 1, 1, 1$ and $1, 1, -1, -1$.

3 What are the eight vectors in the wavelet basis for $\mathbf{R}^8$? They include the long wavelet $(1,1,1,1,-1,-1,-1,-1)$ and the short wavelet $(1,-1,0,0,0,0,0,0)$.

4 The wavelet basis matrix W factors into simpler matrices W_1 and W_2:

$$\begin{bmatrix} 1 & 1 & 1 & 0 \\ 1 & 1 & -1 & 0 \\ 1 & -1 & 0 & 1 \\ 1 & -1 & 0 & -1 \end{bmatrix} = \begin{bmatrix} 1 & 0 & 1 & 0 \\ 1 & 0 & -1 & 0 \\ 0 & 1 & 0 & 1 \\ 0 & 1 & 0 & -1 \end{bmatrix} \begin{bmatrix} 1 & 1 & 0 & 0 \\ 1 & -1 & 0 & 0 \\ 0 & 0 & 1 & 0 \\ 0 & 0 & 0 & 1 \end{bmatrix}.$$

Then $W^{-1} = W_2^{-1} W_1^{-1}$ allows $\boldsymbol{c}$ to be computed in two steps. The first splitting in Example 2 shows $W_1^{-1}\boldsymbol{v}$. Then the second splitting applies W_2^{-1}. Find those inverse matrices W_1^{-1} and W_2^{-1} directly from W_1 and W_2. Apply them to $\boldsymbol{v} = (6, 4, 5, 1)$.

5 The 4 by 4 *Hadamard matrix* is like the wavelet matrix but entirely $+1$ and -1:

$$H = \begin{bmatrix} 1 & 1 & 1 & 1 \\ 1 & -1 & 1 & -1 \\ 1 & 1 & -1 & -1 \\ 1 & -1 & -1 & 1 \end{bmatrix}.$$

Find H^{-1} and write $\boldsymbol{v} = (7, 5, 3, 1)$ as a combination of the columns of H.

6 Suppose we have two bases $\boldsymbol{v}_1, \ldots, \boldsymbol{v}_n$ and $\boldsymbol{w}_1, \ldots, \boldsymbol{w}_n$ for $\mathbf{R}^n$. If a vector has coefficients b_i in one basis and c_i in the other basis, what is the change of basis matrix in $\boldsymbol{b} = M\boldsymbol{c}$? Start from

$$b_1\boldsymbol{v}_1 + \cdots + b_n\boldsymbol{v}_n = V\boldsymbol{b} = c_1\boldsymbol{w}_1 + \cdots + c_n\boldsymbol{w}_n = W\boldsymbol{c}.$$

Your answer represents $T(\boldsymbol{v}) = \boldsymbol{v}$ with input basis of $\boldsymbol{v}$'s and output basis of $\boldsymbol{w}$'s. Because of different bases, the matrix is not I.

7 The dual basis vectors $\boldsymbol{w}_1^*, \ldots, \boldsymbol{w}_n^*$ are the columns of $W^* = (W^{-1})^{\mathrm{T}}$. Show that the original basis $\boldsymbol{w}_1, \ldots, \boldsymbol{w}_n$ is "the dual of the dual." In other words, show that the $\boldsymbol{w}$'s are the rows of $(W^*)^{-1}$. **Hint:** Transpose the equation $WW^{-1} = I$.

DIAGONALIZATION AND THE PSEUDOINVERSE ■ 7.4

This short section combines the ideas from Section 7.2 (matrix of a linear transformation) and Section 7.3 (change of basis). The combination produces a needed result: *the change of matrix due to change of basis.* The matrix depends on the input basis and output basis. We want to produce a better matrix than A, by choosing a better basis than the standard basis.

By reversing the input and output bases, we will find the pseudoinverse A^+. It sends $\mathbf{R}^m$ back to $\mathbf{R}^n$, column space back to row space.

The truth is that all our great factorizations of A can be regarded as a change of basis. But this is a short section, so we concentrate on the two outstanding examples. In both cases the good matrix is *diagonal.* It is either Λ or Σ:

1. $S^{-1}AS = \Lambda$ *when the input and output bases are eigenvectors of* A.

2. $U^{-1}AV = \Sigma$ *when the input and output bases are eigenvectors of* $A^{\mathrm{T}}A$ *and* AA^{T}.

You see immediately the difference between Λ and Σ. In Λ the bases are the same. The matrix A must be square. And some square matrices cannot be diagonalized by any S, because they don't have n independent eigenvectors.

In Σ the input and output bases are different. The matrix A can be rectangular. The bases are *orthonormal* because $A^{\mathrm{T}}A$ and AA^{T} are symmetric. Then $U^{-1} = U^{\mathrm{T}}$ and $V^{-1} = V^{\mathrm{T}}$. Every matrix A is allowed, and can be diagonalized. This is the Singular Value Decomposition (**SVD**) of Section 6.7.

I will just note that the Gram-Schmidt factorization $A = QR$ chooses only *one* new basis. That is the orthogonal output basis given by Q. The input uses the standard basis given by I. We don't reach a diagonal Σ, but we do reach a triangular R. The output basis matrix appears on the left and the input basis appears on the right, in $A = QRI$.

We start with input basis equal to output basis. That will produce S and S^{-1}.

Similar Matrices: A and $S^{-1}AS$ and $W^{-1}AW$

We begin with a square matrix and one basis. The input space $\mathbf{V}$ is $\mathbf{R}^n$ and the output space $\mathbf{W}$ is also $\mathbf{R}^n$. The standard basis vectors are the columns of I. The matrix is n by n, and we call it A. The linear transformation T is "multiplication by A".

Most of this book has been about one fundamental problem—*to make the matrix simple.* We made it triangular in Chapter 2 (by elimination) and Chapter 4 (by Gram-Schmidt). We made it diagonal in Chapter 6 (*by eigenvectors*). Now that change from A to Λ comes from a ***change of basis***.

Here are the main facts in advance. When you change the basis for $\mathbf{V}$, the matrix changes from A to AM. Because $\mathbf{V}$ is the input space, the matrix M goes on the right (to come first). When you change the basis for $\mathbf{W}$, the new matrix is $M^{-1}A$. We are working with the output space so M^{-1} is on the left (to come last). ***If you change both bases in the same way, the new matrix is*** $M^{-1}AM$. The good basis vectors are the eigenvectors of A, in the columns of $M = S$. The matrix becomes $S^{-1}AS = \Lambda$.

7D When the basis contains the eigenvectors $\boldsymbol{x}_1, \ldots, \boldsymbol{x}_n$, the matrix for T becomes Λ.

Reason To find column 1 of the matrix, input the first basis vector $\boldsymbol{x}_1$. The transformation multiplies by A. The output is $A\boldsymbol{x}_1 = \lambda_1\boldsymbol{x}_1$. This is λ_1 times the first basis vector plus zero times the other basis vectors. Therefore the first column of the matrix is $(\lambda_1, 0, \ldots, 0)$. ***In the eigenvector basis, the matrix is diagonal.***

Example 1 Find the diagonal matrix that projects onto the 135° line $y = -x$. The standard basis $(1, 0)$ and $(0, 1)$ is projected to $(.5, -.5)$ and $(-.5, .5)$

$$\textbf{Standard matrix} \qquad A = \begin{bmatrix} .5 & -.5 \\ -.5 & .5 \end{bmatrix}.$$

Solution The eigenvectors for this projection are $\boldsymbol{x}_1 = (1, -1)$ and $\boldsymbol{x}_2 = (1, 1)$. The first eigenvector lies on the 135° line and the second is perpendicular.
Their projections are $\boldsymbol{x}_1$ and $\mathbf{0}$. The eigenvalues are $\lambda_1 = 1$ and $\lambda_2 = 0$. In the eigenvector basis, $P\boldsymbol{x}_1 = \boldsymbol{x}_1$ and $P\boldsymbol{x}_2 = \mathbf{0}$ go into the columns of Λ:

$$\textbf{Diagonalized matrix} \qquad \Lambda = \begin{bmatrix} 1 & 0 \\ 0 & 0 \end{bmatrix}.$$

What if you choose another basis like $\boldsymbol{v}_1 = \boldsymbol{w}_1 = (2, 0)$ and $\boldsymbol{v}_2 = \boldsymbol{w}_2 = (1, 1)$? Since $\boldsymbol{w}_1$ is not an eigenvector, the matrix B in this basis will not be diagonal. The first way to compute B follows the rule of Section 7.2: Find column j of the matrix by writing the output $A\boldsymbol{v}_j$ as a combination of $\boldsymbol{w}$'s.

Apply the projection T to $(2, 0)$. The result is $(1, -1)$ which is $\boldsymbol{w}_1 - \boldsymbol{w}_2$. So the first column of B contains 1 and -1. The second vector $\boldsymbol{w}_2 = (1, 1)$ projects to zero, so the second column of B contains 0 and 0:

$$\text{The matrix is } B = \begin{bmatrix} 1 & 0 \\ -1 & 0 \end{bmatrix} \text{ in the basis } \boldsymbol{w}_1, \boldsymbol{w}_2. \tag{1}$$

The second way to find the same B is more insightful. Use W^{-1} and W to change between the standard basis and the basis of $\boldsymbol{w}$'s. Those change of basis matrices from Section 7.3 are representing the identity transformation. The product of transformations is just ITI, and the product of matrices is $B = W^{-1}AW$. *B is **similar** to A.*

7E For any basis $\boldsymbol{w}_1, \ldots, \boldsymbol{w}_n$ find the matrix B in three steps. Change the input basis to the standard basis with W. The matrix in the standard basis is A. Then change the output basis back to the $\boldsymbol{w}$'s with W^{-1}. The product $B = W^{-1}AW$ represents ITI:

$$B_{\boldsymbol{w}\text{'s to }\boldsymbol{w}\text{'s}} = W^{-1}_{\text{standard to }\boldsymbol{w}\text{'s}} \; A_{\text{standard}} \; W_{\boldsymbol{w}\text{'s to standard}} \tag{2}$$

Example 2 (continuing with the projection) Apply this $W^{-1}AW$ rule to find B, when the basis $(2,0)$ and $(1,1)$ is in the columns of W:

$$W^{-1}AW = \begin{bmatrix} \frac{1}{2} & -\frac{1}{2} \\ 0 & 1 \end{bmatrix} \begin{bmatrix} \frac{1}{2} & -\frac{1}{2} \\ -\frac{1}{2} & \frac{1}{2} \end{bmatrix} \begin{bmatrix} 2 & 1 \\ 0 & 1 \end{bmatrix} = \begin{bmatrix} 1 & 0 \\ -1 & 0 \end{bmatrix}.$$

The $W^{-1}AW$ rule has produced the same B as in equation (1). ***A change of basis produces a similarity transformation in the matrix.*** The matrices A and B are similar. They have the same eigenvalues (1 and 0). And Λ is similar too.

The Singular Value Decomposition (SVD)

Now the input basis $\boldsymbol{v}_1, \ldots, \boldsymbol{v}_n$ can be different from the output basis $\boldsymbol{u}_1, \ldots, \boldsymbol{u}_m$. In fact the input space $\mathbf{R}^n$ can be different from the output space $\mathbf{R}^m$. Again the best matrix is diagonal (now m by n). To achieve this diagonal matrix Σ, each input vector $\boldsymbol{v}_j$ must transform into a multiple of the output vector $\boldsymbol{u}_j$. That multiple is the *singular value* σ_j on the main diagonal of Σ:

$$\textbf{SVD} \qquad A\boldsymbol{v}_j = \begin{cases} \sigma_j \boldsymbol{u}_j & \text{for } j \le r \\ \mathbf{0} & \text{for } j > r \end{cases} \qquad \text{with orthonormal bases.} \tag{3}$$

The singular values are in the order $\sigma_1 \ge \sigma_2 \ge \cdots \ge \sigma_r$. The rank r enters because (by definition) singular values are not zero. The second part of the equation says that $\boldsymbol{v}_j$ is in the nullspace for $j = r+1, \ldots, n$. This gives the correct number $n-r$ of basis vectors for the nullspace.

Let me connect the matrices A and Σ and V and U with the linear transformations they represent. The matrices A and Σ represent the *same transformation*. $A = U\Sigma V^{\mathrm{T}}$ uses the standard bases for $\mathbf{R}^n$ and $\mathbf{R}^m$. The diagonal Σ uses the input basis of $\boldsymbol{v}$'s and the output basis of $\boldsymbol{u}$'s. The orthogonal matrices V and U give the basis changes; they represent the identity transformations (in $\mathbf{R}^n$ and $\mathbf{R}^m$). The product of transformations is ITI, and it is represented in the $\boldsymbol{v}$ and $\boldsymbol{u}$ bases by $U^{-1}AV$ which is Σ:

7F The matrix Σ in the new bases comes from A in the standard bases by $U^{-1}AV$:

$$\Sigma_{\boldsymbol{v}\text{'s to } \boldsymbol{u}\text{'s}} = U^{-1}_{\text{standard to } \boldsymbol{u}\text{'s}} \; A_{\text{standard}} \; V_{\boldsymbol{v}\text{'s to standard}}. \tag{4}$$

The SVD chooses orthonormal bases ($U^{-1} = U^{\mathrm{T}}$ and $V^{-1} = V^{\mathrm{T}}$) that diagonalize A.

The two orthonormal bases in the SVD are the eigenvector bases for $A^{\mathrm{T}}A$ (the $\boldsymbol{v}$'s) and AA^{T} (the $\boldsymbol{u}$'s). Since those are symmetric matrices, their unit eigenvectors are orthonormal. Their eigenvalues are the numbers σ_j^2. Equations (10) and (11) in Section 6.7 proved that those bases diagonalize the standard matrix A to produce Σ.

Polar Decomposition

Every complex number has the polar form $re^{i\theta}$. A nonnegative number r multiplies a number on the unit circle. (Remember that $|e^{i\theta}| = |\cos\theta + i\sin\theta| = 1$.) Thinking of these numbers as 1 by 1 matrices, $r \geq 0$ corresponds to a *positive semidefinite matrix* (call it H) and $e^{i\theta}$ corresponds to an *orthogonal matrix* Q. The ***polar decomposition*** extends this $re^{i\theta}$ factorization to matrices.

7G Every real square matrix can be factored into $A = QH$, where Q is ***orthogonal*** and H is ***symmetric positive semidefinite***. If A is invertible then H is positive definite.

For the proof we just insert $V^{\mathrm{T}}V = I$ into the middle of the SVD:

$$A = U\Sigma V^{\mathrm{T}} = (UV^{\mathrm{T}})(V\Sigma V^{\mathrm{T}}) = (Q)(H). \tag{5}$$

The first factor UV^{T} is Q. The product of orthogonal matrices is orthogonal. The second factor $V\Sigma V^{\mathrm{T}}$ is H. It is positive semidefinite because its eigenvalues are in Σ. If A is invertible then Σ and H are also invertible. ***H is the symmetric positive definite square root of $A^{\mathrm{T}}A$.*** Equation (5) says that $H^2 = V\Sigma^2V^{\mathrm{T}} = A^{\mathrm{T}}A$.

There is also a polar decomposition $A = KQ$ in the reverse order. Q is the same but now $K = U\Sigma U^{\mathrm{T}}$. This is the symmetric positive definite square root of AA^{T}.

Example 3 Find the polar decomposition $A = QH$ from its SVD in Section 6.7:

$$A = \begin{bmatrix} 2 & 2 \\ -1 & 1 \end{bmatrix} = \begin{bmatrix} 0 & 1 \\ 1 & 0 \end{bmatrix} \begin{bmatrix} \sqrt{2} & \\ & 2\sqrt{2} \end{bmatrix} \begin{bmatrix} -1/\sqrt{2} & 1/\sqrt{2} \\ 1/\sqrt{2} & 1/\sqrt{2} \end{bmatrix} = U\Sigma V^{\mathrm{T}}.$$

Solution The orthogonal part is $Q = UV^{\mathrm{T}}$. The positive definite part is $H = V\Sigma V^{\mathrm{T}}$. This is also $H = Q^{-1}A$ which is $Q^{\mathrm{T}}A$:

$$Q = \begin{bmatrix} 0 & 1 \\ 1 & 0 \end{bmatrix} \begin{bmatrix} -1/\sqrt{2} & 1/\sqrt{2} \\ 1/\sqrt{2} & 1/\sqrt{2} \end{bmatrix} = \begin{bmatrix} 1/\sqrt{2} & 1/\sqrt{2} \\ -1/\sqrt{2} & 1/\sqrt{2} \end{bmatrix}$$

$$H = \begin{bmatrix} 1/\sqrt{2} & -1/\sqrt{2} \\ 1/\sqrt{2} & 1/\sqrt{2} \end{bmatrix} \begin{bmatrix} 2 & 2 \\ -1 & 1 \end{bmatrix} = \begin{bmatrix} 3/\sqrt{2} & 1/\sqrt{2} \\ 1/\sqrt{2} & 3/\sqrt{2} \end{bmatrix}.$$

In mechanics, the polar decomposition separates the ***rotation*** (in Q) from the ***stretching*** (in H). The eigenvalues of H are the singular values of A. They give the stretching factors. The eigenvectors of H are the eigenvectors of $A^{\mathrm{T}}A$. They give the stretching directions (the principal axes). Then Q rotates the axes.

The polar decomposition just splits the key equation $A\boldsymbol{v}_i = \sigma_i\boldsymbol{u}_i$ into two steps. The "H" part multiplies $\boldsymbol{v}_i$ by σ_i. The "Q" part swings $\boldsymbol{v}_i$ around into $\boldsymbol{u}_i$.

The Pseudoinverse

By choosing good bases, A multiplies $\boldsymbol{v}_i$ in the row space to give $\sigma_i \boldsymbol{u}_i$ in the column space. A^{-1} must do the opposite! If $A\boldsymbol{v} = \sigma \boldsymbol{u}$ then $A^{-1}\boldsymbol{u} = \boldsymbol{v}/\sigma$. The singular values of A^{-1} are $1/\sigma$, just as the eigenvalues of A^{-1} are $1/\lambda$. The bases are reversed. The $\boldsymbol{u}$'s are in the row space of A^{-1}, the $\boldsymbol{v}$'s are in the column space.

Until this moment we would have added "*if A^{-1} exists.*" Now we don't. A matrix that multiplies $\boldsymbol{u}_i$ to produce $\boldsymbol{v}_i/\sigma_i$ *does* exist. It is the pseudoinverse A^+:

Pseudoinverse

$$A^+ = V\Sigma^+U^{\mathrm{T}} = \underset{n \text{ by } n}{\begin{bmatrix} \boldsymbol{v}_1 \cdots \boldsymbol{v}_r \cdots \boldsymbol{v}_n \end{bmatrix}} \underset{n \text{ by } m}{\begin{bmatrix} \sigma_1^{-1} & & & \\ & \ddots & & \\ & & \sigma_r^{-1} & \\ & & & \end{bmatrix}} \underset{m \text{ by } m}{\begin{bmatrix} \boldsymbol{u}_1 \cdots \boldsymbol{u}_r \cdots \boldsymbol{u}_m \end{bmatrix}^{\mathrm{T}}}$$

The ***pseudoinverse*** A^+ is an n by m matrix. If A^{-1} exists (we said it again), then A^+ is the same as A^{-1}. In that case $m = n = r$ and we are inverting $U\Sigma V^{\mathrm{T}}$ to get $V\Sigma^{-1}U^{\mathrm{T}}$. The new symbol A^+ is needed when $r < m$ or $r < n$. Then A has no two-sided inverse, but it has a *pseudo*inverse A^+ with that same rank r:

$$A^+\boldsymbol{u}_i = \frac{1}{\sigma_i}\boldsymbol{v}_i \quad \text{for } i \le r \qquad \text{and} \qquad A^+\boldsymbol{u}_i = \boldsymbol{0} \quad \text{for } i > r.$$

The vectors $\boldsymbol{u}_1, \ldots, \boldsymbol{u}_r$ in the column space of A go back to the row space. The other vectors $\boldsymbol{u}_{r+1}, \ldots, \boldsymbol{u}_m$ are in the left nullspace, and A^+ sends them to zero. When we know what happens to each basis vector $\boldsymbol{u}_i$, we know A^+.

Notice the pseudoinverse Σ^+ of the diagonal matrix Σ. Each σ is replaced by σ^{-1}. The product $\Sigma^+\Sigma$ is as near to the identity as we can get. We get r 1's. We can't do anything about the zero rows and columns! This example has $\sigma_1 = 2$ and $\sigma_2 = 3$:

$$\Sigma^+\Sigma = \begin{bmatrix} 1/2 & 0 & 0 \\ 0 & 1/3 & 0 \\ 0 & 0 & 0 \end{bmatrix} \begin{bmatrix} 2 & 0 & 0 \\ 0 & 3 & 0 \\ 0 & 0 & 0 \end{bmatrix} = \begin{bmatrix} 1 & 0 & 0 \\ 0 & 1 & 0 \\ 0 & 0 & 0 \end{bmatrix}.$$

7H The pseudoinverse A^+ is the n by m matrix with these two properties:

$AA^+ =$ projection matrix onto the column space of A
$A^+A =$ projection matrix onto the row space of A

Example 4 Find the pseudoinverse of $A = \begin{bmatrix} 2 & 2 \\ 1 & 1 \end{bmatrix}$. This matrix is not invertible. The rank is 1. The only singular value is $\sqrt{10}$. That is inverted to $1/\sqrt{10}$ in Σ^+:

$$A^+ = V\Sigma^+U^{\mathrm{T}} = \frac{1}{\sqrt{2}}\begin{bmatrix} 1 & 1 \\ 1 & -1 \end{bmatrix} \begin{bmatrix} 1/\sqrt{10} & 0 \\ 0 & 0 \end{bmatrix} \frac{1}{\sqrt{5}} \begin{bmatrix} 2 & 1 \\ 1 & -2 \end{bmatrix} = \frac{1}{10}\begin{bmatrix} 2 & 1 \\ 2 & 1 \end{bmatrix}.$$

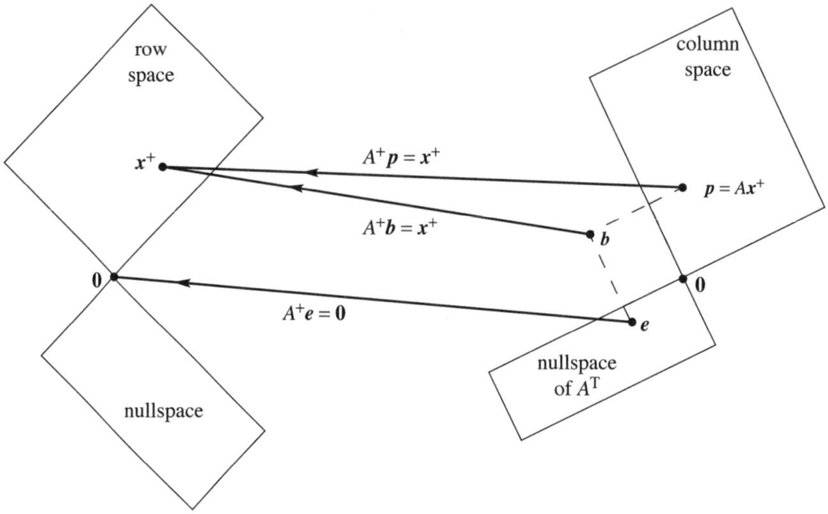

Figure 7.4 A is invertible from row space to column space. A^+ inverts it.

A^+ also has rank 1. Its column space is the row space of A. When A takes $(1, 1)$ in the row space to $(4, 2)$ in the column space, A^+ does the reverse. Every rank one matrix is a column times a row. With unit vectors $\boldsymbol{u}$ and $\boldsymbol{v}$, that is $A = \sigma \boldsymbol{u}\boldsymbol{v}^T$. Then the best inverse of a rank one matrix is $A^+ = \boldsymbol{v}\boldsymbol{u}^T/\sigma$.

The product AA^+ is $\boldsymbol{u}\boldsymbol{u}^T$, the projection onto the line through $\boldsymbol{u}$. The product A^+A is $\boldsymbol{v}\boldsymbol{v}^T$, the projection onto the line through $\boldsymbol{v}$. ***For all matrices,*** AA^+ ***and*** A^+A ***are the projections onto the column space and row space.***

The shortest least squares solution to $A\boldsymbol{x} = \boldsymbol{b}$ is $\boldsymbol{x}^+ = A^+\boldsymbol{b}$. Any other vector that solves $A^TA\widehat{\boldsymbol{x}} = A^T\boldsymbol{b}$ is longer than $\boldsymbol{x}^+$ (Problem 18).

■ REVIEW OF THE KEY IDEAS ■

1. Diagonalization $S^{-1}AS = \Lambda$ is the same as a change to the eigenvector basis.

2. The SVD chooses an input basis of $\boldsymbol{v}$'s and an output basis of $\boldsymbol{u}$'s. Those orthonormal bases diagonalize A. This is $A\boldsymbol{v}_i = \sigma_i \boldsymbol{u}_i$, and $A = U\Sigma V^T$.

3. Polar decomposition factors A into QH, rotation times stretching.

4. The pseudoinverse $A^+ = V\Sigma^+U^T$ transforms the column space of A back to its row space. A^+A is the identity on the row space (and zero on the nullspace).

■ WORKED EXAMPLES ■

7.4 A Start with an m by n matrix A. If its rank is n (full column rank) then it has a *left inverse* $C = (A^{\mathrm{T}}A)^{-1}A^{\mathrm{T}}$. This matrix C gives $CA = I$. Explain why the pseudoinverse is $A^+ = C$ in this case. If A has rank m (full row rank) then it has a *right inverse* B with $B = A^{\mathrm{T}}(AA^{\mathrm{T}})^{-1}$ and $AB = I$. Explain why $A^+ = B$ in this case.

Find B and C if possible and find A^+ for all three matrices:

$$A_1 = \begin{bmatrix} 2 \\ 2 \end{bmatrix} \quad A_2 = \begin{bmatrix} 2 & 2 \end{bmatrix} \quad A_3 = \begin{bmatrix} 2 & 2 \\ 2 & 2 \end{bmatrix}.$$

Solution If A has rank n (independent columns) then $A^{\mathrm{T}}A$ is invertible—this is a key point of Section 4.2. Certainly $C = (A^{\mathrm{T}}A)^{-1}A^{\mathrm{T}}$ multiplies A to give $CA = I$. In the opposite order, $AC = A(A^{\mathrm{T}}A)^{-1}A^{\mathrm{T}}$ is the projection matrix (Section 4.2 again) onto the column space. So C meets the requirements **7H** to be A^+.

If A has rank m (full row rank) then AA^{T} is invertible. Certainly A multiplies $B = A^{\mathrm{T}}(AA^{\mathrm{T}})^{-1}$ to give $AB = I$. In the opposite order, $BA = A^{\mathrm{T}}(AA^{\mathrm{T}})^{-1}A$ is the projection matrix onto the row space. So B is the pseudoinverse A^+.

The example A_1 has full column rank (for C) and A_2 has full row rank (for B):

$$A_1^+ = (A_1^{\mathrm{T}}A_1)^{-1}A_1^{\mathrm{T}} = \frac{1}{\sqrt{8}}\begin{bmatrix} 2 & 2 \end{bmatrix} \qquad A_2^+ = A_2^{\mathrm{T}}(A_2A_2^{\mathrm{T}})^{-1} = \frac{1}{\sqrt{8}}\begin{bmatrix} 2 \\ 2 \end{bmatrix}.$$

Notice $A_1^+A_1 = [1]$ and $A_2A_2^+ = [1]$. But A_3 has no left or right inverse. Its pseudoinverse is $A_3^+ = \sigma_1^{-1}\boldsymbol{v}_1\boldsymbol{u}_1^{\mathrm{T}} = \left[\begin{smallmatrix} 1 & 1 \\ 1 & 1 \end{smallmatrix}\right]/4$.

Problem Set 7.4

Problems 1–6 compute and use the SVD of a particular matrix (not invertible).

1 Compute $A^{\mathrm{T}}A$ and its eigenvalues and unit eigenvectors $\boldsymbol{v}_1$ and $\boldsymbol{v}_2$:

$$A = \begin{bmatrix} 1 & 2 \\ 3 & 6 \end{bmatrix}$$

What is the only singular value σ_1? The rank of A is $r = 1$.

2 (a) Compute AA^{T} and its eigenvalues and unit eigenvectors $\boldsymbol{u}_1$ and $\boldsymbol{u}_2$.

(b) Verify from Problem 1 that $A\boldsymbol{v}_1 = \sigma_1\boldsymbol{u}_1$. Put numbers into the SVD:

$$\begin{bmatrix} 1 & 2 \\ 3 & 6 \end{bmatrix} = \begin{bmatrix} \boldsymbol{u}_1 & \boldsymbol{u}_2 \end{bmatrix}\begin{bmatrix} \sigma_1 & \\ & 0 \end{bmatrix}\begin{bmatrix} \boldsymbol{v}_1 & \boldsymbol{v}_2 \end{bmatrix}^{\mathrm{T}}.$$

3 From the u's and v's write down orthonormal bases for the four fundamental subspaces of this matrix A.

4 Describe all matrices that have those same four subspaces.

5 From U, V, and Σ find the orthogonal matrix $Q = UV^{\mathrm{T}}$ and the symmetric matrix $H = V\Sigma V^{\mathrm{T}}$. Verify the polar decomposition $A = QH$. This H is only semidefinite because ______.

6 Compute the pseudoinverse $A^+ = V\Sigma^+U^{\mathrm{T}}$. The diagonal matrix Σ^+ contains $1/\sigma_1$. Rename the four subspaces (for A) in Figure 7.4 as four subspaces for A^+. Compute A^+A and AA^+.

Problems 7–11 are about the SVD of an invertible matrix.

7 Compute $A^{\mathrm{T}}A$ and its eigenvalues and unit eigenvectors v_1 and v_2. What are the singular values σ_1 and σ_2 for this matrix A?

$$A = \begin{bmatrix} 3 & 3 \\ -1 & 1 \end{bmatrix}.$$

8 AA^{T} has the same eigenvalues σ_1^2 and σ_2^2 as $A^{\mathrm{T}}A$. Find unit eigenvectors u_1 and u_2. Put numbers into the SVD:

$$A = \begin{bmatrix} 3 & 3 \\ -1 & 1 \end{bmatrix} = \begin{bmatrix} u_1 & u_2 \end{bmatrix} \begin{bmatrix} \sigma_1 & \\ & \sigma_2 \end{bmatrix} \begin{bmatrix} v_1 & v_2 \end{bmatrix}^{\mathrm{T}}.$$

9 In Problem 8, multiply columns times rows to show that $A = \sigma_1 u_1 v_1^{\mathrm{T}} + \sigma_2 u_2 v_2^{\mathrm{T}}$. Prove from $A = U\Sigma V^{\mathrm{T}}$ that every matrix of rank r is the sum of r matrices of rank one.

10 From U, V, and Σ find the orthogonal matrix $Q = UV^{\mathrm{T}}$ and the symmetric matrix $K = U\Sigma U^{\mathrm{T}}$. Verify the polar decomposition in the reverse order $A = KQ$.

11 The pseudoinverse of this A is the same as ______ because ______.

Problems 12–13 compute and use the SVD of a 1 by 3 rectangular matrix.

12 Compute $A^{\mathrm{T}}A$ and AA^{T} and their eigenvalues and unit eigenvectors when the matrix is $A = \begin{bmatrix} 3 & 4 & 0 \end{bmatrix}$. What are the singular values of A?

13 Put numbers into the singular value decomposition of A:

$$A = \begin{bmatrix} 3 & 4 & 0 \end{bmatrix} = [u_1] \begin{bmatrix} \sigma_1 & 0 & 0 \end{bmatrix} \begin{bmatrix} v_1 & v_2 & v_3 \end{bmatrix}^{\mathrm{T}}.$$

Put numbers into the pseudoinverse of A. *Compute AA^+ and A^+A*:

$$A^+ = \begin{bmatrix} \\ \\ \\ \end{bmatrix} = \begin{bmatrix} v_1 & v_2 & v_3 \end{bmatrix} \begin{bmatrix} 1/\sigma_1 \\ 0 \\ 0 \end{bmatrix} \begin{bmatrix} u_1 \end{bmatrix}^{\mathrm{T}}.$$

14 What is the only 2 by 3 matrix that has no pivots and no singular values? What is Σ for that matrix? A^+ is the zero matrix, but what shape?

15 If $\det A = 0$ how do you know that $\det A^+ = 0$?

16 When are the factors in $U\Sigma V^{\mathrm{T}}$ the same as in $Q\Lambda Q^{\mathrm{T}}$? The eigenvalues λ_i must be positive, to equal the σ_i. Then A must be _____ and positive _____.

Problems 17–20 bring out the main properties of A^+ and $x^+ = A^+b$.

17 Suppose all matrices have rank one. The vector $\boldsymbol{b}$ is (b_1, b_2).

$$A = \begin{bmatrix} 2 & 2 \\ 1 & 1 \end{bmatrix} \quad A^{\mathrm{T}} = \begin{bmatrix} .2 & .1 \\ .2 & .1 \end{bmatrix} \quad AA^{\mathrm{T}} = \begin{bmatrix} .8 & .4 \\ .4 & .2 \end{bmatrix} \quad A^{\mathrm{T}}A = \begin{bmatrix} .5 & .5 \\ .5 & .5 \end{bmatrix}$$

(a) The equation $A^{\mathrm{T}}A\widehat{\boldsymbol{x}} = A^{\mathrm{T}}\boldsymbol{b}$ has many solutions because $A^{\mathrm{T}}A$ is _____.

(b) Verify that $\boldsymbol{x}^+ = A^+\boldsymbol{b} = (.2b_1 + .1b_2, .2b_1 + .1b_2)$ does solve $A^{\mathrm{T}}A\boldsymbol{x}^+ = A^{\mathrm{T}}\boldsymbol{b}$.

(c) AA^+ projects onto the column space of A. Therefore _____ projects onto the nullspace of A^{T}. Then $A^{\mathrm{T}}(AA^+ - I)\boldsymbol{b} = \boldsymbol{0}$. This gives $A^{\mathrm{T}}A\boldsymbol{x}^+ = A^{\mathrm{T}}\boldsymbol{b}$ and $\widehat{\boldsymbol{x}}$ can be $\boldsymbol{x}^+$.

18 *The vector $\boldsymbol{x}^+$ is the shortest possible solution to $A^{\mathrm{T}}A\widehat{\boldsymbol{x}} = A^{\mathrm{T}}\boldsymbol{b}$.* Reason: The difference $\widehat{\boldsymbol{x}} - \boldsymbol{x}^+$ is in the nullspace of $A^{\mathrm{T}}A$. This is also the nullspace of A. Explain how it follows that

$$\|\widehat{\boldsymbol{x}}\|^2 = \|\boldsymbol{x}^+\|^2 + \|\widehat{\boldsymbol{x}} - \boldsymbol{x}^+\|^2.$$

Any other solution $\widehat{\boldsymbol{x}}$ has greater length than $\boldsymbol{x}^+$.

19 Every $\boldsymbol{b}$ in $\mathbf{R}^m$ is $\boldsymbol{p} + \boldsymbol{e}$. This is the column space part plus the left nullspace part. Every $\boldsymbol{x}$ in $\mathbf{R}^n$ is $\boldsymbol{x}_r + \boldsymbol{x}_n =$ (row space part) + (nullspace part). Then

$AA^+\boldsymbol{p} =$ _____ $\quad AA^+\boldsymbol{e} =$ _____ $\quad A^+A\boldsymbol{x}_r =$ _____ $\quad A^+A\boldsymbol{x}_n =$ _____

20 Find A^+ and A^+A and AA^+ for the 2 by 1 matrix whose SVD is

$$A = \begin{bmatrix} 3 \\ 4 \end{bmatrix} = \begin{bmatrix} .6 & -.8 \\ .8 & .6 \end{bmatrix} \begin{bmatrix} 5 \\ 0 \end{bmatrix} [1].$$

21 A general 2 by 2 matrix A is determined by four numbers. If triangular, it is determined by three. If diagonal, by two. If a rotation, by one. An eigenvector, by one. Check that the total count is four for each factorization of A:

$$LU \quad LDU \quad QR \quad U\Sigma V^{\mathrm{T}} \quad S\Lambda S^{-1}.$$

22 Following Problem 21, check that LDL^{T} and $Q\Lambda Q^{\mathrm{T}}$ are determined by *three* numbers. This is correct because the matrix A is _____.

23 From $A = U\Sigma V^{\mathrm{T}}$ and $A^{+} = V\Sigma^{+}U^{\mathrm{T}}$ explain these splittings into rank 1:

$$A = \sum_{1}^{r} \sigma_i \boldsymbol{u}_i \boldsymbol{v}_i^{\mathrm{T}} \qquad A^{+} = \sum_{1}^{r} \frac{\boldsymbol{v}_i \boldsymbol{u}_i^{\mathrm{T}}}{\sigma_i} \qquad A^{+}A = \sum_{1}^{r} \boldsymbol{v}_i \boldsymbol{v}_i^{\mathrm{T}} \qquad AA^{+} = \sum_{1}^{r} \boldsymbol{u}_i \boldsymbol{u}_i^{\mathrm{T}}$$

24 This problem looks for all matrices A with a given column space in $\mathbf{R}^m$ and a given row space in $\mathbf{R}^n$. Suppose $\widehat{\boldsymbol{u}}_1, \ldots, \widehat{\boldsymbol{u}}_r$ and $\widehat{\boldsymbol{v}}_1, \ldots, \widehat{\boldsymbol{v}}_r$ are bases for those two spaces. Make them columns of $\widehat{U}$ and $\widehat{V}$. Use $A = U\Sigma V^{\mathrm{T}}$ to show that A has the form $\widehat{U}M\widehat{V}^{\mathrm{T}}$ for an r by r invertible matrix M.

25 A pair of singular vectors $\boldsymbol{v}$ and $\boldsymbol{u}$ will satisfy $A\boldsymbol{v} = \sigma\boldsymbol{u}$ and $A^{\mathrm{T}}\boldsymbol{u} = \sigma\boldsymbol{v}$. This means that the double vector $\boldsymbol{x} = \begin{bmatrix} \boldsymbol{u} \\ \boldsymbol{v} \end{bmatrix}$ is an eigenvector of what symmetric matrix? With what eigenvalue?

8

APPLICATIONS

MATRICES IN ENGINEERING ■ 8.1

This section will show how engineering problems produce symmetric matrices K (often positive definite matrices). The "linear algebra reason" for symmetry and positive definiteness is their form $K = A^{\mathrm{T}}A$ and $K = A^{\mathrm{T}}CA$. The "physical reason" is that the expression $\frac{1}{2}\boldsymbol{u}^{\mathrm{T}}K\boldsymbol{u}$ represents *energy*—and energy is never negative.

Our first examples come from mechanical and civil and aeronautical engineering. K is the ***stiffness matrix***, and $K^{-1}\boldsymbol{f}$ is the structure's response to forces $\boldsymbol{f}$ from outside. The next section turns to electrical engineering—the matrices come from networks and circuits. The exercises involve chemical engineering and I could go on! Economics and management and engineering design come later in this chapter (there the key is optimization).

Here we present equilibrium equations $K\boldsymbol{u} = \boldsymbol{f}$. With motion, $Md^2\boldsymbol{u}/dt^2 + K\boldsymbol{u} = \boldsymbol{f}$ becomes dynamic. Then we use eigenvalues, or finite differences between time steps.

Before explaining the physical examples, may I write down the matrices? The tridiagonal K_0 appears many times in this textbook. Now we will see its applications. These matrices are all symmetric, and the first four are positive definite:

$$K_0 = A_0^{\mathrm{T}}A_0 = \begin{bmatrix} \mathbf{2} & -1 & \\ -1 & 2 & -1 \\ & -1 & \mathbf{2} \end{bmatrix} \qquad A_0^{\mathrm{T}}C_0A_0 = \begin{bmatrix} c_1+c_2 & -c_2 & \\ -c_2 & c_2+c_3 & -c_3 \\ & -c_3 & c_3+c_4 \end{bmatrix}$$

$$K_1 = A_1^{\mathrm{T}}A_1 = \begin{bmatrix} \mathbf{2} & -1 & \\ -1 & 2 & -1 \\ & -1 & \mathbf{1} \end{bmatrix} \qquad A_1^{\mathrm{T}}C_1A_1 = \begin{bmatrix} c_1+c_2 & -c_2 & \\ -c_2 & c_2+c_3 & -c_3 \\ & -c_3 & c_3 \end{bmatrix}$$

$$K_{singular} = \begin{bmatrix} \mathbf{1} & -1 & \\ -1 & 2 & -1 \\ & -1 & \mathbf{1} \end{bmatrix} \qquad K_{circular} = \begin{bmatrix} 2 & -1 & \mathbf{-1} \\ -1 & 2 & -1 \\ \mathbf{-1} & -1 & 2 \end{bmatrix}$$

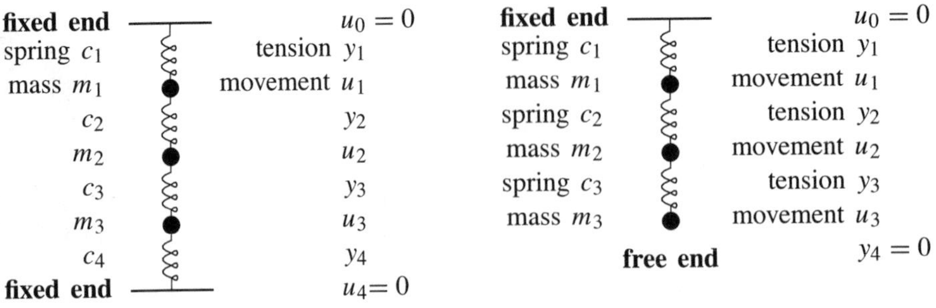

Figure 8.1 Lines of springs and masses with different end conditions: no movement (**fixed-fixed**) and no force at the bottom (**fixed-free**).

The matrices $K_0, K_1, K_{singular}$, and $K_{circular}$ have $C = I$ for simplicity. This means that all the "spring constants" are $c_i = 1$. We included $A_0^T C_0 A_0$ and $A_1^T C_1 A_1$ to show how the spring constants enter the matrix (without changing its positive definiteness). Our first goal is to show where these stiffness matrices come from.

A Line of Springs

Figure 8.1 shows three masses m_1, m_2, m_3 connected by a line of springs. In one case there are four springs, with top and bottom fixed. The fixed-free case has only three springs; the lowest mass hangs freely. The **fixed-fixed** problem will lead to K_0 and $A_0^T C_0 A_0$. The **fixed-free** problem will lead to K_1 and $A_1^T C_1 A_1$. A **free-free** problem, with no support at either end, produces the matrix $K_{singular}$.

We want equations for the mass movements $\boldsymbol{u}$ and the tensions (or compressions) $\boldsymbol{y}$:

$$\boldsymbol{u} = (u_1, u_2, u_3) = \textbf{movements of the masses} \text{ (down or up)}$$
$$\boldsymbol{y} = (y_1, y_2, y_3, y_4) \text{ or } (y_1, y_2, y_3) = \textbf{tensions in the springs}$$

When a mass moves downward, its displacement is positive ($u_i > 0$). For the springs, tension is positive and compression is negative ($y_i < 0$). In tension, the spring is stretched so it pulls the masses inward. Each spring is controlled by its own Hooke's Law $y = c\,e$: (*stretching force*) = (*spring constant*) times (*stretching distance*).

Our job is to link these one-spring equations into a vector equation $K\boldsymbol{u} = \boldsymbol{f}$ for the whole system. The force vector $\boldsymbol{f}$ comes from gravity. The gravitational constant g multiplies each mass to produce $\boldsymbol{f} = (m_1 g, m_2 g, m_3 g)$.

The real problem is to find the stiffness matrix (**fixed-fixed** and **fixed-free**). The best way to create K is in three steps, not one. Instead of connecting the movements $\boldsymbol{u}_i$ directly to the forces $\boldsymbol{f}_i$, it is much better to connect each vector to the next in this list:

$$\begin{aligned}
\boldsymbol{u} &= \textbf{\textit{Movements}} \text{ of } n \text{ masses} &&= (u_1, \ldots, u_n) \\
\boldsymbol{e} &= \textbf{\textit{Elongations}} \text{ of } m \text{ springs} &&= (e_1, \ldots, e_m) \\
\boldsymbol{y} &= \textbf{\textit{Internal forces}} \text{ in } m \text{ springs} &&= (y_1, \ldots, y_m) \\
\boldsymbol{f} &= \textbf{\textit{External forces}} \text{ on } n \text{ masses} &&= (f_1, \ldots, f_n)
\end{aligned}$$

The framework that connects $\boldsymbol{u}$ to $\boldsymbol{e}$ to $\boldsymbol{y}$ to $\boldsymbol{f}$ looks like this:

$$\begin{array}{ccc} \boxed{\boldsymbol{u}} & & \boxed{\boldsymbol{f}} \\ A\downarrow & & \uparrow A^{\mathrm{T}} \\ \boxed{\boldsymbol{e}} & \xrightarrow{C} & \boxed{\boldsymbol{y}} \end{array} \qquad \begin{array}{ll} \boldsymbol{e} = A\boldsymbol{u} & A \text{ is } m \text{ by } n \\ \boldsymbol{y} = C\boldsymbol{e} & C \text{ is } m \text{ by } m \\ \boldsymbol{f} = A^{\mathrm{T}}\boldsymbol{y} & A^{\mathrm{T}} \text{ is } n \text{ by } m \end{array}$$

We will write down the matrices A and C and A^{T} for the two examples, first with fixed ends and then with the lower end free. Forgive the simplicity of these matrices, it is their form that is so important. Especially the appearance of A and A^{T}.

The *elongation* $\boldsymbol{e}$ is the stretching distance—how far the springs are extended. Originally there is no stretching—the system is lying on a table. When it becomes vertical and upright, gravity acts. The masses move down by distances u_1, u_2, u_3. Each spring is stretched or compressed by $e_i = u_i - u_{i-1}$, *the difference in displacements*:

$$\begin{array}{lll} \text{First spring:} & \boldsymbol{e}_1 = \boldsymbol{u}_1 & \text{(the top is fixed so } u_0 = 0) \\ \text{Second spring:} & \boldsymbol{e}_2 = \boldsymbol{u}_2 - \boldsymbol{u}_1 & \\ \text{Third spring:} & \boldsymbol{e}_3 = \boldsymbol{u}_3 - \boldsymbol{u}_2 & \\ \text{Fourth spring:} & \boldsymbol{e}_4 = \quad\;\; - \boldsymbol{u}_3 & \text{(the bottom is fixed so } u_4 = 0) \end{array}$$

If both ends move the same distance, that spring is not stretched: $u_i = u_{i-1}$ and $e_i = 0$. The matrix in those four equations is a 4 by 3 *difference matrix* A, and $\boldsymbol{e} = A\boldsymbol{u}$:

Stretching distances (elongations)

$$\boldsymbol{e} = A\boldsymbol{u} \quad \text{is} \quad \begin{bmatrix} e_1 \\ e_2 \\ e_3 \\ e_4 \end{bmatrix} = \begin{bmatrix} 1 & 0 & 0 \\ -1 & 1 & 0 \\ 0 & -1 & 1 \\ 0 & 0 & -1 \end{bmatrix} \begin{bmatrix} u_1 \\ u_2 \\ u_3 \end{bmatrix}. \tag{1}$$

The next equation $\boldsymbol{y} = C\boldsymbol{e}$ connects spring elongation $\boldsymbol{e}$ with spring tension $\boldsymbol{y}$. *This is Hooke's Law* $y_i = c_i e_i$ *for each separate spring.* It is the "constitutive law" that depends on the material in the spring. A soft spring has small c, so a moderate force y can produce a large stretching e. Hooke's linear law is nearly exact for real springs, before they are overstretched and the material becomes plastic.

Since each spring has its own law, the matrix in $\boldsymbol{y} = C\boldsymbol{e}$ is a diagonal matrix C:

Hooke's Law

$$\begin{array}{rcl} y_1 &=& c_1 e_1 \\ y_2 &=& c_2 e_2 \\ y_3 &=& c_3 e_3 \\ y_4 &=& c_4 e_4 \end{array} \quad \text{is} \quad \begin{bmatrix} y_1 \\ y_2 \\ y_3 \\ y_4 \end{bmatrix} = \begin{bmatrix} c_1 & & & \\ & c_2 & & \\ & & c_3 & \\ & & & c_4 \end{bmatrix} \begin{bmatrix} e_1 \\ e_2 \\ e_3 \\ e_4 \end{bmatrix} = C\boldsymbol{e}. \tag{2}$$

Combining $\boldsymbol{e} = A\boldsymbol{u}$ with $\boldsymbol{y} = C\boldsymbol{e}$, the spring forces are $\boldsymbol{y} = CA\boldsymbol{u}$.

Finally comes the *balance equation*, the most fundamental law of applied mathematics. The internal forces from the springs balance the external forces on the masses. Each mass is pulled or pushed by the spring force y_j above it. From below it feels the spring force y_{j+1} plus f_j from gravity. Thus $y_j = y_{j+1} + f_j$ or $f_j = y_j - y_{j+1}$:

$$\textbf{Force balance}\qquad \begin{aligned} f_1 &= y_1 - y_2 \\ f_2 &= y_2 - y_3 \\ f_3 &= y_3 - y_4 \end{aligned} \quad \text{and} \quad \begin{bmatrix} f_1 \\ f_2 \\ f_3 \end{bmatrix} = \begin{bmatrix} 1 & -1 & 0 & 0 \\ 0 & 1 & -1 & 0 \\ 0 & 0 & 1 & -1 \end{bmatrix} \begin{bmatrix} y_1 \\ y_2 \\ y_3 \\ y_4 \end{bmatrix}. \tag{3}$$

That matrix is A^{T}. *The equation for balance of forces is* $\boldsymbol{f} = A^{\mathrm{T}}\boldsymbol{y}$. Nature transposes the rows and columns of the $\boldsymbol{e} - \boldsymbol{u}$ matrix to produce the $\boldsymbol{f} - \boldsymbol{y}$ matrix. This is the beauty of the framework, that A^{T} appears along with A. The three equations combine into $K\boldsymbol{u} = \boldsymbol{f}$, where the ***stiffness matrix*** is $K = A^{\mathrm{T}}CA$:

$$\left\{ \begin{aligned} \boldsymbol{e} &= A\boldsymbol{u} \\ \boldsymbol{y} &= C\boldsymbol{e} \\ \boldsymbol{f} &= A^{\mathrm{T}}\boldsymbol{y} \end{aligned} \right\} \quad \text{combine into} \quad A^{\mathrm{T}}CA\boldsymbol{u} = \boldsymbol{f} \quad \text{or} \quad K\boldsymbol{u} = \boldsymbol{f}.$$

In the language of elasticity, $\boldsymbol{e} = A\boldsymbol{u}$ is the **kinematic** equation (for displacement). The force balance $\boldsymbol{f} = A^{\mathrm{T}}\boldsymbol{y}$ is the **static** equation (for equilibrium). The **constitutive law** is $\boldsymbol{y} = C\boldsymbol{e}$ (from the material). Then $A^{\mathrm{T}}CA$ is n by n = (n by m)(m by m)(m by n).

Finite element programs spend major effort on assembling $K = A^{\mathrm{T}}CA$ from thousands of smaller pieces. We do it for four springs by multiplying A^{T} times CA:

$$\begin{matrix}\textbf{FIXED}\\ \textbf{FIXED}\end{matrix}\quad \begin{bmatrix} 1 & -1 & 0 & 0 \\ 0 & 1 & -1 & 0 \\ 0 & 0 & 1 & -1 \end{bmatrix} \begin{bmatrix} c_1 & 0 & 0 \\ -c_2 & c_2 & 0 \\ 0 & -c_3 & c_3 \\ 0 & 0 & -c_4 \end{bmatrix} = \begin{bmatrix} c_1 + c_2 & -c_2 & 0 \\ -c_2 & c_2 + c_3 & -c_3 \\ 0 & -c_3 & c_3 + c_4 \end{bmatrix}$$

If all springs are identical, with $c_1 = c_2 = c_3 = c_4 = 1$, then $C = I$. The stiffness matrix reduces to $A^{\mathrm{T}}A$. It becomes the special matrix

$$K_0 = A_0^{\mathrm{T}}A_0 = \begin{bmatrix} 2 & -1 & 0 \\ -1 & 2 & -1 \\ 0 & -1 & 2 \end{bmatrix}. \tag{4}$$

Note the difference between $A^{\mathrm{T}}A$ from engineering and LL^{T} from linear algebra. The matrix A from four springs is 4 by 3. The triangular matrix L from elimination is square. The stiffness matrix K is assembled from $A^{\mathrm{T}}A$, and then broken up into LL^{T}. One step is applied mathematics, the other is computational mathematics. Each K is built from rectangular matrices and factored into square matrices.

May I list some properties of $K = A^{\mathrm{T}}CA$? You know almost all of them:

1. K is **tridiagonal**, because mass 3 is not connected to mass 1.
2. K is **symmetric**, because C is symmetric and A^{T} comes with A.
3. K is **positive definite**, because $c_i > 0$ and A has **independent columns**.
4. K^{-1} is a full matrix in equation (5) with **all positive entries**.

That last property leads to an important fact about $\boldsymbol{u} = K^{-1}\boldsymbol{f}$: *If all forces act downwards* ($f_j > 0$) *then all movements are downwards* ($u_j > 0$). Notice that "positiveness" is different from "positive definiteness". Here K^{-1} is positive (K is not). Both K and K^{-1} are positive definite.

Example 1 Suppose all $c_i = c$ and $m_j = m$. Find the movements $\boldsymbol{u}$ and tensions $\boldsymbol{y}$.

All springs are the same and all masses are the same. But all movements and elongations and tensions will *not* be the same. K^{-1} includes $\frac{1}{c}$ because $A^{\mathrm{T}}CA$ includes c:

$$\boldsymbol{u} = K^{-1}\boldsymbol{f} = \frac{1}{4c}\begin{bmatrix} 3 & 2 & 1 \\ 2 & 4 & 2 \\ 1 & 2 & 3 \end{bmatrix}\begin{bmatrix} mg \\ mg \\ mg \end{bmatrix} = \frac{mg}{c}\begin{bmatrix} \frac{3}{2} \\ 2 \\ \frac{3}{2} \end{bmatrix} \tag{5}$$

The displacement u_2, for the mass in the middle, is greater than u_1 and u_3. The units are correct: the force mg divided by force per unit length c gives a length u. Then

$$\boldsymbol{e} = A\boldsymbol{u} = \begin{bmatrix} 1 & 0 & 0 \\ -1 & 1 & 0 \\ 0 & -1 & 1 \\ 0 & 0 & -1 \end{bmatrix}\frac{mg}{c}\begin{bmatrix} \frac{3}{2} \\ 2 \\ \frac{3}{2} \end{bmatrix} = \frac{mg}{c}\begin{bmatrix} \frac{3}{2} \\ \frac{1}{2} \\ -\frac{1}{2} \\ -\frac{3}{2} \end{bmatrix}.$$

Those elongations add to zero because the ends of the line are fixed. (The sum $u_1 + (u_2 - u_1) + (u_3 - u_2) + (-u_3)$ is certainly zero.) For each spring force y_i we just multiply e_i by c. So y_1, y_2, y_3, y_4 are $\frac{3}{2}mg$, $\frac{1}{2}mg$, $-\frac{1}{2}mg$, $-\frac{3}{2}mg$. The upper two springs are stretched, the lower two springs are compressed.

Notice how $\boldsymbol{u}, \boldsymbol{e}, \boldsymbol{y}$ are computed in that order. We assembled $K = A^{\mathrm{T}}CA$ from rectangular matrices. To find $\boldsymbol{u} = K^{-1}\boldsymbol{f}$, we work with the whole matrix and not its three pieces! The rectangular matrices A and A^{T} do not have (two-sided) inverses.

Warning: *Normally you cannot write* $K^{-1} = A^{-1}C^{-1}(A^{\mathrm{T}})^{-1}$.

The three matrices are mixed together by $A^{\mathrm{T}}CA$, and they cannot easily be untangled. In general, $A^{\mathrm{T}}\boldsymbol{y} = \boldsymbol{f}$ has many solutions. And four equations $A\boldsymbol{u} = \boldsymbol{e}$ would usually have no solution with three unknowns. But $A^{\mathrm{T}}CA$ gives the correct solution to all three equations in the framework. Only when $m = n$ and the matrices are square can we go from $\boldsymbol{y} = (A^{\mathrm{T}})^{-1}\boldsymbol{f}$ to $\boldsymbol{e} = C^{-1}\boldsymbol{y}$ to $\boldsymbol{u} = A^{-1}\boldsymbol{e}$. We will see that now.

Fixed End and Free End

Remove the fourth spring. All matrices become 3 by 3. The pattern does not change! The matrix A loses its fourth row and (of course) A^{T} loses its fourth column. The new stiffness matrix K_1 becomes a product of square matrices:

$$A_1^{\mathrm{T}} C_1 A_1 = \begin{bmatrix} 1 & -1 & 0 \\ 0 & 1 & -1 \\ 0 & 0 & 1 \end{bmatrix} \begin{bmatrix} c_1 & & \\ & c_2 & \\ & & c_3 \end{bmatrix} \begin{bmatrix} 1 & 0 & 0 \\ -1 & 1 & 0 \\ 0 & -1 & 1 \end{bmatrix}.$$

The missing column of A^{T} and row of A multiplied the missing c_4. So the quickest way to find the new $A^{\mathrm{T}}CA$ is to set $c_4 = 0$ in the old one:

FIXED
FREE

$$K_1 = A_1^{\mathrm{T}} C_1 A_1 = \begin{bmatrix} c_1 + c_2 & -c_2 & 0 \\ -c_2 & c_2 + c_3 & -c_3 \\ 0 & -c_3 & c_3 \end{bmatrix}. \qquad (6)$$

If $c_1 = c_2 = c_3 = 1$ and $C = I$, this is the -1, 2, -1 tridiagonal matrix, except the last entry is 1 instead of 2. The spring at the bottom is free.

Example 2 All $c_i = c$ and all $m_j = m$ in the fixed-free hanging line of springs. Then

$$K_1 = c \begin{bmatrix} 2 & -1 & 0 \\ -1 & 2 & -1 \\ 0 & -1 & 1 \end{bmatrix} \quad \text{and} \quad K_1^{-1} = \frac{1}{c} \begin{bmatrix} 1 & 1 & 1 \\ 1 & 2 & 2 \\ 1 & 2 & 3 \end{bmatrix}.$$

The forces mg from gravity are the same. But the movements change from the previous example because the stiffness matrix has changed:

$$\boldsymbol{u} = K_1^{-1} \boldsymbol{f} = \frac{1}{c} \begin{bmatrix} 1 & 1 & 1 \\ 1 & 2 & 2 \\ 1 & 2 & 3 \end{bmatrix} \begin{bmatrix} mg \\ mg \\ mg \end{bmatrix} = \frac{mg}{c} \begin{bmatrix} 3 \\ 5 \\ 6 \end{bmatrix}.$$

Those movements are greater in this fixed-free case. The number 3 appears in u_1 because all three masses are pulling the first spring down. The next mass moves by that 3 plus an additional 2 from the masses below it. The third mass drops even more $(3+2+1=6)$. The elongations $\boldsymbol{e} = A\boldsymbol{u}$ in the springs display those numbers 3, 2, 1:

$$\boldsymbol{e} = \begin{bmatrix} 1 & 0 & 0 \\ -1 & 1 & 0 \\ 0 & -1 & 1 \end{bmatrix} \frac{mg}{c} \begin{bmatrix} 3 \\ 5 \\ 6 \end{bmatrix} = \frac{mg}{c} \begin{bmatrix} 3 \\ 2 \\ 1 \end{bmatrix}.$$

Multiplying by c, the forces $\boldsymbol{y}$ in the three springs are $3mg$ and $2mg$ and mg. And the special point of square matrices is that $\boldsymbol{y}$ can be found directly from $\boldsymbol{f}$! The balance equation $A^{\mathrm{T}}\boldsymbol{y} = \boldsymbol{f}$ determines $\boldsymbol{y}$ immediately, because $m = n$ and A^{T} is square. We are allowed to write $(A^{\mathrm{T}}CA)^{-1} = A^{-1}C^{-1}(A^{\mathrm{T}})^{-1}$:

$$\boldsymbol{y} = (A^{\mathrm{T}})^{-1} \boldsymbol{f} \ \text{is} \ \begin{bmatrix} 1 & 1 & 1 \\ 0 & 1 & 1 \\ 0 & 0 & 1 \end{bmatrix} \begin{bmatrix} mg \\ mg \\ mg \end{bmatrix} = \begin{bmatrix} 3mg \\ 2mg \\ 1mg \end{bmatrix}.$$

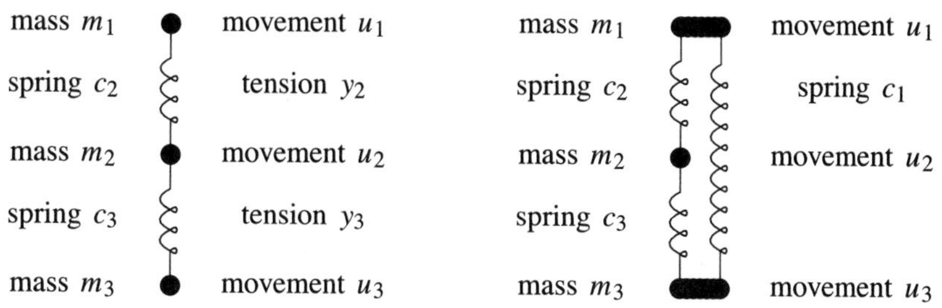

Figure 8.2 **Free-free ends**: A line of springs and a "circle" of springs: *Singular* K's. The masses can move without stretching the springs so $A\boldsymbol{u} = \mathbf{0}$ has nonzero solutions.

Two Free Ends: K is Singular

The first line of springs in Figure 8.2 is free at *both ends*. This means trouble (the whole line can move). The matrix A is 2 by 3, short and wide. Here is $\boldsymbol{e} = A\boldsymbol{u}$:

$$\begin{bmatrix} e_1 \\ e_2 \end{bmatrix} = \begin{bmatrix} u_2 - u_1 \\ u_3 - u_2 \end{bmatrix} = \begin{bmatrix} -1 & 1 & 0 \\ 0 & -1 & 1 \end{bmatrix} \begin{bmatrix} u_1 \\ u_2 \\ u_3 \end{bmatrix}. \qquad (7)$$

Now there is a nonzero solution to $A\boldsymbol{u} = \mathbf{0}$. ***The masses can move with no stretching of the springs***. The whole line can shift by $\boldsymbol{u} = (1, 1, 1)$ and this leaves $\boldsymbol{e} = (0, 0)$. A has *dependent columns* and the vector $(1, 1, 1)$ is in its nullspace:

$$A\boldsymbol{u} = \begin{bmatrix} -1 & 1 & 0 \\ 0 & -1 & 1 \end{bmatrix} \begin{bmatrix} 1 \\ 1 \\ 1 \end{bmatrix} = \begin{bmatrix} 0 \\ 0 \end{bmatrix} = \textbf{no stretching}. \qquad (8)$$

$A\boldsymbol{u} = \mathbf{0}$ certainly leads to $A^{\mathrm{T}}CA\boldsymbol{u} = \mathbf{0}$. So $A^{\mathrm{T}}CA$ is only *positive semidefinite*, without c_1 and c_4. The pivots will be c_2 and c_3 and *no third pivot*:

$$\begin{bmatrix} -1 & 0 \\ 1 & -1 \\ 0 & 1 \end{bmatrix} \begin{bmatrix} c_2 & \\ & c_3 \end{bmatrix} \begin{bmatrix} -1 & 1 & 0 \\ 0 & -1 & 1 \end{bmatrix} = \begin{bmatrix} c_2 & -c_2 & 0 \\ -c_2 & c_2 + c_3 & -c_3 \\ 0 & -c_3 & c_3 \end{bmatrix} \qquad (9)$$

Two eigenvalues will be positive but $\boldsymbol{x} = (1, 1, 1)$ is an eigenvector for $\lambda = 0$. We can solve $A^{\mathrm{T}}CA\boldsymbol{u} = \boldsymbol{f}$ only for special vectors $\boldsymbol{f}$. The forces have to add to $f_1 + f_2 + f_3 = 0$, or the whole line of springs (with both ends free) will take off like a rocket.

Circle of Springs

A third spring will complete the circle from mass 3 back to mass 1. This doesn't make K invertible—the new matrix is still singular. That stiffness matrix $K_{circular}$ is not tridiagonal, but it is symmetric (always) and *semidefinite*:

$$A_{circular}^{\mathrm{T}} A_{circular} = \begin{bmatrix} 1 & -1 & 0 \\ 0 & 1 & -1 \\ -1 & 0 & 1 \end{bmatrix} \begin{bmatrix} 1 & 0 & -1 \\ -1 & 1 & 0 \\ 0 & -1 & 1 \end{bmatrix} = \begin{bmatrix} 2 & -1 & -1 \\ -1 & 2 & -1 \\ -1 & -1 & 2 \end{bmatrix}. \tag{10}$$

The only pivots are 2 and $\frac{3}{2}$. The eigenvalues are 3 and 3 and 0. The determinant is zero. The nullspace still contains $\boldsymbol{x} = (1, 1, 1)$, when all masses move together (nothing is holding them) and the springs are not stretched. This movement vector $(1, 1, 1)$ is in the nullspace of $A_{circular}$ and $K_{circular}$, even after the diagonal matrix C of spring constants is included:

$$(A^{\mathrm{T}} C A)_{circular} = \begin{bmatrix} c_1 + c_2 & -c_2 & -c_1 \\ -c_2 & c_2 + c_3 & -c_3 \\ -c_1 & -c_3 & c_3 + c_1 \end{bmatrix}. \tag{11}$$

Continuous Instead of Discrete

Matrix equations are discrete. Differential equations are continuous. We will see the differential equation that corresponds to the tridiagonal $-1, 2, -1$ matrix $A^{\mathrm{T}}A$. And it is a pleasure to see the boundary conditions that go with K_0 and K_1.

The matrices A and A^{T} correspond to the derivatives d/dx and $-d/dx$! Remember that $\boldsymbol{e} = A\boldsymbol{u}$ took differences $u_i - u_{i-1}$, and $\boldsymbol{f} = A^{\mathrm{T}}\boldsymbol{y}$ took differences $y_i - y_{i+1}$. Now the springs are infinitesimally short, and those differences become derivatives:

$$\frac{u_i - u_{i-1}}{\Delta x} \text{ is like } \frac{du}{dx} \qquad \frac{y_i - y_{i+1}}{\Delta x} \text{ is like } -\frac{dy}{dx}$$

The factor Δx didn't appear earlier—we imagined the distance between masses was 1. To approximate a continuous solid bar, we take many more masses (smaller and closer). Let me jump to the three steps A, C, A^{T} in the continuous model, when there is stretching and Hooke's Law and force balance at every point x:

$$e(x) = Au = \frac{du}{dx} \quad y(x) = c(x)e(x) \quad A^{\mathrm{T}}y = -\frac{dy}{dx} = f(x)$$

Combining those equations into $A^{\mathrm{T}}CAu(x) = f(x)$, we have a differential equation not a matrix equation. The line of springs becomes an elastic bar:

$$\textbf{Solid Elastic Bar} \qquad A^{\mathrm{T}}CAu(x) = f(x) \quad \text{is} \quad -\frac{d}{dx}\left(c(x)\frac{du}{dx}\right) = f(x) \tag{12}$$

A^TA corresponds to a second derivative. A is a "difference matrix" and A^TA is a "second difference matrix". ***The matrix has*** $-1, 2, -1$ ***and the equation has*** $-d^2u/dx^2$:

$$-u_{i+1} + 2u_i - u_{i-1} \text{ is a } \textbf{second difference} \qquad -\frac{d^2u}{dx^2} \text{ is a } \textbf{second derivative}.$$

Now we see why this symmetric matrix is a favorite. When we meet a first derivative du/dx, we have three choices (*forward, backward, and centered differences*):

$$\frac{du}{dx} \simeq \frac{u(x+\Delta x) - u(x)}{\Delta x} \quad \text{or} \quad \frac{u(x) - u(x-\Delta x)}{\Delta x} \quad \text{or} \quad \frac{u(x+\Delta x) - u(x-\Delta x)}{2\Delta x}.$$

When we meet d^2u/dx^2, the natural choice is $u(x+\Delta x) - 2u(x) + u(x-\Delta x)$, divided by $(\Delta x)^2$. *Why reverse these signs to* $-1, 2, -1$? Because the positive definite matrix has $+2$ on the diagonal. First derivatives are *anti*symmetric; the transpose has a minus sign. So second differences are negative definite, and we change to $-d^2u/dx^2$.

We have moved from vectors to functions. Scientific computing moves the other way. It starts with a differential equation like (12). Sometimes there is a formula for the solution $u(x)$, more often not. In reality we *create* the discrete matrix K by approximating the continuous problem. Watch how the boundary conditions on u come in! By missing u_0 we treat it (correctly) as zero:

$$\begin{matrix}\textbf{FIXED}\\ \textbf{FIXED}\end{matrix} \quad A\boldsymbol{u} = \frac{1}{\Delta x}\begin{bmatrix} 1 & 0 & 0 \\ -1 & 1 & 0 \\ 0 & -1 & 1 \\ 0 & 0 & -1 \end{bmatrix}\begin{bmatrix} u_1 \\ u_2 \\ u_3 \end{bmatrix} \approx \frac{du}{dx} \quad \text{with} \quad \begin{matrix}\boldsymbol{u_0 = 0}\\ \boldsymbol{u_4 = 0}\end{matrix} \qquad (13)$$

Fixing the top end gives the boundary condition $u_0 = 0$. What about the free end, when the bar hangs in the air? Row 4 of A is gone and so is u_4. The boundary condition must come from A^T. It is the missing y_4 that we are treating (correctly) as zero:

$$\begin{matrix}\textbf{FIXED}\\ \textbf{FREE}\end{matrix} \quad A^T\boldsymbol{y} = \frac{1}{\Delta x}\begin{bmatrix} 1 & -1 & 0 \\ 0 & 1 & -1 \\ 0 & 0 & 1 \end{bmatrix}\begin{bmatrix} y_1 \\ y_2 \\ y_3 \end{bmatrix} \approx -\frac{dy}{dx} \quad \text{with} \quad \begin{matrix}\boldsymbol{u_0 = 0}\\ \boldsymbol{y_4 = 0}\end{matrix} \qquad (14)$$

The boundary condition $y_4 = 0$ *at the free end becomes* $du/dx = 0$, since $\boldsymbol{y} = A\boldsymbol{u}$ corresponds to du/dx. The force balance $A^T\boldsymbol{y} = \boldsymbol{f}$ at that end (in the air) is $0 = 0$. The last row of $K_1\boldsymbol{u} = \boldsymbol{f}$ has entries $-1, 1$ to reflect this condition $du/dx = 0$.

May I summarize this section? I hope this example will help you turn calculus into linear algebra, replacing differential equations by difference equations. If your step Δx is small enough, you will have a totally satisfactory solution.

$$\textbf{The equation is} \quad -\frac{d}{dx}\left(c(x)\frac{du}{dx}\right) = f(x) \ \textbf{with}\ u(0) = 0 \ \textbf{and} \ \left[u(1) \text{ or } \frac{du}{dx}(1)\right] = 0$$

Divide the bar into N pieces of length Δx. Replace du/dx by $A\boldsymbol{u}$ and $-dy/dx$ by $A^{\mathrm{T}}\boldsymbol{y}$. Now A and A^{T} include $1/\Delta x$. The end conditions are $u_0 = 0$ and $[u_N = 0$ or $y_N = 0]$. The three steps $-d/dx$ and $c(x)$ and d/dx correspond to A^{T} and C and A:

$$\boldsymbol{f} = A^{\mathrm{T}}\boldsymbol{y} \text{ and } \boldsymbol{y} = C\boldsymbol{e} \text{ and } \boldsymbol{e} = A\boldsymbol{u} \text{ give } A^{\mathrm{T}}CA\boldsymbol{u} = \boldsymbol{f}.$$

This is a fundamental example in computational science and engineering. Our book concentrates on Step 3 in that process (linear algebra). Now we have taken Step 2.

1. Model the problem by a differential equation

2. Discretize the differential equation to a difference equation

3. Understand and solve the difference equation (and boundary conditions!)

4. Interpret the solution; visualize it; redesign if needed.

Numerical simulation has become a third branch of science, together with experiment and deduction. Designing the Boeing 777 was much less expensive on a computer than in a wind tunnel. Our discussion still has to move from ordinary to partial differential equations, and from linear to nonlinear. The text *Introduction to Applied Mathematics* (Wellesley-Cambridge Press) develops this whole subject further—see the course page **math.mit.edu/18085**. The principles remain the same, and I hope this book helps you to see the framework behind the computations.

Problem Set 8.1

1 Show that $\det A_0^{\mathrm{T}}C_0A_0 = c_1c_2c_3+c_1c_3c_4+c_1c_2c_4+c_2c_3c_4$. Find also $\det A_1^{\mathrm{T}}C_1A_1$ in the fixed-free example.

2 Invert $A_1^{\mathrm{T}}C_1A_1$ in the fixed-free example by multiplying $A_1^{-1}C_1^{-1}(A_1^{\mathrm{T}})^{-1}$.

3 In the free-free case when $A^{\mathrm{T}}CA$ in equation (9) is singular, add the three equations $A^{\mathrm{T}}CA\boldsymbol{u} = \boldsymbol{f}$ to show that we need $f_1 + f_2 + f_3 = 0$. Find a solution to $A^{\mathrm{T}}CA\boldsymbol{u} = \boldsymbol{f}$ when the forces $\boldsymbol{f} = (-1, 0, 1)$ balance themselves. Find all solutions!

4 Both end conditions for the free-free differential equation are $du/dx = 0$:

$$-\frac{d}{dx}\left(c(x)\frac{du}{dx}\right) = f(x) \quad \text{with} \quad \frac{du}{dx} = 0 \text{ at both ends.}$$

Integrate both sides to show that the force $f(x)$ must balance itself, $\int f(x)\,dx = 0$, or there is no solution. The complete solution is one particular solution $u(x)$ plus any constant. The constant corresponds to $\boldsymbol{u} = (1, 1, 1)$ in the nullspace of $A^{\mathrm{T}}CA$.

5 In the fixed-free problem, the matrix A is square and invertible. We can solve $A^{\mathrm{T}}\boldsymbol{y} = \boldsymbol{f}$ separately from $A\boldsymbol{u} = \boldsymbol{e}$. Do the same for the differential equation:

$$\text{Solve} \quad -\frac{dy}{dx} = f(x) \text{ with } y(1) = 0. \quad \text{Graph } y(x) \text{ if } f(x) = 1.$$

6 The 3 by 3 matrix $K_1 = A_1^{\mathrm{T}} C_1 A_1$ in equation (6) splits into three "element matrices" $c_1E_1 + c_2E_2 + c_3E_3$. Write down those pieces, one for each c. Show how they come from *column times row* multiplication of $A_1^{\mathrm{T}} C_1 A_1$. This is how finite element stiffness matrices are actually assembled.

7 For five springs and four masses with both ends fixed, what are the matrices A and C and K? With $C = I$ solve $K\boldsymbol{u} = \text{ones}(4)$.

8 Compare the solution $\boldsymbol{u} = (u_1, u_2, u_3, u_4)$ in Problem 7 to the solution of the continuous problem $-u'' = 1$ with $u(0) = 0$ and $u(1) = 0$. The parabola $u(x)$ should correspond at $x = \frac{1}{5}, \frac{2}{5}, \frac{3}{5}, \frac{4}{5}$ to $\boldsymbol{u}$—is there a $(\Delta x)^2$ factor to account for?

9 Solve the fixed-free problem $-u'' = mg$ with $u(0) = 0$ and $u'(1) = 0$. Compare $u(x)$ at $x = \frac{1}{3}, \frac{2}{3}, \frac{3}{3}$ with the vector $\boldsymbol{u} = (3mg, 5mg, 6mg)$ in Example 2.

10 (MATLAB) Find the displacements $u(1), \ldots, u(100)$ of 100 masses connected by springs all with $c = 1$. Each force is $f(i) = .01$. Print graphs of $\boldsymbol{u}$ with **fixed-fixed** and **fixed-free** ends. Note that diag(ones(n, 1), d) is a matrix with n ones along diagonal d. This print command will graph a vector u:

```
plot(u, '+'); xlabel('mass number'); ylabel('movement'); print
```

11 (MATLAB) Chemical engineering has a first derivative du/dx from fluid velocity as well as d^2u/dx^2 from diffusion. Replace du/dx by a forward difference and then by a backward difference, with $\Delta x = \frac{1}{8}$. Graph your numerical solutions of

$$-\frac{d^2u}{dx^2} + 10\frac{du}{dx} = 1 \text{ with } u(0) = u(1) = 0.$$

GRAPHS AND NETWORKS ■ 8.2

This chapter is about six selected applications of linear algebra. We had many applications to choose from. Any time you have a connected system, with each part depending on other parts, you have a matrix. Linear algebra deals with interacting systems, provided the laws that govern them are linear. Over the years I have seen one model so often, and found it so basic and useful, that I always put it first. The model consists of ***nodes connected by edges***. This is called a ***graph***.

Graphs of the usual kind display functions $f(x)$. Graphs of this node-edge kind lead to matrices. This section is about the ***incidence matrix*** of a graph—which tells how the n nodes are connected by the m edges. Normally $m > n$, there are more edges than nodes.

For any m by n matrix there are two fundamental subspaces in $\mathbf{R}^n$ and two in $\mathbf{R}^m$. They are the row spaces and nullspaces of A and A^{T}. Their *dimensions* are related by the most important theorem in linear algebra. The second part of that theorem is the *orthogonality* of the subspaces. Our goal is to show how examples from graphs illuminate the Fundamental Theorem of Linear Algebra.

We review the four subspaces (for any matrix). Then we construct a ***directed graph*** and its ***incidence matrix***. The dimensions will be easy to discover. But we want the subspaces themselves—this is where orthogonality helps. It is essential to connect the subspaces to the graph they come from. By specializing to incidence matrices, the laws of linear algebra become Kirchhoff's laws. Please don't be put off by the words "current" and "potential" and "Kirchhoff." These rectangular matrices are the best.

Every entry of an incidence matrix is 0 or 1 or -1. This continues to hold during elimination. All pivots and multipliers are ± 1. Therefore both factors in $A = LU$ also contain $0, 1, -1$. So do the nullspace matrices! All four subspaces have basis vectors with these exceptionally simple components. The matrices are not concocted for a textbook, they come from a model that is absolutely essential in pure and applied mathematics.

Review of the Four Subspaces

Start with an m by n matrix. Its columns are vectors in $\mathbf{R}^m$. Their linear combinations produce the ***column space*** $\boldsymbol{C}(A)$, a subspace of $\mathbf{R}^m$. Those combinations are exactly the matrix-vector products $A\boldsymbol{x}$.

The rows of A are vectors in $\mathbf{R}^n$ (or they would be, if they were column vectors). Their linear combinations produce the ***row space***. To avoid any inconvenience with rows, we transpose the matrix. The row space becomes $\boldsymbol{C}(A^{\mathrm{T}})$, the column space of A^{T}.

The central questions of linear algebra come from these two ways of looking at the same numbers, by columns and by rows.

The ***nullspace*** $N(A)$ contains every $\boldsymbol{x}$ that satisfies $A\boldsymbol{x} = \mathbf{0}$—this is a subspace of $\mathbf{R}^n$. The "***left***" *nullspace* contains all solutions to $A^{\mathrm{T}}\boldsymbol{y} = \mathbf{0}$. Now $\boldsymbol{y}$ has m components, and $N(A^{\mathrm{T}})$ is a subspace of $\mathbf{R}^m$. Written as $\boldsymbol{y}^{\mathrm{T}}A = \mathbf{0}^{\mathrm{T}}$, we are combining rows

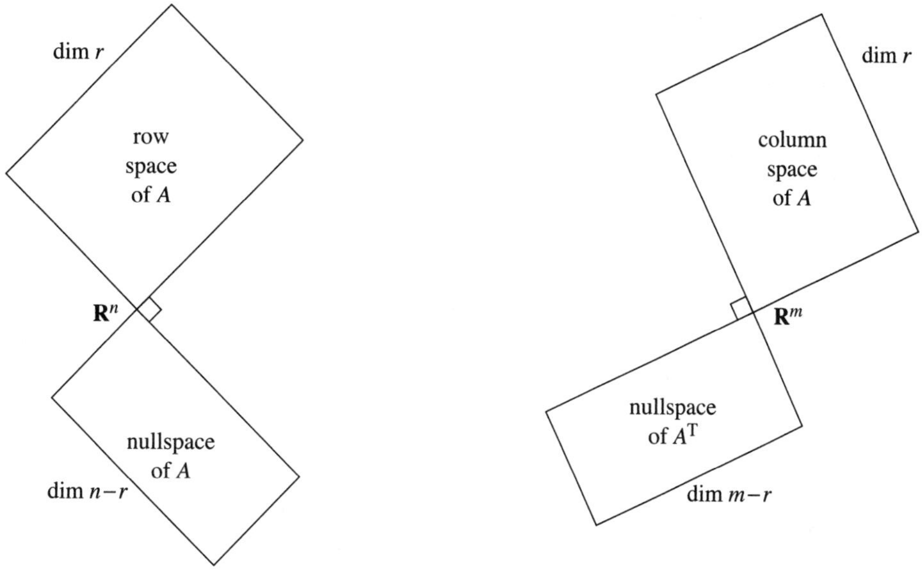

Figure 8.3 The four subspaces with their dimensions and orthogonality.

of A to produce the zero row. The four subspaces are illustrated by Figure 8.3, which shows $\mathbf{R}^n$ on one side and $\mathbf{R}^m$ on the other. The link between them is A.

The information in that figure is crucial. First come the dimensions, which obey the two central laws of linear algebra:

$$\dim \boldsymbol{C}(A) = \dim \boldsymbol{C}(A^{\mathrm{T}}) \quad \text{and} \quad \dim \boldsymbol{C}(A) + \dim \boldsymbol{N}(A) = n.$$

When the row space has dimension r, the nullspace has dimension $n-r$. Elimination leaves these two spaces unchanged, and the echelon form U gives the dimension count. There are r rows and columns with pivots. There are $n-r$ free columns without pivots, and those lead to vectors in the nullspace.

The following incidence matrix A comes from a graph. Its echelon form is U:

$$A = \begin{bmatrix} -1 & 1 & 0 & 0 \\ -1 & 0 & 1 & 0 \\ 0 & -1 & 1 & 0 \\ -1 & 0 & 0 & 1 \\ 0 & -1 & 0 & 1 \\ 0 & 0 & -1 & 1 \end{bmatrix} \quad \text{goes to} \quad U = \begin{bmatrix} -1 & 1 & 0 & 0 \\ 0 & -1 & 1 & 0 \\ 0 & 0 & -1 & 1 \\ 0 & 0 & 0 & 0 \\ 0 & 0 & 0 & 0 \\ 0 & 0 & 0 & 0 \end{bmatrix}.$$

The nullspace of A and U is the line through $\boldsymbol{x} = (1, 1, 1, 1)$. The column spaces of A and U have dimension $r = 3$. The pivot rows are a basis for the row space.

Figure 8.3 shows more—the subspaces are orthogonal. ***Every vector in the nullspace is perpendicular to every vector in the row space.*** This comes directly from the m equations

$Ax = \mathbf{0}$. For A and U above, $\boldsymbol{x} = (1, 1, 1, 1)$ is perpendicular to all rows and thus to the whole row space.

This review of the subspaces applies to any matrix A—only the example was special. Now we concentrate on that example. It is the incidence matrix for a particular graph, and we look to the graph for the meaning of every subspace.

Directed Graphs and Incidence Matrices

Figure 8.4 displays a *graph* with $m = 6$ edges and $n = 4$ nodes, so the matrix A is 6 by 4. It tells which nodes are connected by which edges. The entries -1 and $+1$ also tell the direction of each arrow (this is a *directed* graph). The first row of A gives a record of the first edge:

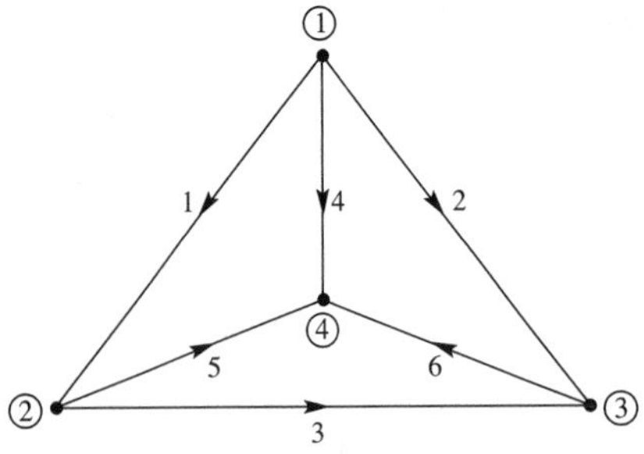

The first edge goes from node 1 to node 2. The first row has −1 in column 1 and +1 in column 2.

$$\begin{array}{c} \textbf{node} \\ ①\ ②\ ③\ ④ \end{array}$$

$$A = \begin{bmatrix} -1 & 1 & 0 & 0 \\ -1 & 0 & 1 & 0 \\ 0 & -1 & 1 & 0 \\ -1 & 0 & 0 & 1 \\ 0 & -1 & 0 & 1 \\ 0 & 0 & -1 & 1 \end{bmatrix} \begin{array}{c} 1 \\ 2 \\ 3 \\ 4 \\ 5 \\ 6 \end{array} \quad \textbf{edge}$$

Figure 8.4a Complete graph with $m = 6$ edges and $n = 4$ nodes.

Row numbers are edge numbers, column numbers are node numbers.

You can write down A immediately by looking at the graph.

The second graph has the same four nodes but only three edges. Its incidence matrix is 3 by 4:

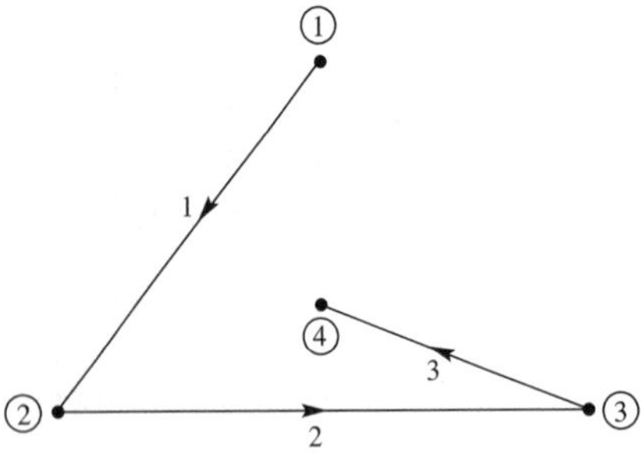

$$\begin{array}{c} \textbf{node} \\ ①\ ②\ ③\ ④ \end{array}$$

$$B = \begin{bmatrix} -1 & 1 & 0 & 0 \\ 0 & -1 & 1 & 0 \\ 0 & 0 & -1 & 1 \end{bmatrix} \begin{array}{c} 1 \\ 2 \\ 3 \end{array} \quad \textbf{edge}$$

Figure 8.4b Tree with 3 edges and 4 nodes and no loops.

The first graph is *complete*—every pair of nodes is connected by an edge. The second graph is a ***tree***—the graph has ***no closed loops***. Those graphs are the two extremes, with the maximum number of edges $m = \frac{1}{2}n(n-1)$ and the minimum number $m = n - 1$. We are assuming that the graph is connected, and it makes no fundamental difference which way the arrows go. On each edge, flow with the arrow is "positive." Flow in the opposite direction counts as negative. The flow might be a current or a signal or a force—or even oil or gas or water.

The rows of B match the nonzero rows of U—the echelon form found earlier. ***Elimination reduces every graph to a tree.*** The loops produce zero rows in U. Look at the loop from edges 1, 2, 3 in the first graph, which leads to a zero row:

$$\begin{bmatrix} -1 & 1 & 0 & 0 \\ -1 & 0 & 1 & 0 \\ 0 & -1 & 1 & 0 \end{bmatrix} \longrightarrow \begin{bmatrix} -1 & 1 & 0 & 0 \\ 0 & -1 & 1 & 0 \\ 0 & -1 & 1 & 0 \end{bmatrix} \longrightarrow \begin{bmatrix} -1 & 1 & 0 & 0 \\ 0 & -1 & 1 & 0 \\ 0 & 0 & 0 & 0 \end{bmatrix}$$

Those steps are typical. When two edges share a node, elimination produces the "shortcut edge" without that node. If the graph already has this shortcut edge, elimination gives a row of zeros. When the dust clears we have a tree.

An idea suggests itself: ***Rows are dependent when edges form a loop***. Independent rows come from trees. This is the key to the row space.

For the column space we look at $A\boldsymbol{x}$, which is a vector of differences:

$$A\boldsymbol{x} = \begin{bmatrix} -1 & 1 & 0 & 0 \\ -1 & 0 & 1 & 0 \\ 0 & -1 & 1 & 0 \\ -1 & 0 & 0 & 1 \\ 0 & -1 & 0 & 1 \\ 0 & 0 & -1 & 1 \end{bmatrix} \begin{bmatrix} x_1 \\ x_2 \\ x_3 \\ x_4 \end{bmatrix} = \begin{bmatrix} x_2 - x_1 \\ x_3 - x_1 \\ x_3 - x_2 \\ x_4 - x_1 \\ x_4 - x_2 \\ x_4 - x_3 \end{bmatrix}. \tag{1}$$

The unknowns x_1, x_2, x_3, x_4 represent ***potentials*** at the nodes. Then $A\boldsymbol{x}$ gives the ***potential differences*** across the edges. It is these differences that cause flows. We now examine the meaning of each subspace.

1 The ***nullspace*** contains the solutions to $A\boldsymbol{x} = \boldsymbol{0}$. All six potential differences are zero. This means: *All four potentials are equal.* Every $\boldsymbol{x}$ in the nullspace is a constant vector (c, c, c, c). The nullspace of A is a line in $\mathbf{R}^n$—its dimension is $n - r = 1$.

The second incidence matrix B has the same nullspace. It contains $(\mathbf{1}, \mathbf{1}, \mathbf{1}, \mathbf{1})$:

$$B\boldsymbol{x} = \begin{bmatrix} -1 & 1 & 0 & 0 \\ 0 & -1 & 1 & 0 \\ 0 & 0 & -1 & 1 \end{bmatrix} \begin{bmatrix} 1 \\ 1 \\ 1 \\ 1 \end{bmatrix} = \begin{bmatrix} 0 \\ 0 \\ 0 \end{bmatrix}.$$

We can raise or lower all potentials by the same amount c, without changing the differences. There is an "arbitrary constant" in the potentials. Compare this with the same statement for functions. We can raise or lower $f(x)$ by the same amount C, without changing its derivative. There is an arbitrary constant C in the integral.

Calculus adds "$+C$" to indefinite integrals. Graph theory adds (c, c, c, c) to the vector $\boldsymbol{x}$ of potentials. Linear algebra adds any vector $\boldsymbol{x}_n$ in the nullspace to one particular solution of $A\boldsymbol{x} = \boldsymbol{b}$.

The "$+C$" disappears in calculus when the integral starts at a known point $x = a$. Similarly the nullspace disappears when we set $x_4 = 0$. The unknown x_4 is removed and so are the fourth columns of A and B. Electrical engineers would say that node 4 has been "grounded."

2 The ***row space*** contains all combinations of the six rows. Its dimension is certainly not six. The equation $r + (n - r) = n$ must be $3 + 1 = 4$. The rank is $r = 3$, as we also saw from elimination. After 3 edges, we start forming loops! The new rows are not independent.

How can we tell if $\boldsymbol{v} = (v_1, v_2, v_3, v_4)$ is in the row space? The slow way is to combine rows. The quick way is by orthogonality:

v is in the row space if and only if it is perpendicular to* (1, 1, 1, 1) *in the nullspace.

The vector $\boldsymbol{v} = (0, 1, 2, 3)$ fails this test—its components add to 6. The vector $(-6, 1, 2, 3)$ passes the test. It lies in the row space because its components add to zero. It equals $6(\text{row } 1) + 5(\text{row } 3) + 3(\text{row } 6)$.

Each row of A adds to zero. This must be true for every vector in the row space.

3 The ***column space*** contains all combinations of the four columns. We expect three independent columns, since there were three independent rows. The first three columns are independent (so are any three). But the four columns add to the zero vector, which says again that $(1, 1, 1, 1)$ is in the nullspace. *How can we tell if a particular vector* $\boldsymbol{b}$ *is in the column space*?

First answer Try to solve $A\boldsymbol{x} = \boldsymbol{b}$. As before, orthogonality gives a better answer. We are now coming to Kirchhoff's two famous laws of circuit theory—the voltage law and current law. Those are natural expressions of "laws" of linear algebra. It is especially pleasant to see the key role of the left nullspace.

Second answer $A\boldsymbol{x}$ is the vector of differences in equation (1). If we add differences around a closed loop in the graph, the cancellation leaves zero. Around the big triangle formed by edges $1, 3, -2$ (the arrow goes backward on edge 2) the differences are

$$(x_2 - x_1) + (x_3 - x_2) - (x_3 - x_1) = 0.$$

This is the ***voltage law***: ***The components of $A\boldsymbol{x}$ add to zero around every loop.*** When $\boldsymbol{b}$ is in the column space, it must obey the same law:

Kirchhoff's Voltage Law: $b_1 + b_3 - b_2 = 0.$

By testing each loop, we decide whether $\boldsymbol{b}$ is in the column space. $A\boldsymbol{x} = \boldsymbol{b}$ can be solved exactly when the components of $\boldsymbol{b}$ satisfy all the same dependencies as the rows of A. Then elimination leads to $0 = 0$, and $A\boldsymbol{x} = \boldsymbol{b}$ is consistent.

4 The ***left nullspace*** contains the solutions to $A^{\mathrm{T}}\boldsymbol{y} = \mathbf{0}$. Its dimension is $m - r = 6 - 3$:

$$A^{\mathrm{T}}\boldsymbol{y} = \begin{bmatrix} -1 & -1 & 0 & -1 & 0 & 0 \\ 1 & 0 & -1 & 0 & -1 & 0 \\ 0 & 1 & 1 & 0 & 0 & -1 \\ 0 & 0 & 0 & 1 & 1 & 1 \end{bmatrix} \begin{bmatrix} y_1 \\ y_2 \\ y_3 \\ y_4 \\ y_5 \\ y_6 \end{bmatrix} = \begin{bmatrix} 0 \\ 0 \\ 0 \\ 0 \end{bmatrix}. \qquad (2)$$

The true number of equations is $r = 3$ and not $n = 4$. Reason: The four equations add to $0 = 0$. The fourth equation follows automatically from the first three.

What do the equations mean? The first equation says that $-y_1 - y_2 - y_4 = 0$. ***The net flow into node* 1 *is zero***. The fourth equation says that $y_4 + y_5 + y_6 = 0$. ***Flow into the node minus flow out is zero***. The equations $A^{\mathrm{T}}\boldsymbol{y} = \mathbf{0}$ are famous and fundamental:

Kirchhoff's Current Law: Flow in equals flow out at each node.

This law deserves first place among the equations of applied mathematics. It expresses "*conservation*" and "*continuity*" and "*balance*." Nothing is lost, nothing is gained. When currents or forces are in equilibrium, the equation to solve is $A^{\mathrm{T}}\boldsymbol{y} = \mathbf{0}$. Notice the beautiful fact that the matrix in this balance equation is the transpose of the incidence matrix A.

What are the actual solutions to $A^{\mathrm{T}}\boldsymbol{y} = \mathbf{0}$? The currents must balance themselves. The easiest way is to **flow around a loop**. If a unit of current goes around the big triangle (forward on edge 1, forward on 3, backward on 2), the vector is $\boldsymbol{y} = (1, -1, 1, 0, 0, 0)$. This satisfies $A^{\mathrm{T}}\boldsymbol{y} = \mathbf{0}$. Every loop current yields a solution $\boldsymbol{y}$, because flow in equals flow out at every node. A smaller loop goes forward on edge 1, forward on 5, back on 4. Then $\boldsymbol{y} = (1, 0, 0, -1, 1, 0)$ is also in the left nullspace.

We expect three independent $\boldsymbol{y}$'s, since $6 - 3 = 3$. The three small loops in the graph are independent. The big triangle seems to give a fourth $\boldsymbol{y}$, but it is the sum of flows around the small loops. The small loops give a basis for the left nullspace.

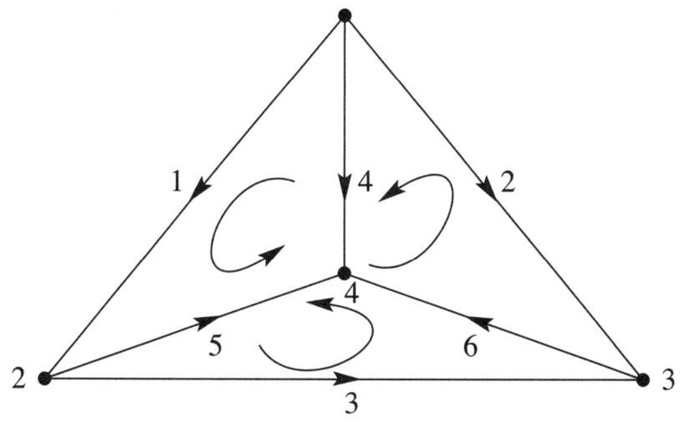

$$\begin{bmatrix} 1 \\ 0 \\ 0 \\ -1 \\ 1 \\ 0 \end{bmatrix} + \begin{bmatrix} 0 \\ 0 \\ 1 \\ 0 \\ -1 \\ 1 \end{bmatrix} + \begin{bmatrix} 0 \\ -1 \\ 0 \\ 1 \\ 0 \\ -1 \end{bmatrix} = \begin{bmatrix} 1 \\ -1 \\ 1 \\ 0 \\ 0 \\ 0 \end{bmatrix}$$

small loops big loop

Summary The incidence matrix A comes from a connected graph with n nodes and m edges. The row space and column space have dimensions $n-1$. The nullspaces have dimension 1 and $m-n+1$:

1 The constant vectors $(c, c, \dots, c)$ make up the nullspace of A.

2 There are $r = n-1$ independent rows, using edges from any tree.

3 ***Voltage law***: The components of Ax add to zero around every loop.

4 ***Current law***: $A^{\mathrm{T}}y = \mathbf{0}$ is solved by loop currents. $N(A^{\mathrm{T}})$ has dimension $m-r$. ***There are $m-r = m-n+1$ independent loops in the graph.***

For every graph in a plane, linear algebra yields ***Euler's formula***:

$$\textit{(number of nodes)} - \textit{(number of edges)} + \textit{(number of small loops)} = \mathbf{1}.$$

This is $\boldsymbol{n - m + (m - n + 1) = 1}$. The graph in our example has $4 - 6 + 3 = 1$.

A single triangle has (3 nodes)−(3 edges)+(1 loop). On a 10-node tree with 9 edges and no loops, Euler's count is $10 - 9 + 0$. All planar graphs lead to the answer 1.

Networks and $A^{\mathrm{T}}CA$

In a real network, the current $\boldsymbol{y}$ along an edge is the product of two numbers. One number is the difference between the potentials $\boldsymbol{x}$ at the ends of the edge. This difference is $A\boldsymbol{x}$ and it drives the flow. The other number is the "***conductance***" c—which measures how easily flow gets through.

In physics and engineering, c is decided by the material. For electrical currents, c is high for metal and low for plastics. For a superconductor, c is nearly infinite. If we consider elastic stretching, c might be low for metal and higher for plastics. In economics, c measures the capacity of an edge or its cost.

To summarize, the graph is known from its "connectivity matrix" A. This tells the connections between nodes and edges. A ***network*** goes further, and assigns a conductance c to each edge. These numbers $c_1, \dots, c_m$ go into the "conductance matrix" C—which is diagonal.

For a network of resistors, the conductance is $c = 1/(\text{resistance})$. In addition to Kirchhoff's laws for the whole system of currents, we have Ohm's law for each particular current. Ohm's law connects the current y_1 on edge 1 to the potential difference $x_2 - x_1$ between the nodes:

Ohm's Law: Current along edge = conductance times potential difference.

Ohm's law for all m currents is $\boldsymbol{y} = -CA\boldsymbol{x}$. The vector $A\boldsymbol{x}$ gives the potential differences, and C multiplies by the conductances. Combining Ohm's law with Kirchhoff's

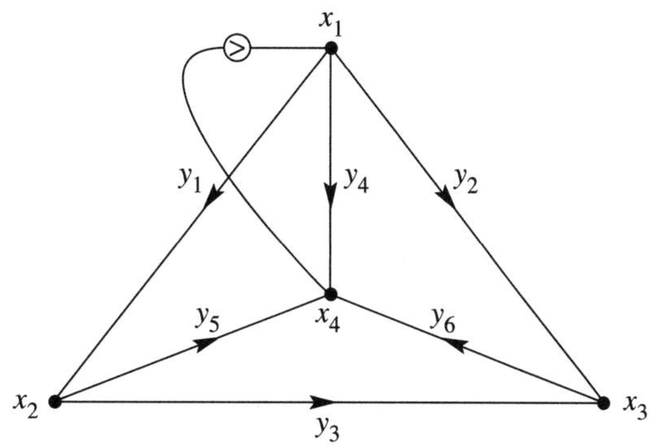

Figure 8.5 The currents in a network with a source S into node 1.

current law $A^{\mathrm{T}}\boldsymbol{y} = \mathbf{0}$, we get $A^{\mathrm{T}}CA\boldsymbol{x} = \mathbf{0}$. This is *almost* the central equation for network flows. The only thing wrong is the zero on the right side! The network needs power from outside—a voltage source or a current source—to make something happen.

Note about signs In circuit theory we change from $A\boldsymbol{x}$ to $-A\boldsymbol{x}$. The flow is from higher potential to lower potential. There is (positive) current from node 1 to node 2 when $x_1 - x_2$ is positive—whereas $A\boldsymbol{x}$ was constructed to yield $x_2 - x_1$. The minus sign in physics and electrical engineering is a plus sign in mechanical engineering and economics. $A\boldsymbol{x}$ versus $-A\boldsymbol{x}$ is a general headache but unavoidable.

Note about applied mathematics Every new application has its own form of Ohm's law. For elastic structures $\boldsymbol{y} = CA\boldsymbol{x}$ is Hooke's law. The stress $\boldsymbol{y}$ is (elasticity C) times (stretching $A\boldsymbol{x}$). For heat conduction, $A\boldsymbol{x}$ is a temperature gradient. For oil flows it is a pressure gradient. There is a similar law for least square regression in statistics. My textbook *Introduction to Applied Mathematics* (Wellesley-Cambridge Press) is practically built on $A^{\mathrm{T}}CA$. This is the key to equilibrium in matrix equations and also in differential equations.

Applied mathematics is more organized than it looks. *I have learned to watch for $A^{\mathrm{T}}CA$.*

We now give an example with a current source. Kirchhoff's law changes from $A^{\mathrm{T}}\boldsymbol{y} = \mathbf{0}$ to $A^{\mathrm{T}}\boldsymbol{y} = \boldsymbol{f}$, to balance the source $\boldsymbol{f}$ from outside. *Flow into each node still equals flow out.* Figure 8.5 shows the network with its conductances $c_1, \ldots, c_6$, and it shows the current source going into node 1. The source comes out at node 4 to keep the balance (in = out). The problem is: ***Find the currents $y_1, \ldots, y_6$ on the six edges.***

Example 1 All conductances are $c = 1$, so that $C = I$. A current y_4 travels directly from node 1 to node 4. Other current goes the long way from node 1 to node 2 to node 4 (this is $y_1 = y_5$). Current also goes from node 1 to node 3 to node 4 (this is $y_2 = y_6$). We can find the six currents by using special rules for symmetry, or we can do it right by using $A^{\mathrm{T}}CA$. Since $C = I$, this matrix is $A^{\mathrm{T}}A$:

$$\begin{bmatrix} -1 & -1 & 0 & -1 & 0 & 0 \\ 1 & 0 & -1 & 0 & -1 & 0 \\ 0 & 1 & 1 & 0 & 0 & -1 \\ 0 & 0 & 0 & 1 & 1 & 1 \end{bmatrix} \begin{bmatrix} -1 & 1 & 0 & 0 \\ -1 & 0 & 1 & 0 \\ 0 & -1 & 1 & 0 \\ -1 & 0 & 0 & 1 \\ 0 & -1 & 0 & 1 \\ 0 & 0 & -1 & 1 \end{bmatrix} = \begin{bmatrix} 3 & -1 & -1 & -1 \\ -1 & 3 & -1 & -1 \\ -1 & -1 & 3 & -1 \\ -1 & -1 & -1 & 3 \end{bmatrix}$$

That last matrix is not invertible! We cannot solve for all potentials because (1, 1, 1, 1) is in the nullspace. One node has to be grounded. Setting $x_4 = 0$ removes the fourth row and column, and this leaves a 3 by 3 invertible matrix. Now we solve $A^{\mathrm{T}}CA\boldsymbol{x} = \boldsymbol{f}$ for the unknown potentials x_1, x_2, x_3, with source S into node 1:

$$\begin{bmatrix} 3 & -1 & -1 \\ -1 & 3 & -1 \\ -1 & -1 & 3 \end{bmatrix} \begin{bmatrix} x_1 \\ x_2 \\ x_3 \end{bmatrix} = \begin{bmatrix} S \\ 0 \\ 0 \end{bmatrix} \quad \text{gives} \quad \begin{bmatrix} x_1 \\ x_2 \\ x_3 \end{bmatrix} = \begin{bmatrix} S/2 \\ S/4 \\ S/4 \end{bmatrix}.$$

Ohm's law $\boldsymbol{y} = -CA\boldsymbol{x}$ yields the six currents. Remember $C = I$ and $x_4 = 0$:

$$\begin{bmatrix} y_1 \\ y_2 \\ y_3 \\ y_4 \\ y_5 \\ y_6 \end{bmatrix} = - \begin{bmatrix} -1 & 1 & 0 & 0 \\ -1 & 0 & 1 & 0 \\ 0 & -1 & 1 & 0 \\ -1 & 0 & 0 & 1 \\ 0 & -1 & 0 & 1 \\ 0 & 0 & -1 & 1 \end{bmatrix} \begin{bmatrix} S/2 \\ S/4 \\ S/4 \\ 0 \end{bmatrix} = \begin{bmatrix} S/4 \\ S/4 \\ 0 \\ S/2 \\ S/4 \\ S/4 \end{bmatrix}.$$

Half the current goes directly on edge 4. That is $y_4 = S/2$. No current crosses from node 2 to node 3. Symmetry indicated $y_3 = 0$ and now the solution proves it.

The same matrix $A^{\mathrm{T}}A$ appears in least squares. Nature distributes the currents to minimize the heat loss. Statistics chooses $\widehat{\boldsymbol{x}}$ to minimize the least squares error.

Problem Set 8.2

Problems 1–7 and 8–14 are about the incidence matrices for these graphs.

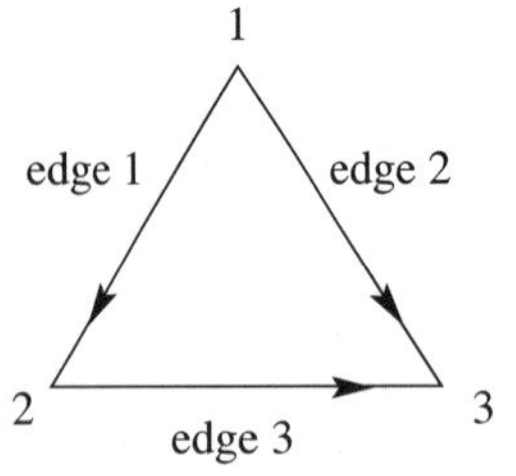

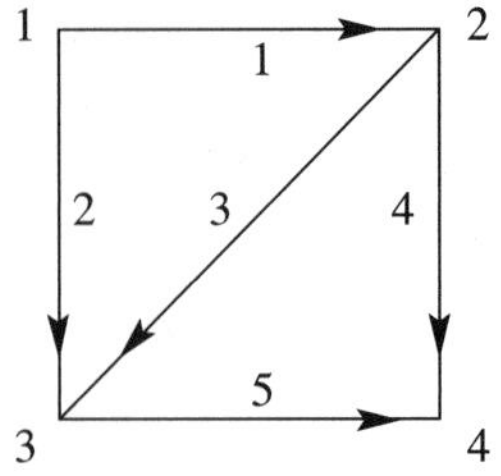

1 Write down the 3 by 3 incidence matrix A for the triangle graph. The first row has -1 in column 1 and $+1$ in column 2. What vectors (x_1, x_2, x_3) are in its nullspace? How do you know that $(1, 0, 0)$ is not in its row space?

2 Write down A^{T} for the triangle graph. Find a vector $\boldsymbol{y}$ in its nullspace. The components of $\boldsymbol{y}$ are currents on the edges—how much current is going around the triangle?

3 Eliminate x_1 and x_2 from the third equation to find the echelon matrix U. What tree corresponds to the two nonzero rows of U?

$$\begin{aligned} -x_1 + x_2 &= b_1 \\ -x_1 + x_3 &= b_2 \\ -x_2 + x_3 &= b_3. \end{aligned}$$

4 Choose a vector (b_1, b_2, b_3) for which $A\boldsymbol{x} = \boldsymbol{b}$ can be solved, and another vector $\boldsymbol{b}$ that allows no solution. How are those $\boldsymbol{b}$'s related to $\boldsymbol{y} = (1, -1, 1)$?

5 Choose a vector (f_1, f_2, f_3) for which $A^{\mathrm{T}}\boldsymbol{y} = \boldsymbol{f}$ can be solved, and a vector $\boldsymbol{f}$ that allows no solution. How are those $\boldsymbol{f}$'s related to $\boldsymbol{x} = (1, 1, 1)$? The equation $A^{\mathrm{T}}\boldsymbol{y} = \boldsymbol{f}$ is Kirchhoff's _____ law.

6 Multiply matrices to find $A^{\mathrm{T}}A$. Choose a vector $\boldsymbol{f}$ for which $A^{\mathrm{T}}A\boldsymbol{x} = \boldsymbol{f}$ can be solved, and solve for $\boldsymbol{x}$. Put those potentials $\boldsymbol{x}$ and the currents $\boldsymbol{y} = -A\boldsymbol{x}$ and current sources $\boldsymbol{f}$ onto the triangle graph. Conductances are 1 because $C = I$.

7 With conductances $c_1 = 1$ and $c_2 = c_3 = 2$, multiply matrices to find $A^{\mathrm{T}}CA$. For $\boldsymbol{f} = (1, 0, -1)$ find a solution to $A^{\mathrm{T}}CA\boldsymbol{x} = \boldsymbol{f}$. Write the potentials $\boldsymbol{x}$ and currents $\boldsymbol{y} = -CA\boldsymbol{x}$ on the triangle graph, when the current source $\boldsymbol{f}$ goes into node 1 and out from node 3.

8 Write down the 5 by 4 incidence matrix A for the square graph with two loops. Find one solution to $A\boldsymbol{x} = \boldsymbol{0}$ and two solutions to $A^{\mathrm{T}}\boldsymbol{y} = \boldsymbol{0}$.

9 Find two requirements on the b's for the five differences $x_2 - x_1, x_3 - x_1, x_3 - x_2, x_4 - x_2, x_4 - x_3$ to equal b_1, b_2, b_3, b_4, b_5. You have found Kirchhoff's _____ law around the two _____ in the graph.

10 Reduce A to its echelon form U. The three nonzero rows give the incidence matrix for what graph? You found one tree in the square graph—find the other seven trees.

11 Multiply matrices to find $A^{\mathrm{T}}A$ and guess how its entries come from the graph:

(a) The diagonal of $A^{\mathrm{T}}A$ tells how many _____ into each node.

(b) The off-diagonals -1 or 0 tell which pairs of nodes are _____.

12 Why is each statement true about $A^{\mathrm{T}}A$? *Answer for $A^{\mathrm{T}}A$ not A.*

(a) Its nullspace contains $(1, 1, 1, 1)$. Its rank is $n-1$.

(b) It is positive semidefinite but not positive definite.

(c) Its four eigenvalues are real and their signs are _____.

13 With conductances $c_1 = c_2 = 2$ and $c_3 = c_4 = c_5 = 3$, multiply the matrices $A^{\mathrm{T}}CA$. Find a solution to $A^{\mathrm{T}}CA\boldsymbol{x} = \boldsymbol{f} = (1, 0, 0, -1)$. Write these potentials $\boldsymbol{x}$ and currents $\boldsymbol{y} = -CA\boldsymbol{x}$ on the nodes and edges of the square graph.

14 The matrix $A^{\mathrm{T}}CA$ is not invertible. What vectors $\boldsymbol{x}$ are in its nullspace? Why does $A^{\mathrm{T}}CA\boldsymbol{x} = \boldsymbol{f}$ have a solution if and only if $f_1 + f_2 + f_3 + f_4 = 0$?

15 A connected graph with 7 nodes and 7 edges has how many loops?

16 For the graph with 4 nodes, 6 edges, and 3 loops, add a new node. If you connect it to one old node, Euler's formula becomes $(\ )-(\ )+(\ ) = 1$. If you connect it to two old nodes, Euler's formula becomes $(\ ) - (\ ) + (\ ) = 1$.

17 Suppose A is a 12 by 9 incidence matrix from a connected (but unknown) graph.

(a) How many columns of A are independent?

(b) What condition on $\boldsymbol{f}$ makes it possible to solve $A^{\mathrm{T}}\boldsymbol{y} = \boldsymbol{f}$?

(c) The diagonal entries of $A^{\mathrm{T}}A$ give the number of edges into each node. What is the sum of those diagonal entries?

18 Why does a complete graph with $n = 6$ nodes have $m = 15$ edges? A tree connecting 6 nodes has _____ edges.

MARKOV MATRICES AND ECONOMIC MODELS ■ 8.3

Early in this book we proposed an experiment. Start with any vector $\boldsymbol{u}_0 = (x, 1-x)$. Multiply it again and again by the "transition matrix" A:

$$A = \begin{bmatrix} .8 & .3 \\ .2 & .7 \end{bmatrix}.$$

The experiment produces $\boldsymbol{u}_1 = A\boldsymbol{u}_0$ and then $\boldsymbol{u}_2 = A\boldsymbol{u}_1$. After k steps we have $A^k\boldsymbol{u}_0$. Unless MATLAB went haywire, the vectors $\boldsymbol{u}_0, \boldsymbol{u}_1, \boldsymbol{u}_2, \boldsymbol{u}_3, \ldots$ approached a "*steady state.*" That limit state is $\boldsymbol{u}_\infty = (.6, .4)$. This final outcome does not depend on the starting vector: ***For every $\boldsymbol{u}_0$ we always converge to*** $(.6, .4)$. The question is why.

At that time we had no good way to answer this question. We knew nothing about eigenvalues. It is true that the steady state equation $A\boldsymbol{u}_\infty = \boldsymbol{u}_\infty$ could be verified:

$$\begin{bmatrix} .8 & .3 \\ .2 & .7 \end{bmatrix}\begin{bmatrix} .6 \\ .4 \end{bmatrix} = \begin{bmatrix} .6 \\ .4 \end{bmatrix}.$$

You would now say that $\boldsymbol{u}_\infty$ is ***an eigenvector with eigenvalue*** 1. *That makes it steady.* Multiplying by A does not change it. But this equation $A\boldsymbol{u}_\infty = \boldsymbol{u}_\infty$ does not explain why all vectors $\boldsymbol{u}_0$ lead to $\boldsymbol{u}_\infty$. Other examples might have a steady state, but it is not necessarily attractive:

$$B = \begin{bmatrix} 1 & 0 \\ 0 & 2 \end{bmatrix} \quad \text{has the steady state} \quad B\begin{bmatrix} 1 \\ 0 \end{bmatrix} = \begin{bmatrix} 1 \\ 0 \end{bmatrix}.$$

In this case, the starting vector $\boldsymbol{u}_0 = (0, 1)$ will give $\boldsymbol{u}_1 = (0, 2)$ and $\boldsymbol{u}_2 = (0, 4)$. The second components are being doubled by the "2" in B. In the language of eigenvalues, B has $\lambda = 1$ but it also has $\lambda = 2$—and an eigenvalue larger than one produces instability. The component of $\boldsymbol{u}$ along that unstable eigenvector is multiplied by λ, and $|\lambda| > 1$ means blowup.

This section is about two special properties of A that guarantee a steady state $\boldsymbol{u}_\infty$. These properties define a ***Markov matrix***, and A above is one particular example:

1. ***Every entry of*** A ***is nonnegative.***

2. ***Every column of*** A ***adds to*** 1.

B did not have Property **2**. When A is a Markov matrix, two facts are immediate:

Multiplying a nonnegative $\boldsymbol{u}_0$ by A produces a nonnegative $\boldsymbol{u}_1 = A\boldsymbol{u}_0$.

If the components of $\boldsymbol{u}_0$ add to 1, so do the components of $\boldsymbol{u}_1 = A\boldsymbol{u}_0$.

Reason: The components of $\boldsymbol{u}_0$ add to 1 when $\begin{bmatrix}1 & \cdots & 1\end{bmatrix}\boldsymbol{u}_0 = 1$. This is true for each column of A by Property 2. Then by matrix multiplication it is true for $A\boldsymbol{u}_0$:

$$\begin{bmatrix}1 & \cdots & 1\end{bmatrix}A\boldsymbol{u}_0 = \begin{bmatrix}1 & \cdots & 1\end{bmatrix}\boldsymbol{u}_0 = 1.$$

The same facts apply to $\boldsymbol{u}_2 = A\boldsymbol{u}_1$ and $\boldsymbol{u}_3 = A\boldsymbol{u}_2$. ***Every vector*** $\boldsymbol{u}_k = A^k\boldsymbol{u}_0$ ***is nonnegative with components adding to*** 1. These are "***probability vectors***." The limit $\boldsymbol{u}_\infty$ is also a probability vector—but we have to prove that there is a limit! The existence of a steady state will follow from **1** and **2** but not so quickly. We must show that $\lambda = 1$ is an eigenvalue of A, and we must estimate the other eigenvalues.

Example 1 The fraction of rental cars in Denver starts at $\frac{1}{50} = .02$. The fraction outside Denver is .98. Every month those fractions (which add to 1) are multiplied by the Markov matrix A:

$$A = \begin{bmatrix} .80 & .05 \\ .20 & .95 \end{bmatrix} \quad \text{leads to} \quad \boldsymbol{u}_1 = A\boldsymbol{u}_0 = A\begin{bmatrix} .02 \\ .98 \end{bmatrix} = \begin{bmatrix} .065 \\ .935 \end{bmatrix}.$$

That is a single step of a ***Markov chain***. In one month, the fraction of cars in Denver is up to .065. The chain of vectors is $\boldsymbol{u}_0, \boldsymbol{u}_1, \boldsymbol{u}_2, \ldots$, and each step multiplies by A:

$$\boldsymbol{u}_1 = A\boldsymbol{u}_0, \qquad \boldsymbol{u}_2 = A^2\boldsymbol{u}_0, \qquad \ldots \quad \text{produces} \quad \boldsymbol{u}_k = A^k\boldsymbol{u}_0.$$

All these vectors are nonnegative because A is nonnegative. Furthermore $.065 + .935 = 1.000$. Each vector $\boldsymbol{u}_k$ will have its components adding to 1. The vector $\boldsymbol{u}_2 = A\boldsymbol{u}_1$ is $(.09875, .90125)$. The first component has grown from .02 to .065 to nearly .099. Cars are moving toward Denver. What happens in the long run?

This section involves powers of matrices. The understanding of A^k was our first and best application of diagonalization. Where A^k can be complicated, the diagonal matrix Λ^k is simple. The eigenvector matrix S connects them: A^k equals $S\Lambda^k S^{-1}$. The new application to Markov matrices follows up on this idea—to use the eigenvalues (in Λ) and the eigenvectors (in S). We will show that $\boldsymbol{u}_\infty$ is an eigenvector corresponding to $\lambda = 1$.

Since every column of A adds to 1, nothing is lost or gained. We are moving rental cars or populations, and no cars or people suddenly appear (or disappear). The fractions add to 1 and the matrix A keeps them that way. The question is how they are distributed after k time periods—which leads us to A^k.

Solution to Example 1 After k steps the fractions in and out of Denver are the components of $A^k\boldsymbol{u}_0$. To study the powers of A we diagonalize it. The eigenvalues are $\lambda = 1$ and $\lambda = .75$. The first eigenvector, with components adding to 1, is $\boldsymbol{x}_1 = (.2, .8)$:

$$|A - \lambda I| = \begin{vmatrix} .80 - \lambda & .05 \\ .20 & .95 - \lambda \end{vmatrix} = \lambda^2 - 1.75\lambda + .75 = (\lambda - 1)(\lambda - .75)$$

$$A\begin{bmatrix} .2 \\ .8 \end{bmatrix} = 1\begin{bmatrix} .2 \\ .8 \end{bmatrix} \quad \text{and} \quad A\begin{bmatrix} -1 \\ 1 \end{bmatrix} = .75\begin{bmatrix} -1 \\ 1 \end{bmatrix}.$$

Those eigenvectors are x_1 and x_2. They are the columns of S. The starting vector u_0 is a combination of x_1 and x_2, in this case with coefficients 1 and .18:

$$u_0 = \begin{bmatrix} .02 \\ .98 \end{bmatrix} = \begin{bmatrix} .2 \\ .8 \end{bmatrix} + .18 \begin{bmatrix} -1 \\ 1 \end{bmatrix}.$$

Now multiply by A to find u_1. The eigenvectors are multiplied by $\lambda_1 = 1$ and $\lambda_2 =$ .75:

$$u_1 = 1 \begin{bmatrix} .2 \\ .8 \end{bmatrix} + (.75)(.18) \begin{bmatrix} -1 \\ 1 \end{bmatrix}.$$

Each time we multiply by A, another .75 multiplies the last vector. The eigenvector x_1 is unchanged:

$$u_k = A^k u_0 = \begin{bmatrix} .2 \\ .8 \end{bmatrix} + (.75)^k (.18) \begin{bmatrix} -1 \\ 1 \end{bmatrix}.$$

This equation reveals what happens. ***The eigenvector*** x_1 ***with*** $\lambda = 1$ ***is the steady state*** u_∞. The other eigenvector x_2 gradually disappears because $|\lambda| < 1$. The more steps we take, the closer we come to $u_\infty = (.2, .8)$. In the limit, $\frac{2}{10}$ of the cars are in Denver and $\frac{8}{10}$ are outside. This is the pattern for Markov chains:

8A If A is a *positive* Markov matrix (entries $a_{ij} > 0$, each column adds to **1**), then $\lambda = 1$ is larger than any other eigenvalue. The eigenvector x_1 is the ***steady state***:

$$u_k = x_1 + c_2(\lambda_2)^k x_2 + \cdots + c_n(\lambda_n)^k x_n \quad \textit{always approaches} \quad u_\infty = x_1.$$

Assume that the components of u_0 add to 1. Then this is true of $u_1, u_2, \ldots$. The key point is that *we approach a multiple of* x_1 *from every starting vector* u_0. If all cars start outside Denver, or all start inside, the limit is still $u_\infty = x_1 = (.2, .8)$.

The first point is to see that $\lambda = 1$ is an eigenvalue of A. *Reason*: Every column of $A - I$ adds to $1 - 1 = 0$. The rows of $A - I$ add up to the zero row. Those rows are linearly dependent, so $A - I$ is singular. Its determinant is zero and $\lambda = 1$ is an eigenvalue. Since the trace of A was 1.75, the other eigenvalue had to be $\lambda_2 = .75$.

The second point is that no eigenvalue can have $|\lambda| > 1$. With such an eigenvalue, the powers A^k would grow. But A^k is also a Markov matrix with nonnegative entries adding to 1—and that leaves no room to get large.

A lot of attention is paid to the possibility that another eigenvalue has $|\lambda| = 1$. Suppose the entries of A or any power A^k are all *positive*—zero is not allowed. In this "regular" case $\lambda = 1$ is strictly bigger than any other eigenvalue. When A and its powers have zero entries, another eigenvalue could be as large as $\lambda_1 = 1$.

Example 2 $A = \begin{bmatrix} 0 & 1 \\ 1 & 0 \end{bmatrix}$ has no steady state because $\lambda_2 = -1$.

This matrix sends all cars from inside Denver to outside, and vice versa. The powers A^k alternate between A and I. The second eigenvector $\boldsymbol{x}_2 = (-1, 1)$ is multiplied by $\lambda_2 = -1$ at every step—and does not become smaller. With a regular Markov matrix, the powers A^k approach the rank one matrix that has the steady state $\boldsymbol{x}_1$ in every column.

Example 3 ("**Everybody moves**") Start with three groups. At each time step, half of group 1 goes to group 2 and the other half goes to group 3. The other groups also *split in half and move*. If the starting populations are p_1, p_2, p_3, then after one step the new populations are

$$\boldsymbol{u}_1 = A\boldsymbol{u}_0 = \begin{bmatrix} 0 & \frac{1}{2} & \frac{1}{2} \\ \frac{1}{2} & 0 & \frac{1}{2} \\ \frac{1}{2} & \frac{1}{2} & 0 \end{bmatrix} \begin{bmatrix} p_1 \\ p_2 \\ p_3 \end{bmatrix} = \begin{bmatrix} \frac{1}{2}p_2 + \frac{1}{2}p_3 \\ \frac{1}{2}p_1 + \frac{1}{2}p_3 \\ \frac{1}{2}p_1 + \frac{1}{2}p_2 \end{bmatrix}.$$

A is a Markov matrix. Nobody is born or lost. It is true that A contains zeros, which gave trouble in Example 2. But after two steps in this new example, the zeros disappear from A^2:

$$\boldsymbol{u}_2 = A^2\boldsymbol{u}_0 = \begin{bmatrix} \frac{1}{2} & \frac{1}{4} & \frac{1}{4} \\ \frac{1}{4} & \frac{1}{2} & \frac{1}{4} \\ \frac{1}{4} & \frac{1}{4} & \frac{1}{2} \end{bmatrix} \begin{bmatrix} p_1 \\ p_2 \\ p_3 \end{bmatrix}.$$

What is the steady state? The eigenvalues of A are $\lambda_1 = 1$ (because A is Markov) and $\lambda_2 = \lambda_3 = -\frac{1}{2}$. ***The eigenvector*** $\boldsymbol{x}_1 = (\frac{1}{3}, \frac{1}{3}, \frac{1}{3})$ ***for*** $\lambda = 1$ ***will be the steady state.*** When three equal populations split in half and move, the final populations are again equal. When the populations start from $\boldsymbol{u}_0 = (8, 16, 32)$, the Markov chain approaches its steady state:

$$\boldsymbol{u}_0 = \begin{bmatrix} 8 \\ 16 \\ 32 \end{bmatrix} \quad \boldsymbol{u}_1 = \begin{bmatrix} 24 \\ 20 \\ 12 \end{bmatrix} \quad \boldsymbol{u}_2 = \begin{bmatrix} 16 \\ 18 \\ 22 \end{bmatrix} \quad \boldsymbol{u}_3 = \begin{bmatrix} 20 \\ 19 \\ 17 \end{bmatrix}.$$

The step to $\boldsymbol{u}_4$ will split some people in half. This cannot be helped. The total population is $8+16+32 = 56$ (and later the total is still $20+19+17 = 56$). The steady state populations $\boldsymbol{u}_\infty$ are 56 times $(\frac{1}{3}, \frac{1}{3}, \frac{1}{3})$. You can see the three populations approaching, but never reaching, their final limits $56/3$.

Linear Algebra in Economics: The Consumption Matrix

A long essay about linear algebra in economics would be out of place here. A short note about one matrix seems reasonable. The ***consumption matrix*** tells how much of each input goes into a unit of output. We have n industries like chemicals, food, and oil. To produce a unit of chemicals may require .2 units of chemicals, .3 units of food, and .4 units of oil. Those numbers go into row 1 of the consumption matrix A:

$$\begin{bmatrix} \text{chemical output} \\ \text{food output} \\ \text{oil output} \end{bmatrix} = \begin{bmatrix} .2 & .3 & .4 \\ .4 & .4 & .1 \\ .5 & .1 & .3 \end{bmatrix} \begin{bmatrix} \text{chemical input} \\ \text{food input} \\ \text{oil input} \end{bmatrix}.$$

Row 2 shows the inputs to produce food—a heavy use of chemicals and food, not so much oil. Row 3 of A shows the inputs consumed to refine a unit of oil. The real consumption matrix for the United States in 1958 contained 83 industries. The models in the 1990's are much larger and more precise. We chose a consumption matrix that has a convenient eigenvector.

Now comes the question: Can this economy meet demands y_1, y_2, y_3 for chemicals, food, and oil? To do that, the inputs p_1, p_2, p_3 will have to be higher—because part of $\boldsymbol{p}$ is consumed in producing $\boldsymbol{y}$. The input is $\boldsymbol{p}$ and the consumption is $A\boldsymbol{p}$, which leaves $\boldsymbol{p} - A\boldsymbol{p}$. This net production is what meets the demand $\boldsymbol{y}$:

Problem Find a vector $\boldsymbol{p}$ such that $\boldsymbol{p} - A\boldsymbol{p} = \boldsymbol{y}$ or $(I - A)\boldsymbol{p} = \boldsymbol{y}$ or $\boldsymbol{p} = (I - A)^{-1}\boldsymbol{y}$.

Apparently the linear algebra question is whether $I - A$ is invertible. But there is more to the problem. The demand vector $\boldsymbol{y}$ is nonnegative, and so is A. *The production levels in $\boldsymbol{p} = (I - A)^{-1}\boldsymbol{y}$ must also be nonnegative.* The real question is:

When is $(I - A)^{-1}$ a nonnegative matrix?

This is the test on $(I - A)^{-1}$ for a productive economy, which can meet any positive demand. If A is small compared to I, then $A\boldsymbol{p}$ is small compared to $\boldsymbol{p}$. There is plenty of output. If A is too large, then production consumes more than it yields. In this case the external demand $\boldsymbol{y}$ cannot be met.

"Small" or "large" is decided by the largest eigenvalue λ_1 of A (which is positive):

If $\lambda_1 > 1$ then $(I - A)^{-1}$ has negative entries
If $\lambda_1 = 1$ then $(I - A)^{-1}$ fails to exist
If $\lambda_1 < 1$ then $(I - A)^{-1}$ is nonnegative as desired.

The main point is that last one. The reasoning makes use of a nice formula for $(I - A)^{-1}$, which we give now. The most important infinite series in mathematics is the ***geometric series*** $1 + x + x^2 + \cdots$. This series adds up to $1/(1 - x)$ provided x is between -1 and 1. (When $x = 1$ the series is $1 + 1 + 1 + \cdots = \infty$. When $|x| \geq 1$ the terms x^n don't go to zero and the series cannot converge.) The nice formula for $(I - A)^{-1}$ is the ***geometric series of matrices***:

$$(I - A)^{-1} = I + A + A^2 + A^3 + \cdots.$$

If you multiply this series by A, you get the same series S except for I. Therefore $S - AS = I$, which is $(I - A)S = I$. The series adds to $S = (I - A)^{-1}$ if it converges. And it converges if $|\lambda_{\max}| < 1$.

In our case $A \geq 0$. All terms of the series are nonnegative. Its sum is $(I-A)^{-1} \geq 0$.

Example 4 $A = \begin{bmatrix} .2 & .3 & .4 \\ .4 & .4 & .1 \\ .5 & .1 & .3 \end{bmatrix}$ has $\boldsymbol{\lambda_1 = .9}$ and $\boldsymbol{(I-A)^{-1} = \frac{1}{93}} \begin{bmatrix} 41 & 25 & 27 \\ 33 & 36 & 24 \\ 34 & 23 & 36 \end{bmatrix}$.

This economy is productive. A is small compared to I, because $\lambda_{\max}$ is .9. To meet the demand $\boldsymbol{y}$, start from $\boldsymbol{p} = (I-A)^{-1}\boldsymbol{y}$. Then $A\boldsymbol{p}$ is consumed in production, leaving $\boldsymbol{p} - A\boldsymbol{p}$. This is $(I-A)\boldsymbol{p} = \boldsymbol{y}$, and the demand is met.

Example 5 $A = \begin{bmatrix} 0 & 4 \\ 1 & 0 \end{bmatrix}$ has $\boldsymbol{\lambda_1 = 2}$ and $\boldsymbol{(I-A)^{-1} = -\frac{1}{3}} \begin{bmatrix} 1 & 4 \\ 1 & 1 \end{bmatrix}$.

This consumption matrix $\boldsymbol{A}$ is too large. Demands can't be met, because production consumes more than it yields. The series $\boldsymbol{I + A + A^2 + \ldots}$ does not converge to $\boldsymbol{(I-A)^{-1}}$. The series is growing while $\boldsymbol{(I-A)^{-1}}$ is actually negative.

Problem Set 8.3

Questions 1–14 are about Markov matrices and their eigenvalues and powers.

1 Find the eigenvalues of this Markov matrix (their sum is the trace):

$$A = \begin{bmatrix} .90 & .15 \\ .10 & .85 \end{bmatrix}.$$

What is the steady state eigenvector for the eigenvalue $\lambda_1 = 1$?

2 Diagonalize the Markov matrix in Problem 1 to $A = S\Lambda S^{-1}$ by finding its other eigenvector:

$$A = \begin{bmatrix} \quad & \quad \\ \quad & \quad \end{bmatrix} \begin{bmatrix} 1 & \\ & .75 \end{bmatrix} \begin{bmatrix} \quad & \quad \\ \quad & \quad \end{bmatrix}.$$

What is the limit of $A^k = S\Lambda^k S^{-1}$ when $\Lambda^k = \begin{bmatrix} 1 & 0 \\ 0 & .75^k \end{bmatrix}$ approaches $\begin{bmatrix} 1 & 0 \\ 0 & 0 \end{bmatrix}$?

3 What are the eigenvalues and the steady state eigenvectors for these Markov matrices?

$$A = \begin{bmatrix} 1 & .2 \\ 0 & .8 \end{bmatrix} \quad A = \begin{bmatrix} .2 & 1 \\ .8 & 0 \end{bmatrix} \quad A = \begin{bmatrix} \frac{1}{2} & \frac{1}{4} & \frac{1}{4} \\ \frac{1}{4} & \frac{1}{2} & \frac{1}{4} \\ \frac{1}{4} & \frac{1}{4} & \frac{1}{2} \end{bmatrix}.$$

4 For every 4 by 4 Markov matrix, what eigenvector of A^{T} corresponds to the (known) eigenvalue $\lambda = 1$?

5 Every year 2% of young people become old and 3% of old people become dead. (No births.) Find the steady state for

$$\begin{bmatrix} \text{young} \\ \text{old} \\ \text{dead} \end{bmatrix}_{k+1} = \begin{bmatrix} .98 & .00 & 0 \\ .02 & .97 & 0 \\ .00 & .03 & 1 \end{bmatrix} \begin{bmatrix} \text{young} \\ \text{old} \\ \text{dead} \end{bmatrix}_k .$$

6 The sum of the components of $\boldsymbol{x}$ equals the sum of the components of $A\boldsymbol{x}$. If $A\boldsymbol{x} = \lambda\boldsymbol{x}$ with $\lambda \neq 1$, prove that the components of this non-steady eigenvector $\boldsymbol{x}$ add to zero.

7 Find the eigenvalues and eigenvectors of A. Factor A into $S\Lambda S^{-1}$:

$$A = \begin{bmatrix} .8 & .3 \\ .2 & .7 \end{bmatrix}.$$

This was a MATLAB example in Chapter 1. There A^{16} was computed by squaring four times. What are the factors in $A^{16} = S\Lambda^{16}S^{-1}$?

8 Explain why the powers A^k in Problem 7 approach this matrix A^{∞}:

$$A^{\infty} = \begin{bmatrix} .6 & .6 \\ .4 & .4 \end{bmatrix}.$$

Challenge problem: Which Markov matrices produce that steady state (.6, .4)?

9 This permutation matrix is also a Markov matrix:

$$P = \begin{bmatrix} 0 & 1 & 0 & 0 \\ 0 & 0 & 1 & 0 \\ 0 & 0 & 0 & 1 \\ 1 & 0 & 0 & 0 \end{bmatrix}.$$

The steady state eigenvector for $\lambda = 1$ is $(\frac{1}{4}, \frac{1}{4}, \frac{1}{4}, \frac{1}{4})$. This is *not* approached when $\boldsymbol{u}_0 = (0, 0, 0, 1)$. What are $\boldsymbol{u}_1$ and $\boldsymbol{u}_2$ and $\boldsymbol{u}_3$ and $\boldsymbol{u}_4$? What are the four eigenvalues of P, which solve $\lambda^4 = 1$?

10 Prove that the square of a Markov matrix is also a Markov matrix.

11 If $A = \left[\begin{smallmatrix} a & b \\ c & d \end{smallmatrix}\right]$ is a Markov matrix, its eigenvalues are 1 and _____. The steady state eigenvector is $\boldsymbol{x}_1 =$ _____.

12 Complete the last row to make A a Markov matrix and find the steady state eigenvector:

$$A = \begin{bmatrix} .7 & .1 & .2 \\ .1 & .6 & .3 \\ \text{—} & \text{—} & \text{—} \end{bmatrix}.$$

When A is a symmetric Markov matrix, why is $\boldsymbol{x}_1 = (1, \ldots, 1)$ its steady state?

13 A Markov differential equation is not $du/dt = Au$ but $du/dt = (A - I)u$. Find the eigenvalues of

$$B = A - I = \begin{bmatrix} -.2 & .3 \\ .2 & -.3 \end{bmatrix}.$$

When $e^{\lambda_1 t}$ multiplies the eigenvector x_1 and $e^{\lambda_2 t}$ multiplies x_2, what is the steady state as $t \to \infty$?

14 The matrix $B = A - I$ for a Markov differential equation has each column adding to _____. The steady state x_1 is the same as for A, but now $\lambda_1 =$ _____ and $e^{\lambda_1 t} =$ _____.

Questions 15–18 are about linear algebra in economics.

15 Each row of the consumption matrix in Example 4 adds to .9. Why does that make $\lambda = .9$ an eigenvalue, and what is the eigenvector?

16 Multiply $I + A + A^2 + A^3 + \cdots$ by $I - A$ to show that the series adds to _____. For $A = \begin{bmatrix} 0 & \frac{1}{2} \\ 1 & 0 \end{bmatrix}$, find A^2 and A^3 and use the pattern to add up the series.

17 For which of these matrices does $I + A + A^2 + \cdots$ yield a nonnegative matrix $(I - A)^{-1}$? Then the economy can meet any demand:

$$A = \begin{bmatrix} 0 & 1 \\ 0 & 0 \end{bmatrix} \quad A = \begin{bmatrix} 0 & 4 \\ .2 & 0 \end{bmatrix} \quad A = \begin{bmatrix} .5 & 1 \\ .5 & 0 \end{bmatrix}.$$

18 If the demands in Problem 17 are $y = (2, 6)$, what are the vectors $p = (I - A)^{-1} y$?

19 (Markov again) This matrix has zero determinant. What are its eigenvalues?

$$A = \begin{bmatrix} .4 & .2 & .3 \\ .2 & .4 & .3 \\ .4 & .4 & .4 \end{bmatrix}.$$

Find the limits of $A^k u_0$ starting from $u_0 = (1, 0, 0)$ and then $u_0 = (100, 0, 0)$.

20 If A is a Markov matrix, does $I + A + A^2 + \cdots$ add up to $(I - A)^{-1}$?

LINEAR PROGRAMMING ■ 8.4

Linear programming is linear algebra plus two new ingredients: ***inequalities*** and ***minimization***. The starting point is still a matrix equation $A\boldsymbol{x} = \boldsymbol{b}$. But the only acceptable solutions are *nonnegative*. We require $\boldsymbol{x} \geq 0$ (meaning that no component of $\boldsymbol{x}$ can be negative). The matrix has $n > m$, more unknowns than equations. If there are any nonnegative solutions to $A\boldsymbol{x} = \boldsymbol{b}$, there are probably a lot. Linear programming picks the solution $\boldsymbol{x}^* \geq 0$ that minimizes the cost:

> ***The cost is $c_1x_1 + \cdots + c_nx_n$. The winning vector x^* is the nonnegative solution of $Ax = b$ that has smallest cost.***

Thus a linear programming problem starts with a matrix A and two vectors $\boldsymbol{b}$ and $\boldsymbol{c}$:

i) A has $n > m$: for example $A = [\,1 \quad 1 \quad 2\,]$

ii) $\boldsymbol{b}$ has m components: for example $\boldsymbol{b} = [\,4\,]$

iii) The cost $\boldsymbol{c}$ has n components: for example $\boldsymbol{c} = [\,5 \quad 3 \quad 8\,]$.

Then the problem is to minimize $\boldsymbol{c} \cdot \boldsymbol{x}$ subject to the requirements $A\boldsymbol{x} = \boldsymbol{b}$ and $\boldsymbol{x} \geq 0$:

$$\textbf{\textit{Minimize}}\ 5x_1 + 3x_2 + 8x_3\ \textbf{\textit{subject to}}\ x_1 + x_2 + 2x_3 = 4\ \textbf{\textit{and}}\ x_1, x_2, x_3 \geq 0.$$

We jumped right into the problem, without explaining where it comes from. Linear programming is actually the most important application of mathematics to management. Development of the fastest algorithm and fastest code is highly competitive. You will see that finding $\boldsymbol{x}^*$ is harder than solving $A\boldsymbol{x} = \boldsymbol{b}$, because of the extra requirements: cost minimization and nonnegativity. We will explain the background, and the famous *simplex method*, after solving the example.

Look first at the "constraints": $A\boldsymbol{x} = \boldsymbol{b}$ and $\boldsymbol{x} \geq 0$. The equation $x_1+x_2+2x_3 = 4$ gives a plane in three dimensions. The nonnegativity $x_1 \geq 0, x_2 \geq 0, x_3 \geq 0$ chops the plane down to a triangle. The solution $\boldsymbol{x}^*$ must lie in the triangle PQR in Figure 8.6. Outside that triangle, some components of $\boldsymbol{x}$ are negative. On the edges of that triangle, one component is zero. At the corners of that triangle, two components are zero. ***The solution x^* will be one of those corners!*** We will now show why.

The triangle contains all vectors $\boldsymbol{x}$ that satisfy $A\boldsymbol{x} = \boldsymbol{b}$ and $\boldsymbol{x} \geq 0$. (Those $\boldsymbol{x}$'s are called *feasible points*, and the triangle is the *feasible set*.) These points are the candidates in the minimization of $\boldsymbol{c} \cdot \boldsymbol{x}$, which is the final step:

> ***Find x^* in the triangle to minimize the cost $5x_1 + 3x_2 + 8x_3$.***

The vectors that have *zero* cost lie on the plane $5x_1+3x_2+8x_3 = 0$. That plane does not meet the triangle. We cannot achieve zero cost, while meeting the requirements on $\boldsymbol{x}$. So increase the cost C until the plane $5x_1 + 3x_2 + 8x_3 = C$ does meet the triangle. This is a family of *parallel planes*, one for each C. As C increases, the planes move toward the triangle.

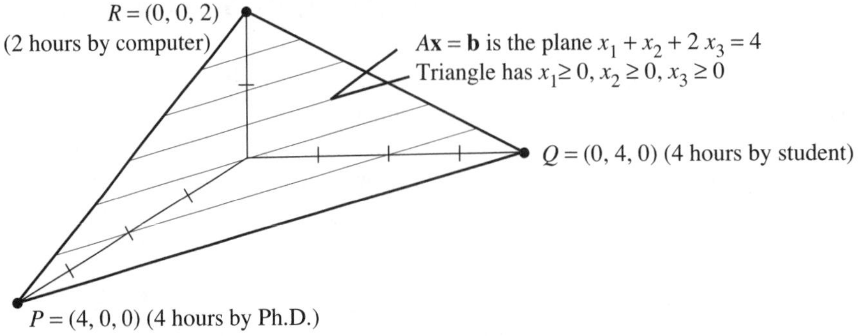

Figure 8.6 The triangle containing nonnegative solutions: $Ax = b$ and $x \geq 0$. The lowest cost solution x^* is one of the corners P, Q, or R.

The first plane to touch the triangle has minimum cost C. *The point where it touches is the solution* x^*. This touching point must be one of the corners P or Q or R. A moving plane could not reach the inside of the triangle before it touches a corner! So check the cost $5x_1 + 3x_2 + 8x_3$ at each corner:

$$P = (4,0,0) \text{ costs } 20 \qquad Q = (0,4,0) \text{ costs } 12 \qquad R = (0,0,2) \text{ costs } 16.$$

The winner is Q. Then $x^* = (0,4,0)$ solves the linear programming problem.

If the cost vector c is changed, the parallel planes are tilted. For small changes, Q is still the winner. For the cost $c \cdot x = 5x_1 + 4x_2 + 7x_3$, the optimum x^* moves to $R = (0,0,2)$. The minimum cost is now $7 \cdot 2 = 14$.

Note 1 Some linear programs *maximize profit* instead of minimizing cost. The mathematics is almost the same. The parallel planes start with a large value of C, instead of a small value. They move toward the origin (instead of away), as C gets smaller. *The first touching point is still a corner.*

Note 2 The requirements $Ax = b$ and $x \geq 0$ could be impossible to satisfy. The equation $x_1 + x_2 + x_3 = -1$ cannot be solved with $x \geq 0$. The feasible set is empty.

Note 3 It could also happen that the feasible set is *unbounded.* If I change the requirement to $x_1 + x_2 - 2x_3 = 4$, the large positive vector $(100, 100, 98)$ is now a candidate. So is the larger vector $(1000, 1000, 998)$. The plane $Ax = b$ is no longer chopped off to a triangle. The two corners P and Q are still candidates for x^*, but the third corner has moved to infinity.

Note 4 With an unbounded feasible set, the minimum cost could be $-\infty$ (*minus infinity*). Suppose the cost is $-x_1 - x_2 + x_3$. Then the vector $(100, 100, 98)$ costs $C = -102$. The vector $(1000, 1000, 998)$ costs $C = -1002$. We are being paid to include x_1 and x_2. Instead of paying a cost for those components. In realistic applications this will not happen. But it is theoretically possible that changes in A, $\boldsymbol{b}$, and $\boldsymbol{c}$ can produce unexpected triangles and costs.

Background to Linear Programming

This first problem is made up to fit the previous example. The unknowns x_1, x_2, x_3 represent hours of work by a Ph.D. and a student and a machine. The costs per hour are \$5, \$3, and \$8. (*I apologize for such low pay.*) The number of hours cannot be negative: $x_1 \geq 0, x_2 \geq 0, x_3 \geq 0$. The Ph.D. and the student get through one homework problem per hour—*the machine solves two problems in one hour*. In principle they can share out the homework, which has four problems to be solved: $x_1 + x_2 + 2x_3 = 4$.

The problem is to finish the four problems at minimum cost.

If all three are working, the job takes one hour: $x_1 = x_2 = x_3 = 1$. The cost is $5 + 3 + 8 = 16$. But certainly the Ph.D. should be put out of work by the student (who is just as fast and costs less—this problem is getting realistic). When the student works two hours and the machine works one, the cost is $6 + 8$ and all four problems get solved. We are on the edge $\boldsymbol{QR}$ of the triangle because the Ph.D. is unemployed: $x_1 = 0$. But the best point is at a corner—all work by student (at $\boldsymbol{Q}$) or all work by machine (at $\boldsymbol{R}$). In this example the student solves four problems in four hours for \$12—the minimum cost.

With only one equation in $A\boldsymbol{x} = \boldsymbol{b}$, the corner $(0, 4, 0)$ has only one nonzero component. When $A\boldsymbol{x} = \boldsymbol{b}$ has m equations, corners have m nonzeros. As in Chapter 3, $n - m$ free variables are set to zero. We solve $A\boldsymbol{x} = \boldsymbol{b}$ for the m basic variables (pivot variables). But unlike Chapter 3, we don't know which m variables to choose as basic. Our choice must minimize the cost.

The number of possible corners is the number of ways to choose m components out of n. This number "n choose m" is heavily involved in gambling and probability. With $n = 20$ unknowns and $m = 8$ equations (still small numbers), the "feasible set" can have $20!/8!12!$ corners. That number is $(20)(19)\cdots(13) = 5{,}079{,}110{,}400$.

Checking three corners for the minimum cost was fine. Checking five billion corners is not the way to go. The simplex method described below is much faster.

The Dual Problem In linear programming, problems come in pairs. There is a minimum problem and a maximum problem—the original and its "dual." The original problem was specified by a matrix A and two vectors $\boldsymbol{b}$ and $\boldsymbol{c}$. The dual problem has the same input, but A is transposed and $\boldsymbol{b}$ and $\boldsymbol{c}$ are switched. Here is the dual to our example:

> **A cheater offers to solve homework problems by looking up the answers.** The charge is y dollars per problem, or $4y$ altogether. (Note how $\boldsymbol{b} = 4$ has gone into the cost.) The cheater must be as cheap as the Ph.D. or student or machine: $y \leq 5$ and $y \leq 3$ and $2y \leq 8$. (Note how $\boldsymbol{c} = (5, 3, 8)$ has gone into inequality constraints). The cheater maximizes the income $4y$.

Dual Problem ***Maximize*** $\boldsymbol{b} \cdot \boldsymbol{y}$ ***subject to*** $A^{\mathrm{T}}\boldsymbol{y} \le \boldsymbol{c}$.

The maximum occurs when $y = 3$. The income is $4y = 12$. The maximum in the dual problem (\$12) equals the minimum in the original (\$12). This is always true:

Duality Theorem If either problem has a best vector ($\boldsymbol{x}^*$ or $\boldsymbol{y}^*$) then so does the other. ***The minimum cost*** $\boldsymbol{c} \cdot \boldsymbol{x}^*$ ***equals the maximum income*** $\boldsymbol{b} \cdot \boldsymbol{y}^*$.

Please note that I personally often look up the answers. It's not cheating.

This book started with a row picture and a column picture. The first "duality theorem" was about rank: The number of independent rows equals the number of independent columns. That theorem, like this one, was easy for small matrices. A proof that minimum cost = maximum income is in our text *Linear Algebra and Its Applications*. Here we establish the easy half of the theorem: ***The cheater's income cannot exceed the honest cost***:

$$\textit{If}\ \ A\boldsymbol{x} = \boldsymbol{b}, \boldsymbol{x} \ge \boldsymbol{0}, A^{\mathrm{T}}\boldsymbol{y} \le \boldsymbol{c} \quad \textit{then} \quad \boldsymbol{b}^{\mathrm{T}}\boldsymbol{y} = (A\boldsymbol{x})^{\mathrm{T}}\boldsymbol{y} = \boldsymbol{x}^{\mathrm{T}}(A^{\mathrm{T}}\boldsymbol{y}) \le \boldsymbol{x}^{\mathrm{T}}\boldsymbol{c}.$$

The full duality theorem says that when $\boldsymbol{b}^{\mathrm{T}}\boldsymbol{y}$ reaches its maximum and $\boldsymbol{x}^{\mathrm{T}}\boldsymbol{c}$ reaches its minimum, they are equal: $\boldsymbol{b} \cdot \boldsymbol{y}^* = \boldsymbol{c} \cdot \boldsymbol{x}^*$.

The Simplex Method

Elimination is the workhorse for linear equations. The simplex method is the workhorse for linear inequalities. We cannot give the simplex method as much space as elimination—but the idea can be briefly described. *The simplex method goes from one corner to a neighboring corner of lower cost.* Eventually (and quite soon in practice) it reaches the corner of minimum cost. This is the solution $\boldsymbol{x}^*$.

A ***corner*** is a vector $\boldsymbol{x} \ge 0$ that satisfies the m equations $A\boldsymbol{x} = \boldsymbol{b}$ with at most m positive components. The other $n - m$ components are *zero*. (Those are the free variables. Back substitution gives the basic variables. All variables must be nonnegative or $\boldsymbol{x}$ is a false corner.) For a *neighboring corner*, one zero component becomes positive and one positive component becomes zero.

The simplex method must decide which component "enters" by becoming positive, and which component "leaves" by becoming zero. That exchange is chosen so as to lower the total cost. This is one step of the simplex method.

Here is the overall plan. Look at each zero component at the current corner. If it changes from 0 to 1, the other nonzeros have to adjust to keep $A\boldsymbol{x} = \boldsymbol{b}$. Find the new $\boldsymbol{x}$ by back substitution and compute the change in the total cost $\boldsymbol{c} \cdot \boldsymbol{x}$. This change is the "reduced cost" r of the new component. The ***entering variable*** is the one that gives the *most negative* r. This is the greatest cost reduction for a single unit of a new variable.

Example 1 Suppose the current corner is $(4, 0, 0)$, with the Ph.D. doing all the work (the cost is \$20). If the student works one hour, the cost of $\boldsymbol{x} = (3, 1, 0)$ is down to \$18. The reduced cost is $r = -2$. If the machine works one hour, then $\boldsymbol{x} = (2, 0, 1)$ also costs \$18. The reduced cost is also $r = -2$. In this case the simplex method can choose either the student or the machine as the entering variable.

Even in this small example, the first step may not go immediately to the best $\boldsymbol{x}^*$. The method chooses the entering variable before it knows how much of that variable to include. We computed r when the entering variable changes from 0 to 1, but one unit may be too much or too little. The method now chooses the leaving variable (the Ph.D.).

The more of the entering variable we include, the lower the cost. This has to stop when one of the positive components (which are adjusting to keep $A\boldsymbol{x} = \boldsymbol{b}$) hits zero. *The **leaving variable** is the first positive x_i to reach zero.* When that happens, a neighboring corner has been found. More of the entering variable would make the leaving variable negative, which is not allowed. We have gone along an edge of the allowed feasible set, from the old corner to the new corner. Then start again (from the new corner) to find the next variables to enter and leave.

When all reduced costs are positive, the current corner is the optimal $\boldsymbol{x}^*$. No zero component can become positive without increasing $\boldsymbol{c} \cdot \boldsymbol{x}$. No new variable should enter. The problem is solved.

Note Generally $\boldsymbol{x}^*$ is reached in αn steps, where α is not large. But examples have been invented which use an exponential number of simplex steps. Eventually a different approach was developed, which is guaranteed to reach $\boldsymbol{x}^*$ in fewer (but more difficult) steps. The new methods travel through the *interior* of the feasible set, to find $\boldsymbol{x}^*$ in polynomial time. Khachian proved this was possible, and Karmarkar made it efficient. There is strong competition between Dantzig's simplex method (traveling around the edges) and Karmarkar's method through the interior.

Example 2 Minimize the cost $\boldsymbol{c} \cdot \boldsymbol{x} = 3x_1 + x_2 + 9x_3 + x_4$. The constraints are $\boldsymbol{x} \geq 0$ and two equations $A\boldsymbol{x} = \boldsymbol{b}$:

$$\begin{aligned} x_1 + 2x_3 + x_4 &= 4 \qquad m = 2 \quad \text{equations} \\ x_2 + x_3 - x_4 &= 2 \qquad n = 4 \quad \text{unknowns.} \end{aligned}$$

A starting corner is $\boldsymbol{x} = (4, 2, 0, 0)$ which costs $\boldsymbol{c} \cdot \boldsymbol{x} = 14$. It has $m = 2$ nonzeros and $n - m = 2$ zeros (x_3 and x_4). The question is whether x_3 or x_4 should enter (become nonzero). Try each of them:

$$\begin{aligned} &\text{If } x_3 = 1 \text{ and } x_4 = 0, \quad \text{then } \boldsymbol{x} = (2, 1, 1, 0) \text{ costs } 16. \\ &\text{If } x_4 = 1 \text{ and } x_3 = 0, \quad \text{then } \boldsymbol{x} = (3, 3, 0, 1) \text{ costs } 13. \end{aligned}$$

Compare those costs with 14. The reduced cost of x_3 is $r = 2$, positive and useless. The reduced cost of x_4 is $r = -1$, negative and helpful. *The entering variable is* x_4.

How much of x_4 can enter? One unit of x_4 made x_1 drop from 4 to 3. Four units will make x_1 drop from 4 to zero (while x_2 increases all the way to 6). *The leaving variable is* x_1. The new corner is $\boldsymbol{x} = (0, 6, 0, 4)$, which costs only $\boldsymbol{c} \cdot \boldsymbol{x} = 10$. This is the optimal $\boldsymbol{x}^*$, but to know that we have to try another simplex step from $(0, 6, 0, 4)$. Suppose x_1 or x_3 tries to enter:

$$\begin{array}{ll} \text{If } x_1 = 1 \text{ and } x_3 = 0, & \text{then } \boldsymbol{x} = (1, 5, 0, 3) \text{ costs } 11. \\ \text{If } x_3 = 1 \text{ and } x_1 = 0, & \text{then } \boldsymbol{x} = (0, 3, 1, 2) \text{ costs } 14. \end{array}$$

Those costs are higher than 10. Both r's are positive—it does not pay to move. The current corner $(0, 6, 0, 4)$ is the solution $\boldsymbol{x}^*$.

These calculations can be streamlined. It turns out that each simplex step solves three linear systems with the same matrix B. (This is the m by m matrix that keeps the m basic columns of A.) When a new column enters and an old column leaves, there is a quick way to update B^{-1}. That is how most computer codes organize the steps of the simplex method.

One final note. We described how to go from one corner to a better neighbor. We did not describe how to find the first corner—which is easy in this example but not always. One way is to create new variables x_5 and x_6, which begin at 4 and 2 (with all the original x's at zero). Then start the simplex method with $x_5 + x_6$ as the cost. Switch to the original problem after x_5 and x_6 reach zero—a starting corner for the original problem has been found.

Problem Set 8.4

1 Draw the region in the xy plane where $x + 2y = 6$ and $x \geq 0$ and $y \geq 0$. Which point in this "feasible set" minimizes the cost $c = x + 3y$? Which point gives maximum cost?

2 Draw the region in the xy plane where $x + 2y \leq 6$, $2x + y \leq 6$, $x \geq 0$, $y \geq 0$. It has four corners. Which corner minimizes the cost $c = 2x - y$?

3 What are the corners of the set $x_1 + 2x_2 - x_3 = 4$ with x_1, x_2, x_3 all ≥ 0? Show that $x_1 + 2x_3$ can be very negative in this set.

4 Start at $\boldsymbol{x} = (0, 0, 2)$ where the machine solves all four problems for \$16. Move to $\boldsymbol{x} = (0, 1, \quad)$ to find the reduced cost r (the savings per hour) for work by the student. Find r for the Ph.D. by moving to $\boldsymbol{x} = (1, 0, \quad)$. Notice that r does not give the *number* of hours or the total savings.

5 Start from $(4, 0, 0)$ with $\boldsymbol{c}$ changed to $[\,5 \;\; 3 \;\; 7\,]$. Show that r is better for the machine but the total cost is lower for the student. The simplex method takes two steps, first to machine and then to student.

6 Choose a different $\boldsymbol{c}$ so the Ph.D. gets the job. Rewrite the dual problem (maximum income to the cheater).

FOURIER SERIES: LINEAR ALGEBRA FOR FUNCTIONS ■ 8.5

This section goes from finite dimensions to *infinite* dimensions. I want to explain linear algebra in infinite-dimensional space, and to show that it still works. First step: look back. This book began with vectors and dot products and linear combinations. We begin by converting those basic ideas to the infinite case—then the rest will follow.

What does it mean for a vector to have infinitely many components? There are two different answers, both good:

1. The vector becomes $\boldsymbol{v} = (v_1, v_2, v_3, \ldots)$. It could be $(1, \frac{1}{2}, \frac{1}{4}, \ldots)$.

2. The vector becomes a function $f(x)$. It could be $\sin x$.

We will go both ways. Then the idea of Fourier series will connect them.

After vectors come *dot products*. The natural dot product of two infinite vectors $(v_1, v_2, \ldots)$ and $(w_1, w_2, \ldots)$ is an infinite series:

$$\boldsymbol{v} \cdot \boldsymbol{w} = v_1 w_1 + v_2 w_2 + \cdots . \tag{1}$$

This brings a new question, which never occurred to us for vectors in $\mathbf{R}^n$. Does this infinite sum add up to a finite number? Does the series converge? Here is the first and biggest difference between finite and infinite.

When $\boldsymbol{v} = \boldsymbol{w} = (1, 1, 1, \ldots)$, the sum certainly does not converge. In that case $\boldsymbol{v} \cdot \boldsymbol{w} = 1 + 1 + 1 + \cdots$ is infinite. Since $\boldsymbol{v}$ equals $\boldsymbol{w}$, we are really computing $\boldsymbol{v} \cdot \boldsymbol{v} = \|\boldsymbol{v}\|^2$ = length squared. The vector $(1, 1, 1, \ldots)$ has infinite length. *We don't want that vector.* Since we are making the rules, we don't have to include it. The only vectors to be allowed are those with finite length:

DEFINITION The vector $(v_1, v_2, \ldots)$ is in our infinite-dimensional "***Hilbert space***" if and only if its length is finite:

$$\|\boldsymbol{v}\|^2 = \boldsymbol{v} \cdot \boldsymbol{v} = v_1^2 + v_2^2 + v_3^2 + \cdots \text{ must add to a finite number.}$$

Example 1 The vector $\boldsymbol{v} = (1, \frac{1}{2}, \frac{1}{4}, \ldots)$ is included in Hilbert space, because its length is $2/\sqrt{3}$. We have a geometric series that adds to $4/3$. The length of $\boldsymbol{v}$ is the square root:

$$\boldsymbol{v} \cdot \boldsymbol{v} = 1 + \tfrac{1}{4} + \tfrac{1}{16} + \cdots = \frac{1}{1 - \frac{1}{4}} = \tfrac{4}{3}.$$

Question If $\boldsymbol{v}$ and $\boldsymbol{w}$ have finite length, how large can their dot product be?

Answer The sum $\boldsymbol{v} \cdot \boldsymbol{w} = v_1 w_1 + v_2 w_2 + \cdots$ also adds to a finite number. The Schwarz inequality is still true:

$$|\boldsymbol{v} \cdot \boldsymbol{w}| \le \|\boldsymbol{v}\| \, \|\boldsymbol{w}\|. \tag{2}$$

The ratio of $\boldsymbol{v} \cdot \boldsymbol{w}$ to $\|\boldsymbol{v}\| \, \|\boldsymbol{w}\|$ is still the cosine of θ (the angle between $\boldsymbol{v}$ and $\boldsymbol{w}$). Even in infinite-dimensional space, $|\cos\theta|$ is not greater than 1.

Now change over to functions. Those are the "vectors." The space of functions $f(x), g(x), h(x), \ldots$ defined for $0 \le x \le 2\pi$ must be somehow bigger than $\mathbf{R}^n$. ***What is the dot product of*** $f(x)$ ***and*** $g(x)$?

Key point in the continuous case: *Sums are replaced by integrals.* Instead of a sum of v_j times w_j, the dot product is an integral of $f(x)$ times $g(x)$. Change the "dot" to parentheses with a comma, and change the words "dot product" to *inner product*:

DEFINITION The ***inner product*** of $f(x)$ and $g(x)$, and the ***length squared***, are

$$(f, g) = \int_0^{2\pi} f(x)g(x)\,dx \quad \text{and} \quad \|f\|^2 = \int_0^{2\pi} \big(f(x)\big)^2\,dx. \tag{3}$$

The interval $[0, 2\pi]$ where the functions are defined could change to a different interval like $[0, 1]$. We chose 2π because our first examples are $\sin x$ and $\cos x$.

Example 2 The length of $f(x) = \sin x$ comes from its inner product with itself:

$$(f, f) = \int_0^{2\pi} (\sin\, x)^2\,dx = \pi. \quad \text{The length of } \sin x \text{ is } \sqrt{\pi}.$$

That is a standard integral in calculus—not part of linear algebra. By writing $\sin^2 x$ as $\frac{1}{2} - \frac{1}{2}\cos 2x$, we see it go above and below its average value $\frac{1}{2}$. Multiply that average by the interval length 2π to get the answer π.

More important: ***The functions*** $\sin x$ ***and*** $\cos x$ ***are orthogonal.*** Their inner product is *zero*:

$$\int_0^{2\pi} \sin x \cos x\,dx = \int_0^{2\pi} \tfrac{1}{2}\sin 2x\,dx = \left[-\tfrac{1}{4}\cos 2x\right]_0^{2\pi} = 0. \tag{4}$$

This zero is no accident. It is highly important to science. The orthogonality goes beyond the two functions $\sin x$ and $\cos x$, to an infinite list of sines and cosines. The list contains $\cos 0x$ (which is 1), $\sin x, \cos x, \sin 2x, \cos 2x, \sin 3x, \cos 3x, \ldots$.

Every function in that list is orthogonal to every other function in the list.

The next step is to look at linear combinations of those sines and cosines.

Fourier Series

The Fourier series of a function $y(x)$ is its expansion into sines and cosines:

$$y(x) = a_0 + a_1 \cos x + b_1 \sin x + a_2 \cos 2x + b_2 \sin 2x + \cdots. \tag{5}$$

We have an orthogonal basis! The vectors in "function space" are combinations of the sines and cosines. On the interval from $x = 2\pi$ to $x = 4\pi$, all our functions repeat what they did from 0 to 2π. They are "*periodic*." The distance between repetitions (the period) is 2π.

Remember: The list is infinite. The Fourier series is an infinite series. Just as we avoided the vector $\boldsymbol{v} = (1, 1, 1, \ldots)$ because its length is infinite, so we avoid a function like $\frac{1}{2} + \cos x + \cos 2x + \cos 3x + \cdots$. (*Note*: This is π times the famous delta function. It is an infinite "spike" above a single point. At $x = 0$ its height $\frac{1}{2}+1+1+\cdots$ is infinite. At all points inside $0 < x < 2\pi$ the series adds in some average way to zero.) The delta function has infinite length, and regretfully it is excluded from our space of functions.

Compute the length of a typical sum $f(x)$:

$$\begin{aligned}(f, f) &= \int_0^{2\pi} (a_0 + a_1 \cos x + b_1 \sin x + a_2 \cos 2x + \cdots)^2\, dx \\ &= \int_0^{2\pi} (a_0^2 + a_1^2 \cos^2 x + b_1^2 \sin^2 x + a_2^2 \cos^2 2x + \cdots)\, dx \\ &= 2\pi a_0^2 + \pi(a_1^2 + b_1^2 + a_2^2 + \cdots).\end{aligned} \tag{6}$$

The step from line 1 to line 2 used orthogonality. All products like $\cos x \cos 2x$ and $\sin x \cos 3x$ integrate to give zero. Line 2 contains what is left—the integrals of each sine and cosine squared. Line 3 evaluates those integrals. Unfortunately the integral of 1^2 is 2π, when all other integrals give π. If we divide by their lengths, our functions become *orthonormal*:

$$\frac{1}{\sqrt{2\pi}}, \frac{\cos x}{\sqrt{\pi}}, \frac{\sin x}{\sqrt{\pi}}, \frac{\cos 2x}{\sqrt{\pi}}, \ldots \textit{ is an orthonormal basis for our function space.}$$

These are unit vectors. We could combine them with coefficients $A_0, A_1, B_1, A_2, \ldots$ to yield a function $F(x)$. Then the 2π and the π's drop out of the formula for length. Equation 6 becomes *function length = vector length*:

$$\|F\|^2 = (F, F) = A_0^2 + A_1^2 + B_1^2 + A_2^2 + \cdots. \tag{7}$$

Here is the important point, for $f(x)$ as well as $F(x)$. *The function has finite length exactly when the vector of coefficients has finite length.* The integral of $\big(F(x)\big)^2$ matches the sum of coefficients squared. Through Fourier series, we have a perfect match between function space and infinite-dimensional Hilbert space. On one side is the function, on the other side are its Fourier coefficients.

8B The function space contains $f(x)$ exactly when the Hilbert space contains the vector $\boldsymbol{v} = (a_0, a_1, b_1, \ldots)$ of Fourier coefficients. Both $f(x)$ and $\boldsymbol{v}$ have finite length.

Example 3 Suppose $f(x)$ is a "square wave," equal to -1 for negative x and $+1$ for positive x. That looks like a step function, not a wave. But remember that $f(x)$

must repeat after each interval of length 2π. We should have said

$$f(x) = -1 \text{ for } \quad -\pi < x < 0 + 1 \text{ for } \quad 0 < x < \pi.$$

The wave goes back to -1 for $\pi < x < 2\pi$. It is an odd function like the sines, and all its cosine coefficients are zero. We will find its Fourier series, containing only sines:

$$f(x) = \frac{4}{\pi}\left[\frac{\sin x}{1} + \frac{\sin 3x}{3} + \frac{\sin 5x}{5} + \cdots\right]. \tag{8}$$

This square wave has length $\sqrt{2\pi}$, because at every point $\big(f(x)\big)^2$ is $(-1)^2$ or $(+1)^2$:

$$\|f\|^2 = \int_0^{2\pi} \big(f(x)\big)^2\,dx = \int_0^{2\pi} 1\,dx = 2\pi.$$

At $x = 0$ the sines are zero and the Fourier series 8 gives zero. This is half way up the jump from -1 to $+1$. The Fourier series is also interesting when $x = \frac{\pi}{2}$. At this point the square wave equals 1, and the sines in equation 8 alternate between $+1$ and -1:

$$1 = \frac{4}{\pi}\left(1 - \frac{1}{3} + \frac{1}{5} - \frac{1}{7} + \cdots\right). \tag{9}$$

Multiply through by π to find a magical formula $4(1 - \frac{1}{3} + \frac{1}{5} - \frac{1}{7} + \cdots)$ for that famous number.

The Fourier Coefficients

How do we find the a's and b's which multiply the cosines and sines? For a given function $f(x)$, we are asking for its Fourier coefficients:

$$f(x) = a_0 + a_1 \cos x + b_1 \sin x + a_2 \cos 2x + \cdots.$$

Here is the way to find a_1. ***Multiply both sides by $\cos x$. Then integrate from 0 to 2π.*** The key is orthogonality! All integrals on the right side are zero, except the integral of $a_1 \cos^2 x$:

$$\int_0^{2\pi} f(x)\cos x\,dx = \int_0^{2\pi} a_1 \cos^2 x\,dx = \pi a_1. \tag{10}$$

Divide by π and you have a_1. To find any other a_k, multiply the Fourier series by $\cos kx$. Integrate from 0 to 2π. Use orthogonality, so only the integral of $a_k \cos^2 kx$ is left. That integral is πa_k, and divide by π:

$$a_k = \frac{1}{\pi}\int_0^{2\pi} f(x)\cos kx\,dx \quad \text{and similarly} \quad b_k = \frac{1}{\pi}\int_0^{2\pi} f(x)\sin kx\,dx. \tag{11}$$

The exception is a_0. This time we multiply by $\cos 0x = 1$. The integral of 1 is 2π:

$$a_0 = \frac{1}{2\pi}\int_0^{2\pi} f(x)\cdot 1\,dx = \text{average value of } f(x). \tag{12}$$

I used those formulas to find the coefficients in 8 for the square wave. The integral of $f(x)\cos kx$ was zero. The integral of $f(x)\sin kx$ was $4/k$ for odd k.

The point to emphasize is how this infinite-dimensional case is so much like the n-dimensional case. Suppose the nonzero vectors $\boldsymbol{v}_1, \ldots, \boldsymbol{v}_n$ are orthogonal. We want to write the vector $\boldsymbol{b}$ as a combination of those $\boldsymbol{v}$'s:

$$\boldsymbol{b} = c_1\boldsymbol{v}_1 + c_2\boldsymbol{v}_2 + \cdots + c_n\boldsymbol{v}_n. \tag{13}$$

Multiply both sides by $\boldsymbol{v}_1^{\mathrm{T}}$. Use orthogonality, so $\boldsymbol{v}_1^{\mathrm{T}}\boldsymbol{v}_2 = 0$. Only the c_1 term is left:

$$\boldsymbol{v}_1^{\mathrm{T}}\boldsymbol{b} = c_1\boldsymbol{v}_1^{\mathrm{T}}\boldsymbol{v}_1 + 0 + \cdots + 0. \quad \text{Therefore } c_1 = \frac{\boldsymbol{v}_1^{\mathrm{T}}\boldsymbol{b}}{\boldsymbol{v}_1^{\mathrm{T}}\boldsymbol{v}_1}. \tag{14}$$

The denominator $\boldsymbol{v}_1^{\mathrm{T}}\boldsymbol{v}_1$ is the length squared, like π in equation (11). The numerator $\boldsymbol{v}_1^{\mathrm{T}}\boldsymbol{b}$ is the inner product like $\int f(x)\cos kx\,dx$. ***Coefficients are easy to find when the basis vectors are orthogonal.*** We are just doing one-dimensional projections, to find the components along each basis vector.

The formulas are even better when the vectors are orthonormal. Then we have unit vectors. The denominators $\boldsymbol{v}_k^{\mathrm{T}}\boldsymbol{v}_k$ are all 1. In this orthonormal case,

$$c_1 = \boldsymbol{v}_1^{\mathrm{T}}\boldsymbol{b} \quad\text{and}\quad c_2 = \boldsymbol{v}_2^{\mathrm{T}}\boldsymbol{b} \quad\text{and}\quad c_n = \boldsymbol{v}_n^{\mathrm{T}}\boldsymbol{b}. \tag{15}$$

You know this in another form. The equation for the c's is

$$c_1\boldsymbol{v}_1 + \cdots + c_n\boldsymbol{v}_n = \boldsymbol{b} \quad\text{or}\quad \begin{bmatrix} & & \\ \boldsymbol{v}_1 & \cdots & \boldsymbol{v}_n \\ & & \end{bmatrix}\begin{bmatrix} c_1 \\ \vdots \\ c_n \end{bmatrix} = \boldsymbol{b}.$$

This is an orthogonal matrix Q! Its inverse is Q^{T}. That gives the c's in (15):

$$Q\boldsymbol{c} = \boldsymbol{b} \quad\text{yields}\quad \boldsymbol{c} = Q^{\mathrm{T}}\boldsymbol{b}. \quad \text{Row by row this is } c_i = \boldsymbol{v}_i^{\mathrm{T}}\boldsymbol{b}.$$

Fourier series is like having a matrix with infinitely many orthogonal columns. Those columns are the basis functions $1, \cos x, \sin x, \ldots$. After dividing by their lengths we have an "infinite orthogonal matrix." Its inverse is its transpose. The formulas for the Fourier coefficients are like (15) when we have unit vectors and like (14) when we don't. Orthogonality is what reduces an infinite series to one single term.

Problem Set 8.5

1 Integrate the trig identity $2\cos jx \cos kx = \cos(j+k)x + \cos(j-k)x$ to show that $\cos jx$ is orthogonal to $\cos kx$, provided $j \neq k$. What is the result when $j = k$?

2 Show that the three functions $1, x$, and $x^2 - \frac{1}{3}$ are orthogonal, when the integration is from $x = -1$ to $x = 1$. Write $f(x) = 2x^2$ as a combination of those orthogonal functions.

3 Find a vector $(w_1, w_2, w_3, \ldots)$ that is orthogonal to $\boldsymbol{v} = (1, \frac{1}{2}, \frac{1}{4}, \ldots)$. Compute its length $\|\boldsymbol{w}\|$.

4 The first three *Legendre polynomials* are $1, x$, and $x^2 - \frac{1}{3}$. Choose the number c so that the fourth polynomial $x^3 - cx$ is orthogonal to the first three. The integrals still go from -1 to 1.

5 For the square wave $f(x)$ in Example 3, show that

$$\int_0^{2\pi} f(x)\cos x\,dx = 0 \qquad \int_0^{2\pi} f(x)\sin x\,dx = 4 \qquad \int_0^{2\pi} f(x)\sin 2x\,dx = 0.$$

Which Fourier coefficients come from those integrals?

6 The square wave has $\|f\|^2 = 2\pi$. This equals what remarkable sum by equation 6?

7 Graph the square wave. Then graph by hand the sum of two sine terms in its series, or graph by machine the sum of two, three, and four terms.

8 Find the lengths of these vectors in Hilbert space:

(a) $\boldsymbol{v} = \left(\frac{1}{\sqrt{1}}, \frac{1}{\sqrt{2}}, \frac{1}{\sqrt{4}}, \ldots\right)$

(b) $\boldsymbol{v} = (1, a, a^2, \ldots)$

(c) $f(x) = 1 + \sin x$.

9 Compute the Fourier coefficients a_k and b_k for $f(x)$ defined from 0 to 2π:

(a) $f(x) = 1$ for $0 \le x \le \pi$, $f(x) = 0$ for $\pi < x < 2\pi$

(b) $f(x) = x$.

10 When $f(x)$ has period 2π, why is its integral from $-\pi$ to π the same as from 0 to 2π? If $f(x)$ is an *odd* function, $f(-x) = -f(x)$, show that $\int_0^{2\pi} f(x)\,dx$ is zero.

11 From trig identities find the only two terms in the Fourier series for $f(x)$:

(a) $f(x) = \cos^2 x$

(b) $f(x) = \cos\left(x + \frac{\pi}{3}\right)$

12 The functions 1, $\cos x$, $\sin x$, $\cos 2x$, $\sin 2x$, . . . are a basis for Hilbert space. Write the derivatives of those first five functions as combinations of the same five functions. What is the 5 by 5 "differentiation matrix" for these functions?

13 Write the complete solution to $dy/dx = \cos x$ as a particular solution plus any solution to $dy/dx = 0$.

COMPUTER GRAPHICS ■ 8.6

Computer graphics deals with three-dimensional images. The images are moved around. Their scale is changed. They are projected into two dimensions. All the main operations are done by matrices—but the shape of these matrices is surprising.

The transformations of three-dimensional space are done with 4 *by* 4 *matrices.* You would expect 3 by 3. The reason for the change is that one of the four key operations cannot be done with a 3 by 3 matrix multiplication. Here are the four operations:

Translation(shift the origin to another point $P_0 = (x_0, y_0, z_0)$)

Rescaling(by c **in all directions or by different factors** c_1, c_2, c_3)

Rotation(around an axis through the origin or an axis through P_0)

Projection(onto a plane through the origin or a plane through P_0).

Translation is the easiest—just add (x_0, y_0, z_0) to every point. But this is not linear! No 3 by 3 matrix can move the origin. So we change the coordinates of the origin to $(0, 0, 0, 1)$. This is why the matrices are 4 by 4. The "*homogeneous coordinates*" of the point (x, y, z) are $(x, y, z, 1)$ and we now show how they work.

1. Translation Shift the whole three-dimensional space along the vector $\boldsymbol{v}_0$. The origin moves to (x_0, y_0, z_0). This vector $\boldsymbol{v}_0$ is added to every point $\boldsymbol{v}$ in $\mathbf{R}^3$. Using homogeneous coordinates, the 4 by 4 matrix T shifts the whole space by $\boldsymbol{v}_0$:

$$\textit{Translation matrix} \qquad T = \begin{bmatrix} 1 & 0 & 0 & 0 \\ 0 & 1 & 0 & 0 \\ 0 & 0 & 1 & 0 \\ x_0 & y_0 & z_0 & 1 \end{bmatrix}.$$

Important: *Computer graphics works with row vectors.* We have row times matrix instead of matrix times column. You can quickly check that $[0\ \ 0\ \ 0\ \ 1]\,T = [x_0\ \ y_0\ \ z_0\ \ 1]$.

To move the points $(0, 0, 0)$ and (x, y, z) by $\boldsymbol{v}_0$, change to homogeneous coordinates $(0, 0, 0, 1)$ and $(x, y, z, 1)$. Then multiply by T. A row vector times T gives a row vector: ***Every*** $\boldsymbol{v}$ ***moves to*** $\boldsymbol{v} + \boldsymbol{v}_0$: $[x\ \ y\ \ z\ \ 1]\,T = [x + x_0\ \ y + y_0\ \ z + z_0\ \ 1]$.

The output tells where any $\boldsymbol{v}$ will move. (It goes to $\boldsymbol{v} + \boldsymbol{v}_0$.) Translation is now achieved by a matrix, which was impossible in $\mathbf{R}^3$.

2. Scaling To make a picture fit a page, we change its width and height. A Xerox copier will rescale a figure by 90%. In linear algebra, we multiply by .9 times the identity matrix. That matrix is normally 2 by 2 for a plane and 3 by 3 for a solid. In computer graphics, with homogeneous coordinates, the matrix is *one size larger:*

$$\textit{Rescale the plane:} \quad S = \begin{bmatrix} .9 & & \\ & .9 & \\ & & 1 \end{bmatrix} \qquad \textit{Rescale a solid:} \quad S = \begin{bmatrix} c & 0 & 0 & 0 \\ 0 & c & 0 & 0 \\ 0 & 0 & c & 0 \\ 0 & 0 & 0 & 1 \end{bmatrix}.$$

Important: S is not cI. We keep the 1 in the lower corner. Then $[x, y, 1]$ times S is the correct answer in homogeneous coordinates. The origin stays in position because $[0\ 0\ 1]S = [0\ 0\ 1]$.

If we change that 1 to c, the result is strange. ***The point*** (cx, cy, cz, c) ***is the same as*** $(x, y, z, 1)$. The special property of homogeneous coordinates is that *multiplying by cI does not move the point.* The origin in $\mathbf{R}^3$ has homogeneous coordinates $(0, 0, 0, 1)$ and $(0, 0, 0, c)$ for every nonzero c. This is the idea behind the word "homogeneous."

Scaling can be different in different directions. To fit a full-page picture onto a half-page, scale the y direction by $\frac{1}{2}$. To create a margin, scale the x direction by $\frac{3}{4}$. The graphics matrix is diagonal but not 2 by 2. It is 3 by 3 to rescale a plane and 4 by 4 to rescale a space:

$$\textbf{\textit{Scaling matrices}} \quad S = \begin{bmatrix} \frac{3}{4} & & \\ & \frac{1}{2} & \\ & & 1 \end{bmatrix} \quad \text{and} \quad S = \begin{bmatrix} c_1 & & & \\ & c_2 & & \\ & & c_3 & \\ & & & 1 \end{bmatrix}.$$

That last matrix S rescales the x, y, z directions by positive numbers c_1, c_2, c_3. The point at the origin doesn't move, because $\begin{bmatrix} 0\ 0\ 0\ 1 \end{bmatrix} S = \begin{bmatrix} 0\ 0\ 0\ 1 \end{bmatrix}$.

Summary The scaling matrix S is the same size as the translation matrix T. They can be multiplied. To translate and then rescale, multiply $\boldsymbol{v}TS$. To rescale and then translate, multiply $\boldsymbol{v}ST$. (Are those different? *Yes.*) The extra column in all these matrices leaves the extra 1 at the end of every vector.

The point (x, y, z) in $\mathbf{R}^3$ has homogeneous coordinates $(x, y, z, 1)$ in $\mathbf{P}^3$. This "projective space" is not the same as $\mathbf{R}^4$. It is still three-dimensional. To achieve such a thing, (cx, cy, cz, c) is the same point as $(x, y, z, 1)$. Those points of $\mathbf{P}^3$ are really lines through the origin in $\mathbf{R}^4$.

Computer graphics uses ***affine*** transformations, *linear plus shift*. An affine transformation T is executed on $\mathbf{P}^3$ by a 4 by 4 matrix with a special fourth column:

$$A = \begin{bmatrix} a_{11} & a_{12} & a_{13} & 0 \\ a_{21} & a_{22} & a_{23} & 0 \\ a_{31} & a_{32} & a_{33} & 0 \\ a_{41} & a_{42} & a_{43} & 1 \end{bmatrix} = \begin{bmatrix} T(1,0,0) & 0 \\ T(0,1,0) & 0 \\ T(0,0,1) & 0 \\ T(0,0,0) & 1 \end{bmatrix}.$$

The usual 3 by 3 matrix tells us three outputs, this tells four. The usual outputs come from the inputs $(1, 0, 0)$ and $(0, 1, 0)$ and $(0, 0, 1)$. When the transformation is linear, three outputs reveal everything. When the transformation is affine, the matrix also contains the output from $(0, 0, 0)$.Then we know the shift.

3. Rotation A rotation in $\mathbf{R}^2$ or $\mathbf{R}^3$ is achieved by an orthogonal matrix Q. The determinant is $+1$. (With determinant -1 we get an extra reflection through a mirror.) Include the extra column when you use homogeneous coordinates!

$$\textbf{\textit{Plane rotation}} \quad Q = \begin{bmatrix} \cos\theta & -\sin\theta \\ \sin\theta & \cos\theta \end{bmatrix} \quad \text{becomes} \quad R = \begin{bmatrix} \cos\theta & -\sin\theta & 0 \\ \sin\theta & \cos\theta & 0 \\ 0 & 0 & 1 \end{bmatrix}.$$

This matrix rotates the plane around the origin. ***How would we rotate around a different point*** (4, 5)? The answer brings out the beauty of homogeneous coordinates. ***Translate*** (4, 5) ***to*** (0, 0)***, then rotate by*** θ***, then translate*** (0, 0) ***back to*** (4, 5):

$$\boldsymbol{v}\,T_-R\,T_+ = \begin{bmatrix} x & y & 1 \end{bmatrix} \begin{bmatrix} 1 & 0 & 0 \\ 0 & 1 & 0 \\ -4 & -5 & 1 \end{bmatrix} \begin{bmatrix} \cos\theta & -\sin\theta & 0 \\ \sin\theta & \cos\theta & 0 \\ 0 & 0 & 1 \end{bmatrix} \begin{bmatrix} 1 & 0 & 0 \\ 0 & 1 & 0 \\ 4 & 5 & 1 \end{bmatrix}.$$

I won't multiply. The point is to apply the matrices one at a time: $\boldsymbol{v}$ translates to $\boldsymbol{v}T_-$, then rotates to $\boldsymbol{v}T_-R$, and translates back to $\boldsymbol{v}T_-RT_+$. Because each point $\begin{bmatrix} x & y & 1 \end{bmatrix}$ is a row vector, T_- acts first. The center of rotation (4, 5)—otherwise known as (4, 5, 1)—moves first to (0, 0, 1). Rotation doesn't change it. Then T_+ moves it back to (4, 5, 1). All as it should be. The point (4, 6, 1) moves to (0, 1, 1), then turns by θ and moves back.

In three dimensions, every rotation Q turns around an axis. The axis doesn't move—it is a line of eigenvectors with $\lambda = 1$. Suppose the axis is in the z direction. The 1 in Q is to leave the z axis alone, the extra 1 in R is to leave the origin alone:

$$Q = \begin{bmatrix} \cos\theta & -\sin\theta & 0 \\ \sin\theta & \cos\theta & 0 \\ 0 & 0 & 1 \end{bmatrix} \quad \text{and} \quad R = \begin{bmatrix} & & & 0 \\ & Q & & 0 \\ & & & 0 \\ 0 & 0 & 0 & 1 \end{bmatrix}.$$

Now suppose the rotation is around the unit vector $\boldsymbol{a} = (a_1, a_2, a_3)$. With this axis $\boldsymbol{a}$, the rotation matrix Q which fits into R has three parts:

$$Q = (\cos\theta)I + (1-\cos\theta) \begin{bmatrix} a_1^2 & a_1a_2 & a_1a_3 \\ a_1a_2 & a_2^2 & a_2a_3 \\ a_1a_3 & a_2a_3 & a_3^2 \end{bmatrix} - \sin\theta \begin{bmatrix} 0 & a_3 & -a_2 \\ -a_3 & 0 & a_1 \\ a_2 & -a_1 & 0 \end{bmatrix}. \tag{1}$$

The axis doesn't move because $\boldsymbol{a}Q = \boldsymbol{a}$. When $\boldsymbol{a} = (0, 0, 1)$ is in the z direction, this Q becomes the previous Q—for rotation around the z axis.

The linear transformation Q always goes in the upper left block of R. Below it we see zeros, because rotation leaves the origin in place. When those are not zeros, the transformation is affine and the origin moves.

4. Projection In a linear algebra course, most planes go through the origin. In real life, most don't. A plane through the origin is a vector space. The other planes are affine spaces, sometimes called "flats." An affine space is what comes from translating a vector space.

We want to project three-dimensional vectors onto planes. Start with a plane through the origin, whose unit normal vector is $\boldsymbol{n}$. (We will keep $\boldsymbol{n}$ as a column vector.) The vectors in the plane satisfy $\boldsymbol{n}^{\mathrm{T}}\boldsymbol{v} = 0$. ***The usual projection onto the plane is the matrix*** $\boldsymbol{I - nn^{\mathrm{T}}}$. To project a vector, multiply by this matrix. The vector $\boldsymbol{n}$ is projected to zero, and the in-plane vectors $\boldsymbol{v}$ are projected onto themselves:

$$(I - \boldsymbol{n}\boldsymbol{n}^{\mathrm{T}})\boldsymbol{n} = \boldsymbol{n} - \boldsymbol{n}(\boldsymbol{n}^{\mathrm{T}}\boldsymbol{n}) = \boldsymbol{0} \quad \text{and} \quad (I - \boldsymbol{n}\boldsymbol{n}^{\mathrm{T}})\boldsymbol{v} = \boldsymbol{v} - \boldsymbol{n}(\boldsymbol{n}^{\mathrm{T}}\boldsymbol{v}) = \boldsymbol{v}.$$

In homogeneous coordinates the projection matrix becomes 4 by 4 (but the origin doesn't move):

$$\textbf{\textit{Projection onto the plane}} \quad \boldsymbol{n}^{\mathrm{T}}\boldsymbol{v} = 0 \qquad P = \begin{bmatrix} & & & 0 \\ & I - \boldsymbol{n}\boldsymbol{n}^{\mathrm{T}} & & 0 \\ & & & 0 \\ 0 & 0 & 0 & 1 \end{bmatrix}.$$

Now project onto a plane $\boldsymbol{n}^{\mathrm{T}}(\boldsymbol{v} - \boldsymbol{v}_0) = 0$ that does *not* go through the origin. One point on the plane is $\boldsymbol{v}_0$. This is an affine space (or a ***flat***). It is like the solutions to $A\boldsymbol{v} = \boldsymbol{b}$ when the right side is not zero. One particular solution $\boldsymbol{v}_0$ is added to the nullspace—to produce a flat.

The projection onto the flat has three steps. Translate $\boldsymbol{v}_0$ to the origin by T_-. Project along the $\boldsymbol{n}$ direction, and translate back along the row vector $\boldsymbol{v}_0$:

$$\textbf{\textit{Projection onto a flat}} \qquad T_- P T_+ = \begin{bmatrix} I & 0 \\ -\boldsymbol{v}_0 & 1 \end{bmatrix} \begin{bmatrix} I - \boldsymbol{n}\boldsymbol{n}^{\mathrm{T}} & 0 \\ 0 & 1 \end{bmatrix} \begin{bmatrix} I & 0 \\ \boldsymbol{v}_0 & 1 \end{bmatrix}.$$

I can't help noticing that T_- and T_+ are inverse matrices: translate and translate back. They are like the elementary matrices of Chapter 2.

The exercises will include reflection matrices, also known as *mirror matrices*. These are the fifth type needed in computer graphics. A reflection moves each point twice as far as a projection—***the reflection goes through the plane and out the other side***. So change the projection $I - \boldsymbol{n}\boldsymbol{n}^{\mathrm{T}}$ to $I - 2\boldsymbol{n}\boldsymbol{n}^{\mathrm{T}}$ for a mirror matrix.

The matrix P gave a "*parallel*" projection. All points move parallel to $\boldsymbol{n}$, until they reach the plane. The other choice in computer graphics is a "*perspective*" projection. This is more popular because it includes foreshortening. With perspective, an object looks larger as it moves closer. Instead of staying parallel to $\boldsymbol{n}$ (and parallel to each other), the lines of projection come *toward the eye*—the center of projection. This is how we perceive depth in a two-dimensional photograph.

The basic problem of computer graphics starts with a scene and a viewing position. Ideally, the image on the screen is what the viewer would see. The simplest image assigns just one bit to every small picture element—called a ***pixel***. It is light or dark. This gives a black and white picture with no shading. You would not approve. In practice, we assign shading levels between 0 and 2^8 for three colors like red, green, and blue. That means $8 \times 3 = 24$ bits for each pixel. Multiply by the number of pixels, and a lot of memory is needed!

Physically, a *raster frame buffer* directs the electron beam. It scans like a television set. The quality is controlled by the number of pixels and the number of bits per pixel. In this area, one standard text is *Computer Graphics: Principles and Practices* by Foley, Van Dam, Feiner, and Hughes (Addison-Wesley, 1990). My best references were notes by Ronald Goldman (Rice University) and by Tony DeRose (University of Washington, now associated with Pixar).

■ REVIEW OF THE KEY IDEAS ■

1. Computer graphics needs shift operations $T(\boldsymbol{v}) = \boldsymbol{v} + \boldsymbol{v}_0$ as well as linear operations $T(\boldsymbol{v}) = A\boldsymbol{v}$.

2. A shift in $\mathbf{R}^n$ can be executed by a matrix of order $n+1$, using homogeneous coordinates.

3. The extra component 1 in $[x\ y\ z\ 1]$ is preserved when all matrices have the numbers $0, 0, 0, 1$ as last column.

Problem Set 8.6

1 A typical point in $\mathbf{R}^3$ is $x\boldsymbol{i} + y\boldsymbol{j} + z\boldsymbol{k}$. The coordinate vectors $\boldsymbol{i}$, $\boldsymbol{j}$, and $\boldsymbol{k}$ are $(1,0,0)$, $(0,1,0)$, $(0,0,1)$. The coordinates of the point are (x, y, z).

This point in computer graphics is $x\boldsymbol{i} + y\boldsymbol{j} + z\boldsymbol{k} +$ **origin**. Its homogeneous coordinates are (, , ,). Other coordinates for the same point are (, , ,).

2 A linear transformation T is determined when we know $T(\boldsymbol{i}), T(\boldsymbol{j}), T(\boldsymbol{k})$. For an affine transformation we also need $T($____$)$. The input point $(x, y, z, 1)$ is transformed to $xT(\boldsymbol{i}) + yT(\boldsymbol{j}) + zT(\boldsymbol{k}) +$ ____.

3 Multiply the 4 by 4 matrix T for translation along $(1,4,3)$ and the matrix T_1 for translation along $(0,2,5)$. The product TT_1 is translation along ____.

4 Write down the 4 by 4 matrix S that scales by a constant c. Multiply ST and also TS, where T is translation by $(1,4,3)$. To blow up the picture around the center point $(1,4,3)$, would you use $\boldsymbol{v}ST$ or $\boldsymbol{v}TS$?

5 What scaling matrix S (in homogeneous coordinates, so 3 by 3) would make this page square?

6 What 4 by 4 matrix would move a corner of a cube to the origin and then multiply all lengths by 2? The corner of the cube is originally at $(1,1,2)$.

7 When the three matrices in equation 1 multiply the unit vector $\boldsymbol{a}$, show that they give $(\cos\theta)\boldsymbol{a}$ and $(1-\cos\theta)\boldsymbol{a}$ and $\mathbf{0}$. Addition gives $\boldsymbol{a}Q = \boldsymbol{a}$ and the rotation axis is not moved.

8 If $\boldsymbol{b}$ is perpendicular to $\boldsymbol{a}$, multiply by the three matrices in 1 to get $(\cos\theta)\boldsymbol{b}$ and $\mathbf{0}$ and a vector perpendicular to $\boldsymbol{b}$. So $Q\boldsymbol{b}$ makes an angle θ with $\boldsymbol{b}$. ***This is rotation.***

9 What is the 3 by 3 projection matrix $I - \boldsymbol{n}\boldsymbol{n}^{\mathrm{T}}$ onto the plane $\frac{2}{3}x + \frac{2}{3}y + \frac{1}{3}z = 0$? In homogeneous coordinates add $0, 0, 0, 1$ as an extra row and column in P.

10 With the same 4 by 4 matrix P, multiply T_-PT_+ to find the projection matrix onto the plane $\frac{2}{3}x+\frac{2}{3}y+\frac{1}{3}z=1$. The translation T_- moves a point on that plane (choose one) to $(0,0,0,1)$. The inverse matrix T_+ moves it back.

11 Project $(3,3,3)$ onto those planes. Use P in Problem 9 and T_-PT_+ in Problem 10.

12 If you project a square onto a plane, what shape do you get?

13 If you project a cube onto a plane, what is the outline of the projection? Make the projection plane perpendicular to a diagonal of the cube.

14 The 3 by 3 mirror matrix that reflects through the plane $\boldsymbol{n}^{\mathrm{T}}\boldsymbol{v}=0$ is $M=I-2\boldsymbol{n}\boldsymbol{n}^{\mathrm{T}}$. Find the reflection of the point $(3,3,3)$ in the plane $\frac{2}{3}x+\frac{2}{3}y+\frac{1}{3}z=0$.

15 Find the reflection of $(3,3,3)$ in the plane $\frac{2}{3}x+\frac{2}{3}y+\frac{1}{3}z=1$. Take three steps T_-MT_+ using 4 by 4 matrices: translate by T_- so the plane goes through the origin, reflect the translated point $(3,3,3,1)T_-$ in that plane, then translate back by T_+.

16 The vector between the origin $(0,0,0,1)$ and the point $(x,y,z,1)$ is the difference $\boldsymbol{v}=$ ______. In homogeneous coordinates, vectors end in ______. So we add a ______ to a point, not a point to a point.

17 If you multiply only the *last* coordinate of each point to get (x,y,z,c), you rescale the whole space by the number ______. This is because the point (x,y,z,c) is the same as $(\ , \ , \ ,1)$.

9

NUMERICAL LINEAR ALGEBRA

GAUSSIAN ELIMINATION IN PRACTICE ■ 9.1

Numerical linear algebra is a struggle for *quick* solutions and also *accurate* solutions. We need efficiency but we have to avoid instability. In Gaussian elimination, the main freedom (always available) is to exchange equations. This section explains when to exchange rows for the sake of speed, and when to do it for the sake of accuracy.

The key to accuracy is to avoid unnecessarily large numbers. Often that requires us to avoid small numbers! A small pivot generally means large multipliers (since we divide by the pivot). Also, a small pivot now means a large pivot later. The product of the pivots is a fixed number (except for its sign). That number is the determinant.

A good plan is to choose the *largest candidate* in each new column as the pivot. This is called "***partial pivoting***." The competitors are in the pivot position and below. We will see why this strategy is built into computer programs.

Other row exchanges are done to save elimination steps. In practice, most large matrices have only a small percentage of nonzero entries. The user probably knows their location. Elimination is generally fastest when the equations are ordered to put those nonzeros close to the diagonal. Then the matrix is as "banded" as possible.

New questions arise for machines with many processors in parallel. Now the problem is communication—to send processors the data they need, when they need it. This is a major research area. The brief comments in this section will try to introduce you to thinking in parallel.

Section 9.2 is about instability that can't be avoided. It is built into the problem, and this sensitivity is measured by the "***condition number***." Then Section 9.3 describes how to solve $A\boldsymbol{x} = \boldsymbol{b}$ by ***iterations***. Instead of direct elimination, the computer solves an easier equation many times. Each answer $\boldsymbol{x}_k$ goes back into the same equation to find the next guess $\boldsymbol{x}_{k+1}$. For good iterations, the $\boldsymbol{x}_k$ converge quickly to $\boldsymbol{x} = A^{-1}\boldsymbol{b}$.

Roundoff Error and Partial Pivoting

Up to now, any pivot (nonzero of course) was accepted. In practice a small pivot is dangerous. A catastrophe can occur when numbers of different sizes are added. Computers keep a fixed number of significant digits (say three decimals, for a very weak machine). The sum $10{,}000 + 1$ is rounded off to 10,000. The "1" is completely lost. Watch how that changes the solution to this problem:

$$\begin{aligned} .0001u + v &= 1 \\ -u + v &= 0 \end{aligned} \qquad \text{starts with coefficient matrix} \qquad A = \begin{bmatrix} .0001 & 1 \\ -1 & 1 \end{bmatrix}.$$

If we accept .0001 as the pivot, elimination adds 10,000 times row 1 to row 2. Roundoff leaves

$$10{,}000v = 10{,}000 \qquad \text{instead of} \qquad 10{,}001v = 10{,}000.$$

The computed answer $v = 1$ is near the true $v = .9999$. But then back substitution leads to

$$.0001\,u + 1 = 1 \qquad \text{instead of} \qquad .0001\,u + .9999 = 1.$$

The first equation gives $u = 0$. The correct answer (look at the second equation) is $u = 1.000$. By losing the "1" in the matrix, we have lost the solution. ***The change from 10,001 to 10,000 has changed the answer from $u = 1$ to $u = 0$*** (100% error!).

If we exchange rows, even this weak computer finds an answer that is correct to three places:

$$\begin{aligned} -u + v &= 0 \\ .0001u + v &= 1 \end{aligned} \quad \longrightarrow \quad \begin{aligned} -u + v &= 0 \\ v &= 1 \end{aligned} \quad \longrightarrow \quad \begin{aligned} u &= 1 \\ v &= 1. \end{aligned}$$

The original pivots were .0001 and 10,000—badly scaled. After a row exchange the exact pivots are -1 and 1.0001—well scaled. The computed pivots -1 and 1 come close to the exact values. Small pivots bring numerical instability, and the remedy is ***partial pivoting***. The kth pivot is decided when we reach and search column k:

Choose the largest number in row k or below. Exchange its row with row k.

The strategy of *complete pivoting* looks also in later columns for the largest pivot. It exchanges columns as well as rows. This expense is seldom justified, and all major codes use partial pivoting. Multiplying a row or column by a scaling constant can also be worthwhile. *If the first equation above is $u + 10{,}000v = 10{,}000$ and we don't rescale, then 1 is the pivot but we are in trouble again.*

For positive definite matrices, row exchanges are *not* required. It is safe to accept the pivots as they appear. Small pivots can occur, but the matrix is not improved by row exchanges. When its condition number is high, the problem is in the matrix and not in the order of elimination steps. In this case the output is unavoidably sensitive to the input.

The reader now understands how a computer actually solves $Ax = b$—***by elimination with partial pivoting***. Compared with the theoretical description—***find*** A^{-1} ***and multiply*** $A^{-1}b$—the details took time. But in computer time, elimination is much faster. I believe this algorithm is also the best approach to the algebra of row spaces and nullspaces.

Operation Counts: Full Matrices and Band Matrices

Here is a practical question about cost. *How many separate operations are needed to solve* $Ax = b$ *by elimination*? This decides how large a problem we can afford.

Look first at A, which changes gradually into U. When a multiple of row 1 is subtracted from row 2, we do n operations. The first is a division by the pivot, to find the multiplier ℓ. For the other $n-1$ entries along the row, the operation is a "multiply-subtract." For convenience, we count this as a single operation. If you regard multiplying by ℓ and subtracting from the existing entry as two separate operations, multiply all our counts by 2.

The matrix A is n by n. The operation count applies to all $n-1$ rows below the first. Thus it requires n times $n-1$ operations, or $n^2 - n$, to produce zeros below the first pivot. *Check*: *All* n^2 *entries are changed, except the n entries in the first row.*

When elimination is down to k equations, the rows are shorter. We need only $k^2 - k$ operations (instead of $n^2 - n$) to clear out the column below the pivot. This is true for $1 \le k \le n$. The last step requires no operations ($1^2 - 1 = 0$), since the pivot is set and forward elimination is complete. The total count to reach U is the sum of $k^2 - k$ over all values of k from 1 to n:

$$(1^2 + \dots + n^2) - (1 + \dots + n) = \frac{n(n+1)(2n+1)}{6} - \frac{n(n+1)}{2} = \frac{n^3 - n}{3}.$$

Those are known formulas for the sum of the first n numbers and the sum of the first n squares. Substituting $n = 1$ into $n^3 - n$ gives zero. Substituting $n = 100$ gives a million minus a hundred—then divide by 3. (That translates into one second on a workstation.) We will ignore the last term n in comparison with the larger term n^3, to reach our main conclusion:

The operation count for forward elimination (A to U) is $\frac{1}{3}n^3$.

That means $\frac{1}{3}n^3$ multiplications and $\frac{1}{3}n^3$ subtractions. Doubling n increases this cost by eight (because n is cubed). 100 equations are OK, 1000 are expensive, 10000 are impossible. We need a faster computer or a lot of zeros or a new idea.

On the right side of the equations, the steps go much faster. We operate on single numbers, not whole rows. ***Each right side needs exactly*** n^2 ***operations***. Remember that we solve two triangular systems, $Lc = b$ forward and $Ux = c$ backward. In back substitution, the last unknown needs only division by the last pivot. The equation above

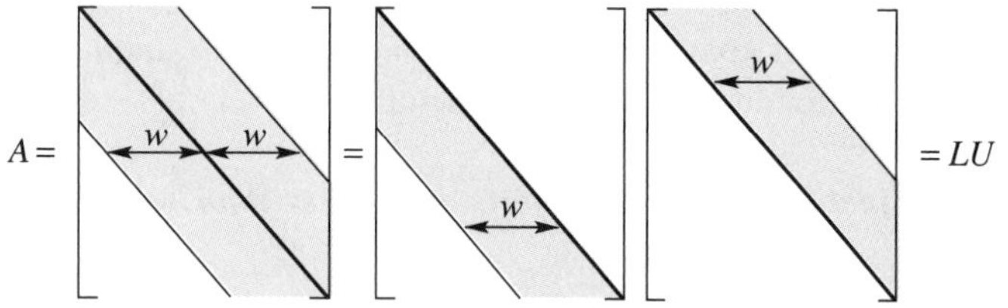

Figure 9.1 $A = LU$ for a band matrix. Good zeros in A stay zero in L and U.

it needs two operations—substituting x_n and dividing by *its* pivot. The kth step needs k operations, and the total for back substitution is

$$1 + 2 + \ldots + n = \frac{n(n+1)}{2} \approx \tfrac{1}{2}n^2 \quad \text{operations.}$$

The forward part is similar. *The n^2 total exactly equals the count for multiplying $A^{-1}\boldsymbol{b}$!* This leaves Gaussian elimination with two big advantages over $A^{-1}\boldsymbol{b}$:

1 Elimination requires $\frac{1}{3}n^3$ operations compared to n^3 for A^{-1}.

2 If A is *banded* so are L and U. But A^{-1} is full of nonzeros.

Band Matrices

These counts are improved when A has "*good zeros.*" A good zero is an entry that remains zero in L and U. The most important good zeros are at the beginning of a row. No elimination steps are required (the multipliers are zero). So we also find those same good zeros in L. That is especially clear for this *tridiagonal matrix* A:

$$\begin{bmatrix} 1 & -1 & & \\ -1 & 2 & -1 & \\ & -1 & 2 & -1 \\ & & -1 & 2 \end{bmatrix} = \begin{bmatrix} 1 & & & \\ -1 & 1 & & \\ & -1 & 1 & \\ & & -1 & 1 \end{bmatrix} \begin{bmatrix} 1 & -1 & & \\ & 1 & -1 & \\ & & 1 & -1 \\ & & & 1 \end{bmatrix}.$$

Rows 3 and 4 of A begin with zeros. No multiplier is needed, so L has the same zeros. Also rows 1 and 2 *end* with zeros. When a multiple of row 1 is subtracted from row 2, no calculation is required beyond the second column. The rows are short. They stay short! Figure 9.1 shows how a band matrix A has band factors L and U.

These zeros lead to a complete change in the operation count, for "half-bandwidth" w:

A band matrix has $a_{ij} = 0$ when $|i - j| > w$.

Thus $w = 1$ for a diagonal matrix and $w = 2$ for a tridiagonal matrix. The length of the pivot row is at most w. There are no more than $w - 1$ nonzeros below any pivot.

Each stage of elimination is complete after $w(w-1)$ operations, and *the band structure survives*. There are n columns to clear out. Therefore:

Forward elimination on a band matrix needs less than $w^2 n$ operations.

For a band matrix, the count is proportional to n instead of n^3. It is also proportional to w^2. A full matrix has $w = n$ and we are back to n^3. For a closer count, remember that the bandwidth drops below w in the lower right corner (not enough space). The exact count to find L and U is

$$\frac{w(w-1)(3n-2w+1)}{3} \quad \text{for a band matrix}$$
$$\frac{n(n-1)(n+1)}{3} = \frac{n^3-n}{3} \quad \text{when} \quad w = n.$$

On the right side, to find $\boldsymbol{x}$ from $\boldsymbol{b}$, the cost is about $2wn$ (compared to the usual n^2). *Main point*: *For a band matrix the operation counts are* ***proportional to*** n. This is extremely fast. A tridiagonal matrix of order 10,000 is very cheap, provided we don't compute A^{-1}. That inverse matrix has no zeros at all:

$$A = \begin{bmatrix} 1 & -1 & 0 & 0 \\ -1 & 2 & -1 & 0 \\ 0 & -1 & 2 & -1 \\ 0 & 0 & -1 & 2 \end{bmatrix} \quad \text{has} \quad A^{-1} = U^{-1}L^{-1} = \begin{bmatrix} 4 & 3 & 2 & 1 \\ 3 & 3 & 2 & 1 \\ 2 & 2 & 2 & 1 \\ 1 & 1 & 1 & 1 \end{bmatrix}.$$

We are actually worse off knowing A^{-1} than knowing L and U. Multiplication by A^{-1} needs the full n^2 steps. Solving $L\boldsymbol{c} = \boldsymbol{b}$ and $U\boldsymbol{x} = \boldsymbol{c}$ needs only $2wn$. Here that means $4n$. A band structure is very common in practice, when the matrix reflects connections between near neighbors. We see $a_{13} = 0$ and $a_{14} = 0$ because 1 is not a neighbor of 3 and 4.

We close with two more operation counts:

1 A^{-1} costs n^3 steps. **2** QR costs $\frac{2}{3}n^3$ steps.

1 Start with $AA^{-1} = I$. The jth column of A^{-1} solves $A\boldsymbol{x}_j = j$th column of I. Normally each of those n right sides needs n^2 operations, making n^3 in all. The left side costs $\frac{1}{3}n^3$ as usual. (This is a one-time cost! L and U are not repeated for each new right side.) This count gives $\frac{4}{3}n^3$, but we can get down to n^3.

The special saving for the jth column of I comes from its first $j-1$ zeros. No work is required on the right side until elimination reaches row j. The forward cost is $\frac{1}{2}(n-j)^2$ instead of $\frac{1}{2}n^2$. Summing over j, the total for forward elimination on the

n right sides is $\frac{1}{6}n^3$. Then the final count of multiplications for A^{-1} (with an equal number of subtractions) is n^3 if we actually want the inverse matrix:

$$\frac{n^3}{3}\ (L \text{ and } U) + \frac{n^3}{6}\ (\text{forward})\ + n\left(\frac{n^2}{2}\right)\ (\text{back substitutions})\ = n^3. \tag{1}$$

2 The Gram-Schmidt process works with columns instead of rows—that is not so important to the count. The key difference from elimination is that *the multiplier is decided by a dot product.* So it takes n operations to find the multiplier, where elimination just divides by the pivot. Then there are n "multiply-subtract" operations to remove from column 2 its projection along column 1. (See Section 4.4 and Problem 4.4.28 for the sequence of projections.) The cost for Gram-Schmidt is $2n$ where for elimination it is n. This factor 2 is the price of orthogonality. We are changing a dot product to zero instead of changing an entry to zero.

Caution To judge a numerical algorithm, it is **not enough** to count the operations. Beyond "flop counting" is a study of stability and the flow of data. Van Loan emphasizes the three levels of linear algebra: linear combinations $c\boldsymbol{u}+\boldsymbol{v}$ (level 1), matrix-vector $A\boldsymbol{u}+\boldsymbol{v}$ (level 2), and matrix-matrix $AB+C$ (level 3). For parallel computing, level 3 is best. AB uses $2n^3$ flops (additions and multiplications) and only $2n^2$ data—a good ratio of work to communication overhead. Solving $UX = B$ for matrices is better than $U\boldsymbol{x} = \boldsymbol{b}$ for vectors. Gauss-Jordan partly wins after all!

Plane Rotations

There are two ways to reach the important factorization $A = QR$. One way works to find Q, the other way works to find R. Gram-Schmidt chose the first way, and the columns of A were orthogonalized to go into Q. (Then R was an afterthought. It was upper triangular because of the order of Gram-Schmidt steps.) Now we look at a method that starts with A and aims directly at R.

Elimination gives $A = LU$, orthogonalization gives $A = QR$. What is the difference, when R and U are both upper triangular? For elimination L is a product of E's—with 1's on the diagonal and the multiplier ℓ_{ij} below. *QR uses orthogonal matrices.* The E's are not allowed. We don't want a triangular L, we want an orthogonal Q.

There are two simple orthogonal matrices to take the place of the E's. The ***reflection matrices*** $I - 2\boldsymbol{u}\boldsymbol{u}^{\mathrm{T}}$ are named after Householder. The ***plane rotation matrices*** are named after Givens. The matrix that rotates the xy plane by θ, and leaves the z direction alone, is Q_{21}:

$$\textbf{Givens Rotation} \qquad Q_{21} = \begin{bmatrix} \cos\theta & -\sin\theta & 0 \\ \sin\theta & \cos\theta & 0 \\ 0 & 0 & 1 \end{bmatrix}.$$

Use Q_{21} the way you used E_{21}, to produce a zero in the $(2, 1)$ position. That determines the angle θ. Here is an example given by Bill Hager in *Applied Numerical Linear Algebra* (Prentice-Hall, 1988):

$$Q_{21}A = \begin{bmatrix} .6 & .8 & 0 \\ -.8 & .6 & 0 \\ 0 & 0 & 1 \end{bmatrix} \begin{bmatrix} 90 & -153 & 114 \\ 120 & -79 & -223 \\ 200 & -40 & 395 \end{bmatrix} = \begin{bmatrix} 150 & -155 & -110 \\ 0 & 75 & -225 \\ 200 & -40 & 395 \end{bmatrix}.$$

The zero came from $-.8(90) + .6(120)$. No need to find θ, what we needed was

$$\cos\theta = \frac{90}{\sqrt{90^2 + 120^2}} \quad \text{and} \quad \sin\theta = \frac{-120}{\sqrt{90^2 + 120^2}}. \tag{2}$$

Now we attack the $(3, 1)$ entry. The rotation will be in rows and columns 3 and 1. The numbers $\cos\theta$ and $\sin\theta$ are determined from 150 and 200, instead of 90 and 120. They happen to be .6 and $-.8$ again:

$$Q_{31}Q_{21}A = \begin{bmatrix} .6 & 0 & .8 \\ 0 & 1 & 0 \\ -.8 & 0 & .6 \end{bmatrix} \begin{bmatrix} 150 & \cdot & \cdot \\ 0 & \cdot & \cdot \\ 200 & \cdot & \cdot \end{bmatrix} = \begin{bmatrix} 250 & -125 & 250 \\ 0 & 75 & -225 \\ 0 & 100 & 325 \end{bmatrix}.$$

One more step to R. The $(3, 2)$ entry has to go. The numbers $\cos\theta$ and $\sin\theta$ now come from 75 and 100. The rotation is now in rows and columns 2 and 3:

$$Q_{32}Q_{31}Q_{21}A = \begin{bmatrix} 1 & 0 & 0 \\ 0 & .6 & .8 \\ 0 & -.8 & .6 \end{bmatrix} \begin{bmatrix} 250 & -125 & \cdot \\ 0 & 75 & \cdot \\ 0 & 100 & \cdot \end{bmatrix} = \begin{bmatrix} 250 & -125 & 250 \\ 0 & 125 & 125 \\ 0 & 0 & 375 \end{bmatrix}.$$

We have reached the upper triangular R. What is Q? Move the plane rotations Q_{ij} to the other side to find $A = QR$—just as you moved the elimination matrices E_{ij} to the other side to find $A = LU$:

$$Q_{32}Q_{31}Q_{21}A = R \quad \text{means} \quad A = (Q_{21}^{-1}Q_{31}^{-1}Q_{32}^{-1})R = QR. \tag{3}$$

The inverse of each Q_{ij} is Q_{ij}^{T} (rotation through $-\theta$). The inverse of E_{ij} was not an orthogonal matrix! E_{ij}^{-1} added back to row i the multiple ℓ_{ij} (times row j) that E_{ij} had subtracted. I hope you see how the big computations of linear algebra—LU and QR—are similar but not the same.

There is a third big computation—***eigenvalues and eigenvectors***. If we can make A triangular, we can see its eigenvalues on the diagonal. But we can't use U and we can't use R. To preserve the eigenvalues, the allowed step is not $Q_{21}A$ but $Q_{21}AQ_{21}^{-1}$. That extra factor Q_{21}^{-1} for a similar matrix wipes out the zero that Q_{21} created!

There are two ways to go. Neither one gives the eigenvalues in a fixed number of steps. (That is impossible. The calculation of $\cos\theta$ and $\sin\theta$ involved only a square root. The nth degree equation $\det(A - \lambda I) = 0$ cannot be solved by a succession of square roots.) But still the rotations Q_{ij} are useful:

Method 1 Produce a zero in the $(3, 1)$ entry of $Q_{21}A$, instead of $(2, 1)$. That zero is not destroyed when Q_{21}^{-1} multiplies on the right. We are leaving a diagonal of nonzeros under the main diagonal, so we can't read off the eigenvalues. But this "Hessenberg matrix" with the extra diagonal of nonzeros still has a lot of good zeros.

Method 2 Choose a different Q_{21}, which does produce a zero in the $(2, 1)$ position of $Q_{21}AQ_{21}^{-1}$. This is just a 2 by 2 eigenvalue problem, for the matrix in the upper left corner of A. The column $(\cos\theta, -\sin\theta)$ is an eigenvector of that matrix. This is the first step in "Jacobi's method."

The problem of destroying zeros will not go away. The second step chooses Q_{31} so that $Q_{31}Q_{21}AQ_{21}^{-1}Q_{31}^{-1}$ has a zero in the $(3, 1)$ position. But it loses the zero in the $(2, 1)$ position. Jacobi solves 2 by 2 eigenvalue problems to find his Q_{ij}, but earlier nonzeros keep coming back. In general those nonzeros are smaller each time, and after several loops through the matrix the lower triangular part is substantially reduced. Then the eigenvalues gradually appear on the diagonal.

What you should remember is this. The Q's are orthogonal matrices—their columns with $(\cos\theta, \sin\theta)$ and $(-\sin\theta, \cos\theta)$ are orthogonal unit vectors. Computations with the Q's are very stable. The angle θ can be chosen to make a particular entry zero. This is a step toward the final goal of a triangular matrix. That was the goal at the beginning of the book, and it still is.

Problem Set 9.1

1 Find the two pivots with and without partial pivoting for

$$A = \begin{bmatrix} .001 & 0 \\ 1 & 1000 \end{bmatrix}.$$

With partial pivoting, why are no entries of L larger than 1? Find a 3 by 3 matrix A with all $|a_{ij}| \le 1$ and $|\ell_{ij}| \le 1$ but third pivot $= 4$.

2 Compute the exact inverse of the Hilbert matrix A by elimination. Then compute A^{-1} again by rounding all numbers to three figures:

$$A = \mathsf{hilb}(3) = \begin{bmatrix} 1 & \frac{1}{2} & \frac{1}{3} \\ \frac{1}{2} & \frac{1}{3} & \frac{1}{4} \\ \frac{1}{3} & \frac{1}{4} & \frac{1}{5} \end{bmatrix}.$$

3 For the same A compute $\boldsymbol{b} = A\boldsymbol{x}$ for $\boldsymbol{x} = (1, 1, 1)$ and $\boldsymbol{x} = (0, 6, -3.6)$. A small change $\Delta\boldsymbol{b}$ produces a large change $\Delta\boldsymbol{x}$.

4 Find the eigenvalues (by computer) of the 8 by 8 Hilbert matrix $a_{ij} = 1/(i + j - 1)$. In the equation $A\boldsymbol{x} = \boldsymbol{b}$ with $\|\boldsymbol{b}\| = 1$, how large can $\|\boldsymbol{x}\|$ be? If $\boldsymbol{b}$ has roundoff error less than 10^{-16}, how large an error can this cause in $\boldsymbol{x}$?

5 For back substitution with a band matrix (width w), show that the number of multiplications to solve $U\boldsymbol{x} = \boldsymbol{c}$ is approximately wn.

6 If you know L and U and Q and R, is it faster to solve $LU\boldsymbol{x} = \boldsymbol{b}$ or $QR\boldsymbol{x} = \boldsymbol{b}$?

7 Show that the number of multiplications to invert an upper triangular n by n matrix is about $\frac{1}{6}n^3$. Use back substitution on the columns of I, upward from 1's.

8 Choosing the largest available pivot in each column (partial pivoting), factor each A into $PA = LU$:

$$A = \begin{bmatrix} 1 & 0 \\ 2 & 2 \end{bmatrix} \quad \text{and} \quad A = \begin{bmatrix} 1 & 0 & 1 \\ 2 & 2 & 0 \\ 0 & 2 & 0 \end{bmatrix}.$$

9 Put 1's on the three central diagonals of a 4 by 4 tridiagonal matrix. Find the cofactors of the six zero entries. Those entries are nonzero in A^{-1}.

10 (Suggested by C. Van Loan.) Find the LU factorization of $A = \left[\begin{smallmatrix} \varepsilon & 1 \\ 1 & 1 \end{smallmatrix}\right]$. On your computer solve by elimination when $\varepsilon = 10^{-3}, 10^{-6}, 10^{-9}, 10^{-12}, 10^{-15}$:

$$\begin{bmatrix} \varepsilon & 1 \\ 1 & 1 \end{bmatrix} \begin{bmatrix} x_1 \\ x_2 \end{bmatrix} = \begin{bmatrix} 1+\varepsilon \\ 2 \end{bmatrix}.$$

The true $\boldsymbol{x}$ is $(1, 1)$. Make a table to show the error for each ε. Exchange the two equations and solve again—the errors should almost disappear.

11 Choose $\sin\theta$ and $\cos\theta$ to triangularize A, and find R:

$$Q_{21}A = \begin{bmatrix} \cos\theta & -\sin\theta \\ \sin\theta & \cos\theta \end{bmatrix} \begin{bmatrix} 1 & -1 \\ 3 & 5 \end{bmatrix} = \begin{bmatrix} * & * \\ 0 & * \end{bmatrix} = R.$$

12 Choose $\sin\theta$ and $\cos\theta$ to make $Q_{21}AQ_{21}^{-1}$ triangular (same A). What are the eigenvalues?

13 When A is multiplied by Q_{ij}, which of the n^2 entries of A are changed? When $Q_{ij}A$ is multiplied on the right by Q_{ij}^{-1}, which entries are changed now?

14 How many multiplications and how many additions are used to compute $Q_{ij}A$? (A careful organization of the whole sequence of rotations gives $\frac{2}{3}n^3$ multiplications and $\frac{2}{3}n^3$ additions—the same as for QR by reflectors and twice as many as for LU.)

15 (**Turning a robot hand**) The robot produces any 3 by 3 rotation A from plane rotations around the x, y, z axes. Then $Q_{32}Q_{31}Q_{21}A = R$, where A is orthogonal so R is I! The three robot turns are in $A = Q_{21}^{-1}Q_{31}^{-1}Q_{32}^{-1}$. The three angles are "Euler angles" and det $Q = 1$ to avoid reflection. Start by choosing $\cos\theta$ and $\sin\theta$ so that

$$Q_{21}A = \begin{bmatrix} \cos\theta & -\sin\theta & 0 \\ \sin\theta & \cos\theta & 0 \\ 0 & 0 & 1 \end{bmatrix} \frac{1}{3} \begin{bmatrix} -1 & 2 & 2 \\ 2 & -1 & 2 \\ 2 & 2 & -1 \end{bmatrix} \quad \text{is zero in the (2, 1) position.}$$

NORMS AND CONDITION NUMBERS ■ 9.2

How do we measure the size of a matrix? For a vector, the length is $\|x\|$. For a matrix, ***the norm is*** $\|A\|$. This word "*norm*" is sometimes used for vectors, instead of length. It is always used for matrices, and there are many ways to measure $\|A\|$. We look at the requirements on all "matrix norms", and then choose one.

Frobenius squared all the entries of A and added; his norm $\|A\|_F$ is the square root. This treats the matrix like a long vector. It is better to treat the matrix as a matrix.

Start with a vector norm: $\|x + y\|$ is not greater than $\|x\| + \|y\|$. This is the triangle inequality: $x + y$ is the third side of the triangle. Also for vectors, the length of $2x$ or $-2x$ is doubled to $2\|x\|$. The same rules apply to matrix norms:

$$\|A + B\| \leq \|A\| + \|B\| \quad \text{and} \quad \|cA\| = |c|\,\|A\|. \tag{1}$$

The second requirements for a norm are new for matrices—because matrices multiply. The size of Ax and the size of AB must stay under control. For all matrices and all vectors, we want

$$\|Ax\| \leq \|A\|\,\|x\| \quad \text{and} \quad \|AB\| \leq \|A\|\,\|B\|. \tag{2}$$

This leads to a natural way to define $\|A\|$. Except for the zero matrix, the norm is a positive number. The following choice satisfies all requirements:

DEFINITION ***The norm of a matrix A is the largest ratio*** $\|Ax\|/\|x\|$:

$$\|A\| = \max_{x \neq 0} \frac{\|Ax\|}{\|x\|}. \tag{3}$$

$\|Ax\|/\|x\|$ is never larger than $\|A\|$ (its maximum). This says that $\|Ax\| \leq \|A\|\,\|x\|$.

Example 1 If A is the identity matrix I, the ratios are always $\|x\|/\|x\|$. Therefore $\|I\| = 1$. If A is an orthogonal matrix Q, then again lengths are preserved: $\|Qx\| = \|x\|$ for every x. The ratios again give $\|Q\| = 1$.

Example 2 The norm of a diagonal matrix is its largest entry (using absolute values):

$$\text{The norm of} \quad A = \begin{bmatrix} 2 & 0 \\ 0 & 3 \end{bmatrix} \quad \text{is} \quad \|A\| = 3.$$

The ratio is $\|A\boldsymbol{x}\| = \sqrt{2^2x_1^2 + 3^2x_2^2}$ divided by $\|\boldsymbol{x}\| = \sqrt{x_1^2 + x_2^2}$. That is a maximum when $x_1 = 0$ and $x_2 = 1$. This vector $\boldsymbol{x} = (0, 1)$ is an eigenvector with $A\boldsymbol{x} = (0, 3)$. The eigenvalue is 3. This is the largest eigenvalue of A and it equals the norm.

For a positive definite symmetric matrix the norm is $\|A\| = \lambda_{\max}$.

Choose $\boldsymbol{x}$ to be the eigenvector with maximum eigenvalue: $A\boldsymbol{x} = \lambda_{\max}\boldsymbol{x}$. Then $\|A\boldsymbol{x}\|/\|\boldsymbol{x}\|$ equals $\lambda_{\max}$. The point is that no other vector $\boldsymbol{x}$ can make the ratio larger. The matrix is $A = Q\Lambda Q^{\mathrm{T}}$, and the orthogonal matrices Q and Q^{T} leave lengths unchanged. So the ratio to maximize is really $\|\Lambda\boldsymbol{x}\|/\|\boldsymbol{x}\|$. The norm $\lambda_{\max}$ is the largest eigenvalue in the diagonal matrix Λ.

Symmetric matrices Suppose A is symmetric but not positive definite—some eigenvalues of A are negative or zero. Then the norm $\|A\|$ is the largest of $|\lambda_1|, |\lambda_2|, \ldots, |\lambda_n|$. We take absolute values of the λ's, because the norm is only concerned with length. For an eigenvector we have $\|A\boldsymbol{x}\| = \|\lambda\boldsymbol{x}\|$, which is $|\lambda|$ times $\|\boldsymbol{x}\|$. Dividing by $\|\boldsymbol{x}\|$ leaves $|\lambda|$. The $\boldsymbol{x}$ that gives the maximum ratio is the eigenvector for the maximum $|\lambda|$.

Unsymmetric matrices If A is not symmetric, its eigenvalues may not measure its true size. The norm can be large when the eigenvalues are small. *Thus the norm is generally larger than* $|\lambda|_{\max}$. A very unsymmetric example has $\lambda_1 = \lambda_2 = 0$ but its norm is not zero:

$$A = \begin{bmatrix} 0 & 2 \\ 0 & 0 \end{bmatrix} \quad \text{has norm} \quad \|A\| = \max_{\boldsymbol{x} \neq \boldsymbol{0}} \frac{\|A\boldsymbol{x}\|}{\|\boldsymbol{x}\|} = 2.$$

The vector $\boldsymbol{x} = (0, 1)$ gives $A\boldsymbol{x} = (2, 0)$. The ratio of lengths is $2/1$. This is the maximum ratio $\|A\|$, even though $\boldsymbol{x}$ is not an eigenvector.

It is the symmetric matrix $A^{\mathrm{T}}A$, not the unsymmetric A, that has $\boldsymbol{x} = (0, 1)$ as its eigenvector. The norm is really decided by *the largest eigenvalue of* $A^{\mathrm{T}}A$, as we now prove.

9A ***The norm of*** A (symmetric or not) ***is the square root of*** $\lambda_{\max}(A^{\mathrm{T}}A)$:

$$\|A\|^2 = \max_{\boldsymbol{x} \neq \boldsymbol{0}} \frac{\|A\boldsymbol{x}\|^2}{\|\boldsymbol{x}\|^2} = \max_{\boldsymbol{x} \neq \boldsymbol{0}} \frac{\boldsymbol{x}^{\mathrm{T}}A^{\mathrm{T}}A\boldsymbol{x}}{\boldsymbol{x}^{\mathrm{T}}\boldsymbol{x}} = \lambda_{\max}(A^{\mathrm{T}}A). \tag{4}$$

Proof Choose $\boldsymbol{x}$ to be the eigenvector of $A^{\mathrm{T}}A$ corresponding to its largest eigenvalue $\lambda_{\max}$. The ratio in equation (1) is then $\boldsymbol{x}^{\mathrm{T}}A^{\mathrm{T}}A\boldsymbol{x} = \boldsymbol{x}^{\mathrm{T}}(\lambda_{\max})\boldsymbol{x}$ divided by $\boldsymbol{x}^{\mathrm{T}}\boldsymbol{x}$. For this particular $\boldsymbol{x}$, the ratio equals $\lambda_{\max}$.

No other $\boldsymbol{x}$ can give a larger ratio. The symmetric matrix $A^{\mathrm{T}}A$ has orthonormal eigenvectors $\boldsymbol{q}_1, \boldsymbol{q}_2, \ldots, \boldsymbol{q}_n$. Every $\boldsymbol{x}$ is a combination of those vectors. Try this combination in the ratio and remember that $\boldsymbol{q}_i^{\mathrm{T}}\boldsymbol{q}_j = 0$:

$$\frac{\boldsymbol{x}^{\mathrm{T}}A^{\mathrm{T}}A\boldsymbol{x}}{\boldsymbol{x}^{\mathrm{T}}\boldsymbol{x}} = \frac{(c_1\boldsymbol{q}_1 + \cdots + c_n\boldsymbol{q}_n)^{\mathrm{T}}(c_1\lambda_1\boldsymbol{q}_1 + \cdots + c_n\lambda_n\boldsymbol{q}_n)}{(c_1\boldsymbol{q}_1 + \cdots + c_n\boldsymbol{q}_n)^{\mathrm{T}}(c_1\boldsymbol{q}_1 + \cdots + c_n\boldsymbol{q}_n)} = \frac{c_1^2\lambda_1 + \cdots + c_n^2\lambda_n}{c_1^2 + \cdots + c_n^2}. \tag{5}$$

That last ratio cannot be larger than $\lambda_{\max}$. The maximum ratio is when all c's are zero, except the one that multiplies $\lambda_{\max}$.

Note 1 The ratio in (5) is known as the *Rayleigh quotient* for the matrix $A^{\mathrm{T}}A$. The maximum is the largest eigenvalue $\lambda_{\max}(A^{\mathrm{T}}A)$. The minimum is $\lambda_{\min}(A^{\mathrm{T}}A)$. If you substitute any vector $\boldsymbol{x}$ into the Rayleigh quotient $\boldsymbol{x}^{\mathrm{T}}A^{\mathrm{T}}A\boldsymbol{x}/\boldsymbol{x}^{\mathrm{T}}\boldsymbol{x}$, you are guaranteed to get a number between $\lambda_{\min}$ and $\lambda_{\max}$.

Note 2 The norm $\|A\|$ equals the largest *singular value* $\sigma_{\max}$ of A. The singular values $\sigma_1, \ldots, \sigma_r$ are the square roots of the positive eigenvalues of $A^{\mathrm{T}}A$. So certainly $\sigma_{\max} = (\lambda_{\max})^{1/2}$. This is the norm of A.

Note 3 Check that the unsymmetric example in equation (3) has $\lambda_{\max}(A^{\mathrm{T}}A) = 4$:

$$A = \begin{bmatrix} 0 & 2 \\ 0 & 0 \end{bmatrix} \text{ leads to } A^{\mathrm{T}}A = \begin{bmatrix} 0 & 0 \\ 0 & 4 \end{bmatrix} \text{ with } \lambda_{\max} = 4. \text{ So the norm is } \|A\| = \sqrt{4}.$$

The Condition Number of A

Section 9.1 showed that roundoff error can be serious. Some systems are sensitive, others are not so sensitive. The sensitivity to error is measured by the ***condition number***. This is the first chapter in the book which intentionally introduces errors. We want to estimate how much they change $\boldsymbol{x}$.

The original equation is $A\boldsymbol{x} = \boldsymbol{b}$. Suppose the right side is changed to $\boldsymbol{b} + \Delta\boldsymbol{b}$ because of roundoff or measurement error. The solution is then changed to $\boldsymbol{x} + \Delta\boldsymbol{x}$. Our goal is to estimate the change $\Delta\boldsymbol{x}$ in the solution from the change $\Delta\boldsymbol{b}$ in the equation. Subtraction gives the *error equation* $A(\Delta\boldsymbol{x}) = \Delta\boldsymbol{b}$:

$$\text{Subtract } A\boldsymbol{x} = \boldsymbol{b} \text{ from } A(\boldsymbol{x} + \Delta\boldsymbol{x}) = \boldsymbol{b} + \Delta\boldsymbol{b} \quad \text{to find} \quad A(\Delta\boldsymbol{x}) = \Delta\boldsymbol{b}. \tag{6}$$

The error is $\Delta\boldsymbol{x} = A^{-1}\Delta\boldsymbol{b}$. It is large when A^{-1} is large (then A is nearly singular). The error $\Delta\boldsymbol{x}$ is especially large when $\Delta\boldsymbol{b}$ points in the worst direction—which is amplified most by A^{-1}. The worst error has $\|\Delta\boldsymbol{x}\| = \|A^{-1}\| \, \|\Delta\boldsymbol{b}\|$. That is the largest possible output error $\Delta\boldsymbol{x}$.

This error bound $\|A^{-1}\|$ has one serious drawback. If we multiply A by 1000, then A^{-1} is divided by 1000. The matrix looks a thousand times better. But a simple rescaling cannot change the reality of the problem. It is true that $\Delta \boldsymbol{x}$ will be divided by 1000, but so will the exact solution $\boldsymbol{x} = A^{-1}\boldsymbol{b}$. The ***relative error*** $\|\Delta \boldsymbol{x}\|/\|\boldsymbol{x}\|$ will stay the same. It is this relative change in $\boldsymbol{x}$ that should be compared to the relative change in $\boldsymbol{b}$.

Comparing relative errors will now lead to the "condition number" $c = \|A\|\,\|A^{-1}\|$. Multiplying A by 1000 does not change this number, because A^{-1} is divided by 1000 and the product c stays the same.

9B ***The solution error is less than*** $c = \|A\|\,\|A^{-1}\|$ ***times the problem error:***

$$\frac{\|\Delta \boldsymbol{x}\|}{\|\boldsymbol{x}\|} \le c\frac{\|\Delta \boldsymbol{b}\|}{\|\boldsymbol{b}\|}. \tag{7}$$

If the problem error is ΔA (error in the matrix instead of in $\boldsymbol{b}$), ***this changes to***

$$\frac{\|\Delta \boldsymbol{x}\|}{\|\boldsymbol{x} + \Delta \boldsymbol{x}\|} \le c\frac{\|\Delta A\|}{\|A\|}. \tag{8}$$

Proof The original equation is $\boldsymbol{b} = A\boldsymbol{x}$. The error equation (6) is $\Delta \boldsymbol{x} = A^{-1}\Delta \boldsymbol{b}$. Apply the key property (2) of matrix norms:

$$\|\boldsymbol{b}\| \le \|A\|\,\|\boldsymbol{x}\| \qquad \text{and} \qquad \|\Delta \boldsymbol{x}\| \le \|A^{-1}\|\,\|\Delta \boldsymbol{b}\|.$$

Multiply the left sides to get $\|\boldsymbol{b}\|\,\|\Delta \boldsymbol{x}\|$, and also multiply the right sides. Divide both sides by $\|\boldsymbol{b}\|\,\|\boldsymbol{x}\|$. The left side is now the relative error $\|\Delta \boldsymbol{x}\|/\|\boldsymbol{x}\|$. The right side is now the upper bound in equation (7).

The same condition number $c = \|A\|\,\|A^{-1}\|$ appears when the error is in the matrix. We have ΔA instead of $\Delta \boldsymbol{b}$:

Subtract $A\boldsymbol{x} = \boldsymbol{b}$ from $(A + \Delta A)(\boldsymbol{x} + \Delta \boldsymbol{x}) = \boldsymbol{b}$ to find $A(\Delta \boldsymbol{x}) = -(\Delta A)(\boldsymbol{x} + \Delta \boldsymbol{x})$.

Multiply the last equation by A^{-1} and take norms to reach equation (8):

$$\|\Delta \boldsymbol{x}\| \le \|A^{-1}\|\,\|\Delta A\|\,\|\boldsymbol{x} + \Delta \boldsymbol{x}\| \quad \text{or} \quad \frac{\|\Delta \boldsymbol{x}\|}{\|\boldsymbol{x} + \Delta \boldsymbol{x}\|} \le \|A\|\,\|A^{-1}\|\frac{\|\Delta A\|}{\|A\|}.$$

Conclusion Errors enter in two ways. They begin with an error ΔA or $\Delta \boldsymbol{b}$—a wrong matrix or a wrong $\boldsymbol{b}$. This problem error is amplified (a lot or a little) into the solution error $\Delta \boldsymbol{x}$. That error is bounded, relative to $\boldsymbol{x}$ itself, by the condition number c.

The error $\Delta \boldsymbol{b}$ depends on computer roundoff and on the original measurements of $\boldsymbol{b}$. The error ΔA also depends on the elimination steps. Small pivots tend to produce large errors in L and U. Then $L + \Delta L$ times $U + \Delta U$ equals $A + \Delta A$. When ΔA or the condition number is very large, the error $\Delta \boldsymbol{x}$ can be unacceptable.

Example 3 When A is symmetric, $c = \|A\|\,\|A^{-1}\|$ comes from the eigenvalues:

$$A = \begin{bmatrix} 6 & 0 \\ 0 & 2 \end{bmatrix} \text{ has norm } 6. \qquad A^{-1} = \begin{bmatrix} \frac{1}{6} & 0 \\ 0 & \frac{1}{2} \end{bmatrix} \text{ has norm } \tfrac{1}{2}.$$

This A is symmetric positive definite. Its norm is $\lambda_{\max} = 6$. The norm of A^{-1} is $1/\lambda_{\min} = \frac{1}{2}$. Multiplying those norms gives the *condition number*:

$$c = \frac{\lambda_{\max}}{\lambda_{\min}} = \frac{6}{2} = 3.$$

Example 4 Keep the same A, with eigenvalues 6 and 2. To make $\boldsymbol{x}$ small, choose $\boldsymbol{b}$ along the first eigenvector $(1, 0)$. To make $\Delta\boldsymbol{x}$ large, choose $\Delta\boldsymbol{b}$ along the second eigenvector $(0, 1)$. Then $\boldsymbol{x} = \frac{1}{6}\boldsymbol{b}$ and $\Delta\boldsymbol{x} = \frac{1}{2}\boldsymbol{b}$. The ratio $\|\Delta\boldsymbol{x}\|/\|\boldsymbol{x}\|$ is exactly $c = 3$ times the ratio $\|\Delta\boldsymbol{b}\|/\|\boldsymbol{b}\|$.

This shows that the worst error allowed by the condition number can actually happen. Here is a useful rule of thumb, experimentally verified for Gaussian elimination: *The computer can lose* $\log c$ *decimal places to roundoff error.*

Problem Set 9.2

1 Find the norms $\lambda_{\max}$ and condition numbers $\lambda_{\max}/\lambda_{\min}$ of these positive definite matrices:

$$\begin{bmatrix} .5 & 0 \\ 0 & 2 \end{bmatrix} \quad \begin{bmatrix} 2 & 1 \\ 1 & 2 \end{bmatrix} \quad \begin{bmatrix} 3 & 1 \\ 1 & 1 \end{bmatrix}.$$

2 Find the norms and condition numbers from the square roots of $\lambda_{\max}(A^{\mathrm{T}}A)$ and $\lambda_{\min}(A^{\mathrm{T}}A)$:

$$\begin{bmatrix} -2 & 0 \\ 0 & 2 \end{bmatrix} \quad \begin{bmatrix} 1 & 1 \\ 0 & 0 \end{bmatrix} \quad \begin{bmatrix} 1 & 1 \\ -1 & 1 \end{bmatrix}.$$

3 Explain these two inequalities from the definitions of $\|A\|$ and $\|B\|$:

$$\|AB\boldsymbol{x}\| \le \|A\|\,\|B\boldsymbol{x}\| \le \|A\|\,\|B\|\,\|\boldsymbol{x}\|.$$

From the ratio that gives $\|AB\|$, deduce that $\|AB\| \le \|A\|\,\|B\|$. This is the key to using matrix norms.

4 Use $\|AB\| \le \|A\|\,\|B\|$ to prove that the condition number of any matrix A is at least 1.

5 Why is I the only symmetric positive definite matrix that has $\lambda_{\max} = \lambda_{\min} = 1$? Then the only matrices with $\|A\| = 1$ and $\|A^{-1}\| = 1$ must have $A^{\mathrm{T}}A = I$. They are _____ matrices.

6 Orthogonal matrices have norm $\|Q\| = 1$. If $A = QR$ show that $\|A\| \le \|R\|$ and also $\|R\| \le \|A\|$. Then $\|A\| = \|R\|$. Find an example of $A = LU$ with $\|A\| < \|L\|\|U\|$.

7 (a) Which famous inequality gives $\|(A + B)\boldsymbol{x}\| \le \|A\boldsymbol{x}\| + \|B\boldsymbol{x}\|$ for every $\boldsymbol{x}$?

(b) Why does the definition (4) of matrix norms lead to $\|A+B\| \le \|A\|+\|B\|$?

8 Show that if λ is any eigenvalue of A, then $|\lambda| \le \|A\|$. Start from $A\boldsymbol{x} = \lambda\boldsymbol{x}$.

9 The "*spectral radius*" $\rho(A) = |\lambda_{\max}|$ is the largest absolute value of the eigenvalues. Show with 2 by 2 examples that $\rho(A+B) \le \rho(A)+\rho(B)$ and $\rho(AB) \le \rho(A)\rho(B)$ can both be *false*. The spectral radius is not acceptable as a norm.

10 (a) Explain why A and A^{-1} have the same condition number.

(b) Explain why A and A^{T} have the same norm.

11 Estimate the condition number of the ill-conditioned matrix $A = \begin{bmatrix} 1 & 1 \\ 1 & 1.0001 \end{bmatrix}$.

12 Why is the determinant of A no good as a norm? Why is it no good as a condition number?

13 (Suggested by C. Moler and C. Van Loan.) Compute $\boldsymbol{b} - A\boldsymbol{y}$ and $\boldsymbol{b} - A\boldsymbol{z}$ when

$$\boldsymbol{b} = \begin{bmatrix} .217 \\ .254 \end{bmatrix} \quad A = \begin{bmatrix} .780 & .563 \\ .913 & .659 \end{bmatrix} \quad \boldsymbol{y} = \begin{bmatrix} .341 \\ -.087 \end{bmatrix} \quad \boldsymbol{z} = \begin{bmatrix} .999 \\ -1.0 \end{bmatrix}.$$

Is $\boldsymbol{y}$ closer than $\boldsymbol{z}$ to solving $A\boldsymbol{x} = \boldsymbol{b}$? Answer in two ways: Compare the *residual* $\boldsymbol{b} - A\boldsymbol{y}$ to $\boldsymbol{b} - A\boldsymbol{z}$. Then compare $\boldsymbol{y}$ and $\boldsymbol{z}$ to the true $\boldsymbol{x} = (1, -1)$. Both answers can be right. Sometimes we want a small residual, sometimes a small $\Delta\boldsymbol{x}$.

14 (a) Compute the determinant of A in Problem 13. Compute A^{-1}.

(b) If possible compute $\|A\|$ and $\|A^{-1}\|$ and show that $c > 10^6$.

Problems 15–19 are about vector norms other than the usual $\|\boldsymbol{x}\| = \sqrt{\boldsymbol{x} \cdot \boldsymbol{x}}$.

15 The "l^1 norm" and the "l^∞ norm" of $\boldsymbol{x} = (x_1, \ldots, x_n)$ are

$$\|\boldsymbol{x}\|_1 = |x_1| + \cdots + |x_n| \quad \text{and} \quad \|\boldsymbol{x}\|_\infty = \max_{1 \le i \le n} |x_i|.$$

Compute the norms $\|\boldsymbol{x}\|$ and $\|\boldsymbol{x}\|_1$ and $\|\boldsymbol{x}\|_\infty$ of these two vectors in $\mathbf{R}^5$:

$$\boldsymbol{x} = (1, 1, 1, 1, 1) \qquad \boldsymbol{x} = (.1, .7, .3, .4, .5).$$

16 Prove that $\|\boldsymbol{x}\|_\infty \le \|\boldsymbol{x}\| \le \|\boldsymbol{x}\|_1$. Show from the Schwarz inequality that the ratios $\|\boldsymbol{x}\|/\|\boldsymbol{x}\|_\infty$ and $\|\boldsymbol{x}\|_1/\|\boldsymbol{x}\|$ are never larger than $\sqrt{n}$. Which vector $(x_1, \ldots, x_n)$ gives ratios equal to $\sqrt{n}$?

17 All vector norms must satisfy the *triangle inequality*. Prove that

$$\|\boldsymbol{x}+\boldsymbol{y}\|_\infty \le \|\boldsymbol{x}\|_\infty + \|\boldsymbol{y}\|_\infty \qquad \text{and} \qquad \|\boldsymbol{x}+\boldsymbol{y}\|_1 \le \|\boldsymbol{x}\|_1 + \|\boldsymbol{y}\|_1.$$

18 Vector norms must also satisfy $\|c\boldsymbol{x}\| = |c|\,\|\boldsymbol{x}\|$. The norm must be positive except when $\boldsymbol{x} = \boldsymbol{0}$. Which of these are norms for (x_1, x_2)?

$$\|\boldsymbol{x}\|_A = |x_1| + 2|x_2| \qquad \|\boldsymbol{x}\|_B = \min |x_i|$$

$$\|\boldsymbol{x}\|_C = \|\boldsymbol{x}\| + \|\boldsymbol{x}\|_\infty \qquad \|\boldsymbol{x}\|_D = \|A\boldsymbol{x}\| \quad \text{(answer depends on } A\text{)}.$$

ITERATIVE METHODS FOR LINEAR ALGEBRA ■ 9.3

Up to now, our approach to $A\boldsymbol{x} = \boldsymbol{b}$ has been "direct." We accepted A as it came. We attacked it with Gaussian elimination. This section is about ***iterative methods***, which replace A by a simpler matrix S. The difference $T = S - A$ is moved over to the right side of the equation. The problem becomes easier to solve, with S instead of A. But there is a price—*the simpler system has to be solved over and over.*

An iterative method is easy to invent. Just split A into $S - T$. Then $A\boldsymbol{x} = \boldsymbol{b}$ is the same as

$$S\boldsymbol{x} = T\boldsymbol{x} + \boldsymbol{b}. \tag{1}$$

The novelty is to solve (1) iteratively. Each guess $\boldsymbol{x}_k$ leads to the next $\boldsymbol{x}_{k+1}$:

$$S\boldsymbol{x}_{k+1} = T\boldsymbol{x}_k + \boldsymbol{b}. \tag{2}$$

Start with any $\boldsymbol{x}_0$. Then solve $S\boldsymbol{x}_1 = T\boldsymbol{x}_0 + \boldsymbol{b}$. Continue to the second iteration $S\boldsymbol{x}_2 = T\boldsymbol{x}_1 + \boldsymbol{b}$. A hundred iterations are very common—maybe more. Stop when (and if!) the new vector $\boldsymbol{x}_{k+1}$ is sufficiently close to $\boldsymbol{x}_k$—or when the residual $A\boldsymbol{x}_k - \boldsymbol{b}$ is near zero. We can choose the stopping test. Our hope is to get near the true solution, more quickly than by elimination. When the sequence $\boldsymbol{x}_k$ converges, its limit $\boldsymbol{x} = \boldsymbol{x}_\infty$ does solve equation (1). The proof is to let $k \to \infty$ in equation (2).

The two goals of the splitting $A = S - T$ are ***speed per step*** and ***fast convergence of the*** $\boldsymbol{x}_k$. The speed of each step depends on S and the speed of convergence depends on $S^{-1}T$:

1 Equation (2) should be easy to solve for $\boldsymbol{x}_{k+1}$. The "***preconditioner***" S could be diagonal or triangular. When its LU factorization is known, each iteration step is fast.

2 The difference $\boldsymbol{x} - \boldsymbol{x}_k$ (this is the error $\boldsymbol{e}_k$) should go quickly to zero. Subtracting equation (2) from (1) cancels $\boldsymbol{b}$, and it leaves the ***error equation***:

$$S\boldsymbol{e}_{k+1} = T\boldsymbol{e}_k \text{ which means } \boldsymbol{e}_{k+1} = S^{-1}T\boldsymbol{e}_k. \tag{3}$$

At every step the error is multiplied by $S^{-1}T$. If $S^{-1}T$ is small, its powers go quickly to zero. But what is "small"?

The extreme splitting is $S = A$ and $T = 0$. Then the first step of the iteration is the original $A\boldsymbol{x} = \boldsymbol{b}$. Convergence is perfect and $S^{-1}T$ is zero. But the cost of that step is what we wanted to avoid. The choice of S is a battle between speed per step (a simple S) and fast convergence (S close to A). Here are some popular choices:

J $S =$ diagonal part of A (the iteration is called *Jacobi's method*)

GS $S =$ lower triangular part of A (*Gauss-Seidel method*)

SOR $S =$ combination of Jacobi and Gauss-Seidel (*successive overrelaxation*)

ILU $S =$ approximate L times approximate U (*incomplete LU method*).

Our first question is pure linear algebra: ***When do the x_k's converge to x?*** The answer uncovers the number $|\lambda|_{\max}$ that controls convergence. In examples of **J** and **GS** and **SOR**, we will compute this "*spectral radius*" $|\lambda|_{\max}$. It is the largest eigenvalue of the iteration matrix $S^{-1}T$.

The Spectral Radius Controls Convergence

Equation (3) is $e_{k+1} = S^{-1}Te_k$. Every iteration step multiplies the error by the same matrix $B = S^{-1}T$. The error after k steps is $e_k = B^k e_0$. ***The error approaches zero if the powers of $B = S^{-1}T$ approach zero.*** It is beautiful to see how the eigenvalues of B—the largest eigenvalue in particular—control the matrix powers B^k.

9C Convergence The powers B^k approach zero if and only if every eigenvalue of B satisfies $|\lambda| < 1$. ***The rate of convergence is controlled by the spectral radius $|\lambda|_{\max}$.***

The test for convergence is $|\lambda|_{\max} < 1$. Real eigenvalues must lie between -1 and 1. Complex eigenvalues $\lambda = a+ib$ must lie inside the unit circle in the complex plane. In that case the absolute value $|\lambda|$ is the square root of a^2+b^2—Chapter 10 will discuss complex numbers. In every case the spectral radius is the largest distance from the origin 0 to the eigenvalues $\lambda_1, \ldots, \lambda_n$. Those are eigenvalues of the iteration matrix $B = S^{-1}T$.

To see why $|\lambda|_{\max} < 1$ is necessary, suppose the starting error e_0 happens to be an eigenvector of B. After one step the error is $Be_0 = \lambda e_0$. After k steps the error is $B^k e_0 = \lambda^k e_0$. If we start with an eigenvector, we continue with that eigenvector—and it grows or decays with the powers λ^k. *This factor λ^k goes to zero when* $|\lambda| < 1$. Since this condition is required of every eigenvalue, we need $|\lambda|_{\max} < 1$.

To see why $|\lambda|_{\max} < 1$ is sufficient for the error to approach zero, suppose e_0 is a combination of eigenvectors:

$$e_0 = c_1 x_1 + \cdots + c_n x_n \quad \text{leads to} \quad e_k = c_1(\lambda_1)^k x_1 + \cdots + c_n(\lambda_n)^k x_n. \tag{4}$$

This is the point of eigenvectors! They grow independently, each one controlled by its eigenvalue. When we multiply by B, the eigenvector x_i is multiplied by λ_i. If all $|\lambda_i| < 1$ then equation (4) ensures that e_k goes to zero.

Example 1 $B = \begin{bmatrix} .6 & .5 \\ .6 & .5 \end{bmatrix}$ has $\lambda_{\max} = 1.1$ $B' = \begin{bmatrix} .6 & 1.1 \\ 0 & .5 \end{bmatrix}$ has $\lambda_{\max} = .6$ B^2 is 1.1 times B. Then B^3 is $(1.1)^2$ times B. The powers of B blow up. Contrast with the powers of B'. The matrix $(B')^k$ has $(.6)^k$ and $(.5)^k$ on its diagonal. The off-diagonal entries also involve $(.6)^k$, which sets the speed of convergence.

Note There is a technical difficulty when B does not have n independent eigenvectors. (To produce this effect in B', change .5 to .6.) The starting error $\boldsymbol{e}_0$ may not be a combination of eigenvectors—there are too few for a basis. Then diagonalization is impossible and equation (4) is not correct. We turn to the *Jordan form*:

$$B = SJS^{-1} \quad \text{and} \quad B^k = SJ^kS^{-1}. \tag{5}$$

Section 6.6 shows how J and J^k are made of "blocks" with one repeated eigenvalue:

$$\text{The powers of a 2 by 2 block are} \quad \begin{bmatrix} \lambda & 1 \\ 0 & \lambda \end{bmatrix}^k = \begin{bmatrix} \lambda^k & k\lambda^{k-1} \\ 0 & \lambda^k \end{bmatrix}.$$

If $|\lambda| < 1$ then these powers approach zero. The extra factor k from a double eigenvalue is overwhelmed by the decreasing factor λ^{k-1}. This applies to all Jordan blocks. A larger block has $k^2\lambda^{k-2}$ in J^k, which also approaches zero when $|\lambda| < 1$.

If all $|\lambda| < 1$ then $J^k \to 0$. This proves **9C:** ***Convergence requires*** $|\lambda|_{\max} < 1$**.**

Jacobi versus Seidel

We now solve a specific 2 by 2 problem. The theory of iteration says that the key number is the spectral radius of $B = S^{-1}T$. Watch for that number $|\lambda|_{\max}$. It is also written $\rho(B)$—the Greek letter "rho" stands for the spectral radius:

$$\begin{aligned} 2u - v &= 4 \\ -u + 2v &= -2 \end{aligned} \quad \text{has the solution} \quad \begin{bmatrix} u \\ v \end{bmatrix} = \begin{bmatrix} 2 \\ 0 \end{bmatrix}. \tag{6}$$

The first splitting is ***Jacobi's method***. Keep the *diagonal terms* on the left side (this is S). Move the off-diagonal part of A to the right side (this is T). Then iterate:

$$\begin{aligned} 2u_{k+1} &= v_k + 4 \\ 2v_{k+1} &= u_k - 2. \end{aligned}$$

Start the iteration from $u_0 = v_0 = 0$. The first step goes to $u_1 = 2, v_1 = -1$. Keep going:

$$\begin{bmatrix} 0 \\ 0 \end{bmatrix} \begin{bmatrix} 2 \\ -1 \end{bmatrix} \begin{bmatrix} 3/2 \\ 0 \end{bmatrix} \begin{bmatrix} 2 \\ -1/4 \end{bmatrix} \begin{bmatrix} 15/8 \\ 0 \end{bmatrix} \begin{bmatrix} 2 \\ -1/16 \end{bmatrix} \quad \text{approaches} \quad \begin{bmatrix} 2 \\ 0 \end{bmatrix}.$$

This shows convergence. At steps 1, 3, 5 the second component is $-1, -1/4, -1/16$. The error is multiplied by $\frac{1}{4}$ every two steps. So is the error in the first component. The values $0, 3/2, 15/8$ have errors $2, \frac{1}{2}, \frac{1}{8}$. Those also drop by 4 in each two steps. *The error equation is* $S\boldsymbol{e}_{k+1} = T\boldsymbol{e}_k$:

$$\begin{bmatrix} 2 & 0 \\ 0 & 2 \end{bmatrix} \boldsymbol{e}_{k+1} = \begin{bmatrix} 0 & 1 \\ 1 & 0 \end{bmatrix} \boldsymbol{e}_k \quad \text{or} \quad \boldsymbol{e}_{k+1} = \begin{bmatrix} 0 & \frac{1}{2} \\ \frac{1}{2} & 0 \end{bmatrix} \boldsymbol{e}_k. \tag{7}$$

That last matrix is $S^{-1}T$. Its eigenvalues are $\frac{1}{2}$ and $-\frac{1}{2}$. So its spectral radius is $\frac{1}{2}$:

$$B = S^{-1}T = \begin{bmatrix} 0 & \frac{1}{2} \\ \frac{1}{2} & 0 \end{bmatrix} \quad \text{has } |\lambda|_{\max} = \tfrac{1}{2} \quad \text{and} \quad \begin{bmatrix} 0 & \frac{1}{2} \\ \frac{1}{2} & 0 \end{bmatrix}^2 = \begin{bmatrix} \frac{1}{4} & 0 \\ 0 & \frac{1}{4} \end{bmatrix}.$$

Two steps multiply the error by $\frac{1}{4}$ exactly, in this special example. The important message is this: Jacobi's method works well when the main diagonal of A is large compared to the off-diagonal part. The diagonal part is S, the rest is $-T$. We want the diagonal to dominate and $S^{-1}T$ to be small.

The eigenvalue $\lambda = \frac{1}{2}$ is unusually small. Ten iterations reduce the error by $2^{10} = 1024$. Twenty iterations reduce $\boldsymbol{e}$ by $(1024)^2$. More typical and more expensive is $|\lambda|_{\max} = .99$ or $.999$.

The ***Gauss-Seidel method*** keeps the whole lower triangular part of A on the left side as S:

$$\begin{aligned} 2u_{k+1} \qquad\quad &= v_k + 4 \\ -u_{k+1} + 2v_{k+1} &= \quad -2 \end{aligned} \qquad \text{or} \qquad \begin{aligned} u_{k+1} &= \tfrac{1}{2}v_k \quad + 2 \\ v_{k+1} &= \tfrac{1}{2}u_{k+1} - 1. \end{aligned} \tag{8}$$

Notice the change. The new u_{k+1} from the first equation is used *immediately* in the second equation. With Jacobi, we saved the old u_k until the whole step was complete. With Gauss-Seidel, the new values enter right away and the old u_k is destroyed. This cuts the storage in half! It also speeds up the iteration (usually). And it costs no more than the Jacobi method.

Starting from $(0, 0)$, the exact answer $(2, 0)$ is reached in one step. That is an accident I did not expect. Test the iteration from another start $u_0 = 0$ and $v_0 = -1$:

$$\begin{bmatrix} 0 \\ -1 \end{bmatrix} \quad \begin{bmatrix} 3/2 \\ -1/4 \end{bmatrix} \quad \begin{bmatrix} 15/8 \\ -1/16 \end{bmatrix} \quad \begin{bmatrix} 63/32 \\ -1/64 \end{bmatrix} \quad \text{approaches} \quad \begin{bmatrix} 2 \\ 0 \end{bmatrix}.$$

The errors in the first component are 2, 1/2, 1/8, 1/32. The errors in the second component are -1, $-1/4$, $-1/16$, $-1/32$. We divide by 4 in *one* step not two steps. ***Gauss-Seidel is twice as fast as Jacobi.***

This is true for every positive definite tridiagonal matrix; $|\lambda|_{\max}$ for Gauss-Seidel is the *square* of $|\lambda|_{\max}$ for Jacobi. This holds in many other applications—but not for every matrix. Anything is possible when A is strongly nonsymmetric—Jacobi is sometimes better, and both methods might fail. Our example is small:

$$S = \begin{bmatrix} 2 & 0 \\ -1 & 2 \end{bmatrix} \quad \text{and} \quad T = \begin{bmatrix} 0 & 1 \\ 0 & 0 \end{bmatrix} \quad \text{and} \quad S^{-1}T = \begin{bmatrix} 0 & \frac{1}{2} \\ 0 & \frac{1}{4} \end{bmatrix}.$$

The Gauss-Seidel eigenvalues are 0 and $\frac{1}{4}$. Compare with $\frac{1}{2}$ and $-\frac{1}{2}$ for Jacobi.

With a small push we can explain the ***successive overrelaxation method*** (**SOR**). The new idea is to introduce a parameter ω (omega) into the iteration. Then choose this number ω to make the spectral radius of $S^{-1}T$ as small as possible.

Rewrite $Ax = b$ as $\omega Ax = \omega b$. The matrix S in **SOR** has the diagonal of the original A, but below the diagonal we use ωA. The matrix T on the right side is $S - \omega A$:

$$\begin{aligned} 2u_{k+1} \qquad &= (2-2\omega)u_k + \qquad \omega v_k + 4\omega \\ -\omega u_{k+1} + 2v_{k+1} &= \qquad (2-2\omega)v_k - 2\omega. \end{aligned} \tag{9}$$

This looks more complicated to us, but the computer goes as fast as ever. Each new u_{k+1} from the first equation is used immediately to find v_{k+1} in the second equation. This is like Gauss-Seidel, with an adjustable number ω. The key matrix is always $S^{-1}T$:

$$S^{-1}T = \begin{bmatrix} 1-\omega & \frac{1}{2}\omega \\ \frac{1}{2}\omega(1-\omega) & 1-\omega+\frac{1}{4}\omega^2 \end{bmatrix}. \tag{10}$$

The determinant is $(1-\omega)^2$. At the best ω, both eigenvalues turn out to equal $7-4\sqrt{3}$, which is close to $(\frac{1}{4})^2$. Therefore **SOR** is twice as fast as Gauss-Seidel in this example. In other examples **SOR** can converge ten or a hundred times as fast.

I will put on record the most valuable test matrix of order n. It is our favorite -1, 2, -1 tridiagonal matrix. The diagonal is $2I$. Below and above are -1's. Our example had $n = 2$, which leads to $\cos\frac{\pi}{3} = \frac{1}{2}$ as the Jacobi eigenvalue. (We found that $\frac{1}{2}$ above.) Notice especially that this eigenvalue is squared for Gauss-Seidel:

9D The splittings of the -1, 2, -1 matrix of order n yield these eigenvalues of B:

Jacobi ($S = 0, 2, 0$ matrix): $\quad S^{-1}T$ has $|\lambda|_{\max} = \cos\dfrac{\pi}{n+1}$

Gauss-Seidel ($S = -1, 2, 0$ matrix): $\quad S^{-1}T$ has $|\lambda|_{\max} = \left(\cos\dfrac{\pi}{n+1}\right)^2$

SOR (with the best ω): $\quad S^{-1}T$ has $|\lambda|_{\max} = \left(\cos\dfrac{\pi}{n+1}\right)^2 \Big/ \left(1+\sin\dfrac{\pi}{n+1}\right)^2$.

Let me be clear: For the -1, 2, -1 matrix you should not use any of these iterations! Elimination is very fast (exact LU). Iterations are intended for large sparse matrices—when a high percentage of the zero entries are "not good." The not good zeros are inside the band, which is wide. They become nonzero in the exact L and U, which is why elimination becomes expensive.

We mention one more splitting. It is associated with the words "***incomplete*** LU." The idea is to set the small nonzeros in L and U *back to zero*. This leaves triangular matrices L_0 and U_0 which are again sparse. That allows fast computations.

The splitting has $S = L_0U_0$ on the left side. Each step is quick:

$$L_0U_0x_{k+1} = (A - L_0U_0)x_k + b.$$

On the right side we do sparse matrix-vector multiplications. Don't multiply L_0 times U_0—those are matrices. Multiply x_k by U_0 and then multiply that vector by L_0. On the left side we do forward and back substitutions. If L_0U_0 is close to A, then $|\lambda|_{\max}$ is small. A few iterations will give a close answer.

The difficulty with all four of these splittings is that a single large eigenvalue in $S^{-1}T$ would spoil everything. There is a safer iteration—the ***conjugate gradient method***—which avoids this difficulty. Combined with a good preconditioner S (from the splitting $A = S-T$), this produces one of the most popular and powerful algorithms in numerical linear algebra.[1]

Iterative Methods for Eigenvalues

We move from $A\boldsymbol{x} = \boldsymbol{b}$ to $A\boldsymbol{x} = \lambda\boldsymbol{x}$. Iterations are an option for linear equations. They are a necessity for eigenvalue problems. The eigenvalues of an n by n matrix are the roots of an nth degree polynomial. The determinant of $A - \lambda I$ starts with $(-\lambda)^n$. This book must not leave the impression that eigenvalues should be computed from this polynomial. The determinant of $A - \lambda I$ is a very poor approach—except when n is small.

For $n > 4$ there is no formula to solve $\det(A - \lambda I) = 0$. Worse than that, the λ's can be very unstable and sensitive. It is much better to work with A itself, gradually making it diagonal or triangular. (Then the eigenvalues appear on the diagonal.) Good computer codes are available in the LAPACK library—individual routines are free on **www.netlib.org**. This library combines the earlier LINPACK and EISPACK, with improvements. It is a collection of Fortran 77 programs for linear algebra on high-performance computers. (The email message **send index from lapack** brings information.) For your computer and mine, the same efficiency is achieved by high quality matrix packages like MATLAB.

We will briefly discuss the power method and the QR method for computing eigenvalues. It makes no sense to give full details of the codes.

1 Power methods and inverse power methods. Start with any vector $\boldsymbol{u}_0$. Multiply by A to find $\boldsymbol{u}_1$. Multiply by A again to find $\boldsymbol{u}_2$. If $\boldsymbol{u}_0$ is a combination of the eigenvectors, then A multiplies each eigenvector $\boldsymbol{x}_i$ by λ_i. After k steps we have $(\lambda_i)^k$:

$$\boldsymbol{u}_k = A^k\boldsymbol{u}_0 = c_1(\lambda_1)^k\boldsymbol{x}_1 + \cdots + c_n(\lambda_n)^k\boldsymbol{x}_n. \tag{11}$$

As the power method continues, ***the largest eigenvalue begins to dominate***. The vectors $\boldsymbol{u}_k$ point toward that dominant eigenvector. We saw this for Markov matrices in Chapter 8:

$$A = \begin{bmatrix} .9 & .3 \\ .1 & .7 \end{bmatrix} \quad \text{has} \quad \lambda_{\max} = 1 \quad \text{with eigenvector} \quad \begin{bmatrix} .75 \\ .25 \end{bmatrix}.$$

Start with $\boldsymbol{u}_0$ and multiply at every step by A:

$$\boldsymbol{u}_0 = \begin{bmatrix} 1 \\ 0 \end{bmatrix}, \quad \boldsymbol{u}_1 = \begin{bmatrix} .9 \\ .1 \end{bmatrix}, \quad \boldsymbol{u}_2 = \begin{bmatrix} .84 \\ .16 \end{bmatrix} \quad \text{is approaching} \quad \boldsymbol{u}_\infty = \begin{bmatrix} .75 \\ .25 \end{bmatrix}.$$

[1] Conjugate gradients are described in the author's book *Introduction to Applied Mathematics* and in greater detail by Golub-Van Loan and by Trefethen-Bau.

The speed of convergence depends on the *ratio* of the second largest eigenvalue λ_2 to the largest λ_1. We don't want λ_1 to be small, we want λ_2/λ_1 to be small. Here $\lambda_2/\lambda_1 = .6/1$ and the speed is reasonable. For large matrices it often happens that $|\lambda_2/\lambda_1|$ is very close to 1. Then the power method is too slow.

Is there a way to find the *smallest* eigenvalue—which is often the most important in applications? Yes, by the *inverse* power method: Multiply $\boldsymbol{u}_0$ by A^{-1} instead of A. Since we never want to compute A^{-1}, we actually solve $A\boldsymbol{u}_1 = \boldsymbol{u}_0$. By saving the LU factors, the next step $A\boldsymbol{u}_2 = \boldsymbol{u}_1$ is fast. Eventually

$$\boldsymbol{u}_k = A^{-k}\boldsymbol{u}_0 = \frac{c_1\boldsymbol{x}_1}{(\lambda_1)^k} + \cdots + \frac{c_n\boldsymbol{x}_n}{(\lambda_n)^k}. \tag{12}$$

Now the *smallest* eigenvalue λ_{min} is in control. When it is very small, the factor $1/\lambda_{\text{min}}^k$ is large. For high speed, we make λ_{min} even smaller by shifting the matrix to $A-\lambda^*I$. If λ^* is close to λ_{min} then $A-\lambda^*I$ has the very small eigenvalue $\lambda_{\text{min}}-\lambda^*$. Each *shifted inverse power step* divides the eigenvector by this number, and that eigenvector quickly dominates.

2 The QR Method This is a major achievement in numerical linear algebra. Fifty years ago, eigenvalue computations were slow and inaccurate. We didn't even realize that solving $\det(A-\lambda I) = 0$ was a terrible method. Jacobi had suggested earlier that A should gradually be made triangular—then the eigenvalues appear automatically on the diagonal. He used 2 by 2 rotations to produce off-diagonal zeros. (Unfortunately the previous zeros can become nonzero again. But Jacobi's method made a partial comeback with parallel computers.) At present the QR method is the leader in eigenvalue computations and we describe it briefly.

The basic step is to factor A, whose eigenvalues we want, into QR. Remember from Gram-Schmidt (Section 4.4) that Q has orthonormal columns and R is triangular. For eigenvalues the key idea is: ***Reverse Q and R***. The new matrix is RQ. Since $A_1 = RQ$ is similar to $A = QR$, *the eigenvalues are not changed*:

$$QRx = \lambda x \quad \text{gives} \quad RQ(Q^{-1}x) = \lambda(Q^{-1}x). \tag{13}$$

This process continues. Factor the new matrix A_1 into Q_1R_1. Then reverse the factors to R_1Q_1. This is the next matrix A_2, and again no change in the eigenvalues. Amazingly, those eigenvalues begin to show up on the diagonal. Often the last entry of A_4 holds an accurate eigenvalue. In that case we remove the last row and column and continue with a smaller matrix to find the next eigenvalue.

Two extra ideas make this method a success. One is to shift the matrix by a multiple of I, before factoring into QR. Then RQ is shifted back:

$$\text{Factor } A_k - c_kI \text{ into } Q_kR_k. \text{ The next matrix is } A_{k+1} = R_kQ_k + c_kI.$$

A_{k+1} has the same eigenvalues as A_k, and the same as the original $A_0 = A$. A good shift chooses c near an (unknown) eigenvalue. That eigenvalue appears more accurately on the diagonal of A_{k+1}—which tells us a better c for the next step to A_{k+2}.

The other idea is to obtain off-diagonal zeros before the QR method starts. Change A to the similar matrix $L^{-1}AL$ (no change in the eigenvalues):

$$L^{-1}AL = \begin{bmatrix} 1 & & \\ & 1 & \\ & -1 & 1 \end{bmatrix} \begin{bmatrix} 1 & 2 & 3 \\ 1 & 4 & 5 \\ 1 & 6 & 7 \end{bmatrix} \begin{bmatrix} 1 & & \\ & 1 & \\ & 1 & 1 \end{bmatrix} = \begin{bmatrix} 1 & 5 & 3 \\ 1 & 9 & 5 \\ 0 & 4 & 2 \end{bmatrix}.$$

L^{-1} subtracted row 2 from row 3 to produce the zero in column 1. Then L added column 3 to column 2 and left the zero alone. If I try for another zero (too ambitious), I will fail. Subtracting row 1 from row 2 produces a zero. But now L adds column 2 to column 1 and destroys it.

We must leave those nonzeros 1 and 4 along one subdiagonal. This is a "*Hessenberg matrix*", which is reachable in a fixed number of steps. The zeros in the lower left corner will stay zero through the QR method. The operation count for each QR factorization drops from $O(n^3)$ to $O(n^2)$.

Golub and Van Loan give this example of one shifted QR step on a Hessenberg matrix A. The shift is $cI = 7I$:

$$A = \begin{bmatrix} 1 & 2 & 3 \\ 4 & 5 & 6 \\ 0 & .001 & 7 \end{bmatrix} \quad \text{leads to} \quad A_1 = \begin{bmatrix} -.54 & 1.69 & 0.835 \\ .31 & 6.53 & -6.656 \\ 0 & .00002 & 7.012 \end{bmatrix}.$$

Factoring $A - 7I$ into QR produced $A_1 = RQ + 7I$. Notice the very small number .00002. The diagonal entry 7.012 is almost an exact eigenvalue of A_1, and therefore of A. Another QR step with shift by $7.012I$ would give terrific accuracy.

Problem Set 9.3

Problems 1–12 are about iterative methods for $Ax = b$.

1 Change $Ax = b$ to $x = (I - A)x + b$. What are S and T for this splitting? What matrix $S^{-1}T$ controls the convergence of $x_{k+1} = (I - A)x_k + b$?

2 If λ is an eigenvalue of A, then ______ is an eigenvalue of $B = I - A$. The real eigenvalues of B have absolute value less than 1 if the real eigenvalues of A lie between ______ and ______.

3 Show why the iteration $x_{k+1} = (I - A)x_k + b$ does not converge for $A = \begin{bmatrix} 2 & -1 \\ -1 & 2 \end{bmatrix}$.

4 Why is the norm of B^k never larger than $\|B\|^k$? Then $\|B\| < 1$ guarantees that the powers B^k approach zero (convergence). No surprise since $|\lambda|_{\max}$ is below $\|B\|$.

5 If A is singular then all splittings $A = S - T$ must fail. From $Ax = 0$ show that $S^{-1}Tx = x$. So this matrix $B = S^{-1}T$ has $\lambda = 1$ and fails.

6 Change the 2's to 3's and find the eigenvalues of $S^{-1}T$ for Jacobi's method:

$$S\boldsymbol{x}_{k+1} = T\boldsymbol{x}_k + \boldsymbol{b} \quad \text{is} \quad \begin{bmatrix} 3 & 0 \\ 0 & 3 \end{bmatrix} \boldsymbol{x}_{k+1} = \begin{bmatrix} 0 & 1 \\ 1 & 0 \end{bmatrix} \boldsymbol{x}_k + \boldsymbol{b}.$$

7 Find the eigenvalues of $S^{-1}T$ for the Gauss-Seidel method applied to Problem 6:

$$\begin{bmatrix} 3 & 0 \\ -1 & 3 \end{bmatrix} \boldsymbol{x}_{k+1} = \begin{bmatrix} 0 & 1 \\ 0 & 0 \end{bmatrix} \boldsymbol{x}_k + \boldsymbol{b}.$$

Does $|\lambda|_{\max}$ for Gauss-Seidel equal $|\lambda|^2_{\max}$ for Jacobi?

8 For any 2 by 2 matrix $\left[\begin{smallmatrix} \mathbf{a} & \mathbf{b} \\ \mathbf{c} & \mathbf{d} \end{smallmatrix}\right]$ show that $|\lambda|_{\max}$ equals $|bc/ad|$ for Gauss-Seidel and $|bc/ad|^{1/2}$ for Jacobi. We need $ad \neq 0$ for the matrix S to be invertible.

9 The best ω produces two equal eigenvalues for $S^{-1}T$ in the **SOR** method. Those eigenvalues are $\omega - 1$ because the determinant is $(\omega - 1)^2$. Set the trace in equation (10) equal to $(\omega - 1) + (\omega - 1)$ and find this optimal ω.

10 Write a computer code (MATLAB or other) for the Gauss-Seidel method. You can define S and T from A, or set up the iteration loop directly from the entries a_{ij}. Test it on the -1, 2, -1 matrices A of order 10, 20, 50 with $\boldsymbol{b} = (1, 0, \ldots, 0)$.

11 The Gauss-Seidel iteration at component i is

$$x_i^{\text{new}} = x_i^{\text{old}} + \frac{1}{a_{ii}} \Big(b_i - \sum_{j=1}^{i-1} a_{ij} x_j^{\text{new}} - \sum_{j=i}^{n} a_{ij} x_j^{\text{old}} \Big).$$

If every $x_i^{\text{new}} = x_i^{\text{old}}$ how does this show that the solution $\boldsymbol{x}$ is correct? How does the formula change for Jacobi's method? For **SOR** insert ω outside the parentheses.

12 The **SOR** splitting matrix S is the same as for Gauss-Seidel except that the diagonal is divided by ω. Write a program for **SOR** on an n by n matrix. Apply it with $\omega = 1$, 1.4, 1.8, 2.2 when A is the -1, 2, -1 matrix of order $n = 10$.

13 Divide equation (11) by λ_1^k and explain why $|\lambda_2/\lambda_1|$ controls the convergence of the power method. Construct a matrix A for which this method *does not converge*.

14 The Markov matrix $A = \left[\begin{smallmatrix} .9 & .3 \\ .1 & .7 \end{smallmatrix}\right]$ has $\lambda = 1$ and .6, and the power method $\boldsymbol{u}_k = A^k \boldsymbol{u}_0$ converges to $\left[\begin{smallmatrix} .75 \\ .25 \end{smallmatrix}\right]$. Find the eigenvectors of A^{-1}. What does the inverse power method $\boldsymbol{u}_{-k} = A^{-k}\boldsymbol{u}_0$ converge to (after you multiply by $.6^k$)?

15 Show that the n by n matrix with diagonals -1, 2, -1 has the eigenvector $\boldsymbol{x}_1 = (\sin \frac{\pi}{n+1}, \sin \frac{2\pi}{n+1}, \ldots, \sin \frac{n\pi}{n+1})$. Find the eigenvalue λ_1 by multiplying $A\boldsymbol{x}_1$. Note: For the other eigenvectors and eigenvalues of this matrix, change π to $j\pi$ in $\boldsymbol{x}_1$ and λ_1.

16 For $A = \begin{bmatrix} 2 & -1 \\ -1 & 2 \end{bmatrix}$ apply the power method $\boldsymbol{u}_{k+1} = A\boldsymbol{u}_k$ three times starting with $\boldsymbol{u}_0 = \begin{bmatrix} 1 \\ 0 \end{bmatrix}$. What eigenvector is the power method converging to?

17 In Problem 11 apply the *inverse* power method $\boldsymbol{u}_{k+1} = A^{-1}\boldsymbol{u}_k$ three times with the same $\boldsymbol{u}_0$. What eigenvector are the $\boldsymbol{u}_k$'s approaching?

18 In the QR method for eigenvalues, show that the 2, 1 entry drops from $\sin\theta$ in $A = QR$ to $-\sin^3\theta$ in RQ. (*Compute R and RQ.*) This "cubic convergence" makes the method a success:

$$A = \begin{bmatrix} \cos\theta & \sin\theta \\ \sin\theta & 0 \end{bmatrix} = QR = \begin{bmatrix} \cos\theta & -\sin\theta \\ \sin\theta & \cos\theta \end{bmatrix} \begin{bmatrix} 1 & ? \\ 0 & ? \end{bmatrix}.$$

19 If A is an orthogonal matrix, its QR factorization has $Q =$ _____ and $R =$ _____. Therefore $RQ =$ _____. These are among the rare examples when the QR method fails.

20 The shifted QR method factors $A - cI$ into QR. Show that the next matrix $A_1 = RQ + cI$ equals $Q^{-1}AQ$. Therefore A_1 has the _____ eigenvalues as A (but is closer to triangular).

21 When $A = A^{\mathrm{T}}$, the "*Lanczos method*" finds a's and b's and orthonormal $\boldsymbol{q}$'s so that $A\boldsymbol{q}_j = b_{j-1}\boldsymbol{q}_{j-1} + a_j\boldsymbol{q}_j + b_j\boldsymbol{q}_{j+1}$ (with $\boldsymbol{q}_0 = \boldsymbol{0}$). Multiply by $\boldsymbol{q}_j^{\mathrm{T}}$ to find a formula for a_j. The equation says that $AQ = QT$ where T is a _____ matrix.

22 The equation in Problem 21 develops from this loop with $b_0 = 1$ and $\boldsymbol{r}_0 =$ any $\boldsymbol{q}_1$:

$$\boldsymbol{q}_{j+1} = \boldsymbol{r}_j/b_j;\ j = j+1;\ a_j = \boldsymbol{q}_j^{\mathrm{T}}A\boldsymbol{q}_j;\ \boldsymbol{r}_j = A\boldsymbol{q}_j - b_{j-1}\boldsymbol{q}_{j-1} - a_j\boldsymbol{q}_j;\ b_j = \|\boldsymbol{r}_j\|.$$

Write a computer program. Test on the $-1, 2, -1$ matrix A. $Q^{\mathrm{T}}Q$ should be I.

23 Suppose A is *tridiagonal and symmetric in the* QR *method.* From $A_1 = Q^{-1}AQ$ show that A_1 is symmetric. Then change Then change to $A_1 = RAR^{-1}$ and show that A_1 is also tridiagonal. (If the lower part of A_1 is proved tridiagonal then by symmetry the upper part is too.) Symmetric tridiagonal matrices are at the heart of the QR method.

Questions 24–26 are about quick ways to estimate the location of the eigenvalues.

24 If the sum of $|a_{ij}|$ along every row is less than 1, prove that $|\lambda| < 1$. (If $|x_i|$ is larger than the other components of $\boldsymbol{x}$, why is $|\Sigma a_{ij}x_j|$ less than $|x_i|$? That means $|\lambda x_i| < |x_i|$ so $|\lambda| < 1$.)

(Gershgorin circles) Every eigenvalue of A is in a circle centered at a diagonal entry a_{ii} with radius $r_i = \Sigma_{j \neq i}|a_{ij}|$. *This follows from* $(\lambda - a_{ii})x_i = \Sigma_{j \neq i}a_{ij}x_j$.

If $|x_i|$ is larger than the other components of $\boldsymbol{x}$, this sum is at most $r_i|x_i|$. Dividing by $|x_i|$ leaves $|\lambda - a_{ii}| \leq r_i$.

25 What bound on $|\lambda|_{\max}$ does Problem 24 give for these matrices? What are the three Gershgorin circles that contain all the eigenvalues?

$$A = \begin{bmatrix} .3 & .3 & .2 \\ .3 & .2 & .4 \\ .2 & .4 & .1 \end{bmatrix} \qquad A = \begin{bmatrix} 2 & -1 & 0 \\ -1 & 2 & -1 \\ 0 & -1 & 2 \end{bmatrix}.$$

26 These matrices are diagonally dominant because each $a_{ii} > r_i =$ absolute sum along the rest of row i. From the Gershgorin circles containing all λ's, show that diagonally dominant matrices are invertible.

$$A = \begin{bmatrix} 1 & .3 & .4 \\ .3 & 1 & .5 \\ .4 & .5 & 1 \end{bmatrix} \qquad A = \begin{bmatrix} 4 & 2 & 1 \\ 1 & 3 & 1 \\ 2 & 2 & 5 \end{bmatrix}.$$

The key point for large matrices is that matrix-vector multiplication is much faster than matrix-matrix multiplication. A crucial construction starts with a vector $\boldsymbol{b}$ and computes $A\boldsymbol{b}$, $A^2\boldsymbol{b}, \ldots$ (but never A^2!). The first N vectors span the ***Nth Krylov subspace***. They are the columns of the ***Krylov matrix*** K_N:

$$K_N = \begin{bmatrix} \boldsymbol{b} & A\boldsymbol{b} & A^2\boldsymbol{b} & \cdots & A^{N-1}\boldsymbol{b} \end{bmatrix}.$$

Here in "pseudocode" are two of the most important algorithms in numerical linear algebra:

Arnoldi Iteration	**Conjugate Gradient Iteration for Positive Definite** A	
$\boldsymbol{q}_1 = \boldsymbol{b}/\|\boldsymbol{b}\|$	$x_0 = 0, \boldsymbol{r}_0 = \boldsymbol{b}, \boldsymbol{p}_0 = \boldsymbol{r}_0$	
for $n = 1$ **to** $N-1$	**for** $n = 1$ **to** N	
$\boldsymbol{v} = A\boldsymbol{q}_n$	$\alpha_n = (\boldsymbol{r}_{n-1}^{\mathrm{T}}\boldsymbol{r}_{n-1})/(\boldsymbol{p}_{n-1}^{\mathrm{T}}A\boldsymbol{p}_{n-1})$	step length $\boldsymbol{x}_{n-1}$ to $\boldsymbol{x}_n$
for $j = 1$ **to** n	$\boldsymbol{x}_n = \boldsymbol{x}_{n-1} + \alpha_n \boldsymbol{p}_{n-1}$	approximate solution
$h_{jn} = \boldsymbol{q}_j^{\mathrm{T}}\boldsymbol{v}$	$\boldsymbol{r}_n = \boldsymbol{r}_{n-1} - \alpha_n A\boldsymbol{p}_{n-1}$	new residual $\boldsymbol{b} - A\boldsymbol{x}_n$
$\boldsymbol{v} = \boldsymbol{v} - h_{jn}\boldsymbol{q}_j$	$\beta_n = (\boldsymbol{r}_n^{\mathrm{T}}\boldsymbol{r}_n)/(\boldsymbol{r}_{n-1}^{\mathrm{T}}\boldsymbol{r}_{n-1})$	improvement this step
$h_{n+1,n} = \|\boldsymbol{v}\|$	$\boldsymbol{p}_n = \boldsymbol{r}_n + \beta_n \boldsymbol{p}_{n-1}$	next search direction
$\boldsymbol{q}_{n+1} = \boldsymbol{v}/h_{n+1,n}$	*% Notice: only* 1 *matrix-vector multiplication* $A\boldsymbol{q}$ *and* $A\boldsymbol{p}$	

27 In Arnoldi show that $\boldsymbol{q}_2$ is orthogonal to $\boldsymbol{q}_1$. The Arnoldi method is Gram-Schmidt orthogonalization applied to the Krylov matrix: $K_N = Q_N R_N$. The eigenvalues of $Q_N^{\mathrm{T}} A Q_N$ are often very close to those of A even for $N \ll n$. The *Lanczos iteration* is Arnoldi for symmetric matrices (all coded in ARPACK).

28 In Conjugate Gradients show that $\boldsymbol{r}_1$ is orthogonal to $\boldsymbol{r}_0$ (orthogonal residuals) and $\boldsymbol{p}_1^{\mathrm{T}} A \boldsymbol{p}_0 = 0$ (search directions are A-orthogonal). The iteration solves $A\boldsymbol{x} = \boldsymbol{b}$ by minimizing the error $\boldsymbol{e}^{\mathrm{T}} A \boldsymbol{e}$ in the Krylov subspace. It is a fantastic algorithm.

10

COMPLEX VECTORS AND MATRICES

COMPLEX NUMBERS ■ 10.1

A complete theory of linear algebra must include complex numbers. Even when the matrix is real, the eigenvalues and eigenvectors are often complex. Example: A 2 by 2 rotation matrix has no real eigenvectors. Every vector turns by θ—the direction is changed. But there are complex eigenvectors $(1, i)$ and $(1, -i)$. The eigenvalues are also complex numbers $e^{i\theta}$ and $e^{-i\theta}$. If we insist on staying with real numbers, the theory of eigenvalues will be left in midair.

The second reason for allowing complex numbers goes beyond λ and $\boldsymbol{x}$ to the matrix A. ***The matrix itself may be complex***. We will devote a whole section to the most important example—*the Fourier matrix*. Engineering and science and music and economics all use Fourier series. In reality the series is finite, not infinite. Computing the coefficients in $c_1 e^{ix} + c_2 e^{i2x} + \cdots + c_n e^{inx}$ is a linear algebra problem.

This section gives the main facts about complex numbers. It is a review for some students and a reference for everyone. The underlying fact is that $i^2 = -1$. Everything comes from that. We will get as far as the amazing formula $e^{2\pi i} = 1$.

Adding and Multiplying Complex Numbers

Start with the imaginary number i. Everybody knows that $x^2 = -1$ has no real solution. When you square a real number, the answer is never negative. So the world has agreed on a solution called i. (Except that electrical engineers call it j.) Imaginary numbers follow the normal rules of addition and multiplication, with one difference. *Whenever i^2 appears it is replaced by -1.*

10A ***A complex number*** (say $3+2i$) ***is the sum of a real number*** (3) ***and a pure imaginary number*** ($2i$). Addition keeps the real and imaginary parts separate. Multiplication uses $i^2=-1$:

$$\textbf{Add:}\quad (3+2i)+(3+2i)=6+4i$$
$$\textbf{Multiply:}\quad (3+2i)(1-i)=3+2i-3i-2i^2=5-i.$$

If I add $3+2i$ to $1-i$, the answer is $4+i$. The real numbers $3+1$ stay separate from the imaginary numbers $2i-i$. We are adding the vectors $(3,2)$ and $(1,-1)$.

The number $(1-i)^2$ is $1-i$ times $1-i$. The rules give the surprising answer $-2i$:

$$(1-i)(1-i)=1-i-i+i^2=-2i.$$

In the complex plane, $1-i$ is at an angle of -45°. When we square it to get $-2i$, the angle doubles to -90°. If we square again, the answer is $(-2i)^2=-4$. The -90° angle has become -180°, which is the direction of a negative real number.

A real number is just a complex number $z=a+bi$, with zero imaginary part: $b=0$. A pure imaginary number has $a=0$:

The ***real part*** is $a=\text{Re }(a+bi)$. The ***imaginary part*** is $b=\text{Im }(a+bi)$.

The Complex Plane

Complex numbers correspond to points in a plane. Real numbers go along the x axis. Pure imaginary numbers are on the y axis. ***The complex number* $3+2i$ *is at the point with coordinates* $(3,2)$.** The number zero, which is $0+0i$, is at the origin.

Adding and subtracting complex numbers is like adding and subtracting vectors in the plane. The real component stays separate from the imaginary component. The vectors go head-to-tail as usual. The complex plane $\mathbf{C}^1$ is like the ordinary two-dimensional plane $\mathbf{R}^2$, except that we multiply complex numbers and we didn't multiply vectors.

Now comes an important idea. ***The complex conjugate of* $3+2i$ *is* $3-2i$.** The complex conjugate of $z=1-i$ is $\bar{z}=1+i$. In general the conjugate of $z=a+bi$ is $\bar{z}=a-bi$. (Notice the "*bar*" on the number to indicate the conjugate.) The imaginary parts of z and "z bar" have opposite signs. In the complex plane, $\bar{z}$ is the image of z on the other side of the real axis.

Two useful facts. ***When we multiply conjugates* $\bar{z}_1$ *and* $\bar{z}_2$*, we get the conjugate of* z_1z_2.** When we add $\bar{z}_1$ and $\bar{z}_2$, we get the conjugate of z_1+z_2:

$$\bar{z}_1+\bar{z}_2=(3-2i)+(1+i)=4-i=\text{ conjugate of } z_1+z_2.$$
$$\bar{z}_1\times\bar{z}_2=(3-2i)\times(1+i)=5+i=\text{ conjugate of } z_1\times z_2.$$

Adding and multiplying is exactly what linear algebra needs. By taking conjugates of $Ax=\lambda x$, when A is real, we have another eigenvalue $\bar{\lambda}$ and its eigenvector $\bar{x}$:

$$\textit{If } Ax=\lambda x \textit{ and } A \textit{ is real then } A\bar{x}=\bar{\lambda}\bar{x}. \tag{1}$$

Complex plane

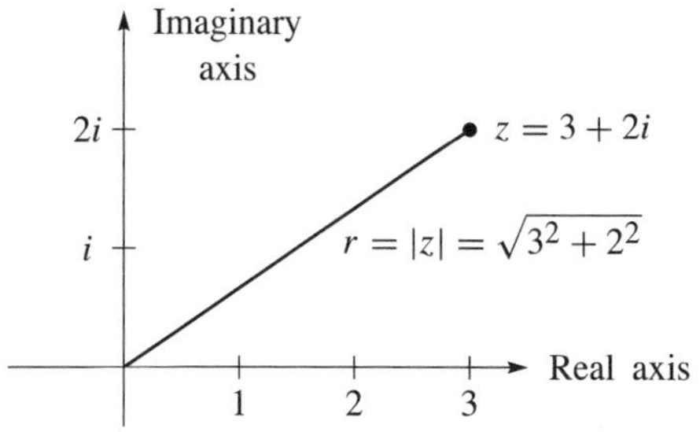

Figure 10.1 $z = a + bi$ corresponds to the point (a, b) and the vector $\begin{bmatrix} a \\ b \end{bmatrix}$.

Something special happens when $z = 3+2i$ combines with its own complex conjugate $\bar{z} = 3 - 2i$. The result from adding $z + \bar{z}$ or multiplying $z\bar{z}$ is always real:

$$(3 + 2i) + (3 - 2i) = 6 \quad \textbf{(real)}$$
$$(3 + 2i) \times (3 - 2i) = 9 + 6i - 6i - 4i^2 = 13 \quad \textbf{(real)}.$$

The sum of $z = a+bi$ and its conjugate $\bar{z} = a-bi$ is the real number $2a$. The product of z times $\bar{z}$ is the real number $a^2 + b^2$:

$$z\bar{z} = (a + bi)(a - bi) = a^2 + b^2. \tag{2}$$

The next step with complex numbers is division. The best idea is to multiply the denominator by its conjugate to produce $a^2 + b^2$ which is real:

$$\frac{1}{a+ib} = \frac{1}{a+ib}\,\frac{a-ib}{a-ib} = \frac{a-ib}{a^2+b^2} \qquad \frac{1}{3+2i} = \frac{1}{3+2i}\,\frac{3-2i}{3-2i} = \frac{3-2i}{13}.$$

In case $a^2 + b^2 = 1$, this says that $(a + ib)^{-1}$ is $a - ib$. ***On the unit circle, $1/z$ is $\bar{z}$.*** Later we will say: $1/e^{i\theta}$ is $e^{-i\theta}$ (the conjugate). A better way to multiply and divide is to use the polar form with distance r and angle θ.

The Polar Form

The square root of a^2+b^2 is $|z|$. This is the ***absolute value*** (or ***modulus***) of the number $z = a + ib$. The same square root is also written r, because it is the distance from 0 to the complex number. The number r in the polar form gives the size of z:

The absolute value of $z = a + ib$ is $|z| = \sqrt{a^2 + b^2}$. This is also called r.

The absolute value of $z = 3 + 2i$ is $|z| = \sqrt{3^2 + 2^2}$. This is $r = \sqrt{13}$.

The other part of the polar form is the angle θ. The angle for $z = 5$ is $\theta = 0$ (because this z is real and positive). The angle for $z = 3i$ is $\pi/2$ radians. The angle for $z = -9$

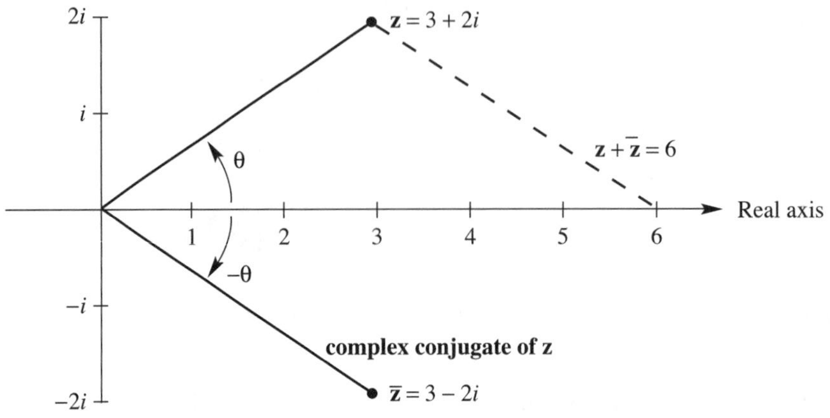

Figure 10.2 Conjugates give the mirror image of the previous figure: $z + \bar{z}$ is real.

is π radians. ***The angle doubles when the number is squared.*** This is one reason why the polar form is good for multiplying complex numbers (not so good for addition).

When the distance is r and the angle is θ, trigonometry gives the other two sides of the triangle. The real part (along the bottom) is $a = r\cos\theta$. The imaginary part (up or down) is $b = r\sin\theta$. Put those together, and the rectangular form becomes the polar form:

$$\textbf{\textit{The number}} \quad z = a + ib \quad \textbf{\textit{is also}} \quad z = r\cos\theta + ir\sin\theta.$$

Note: $\cos\theta + i\sin\theta$ ***has absolute value*** $r = 1$ ***because*** $\cos^2\theta + \sin^2\theta = 1$. Thus $\cos\theta + i\sin\theta$ lies on the circle of radius 1—*the unit circle.*

Example 1 Find r and θ for $z = 1 + i$ and also for the conjugate $\bar{z} = 1 - i$.

Solution The absolute value is the same for z and $\bar{z}$. Here it is $r = \sqrt{1+1} = \sqrt{2}$:

$$|z|^2 = 1^2 + 1^2 = 2 \qquad \text{and also} \qquad |\bar{z}|^2 = 1^2 + (-1)^2 = 2.$$

The distance from the center is $\sqrt{2}$. What about the angle? The number $1 + i$ is at the point $(1, 1)$ in the complex plane. The angle to that point is $\pi/4$ radians or 45°. The cosine is $1/\sqrt{2}$ and the sine is $1/\sqrt{2}$. Combining r and θ brings back $z = 1 + i$:

$$r\cos\theta + ir\sin\theta = \sqrt{2}\left(\frac{1}{\sqrt{2}}\right) + i\sqrt{2}\left(\frac{1}{\sqrt{2}}\right) = 1 + i.$$

The angle to the conjugate $1 - i$ can be positive or negative. We can go to $7\pi/4$ radians which is 315°. Or we can go *backwards through a negative angle*, to $-\pi/4$ radians or −45°. ***If z is at angle θ, its conjugate z̄ is at 2π − θ and also at −θ.***

We can freely add 2π or 4π or -2π to any angle! Those go full circles so the final point is the same. This explains why there are infinitely many choices of θ. Often we select the angle between zero and 2π radians. But $-\theta$ is very useful for the conjugate $\bar{z}$.

Powers and Products: Polar Form

Computing $(1+i)^2$ and $(1+i)^8$ is quickest in polar form. That form has $r = \sqrt{2}$ and $\theta = \pi/4$ (or 45°). If we square the absolute value to get $r^2 = 2$, and double the angle to get $2\theta = \pi/2$ (or 90°), we have $(1+i)^2$. For the eighth power we need r^8 and 8θ:

$$r^8 = 2 \cdot 2 \cdot 2 \cdot 2 = 16 \quad \text{and} \quad 8\theta = 8 \cdot \frac{\pi}{4} = 2\pi.$$

This means: $(1+i)^8$ has absolute value 16 and angle 2π. *The eighth power of* $1+i$ *is the real number* 16.

Powers are easy in polar form. So is multiplication of complex numbers.

10B The polar form of z^n has absolute value r^n. The angle is n times θ:

The nth power of $\boldsymbol{z = r(\cos\theta + i\sin\theta)}$ ***is*** $\boldsymbol{z^n = r^n(\cos n\theta + i\sin n\theta)}$. (3)

In that case z multiplies itself. In all cases, ***multiply r's and add angles***:

$$r(\cos\theta + i\sin\theta) \text{ times } r'(\cos\theta' + i\sin\theta') = rr'\big(\cos(\theta+\theta') + i\sin(\theta+\theta')\big). \tag{4}$$

One way to understand this is by trigonometry. Concentrate on angles. Why do we get the double angle 2θ for z^2?

$$(\cos\theta + i\sin\theta) \times (\cos\theta + i\sin\theta) = \cos^2\theta + i^2\sin^2\theta + 2i\sin\theta\cos\theta.$$

The real part $\cos^2\theta - \sin^2\theta$ is $\cos 2\theta$. The imaginary part $2\sin\theta\cos\theta$ is $\sin 2\theta$. Those are the "double angle" formulas. They show that θ in z becomes 2θ in in z^2.

When the angles θ and θ' are different, use the "addition formulas" instead:

$$(\cos\theta + i\sin\theta)(\cos\theta' + i\sin\theta') = [\cos\theta\cos\theta' - \sin\theta\sin\theta'] + i[\sin\theta\cos\theta' + \cos\theta\sin\theta']$$

In those brackets, trigonometry sees the cosine and sine of $\theta + \theta'$. This confirms equation (4), that angles add when you multiply complex numbers.

There is a second way to understand the rule for z^n. It uses the only amazing formula in this section. Remember that $\cos\theta + i\sin\theta$ has absolute value 1. The cosine is made up of even powers, starting with $1 - \frac{1}{2}\theta^2$. The sine is made up of odd powers, starting with $\theta - \frac{1}{6}\theta^3$. The beautiful fact is that $e^{i\theta}$ combines both of those series into $\cos\theta + i\sin\theta$:

$$e^x = 1 + x + \frac{1}{2}x^2 + \frac{1}{6}x^3 + \cdots \quad \text{becomes} \quad e^{i\theta} = 1 + i\theta + \frac{1}{2}i^2\theta^2 + \frac{1}{6}i^3\theta^3 + \cdots$$

Write -1 for i^2. The real part $1 - \frac{1}{2}\theta^2 + \cdots$ is exactly $\cos\theta$. The imaginary part $\theta - \frac{1}{6}\theta^3 + \cdots$ is $\sin\theta$. ***The whole right side is*** $\boldsymbol{\cos\theta + i\sin\theta}$***:***

Euler's Formula $e^{i\theta} = \cos\theta + i\sin\theta.$ (5)

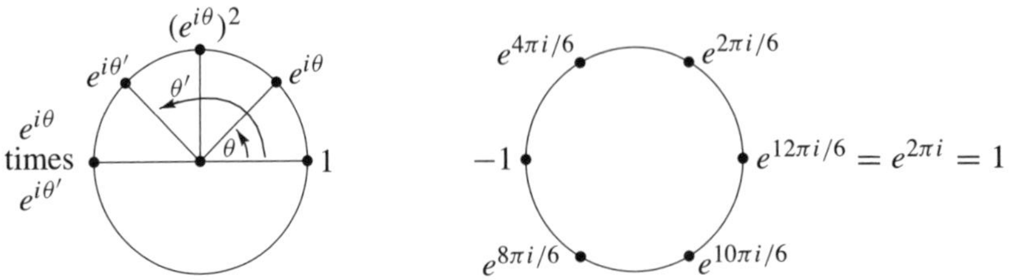

Figure 10.3 (a) Multiplying $e^{i\theta}$ times $e^{i\theta'}$. (b) The 6th power of $e^{2\pi i/6}$ is $e^{2\pi i} = 1$.

The special choice $\theta = 2\pi$ gives $\cos 2\pi + i \sin 2\pi$ which is 1. Somehow the infinite series $e^{2\pi i} = 1 + 2\pi i + \frac{1}{2}(2\pi i)^2 + \cdots$ adds up to 1.

Now multiply $e^{i\theta}$ times $e^{i\theta'}$. Angles add for the same reason that exponents add:

$$\begin{aligned} e^2 \text{ times } e^3 \text{ is } e^5 \text{ because } (e)(e) \times (e)(e)(e) &= (e)(e)(e)(e)(e) \\ e^{i\theta} \text{ times } e^{i\theta} \text{ is } e^{2i\theta} \qquad e^{i\theta} \text{ times } e^{i\theta'} &\text{ is } e^{i(\theta+\theta')}. \end{aligned} \tag{6}$$

Every complex number $a + ib = r\cos\theta + ir\sin\theta$ now goes into its best possible form. That form is $re^{i\theta}$.

The powers $(re^{i\theta})^n$ are equal to $r^n e^{in\theta}$. They stay on the unit circle when $r = 1$ and $r^n = 1$. Then we find n different numbers whose nth powers equal 1:

Set $w = e^{2\pi i/n}$. The nth powers of $1, w, w^2, \ldots, w^{n-1}$ all equal 1.

Those are the "nth roots of 1." They solve the equation $z^n = 1$. They are equally spaced around the unit circle in Figure 10.3b, where the full 2π is divided by n. Multiply their angles by n to take nth powers. That gives $w^n = e^{2\pi i}$ which is 1. Also $(w^2)^n = e^{4\pi i} = 1$. Each of those numbers, to the nth power, comes around the unit circle to 1.

These roots of 1 are the key numbers for signal processing. A real digital computer uses only 0 and 1. The complex Fourier transform uses w and its powers. The last section of the book shows how to decompose a vector (a signal) into n frequencies by the Fast Fourier Transform.

Problem Set 10.1

Questions 1–8 are about operations on complex numbers.

1 Add and multiply each pair of complex numbers:

(a) $2+i, 2-i$ (b) $-1+i, -1+i$ (c) $\cos\theta + i\sin\theta, \cos\theta - i\sin\theta$

2 Locate these points on the complex plane. Simplify them if necessary:

(a) $2+i$ (b) $(2+i)^2$ (c) $\frac{1}{2+i}$ (d) $|2+i|$

3 Find the absolute value $r = |z|$ of these four numbers. If θ is the angle for $6-8i$, what are the angles for the other three numbers?

(a) $6-8i$ (b) $(6-8i)^2$ (c) $\frac{1}{6-8i}$ (d) $(6+8i)^2$

4 If $|z| = 2$ and $|w| = 3$ then $|z \times w| =$ ____ and $|z+w| \le$ ____ and $|z/w| =$ ____ and $|z-w| \le$ ____.

5 Find $a+ib$ for the numbers at angles $30°, 60°, 90°, 120°$ on the unit circle. If w is the number at $30°$, check that w^2 is at $60°$. What power of w equals 1?

6 If $z = r\cos\theta + ir\sin\theta$ then $1/z$ has absolute value ____ and angle ____. Its polar form is ____. Multiply $z \times 1/z$ to get 1.

7 The 1 by 1 complex multiplication $M = (a+bi)(c+di)$ is a 2 by 2 real multiplication

$$\begin{bmatrix} a & -b \\ b & a \end{bmatrix} \begin{bmatrix} c \\ d \end{bmatrix} = \begin{bmatrix} \ \end{bmatrix}.$$

The right side contains the real and imaginary parts of M. Test $M = (1+3i)(1-3i)$.

8 $A = A_1 + iA_2$ is a complex n by n matrix and $\boldsymbol{b} = \boldsymbol{b}_1 + i\boldsymbol{b}_2$ is a complex vector. The solution to $A\boldsymbol{x} = \boldsymbol{b}$ is $\boldsymbol{x}_1 + i\boldsymbol{x}_2$. Write $A\boldsymbol{x} = \boldsymbol{b}$ as a real system of size $2n$:

$$\begin{bmatrix} \ & \ \\ \ & \ \end{bmatrix} \begin{bmatrix} \boldsymbol{x}_1 \\ \boldsymbol{x}_2 \end{bmatrix} = \begin{bmatrix} \boldsymbol{b}_1 \\ \boldsymbol{b}_2 \end{bmatrix}.$$

Questions 9–16 are about the conjugate $\bar{z} = a - ib = re^{-i\theta}$ of the number $z = a + ib = re^{i\theta}$.

9 Write down the complex conjugate of each number by changing i to $-i$:

(a) $2-i$ (b) $(2-i)(1-i)$ (c) $e^{i\pi/2}$ (which is i)

(d) $e^{i\pi} = -1$ (e) $\frac{1+i}{1-i}$ (which is also i) (f) $i^{103} =$ ____.

10 The sum $z + \bar{z}$ is always ____. The difference $z - \bar{z}$ is always ____. Assume $z \neq 0$. The product $z \times \bar{z}$ is always ____. The ratio $z/\bar{z}$ always has absolute value ____.

11 For a real 3 by 3 matrix, the numbers a_2, a_1, a_0 from the determinant are real:

$$\det(A - \lambda I) = -\lambda^3 + a_2\lambda^2 + a_1\lambda + a_0 = 0.$$

Each root λ is an eigenvalue. Taking conjugates gives $-\bar{\lambda}^3 + a_2\bar{\lambda}^2 + a_1\bar{\lambda} + a_0 = 0$, so $\bar{\lambda}$ is also an eigenvalue. For the matrix with $a_{ij} = i - j$, find $\det(A - \lambda I)$ and the three eigenvalues.

Note The conjugate of $A\boldsymbol{x} = \lambda\boldsymbol{x}$ is $A\bar{\boldsymbol{x}} = \bar{\lambda}\bar{\boldsymbol{x}}$. This proves two things: $\bar{\lambda}$ is an eigenvalue and $\bar{\boldsymbol{x}}$ is its eigenvector. Problem 11 only proves that $\bar{\lambda}$ is an eigenvalue.

12 The eigenvalues of a real 2 by 2 matrix come from the quadratic formula:

$$\begin{vmatrix} a - \lambda & b \\ c & d - \lambda \end{vmatrix} = \lambda^2 - (a + d)\lambda + (ad - bc) = 0$$

gives the two eigenvalues (notice the $\pm$ symbol):

$$\lambda = \frac{a + d \pm \sqrt{(a + d)^2 - 4(ad - bc)}}{2}.$$

(a) If $a = b = d = 1$, the eigenvalues are complex when c is ____.

(b) What are the eigenvalues when $ad = bc$?

(c) The two eigenvalues (plus sign and minus sign) are not always conjugates of each other. Why not?

13 In Problem 12 the eigenvalues are not real when $(\text{trace})^2 = (a + d)^2$ is smaller than ____. Show that the λ's *are* real when $bc > 0$.

14 Find the eigenvalues and eigenvectors of this permutation matrix:

$$P_4 = \begin{bmatrix} 0 & 0 & 0 & 1 \\ 1 & 0 & 0 & 0 \\ 0 & 1 & 0 & 0 \\ 0 & 0 & 1 & 0 \end{bmatrix} \quad \text{has} \quad \det(P_4 - \lambda I) = \underline{\qquad}.$$

15 Extend P_4 above to P_6 (five 1's below the diagonal and one in the corner). Find $\det(P_6 - \lambda I)$ and the six eigenvalues in the complex plane.

16 A real skew-symmetric matrix ($A^{\mathrm{T}} = -A$) has pure imaginary eigenvalues. First proof: If $A\boldsymbol{x} = \lambda\boldsymbol{x}$ then block multiplication gives

$$\begin{bmatrix} 0 & A \\ -A & 0 \end{bmatrix} \begin{bmatrix} \boldsymbol{x} \\ i\boldsymbol{x} \end{bmatrix} = i\lambda \begin{bmatrix} \boldsymbol{x} \\ i\boldsymbol{x} \end{bmatrix}.$$

This block matrix is symmetric. Its eigenvalues must be ____! So λ is ____.

Questions 17–24 are about the form $re^{i\theta}$ of the complex number $r\cos\theta + ir\sin\theta$.

17 Write these numbers in Euler's form $re^{i\theta}$. Then square each number:

(a) $1+\sqrt{3}i$ (b) $\cos 2\theta + i\sin 2\theta$ (c) $-7i$ (d) $5-5i$.

18 Find the absolute value and the angle for $z = \sin\theta + i\cos\theta$ (careful). Locate this z in the complex plane. Multiply z by $\cos\theta + i\sin\theta$ to get _____.

19 Draw all eight solutions of $z^8 = 1$ in the complex plane. What are the rectangular forms $a+ib$ of these eight numbers?

20 Locate the cube roots of 1 in the complex plane. Locate the cube roots of -1. Together these are the sixth roots of _____.

21 By comparing $e^{3i\theta} = \cos 3\theta + i\sin 3\theta$ with $(e^{i\theta})^3 = (\cos\theta + i\sin\theta)^3$, find the "triple angle" formulas for $\cos 3\theta$ and $\sin 3\theta$ in terms of $\cos\theta$ and $\sin\theta$.

22 Suppose the conjugate $\overline{z}$ is equal to the reciprocal $1/z$. What are all possible z's?

23 (a) Why do e^i and i^e both have absolute value 1?

(b) In the complex plane put stars near the points e^i and i^e.

(c) The number i^e could be $(e^{i\pi/2})^e$ or $(e^{5i\pi/2})^e$. Are those equal?

24 Draw the paths of these numbers from $t=0$ to $t=2\pi$ in the complex plane:

(a) e^{it} (b) $e^{(-1+i)t} = e^{-t}e^{it}$ (c) $(-1)^t = e^{t\pi i}$.

HERMITIAN AND UNITARY MATRICES ■ 10.2

The main message of this section can be presented in the first sentence: ***When you transpose a complex vector z or a matrix A, take the complex conjugate too.*** Don't stop at z^{T} or A^{T}. Reverse the signs of all imaginary parts. Starting from a column vector with components $z_j = a_j + ib_j$, the good row vector is the *conjugate transpose* with components $a_j - ib_j$:

$$\bar{z}^{\mathrm{T}} = \begin{bmatrix} \bar{z}_1 & \cdots & \bar{z}_n \end{bmatrix} = \begin{bmatrix} a_1 - ib_1 & \cdots & a_n - ib_n \end{bmatrix}. \tag{1}$$

Here is one reason to go to $\bar{z}$. The length squared of a real vector is $x_1^2 + \cdots + x_n^2$. The length squared of a complex vector is *not* $z_1^2 + \cdots + z_n^2$. With that wrong definition, the length of $(1, i)$ would be $1^2 + i^2 = 0$. A nonzero vector would have zero length—not good. Other vectors would have complex lengths. Instead of $(a+bi)^2$ we want $a^2 + b^2$, the *absolute value squared.* This is $(a + bi)$ times $(a - bi)$.

For each component we want z_j times $\bar{z}_j$, which is $|z_j|^2 = a^2 + b^2$. That comes when the components of z multiply the components of $\bar{z}$:

$$\begin{bmatrix} \bar{z}_1 & \cdots & \bar{z}_n \end{bmatrix} \begin{bmatrix} z_1 \\ \vdots \\ z_n \end{bmatrix} = |z_1|^2 + \cdots + |z_n|^2. \quad \text{This is} \quad \bar{z}^{\mathrm{T}} z = \|z\|^2. \tag{2}$$

Now the squared length of $(1, i)$ is $1^2 + |i|^2 = 2$. The length is $\sqrt{2}$ and not zero. The squared length of $(1+i, 1-i)$ is 4. The only vectors with zero length are zero vectors.

DEFINITION ***The length $\|z\|$ is the square root of*** $\|z\|^2 = \bar{z}^{\mathrm{T}} z = |z_1|^2 + \cdots + |z_n|^2$

Before going further we replace two symbols by one symbol. Instead of a bar for the conjugate and T for the transpose, we just use a superscript H. Thus $\bar{z}^{\mathrm{T}} = z^{\mathrm{H}}$. This is "$z$ Hermitian," the *conjugate transpose* of z. The new word is pronounced "Hermeeshan." The new symbol applies also to matrices: The conjugate transpose of a matrix A is A^{H}.

Notation The vector z^{H} is $\bar{z}^{\mathrm{T}}$. The matrix A^{H} is $\bar{A}^{\mathrm{T}}$, the conjugate transpose of A:

$$\text{If} \quad A = \begin{bmatrix} 1 & i \\ 0 & 1+i \end{bmatrix} \quad \text{then} \quad A^{\mathrm{H}} = \begin{bmatrix} 1 & 0 \\ -i & 1-i \end{bmatrix} = \text{"}A \text{ Hermitian."}$$

Complex Inner Products

For real vectors, the length squared is $\boldsymbol{x}^{\mathrm{T}}\boldsymbol{x}$—*the inner product of $\boldsymbol{x}$ with itself.* For complex vectors, the length squared is $z^{\mathrm{H}}z$. It will be very desirable if this is the inner product of z with itself. To make that happen, the complex inner product should use the conjugate transpose (not just the transpose). There will be no effect when the vectors are real, but there is a definite effect when they are complex:

DEFINITION The inner product of real or complex vectors $\boldsymbol{u}$ and $\boldsymbol{v}$ is $\boldsymbol{u}^{\mathrm{H}}\boldsymbol{v}$:

$$\boldsymbol{u}^{\mathrm{H}}\boldsymbol{v} = \begin{bmatrix} \overline{u}_1 & \cdots & \overline{u}_n \end{bmatrix} \begin{bmatrix} v_1 \\ \vdots \\ v_n \end{bmatrix} = \overline{u}_1 v_1 + \cdots + \overline{u}_n v_n. \tag{3}$$

With complex vectors, $\boldsymbol{u}^{\mathrm{H}}\boldsymbol{v}$ is different from $\boldsymbol{v}^{\mathrm{H}}\boldsymbol{u}$. *The order of the vectors is now important.* In fact $\boldsymbol{v}^{\mathrm{H}}\boldsymbol{u} = \overline{v}_1 u_1 + \cdots + \overline{v}_n u_n$ is the complex conjugate of $\boldsymbol{u}^{\mathrm{H}}\boldsymbol{v}$. We have to put up with a few inconveniences for the greater good.

Example 1 The inner product of $\boldsymbol{u} = \begin{bmatrix} 1 \\ i \end{bmatrix}$ with $\boldsymbol{v} = \begin{bmatrix} i \\ 1 \end{bmatrix}$ is $\begin{bmatrix} 1 & -i \end{bmatrix} \begin{bmatrix} i \\ 1 \end{bmatrix} = 0$. *Not* $2i$.

Example 2 The inner product of $\boldsymbol{u} = \begin{bmatrix} 1+i \\ 0 \end{bmatrix}$ with $\boldsymbol{v} = \begin{bmatrix} 2 \\ i \end{bmatrix}$ is $\boldsymbol{u}^{\mathrm{H}}\boldsymbol{v} = 2 - 2i$.

Example 1 is surprising. Those vectors $(1, i)$ and $(i, 1)$ don't look perpendicular. But they are. ***A zero inner product still means that the*** (complex) ***vectors are orthogonal.*** Similarly the vector $(1, i)$ is orthogonal to the vector $(1, -i)$. Their inner product is $1 - 1 = 0$. We are correctly getting zero for the inner product—where we would be incorrectly getting zero for the length of $(1, i)$ if we forgot to take the conjugate.

Note We have chosen to conjugate the first vector $\boldsymbol{u}$. Some authors choose the second vector $\boldsymbol{v}$. Their complex inner product would be $\boldsymbol{u}^{\mathrm{T}}\overline{\boldsymbol{v}}$. It is a free choice, as long as we stick to one or the other. We wanted to use the single symbol ^{H} in the next formula too:

The inner product of Au ***with*** $\boldsymbol{v}$ ***equals the inner product of*** $\boldsymbol{u}$ ***with*** $A^{\mathrm{H}}\boldsymbol{v}$:

$$(A\boldsymbol{u})^{\mathrm{H}}\boldsymbol{v} = \boldsymbol{u}^{\mathrm{H}}(A^{\mathrm{H}}\boldsymbol{v}). \tag{4}$$

The conjugate of $A\boldsymbol{u}$ is $\overline{A\boldsymbol{u}}$. Transposing it gives $\overline{\boldsymbol{u}}^{\mathrm{T}}\overline{A}^{\mathrm{T}}$ as usual. This is $\boldsymbol{u}^{\mathrm{H}}A^{\mathrm{H}}$. Everything that should work, does work. The rule for ^{H} comes from the rule for ^{T}. That applies to products of matrices:

10C ***The conjugate transpose of*** AB ***is*** $(AB)^{\mathrm{H}} = B^{\mathrm{H}}A^{\mathrm{H}}$.

We are constantly using the fact that $(a-ib)(c-id)$ is the conjugate of $(a+ib)(c+id)$.

Among real matrices, the *symmetric matrices* form the most important special class: $A = A^{\mathrm{T}}$. They have real eigenvalues and a full set of orthogonal eigenvectors. The diagonalizing matrix S is an orthogonal matrix Q. Every symmetric matrix can be written as $A = Q\Lambda Q^{-1}$ and also as $A = Q\Lambda Q^{\mathrm{T}}$ (because $Q^{-1} = Q^{\mathrm{T}}$). All this follows from $a_{ij} = a_{ji}$, when A is real.

Among complex matrices, the special class consists of the ***Hermitian matrices***: $A = A^{\mathrm{H}}$. The condition on the entries is now $a_{ij} = \overline{a_{ji}}$. In this case we say that "*A is* Hermitian." *Every real symmetric matrix is Hermitian*, because taking its conjugate has no effect. The next matrix is also Hermitian:

$$A = \begin{bmatrix} 2 & 3-3i \\ 3+3i & 5 \end{bmatrix} \qquad \begin{array}{l} \text{The main diagonal is real since } a_{ii} = \overline{a_{ii}}. \\ \text{Across it are conjugates } 3+3i \text{ and } 3-3i. \end{array}$$

This example will illustrate the three crucial properties of all Hermitian matrices.

10D ***If*** $\boldsymbol{A = A^{\mathrm{H}}}$ ***and*** $\boldsymbol{z}$ ***is any vector, the number*** $\boldsymbol{z^{\mathrm{H}}Az}$ ***is real.***

Quick proof: $z^{\mathrm{H}}Az$ is certainly 1 by 1. Take its conjugate transpose:

$$(z^{\mathrm{H}}Az)^{\mathrm{H}} = z^{\mathrm{H}}A^{\mathrm{H}}(z^{\mathrm{H}})^{\mathrm{H}} \quad \text{which is } z^{\mathrm{H}}Az \text{ again.}$$

Reversing the order has produced the same 1 by 1 matrix (this used $A = A^{\mathrm{H}}$!) For 1 by 1 matrices, the conjugate transpose is simply the conjugate. So the number $z^{\mathrm{H}}Az$ equals its conjugate and must be real. Here is $z^{\mathrm{H}}Az$ in our example:

$$\begin{bmatrix} \bar{z}_1 & \bar{z}_2 \end{bmatrix} \begin{bmatrix} 2 & 3-3i \\ 3+3i & 5 \end{bmatrix} \begin{bmatrix} z_1 \\ z_2 \end{bmatrix} = 2\bar{z}_1 z_1 + 5\bar{z}_2 z_2 + (3-3i)\bar{z}_1 z_2 + (3+3i) z_1 \bar{z}_2.$$

The terms $2|z_1|^2$ and $5|z_2|^2$ from the diagonal are both real. The off-diagonal terms are conjugates of each other—so their sum is real. (The imaginary parts cancel when we add.) The whole expression $z^{\mathrm{H}}Az$ is real.

10E ***Every eigenvalue of a Hermitian matrix is real.***

Proof Suppose $Az = \lambda z$. *Multiply both sides by* z^{H} *to get* $z^{\mathrm{H}}Az = \lambda z^{\mathrm{H}}z$. On the left side, $z^{\mathrm{H}}Az$ is real by **10D**. On the right side, $z^{\mathrm{H}}z$ is the length squared, real and positive. So the ratio $\lambda = z^{\mathrm{H}}Az/z^{\mathrm{H}}z$ is a real number. Q.E.D.

The example above has real eigenvalues $\lambda = 8$ and $\lambda = -1$. Take the determinant of $A - \lambda I$ to get $(d-8)(d+1)$:

$$\begin{vmatrix} 2-\lambda & 3-3i \\ 3+3i & 5-\lambda \end{vmatrix} = \lambda^2 - 7\lambda + 10 - |3+3i|^2$$
$$= \lambda^2 - 7\lambda + 10 - 18 = (\lambda - 8)(\lambda + 1).$$

10F ***The eigenvectors of a Hermitian matrix are orthogonal*** (provided they correspond to different eigenvalues). If $Az = \lambda z$ and $Ay = \beta y$ then $y^{\mathrm{H}}z = 0$.

Proof Multiply $Az = \lambda z$ on the left by y^H. Multiply $y^H A^H = \beta y^H$ on the right by z:

$$y^H A z = \lambda y^H z \quad \text{and} \quad y^H A^H z = \beta y^H z. \tag{5}$$

The left sides are equal because $A = A^H$. Therefore the right sides are equal. Since β is different from λ, the other factor $y^H z$ must be zero. The eigenvectors are orthogonal, as in the example with $\lambda = 8$ and $\beta = -1$:

$$(A - 8I)z = \begin{bmatrix} -6 & 3-3i \\ 3+3i & -3 \end{bmatrix} \begin{bmatrix} z_1 \\ z_2 \end{bmatrix} = \begin{bmatrix} 0 \\ 0 \end{bmatrix} \quad \text{and} \quad z = \begin{bmatrix} 1 \\ 1+i \end{bmatrix}$$

$$(A + I)y = \begin{bmatrix} 3 & 3-3i \\ 3+3i & 6 \end{bmatrix} \begin{bmatrix} y_1 \\ y_2 \end{bmatrix} = \begin{bmatrix} 0 \\ 0 \end{bmatrix} \quad \text{and} \quad y = \begin{bmatrix} 1-i \\ -1 \end{bmatrix}.$$

Take the inner product of those eigenvectors y and z:

$$y^H z = \begin{bmatrix} 1+i & -1 \end{bmatrix} \begin{bmatrix} 1 \\ 1+i \end{bmatrix} = 0 \qquad \text{(orthogonal eigenvectors).}$$

These eigenvectors have squared length $1^2 + 1^2 + 1^2 = 3$. After division by $\sqrt{3}$ they are unit vectors. They were orthogonal, now they are ***orthonormal***. They go into the columns of the *eigenvector matrix* S, which diagonalizes A.

When A is real and symmetric, it has real orthogonal eigenvectors. Then S is Q—an orthogonal matrix. Now A is complex and Hermitian. Its eigenvectors are complex and orthonormal. ***The eigenvector matrix S is like Q, but complex.*** We now assign a new name and a new letter to a complex orthogonal matrix.

Unitary Matrices

A ***unitary matrix*** is a (complex) square matrix with ***orthonormal columns***. It is denoted by U—the complex equivalent of Q. The eigenvectors above, divided by $\sqrt{3}$ to become unit vectors, are a perfect example:

$$U = \frac{1}{\sqrt{3}} \begin{bmatrix} 1 & 1-i \\ 1+i & -1 \end{bmatrix} \quad \text{is a unitary matrix.}$$

This U is also a Hermitian matrix. I didn't expect that! The example is almost too perfect. Its second column could be multiplied by -1, or even by i, and the matrix of eigenvectors would still be unitary:

$$U = \frac{1}{\sqrt{3}} \begin{bmatrix} 1 & -1+i \\ 1+i & 1 \end{bmatrix} \quad \text{is also a unitary matrix.}$$

The matrix test for real orthonormal columns was $Q^T Q = I$. When Q^T multiplies Q, the zero inner products appear off the diagonal. In the complex case, Q becomes U and the symbol ^{T} becomes ^{H}. The columns show themselves as orthonormal when U^H multiplies U. The inner products of the columns are again 1 and 0, and they fill up $U^H U = I$:

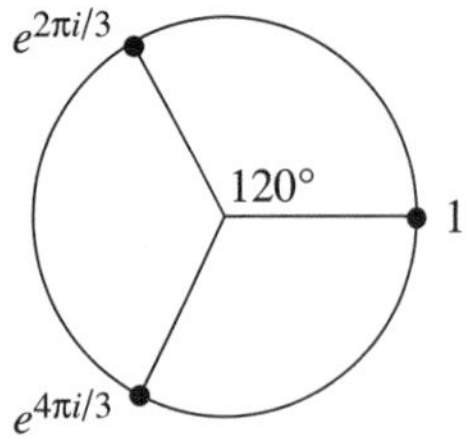

Fourier matrix $$F = \frac{1}{\sqrt{3}} \begin{bmatrix} 1 & 1 & 1 \\ 1 & e^{2\pi i/3} & e^{4\pi i/3} \\ 1 & e^{4\pi i/3} & e^{2\pi i/3} \end{bmatrix}.$$

Figure 10.4 The cube roots of 1 go into the Fourier matrix $F = F_3$.

10G ***The matrix U has orthonormal columns when $U^{\mathrm{H}}U = I$.***
If U is square, it is a ***unitary matrix***. Then $U^{\mathrm{H}} = U^{-1}$.

$$U^{\mathrm{H}}U = \frac{1}{\sqrt{3}} \begin{bmatrix} 1 & 1-i \\ 1+i & -1 \end{bmatrix} \frac{1}{\sqrt{3}} \begin{bmatrix} 1 & 1-i \\ 1+i & -1 \end{bmatrix} = \begin{bmatrix} 1 & 0 \\ 0 & 1 \end{bmatrix}. \tag{6}$$

Suppose U (with orthogonal column) multiplies any z. The vector length stays the same, because $z^{\mathrm{H}}U^{\mathrm{H}}Uz = z^{\mathrm{H}}z$. If z is an eigenvector, we learn something more: ***The eigenvalues of unitary (and orthogonal) matrices all have absolute value*** $|\lambda| = 1$.

10H ***If U is unitary then*** $\|Uz\| = \|z\|$. Therefore $Uz = \lambda z$ leads to $|\lambda| = 1$.

Our 2 by 2 example is both Hermitian ($U = U^{\mathrm{H}}$) and unitary ($U^{-1} = U^{\mathrm{H}}$). That means real eigenvalues ($\lambda = \bar{\lambda}$), and it means absolute value one ($\lambda^{-1} = \bar{\lambda}$). A real number with absolute value 1 has only two possibilities: *The eigenvalues are* 1 *or* -1.

One thing more about the example: The diagonal of U adds to zero. The trace is zero. So one eigenvalue is $\lambda = 1$, the other is $\lambda = -1$. The determinant must be 1 times -1, the product of the λ's.

Example 3 The 3 by 3 ***Fourier matrix*** is in Figure 10.4. Is it Hermitian? Is it unitary? The Fourier matrix is certainly symmetric. It equals its transpose. But it doesn't equal its conjugate transpose—*it is not Hermitian.* If you change i to $-i$, you get a different matrix.

Is F unitary? *Yes.* The squared length of every column is $\frac{1}{3}(1+1+1)$. The columns are unit vectors. The first column is orthogonal to the second column because $1 + e^{2\pi i/3} + e^{4\pi i/3} = 0$. This is the sum of the three numbers marked in Figure 10.4.

Notice the symmetry of the figure. If you rotate it by 120°, the three points are in the same position. Therefore their sum S also stays in the same position! The only possible sum is $S = 0$, because this is the only point that is in the same position after 120° rotation.

Is column 2 of F orthogonal to column 3? Their dot product looks like

$$\tfrac{1}{3}(1 + e^{6\pi i/3} + e^{6\pi i/3}) = \tfrac{1}{3}(1 + 1 + 1).$$

This is not zero. That is because we forgot to take complex conjugates! The complex inner product uses ^{H} not ^{T}:

$$\begin{aligned}(\text{column } 2)^{\mathrm{H}}(\text{column } 3) &= \tfrac{1}{3}(1 \cdot 1 + e^{-2\pi i/3}e^{4\pi i/3} + e^{-4\pi i/3}e^{2\pi i/3}) \\ &= \tfrac{1}{3}(1 + e^{2\pi i/3} + e^{-2\pi i/3}) = 0.\end{aligned}$$

So we do have orthogonality. ***Conclusion: F is a unitary matrix.***

The next section will study the n by n Fourier matrices. Among all complex unitary matrices, these are the most important. When we multiply a vector by F, we are computing its ***discrete Fourier transform***. When we multiply by F^{-1}, we are computing the *inverse transform*. The special property of unitary matrices is that $F^{-1} = F^{\mathrm{H}}$. The inverse transform only differs by changing i to $-i$:

$$F^{-1} = F^{\mathrm{H}} = \frac{1}{\sqrt{3}}\begin{bmatrix} 1 & 1 & 1 \\ 1 & e^{-2\pi i/3} & e^{-4\pi i/3} \\ 1 & e^{-4\pi i/3} & e^{-2\pi i/3} \end{bmatrix}.$$

Everyone who works with F recognizes its value. The last section of the book will bring together Fourier analysis and linear algebra.

This section ends with a table to translate between real and complex—for vectors and for matrices:

Real versus Complex

$\mathbf{R}^n$: vectors with n real components	$\leftrightarrow$	$\mathbf{C}^n$: vectors with n complex components
length: $\|\boldsymbol{x}\|^2 = x_1^2 + \cdots + x_n^2$	$\leftrightarrow$	length: $\|\boldsymbol{z}\|^2 = \lvert z_1\rvert^2 + \cdots + \lvert z_n\rvert^2$
$(A^{\mathrm{T}})_{ij} = A_{ji}$	$\leftrightarrow$	$(A^{\mathrm{H}})_{ij} = \overline{A_{ji}}$
$(AB)^{\mathrm{T}} = B^{\mathrm{T}}A^{\mathrm{T}}$	$\leftrightarrow$	$(AB)^{\mathrm{H}} = B^{\mathrm{H}}A^{\mathrm{H}}$
dot product: $\boldsymbol{x}^{\mathrm{T}}\boldsymbol{y} = x_1y_1 + \cdots + x_ny_n$	$\leftrightarrow$	inner product: $\boldsymbol{u}^{\mathrm{H}}\boldsymbol{v} = \overline{u}_1v_1 + \cdots + \overline{u}_nv_n$
$(A\boldsymbol{x})^{\mathrm{T}}\boldsymbol{y} = \boldsymbol{x}^{\mathrm{T}}(A^{\mathrm{T}}\boldsymbol{y})$	$\leftrightarrow$	$(A\boldsymbol{u})^{\mathrm{H}}\boldsymbol{v} = \boldsymbol{u}^{\mathrm{H}}(A^{\mathrm{H}}\boldsymbol{v})$
orthogonality: $\boldsymbol{x}^{\mathrm{T}}\boldsymbol{y} = 0$	$\leftrightarrow$	orthogonality: $\boldsymbol{u}^{\mathrm{H}}\boldsymbol{v} = 0$
symmetric matrices: $A = A^{\mathrm{T}}$	$\leftrightarrow$	Hermitian matrices: $A = A^{\mathrm{H}}$
$A = Q\Lambda Q^{-1} = Q\Lambda Q^{\mathrm{T}}(\text{real } \Lambda)$	$\leftrightarrow$	$A = U\Lambda U^{-1} = U\Lambda U^{\mathrm{H}}$ (real Λ)
skew-symmetric matrices: $K^{\mathrm{T}} = -K$	$\leftrightarrow$	skew-Hermitian matrices $K^{\mathrm{H}} = -K$
orthogonal matrices: $Q^{\mathrm{T}} = Q^{-1}$	$\leftrightarrow$	unitary matrices: $U^{\mathrm{H}} = U^{-1}$
orthonormal columns: $Q^{\mathrm{T}}Q = I$	$\leftrightarrow$	orthonormal columns: $U^{\mathrm{H}}U = I$
$(Q\boldsymbol{x})^{\mathrm{T}}(Q\boldsymbol{y}) = \boldsymbol{x}^{\mathrm{T}}\boldsymbol{y}$ and $\|Q\boldsymbol{x}\| = \|\boldsymbol{x}\|$	$\leftrightarrow$	$(U\boldsymbol{x})^{\mathrm{H}}(U\boldsymbol{y}) = \boldsymbol{x}^{\mathrm{H}}\boldsymbol{y}$ and $\|U\boldsymbol{z}\| = \|\boldsymbol{z}\|$

The columns and also the eigenvectors of Q and U are orthonormal. Every $|\lambda| = 1$.

Problem Set 10.2

1 Find the lengths of $\boldsymbol{u} = (1+i, 1-i, 1+2i)$ and $\boldsymbol{v} = (i, i, i)$. Also find $\boldsymbol{u}^{\mathrm{H}}\boldsymbol{v}$ and $\boldsymbol{v}^{\mathrm{H}}\boldsymbol{u}$.

2 Compute $A^{\mathrm{H}}A$ and AA^{H}. Those are both _____ matrices:

$$A = \begin{bmatrix} i & 1 & i \\ 1 & i & i \end{bmatrix}.$$

3 Solve $Az = \mathbf{0}$ to find a vector in the nullspace of A in Problem 2. Show that z is orthogonal to the columns of A^{H}. Show that z is *not* orthogonal to the columns of A^{T}.

4 Problem 3 indicates that the four fundamental subspaces are $\boldsymbol{C}(A)$ and $\boldsymbol{N}(A)$ and _____ and _____. Their dimensions are still r and $n-r$ and r and $m-r$. They are still orthogonal subspaces. *The symbol* ^{H} *takes the place of* ^{T}.

5 (a) Prove that $A^{\mathrm{H}}A$ is always a Hermitian matrix.

(b) If $Az = \mathbf{0}$ then $A^{\mathrm{H}}Az = \mathbf{0}$. If $A^{\mathrm{H}}Az = \mathbf{0}$, multiply by z^{H} to prove that $Az = \mathbf{0}$. The nullspaces of A and $A^{\mathrm{H}}A$ are _____. Therefore $A^{\mathrm{H}}A$ is an invertible Hermitian matrix when the nullspace of A contains only $z =$ _____.

6 True or false (give a reason if true or a counterexample if false):

(a) If A is a real matrix then $A + iI$ is invertible.

(b) If A is a Hermitian matrix then $A + iI$ is invertible.

(c) If U is a unitary matrix then $A + iI$ is invertible.

7 When you multiply a Hermitian matrix by a real number c, is cA still Hermitian? If $c = i$ show that iA is skew-Hermitian. The 3 by 3 Hermitian matrices are a subspace provided the "scalars" are real numbers.

8 Which classes of matrices does P belong to: orthogonal, invertible, Hermitian, unitary, factorizable into LU, factorizable into QR?

$$P = \begin{bmatrix} 0 & 1 & 0 \\ 0 & 0 & 1 \\ 1 & 0 & 0 \end{bmatrix}.$$

9 Compute P^2, P^3, and P^{100} in Problem 8. What are the eigenvalues of P?

10 Find the unit eigenvectors of P in Problem 8, and put them into the columns of a unitary matrix F. What property of P makes these eigenvectors orthogonal?

11 Write down the 3 by 3 circulant matrix $C = 2I + 5P + 4P^2$. It has the same eigenvectors as P in Problem 8. Find its eigenvalues.

12 If U and V are unitary matrices, show that U^{-1} is unitary and also UV is unitary. Start from $U^{\mathrm{H}}U = I$ and $V^{\mathrm{H}}V = I$.

13 How do you know that the determinant of every Hermitian matrix is real?

14 The matrix $A^{\mathrm{H}}A$ is not only Hermitian but also positive definite, when the columns of A are independent. Proof: $z^{\mathrm{H}}A^{\mathrm{H}}Az$ is positive if z is nonzero because _____.

15 Diagonalize this Hermitian matrix to reach $A = U\Lambda U^{\mathrm{H}}$:

$$A = \begin{bmatrix} 0 & 1-i \\ i+1 & 1 \end{bmatrix}.$$

16 Diagonalize this skew-Hermitian matrix to reach $K = U\Lambda U^{\mathrm{H}}$. All λ's are _____:

$$K = \begin{bmatrix} 0 & -1+i \\ 1+i & i \end{bmatrix}.$$

17 Diagonalize this orthogonal matrix to reach $Q = U\Lambda U^{\mathrm{H}}$. Now all λ's are _____:

$$Q = \begin{bmatrix} \cos\theta & -\sin\theta \\ \sin\theta & \cos\theta \end{bmatrix}.$$

18 Diagonalize this unitary matrix V to reach $V = U\Lambda U^{\mathrm{H}}$. Again all λ's are _____:

$$V = \frac{1}{\sqrt{3}}\begin{bmatrix} 1 & 1-i \\ 1+i & -1 \end{bmatrix}.$$

19 If $\boldsymbol{v}_1, \ldots, \boldsymbol{v}_n$ is an orthogonal basis for $\mathbf{C}^n$, the matrix with those columns is a _____ matrix. Show that any vector z equals $(\boldsymbol{v}_1^{\mathrm{H}}z)\boldsymbol{v}_1 + \cdots + (\boldsymbol{v}_n^{\mathrm{H}}z)\boldsymbol{v}_n$.

20 The functions e^{-ix} and e^{ix} are orthogonal on the interval $0 \le x \le 2\pi$ because their inner product is $\int_0^{2\pi}$ _____ $= 0$.

21 The vectors $\boldsymbol{v} = (1, i, 1)$, $\boldsymbol{w} = (i, 1, 0)$ and $z =$ _____ are an orthogonal basis for _____.

22 If $A = R + iS$ is a Hermitian matrix, are its real and imaginary parts symmetric?

23 The (complex) dimension of $\mathbf{C}^n$ is _____. Find a non-real basis for $\mathbf{C}^n$.

24 Describe all 1 by 1 Hermitian matrices and unitary matrices. Do the same for 2 by 2.

25 How are the eigenvalues of A^{H} related to the eigenvalues of the square complex matrix A?

26 If $\boldsymbol{u}^{\mathrm{H}}\boldsymbol{u} = 1$ show that $I - 2\boldsymbol{u}\boldsymbol{u}^{\mathrm{H}}$ is Hermitian and also unitary. The rank-one matrix $\boldsymbol{u}\boldsymbol{u}^{\mathrm{H}}$ is the projection onto what line in $\mathbf{C}^n$?

27 If $A + iB$ is a unitary matrix (A and B are real) show that $Q = \begin{bmatrix} \mathbf{A} & -\mathbf{B} \\ \mathbf{B} & \mathbf{A} \end{bmatrix}$ is an orthogonal matrix.

28 If $A + iB$ is a Hermitian matrix (A and B are real) show that $\begin{bmatrix} \mathbf{A} & -\mathbf{B} \\ \mathbf{B} & \mathbf{A} \end{bmatrix}$ is symmetric.

29 Prove that the inverse of a Hermitian matrix is a Hermitian matrix.

30 Diagonalize this matrix by constructing its eigenvalue matrix Λ and its eigenvector matrix S:

$$A = \begin{bmatrix} 2 & 1-i \\ 1+i & 3 \end{bmatrix} = A^{\mathrm{H}}.$$

31 A matrix with orthonormal eigenvectors has the form $A = U\Lambda U^{-1} = U\Lambda U^{\mathrm{H}}$. *Prove that* $AA^{\mathrm{H}} = A^{\mathrm{H}}A$. These are exactly the *normal matrices*.

THE FAST FOURIER TRANSFORM ■ 10.3

Many applications of linear algebra take time to develop. It is not easy to explain them in an hour. The teacher and the author must choose between completing the theory and adding new applications. Generally the theory wins, because this course is the best chance to make it clear—and the importance of any one application seems limited. This section is almost an exception, because the importance of Fourier transforms is almost unlimited.

More than that, the algebra is basic. ***We want to multiply quickly by F and F^{-1}, the Fourier matrix and its inverse.*** This is achieved by the Fast Fourier Transform—the most valuable numerical algorithm in our lifetime.

The FFT has revolutionized signal processing. Whole industries are speeded up by this one idea. Electrical engineers are the first to know the difference—they take your Fourier transform as they meet you (if you are a function). Fourier's idea is to represent f as a sum of harmonics $c_k e^{ikx}$. The function is seen in *frequency space* through the coefficients c_k, instead of *physical space* through its values $f(x)$. The passage backward and forward between c's and f's is by the Fourier transform. Fast passage is by the FFT.

An ordinary product Fc uses n^2 multiplications (the matrix has n^2 nonzero entries). The Fast Fourier Transform needs only n times $\frac{1}{2}\log_2 n$. We will see how.

Roots of Unity and the Fourier Matrix

Quadratic equations have two roots (or one repeated root). Equations of degree n have n roots (counting repetitions). This is the Fundamental Theorem of Algebra, and to make it true we must allow complex roots. This section is about the very special equation $z^n = 1$. The solutions z are the "nth roots of unity." They are n evenly spaced points around the unit circle in the complex plane.

Figure 10.5 shows the eight solutions to $z^8 = 1$. Their spacing is $\frac{1}{8}(360°) = 45°$. The first root is at 45° or $\theta = 2\pi/8$ radians. ***It is the complex number $w = e^{i\theta} = e^{i2\pi/8}$.*** We call this number w_8 to emphasize that it is an 8th root. You could write it in terms of $\cos\frac{2\pi}{8}$ and $\sin\frac{2\pi}{8}$, but don't do it. The seven other 8th roots are $w^2, w^3, \ldots, w^8$, going around the circle. Powers of w are best in polar form, because we work only with the angle.

The fourth roots of 1 are also in the figure. They are $i, -1, -i, 1$. The angle is now $2\pi/4$ or 90°. The first root $w_4 = e^{2\pi i/4}$ is nothing but i. Even the square roots of 1 are seen, with $w_2 = e^{i2\pi/2} = -1$. Do not despise those square roots 1 and -1. The idea behind the FFT is to go from an 8 by 8 Fourier matrix (containing powers of w_8) to the 4 by 4 matrix below (with powers of $w_4 = i$). The same idea goes from 4 to 2. By exploiting the connections of F_8 down to F_4 and up to F_{16} (and beyond), the FFT makes multiplication by F_{1024} very quick.

We describe the *Fourier matrix*, first for $n = 4$. Its rows contain powers of 1 and w and w^2 and w^3. These are the fourth roots of 1, and their powers come in a special order:

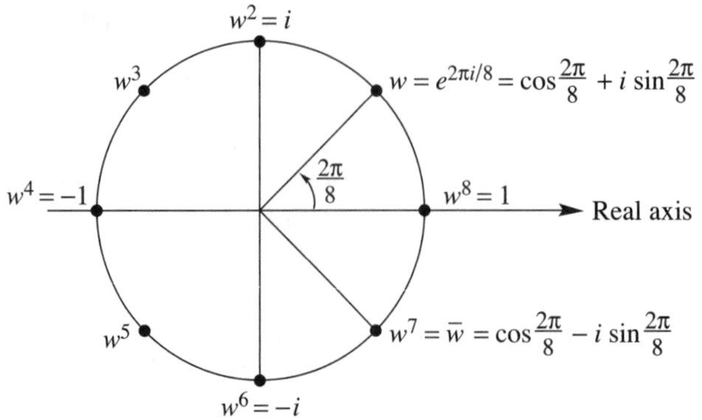

Figure 10.5 The eight solutions to $z^8 = 1$ are $1, w, w^2, \ldots, w^7$ with $w = (1+i)/\sqrt{2}$.

$$F_4 = \begin{bmatrix} 1 & 1 & 1 & 1 \\ 1 & w & w^2 & w^3 \\ 1 & w^2 & w^4 & w^6 \\ 1 & w^3 & w^6 & w^9 \end{bmatrix} = \begin{bmatrix} 1 & 1 & 1 & 1 \\ 1 & i & i^2 & i^3 \\ 1 & i^2 & i^4 & i^6 \\ 1 & i^3 & i^6 & i^9 \end{bmatrix}.$$

The matrix is symmetric ($F = F^{\mathrm{T}}$). It is *not* Hermitian. Its main diagonal is not real. But $\frac{1}{2}F$ is a ***unitary matrix***, which means that $(\frac{1}{2}F^{\mathrm{H}})(\frac{1}{2}F) = I$:

The columns of F give $F^{\mathrm{H}}F = 4I$. The inverse of F is $\frac{1}{4}F^{\mathrm{H}}$ which is $\frac{1}{4}\overline{F}$.

The inverse changes from $w = i$ to $\overline{w} = -i$. That takes us from F to $\overline{F}$. When the Fast Fourier Transform gives a quick way to multiply by F_4, it does the same for the inverse.

The unitary matrix is $U = F/\sqrt{n}$. We prefer to avoid that $\sqrt{n}$ and just put $\frac{1}{n}$ outside F^{-1}. The main point is to multiply the matrix F times the coefficients in the Fourier series $c_0 + c_1e^{ix} + c_2e^{2ix} + c_3e^{3ix}$:

$$F\boldsymbol{c} = \begin{bmatrix} 1 & 1 & 1 & 1 \\ 1 & w & w^2 & w^3 \\ 1 & w^2 & w^4 & w^6 \\ 1 & w^3 & w^6 & w^9 \end{bmatrix} \begin{bmatrix} c_0 \\ c_1 \\ c_2 \\ c_3 \end{bmatrix} = \begin{bmatrix} c_0 + c_1 + c_2 + c_3 \\ c_0 + c_1w + c_2w^2 + c_3w^3 \\ c_0 + c_1w^2 + c_2w^4 + c_3w^6 \\ c_0 + c_1w^3 + c_2w^6 + c_3w^9 \end{bmatrix}. \quad (1)$$

The input is four complex coefficients c_0, c_1, c_2, c_3. The output is four function values y_0, y_1, y_2, y_3. The first output $y_0 = c_0 + c_1 + c_2 + c_3$ is the value of the Fourier series at $x = 0$. *The second output is the value of that series* $\sum c_ke^{ikx}$ *at* $x = 2\pi/4$:

$$y_1 = c_0 + c_1e^{i2\pi/4} + c_2e^{i4\pi/4} + c_3e^{i6\pi/4} = c_0 + c_1w + c_2w^2 + c_3w^3.$$

The third and fourth outputs y_2 and y_3 are the values of $\sum c_ke^{ikx}$ at $x = 4\pi/4$ and $x = 6\pi/4$. These are *finite* Fourier series! They contain $n = 4$ terms and they are evaluated at $n = 4$ points. Those points $x = 0, 2\pi/4, 4\pi/4, 6\pi/4$ are equally spaced.

The next point would be $x = 8\pi/4$ which is 2π. Then the series is back to y_0, because $e^{2\pi i}$ is the same as $e^0 = 1$. Everything cycles around with period 4. In this world $2 + 2$ is 0 because $(w^2)(w^2) = w^0 = 1$. In matrix shorthand, F times $\boldsymbol{c}$ gives a column vector $\boldsymbol{y}$. The four y's come from evaluating the series at the four x's with spacing $2\pi/4$:

$$\boldsymbol{y} = F\boldsymbol{c} \text{ produces } y_j = \sum_{k=0}^{3} c_k e^{ik(2\pi j/4)} = \text{ the value of the series at } x = \frac{2\pi j}{4}.$$

We will follow the convention that j and k go from 0 to $n-1$ (instead of 1 to n). The "zeroth row" and "zeroth column" of F contain all ones.

The n by n Fourier matrix contains powers of $w = e^{2\pi i/n}$:

$$F_n\boldsymbol{c} = \begin{bmatrix} 1 & 1 & 1 & \cdot & 1 \\ 1 & w & w^2 & \cdot & w^{n-1} \\ 1 & w^2 & w^4 & \cdot & w^{2(n-1)} \\ \cdot & \cdot & \cdot & \cdot & \cdot \\ 1 & w^{n-1} & w^{2(n-1)} & \cdot & w^{(n-1)^2} \end{bmatrix} \begin{bmatrix} c_0 \\ c_1 \\ c_2 \\ \cdot \\ c_{n-1} \end{bmatrix} = \begin{bmatrix} y_0 \\ y_1 \\ y_2 \\ \cdot \\ y_{n-1} \end{bmatrix}. \qquad (2)$$

F_n is symmetric but not Hermitian. ***Its columns are orthogonal***, and $F_n\overline{F}_n = nI$. *Then* F_n^{-1} *is* $\overline{F}_n/n$. The inverse contains powers of $\overline{w}_n = e^{-2\pi i/n}$. Look at the pattern in F:

The entry in row j, column k ***is*** w^{jk}. ***Row zero and column zero contain*** $w^0 = 1$.

The zeroth output is $y_0 = c_0 + c_1 + \cdots + c_{n-1}$. This is the series $\sum c_k e^{ikx}$ at $x = 0$. When we multiply $\boldsymbol{c}$ by F_n, we sum the series at n points. *When we multiply* $\boldsymbol{y}$ *by* F_n^{-1}, *we find the coefficients* $\boldsymbol{c}$ *from the function values* $\boldsymbol{y}$. The matrix F passes from "frequency space" to "physical space." F^{-1} returns from the function values $\boldsymbol{y}$ to the Fourier coefficients $\boldsymbol{c}$.

One Step of the Fast Fourier Transform

We want to multiply F times $\boldsymbol{c}$ as quickly as possible. Normally a matrix times a vector takes n^2 separate multiplications—the matrix has n^2 entries. You might think it is impossible to do better. (If the matrix has zero entries then multiplications can be skipped. But the Fourier matrix has no zeros!) By using the special pattern w^{jk} for its entries, F can be factored in a way that produces many zeros. This is the **FFT**.

The key idea is to connect $\boldsymbol{F_n}$ ***with the half-size Fourier matrix*** $\boldsymbol{F_{n/2}}$. Assume that n is a power of 2 (say $n = 2^{10} = 1024$). We will connect F_{1024} to F_{512}—or rather to ***two copies of*** F_{512}. When $n = 4$, the key is in the relation between the matrices

$$F_4 = \begin{bmatrix} 1 & 1 & 1 & 1 \\ 1 & i & i^2 & i^3 \\ 1 & i^2 & i^4 & i^6 \\ 1 & i^3 & i^6 & i^9 \end{bmatrix} \quad \text{and} \quad \begin{bmatrix} F_2 & \\ & F_2 \end{bmatrix} = \begin{bmatrix} 1 & 1 & & \\ 1 & i^2 & & \\ & & 1 & 1 \\ & & 1 & i^2 \end{bmatrix}.$$

On the left is F_4, with no zeros. On the right is a matrix that is half zero. The work is cut in half. But wait, those matrices are not the same. The block matrix with two copies of the half-size F is one piece of the picture but not the only piece. Here is the factorization of F_4 with many zeros:

$$F_4 = \begin{bmatrix} 1 & & 1 & \\ & 1 & & i \\ 1 & & -1 & \\ & 1 & & -i \end{bmatrix} \begin{bmatrix} 1 & 1 & & \\ 1 & i^2 & & \\ & & 1 & 1 \\ & & 1 & i^2 \end{bmatrix} \begin{bmatrix} 1 & & & \\ & & 1 & \\ & 1 & & \\ & & & 1 \end{bmatrix}. \tag{3}$$

The matrix on the right is a permutation. It puts the even c's (c_0 and c_2) ahead of the odd c's (c_1 and c_3). The middle matrix performs separate half-size transforms on the evens and odds. The matrix at the left combines the two half-size outputs—in a way that produces the correct full-size output $\boldsymbol{y} = F_4\boldsymbol{c}$. You could multiply those three matrices to see that their product is F_4.

The same idea applies when $n = 1024$ and $m = \frac{1}{2}n = 512$. The number w is $e^{2\pi i/1024}$. It is at the angle $\theta = 2\pi/1024$ on the unit circle. The Fourier matrix F_{1024} is full of powers of w. The first stage of the FFT is the great factorization discovered by Cooley and Tukey (and foreshadowed in 1805 by Gauss):

$$F_{1024} = \begin{bmatrix} I_{512} & D_{512} \\ I_{512} & -D_{512} \end{bmatrix} \begin{bmatrix} F_{512} & \\ & F_{512} \end{bmatrix} \begin{bmatrix} \text{even-odd} \\ \text{permutation} \end{bmatrix}. \tag{4}$$

I_{512} is the identity matrix. D_{512} is the diagonal matrix with entries $(1, w, \dots, w^{511})$. The two copies of F_{512} are what we expected. Don't forget that they use the 512th root of unity (which is nothing but w^2!!) The permutation matrix separates the incoming vector $\boldsymbol{c}$ into its even and odd parts $\boldsymbol{c}' = (c_0, c_2, \dots, c_{1022})$ and $\boldsymbol{c}'' = (c_1, c_3, \dots, c_{1023})$.

Here are the algebra formulas which say the same thing as the factorization of F_{1024}:

10I (FFT) Set $m = \frac{1}{2}n$. The first m and last m components of $\boldsymbol{y} = F_n\boldsymbol{c}$ are combinations of the half-size transforms $\boldsymbol{y}' = F_m\boldsymbol{c}'$ and $\boldsymbol{y}'' = F_m\boldsymbol{c}''$. Equation (4) shows $I\boldsymbol{y}' + D\boldsymbol{y}''$ and $I\boldsymbol{y}' - D\boldsymbol{y}''$:

$$\begin{aligned} y_j &= y'_j + w_n^j y''_j, \quad j = 0, \dots, m-1 \\ y_{j+m} &= y'_j - w_n^j y''_j, \quad j = 0, \dots, m-1. \end{aligned} \tag{5}$$

Thus the three steps are: split $\boldsymbol{c}$ into $\boldsymbol{c}'$ and $\boldsymbol{c}''$, transform them by F_m into $\boldsymbol{y}'$ and $\boldsymbol{y}''$, and reconstruct $\boldsymbol{y}$ from equation (5).

You might like the flow graph in Figure 10.6 better than these formulas. The graph for $n = 4$ shows $\boldsymbol{c}'$ and $\boldsymbol{c}''$ going through the half-size F_2. Those steps are called "*butterflies*," from their shape. Then the outputs from the F_2's are combined using the I and D matrices to produce $\boldsymbol{y} = F_4\boldsymbol{c}$:

This reduction from F_n to two F_m's almost cuts the work in half—you see the zeros in the matrix factorization. That reduction is good but not great. The full idea of the **FFT** is much more powerful. It saves much more than half the time.

The Full FFT by Recursion

If you have read this far, you have probably guessed what comes next. We reduced F_n to $F_{n/2}$. ***Keep going to*** $F_{n/4}$. The matrices F_{512} lead to F_{256} (in four copies). Then 256 leads to 128. *That is recursion.* It is a basic principle of many fast algorithms, and here is the second stage with four copies of $F = F_{256}$ and $D = D_{256}$:

$$\begin{bmatrix} F_{512} & \\ & F_{512} \end{bmatrix} = \begin{bmatrix} I & D & & \\ I & -D & & \\ & & I & D \\ & & I & -D \end{bmatrix} \begin{bmatrix} F & & & \\ & F & & \\ & & F & \\ & & & F \end{bmatrix} \begin{bmatrix} \text{pick} & 0, 4, 8, \cdots \\ \text{pick} & 2, 6, 10, \cdots \\ \text{pick} & 1, 5, 9, \cdots \\ \text{pick} & 3, 7, 11, \cdots \end{bmatrix}.$$

We will count the individual multiplications, to see how much is saved. Before the **FFT** was invented, the count was the usual $n^2 = (1024)^2$. This is about a million multiplications. I am not saying that they take a long time. The cost becomes large when we have many, many transforms to do—which is typical. Then the saving by the **FFT** is also large:

The final count for size $n = 2^l$ ***is reduced from*** n^2 ***to*** $\frac{1}{2}nl$.

The number 1024 is 2^{10}, so $l = 10$. The original count of $(1024)^2$ is reduced to (5)(1024). The saving is a factor of 200. A million is reduced to five thousand. That is why the **FFT** has revolutionized signal processing.

Here is the reasoning behind $\frac{1}{2}nl$. There are l levels, going from $n = 2^l$ down to $n = 1$. Each level has $\frac{1}{2}n$ multiplications from the diagonal D's, to reassemble the half-size outputs from the lower level. This yields the final count $\frac{1}{2}nl$, which is $\frac{1}{2}n \log_2 n$.

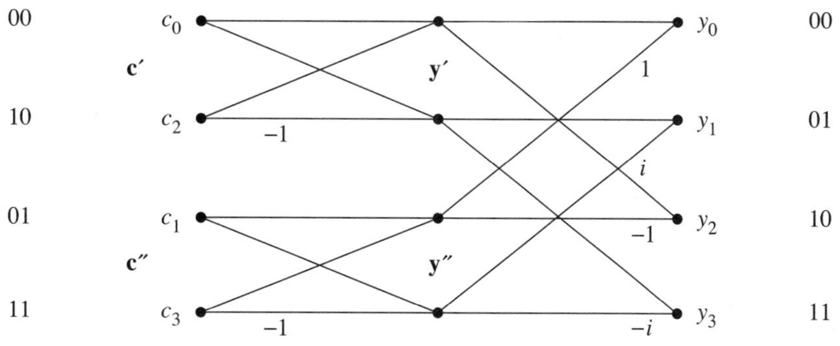

Figure 10.6 Flow graph for the Fast Fourier Transform with $n = 4$.

One last note about this remarkable algorithm. There is an amazing rule for the order that the c's enter the **FFT**, after all the even-odd permutations. Write the numbers 0 to $n-1$ in binary (base 2). *Reverse the order of their digits.* The complete picture shows the bit-reversed order at the start, the $l = \log_2 n$ steps of the recursion, and the final output $y_0, \ldots, y_{n-1}$ which is F_n times $\boldsymbol{c}$. The book ends with that very fundamental idea, a matrix multiplying a vector.

Thank you for studying linear algebra. I hope you enjoyed it, and I very much hope you will use it. It was a pleasure to write about this neat subject.

Problem Set 10.3

1 Multiply the three matrices in equation (3) and compare with F. In which six entries do you need to know that $i^2 = -1$?

2 Invert the three factors in equation (3) to find a fast factorization of F^{-1}.

3 F is symmetric. So transpose equation (3) to find a new Fast Fourier Transform!

4 All entries in the factorization of F_6 involve powers of $w =$ sixth root of 1:

$$F_6 = \begin{bmatrix} I & D \\ I & -D \end{bmatrix} \begin{bmatrix} F_3 & \\ & F_3 \end{bmatrix} \begin{bmatrix} & P & \end{bmatrix}.$$

Write down these three factors with $1, w, w^2$ in D and powers of w^2 in F_3. Multiply!

5 If $\boldsymbol{v} = (1, 0, 0, 0)$ and $\boldsymbol{w} = (1, 1, 1, 1)$, show that $F\boldsymbol{v} = \boldsymbol{w}$ and $F\boldsymbol{w} = 4\boldsymbol{v}$. Therefore $F^{-1}\boldsymbol{w} = \boldsymbol{v}$ and $F^{-1}\boldsymbol{v} =$ ______.

6 What is F^2 and what is F^4 for the 4 by 4 Fourier matrix?

7 Put the vector $\boldsymbol{c} = (1, 0, 1, 0)$ through the three steps of the **FFT** to find $\boldsymbol{y} = F\boldsymbol{c}$. Do the same for $\boldsymbol{c} = (0, 1, 0, 1)$.

8 Compute $\boldsymbol{y} = F_8\boldsymbol{c}$ by the three **FFT** steps for $\boldsymbol{c} = (1, 0, 1, 0, 1, 0, 1, 0)$. Repeat the computation for $\boldsymbol{c} = (0, 1, 0, 1, 0, 1, 0, 1)$.

9 If $w = e^{2\pi i/64}$ then w^2 and $\sqrt{w}$ are among the ______ and ______ roots of 1.

10 (a) Draw all the sixth roots of 1 on the unit circle. Prove they add to zero.

(b) What are the three cube roots of 1? Do they also add to zero?

11 The columns of the Fourier matrix F are the *eigenvectors* of the cyclic permutation P. Multiply PF to find the eigenvalues λ_1 to λ_4:

$$\begin{bmatrix} 0 & 1 & 0 & 0 \\ 0 & 0 & 1 & 0 \\ 0 & 0 & 0 & 1 \\ 1 & 0 & 0 & 0 \end{bmatrix} \begin{bmatrix} 1 & 1 & 1 & 1 \\ 1 & i & i^2 & i^3 \\ 1 & i^2 & i^4 & i^6 \\ 1 & i^3 & i^6 & i^9 \end{bmatrix} = \begin{bmatrix} 1 & 1 & 1 & 1 \\ 1 & i & i^2 & i^3 \\ 1 & i^2 & i^4 & i^6 \\ 1 & i^3 & i^6 & i^9 \end{bmatrix} \begin{bmatrix} \lambda_1 & & & \\ & \lambda_2 & & \\ & & \lambda_3 & \\ & & & \lambda_4 \end{bmatrix}.$$

This is $PF = F\Lambda$ or $P = F\Lambda F^{-1}$. The eigenvector matrix (usually S) is F.

12 The equation $\det(P - \lambda I) = 0$ is $\lambda^4 = 1$. This shows again that the eigenvalue matrix Λ is ______. Which permutation P has eigenvalues = cube roots of 1?

13 (a) Two eigenvectors of C are $(1, 1, 1, 1)$ and $(1, i, i^2, i^3)$. What are the eigenvalues?

$$\begin{bmatrix} c_0 & c_1 & c_2 & c_3 \\ c_3 & c_0 & c_1 & c_2 \\ c_2 & c_3 & c_0 & c_1 \\ c_1 & c_2 & c_3 & c_0 \end{bmatrix} \begin{bmatrix} 1 \\ 1 \\ 1 \\ 1 \end{bmatrix} = e_1 \begin{bmatrix} 1 \\ 1 \\ 1 \\ 1 \end{bmatrix} \quad \text{and} \quad C \begin{bmatrix} 1 \\ i \\ i^2 \\ i^3 \end{bmatrix} = e_2 \begin{bmatrix} 1 \\ i \\ i^2 \\ i^3 \end{bmatrix}.$$

(b) $P = F\Lambda F^{-1}$ immediately gives $P^2 = F\Lambda^2 F^{-1}$ and $P^3 = F\Lambda^3 F^{-1}$. Then $C = c_0 I + c_1 P + c_2 P^2 + c_3 P^3 = F(c_0 I + c_1 \Lambda + c_2 \Lambda^2 + c_3 \Lambda^3)F^{-1} = FEF^{-1}$. That matrix E in parentheses is diagonal. It contains the ______ of C.

14 Find the eigenvalues of the "periodic" $-1, 2, -1$ matrix from $E = 2I - \Lambda - \Lambda^3$, with the eigenvalues of P in Λ. The -1's in the corners make this matrix periodic:

$$C = \begin{bmatrix} 2 & -1 & 0 & -1 \\ -1 & 2 & -1 & 0 \\ 0 & -1 & 2 & -1 \\ -1 & 0 & -1 & 2 \end{bmatrix} \quad \text{has } c_0 = 2, c_1 = -1, c_2 = 0, c_3 = -1.$$

15 To multiply C times a vector $\boldsymbol{x}$, we can multiply $F(E(F^{-1}\boldsymbol{x}))$ instead. The direct way uses n^2 separate multiplications. Knowing E and F, the second way uses only $n \log_2 n + n$ multiplications. How many of those come from E, how many from F, and how many from F^{-1}?

16 How could you quickly compute these four components of $F\boldsymbol{c}$ starting from $c_0 + c_2, c_0 - c_2, c_1 + c_3, c_1 - c_3$? You are finding the Fast Fourier Transform!

$$F\boldsymbol{c} = \begin{bmatrix} c_0 + c_1 + c_2 + c_3 \\ c_0 + ic_1 + i^2 c_2 + i^3 c_3 \\ c_0 + i^2 c_1 + i^4 c_2 + i^6 c_3 \\ c_0 + i^3 c_1 + i^6 c_2 + i^9 c_3 \end{bmatrix}.$$

SOLUTIONS TO SELECTED EXERCISES

Problem Set 1.1, page 7

4 $3\boldsymbol{v} + \boldsymbol{w} = (7, 5)$ and $\boldsymbol{v} - 3\boldsymbol{w} = (-1, -5)$ and $c\boldsymbol{v} + d\boldsymbol{w} = (2c + d, c + 2d)$.

6 The components of every $c\boldsymbol{v}+d\boldsymbol{w}$ add to zero. Choose $c = 4$ and $d = 10$ to get $(4, 2, -6)$.

9 The fourth corner can be $(4, 4)$ or $(4, 0)$ or $(-2, 2)$.

11 Five more corners $(0, 0, 1), (1, 1, 0), (1, 0, 1), (0, 1, 1), (1, 1, 1)$. The center point is $(\frac{1}{2}, \frac{1}{2}, \frac{1}{2})$. The centers of the six faces are $(\frac{1}{2}, \frac{1}{2}, 0), (\frac{1}{2}, \frac{1}{2}, 1)$ and $(0, \frac{1}{2}, \frac{1}{2}), (1, \frac{1}{2}, \frac{1}{2})$ and $(\frac{1}{2}, 0, \frac{1}{2}), (\frac{1}{2}, 1, \frac{1}{2})$.

12 A four-dimensional cube has $2^4 = 16$ corners and $2 \cdot 4 = 8$ three-dimensional sides and 24 two-dimensional faces and 32 one-dimensional edges. See Worked Example **2.4 A**.

13 sum = zero vector; sum = −4:00 vector; 1:00 is 60° from horizontal $= (\cos\frac{\pi}{3}, \sin\frac{\pi}{3}) = (\frac{1}{2}, \frac{\sqrt{3}}{2})$.

16 All combinations with $c+d = 1$ are on the line through $\boldsymbol{v}$ and $\boldsymbol{w}$. The point $\boldsymbol{V} = -\boldsymbol{v}+2\boldsymbol{w}$ is on that line beyond $\boldsymbol{w}$.

17 The vectors $c\boldsymbol{v}+c\boldsymbol{w}$ fill out the line passing through $(0, 0)$ and $\boldsymbol{u} = \frac{1}{2}\boldsymbol{v}+\frac{1}{2}\boldsymbol{w}$. It continues beyond $\boldsymbol{v} + \boldsymbol{w}$ and $(0, 0)$. With $c \geq 0$, half this line is removed and the "ray" starts at $(0, 0)$.

20 (a) $\frac{1}{3}\boldsymbol{u} + \frac{1}{3}\boldsymbol{v} + \frac{1}{3}\boldsymbol{w}$ is the center of the triangle between $\boldsymbol{u}, \boldsymbol{v}$ and $\boldsymbol{w}$; $\frac{1}{2}\boldsymbol{u} + \frac{1}{2}\boldsymbol{w}$ is the center of the edge between $\boldsymbol{u}$ and $\boldsymbol{w}$ (b) To fill in the triangle keep $c \geq 0$, $d \geq 0$, $e \geq 0$, and $c + d + e = 1$.

22 The vector $\frac{1}{2}(\boldsymbol{u} + \boldsymbol{v} + \boldsymbol{w})$ is *outside* the pyramid because $c + d + e = \frac{1}{2} + \frac{1}{2} + \frac{1}{2} > 1$.

25 (a) Choose $\boldsymbol{u} = \boldsymbol{v} = \boldsymbol{w} =$ any nonzero vector (b) Choose $\boldsymbol{u}$ and $\boldsymbol{v}$ in different directions, and $\boldsymbol{w}$ to be a combination like $\boldsymbol{u} + \boldsymbol{v}$.

28 An example is $(a, b) = (3, 6)$ and $(c, d) = (1, 2)$. The ratios a/c and b/d are equal. Then $ad = bc$. Then (divide by bd) the ratios a/b and c/d are equal!

Problem Set 1.2, page 17

3 Unit vectors $\boldsymbol{v}/\|\boldsymbol{v}\| = (\frac{3}{5}, \frac{4}{5}) = (.6, .8)$ and $\boldsymbol{w}/\|\boldsymbol{w}\| = (\frac{4}{5}, \frac{3}{5}) = (.8, .6)$. The cosine of θ is $\frac{\boldsymbol{v}}{\|\boldsymbol{v}\|} \cdot \frac{\boldsymbol{w}}{\|\boldsymbol{w}\|} = \frac{24}{25}$. The vectors $\boldsymbol{w}, \boldsymbol{u}, -\boldsymbol{w}$ make $0°, 90°, 180°$ angles with $\boldsymbol{w}$.

5 (a) $\boldsymbol{v}\cdot(-\boldsymbol{v}) = -1$ (b) $(\boldsymbol{v}+\boldsymbol{w})\cdot(\boldsymbol{v}-\boldsymbol{w}) = \boldsymbol{v}\cdot\boldsymbol{v}+\boldsymbol{w}\cdot\boldsymbol{v}-\boldsymbol{v}\cdot\boldsymbol{w}-\boldsymbol{w}\cdot\boldsymbol{w} = 1+(\ \ )-(\ \ )-1 = 0$ so $\theta = 90°$ (c) $(\boldsymbol{v}-2\boldsymbol{w})\cdot(\boldsymbol{v}+2\boldsymbol{w}) = \boldsymbol{v}\cdot\boldsymbol{v} - 4\boldsymbol{w}\cdot\boldsymbol{w} = -3$

7 All vectors $\boldsymbol{w} = (c, 2c)$; all vectors (x, y, z) with $x+y+z = 0$ lie on a *plane*; all vectors perpendicular to $(1,1,1)$ and $(1,2,3)$ lie on a *line*.

9 If $v_2w_2/v_1w_1 = -1$ then $v_2w_2 = -v_1w_1$ or $v_1w_1 + v_2w_2 = 0$.

11 $\boldsymbol{v}\cdot\boldsymbol{w} < 0$ means angle $> 90°$; this is half of the plane.

12 $(1,1)$ perpendicular to $(1,5)-c(1,1)$ if $6-2c = 0$ or $c = 3$; $\boldsymbol{v}\cdot(\boldsymbol{w}-c\boldsymbol{v}) = 0$ if $c = \boldsymbol{v}\cdot\boldsymbol{w}/\boldsymbol{v}\cdot\boldsymbol{v}$.

15 $\frac{1}{2}(x+y) = 5$; $\cos\theta = 2\sqrt{16}/\sqrt{10}\sqrt{10} = .8$.

17 $\cos\alpha = 1/\sqrt{2}$, $\cos\beta = 0$, $\cos\gamma = -1/\sqrt{2}$, $\cos^2\alpha+\cos^2\beta+\cos^2\gamma = (v_1^2+v_2^2+v_3^2)/\|\boldsymbol{v}\|^2 = 1$.

21 $2\boldsymbol{v}\cdot\boldsymbol{w} \le 2\|\boldsymbol{v}\|\|\boldsymbol{w}\|$ leads to $\|\boldsymbol{v}+\boldsymbol{w}\|^2 = \boldsymbol{v}\cdot\boldsymbol{v}+2\boldsymbol{v}\cdot\boldsymbol{w}+\boldsymbol{w}\cdot\boldsymbol{w} \le \|\boldsymbol{v}\|^2+2\|\boldsymbol{v}\|\|\boldsymbol{w}\|+\|\boldsymbol{w}\|^2 = (\|\boldsymbol{v}\|+\|\boldsymbol{w}\|)^2$.

23 $\cos\beta = w_1/\|\boldsymbol{w}\|$ and $\sin\beta = w_2/\|\boldsymbol{w}\|$. Then $\cos(\beta - a) = \cos\beta\cos\alpha + \sin\beta\sin\alpha = v_1w_1/\|\boldsymbol{v}\|\|\boldsymbol{w}\| + v_2w_2/\|\boldsymbol{v}\|\|\boldsymbol{w}\| = \boldsymbol{v}\cdot\boldsymbol{w}/\|\boldsymbol{v}\|\|\boldsymbol{w}\|$.

25 (a) $v_1^2w_1^2+2v_1w_1v_2w_2+v_2^2w_2^2 \le v_1^2w_1^2+v_1^2w_2^2+v_2^2w_1^2+v_2^2w_2^2$ is true because the difference is $v_1^2w_2^2+v_2^2w_1^2-2v_1w_1v_2w_2$ which is $(v_1w_2-v_2w_1)^2 \ge 0$.

26 Example 6 gives $|u_1||U_1| \le \frac{1}{2}(u_1^2+U_1^2)$ and $|u_2||U_2| \le \frac{1}{2}(u_2^2+U_2^2)$. The whole line becomes $.96 \le (.6)(.8)+(.8)(.6) \le \frac{1}{2}(.6^2+.8^2)+\frac{1}{2}(.8^2+.6^2) = 1$.

28 Try $\boldsymbol{v} = (1,2,-3)$ and $\boldsymbol{w} = (-3,1,2)$ with $\cos\theta = \frac{-7}{14}$ and $\theta = 120°$. Write $\boldsymbol{v}\cdot\boldsymbol{w} = xz+yz+xy$ as $\frac{1}{2}(x+y+z)^2 - \frac{1}{2}(x^2+y^2+z^2)$. If $x+y+z = 0$ this is $-\frac{1}{2}(x^2+y^2+z^2)$, so $\boldsymbol{v}\cdot\boldsymbol{w}/\|\boldsymbol{v}\|\|\boldsymbol{w}\| = -\frac{1}{2}$.

31 Three vectors in the plane could make angles $> 90°$ with each other: $(1,0), (-1,4), (-1,-4)$. Four vectors could not do this ($360°$ total angle). How many can do this in $\mathbf{R}^3$ or $\mathbf{R}^n$?

Problem Set 2.1, page 30

2 The columns are $\boldsymbol{i} = (1,0,0)$ and $\boldsymbol{j} = (0,1,0)$ and $\boldsymbol{k} = (0,0,1)$ and $\boldsymbol{b} = (2,3,4) = 2\boldsymbol{i}+3\boldsymbol{j}+4\boldsymbol{k}$.

3 The planes are the same: $2x = 4$ is $x = 2$, $3y = 9$ is $y = 3$, and $4z = 16$ is $z = 4$. The solution is the same intersection point. The columns are changed; but same combination $\widehat{\boldsymbol{x}} = \boldsymbol{x}$.

5 If $z = 2$ then $x + y = 0$ and $x - y = z$ give the point $(1, -1, 2)$. If $z = 0$ then $x + y = 6$ and $x - y = 4$ give the point $(5, 1, 0)$. Halfway between is $(3, 0, 1)$.

7 Equation 1 + equation 2 − equation 3 is now $0 = -4$. Line misses plane; *no solution.*

9 Four planes in 4-dimensional space normally meet at a *point*. The solution to $A\boldsymbol{x} = (3, 3, 3, 2)$ is $\boldsymbol{x} = (0, 0, 1, 2)$ if A has columns $(1, 0, 0, 0)$, $(1, 1, 0, 0)$, $(1, 1, 1, 0)$, $(1, 1, 1, 1)$. The equations are $x + y + z + t = 3$, $y + z + t = 3$, $z + t = 3$, $t = 2$.

15 $2x + 3y + z + 5t = 8$ is $A\boldsymbol{x} = \boldsymbol{b}$ with the 1 by 4 matrix $A = [\,2 \;\; 3 \;\; 1 \;\; 5\,]$. The solutions $\boldsymbol{x}$ fill a 3D "plane" in 4 dimensions.

17 $R = \begin{bmatrix} 0 & 1 \\ -1 & 0 \end{bmatrix}$, 180° rotation from $R^2 = \begin{bmatrix} -1 & 0 \\ 0 & -1 \end{bmatrix} = -I$.

19 $E = \begin{bmatrix} 1 & 0 \\ -1 & 1 \end{bmatrix}$, $E = \begin{bmatrix} 1 & 0 & 0 \\ -1 & 1 & 0 \\ 0 & 0 & 1 \end{bmatrix}$.

23 The dot product $[\,1 \;\; 4 \;\; 5\,]\begin{bmatrix} x \\ y \\ z \end{bmatrix} = (1 \text{ by } 3)(3 \text{ by } 1)$ is zero for points (x, y, z) on a plane in three dimensions. The columns of A are one-dimensional vectors.

24 $A = [\,1 \;\; 2 \;\; ; \;\; 3 \;\; 4\,]$ and $\boldsymbol{x} = [\,5 \;\; -2\,]'$ and $\boldsymbol{b} = [\,1 \;\; 7\,]'$. $\boldsymbol{r} = \boldsymbol{b} - A * \boldsymbol{x}$ prints as zero.

26 $\text{ones}(4, 4) * \text{ones}(4, 1) = [\,4 \;\; 4 \;\; 4 \;\; 4\,]'$; $B * \boldsymbol{w} = [\,10 \;\; 10 \;\; 10 \;\; 10\,]'$.

29 The row picture shows four *lines*. The column picture is in *four*-dimensional space. No solution unless the right side is a combination of *the two columns.*

31 $\boldsymbol{u}_7, \boldsymbol{v}_7, \boldsymbol{w}_7$ are all close to $(.6, .4)$. Their components still add to 1.

32 $\begin{bmatrix} .8 & .3 \\ .2 & .7 \end{bmatrix}\begin{bmatrix} .6 \\ .4 \end{bmatrix} = \begin{bmatrix} .6 \\ .4 \end{bmatrix} =$ *steady state* $\boldsymbol{s}$. No change when multiplied by $\begin{bmatrix} .8 & .3 \\ .2 & .7 \end{bmatrix}$.

34 $M = \begin{bmatrix} 8 & 3 & 4 \\ 1 & 5 & 9 \\ 6 & 7 & 2 \end{bmatrix} = \begin{bmatrix} 5+u & 5-u+v & 5-v \\ 5-u-v & 5 & 5+u+v \\ 5+v & 5+u-v & 5-u \end{bmatrix}$; $M_3(1, 1, 1) = (15, 15, 15)$;
$M_4(1, 1, 1, 1) = (34, 34, 34, 34)$ because the numbers 1 to 16 add to 136 which is 4(34).

Problem Set 2.2, page 41

3 Subtract $-\frac{1}{2}$ times equation 1 (or add $\frac{1}{2}$ times equation 1). The new second equation is $3y = 3$. Then $y = 1$ and $x = 5$. If the right side changes sign, so does the solution: $(x, y) = (-5, -1)$.

4 Subtract $l = \frac{c}{a}$ times equation 1. The new second pivot multiplying y is $d - (cb/a)$ or $(ad - bc)/a$. Then $y = (ag - cf)/(ad - bc)$.

6 Singular system if $b = 4$, because $4x + 8y$ is 2 times $2x + 4y$. Then $g = 2 \cdot 16 = 32$ makes the system solvable. The lines become the *same*: infinitely many solutions like $(8, 0)$ and $(0, 4)$.

8 If $k = 3$ elimination must fail: no solution. If $k = -3$, elimination gives $0 = 0$ in equation 2: infinitely many solutions. If $k = 0$ a row exchange is needed: one solution.

13 Subtract 2 times row 1 from row 2 to reach $(d - 10)y - z = 2$. Equation (3) is $y - z = 3$. If $d = 10$ exchange rows 2 and 3. If $d = 11$ the system is singular; third pivot is missing.

14 The second pivot position will contain $-2 - b$. If $b = -2$ we exchange with row 3. If $b = -1$ (singular case) the second equation is $-y - z = 0$. A solution is $(1, 1, -1)$.

16 If row 1 = row 2, then row 2 is zero after the first step; exchange the zero row with row 3 and there is no *third* pivot. If column 1 = column 2 there is no *second* pivot.

18 Row 2 becomes $3y - 4z = 5$, then row 3 becomes $(q + 4)z = t - 5$. If $q = -4$ the system is singular — no third pivot. Then if $t = 5$ the third equation is $0 = 0$. Choosing $z = 1$ the equation $3y - 4z = 5$ gives $y = 3$ and equation 1 gives $x = -9$.

20 Singular if row 3 is a combination of rows 1 and 2. From the end view, the three planes form a triangle. This happens if rows $1+2$ = row 3 on the left side but not the right side: for example $x + y + z = 0$, $x - 2y - z = 1$, $2x - y = 1$. No parallel planes but still no solution.

24 $A = \begin{bmatrix} 1 & 1 & 1 \\ a & a+1 & a+1 \\ b & b+c & b+c+3 \end{bmatrix}$ for any a, b, c leads to $U = \begin{bmatrix} 1 & 1 & 1 \\ 0 & 1 & 1 \\ 0 & 0 & 3 \end{bmatrix}$.

26 $a = 2$ (equal columns), $a = 4$ (equal rows), $a = 0$ (zero column).

29 $A(2, :) = A(2, :) - 3 * A(1, :)$ Subtracts 3 times row 1 from row 2.

30 The average pivots for rand(3) *without* row exchanges were $\frac{1}{2}, 5, 10$ in one experiment— but pivots 2 and 3 can be arbitrarily large. Their averages are actually infinite! *With* row exchanges in MATLAB's lu code, the averages .75 and .50 and .365 are much more stable (and should be predictable, also for randn with normal instead of uniform probability distribution).

Problem Set 2.3, page 53

1 $E_{21} = \begin{bmatrix} 1 & 0 & 0 \\ -5 & 1 & 0 \\ 0 & 0 & 1 \end{bmatrix}$, $E_{32} = \begin{bmatrix} 1 & 0 & 0 \\ 0 & 1 & 0 \\ 0 & 7 & 1 \end{bmatrix}$, $P = \begin{bmatrix} 1 & 0 & 0 \\ 0 & 0 & 1 \\ 0 & 1 & 0 \end{bmatrix} \begin{bmatrix} 0 & 1 & 0 \\ 1 & 0 & 0 \\ 0 & 0 & 1 \end{bmatrix} = \begin{bmatrix} 0 & 1 & 0 \\ 0 & 0 & 1 \\ 1 & 0 & 0 \end{bmatrix}$.

3 $\begin{bmatrix} 1 & 0 & 0 \\ -4 & 1 & 0 \\ 0 & 0 & 1 \end{bmatrix}, \begin{bmatrix} 1 & 0 & 0 \\ 0 & 1 & 0 \\ 2 & 0 & 1 \end{bmatrix}, \begin{bmatrix} 1 & 0 & 0 \\ 0 & 1 & 0 \\ 0 & -2 & 1 \end{bmatrix} \begin{bmatrix} \\ M = E_{32}E_{31}E_{21} \\ \\ \end{bmatrix} = \begin{bmatrix} 1 & 0 & 0 \\ -4 & 1 & 0 \\ 10 & -2 & 1 \end{bmatrix}$.

5 Changing a_{33} from 7 to 11 will change the third pivot from 5 to 9. Changing a_{33} from 7 to 2 will change the pivot from 5 to *no pivot*.

7 To reverse E_{31}, *add* 7 times row 1 to row 3. The matrix is $R_{31} = \begin{bmatrix} 1 & 0 & 0 \\ 0 & 1 & 0 \\ 7 & 0 & 1 \end{bmatrix}$.

9 $M = \begin{bmatrix} 1 & 0 & 0 \\ 0 & 0 & 1 \\ -1 & 1 & 0 \end{bmatrix}$. After the exchange, we need E_{31} (not E_{21}) to act on the new row 3.

10 $E_{13} = \begin{bmatrix} 1 & 0 & 1 \\ 0 & 1 & 0 \\ 0 & 0 & 1 \end{bmatrix}$; $\begin{bmatrix} 1 & 0 & 1 \\ 0 & 1 & 0 \\ 1 & 0 & 1 \end{bmatrix}$; $E_{31}E_{13} = \begin{bmatrix} 2 & 0 & 1 \\ 0 & 1 & 0 \\ 1 & 0 & 1 \end{bmatrix}$. Test on the identity matrix!

12 $\begin{bmatrix} 9 & 8 & 7 \\ 6 & 5 & 4 \\ 3 & 2 & 1 \end{bmatrix}$, $\begin{bmatrix} 1 & 2 & 3 \\ 0 & 1 & -2 \\ 0 & 2 & -3 \end{bmatrix}$.

14 E_{21} has $\ell_{21} = -\frac{1}{2}$, E_{32} has $\ell_{32} = -\frac{2}{3}$, E_{43} has $\ell_{43} = -\frac{3}{4}$. Otherwise the E's match I.

18 $EF = \begin{bmatrix} 1 & 0 & 0 \\ a & 1 & 0 \\ b & c & 1 \end{bmatrix}$, $FE = \begin{bmatrix} 1 & 0 & 0 \\ a & 1 & 0 \\ b+ac & c & 1 \end{bmatrix}$, $E^2 = \begin{bmatrix} 1 & 0 & 0 \\ 2a & 1 & 0 \\ 2b & 0 & 1 \end{bmatrix}$, $F^3 = \begin{bmatrix} 1 & 0 & 0 \\ 0 & 1 & 0 \\ 0 & 3c & 1 \end{bmatrix}$.

22 (a) $\sum a_{3j}x_j$ (b) $a_{21} - a_{11}$ (c) $a_{21} - 2a_{11}$ (d) $(EA\boldsymbol{x})_1 = (A\boldsymbol{x})_1 = \sum a_{1j}x_j$.

25 The last equation becomes $0 = 3$. Change the original 6 to 3. Then row 1 + row 2 = row 3.

27 (a) No solution if $d = 0$ and $c \neq 0$ (b) Infinitely many solutions if $d = 0$ and $c = 0$. No effect from a and b.

28 $A = AI = A(BC) = (AB)C = IC = C$.

29 Given positive integers with $ad - bc = 1$. Certainly $c < a$ and $b < d$ would be impossible. Also $c > a$ and $b > d$ would be impossible with integers. This leaves row 1 < row 2 OR row 2 < row 1. An example is $M = \begin{bmatrix} 3 & 4 \\ 2 & 3 \end{bmatrix}$. Multiply by $\begin{bmatrix} 1 & -1 \\ 0 & 1 \end{bmatrix}$ to get $\begin{bmatrix} 1 & 1 \\ 2 & 3 \end{bmatrix}$, then multiply twice by $\begin{bmatrix} 1 & 0 \\ -1 & 1 \end{bmatrix}$ to get $\begin{bmatrix} 1 & 1 \\ 0 & 1 \end{bmatrix}$. This shows that $M = \begin{bmatrix} 1 & 1 \\ 0 & 1 \end{bmatrix}\begin{bmatrix} 1 & 0 \\ 1 & 1 \end{bmatrix}\begin{bmatrix} 1 & 0 \\ 1 & 1 \end{bmatrix}\begin{bmatrix} 1 & 1 \\ 0 & 1 \end{bmatrix}$.

30 $E = \begin{bmatrix} 1 & 0 & 0 & 0 \\ -1 & 1 & 0 & 0 \\ 0 & -1 & 1 & 0 \\ 0 & 0 & -1 & 1 \end{bmatrix}$. Eventually $M = \begin{bmatrix} 1 & 0 & 0 & 0 \\ -1 & 1 & 0 & 0 \\ 1 & -2 & 1 & 0 \\ -1 & 3 & -3 & 1 \end{bmatrix}$ = "inverse of Pascal" reduces Pascal to I.

Problem Set 2.4, page 65

2 (a) A (column 3 of B) (b) (Row 1 of A) B (c) (Row 3 of A)(column 4 of B) (d) (Row 1 of C)D(column 1 of E).

5 $A^n = \begin{bmatrix} 1 & bn \\ 0 & 1 \end{bmatrix}$ and $A^n = \begin{bmatrix} 2^n & 2^n \\ 0 & 0 \end{bmatrix}$.

7 (a) True (b) False (c) True (d) False.

9 $AF = \begin{bmatrix} a & a+b \\ c & c+d \end{bmatrix}$ and $E(AF)$ equals $(EA)F$ because matrix multiplication is *associative*.

11 (a) $B = 4I$ (b) $B = 0$ (c) $B = \begin{bmatrix} 0 & 0 & 1 \\ 0 & 1 & 0 \\ 1 & 0 & 0 \end{bmatrix}$ (d) Every row of B is $1, 0, 0$.

15 (a) mn (every entry) (b) mnp (c) n^3 (this is n^2 dot products).

17 (a) Use only column 2 of B (b) Use only row 2 of A (c)–(d) Use row 2 of first A.

19 Diagonal matrix, lower triangular, symmetric, all rows equal. Zero matrix.

20 (a) a_{11} (b) $\ell_{31} = a_{31}/a_{11}$ (c) $a_{32} - (\frac{a_{31}}{a_{11}})a_{12}$ (d) $a_{22} - (\frac{a_{21}}{a_{11}})a_{12}$.

23 $A = \begin{bmatrix} 0 & 1 \\ -1 & 0 \end{bmatrix}$ has $A^2 = -I$; $BC = \begin{bmatrix} 1 & -1 \\ 1 & -1 \end{bmatrix}\begin{bmatrix} 1 & 1 \\ 1 & 1 \end{bmatrix} = \begin{bmatrix} 0 & 0 \\ 0 & 0 \end{bmatrix}$;

$DE = \begin{bmatrix} 0 & 1 \\ 1 & 0 \end{bmatrix}\begin{bmatrix} 0 & 1 \\ -1 & 0 \end{bmatrix} = \begin{bmatrix} -1 & 0 \\ 0 & 1 \end{bmatrix} = -ED.$

25 $A_1^n = \begin{bmatrix} 2^n & 2^n - 1 \\ 0 & 1 \end{bmatrix}$, $A_2^n = 2^{n-1}\begin{bmatrix} 1 & 1 \\ 1 & 1 \end{bmatrix}$, $A_3^n = \begin{bmatrix} a^n & a^{n-1}b \\ 0 & 0 \end{bmatrix}$.

27 (a) (Row 3 of A)·(column 1 of B) and (Row 3 of A)·(column 2 of B) are both zero.

(b) $\begin{bmatrix} x \\ x \\ 0 \end{bmatrix}\begin{bmatrix} 0 & x & x \end{bmatrix} = \begin{bmatrix} 0 & x & x \\ 0 & x & x \\ 0 & 0 & 0 \end{bmatrix}$ and $\begin{bmatrix} x \\ x \\ x \end{bmatrix}\begin{bmatrix} 0 & 0 & x \end{bmatrix} = \begin{bmatrix} 0 & 0 & x \\ 0 & 0 & x \\ 0 & 0 & x \end{bmatrix}$: upper triangular!

28 A times B is $A\begin{bmatrix} | & | & | & | \end{bmatrix}$, $[\text{——}]B$, $[\text{——}]\begin{bmatrix} | & | & | & | \end{bmatrix}$, $\begin{bmatrix} | & | & | \end{bmatrix}\begin{bmatrix} \text{——} \\ \text{——} \\ \text{——} \end{bmatrix}$

31 In Problem 30, $\boldsymbol{c} = \begin{bmatrix} -2 \\ 8 \end{bmatrix}$, $D = \begin{bmatrix} 0 & 1 \\ 5 & 3 \end{bmatrix}$, $D - \boldsymbol{cb}/a = \begin{bmatrix} 1 & 1 \\ 1 & 3 \end{bmatrix}$ in lower corner of EA.

33 A times $X = [\,\boldsymbol{x}_1 \;\; \boldsymbol{x}_2 \;\; \boldsymbol{x}_3\,]$ will be the identity matrix $I = [\,A\boldsymbol{x}_1 \;\; A\boldsymbol{x}_2 \;\; A\boldsymbol{x}_3\,]$.

34 The solution for $\boldsymbol{b} = \begin{bmatrix} 3 \\ 5 \\ 8 \end{bmatrix}$ is $\boldsymbol{x} = 3\boldsymbol{x}_1 + 5\boldsymbol{x}_2 + 8\boldsymbol{x}_3 = \begin{bmatrix} 3 \\ 8 \\ 16 \end{bmatrix}$; $A = \begin{bmatrix} 1 & 0 & 0 \\ -1 & 1 & 0 \\ 0 & -1 & 1 \end{bmatrix}$ will produce those $\boldsymbol{x}_1 = (1, 1, 1)$, $\boldsymbol{x}_2 = (0, 1, 1)$, $\boldsymbol{x}_3 = (0, 0, 1)$ as columns of its "inverse".

37 $A = \begin{bmatrix} 0 & 1 & 0 & 0 & 1 \\ 1 & 0 & 1 & 0 & 0 \\ 0 & 1 & 0 & 1 & 0 \\ 0 & 0 & 1 & 0 & 1 \\ 1 & 0 & 0 & 1 & 0 \end{bmatrix}$, $A^2 = \begin{bmatrix} 2 & 0 & 1 & 1 & 0 \\ 0 & 2 & 0 & 1 & 1 \\ 1 & 0 & 2 & 0 & 1 \\ 1 & 1 & 0 & 2 & 0 \\ 0 & 1 & 1 & 0 & 2 \end{bmatrix}$, $A^3 = \begin{bmatrix} 0 & 3 & 1 & 1 & 3 \\ 3 & 0 & 3 & 1 & 1 \\ 1 & 3 & 0 & 3 & 1 \\ 1 & 1 & 3 & 0 & 3 \\ 3 & 1 & 1 & 3 & 0 \end{bmatrix}$, $A^3 + A^2$ no zeros so diameter 3

39 If A is "northwest" and B is "southeast", AB is upper triangular and BA is lower triangular. Row i of A ends with $i-1$ zeros. Column j of B starts with $n-j$ zeros. If $i > j$ then (row i of A) $\cdot$ (column j of B) $= 0$. So AB is upper triangular. Similarly BA is lower triangular. Problem 2.7.40 asks about inverses and transposes and permutations of a northwest A and a southeast B.

Problem Set 2.5, page 78

1 $A^{-1} = \begin{bmatrix} 0 & \frac{1}{4} \\ \frac{1}{3} & 0 \end{bmatrix}$, $B^{-1} = \begin{bmatrix} \frac{1}{2} & 0 \\ -1 & \frac{1}{2} \end{bmatrix}$, $C^{-1} = \begin{bmatrix} 7 & -4 \\ -5 & 3 \end{bmatrix}$.

7 (a) In $Ax = (1,0,0)$, equation 1 + equation 2 − equation 3 is $0 = 1$ (b) The right sides must satisfy $b_1 + b_2 = b_3$ (c) Row 3 becomes a row of zeros—no third pivot.

8 (a) The vector $x = (1,1,-1)$ solves $Ax = \mathbf{0}$ (b) Elimination keeps columns $1 + 2 =$ column 3. When columns 1 and 2 end in zeros so does column 3: no third pivot.

12 $C = AB$ gives $C^{-1} = B^{-1}A^{-1}$ so $A^{-1} = BC^{-1}$.

14 $B^{-1} = A^{-1}\begin{bmatrix} 1 & 0 \\ 1 & 1 \end{bmatrix}^{-1} = A^{-1}\begin{bmatrix} 1 & 0 \\ -1 & 1 \end{bmatrix}$: subtract column 2 of A^{-1} from column 1.

16 $\begin{bmatrix} a & b \\ c & d \end{bmatrix}\begin{bmatrix} d & -b \\ -c & a \end{bmatrix} = \begin{bmatrix} ad-bc & 0 \\ 0 & ad-bc \end{bmatrix} = (ad-bc)I$. The inverse of one matrix is the other divided by $ad - bc$.

18 $A^2B = I$ can be written as $A(AB) = I$. Therefore A^{-1} is AB.

21 6 of the 16 are invertible, including all four with three 1's.

22 $\begin{bmatrix} 1 & 3 & 1 & 0 \\ 2 & 7 & 0 & 1 \end{bmatrix} \to \begin{bmatrix} 1 & 3 & 1 & 0 \\ 0 & 1 & -2 & 1 \end{bmatrix} \to \begin{bmatrix} 1 & 0 & 7 & -3 \\ 0 & 1 & -2 & 1 \end{bmatrix} = \begin{bmatrix} I & A^{-1} \end{bmatrix}$;

$\begin{bmatrix} 1 & 3 & 1 & 0 \\ 3 & 8 & 0 & 1 \end{bmatrix} \to \begin{bmatrix} 1 & 0 & -8 & 3 \\ 0 & 1 & 3 & -1 \end{bmatrix} = \begin{bmatrix} I & A^{-1} \end{bmatrix}$.

24 $\begin{bmatrix} 1 & a & b & 1 & 0 & 0 \\ 0 & 1 & c & 0 & 1 & 0 \\ 0 & 0 & 1 & 0 & 0 & 1 \end{bmatrix} \to \begin{bmatrix} 1 & a & 0 & 1 & 0 & -b \\ 0 & 1 & 0 & 0 & 1 & -c \\ 0 & 0 & 1 & 0 & 0 & 1 \end{bmatrix} \to \begin{bmatrix} 1 & 0 & 0 & 1 & -a & ac-b \\ 0 & 1 & 0 & 0 & 1 & -c \\ 0 & 0 & 1 & 0 & 0 & 1 \end{bmatrix}$.

27 $A^{-1} = \begin{bmatrix} 1 & 0 & 0 \\ -2 & 1 & -3 \\ 0 & 0 & 1 \end{bmatrix}$ (notice the pattern); $A^{-1} = \begin{bmatrix} 2 & -1 & 0 \\ -1 & 2 & -1 \\ 0 & -1 & 1 \end{bmatrix}$.

31 Elimination produces the pivots a and $a-b$ and $a-b$. $A^{-1} = \dfrac{1}{a(a-b)}\begin{bmatrix} a & 0 & -b \\ -a & a & 0 \\ 0 & -a & a \end{bmatrix}$.

34 $x = (1, 1, \ldots, 1)$ has $Px = Qx$ so $(P - Q)x = \mathbf{0}$.

35 $\begin{bmatrix} I & 0 \\ -C & I \end{bmatrix}$ and $\begin{bmatrix} A^{-1} & 0 \\ -D^{-1}CA^{-1} & D^{-1} \end{bmatrix}$ and $\begin{bmatrix} -D & I \\ I & 0 \end{bmatrix}$.

37 A can be invertible but B is always singular. Each row of B will add to zero, from $0+1+2-3$, so the vector $\boldsymbol{x}=(1,1,1,1)$ will give $B\boldsymbol{x}=\mathbf{0}$. I thought A would be invertible as long as you put the 3's on its main diagonal, but that's wrong:

$$A\boldsymbol{x}=\begin{bmatrix} 3 & 0 & 1 & 2 \\ 0 & 3 & 1 & 2 \\ 1 & 2 & 3 & 0 \\ 1 & 2 & 0 & 3 \end{bmatrix}\begin{bmatrix} 1 \\ 1 \\ -1 \\ -1 \end{bmatrix}=\mathbf{0} \quad \text{but} \quad A=\begin{bmatrix} 0 & 1 & 2 & 3 \\ 3 & 0 & 1 & 2 \\ 2 & 3 & 0 & 1 \\ 1 & 2 & 3 & 0 \end{bmatrix} \quad \text{is invertible}$$

40 The three Pascal matrices have $S=LU=LL^{\mathrm{T}}$ and then $\mathsf{inv}(S)=\mathsf{inv}(L^{\mathrm{T}})\mathsf{inv}(L)$. Note that the triangular L is $\mathsf{abs}(\mathsf{pascal}(n,1))$ in MATLAB.

42 If $AC=I$ for square matrices then $C=A^{-1}$ (it is proved in **2I** that $CA=I$ will also be true). The same will be true for C^*. But a square matrix has only one inverse so $C=C^*$.

43
$$\begin{aligned} MM^{-1} &= (I_n-UV)\,(I_n+U(I_m-VU)^{-1}V) \\ &= I_n-UV+U(I_m-VU)^{-1}V-UVU(I_m-VU)^{-1}V \\ &= I_n-UV+U(I_m-VU)(I_m-VU)^{-1}V=I_n \quad \text{(formulas } \mathbf{1,2,4} \text{ are similar)} \end{aligned}$$

Problem Set 2.6, page 91

2 $\ell_{31}=1$ and $\ell_{32}=2$ (and $\ell_{33}=1$): reverse the steps to recover $x+3y+6z=11$ from $U\boldsymbol{x}=\boldsymbol{c}$:

1 times $(x+y+z=5)+2$ times $(y+2z=2)+1$ times $(z=2)$ gives $x+3y+6z=11$.

4 $Lc=\begin{bmatrix} 1 & & \\ 1 & 1 & \\ 1 & 2 & 1 \end{bmatrix}\begin{bmatrix} \\ \boldsymbol{c} \\ \\ \end{bmatrix}=\begin{bmatrix} 5 \\ 7 \\ 11 \end{bmatrix}$; $\boldsymbol{c}=\begin{bmatrix} 5 \\ 2 \\ 2 \end{bmatrix}$. $U\boldsymbol{x}=\begin{bmatrix} 1 & 1 & 1 \\ & 1 & 2 \\ & & 1 \end{bmatrix}\begin{bmatrix} \\ \boldsymbol{x} \\ \\ \end{bmatrix}=\begin{bmatrix} 5 \\ 2 \\ 2 \end{bmatrix}$; $\boldsymbol{x}=\begin{bmatrix} 5 \\ -2 \\ 2 \end{bmatrix}$.

6 $\begin{bmatrix} 1 & & \\ 0 & 1 & \\ 0 & -2 & 1 \end{bmatrix}\begin{bmatrix} 1 & & \\ -2 & 1 & \\ 0 & 0 & 1 \end{bmatrix}A=\begin{bmatrix} 1 & 1 & 1 \\ 0 & 2 & 3 \\ 0 & 0 & -6 \end{bmatrix}=U$. Then $A=\begin{bmatrix} 1 & 0 & 0 \\ 2 & 1 & 0 \\ 0 & 2 & 1 \end{bmatrix}$

$U=E_{21}^{-1}E_{32}^{-1}U=LU$.

10 $c=2$ leads to zero in the second pivot position: exchange rows and the matrix will be OK. $c=1$ leads to zero in the third pivot position. In this case the matrix is *singular*.

12 $A=\begin{bmatrix} 2 & 4 \\ 4 & 11 \end{bmatrix}=\begin{bmatrix} 1 & 0 \\ 2 & 1 \end{bmatrix}\begin{bmatrix} 2 & 4 \\ 0 & 3 \end{bmatrix}=\begin{bmatrix} 1 & 0 \\ 2 & 1 \end{bmatrix}\begin{bmatrix} 2 & 0 \\ 0 & 3 \end{bmatrix}\begin{bmatrix} 1 & 2 \\ 0 & 1 \end{bmatrix}=LDU$; notice U is L^{T}

$$A=\begin{bmatrix} 1 & & \\ 4 & 1 & \\ 0 & -1 & 1 \end{bmatrix}\begin{bmatrix} 1 & 4 & 0 \\ 0 & -4 & 4 \\ 0 & 0 & 4 \end{bmatrix}=\begin{bmatrix} 1 & & \\ 4 & 1 & \\ 0 & -1 & 1 \end{bmatrix}\begin{bmatrix} 1 & & \\ & -4 & \\ & & 4 \end{bmatrix}\begin{bmatrix} 1 & 4 & 0 \\ 0 & 1 & -1 \\ 0 & 0 & 1 \end{bmatrix}$$

$=LDL^{\mathrm{T}}$.

14 $\begin{bmatrix} a & r & r & r \\ a & b & s & s \\ a & b & c & t \\ a & b & c & d \end{bmatrix} = \begin{bmatrix} 1 & & & \\ 1 & 1 & & \\ 1 & 1 & 1 & \\ 1 & 1 & 1 & 1 \end{bmatrix} \begin{bmatrix} a & r & r & r \\ & b-r & s-r & s-r \\ & & c-s & t-s \\ & & & d-t \end{bmatrix}$. Need $\begin{matrix} a \neq 0 \\ b \neq r \\ c \neq s \\ d \neq t \end{matrix}$

15 $\begin{bmatrix} 1 & 0 \\ 4 & 1 \end{bmatrix} \boldsymbol{c} = \begin{bmatrix} 2 \\ 11 \end{bmatrix}$ gives $\boldsymbol{c} = \begin{bmatrix} 2 \\ 3 \end{bmatrix}$. Then $\begin{bmatrix} 2 & 4 \\ 0 & 1 \end{bmatrix} \boldsymbol{x} = \begin{bmatrix} 2 \\ 3 \end{bmatrix}$ gives $\boldsymbol{x} = \begin{bmatrix} -5 \\ 3 \end{bmatrix}$.

Check that $A = LU = \begin{bmatrix} 2 & 4 \\ 8 & 17 \end{bmatrix}$ times $\boldsymbol{x}$ is $\boldsymbol{b} = \begin{bmatrix} 2 \\ 11 \end{bmatrix}$.

18 (a) Multiply $LDU = L_1 D_1 U_1$ by inverses to get $L_1^{-1} L D = D_1 U_1 U^{-1}$. The left side is lower triangular, the right side is upper triangular $\Rightarrow$ both sides are diagonal.

(b) Since L, U, L_1, U_1 have diagonals of 1's we get $D = D_1$. Then $L_1^{-1} L$ is I and $U_1 U^{-1}$ is I.

20 A tridiagonal T has 2 nonzeros in the pivot row and only one nonzero below the pivot (so 1 operation to find the multiplier and 1 to find the new pivot!). $T =$ bidiagonal L times U:

$$T = \begin{bmatrix} 1 & 2 & 0 & 0 \\ 2 & 3 & 1 & 0 \\ 0 & 1 & 2 & 3 \\ 0 & 0 & 3 & 4 \end{bmatrix} \longrightarrow U = \begin{bmatrix} 1 & 2 & 0 & 0 \\ 0 & -1 & 1 & 0 \\ 0 & 0 & 3 & 3 \\ 0 & 0 & 0 & 1 \end{bmatrix}. \text{ Reverse steps by } L = \begin{bmatrix} 1 & 0 & 0 & 0 \\ 2 & 1 & 0 & 0 \\ 0 & -1 & 1 & 0 \\ 0 & 0 & 1 & 1 \end{bmatrix}.$$

22 $\begin{bmatrix} x & x & x \\ x & x & x \\ x & x & x \end{bmatrix} = \begin{bmatrix} 1 & 0 & 0 \\ * & 1 & 0 \\ * & & 1 \end{bmatrix} \begin{bmatrix} * & * & * \\ 0 & & \\ 0 & & \end{bmatrix}$ ($*$'s are all known after the first pivot is used).

25 The 2 by 2 upper submatrix B has the first two pivots $2, 7$. Reason: Elimination on A starts in the upper left corner with elimination on B.

27 $\begin{bmatrix} 1 & 1 & 1 & 1 & 1 \\ 1 & 2 & 3 & 4 & 5 \\ 1 & 3 & 6 & 10 & 15 \\ 1 & 4 & 10 & 20 & 35 \\ 1 & 5 & 15 & 35 & 70 \end{bmatrix} = \begin{bmatrix} 1 & & & & \\ 1 & 1 & & & \\ 1 & 2 & 1 & & \\ 1 & 3 & 3 & 1 & \\ 1 & 4 & 6 & 4 & 1 \end{bmatrix} \begin{bmatrix} 1 & 1 & 1 & 1 & 1 \\ & 1 & 2 & 3 & 4 \\ & & 1 & 3 & 6 \\ & & & 1 & 4 \\ & & & & 1 \end{bmatrix}$. *Pascal's triangle* in L and U. MATLAB's lu code will wreck the pattern. chol does no row exchanges for symmetric matrices with positive pivots.

32 inv(A) $* \boldsymbol{b}$ should take 3 times as long as $A \backslash \boldsymbol{b}$ (n^3 for A^{-1} *vs* $n^3/3$ multiplications for LU).

36 This L comes from the $-1, 2, -1$ tridiagonal $A = LDL^{\mathrm{T}}$. (Row i of L) $\cdot$ (Column j of L^{-1}) $= \left(\frac{1-i}{i}\right)\left(\frac{j}{i-1}\right) + (1)\left(\frac{j}{i}\right) = 0$ for $i > j$ so $LL^{-1} = I$. Then L^{-1} leads to $A^{-1} = (L^{-1})^{\mathrm{T}} D^{-1} L^{-1}$. ***The*** $-1, 2, -1$ ***matrix has inverse*** $A_{ij}^{-1} = j(n-i+1)/(n+1)$ for $i \geq j$ (reverse for $i \leq j$).

Problem Set 2.7, page 104

2 $(AB)^{\mathrm{T}}$ is not $A^{\mathrm{T}}B^{\mathrm{T}}$ *except when* $AB = BA$. In that case transpose to find: $B^{\mathrm{T}}A^{\mathrm{T}} = A^{\mathrm{T}}B^{\mathrm{T}}$.

4 $A = \begin{bmatrix} 0 & 1 \\ 0 & 0 \end{bmatrix}$ has $A^2 = 0$. But the diagonal entries of $A^{\mathrm{T}}A$ are dot products of columns of A with themselves. If $A^{\mathrm{T}}A = 0$, zero dot products $\Rightarrow$ zero columns $\Rightarrow A =$ zero matrix.

6 $M^{\mathrm{T}} = \begin{bmatrix} A^{\mathrm{T}} & C^{\mathrm{T}} \\ B^{\mathrm{T}} & D^{\mathrm{T}} \end{bmatrix}$; $M^{\mathrm{T}} = M$ needs $A^{\mathrm{T}} = A$, $B^{\mathrm{T}} = C$, $D^{\mathrm{T}} = D$.

8 The 1 in row 1 has n choices; then the 1 in row 2 has $n-1$ choices ... ($n!$ choices overall).

10 $(3, 1, 2, 4)$ and $(2, 3, 1, 4)$ keep only 4 in position; 6 more even P's keep 1 or 2 or 3 in position; $(2, 1, 4, 3)$ and $(3, 4, 1, 2)$ exchange 2 pairs. Then $(1, 2, 3, 4)$ and $(4, 3, 2, 1)$ make 12 even P's.

14 There are $n!$ permutation matrices of order n. Eventually two powers of P *must be the same*: If $P^r = P^s$ then $P^{r-s} = I$. Certainly $r - s \le n!$

$$P = \begin{bmatrix} P_2 & \\ & P_3 \end{bmatrix} \text{ is 5 by 5 with } P_2 = \begin{bmatrix} 0 & 1 \\ 1 & 0 \end{bmatrix} \text{ and } P_3 = \begin{bmatrix} 0 & 1 & 0 \\ 0 & 0 & 1 \\ 1 & 0 & 0 \end{bmatrix} \text{ and } P^6 = I.$$

18 (a) $5+4+3+2+1 = 15$ independent entries if $A = A^{\mathrm{T}}$ (b) L has 10 and D has 5: total 15 in LDL^{T} (c) Zero diagonal if $A^{\mathrm{T}} = -A$, leaving $4+3+2+1 = 10$ choices.

20 $$\begin{bmatrix} 1 & 3 \\ 3 & 2 \end{bmatrix} = \begin{bmatrix} 1 & 0 \\ 3 & 1 \end{bmatrix}\begin{bmatrix} 1 & 0 \\ 0 & -7 \end{bmatrix}\begin{bmatrix} 1 & 3 \\ 0 & 1 \end{bmatrix}; \quad \begin{bmatrix} 1 & b \\ b & c \end{bmatrix} = \begin{bmatrix} 1 & 0 \\ b & 1 \end{bmatrix}\begin{bmatrix} 1 & 0 \\ 0 & c-b^2 \end{bmatrix}\begin{bmatrix} 1 & b \\ 0 & 1 \end{bmatrix} = LDL^{\mathrm{T}}.$$

$$\begin{bmatrix} 2 & -1 & 0 \\ -1 & 2 & -1 \\ 0 & -1 & 2 \end{bmatrix} = \begin{bmatrix} 1 & & \\ -\frac{1}{2} & 1 & \\ 0 & -\frac{2}{3} & 1 \end{bmatrix}\begin{bmatrix} 2 & & \\ & \frac{3}{2} & \\ & & \frac{4}{3} \end{bmatrix}\begin{bmatrix} 1 & -\frac{1}{2} & 0 \\ & 1 & -\frac{2}{3} \\ & & 1 \end{bmatrix}.$$

22 $$\begin{bmatrix} 0 & 1 & \\ 1 & 0 & \\ & & 1 \end{bmatrix} A = \begin{bmatrix} 1 & & \\ 0 & 1 & \\ 2 & 3 & 1 \end{bmatrix}\begin{bmatrix} 1 & 0 & 1 \\ & 1 & 1 \\ & & -1 \end{bmatrix}; \quad \begin{bmatrix} 1 & & \\ & 0 & 1 \\ & 1 & 0 \end{bmatrix} A = \begin{bmatrix} 1 & & \\ 1 & 1 & \\ 2 & 0 & 1 \end{bmatrix}\begin{bmatrix} 1 & 2 & 0 \\ & -1 & 1 \\ & & 1 \end{bmatrix}$$

24 $PA = LU$ is $\begin{bmatrix} & & 1 \\ & 1 & \\ 1 & & \end{bmatrix}\begin{bmatrix} 0 & 1 & 2 \\ 0 & 3 & 8 \\ 2 & 1 & 1 \end{bmatrix} = \begin{bmatrix} 1 & & \\ 0 & 1 & \\ 0 & 1/3 & 1 \end{bmatrix}\begin{bmatrix} 2 & 1 & 1 \\ & 3 & 8 \\ & & -2/3 \end{bmatrix}$. If we wait to exchange and use a_{12} as pivot then $A = L_1P_1U_1 = \begin{bmatrix} 1 & & \\ 3 & 1 & \\ & & 1 \end{bmatrix}\begin{bmatrix} & & 1 \\ 1 & & \\ & 1 & \end{bmatrix}\begin{bmatrix} 2 & 1 & 1 \\ 0 & 1 & 2 \\ 0 & 0 & 2 \end{bmatrix}$.

29 One way to decide even vs. odd is to count all pairs that P has in the wrong order. Then P is even or odd when that count is even or odd. Hard step: show that an exchange always reverses that count! Then 3 or 5 exchanges will leave that count odd.

32 Inputs $\begin{bmatrix} 1 & 50 \\ 40 & 1000 \\ 2 & 50 \end{bmatrix}\begin{bmatrix} x_1 \\ x_2 \end{bmatrix} = A\boldsymbol{x}$; $A^{\mathrm{T}}\boldsymbol{y} = \begin{bmatrix} 1 & 40 & 2 \\ 50 & 1000 & 50 \end{bmatrix}\begin{bmatrix} 700 \\ 3 \\ 3000 \end{bmatrix} = \begin{bmatrix} 6820 \\ 188000 \end{bmatrix}$ 1 truck 1 plane

33 $A\boldsymbol{x} \cdot \boldsymbol{y}$ is the *cost* of inputs while $\boldsymbol{x} \cdot A^{\mathrm{T}}\boldsymbol{y}$ is the *value* of outputs.

34 $P^3 = I$ so three rotations for 360°; P rotates around $(1, 1, 1)$ by 120°.

37 These are groups: Lower triangular with diagonal 1's, diagonal invertible D, permutations P, orthogonal matrices with $Q^{\mathrm{T}} = Q^{-1}$.

40 Certainly B^{T} is northwest. B^2 is a full matrix! B^{-1} is southeast: $\begin{bmatrix} 1 & 1 \\ 1 & 0 \end{bmatrix}^{-1} = \begin{bmatrix} 0 & 1 \\ 1 & -1 \end{bmatrix}$. The rows of B are in reverse order from a lower triangular L, so $B = PL$. Then $B^{-1} = L^{-1}P^{-1}$ has the *columns* in reverse order from L^{-1}. So B^{-1} is *southeast*. Northwest times southeast is upper triangular! $B = PL$ and $C = PU$ give $BC = (PLP)U =$ upper times upper.

41 The i, j entry of PAP is the $n - i + 1, n - j + 1$ entry of A. The main diagonal reverses order.

Problem Set 3.1, page 118

1 $\boldsymbol{x} + \boldsymbol{y} \neq \boldsymbol{y} + \boldsymbol{x}$ and $\boldsymbol{x} + (\boldsymbol{y} + \boldsymbol{z}) \neq (\boldsymbol{x} + \boldsymbol{y}) + \boldsymbol{z}$ and $(c_1 + c_2)\boldsymbol{x} \neq c_1\boldsymbol{x} + c_2\boldsymbol{x}$.

3 (a) $c\boldsymbol{x}$ may not be in our set: not closed under scalar multiplication. Also no $\mathbf{0}$ and no $-\boldsymbol{x}$

(b) $c(\boldsymbol{x} + \boldsymbol{y})$ is the usual $(xy)^c$, while $c\boldsymbol{x} + c\boldsymbol{y}$ is the usual $(x^c)(y^c)$. Those are equal. With $c = 3, x = 2, y = 1$ they equal 8. This is $3(\mathbf{2} + \mathbf{1})$!! The zero vector is the number 1.

5 (a) One possibility: The matrices cA form a subspace not containing B (b) Yes: the subspace must contain $A - B = I$ (c) All matrices whose main diagonal is all zero.

9 (a) The vectors with integer components allow addition, but not multiplication by $\frac{1}{2}$

(b) Remove the x axis from the xy plane (but leave the origin). Multiplication by any c is allowed but not all vector additions.

11 (a) All matrices $\begin{bmatrix} a & b \\ 0 & 0 \end{bmatrix}$ (b) All matrices $\begin{bmatrix} a & a \\ 0 & 0 \end{bmatrix}$ (c) All diagonal matrices.

15 (a) Two planes through $(0, 0, 0)$ probably intersect in a line through $(0, 0, 0)$ (b) The plane and line probably intersect in the point $(0, 0, 0)$ (c) Suppose $\boldsymbol{x}$ is in $\boldsymbol{S} \cap \boldsymbol{T}$ and $\boldsymbol{y}$ is in $\boldsymbol{S} \cap \boldsymbol{T}$. Both vectors are in both subspaces, so $\boldsymbol{x} + \boldsymbol{y}$ and $c\boldsymbol{x}$ are in both subspaces.

20 (a) Solution only if $b_2 = 2b_1$ and $b_3 = -b_1$ (b) Solution only if $b_3 = -b_1$.

23 The extra column $\boldsymbol{b}$ enlarges the column space unless $\boldsymbol{b}$ is *already in* the column space of A: $[A \;\; \boldsymbol{b}] = \begin{bmatrix} 1 & 0 & 1 \\ 0 & 0 & 1 \end{bmatrix}$ (larger column space) (no solution to $A\boldsymbol{x} = \boldsymbol{b}$) $\begin{bmatrix} 1 & 0 & 1 \\ 0 & 1 & 1 \end{bmatrix}$ ($\boldsymbol{b}$ already in column space) ($A\boldsymbol{x} = \boldsymbol{b}$ has a solution)

25 The solution to $A\boldsymbol{z} = \boldsymbol{b} + \boldsymbol{b}^*$ is $\boldsymbol{z} = \boldsymbol{x} + \boldsymbol{y}$. If $\boldsymbol{b}$ and $\boldsymbol{b}^*$ are in the column space so is $\boldsymbol{b} + \boldsymbol{b}^*$.

Problem Set 3.2, page 130

2 (a) Free variables x_2, x_4, x_5 and solutions $(-2,1,0,0,0)$, $(0,0,-2,1,0)$, $(0,0,-3,0,1)$

(b) Free variable x_3: solution $(1,-1,1)$.

4 $R = \begin{bmatrix} 1 & 2 & 0 & 0 & 0 \\ 0 & 0 & 1 & 2 & 3 \\ 0 & 0 & 0 & 0 & 0 \end{bmatrix}$, $R = \begin{bmatrix} 1 & 0 & -1 \\ 0 & 1 & 1 \\ 0 & 0 & 0 \end{bmatrix}$, R has the same nullspace as U and A.

6 (a) Special solutions $(3,1,0)$ and $(5,0,1)$ (b) $(3,1,0)$. Total count of pivot and free is n.

8 $R = \begin{bmatrix} 1 & -3 & -5 \\ 0 & 0 & 0 \end{bmatrix}$ with $I = [\,1\,]$; $R = \begin{bmatrix} 1 & -3 & 0 \\ 0 & 0 & 1 \end{bmatrix}$ with $I = \begin{bmatrix} 1 & 0 \\ 0 & 1 \end{bmatrix}$.

10 (a) Impossible above diagonal (b) $A = \text{invertible} = \begin{bmatrix} 1 & 1 & 1 \\ 1 & 2 & 1 \\ 1 & 1 & 2 \end{bmatrix}$ (c) $A = \begin{bmatrix} 1 & 1 & 1 \\ 1 & 1 & 1 \\ 1 & 1 & 1 \end{bmatrix}$

(d) $A = 2I, U = 2I, R = I$.

14 If column 1 = column 5 then x_5 is a free variable. Its special solution is $(-1,0,0,0,1)$.

16 The nullspace contains only $\boldsymbol{x} = \mathbf{0}$ when A has 5 pivots. Also the column space is $\mathbf{R}^5$, because we can solve $A\boldsymbol{x} = \boldsymbol{b}$ and every $\boldsymbol{b}$ is in the column space.

20 Column 5 is sure to have no pivot since it is a combination of earlier columns. With 4 pivots in the other columns, the special solution is $\boldsymbol{s} = (1,0,1,0,1)$. The nullspace contains all multiples of $\boldsymbol{s}$ (a line in $\mathbf{R}^5$).

24 This construction is impossible: 2 pivot columns, 2 free variables, only 3 columns.

26 $A = \begin{bmatrix} 0 & 1 \\ 0 & 0 \end{bmatrix}$.

30 $A = \begin{bmatrix} 0 & 1 \\ 0 & 0 \end{bmatrix}$ shows that (a)(b)(c) are all false. Notice $\text{rref}(A^{\mathrm{T}}) = \begin{bmatrix} 1 & 0 \\ 0 & 0 \end{bmatrix}$.

32 Any zero rows come after these rows: $R = [\,1 \;\; -2 \;\; -3\,]$, $R = \begin{bmatrix} 1 & 0 & 0 \\ 0 & 1 & 0 \end{bmatrix}$, $R = I$.

33 (a) $\begin{bmatrix} 1 & 0 \\ 0 & 1 \end{bmatrix}, \begin{bmatrix} 1 & 0 \\ 0 & 0 \end{bmatrix}, \begin{bmatrix} 1 & 1 \\ 0 & 0 \end{bmatrix}, \begin{bmatrix} 0 & 1 \\ 0 & 0 \end{bmatrix}, \begin{bmatrix} 0 & 0 \\ 0 & 0 \end{bmatrix}$ (b) All 8 matrices are R's!

Problem Set 3.3, page 141

1 (a) and (c) are correct; (d) is false because R might happen to have 1's in nonpivot columns.

3 $R_A = \begin{bmatrix} 1 & 2 & 0 \\ 0 & 0 & 1 \\ 0 & 0 & 0 \end{bmatrix}$ $\quad R_B = \begin{bmatrix} R_A & R_A \end{bmatrix}$ $\quad R_C \longrightarrow \begin{bmatrix} R_A & 0 \\ 0 & R_A \end{bmatrix} \longrightarrow$ Zero row in the upper R moves all the way to the bottom.

5 I think this is true.

7 Special solutions are columns of $N = [\,-2\ \ -1\ \ 1\ \ 0;\ \ -3\ \ -5\ \ 0\ \ 1\,]$ and $[\,1\ \ 0\ \ 0;\ \ 0\ \ -2\ \ 1\,]$.

13 P has rank r (the same as A) because elimination produces the same pivot columns.

14 The rank of R^{T} is also r, and the example matrix A has rank 2:

$$P = \begin{bmatrix} 1 & 3 \\ 2 & 6 \\ 2 & 7 \end{bmatrix} \qquad P^{\mathrm{T}} = \begin{bmatrix} 1 & 2 & 2 \\ 3 & 6 & 7 \end{bmatrix} \qquad S^{\mathrm{T}} = \begin{bmatrix} 1 & 2 \\ 3 & 7 \end{bmatrix} \qquad S = \begin{bmatrix} 1 & 3 \\ 2 & 7 \end{bmatrix}.$$

16 $(\boldsymbol{u}\boldsymbol{v}^{\mathrm{T}})(\boldsymbol{w}\boldsymbol{z}^{\mathrm{T}}) = \boldsymbol{u}(\boldsymbol{v}^{\mathrm{T}}\boldsymbol{w})\boldsymbol{z}^{\mathrm{T}}$ has rank one unless $\boldsymbol{v}^{\mathrm{T}}\boldsymbol{w} = 0$.

18 If we know that $\operatorname{rank}(B^{\mathrm{T}}A^{\mathrm{T}}) \le \operatorname{rank}(A^{\mathrm{T}})$, then since rank stays the same for transposes, we have $\operatorname{rank}(AB) \le \operatorname{rank}(A)$.

20 Certainly A and B have at most rank 2. Then their product AB has at most rank 2. Since BA is 3 by 3, it cannot be I even if $AB = I$.

21 (a) A and B will both have the same nullspace and row space as R (same R for both matrices). (b) A equals an *invertible* matrix times B, when they share the same R. A key fact!

23 $$A = (\text{pivot columns})(\text{nonzero rows of } R) = \begin{bmatrix} 1 & 0 \\ 1 & 4 \\ 1 & 8 \end{bmatrix} \begin{bmatrix} 1 & 1 & 0 \\ 0 & 0 & 1 \end{bmatrix} = \begin{bmatrix} 1 & 1 & 0 \\ 1 & 1 & 0 \\ 1 & 1 & 0 \end{bmatrix} + \begin{bmatrix} 0 & 0 & 0 \\ 0 & 0 & 4 \\ 0 & 0 & 8 \end{bmatrix}.$$

24 The m by n matrix Z has r ones at the start of its main diagonal. Otherwise Z is all zeros.

Problem Set 3.4, page 152

2 $$\begin{bmatrix} 2 & 1 & 3 & b_1 \\ 6 & 3 & 9 & b_2 \\ 4 & 2 & 6 & b_3 \end{bmatrix} \to \begin{bmatrix} 2 & 1 & 3 & b_1 \\ 0 & 0 & 0 & b_2 - 3b_1 \\ 0 & 0 & 0 & b_3 - 2b_1 \end{bmatrix} \quad \text{Then } [\,R \ \ \boldsymbol{d}\,] = \begin{bmatrix} 1 & 1/2 & 3/2 & \mathbf{5} \\ 0 & 0 & 0 & \mathbf{0} \\ 0 & 0 & 0 & \mathbf{0} \end{bmatrix}$$

$A\boldsymbol{x} = \boldsymbol{b}$ has a solution when $b_2 - 3b_1 = 0$ and $b_3 - 2b_1 = 0$; the column space is the line through $(2, 6, 4)$ which is the intersection of the planes $b_2 - 3b_1 = 0$ and $b_3 - 2b_1 = 0$; the nullspace contains all combinations of $\boldsymbol{s}_1 = (-1/2, 1, 0)$ and $\boldsymbol{s}_2 = (-3/2, 0, 1)$; particular solution $\boldsymbol{x}_p = \boldsymbol{d} = (5, 0, 0)$ and complete solution $\boldsymbol{x}_p + c_1\boldsymbol{s}_1 + c_2\boldsymbol{s}_2$.

4 $\boldsymbol{x}_{complete} = (\frac{1}{2}, 0, \frac{1}{2}, 0) + x_2(-3, 1, 0, 0) + x_4(0, 0, -2, 1)$.

6 (a) Solvable if $b_2 = 2b_1$ and $3b_1 - 3b_3 + b_4 = 0$. Then $\boldsymbol{x} = \begin{bmatrix} 5b_1 - 2b_3 \\ b_3 - 2b_1 \end{bmatrix}$ (no free variables)

(b) Solvable if $b_2 = 2b_1$ and $3b_1 - 3b_3 + b_4 = 0$. Then $\boldsymbol{x} = \begin{bmatrix} 5b_1 - 2b_3 \\ b_3 - 2b_1 \\ 0 \end{bmatrix} + x_3 \begin{bmatrix} -1 \\ -1 \\ 1 \end{bmatrix}$.

8 (a) Every $\boldsymbol{b}$ is in the column space: *independent rows.* (b) Need $b_3 = 2b_2$. Row 3 − 2 row 2 = 0.

12 (a) $\boldsymbol{x}_1 - \boldsymbol{x}_2$ and $\boldsymbol{0}$ solve $A\boldsymbol{x} = \boldsymbol{0}$ (b) $2\boldsymbol{x}_1 - 2\boldsymbol{x}_2$ solves $A\boldsymbol{x} = \boldsymbol{0}$; $2\boldsymbol{x}_1 - \boldsymbol{x}_2$ solves $A\boldsymbol{x} = \boldsymbol{b}$.

13 (a) The particular solution $\boldsymbol{x}_p$ is always multiplied by 1 (b) Any solution can be the particular solution (c) $\begin{bmatrix} 3 & 3 \\ 3 & 3 \end{bmatrix}\begin{bmatrix} x \\ y \end{bmatrix} = \begin{bmatrix} 6 \\ 6 \end{bmatrix}$. Then $\begin{bmatrix} 1 \\ 1 \end{bmatrix}$ is shorter (length $\sqrt{2}$) than $\begin{bmatrix} 2 \\ 0 \end{bmatrix}$
(d) The "homogeneous" solution in the nullspace is $\boldsymbol{x}_n = \boldsymbol{0}$ when A is invertible.

14 If column 5 has no pivot, x_5 is a free variable. The zero vector *is not* the only solution to $A\boldsymbol{x} = \boldsymbol{0}$. If $A\boldsymbol{x} = \boldsymbol{b}$ has a solution, it has *infinitely many* solutions.

16 The largest rank is 3. Then there is a pivot in every *row*. The solution *always exists*. The column space is $\mathbf{R}^3$. An example is $A = [\,I \quad F\,]$ for any 3 by 2 matrix F.

18 Rank = 3; rank = 3 unless $q = 2$ (then rank = 2).

25 (a) $r < m$, always $r \le n$ (b) $r = m$, $r < n$ (c) $r < m$, $r = n$ (d) $r = m = n$.

28 $\begin{bmatrix} 1 & 2 & 3 & \mathbf{0} \\ 0 & 0 & 4 & \mathbf{0} \end{bmatrix} \to \begin{bmatrix} 1 & 2 & 0 & \mathbf{0} \\ 0 & 0 & 1 & \mathbf{0} \end{bmatrix}$; $\boldsymbol{x}_n = \begin{bmatrix} -2 \\ 1 \\ 0 \end{bmatrix}$; $\begin{bmatrix} 1 & 2 & 3 & \mathbf{5} \\ 0 & 0 & 4 & \mathbf{8} \end{bmatrix} \to \begin{bmatrix} 1 & 2 & 0 & \mathbf{-1} \\ 0 & 0 & 1 & \mathbf{2} \end{bmatrix} \boldsymbol{x}_p = \begin{bmatrix} -1 \\ 0 \\ 2 \end{bmatrix}$.

The pivot columns contain I so -1 and 2 go into $\boldsymbol{x}_p$.

30 $\begin{bmatrix} 1 & 0 & 2 & 3 & \mathbf{2} \\ 1 & 3 & 2 & 0 & \mathbf{5} \\ 2 & 0 & 4 & 9 & \mathbf{10} \end{bmatrix} \to \begin{bmatrix} 1 & 0 & 2 & 3 & \mathbf{2} \\ 0 & 3 & 0 & -3 & \mathbf{3} \\ 0 & 0 & 0 & 3 & \mathbf{6} \end{bmatrix} \to \begin{bmatrix} 1 & 0 & 2 & 0 & \mathbf{-4} \\ 0 & 1 & 0 & 0 & \mathbf{3} \\ 0 & 0 & 0 & 1 & \mathbf{2} \end{bmatrix}$; $\boldsymbol{x}_p = \begin{bmatrix} -4 \\ 3 \\ 0 \\ 2 \end{bmatrix}$

and $\boldsymbol{x}_n = x_3 \begin{bmatrix} -2 \\ 0 \\ 1 \\ 0 \end{bmatrix}$.

Problem Set 3.5, page 167

2 $\boldsymbol{v}_1, \boldsymbol{v}_2, \boldsymbol{v}_3$ are independent. All six vectors are on the plane $(1,1,1,1) \cdot \boldsymbol{v} = 0$ so no four of these six vectors can be independent.

3 If $a = 0$ then column 1 = $\boldsymbol{0}$; if $d = 0$ then b(column 1) − a(column 2) = $\boldsymbol{0}$; if $f = 0$ then all columns end in zero (all are perpendicular to $(0,0,1)$, all in the xy plane, must be dependent).

6 Columns 1, 2, 4 are independent. Also 1, 3, 4 and 2, 3, 4 and others (but not 1, 2, 3). Same column numbers (not same columns!) for A.

8 If $c_1(\boldsymbol{w}_2+\boldsymbol{w}_3)+c_2(\boldsymbol{w}_1+\boldsymbol{w}_3)+c_3(\boldsymbol{w}_1+\boldsymbol{w}_2) = \boldsymbol{0}$ then $(c_2+c_3)\boldsymbol{w}_1+(c_1+c_3)\boldsymbol{w}_2+(c_1+c_2)\boldsymbol{w}_3 = \boldsymbol{0}$. Since the $\boldsymbol{w}$'s are independent this requires $c_2 + c_3 = 0, c_1 + c_3 = 0, c_1 + c_2 = 0$. The only solution is $c_1 = c_2 = c_3 = 0$. Only this combination of $\boldsymbol{v}_1, \boldsymbol{v}_2, \boldsymbol{v}_3$ gives zero.

11 (a) Line in $\mathbf{R}^3$ (b) Plane in $\mathbf{R}^3$ (c) Plane in $\mathbf{R}^3$ (d) All of $\mathbf{R}^3$.

12 $\boldsymbol{b}$ is in the column space when there is a solution to $A\boldsymbol{x} = \boldsymbol{b}$; $\boldsymbol{c}$ is in the row space when there is a solution to $A^{\mathrm{T}}\boldsymbol{y} = \boldsymbol{c}$. *False.* The zero vector is always in the row space.

14 The dimension of $\mathbf{S}$ is (a) zero when $\boldsymbol{x} = \mathbf{0}$ (b) one when $\boldsymbol{x} = (1, 1, 1, 1)$ (c) three when $\boldsymbol{x} = (1, 1, -1, -1)$ because all rearrangements of this $\boldsymbol{x}$ are perpendicular to $(1, 1, 1, 1)$ (d) four when the $\boldsymbol{x}$'s are not equal and don't add to zero. **No $\boldsymbol{x}$ gives** $\dim \mathbf{S} = 2$.

16 The n independent vectors span a space of dimension n. They are a *basis* for that space. If they are the columns of A then m is *not less* than n $(m \geq n)$.

19 (a) The 6 vectors *might not* span $\mathbf{R}^4$ (b) The 6 vectors *are not* independent

(c) Any four *might be* a basis.

21 One basis is $(2, 1, 0)$, $(-3, 0, 1)$. The vector $(2, 1, 0)$ is a basis for the intersection with the xy plane. The normal vector $(1, -2, 3)$ is a basis for the line perpendicular to the plane.

23 (a) True (b) False because the basis vectors may not be in $\mathbf{S}$.

26 Rank 2 if $c = 0$ and $d = 2$; rank 2 except when $c = d$ or $c = -d$.

29 $\begin{bmatrix} 1 & 0 & 0 \\ -1 & 0 & 0 \end{bmatrix}, \begin{bmatrix} 0 & 1 & 0 \\ 0 & -1 & 0 \end{bmatrix}, \begin{bmatrix} 0 & 0 & 1 \\ 0 & 0 & -1 \end{bmatrix}; \begin{bmatrix} 1 & -1 & 0 \\ -1 & 1 & 0 \end{bmatrix}$ and $\begin{bmatrix} 1 & 0 & -1 \\ -1 & 0 & 1 \end{bmatrix}$.

30 $-\begin{bmatrix} 1 & & \\ & 1 & \\ & & 1 \end{bmatrix} + \begin{bmatrix} & 1 & \\ 1 & & \\ & & 1 \end{bmatrix} - \begin{bmatrix} & 1 & \\ & & 1 \\ 1 & & \end{bmatrix} + \begin{bmatrix} & & 1 \\ & 1 & \\ 1 & & \end{bmatrix} + \begin{bmatrix} 1 & & \\ & & 1 \\ & 1 & \end{bmatrix} - \begin{bmatrix} & & 1 \\ 1 & & \\ & 1 & \end{bmatrix} = 0$

34 $y(0) = 0$ requires $A + B + C = 0$. One basis is $\cos x - \cos 2x$ and $\cos x - \cos 3x$.

36 $y_1(x), y_2(x), y_3(x)$ can be $x, 2x, 3x$ (dim 1) or $x, 2x, x^2$ (dim 2) or x, x^2, x^3 (dim 3).

40 The subspace of matrices that have $AS = SA$ has dimension *three*.

42 If the 5 by 5 matrix $[\,A \;\; \boldsymbol{b}\,]$ is invertible, $\boldsymbol{b}$ is not a combination of the columns of A. If $[\,A \;\; \boldsymbol{b}\,]$ is singular, and the 4 columns of A are independent, $\boldsymbol{b}$ *is* a combination of those columns.

Problem Set 3.6, page 180

1 (a) Row and column space dimensions $= 5$, nullspace dimension $= 4$, left nullspace dimension $= 2$ sum $= 16 = m + n$ (b) Column space is $\mathbf{R}^3$; left nullspace contains only $\mathbf{0}$.

4 (a) $\begin{bmatrix} 1 & 0 \\ 1 & 0 \\ 0 & 1 \end{bmatrix}$ (b) Impossible: $r + (n - r)$ must be 3 (c) $[\,1 \;\; 1\,]$ (d) $\begin{bmatrix} -9 & -3 \\ 3 & 1 \end{bmatrix}$

(e) Row space $=$ column space requires $m = n$. Then $m - r = n - r$; nullspaces have the same dimension and actually the same vectors ($A\boldsymbol{x} = \mathbf{0}$ means $\boldsymbol{x} \perp$ row space, $A^{\mathrm{T}}\boldsymbol{x} = \mathbf{0}$ means $\boldsymbol{x} \perp$ column space).

6 A: Row space $(0,3,3,3)$ and $(0,1,0,1)$; column space $(3,0,1)$ and $(3,0,0)$; nullspace $(1,0,0,0)$ and $(0,-1,0,1)$; left nullspace $(0,1,0)$. B: Row space (1), column space $(1,4,5)$, nullspace: empty basis, left nullspace $(-4,1,0)$ and $(-5,0,1)$.

9 (a) Same row space and nullspace. Therefore rank (dimension of row space) is the same (b) Same column space and left nullspace. Same rank (dimension of column space).

11 (a) No solution means that $r < m$. Always $r \le n$. Can't compare m and n (b) If $m - r > 0$, the left nullspace contains a nonzero vector.

12 $\begin{bmatrix} 1 & 1 \\ 0 & 2 \\ 1 & 0 \end{bmatrix} \begin{bmatrix} 1 & 0 & 1 \\ 1 & 2 & 0 \end{bmatrix} = \begin{bmatrix} 2 & 2 & 1 \\ 2 & 4 & 0 \\ 1 & 0 & 1 \end{bmatrix}$; $r + (n - r) = n = 3$ but $2 + 2$ is 4.

16 If $A\boldsymbol{v} = \mathbf{0}$ and $\boldsymbol{v}$ is a row of A then $\boldsymbol{v} \cdot \boldsymbol{v} = 0$.

18 Row $3 - 2$ row $2 +$ row $1 =$ zero row so the vectors $c(1,-2,1)$ are in the left nullspace. The same vectors happen to be in the nullspace.

20 (a) All combinations of $(-1,2,0,0)$ and $(-\frac{1}{4},0,-3,1)$ (b) One (c) $(1,2,3)$, $(0,1,4)$.

21 (a) $\boldsymbol{u}$ and $\boldsymbol{w}$ (b) $\boldsymbol{v}$ and $\boldsymbol{z}$ (c) rank < 2 if $\boldsymbol{u}$ and $\boldsymbol{w}$ are dependent or $\boldsymbol{v}$ and $\boldsymbol{z}$ are dependent (d) The rank of $\boldsymbol{u}\boldsymbol{v}^{\mathrm{T}} + \boldsymbol{w}\boldsymbol{z}^{\mathrm{T}}$ is 2.

24 $A^{\mathrm{T}}\boldsymbol{y} = \boldsymbol{d}$ puts $\boldsymbol{d}$ in the *row space* of A; unique solution if the *left nullspace* (nullspace of A^{T}) contains only $\boldsymbol{y} = \mathbf{0}$.

26 The rows of $AB = C$ are combinations of the rows of B. So rank $C \le$ rank B. Also rank $C \le$ rank A. (The columns of C are combinations of the columns of A).

29 $a_{11} = 1, a_{12} = 0, a_{13} = 1, a_{22} = 0, a_{32} = 1, a_{31} = 0, a_{23} = 1, a_{33} = 0, a_{21} = 1$ (not unique).

Problem Set 4.1, page 191

1 Both nullspace vectors are orthogonal to the row space vector in $\mathbf{R}^3$. Column space is perpendicular to the nullspace of A^{T} in $\mathbf{R}^2$.

3 (a) $\begin{bmatrix} 1 & 2 & -3 \\ 2 & -3 & 1 \\ -3 & 5 & -2 \end{bmatrix}$ (b) Impossible, $\begin{bmatrix} 2 \\ -3 \\ 5 \end{bmatrix}$ not orthogonal to $\begin{bmatrix} 1 \\ 1 \\ 1 \end{bmatrix}$ (c) $\begin{bmatrix} 1 \\ 1 \\ 1 \end{bmatrix}$ in $\boldsymbol{C}(A)$ and $\begin{bmatrix} 1 \\ 0 \\ 0 \end{bmatrix}$ in $\boldsymbol{N}(A^{\mathrm{T}})$ is impossible: not perpendicular (d) This asks for $A^2 = 0$; take $A = \left[\begin{smallmatrix} 1 & -1 \\ 1 & -1 \end{smallmatrix}\right]$ (e) $(1,1,1)$ will be in the nullspace and row space; no such matrix.

6 Multiply the equations by $y_1 = 1$, $y_2 = 1$, $y_3 = -1$. They add to $0 = 1$ so no solution: $\boldsymbol{y} = (1,1,-1)$ is in the left nullspace. Can't have $0 = (\boldsymbol{y}^{\mathrm{T}}A)\boldsymbol{x} = \boldsymbol{y}^{\mathrm{T}}\boldsymbol{b} = 1$.

8 $x = x_r + x_n$, where x_r is in the row space and x_n is in the nullspace. Then $Ax_n = \mathbf{0}$ and $Ax = Ax_r + Ax_n = Ax_r$. All vectors Ax are combinations of the columns of A.

9 Ax is always in the *column space* of A. If $A^{\mathrm{T}}Ax = \mathbf{0}$ then Ax is also in the nullspace of A^{T}. Perpendicular to itself, so $Ax = \mathbf{0}$.

10 (a) For a symmetric matrix the column space and row space are the same (b) x is in the nullspace and z is in the column space = row space: so these "eigenvectors" have $x^{\mathrm{T}}z = 0$.

12 x splits into $x_r + x_n = (1,-1) + (1,1) = (2,0)$.

13 $V^{\mathrm{T}}W$ = zero matrix makes each basis vector for V orthogonal to each basis vector for W. Then every v in V is orthogonal to every w in W (they are combinations of the basis vectors).

14 $Ax = B\widehat{x}$ means that $[\,A \;\; B\,]\left[\begin{smallmatrix} x \\ -\widehat{x} \end{smallmatrix}\right] = \mathbf{0}$. Three homogeneous equations in four unknowns always have a nonzero solution. Here $x = (3,1)$ and $\widehat{x} = (1,0)$ and $Ax = B\widehat{x} = (5,6,5)$ is in both column spaces. Two planes in $\mathbf{R}^3$ must intersect in a line at least!

16 $A^{\mathrm{T}}y = \mathbf{0} \Rightarrow (Ax)^{\mathrm{T}}y = x^{\mathrm{T}}A^{\mathrm{T}}y = 0$. Then $y \perp Ax$ and $N(A^{\mathrm{T}}) \perp C(A)$.

18 $S^{\perp}$ is the nullspace of $A = \begin{bmatrix} 1 & 5 & 1 \\ 2 & 2 & 2 \end{bmatrix}$. Therefore $S^{\perp}$ is a *subspace* even if S is not.

21 For example $(-5,0,1,1)$ and $(0,1,-1,0)$ span $S^{\perp}$ = nullspace of $A = [1 \;\; 2 \;\; 2 \;\; 3; \;\; 1 \;\; 3 \;\; 3 \;\; 2]$.

23 x in $V^{\perp}$ is perpendicular to any vector in V. Since V contains all the vectors in S, x is also perpendicular to any vector in S. So every x in $V^{\perp}$ is also in $S^{\perp}$.

28 (a) $(1,-1,0)$ is in both planes. Normal vectors are perpendicular, but planes still intersect!
(b) Need *three* orthogonal vectors to span the whole orthogonal complement.
(c) Lines can meet without being orthogonal.

30 When $AB = 0$, the column space of B is contained in the nullspace of A. Therefore the dimension of $C(B) \le$ dimension of $N(A)$. This means $\operatorname{rank}(B) \le 4 - \operatorname{rank}(A)$.

31 null(N') produces a basis for the *row space* of A (perpendicular to $N(A)$).

Problem Set 4.2, page 202

1 (a) $a^{\mathrm{T}}b/a^{\mathrm{T}}a = 5/3$; $p = (5/3, 5/3, 5/3)$; $e = (-2/3, 1/3, 1/3)$
(b) $a^{\mathrm{T}}b/a^{\mathrm{T}}a = -1$; $p = (1,3,1)$; $e = (0,0,0)$.

3 $P_1 = \dfrac{1}{3}\begin{bmatrix} 1 & 1 & 1 \\ 1 & 1 & 1 \\ 1 & 1 & 1 \end{bmatrix}$ and $P_1 b = \dfrac{1}{3}\begin{bmatrix} 5 \\ 5 \\ 5 \end{bmatrix}$ and $P_1^2 = P_1$. $P_2 = \dfrac{1}{11}\begin{bmatrix} 1 & 3 & 1 \\ 3 & 9 & 3 \\ 1 & 3 & 1 \end{bmatrix}$ and $P_2 b = \begin{bmatrix} 1 \\ 3 \\ 1 \end{bmatrix}$.

6 $p_1 = (\frac{1}{9}, -\frac{2}{9}, -\frac{2}{9})$ and $p_2 = (\frac{4}{9}, \frac{4}{9}, -\frac{2}{9})$ and $p_3 = (\frac{4}{9}, -\frac{2}{9}, \frac{4}{9})$. Then $p_1 + p_2 + p_3 = (1,0,0) = b$.

9 Since A is invertible, $P = A(A^{\mathrm{T}}A)^{-1}A^{\mathrm{T}} = AA^{-1}(A^{\mathrm{T}})^{-1}A^{\mathrm{T}} = I$: project onto all of $\mathbf{R}^2$.

11 (a) $\boldsymbol{p} = A(A^{\mathrm{T}}A)^{-1}A^{\mathrm{T}}\boldsymbol{b} = (2,3,0)$ and $\boldsymbol{e} = (0,0,4)$ (b) $\boldsymbol{p} = (4,4,6)$ and $\boldsymbol{e} = (0,0,0)$.

15 The column space of $2A$ is the same as the column space of A. $\widehat{\boldsymbol{x}}$ for $2A$ is *half* of $\widehat{\boldsymbol{x}}$ for A.

16 $\frac{1}{2}(1,2,-1) + \frac{3}{2}(1,0,1) = (2,1,1)$. Therefore $\boldsymbol{b}$ is in the plane. Projection shows $P\boldsymbol{b} = \boldsymbol{b}$.

18 (a) $I - P$ is the projection matrix onto $(1,-1)$ in the perpendicular direction to $(1,1)$

(b) $I - P$ is the projection matrix onto the plane $x + y + z = 0$ perpendicular to $(1,1,1)$.

20 $$\boldsymbol{e} = \begin{bmatrix} 1 \\ -1 \\ -2 \end{bmatrix}, \quad Q = \boldsymbol{e}\boldsymbol{e}^{\mathrm{T}}/\boldsymbol{e}^{\mathrm{T}}\boldsymbol{e} = \begin{bmatrix} 1/6 & -1/6 & -1/3 \\ -1/6 & 1/6 & 1/3 \\ -1/3 & 1/3 & 2/3 \end{bmatrix}, \quad P = I - Q = \begin{bmatrix} 5/6 & 1/6 & 1/3 \\ 1/6 & 5/6 & -1/3 \\ 1/3 & -1/3 & 1/3 \end{bmatrix}.$$

21 $\left(A(A^{\mathrm{T}}A)^{-1}A^{\mathrm{T}}\right)^2 = A(A^{\mathrm{T}}A)^{-1}(A^{\mathrm{T}}A)(A^{\mathrm{T}}A)^{-1}A^{\mathrm{T}} = A(A^{\mathrm{T}}A)^{-1}A^{\mathrm{T}}$. Therefore $P^2 = P$. $P\boldsymbol{b}$ is always in the column space (where P projects). Therefore its projection $P(P\boldsymbol{b})$ is $P\boldsymbol{b}$.

24 The nullspace of A^{T} is *orthogonal* to the column space $\boldsymbol{C}(A)$. So if $A^{\mathrm{T}}\boldsymbol{b} = \mathbf{0}$, the projection of $\boldsymbol{b}$ onto $\boldsymbol{C}(A)$ should be $\boldsymbol{p} = \mathbf{0}$. Check $P\boldsymbol{b} = A(A^{\mathrm{T}}A)^{-1}A^{\mathrm{T}}\boldsymbol{b} = A(A^{\mathrm{T}}A)^{-1}\mathbf{0} = \mathbf{0}$.

28 $P^2 = P = P^{\mathrm{T}}$ give $P^{\mathrm{T}}P = P$. Then the $(2,2)$ entry of P equals the $(2,2)$ entry of $P^{\mathrm{T}}P$ which is the length squared of column 2.

29 Set $A = B^{\mathrm{T}}$. Then A has independent columns. By **4G**, $A^{\mathrm{T}}A = BB^{\mathrm{T}}$ is invertible.

30 (a) The column space is the line through $\boldsymbol{a} = \begin{bmatrix} 3 \\ 4 \end{bmatrix}$ so $P_C = \dfrac{\boldsymbol{a}\boldsymbol{a}^{\mathrm{T}}}{\boldsymbol{a}^{\mathrm{T}}\boldsymbol{a}} = \dfrac{1}{25}\begin{bmatrix} 9 & 12 \\ 12 & 25 \end{bmatrix}$. We can't use $(A^{\mathrm{T}}A)^{-1}$ because A has dependent columns. (b) The row space is the line through $\boldsymbol{v} = (1,2,2)$ and $P_R = \boldsymbol{v}\boldsymbol{v}^{\mathrm{T}}/\boldsymbol{v}^{\mathrm{T}}\boldsymbol{v}$. Always $P_C A = A$ and $AP_R = A$ and then $P_C A P_R = A$!

Problem Set 4.3, page 215

1 $$A = \begin{bmatrix} 1 & 0 \\ 1 & 1 \\ 1 & 3 \\ 1 & 4 \end{bmatrix} \text{ and } \boldsymbol{b} = \begin{bmatrix} 0 \\ 8 \\ 8 \\ 20 \end{bmatrix} \text{ give } A^{\mathrm{T}}A = \begin{bmatrix} 4 & 8 \\ 8 & 26 \end{bmatrix} \text{ and } A^{\mathrm{T}}\boldsymbol{b} = \begin{bmatrix} 36 \\ 112 \end{bmatrix}.$$

$$A^{\mathrm{T}}A\widehat{\boldsymbol{x}} = A^{\mathrm{T}}\boldsymbol{b} \text{ gives } \widehat{\boldsymbol{x}} = \begin{bmatrix} 1 \\ 4 \end{bmatrix} \text{ and } \boldsymbol{p} = A\widehat{\boldsymbol{x}} = \begin{bmatrix} 1 \\ 5 \\ 13 \\ 17 \end{bmatrix} \text{ and } \boldsymbol{e} = \boldsymbol{b} - \boldsymbol{p} = \begin{bmatrix} -1 \\ 3 \\ -5 \\ 3 \end{bmatrix}.$$

$E = \|\boldsymbol{e}\|^2 = 44$.

5 $E = (C-0)^2 + (C-8)^2 + (C-8)^2 + (C-20)^2$. $A^{\mathrm{T}} = [\,1 \;\; 1 \;\; 1 \;\; 1\,]$, $A^{\mathrm{T}}A = [\,4\,]$ and $A^{\mathrm{T}}\boldsymbol{b} = [\,36\,]$ and $(A^{\mathrm{T}}A)^{-1}A^{\mathrm{T}}\boldsymbol{b} = 9 =$ best height C. Errors $\boldsymbol{e} = (-9,-1,-1,11)$.

7 $A = [\,0 \;\; 1 \;\; 3 \;\; 4\,]^{\mathrm{T}}$, $A^{\mathrm{T}}A = [\,26\,]$ and $A^{\mathrm{T}}\boldsymbol{b} = [\,112\,]$. Best $D = 112/26 = 56/13$.

8 $\widehat{x} = 56/13$, $\boldsymbol{p} = (56/13)(0,1,3,4)$. $C = 9$, $D = 56/13$ don't match $(C, D) = (1,4)$; the columns of A were not perpendicular so we can't project separately to find $C = 1$ and $D = 4$.

9 Closest parabola: Projecting $\boldsymbol{b}$ onto a 3-dimensional column space
$$\begin{bmatrix} 1 & 0 & 0 \\ 1 & 1 & 1 \\ 1 & 3 & 9 \\ 1 & 4 & 16 \end{bmatrix} \begin{bmatrix} C \\ D \\ E \end{bmatrix} = \begin{bmatrix} 0 \\ 8 \\ 8 \\ 20 \end{bmatrix}.$$
$$A^{\mathrm{T}} A\widehat{x} = \begin{bmatrix} 4 & 8 & 26 \\ 8 & 26 & 92 \\ 26 & 92 & 338 \end{bmatrix} \begin{bmatrix} C \\ D \\ E \end{bmatrix} = \begin{bmatrix} 36 \\ 112 \\ 400 \end{bmatrix}.$$

11 (a) The best line is $x = 1 + 4t$, which goes through the center point $(\widehat{t}, \widehat{\boldsymbol{b}}) = (2, 9)$

(b) From the first equation: $C \cdot m + D \cdot \sum_{i=1}^{m} t_i = \sum_{i=1}^{m} b_i$. Divide by m to get $C + D\widehat{t} = \widehat{\boldsymbol{b}}$.

13 $(A^{\mathrm{T}}A)^{-1}A^{\mathrm{T}}(\boldsymbol{b} - A\boldsymbol{x}) = \widehat{\boldsymbol{x}} - \boldsymbol{x}$. Errors $\boldsymbol{b} - A\boldsymbol{x} = (\pm 1, \pm 1, \pm 1)$ add to $\boldsymbol{0}$, so the $\widehat{\boldsymbol{x}} - \boldsymbol{x}$ add to $\boldsymbol{0}$.

14 $(\widehat{\boldsymbol{x}} - \boldsymbol{x})(\widehat{\boldsymbol{x}} - \boldsymbol{x})^{\mathrm{T}} = (A^{\mathrm{T}}A)^{-1}A^{\mathrm{T}}(\boldsymbol{b} - A\boldsymbol{x})(\boldsymbol{b} - A\boldsymbol{x})^{\mathrm{T}}A(A^{\mathrm{T}}A)^{-1}$. Average $(\boldsymbol{b} - A\boldsymbol{x})(\boldsymbol{b} - A\boldsymbol{x})^{\mathrm{T}} = \sigma^2 I$ gives the *covariance matrix* $(A^{\mathrm{T}}A)^{-1}A^{\mathrm{T}}\sigma^2 A(A^{\mathrm{T}}A)^{-1}$ which simplifies to $\sigma^2(A^{\mathrm{T}}A)^{-1}$.

16 $\dfrac{1}{10}b_{10} + \dfrac{\mathbf{9}}{\mathbf{10}}\widehat{x}_9 = \dfrac{1}{10}(b_1 + \cdots + b_{10})$.

18 $\boldsymbol{p} = A\widehat{\boldsymbol{x}} = (5, 13, 17)$ gives the heights of the closest line. The error is $\boldsymbol{b} - \boldsymbol{p} = (2, -6, 4)$.

21 $\boldsymbol{e}$ is in $\boldsymbol{N}(A^{\mathrm{T}})$; $\boldsymbol{p}$ is in $\boldsymbol{C}(A)$; $\widehat{\boldsymbol{x}}$ is in $\boldsymbol{C}(A^{\mathrm{T}})$; $\boldsymbol{N}(A) = \{\boldsymbol{0}\} =$ zero vector.

23 The square of the distance between points on two lines is $E = (y-x)^2 + (3y-x)^2 + (1+x)^2$. Set $\frac{1}{2}\partial E/\partial x = -(y-x) - (3y-x) + (x+1) = 0$ and $\frac{1}{2}\partial E/\partial y = (y-x) + 3(3y-x) = 0$. The solution is $x = -5/7$, $y = -2/7$; $E = 2/7$, and the minimal distance is $\sqrt{2/7}$.

26 Direct approach to 3 points on a line: *Equal slopes* $(b_2 - b_1)/(t_2 - t_1) = (b_3 - b_2)/(t_3 - t_2)$. Linear algebra approach: If $\boldsymbol{y}$ is orthogonal to the columns $(1, 1, 1)$ and (t_1, t_2, t_3) and $\boldsymbol{b}$ is in the column space then $\boldsymbol{y}^{\mathrm{T}}\boldsymbol{b} = 0$. This $\boldsymbol{y} = (t_2 - t_3, t_3 - t_1, t_1 - t_2)$ is in the left nullspace. Then $\boldsymbol{y}^{\mathrm{T}}\boldsymbol{b} = 0$ is the same equal slopes condition written as $(b_2 - b_1)(t_3 - t_2) = (b_3 - b_2)(t_2 - t_1)$.

Problem Set 4.4, page 228

3 (a) $A^{\mathrm{T}}A = 16I$ (b) $A^{\mathrm{T}}A$ is diagonal with entries 1, 4, 9.

6 If Q_1 and Q_2 are orthogonal matrices then $(Q_1Q_2)^{\mathrm{T}}Q_1Q_2 = Q_2^{\mathrm{T}}Q_1^{\mathrm{T}}Q_1Q_2 = Q_2^{\mathrm{T}}Q_2 = I$ which means that Q_1Q_2 is orthogonal also.

8 If $\boldsymbol{q}_1$ and $\boldsymbol{q}_2$ are *orthonormal* vectors in $\mathbf{R}^5$ then $(\boldsymbol{q}_1^{\mathrm{T}}\boldsymbol{b})\boldsymbol{q}_1 + (\boldsymbol{q}_2^{\mathrm{T}}\boldsymbol{b})\boldsymbol{q}_2$ is closest to $\boldsymbol{b}$.

11 (a) Two *orthonormal* vectors are $\frac{1}{10}(1, 3, 4, 5, 7)$ and $\frac{1}{10}(7, -3, -4, 5, -1)$ (b) The closest vector in the plane is the *projection* $QQ^{\mathrm{T}}(1, 0, 0, 0, 0) = (0.5, -0.18, -0.24, 0.4, 0)$.

13 The multiple to subtract is $a^{\mathrm{T}}b/a^{\mathrm{T}}a$. Then $B = b - \dfrac{a^{\mathrm{T}}b}{a^{\mathrm{T}}a}a = (4,0) - 2\cdot(1,1) = (2,-2)$.

14 $\begin{bmatrix}1 & 4\\ 1 & 0\end{bmatrix} = \begin{bmatrix}q_1 & q_2\end{bmatrix}\begin{bmatrix}\|a\| & q_1^{\mathrm{T}}b\\ 0 & \|B\|\end{bmatrix} = \begin{bmatrix}1/\sqrt{2} & 1/\sqrt{2}\\ 1/\sqrt{2} & -1/\sqrt{2}\end{bmatrix}\begin{bmatrix}\sqrt{2} & 2\sqrt{2}\\ 0 & 2\sqrt{2}\end{bmatrix} = QR.$

15 (a) $q_1 = \frac{1}{3}(1,2,-2)$, $q_2 = \frac{1}{3}(2,1,2)$, $q_3 = \frac{1}{3}(2,-2,-1)$ (b) The nullspace of A^{T} contains q_3 (c) $\widehat{x} = (A^{\mathrm{T}}A)^{-1}A^{\mathrm{T}}(1,2,7) = (1,2)$.

16 The projection $p = (a^{\mathrm{T}}b/a^{\mathrm{T}}a)a = 14a/49 = 2a/7$ is closest to b; $q_1 = a/\|a\| = a/7$ is $(4,5,2,2)/7$. $B = b - p = (-1,4,-4,-4)/7$ has $\|B\| = 1$ so $q_2 = B$.

18 $A = a = (1,-1,0,0)$; $B = b - p = (\frac{1}{2}, \frac{1}{2}, -1, 0)$; $C = c - p_A - p_B = (\frac{1}{3}, \frac{1}{3}, \frac{1}{3}, -1)$. Notice the pattern in those orthogonal vectors A, B, C.

20 (a) True (b) True. $Qx = x_1q_1 + x_2q_2$. $\|Qx\|^2 = x_1^2 + x_2^2$ because $q_1 \cdot q_2 = 0$.

21 The orthonormal vectors are $q_1 = (1,1,1,1)/2$ and $q_2 = (-5,-1,1,5)/\sqrt{52}$. Then $b = (-4,-3,3,0)$ projects to $p = (-7,-3,-1,3)/2$. Check that $b - p = (-1,-3,7,-3)/2$ is orthogonal to both q_1 and q_2.

22 $A = (1,1,2)$, $B = (1,-1,0)$, $C = (-1,-1,1)$. Not yet orthonormal.

26 $(q_2^{\mathrm{T}}C^*)q_2 = \dfrac{B^{\mathrm{T}}c}{B^{\mathrm{T}}B}B$ because $q_2 = \dfrac{B}{\|B\|}$ and the extra q_1 in C^* is orthogonal to q_2.

29 There are mn multiplications in (11) and $\frac{1}{2}m^2n$ multiplications in each part of (12).

30 The columns of the wavelet matrix W are orthonormal. Then $W^{-1} = W^{\mathrm{T}}$. See Section 7.3 for more about wavelets.

33 $Q_1 = \begin{bmatrix}1 & 0\\ 0 & -1\end{bmatrix}$ reflects across x axis, $Q_2 = \begin{bmatrix}1 & 0 & 0\\ 0 & 0 & -1\\ 0 & -1 & 0\end{bmatrix}$ across plane $y + z = 0$.

36 Orthogonal and lower triangular $\Rightarrow \pm 1$ on the main diagonal, 0 elsewhere.

Problem Set 5.1, page 240

1 $\det(2A) = 8$ and $\det(-A) = (-1)^4 \det A = \frac{1}{2}$ and $\det(A^2) = \frac{1}{4}$ and $\det(A^{-1}) = 2$.

5 $|J_5| = 1$, $|J_6| = -1$, $|J_7| = -1$. The determinants are 1, 1, -1, -1 repeating, so $|J_{101}| = 1$.

8 $Q^{\mathrm{T}}Q = I \Rightarrow |Q|^2 = 1 \Rightarrow |Q| = \pm 1$; Q^n stays orthogonal so can't blow up. Same for Q^{-1}.

10 If the entries in every row add to zero, then $(1,1,\ldots,1)$ is in the nullspace: singular A has det $= 0$. (The columns add to the zero column so they are linearly dependent.) If every row adds to one, then rows of $A - I$ add to zero (not necessarily $\det A = 1$).

11 $CD = -DC \Rightarrow |CD| = (-1)^n|DC|$ and *not* $-|DC|$. If n is even we can have $|CD| \neq 0$.

14 $\det(A) = 24$ and $\det(A) = 5$.

15 det $= 0$ and det $= 1 - 2t^2 + t^4 = (1-t^2)^2$.

17 Any 3 by 3 skew-symmetric K has $\det(K^{\mathrm{T}}) = \det(-K) = (-1)^3\det(K)$. This is $-\det(K)$. But also $\det(K^{\mathrm{T}}) = \det(K)$, so we must have $\det(K) = 0$.

21 Rules 5 and 3 give Rule 2. (Since Rules 4 and 3 give 5, they also give Rule 2.)

23 $\det(A) = 10$, $A^2 = \begin{bmatrix} 18 & 7 \\ 14 & 11 \end{bmatrix}$, $\det(A^2) = 100$, $A^{-1} = \frac{1}{10}\begin{bmatrix} 3 & -1 \\ -2 & 4 \end{bmatrix}$, $\det(A^{-1}) = \frac{1}{10}$.
$\det(A - \lambda I) = \lambda^2 - 7\lambda + 10 = 0$ when $\lambda = 2$ or $\lambda = 5$.

27 $\det A = abc$, $\det B = -abcd$, $\det C = a(b-a)(c-b)$.

30 $$\begin{bmatrix} \partial f/\partial a & \partial f/\partial c \\ \partial f/\partial b & \partial f/\partial d \end{bmatrix} = \begin{bmatrix} \frac{d}{ad-bc} & \frac{-b}{ad-bc} \\ \frac{-c}{ad-bc} & \frac{a}{ad-bc} \end{bmatrix} = \frac{1}{ad-bc}\begin{bmatrix} d & -b \\ -c & a \end{bmatrix} = A^{-1}.$$

32 Typical determinants of rand(n) are $10^6, 10^{25}, 10^{79}, 10^{218}$ for $n = 50, 100, 200, 400$. Using randn(n) with normal bell-shaped probabilities these are $10^{31}, 10^{78}, 10^{186}$, Inf means $\geq 2^{1024}$. MATLAB computes $1.999999999999999 \times 2^{1023} \approx 1.8 \times 10^{308}$ but one more 9 gives Inf!

34 Reduce B to [row 3 : row 2; row 1]. Then $\det B = -6$.

Problem Set 5.2, page 253

2 $\det A = -2$, independent; $\det B = 0$, dependent; $\det C = (-2)(0)$, dependent.

4 (a) The last three rows must be dependent (b) In each of the 120 terms: Choices from the last 3 rows must use 3 columns; at least one choice will be zero.

5 $a_{11}a_{23}a_{32}a_{44}$ gives -1, $a_{14}a_{23}a_{32}a_{41}$ gives $+1$ so $\det A = 0$;
$\det B = 2 \cdot 4 \cdot 4 \cdot 2 - 1 \cdot 4 \cdot 4 \cdot 1 = 48$.

7 (a) If $a_{11} = a_{22} = a_{33} = 0$ then 4 terms are sure zeros (b) 15 terms are certainly zero.

9 Some term $a_{1\alpha}a_{2\beta}\cdots a_{n\omega}$ is not zero! Move rows $1, 2, \ldots, n$ into rows $\alpha, \beta, \ldots, \omega$. Then these nonzero a's will be on the main diagonal.

10 To get $+1$ for the even permutations the matrix needs an *even* number of -1's. For the odd P's the matrix needs an *odd* number of -1's. So six 1's and det $= 6$ are impossible: max(det) $= 4$.

12 $C = \begin{bmatrix} 6 & -3 \\ -1 & 2 \end{bmatrix}$. $C = \begin{bmatrix} 0 & 42 & -35 \\ 0 & -21 & 14 \\ -3 & 6 & -3 \end{bmatrix}$. $\det B = 1(0) + 2(42) + 3(-35) = -21$.

13 $C = \begin{bmatrix} 3 & 2 & 1 \\ 2 & 4 & 2 \\ 1 & 2 & 3 \end{bmatrix}$ and $AC^{\mathrm{T}} = \begin{bmatrix} 4 & 0 & 0 \\ 0 & 4 & 0 \\ 0 & 0 & 4 \end{bmatrix}$. Therefore $A^{-1} = \frac{1}{4}C^{\mathrm{T}}$.

15 (a) $C_1 = 0$, $C_2 = -1$, $C_3 = 0$, $C_4 = 1$ (b) $C_n = -C_{n-2}$ by cofactors of row 1 then cofactors of column 1. Therefore $C_{10} = -C_8 = C_6 = -C_4 = -1$.

17 The $1,1$ cofactor is E_{n-1}. The $1,2$ cofactor has a single 1 in its first column, with cofactor E_{n-2}. Signs give $E_n = E_{n-1} - E_{n-2}$. Then $1, 0, -1, -1, 0, 1$ repeats by sixes; $E_{100} = -1$.

18 The $1,1$ cofactor is F_{n-1}. The $1,2$ cofactor has a 1 in column 1, with cofactor F_{n-2}. Multiply by $(-1)^{1+2}$ and also (-1) from the $1,2$ entry to find $F_n = F_{n-1} + F_{n-2}$ (so Fibonacci).

20 Since x, x^2, x^3 are all in the same row, they are never multiplied in $\det V_4$. The determinant is zero at $x = a$ or b or c, so $\det V$ has factors $(x-a)(x-b)(x-c)$. Multiply by the cofactor V_3. Any Vandermonde matrix $V_{ij} = (c_i)^{j-1}$ has $\det V =$ product of all $(c_l - c_k)$ for $l > k$.

21 $G_2 = -1$, $G_3 = 2$, $G_4 = -3$, and $G_n = (-1)^{n-1}(n-1) =$ (product of the n eigenvalues!)

23 The problem asks us to show that $F_{2n+2} = 3F_{2n} - F_{2n-2}$. Keep using the Fibonacci rule:

$$F_{2n+2} = F_{2n+1} + F_{2n} = F_{2n} + F_{2n-1} + F_{2n} = F_{2n} + (F_{2n} - F_{2n-2}) + F_{2n} = 3F_{2n} - F_{2n-2}.$$

26 (a) All L's have $\det = 1$; $\det U_k = \det A_k = 2, 6, -6$ for $k = 1, 2, 3$ (b) Pivots $2, \frac{3}{2}, -\frac{1}{3}$.

27 Problem 25 gives $\det \begin{bmatrix} I & 0 \\ -CA^{-1} & I \end{bmatrix} = 1$ and $\det \begin{bmatrix} A & B \\ C & D \end{bmatrix} = |A|$ times $|D - CA^{-1}B|$ which is $|AD - ACA^{-1}B|$. If $AC = CA$ this is $|AD - CAA^{-1}B| = \det(AD - CB)$.

29 (a) $\det A = a_{11}C_{11} + \cdots + a_{1n}C_{1n}$. The derivative with respect to a_{11} is the cofactor C_{11}.

31 There are five nonzero products, all 1's with a plus or minus sign. Here are the (row, column) numbers and the signs: $+(1,1)(2,2)(3,3)(4,4) + (1,2)(2,1)(3,4)(4,3) - (1,2)(2,1)$ $(3,3)(4,4) - (1,1)(2,2)(3,4)(4,3) - (1,1)(2,3)(3,2)(4,4)$. Total $1 + 1 - 1 - 1 - 1 = -1$.

34 With $a_{11} = 1$, the $-1, 2, -1$ matrix has $\det = 1$ and inverse $(A^{-1})_{ij} = n + 1 - \max(i, j)$.

35 With $a_{11} = 2$, the $-1, 2, -1$ matrix has $\det = n + 1$ and $(n+1)(A^{-1})_{ij} = i(n - j + 1)$ for $i \le j$ and symmetrically $(n+1)(A^{-1})_{ij} = j(n - i + 1)$ for $i \ge j$.

Problem Set 5.3, page 269

2 (a) $y = -c/(ad - bc)$ (b) $y = (fg - id)/D$.

3 (a) $x_1 = 3/0$ and $x_2 = -2/0$: no solution (b) $x_1 = 0/0$ and $x_2 = 0/0$: *undetermined*.

4 (a) $x_1 = \det([\boldsymbol{b}\ \ \boldsymbol{a}_2\ \ \boldsymbol{a}_3])/\det A$, if $\det A \neq 0$ (b) The determinant is linear in its first column so $x_1|\boldsymbol{a}_1\ \boldsymbol{a}_2\ \boldsymbol{a}_3| + x_2|\boldsymbol{a}_2\ \boldsymbol{a}_2\ \boldsymbol{a}_3| + x_3|\boldsymbol{a}_3\ \boldsymbol{a}_2\ \boldsymbol{a}_3|$. The last two determinants are zero.

6 (a) $\begin{bmatrix} 1 & -\frac{2}{3} & 0 \\ 0 & \frac{1}{3} & 0 \\ 0 & -\frac{4}{3} & 1 \end{bmatrix}$ (b) $\dfrac{1}{4}\begin{bmatrix} 3 & 2 & 1 \\ 2 & 1 & 2 \\ 1 & 2 & 3 \end{bmatrix}$. The inverse of a symmetric matrix is symmetric.

8 $C = \begin{bmatrix} 6 & -3 & 0 \\ 3 & 1 & -1 \\ -6 & 2 & 1 \end{bmatrix}$ and $AC^{\mathrm{T}} = \begin{bmatrix} 3 & 0 & 0 \\ 0 & 3 & 0 \\ 0 & 0 & 3 \end{bmatrix}$. Therefore $\det A = 3$. Cofactor of 100 is 0.

9 If we know the cofactors and $\det A = 1$ then $C^{\mathrm{T}} = A^{-1}$ and $\det A^{-1} = 1$. Now A is the inverse of A^{-1}, so A is the cofactor matrix for C.

11 We find $\det A = (\det C)^{\frac{1}{n-1}}$ with $n = 4$. Then $\det A^{-1}$ is $1/\det A$. Construct A^{-1} using the cofactors. Invert to find A.

12 The cofactors of A are integers. Division by $\det A = \pm 1$ gives integer entries in A^{-1}.

16 For $n = 5$ the matrix C contains *25* cofactors and each 4 by 4 cofactor contains *24* terms and each term needs *3* multiplications: total 1800 multiplications vs. 125 for Gauss-Jordan.

18 Volume $= \begin{vmatrix} 3 & 1 & 1 \\ 1 & 3 & 1 \\ 1 & 1 & 3 \end{vmatrix} = 20$. Area of faces $=$ length of cross product $\begin{vmatrix} \boldsymbol{i} & \boldsymbol{j} & \boldsymbol{k} \\ 3 & 1 & 1 \\ 1 & 3 & 1 \end{vmatrix} = -2\boldsymbol{i} - 2\boldsymbol{j} + 8\boldsymbol{k} = 6\sqrt{2}$.

19 (a) Area $\frac{1}{2}\begin{vmatrix} 2 & 1 & 1 \\ 3 & 4 & 1 \\ 0 & 5 & 1 \end{vmatrix} = 5$ (b) $5 +$ new triangle area $\frac{1}{2}\begin{vmatrix} 2 & 1 & 1 \\ 0 & 5 & 1 \\ -1 & 0 & 1 \end{vmatrix} = 5 + 7 = 12$.

22 The maximum volume is $L_1L_2L_3L_4$ reached when the four edges are orthogonal in $\mathbf{R}^4$. With entries 1 and -1 all lengths are $\sqrt{1+1+1+1} = 2$. The maximum determinant is $2^4 = 16$, achieved by Hadamard above. For a 3 by 3 matrix, $\det A = (\sqrt{3})^3$ can't be achieved.

24 $A^{\mathrm{T}}A = \begin{bmatrix} \boldsymbol{a}^{\mathrm{T}} \\ \boldsymbol{b}^{\mathrm{T}} \\ \boldsymbol{c}^{\mathrm{T}} \end{bmatrix} \begin{bmatrix} \boldsymbol{a} & \boldsymbol{b} & \boldsymbol{c} \end{bmatrix} = \begin{bmatrix} \boldsymbol{a}^{\mathrm{T}}\boldsymbol{a} & 0 & 0 \\ 0 & \boldsymbol{b}^{\mathrm{T}}\boldsymbol{b} & 0 \\ 0 & 0 & \boldsymbol{c}^{\mathrm{T}}\boldsymbol{c} \end{bmatrix}$ has $\begin{array}{lcl} \det A^{\mathrm{T}}A & = & (\|a\|\,\|b\|\,\|c\|)^2 \\ \det A & = & \pm\|a\|\,\|b\|\,\|c\| \end{array}$.

26 The n-dimensional cube has 2^n corners, $n2^{n-1}$ edges and $2n$ $(n-1)$-dimensional faces. Coefficients from $(2+x)^n$ in Worked Example **2.4A**. The cube from $2I$ has volume 2^n.

27 The pyramid has volume $\frac{1}{6}$. The 4-dimensional pyramid has volume $\frac{1}{24}$.

32 Base area 10, height 2, volume 20.

36 $S = (2, 1, -1)$. The area is $\|PQ \times PS\| = \|(-2, -2, -1)\| = 3$. The other four corners could be $(0,0,0)$, $(0,0,2)$, $(1,2,2)$, $(1,1,0)$. The volume of the tilted box is $|det| = 1$.

Problem Set 6.1, page 283

1 A and A^2 and A^∞ all have the same eigenvectors. The eigenvalues are 1 and 0.5 for A, 1 and 0.25 for A^2, 1 and 0 for A^∞. Therefore A^2 is halfway between A and A^∞.

Exchanging the rows of A changes the eigenvalues to 1 and -0.5 (it is still a Markov matrix with eigenvalue 1, and the trace is now $0.2+0.3$—so the other eigenvalue is -0.5).

Singular matrices stay singular during elimination, so $\lambda = 0$ does not change.

3 A has $\lambda_1 = 4$ and $\lambda_2 = -1$ (check trace and determinant) with $\boldsymbol{x}_1 = (1, 2)$ and $\boldsymbol{x}_2 = (2, -1)$. A^{-1} has the same eigenvectors as A, with eigenvalues $1/\lambda_1 = 1/4$ and $1/\lambda_2 = -1$.

6 A and B have $\lambda_1 = 1$ and $\lambda_2 = 1$. AB and BA have $\lambda = \frac{1}{2}(3 \pm \sqrt{5})$. Eigenvalues of AB *are not equal* to eigenvalues of A times eigenvalues of B. Eigenvalues of AB and BA ***are equal***.

8 (a) Multiply $A\boldsymbol{x}$ to see $\lambda\boldsymbol{x}$ which reveals λ (b) Solve $(A - \lambda I)\boldsymbol{x} = \mathbf{0}$ to find $\boldsymbol{x}$.

10 A has $\lambda_1 = 1$ and $\lambda_2 = .4$ with $\boldsymbol{x}_1 = (1, 2)$ and $\boldsymbol{x}_2 = (1, -1)$. A^∞ has $\lambda_1 = 1$ and $\lambda_2 = 0$ (same eigenvectors). A^{100} has $\lambda_1 = 1$ and $\lambda_2 = (.4)^{100}$ which is near zero. So A^{100} is very near A^∞.

11 $M = (A - \lambda_2 I)(A - \lambda_1 I) =$ zero matrix so the columns of $A - \lambda_1 I$ are in the nullspace of $A - \lambda_2 I$. This "Cayley-Hamilton Theorem" $M = 0$ in Problem 6.2.35 has a short proof: by Problem 9, M has eigenvalues $(\lambda_1 - \lambda_2)(\lambda_1 - \lambda_1) = 0$ and $(\lambda_2 - \lambda_2)(\lambda_2 - \lambda_1) = 0$. Same $\boldsymbol{x}_1, \boldsymbol{x}_2$.

13 (a) $P\boldsymbol{u} = (\boldsymbol{u}\boldsymbol{u}^{\mathrm{T}})\boldsymbol{u} = \boldsymbol{u}(\boldsymbol{u}^{\mathrm{T}}\boldsymbol{u}) = \boldsymbol{u}$ so $\lambda = 1$ (b) $P\boldsymbol{v} = (\boldsymbol{u}\boldsymbol{u}^{\mathrm{T}})\boldsymbol{v} = \boldsymbol{u}(\boldsymbol{u}^{\mathrm{T}}\boldsymbol{v}) = \mathbf{0}$ so $\lambda = 0$
(c) $\boldsymbol{x}_1 = (-1, 1, 0, 0)$, $\boldsymbol{x}_2 = (-3, 0, 1, 0)$, $\boldsymbol{x}_3 = (-5, 0, 0, 1)$ are eigenvectors with $\lambda = 0$.

15 $\lambda = \frac{1}{2}(-1 \pm i\sqrt{3})$; the three eigenvalues are $1, 1, -1$.

16 Set $\lambda = 0$ to find $\det A = (\lambda_1)(\lambda_2)\cdots(\lambda_n)$.

17 If A has $\lambda_1 = 3$ and $\lambda_2 = 4$ then $\det(A - \lambda I) = (\lambda - 3)(\lambda - 4) = \lambda^2 - 7\lambda + 12$. Always $\lambda_1 = \frac{1}{2}(a + d + \sqrt{(a-d)^2 + 4bc})$ and $\lambda_2 = \frac{1}{2}(a + d - \sqrt{\quad})$. Their sum is $a + d$.

19 (a) rank $= 2$ (b) $\det(B^{\mathrm{T}}B) = 0$ (d) eigenvalues of $(B + I)^{-1}$ are $1, \frac{1}{2}, \frac{1}{3}$.

21 $a = 0$, $b = 9$, $c = 0$ multiply $1, \lambda, \lambda^2$ in $\det(A - \lambda I) = 9\lambda - \lambda^3$: $A =$ *companion matrix*.

23 $\lambda = 1$ (for Markov), 0 (for singular), $-\frac{1}{2}$ (so sum of eigenvalues $=$ trace $= \frac{1}{2}$).

24 $\begin{bmatrix} 0 & 0 \\ 1 & 0 \end{bmatrix}, \begin{bmatrix} 0 & 1 \\ 0 & 0 \end{bmatrix}, \begin{bmatrix} -1 & 1 \\ -1 & 1 \end{bmatrix}$. Always $A^2 =$ zero matrix if $\lambda = 0, 0$ (Cayley-Hamilton 6.2.35).

27 $\lambda = 1, 2, 5, 7$.

29 B has $\lambda = -1, -1, -1, 3$ so $\det B = -3$. The 5 by 5 matrix A has $\lambda = 0, 0, 0, 0, 5$ and $B = A - I$ has $\lambda = -1, -1, -1, -1, 4$.

33 (a) $\boldsymbol{u}$ is a basis for the nullspace, $\boldsymbol{v}$ and $\boldsymbol{w}$ give a basis for the column space
(b) $\boldsymbol{x} = (0, \frac{1}{3}, \frac{1}{5})$ is a particular solution. Add any $c\boldsymbol{u}$ from the nullspace
(c) If $A\boldsymbol{x} = \boldsymbol{u}$ had a solution, $\boldsymbol{u}$ would be in the column space, giving dimension 3.

34 With $\lambda_1 = e^{2\pi i/3}$ and $\lambda_2 = e^{-2\pi i/3}$, the determinant is $\lambda_1\lambda_2 = 1$ and the trace is $\lambda_1 + \lambda_2 = -1$:

$$e^{2\pi i/3} + e^{-2\pi i/3} = \cos\frac{2\pi}{3} + i\sin\frac{2\pi}{3} + \cos\frac{2\pi}{3} - i\sin\frac{2\pi}{3} = -1. \text{ Also } \lambda_1^3 = \lambda_2^3 = 1.$$

$A = \left[\begin{smallmatrix} -1 & 1 \\ -1 & 0 \end{smallmatrix}\right]$ has this trace -1 and determinant 1. Then $A^3 = I$ and every $(M^{-1}AM)^3 = I$.
Choosing $\lambda_1 = \lambda_2 = 1$ leads to I or else to a matrix like $A = \left[\begin{smallmatrix} 1 & 1 \\ 0 & 1 \end{smallmatrix}\right]$ that has $A^3 \neq I$.

35 $\det(P-\lambda I)=0$ gives the equation $\lambda^3=1$. This reflects the fact that $P^3=I$. The solutions of $\lambda^3=1$ are $\lambda=1$ (real) and $\lambda=e^{2\pi i/3}, \lambda=e^{-2\pi i/3}$ (complex conjugates). The real eigenvector $\boldsymbol{x}_1=(1,1,1)$ is not changed by the permutation P. The complex eigenvectors are $\boldsymbol{x}_2=(1,e^{-2\pi i/3},e^{-4\pi i/3})$ and $\boldsymbol{x}_3=(1,e^{2\pi i/3},e^{4\pi i/3})=\overline{\boldsymbol{x}}_2$.

Problem Set 6.2, page 298

1 $\begin{bmatrix}1&2\\0&3\end{bmatrix}=\begin{bmatrix}1&1\\0&1\end{bmatrix}\begin{bmatrix}1&0\\0&3\end{bmatrix}\begin{bmatrix}1&-1\\0&1\end{bmatrix}$; $\begin{bmatrix}1&1\\2&2\end{bmatrix}=\begin{bmatrix}1&1\\-1&2\end{bmatrix}\begin{bmatrix}0&0\\0&3\end{bmatrix}\begin{bmatrix}\frac{2}{3}&-\frac{1}{3}\\\frac{1}{3}&\frac{1}{3}\end{bmatrix}$.

4 If $A=S\Lambda S^{-1}$ then the eigenvalue matrix for $A+2I$ is $\Lambda+2I$ and the eigenvector matrix is still S. $A+2I=S(\Lambda+2I)S^{-1}=S\Lambda S^{-1}+S(2I)S^{-1}=A+2I$.

5 (a) False: don't know λ's (b) True (c) True (d) False: need eigenvectors of S!.

7 The columns of S are nonzero multiples of $(2,1)$ and $(0,1)$ in either order. Same for A^{-1}.

9 $A^2=\begin{bmatrix}2&1\\1&1\end{bmatrix}$, $A^3=\begin{bmatrix}3&2\\2&1\end{bmatrix}$, $A^4=\begin{bmatrix}5&3\\3&2\end{bmatrix}$; $F_{20}=6765$.

10 (a) $A=\begin{bmatrix}.5&.5\\1&0\end{bmatrix}$ has $\lambda_1=1,\ \lambda_2=-\frac{1}{2}$ with $\boldsymbol{x}_1=(1,1),\ \boldsymbol{x}_2=(1,-2)$

(b) $A^n=\begin{bmatrix}1&1\\1&-2\end{bmatrix}\begin{bmatrix}1^n&0\\0&(-.5)^n\end{bmatrix}\begin{bmatrix}\frac{2}{3}&\frac{1}{3}\\\frac{1}{3}&-\frac{1}{3}\end{bmatrix}\to A^\infty=\begin{bmatrix}\frac{2}{3}&\frac{1}{3}\\\frac{2}{3}&\frac{1}{3}\end{bmatrix}$

11 $A=S\Lambda S^{-1}=\begin{bmatrix}1&1\\1&0\end{bmatrix}=\dfrac{1}{\lambda_1-\lambda_2}\begin{bmatrix}\lambda_1&\lambda_2\\1&1\end{bmatrix}\begin{bmatrix}\lambda_1&0\\0&\lambda_2\end{bmatrix}\begin{bmatrix}1&-\lambda_2\\-1&\lambda_1\end{bmatrix}$.

$S\Lambda^kS^{-1}=\dfrac{1}{\lambda_1-\lambda_2}\begin{bmatrix}\lambda_1&\lambda_2\\1&1\end{bmatrix}\begin{bmatrix}\lambda_1^k&0\\0&\lambda_2^k\end{bmatrix}\begin{bmatrix}1&-\lambda_2\\-1&\lambda_1\end{bmatrix}\begin{bmatrix}1\\0\end{bmatrix}=\begin{bmatrix}-\\(\lambda_1^k-\lambda_2^k)/(\lambda_1-\lambda_2)\end{bmatrix}$.

13 Direct computation gives $L_0,\ldots,L_{10}$ as $2,1,3,4,7,11,18,29,47,76,123$. My calculator gives $\lambda_1^{10}=(1.618\ldots)^{10}=122.991\ldots$.

16 (a) False: don't know λ (b) True: missing an eigenvector (c) True.

17 $A=\begin{bmatrix}8&3\\-3&2\end{bmatrix}$ (or other), $A=\begin{bmatrix}9&4\\-4&1\end{bmatrix}$, $A=\begin{bmatrix}10&5\\-5&0\end{bmatrix}$; only eigenvectors are $(c,-c)$.

19 $S\Lambda^kS^{-1}$ approaches zero if and only if every $|\lambda|<1$; $B^k\to 0$.

21 $\Lambda=\begin{bmatrix}.9&0\\0&.3\end{bmatrix}$, $S=\begin{bmatrix}3&-3\\1&1\end{bmatrix}$; $B^{10}\begin{bmatrix}3\\1\end{bmatrix}=(.9)^{10}\begin{bmatrix}3\\1\end{bmatrix}$, $B^{10}\begin{bmatrix}3\\-1\end{bmatrix}=(.3)^{10}\begin{bmatrix}3\\-1\end{bmatrix}$, $B^{10}\begin{bmatrix}6\\0\end{bmatrix}=$ sum of those two.

23 $B^k=\begin{bmatrix}1&1\\0&-1\end{bmatrix}\begin{bmatrix}3&0\\0&2\end{bmatrix}^k\begin{bmatrix}1&1\\0&-1\end{bmatrix}=\begin{bmatrix}3^k&3^k-2^k\\0&2^k\end{bmatrix}$.

25 trace $AB=(aq+bs)+(cr+dt)=(qa+rc)+(sb+td)=$ trace BA. Proof for diagonalizable case: the trace of $S\Lambda S^{-1}$ is the trace of $(\Lambda S^{-1})S=\Lambda$ which is *the sum of* the λ's.

28 The A's form a subspace since cA and $A_1 + A_2$ have the same S. When $S = I$ the A's give the subspace of diagonal matrices. Dimension 4.

30 Two problems: The nullspace and column space can overlap, so $\boldsymbol{x}$ could be in both. There may not be r independent eigenvectors in the column space.

31 $R = S\sqrt{\Lambda}S^{-1} = \begin{bmatrix} 2 & 1 \\ 1 & 2 \end{bmatrix}$ has $R^2 = A$. $\sqrt{B}$ would have $\lambda = \sqrt{9}$ and $\lambda = \sqrt{-1}$ so its trace is not real. Note $\begin{bmatrix} -1 & 0 \\ 0 & -1 \end{bmatrix}$ can have $\sqrt{-1} = i$ and $-i$, and real square root $\begin{bmatrix} 0 & 1 \\ -1 & 0 \end{bmatrix}$.

32 $A^{\mathrm{T}} = A$ gives $\boldsymbol{x}^{\mathrm{T}}AB\boldsymbol{x} = (A\boldsymbol{x})^{\mathrm{T}}(B\boldsymbol{x}) \leq \|A\boldsymbol{x}\|\|B\boldsymbol{x}\|$ by the Schwarz inequality. $B^{\mathrm{T}} = -B$ gives $-\boldsymbol{x}^{\mathrm{T}}BA\boldsymbol{x} = (B\boldsymbol{x})^{\mathrm{T}}A\boldsymbol{x} \leq \|A\boldsymbol{x}\|\|B\boldsymbol{x}\|$. Add these to get Heisenberg when $AB - BA = I$.

35 If $A = S\Lambda S^{-1}$ then the product $(A - \lambda_1 I) \cdots (A - \lambda_n I)$ equals $S(\Lambda - \lambda_1 I) \cdots (\Lambda - \lambda_n I)S^{-1}$. The factor $\Lambda - \lambda_j I$ is zero in row j. *The product is zero in all rows* = zero matrix.

38 (a) The eigenvectors for $\lambda = 0$ always span the nullspace (b) The eigenvectors for $\lambda \neq 0$ span the column space if there are r independent eigenvectors: then algebraic multiplicity = geometric multiplicity for each nonzero λ.

39 The eigenvalues $2, -1, 0$ and their eigenvectors are in Λ and S. Then $A^k = S\Lambda^k S^{-1}$ is

$$\begin{bmatrix} 2 & 1 & 0 \\ 1 & -1 & 1 \\ 1 & -1 & -1 \end{bmatrix} \begin{bmatrix} 2^k & & \\ & (-1)^k & \\ & & 0^k \end{bmatrix} \frac{1}{6} \begin{bmatrix} 4 & 1 & 1 \\ 2 & -2 & -2 \\ 0 & 3 & -3 \end{bmatrix} = \frac{2^k}{6} \begin{bmatrix} 4 & 2 & 2 \\ 2 & 1 & 1 \\ 2 & 1 & 1 \end{bmatrix} + \frac{(-1)^k}{3} \begin{bmatrix} 1 & -1 & -1 \\ -1 & 1 & 1 \\ -1 & 1 & 1 \end{bmatrix}$$

Check $k = 1$! The $(2,2)$ entry of A^4 is $2^4/6 + (-1)^4/3 = 18/6 = 3$. The 4-step paths that begin and end at node 2 are 2 to 1 to 1 to 1 to 2, 2 to 1 to 2 to 1 to 2, and 2 to 1 to 3 to 1 to 2. Harder to find the eleven 4-step paths that start and end at node 1.

41 $AB = BA$ always has the solution $B = A$. (In case $A = 0$ every B is a solution.)

42 B has $\lambda = i$ and $-i$, so B^4 has $\lambda^4 = 1$ and 1; C has $\lambda = (1 \pm \sqrt{3}i)/2 = \exp(\pm\pi i/3)$ so $\lambda^3 = -1$ and -1. Then $C^3 = -I$ and $C^{1024} = -C$.

Problem Set 6.3, page 315

1 $\boldsymbol{u}_1 = e^{4t}\begin{bmatrix} 1 \\ 0 \end{bmatrix}$, $\boldsymbol{u}_2 = e^{t}\begin{bmatrix} 1 \\ -1 \end{bmatrix}$. If $\boldsymbol{u}(0) = (5, -2)$, then $\boldsymbol{u}(t) = 3e^{4t}\begin{bmatrix} 1 \\ 0 \end{bmatrix} + 2e^{t}\begin{bmatrix} 1 \\ -1 \end{bmatrix}$.

4 $\begin{bmatrix} 6 & -2 \\ 2 & 1 \end{bmatrix}$ has $\lambda_1 = 5$, $\boldsymbol{x}_1 = \begin{bmatrix} 2 \\ 1 \end{bmatrix}$, $\lambda_2 = 2$, $\boldsymbol{x}_2 = \begin{bmatrix} 1 \\ 2 \end{bmatrix}$; rabbits $r(t) = 20e^{5t} + 10e^{2t}$, $w(t) = 10e^{5t} + 20e^{2t}$. The ratio of rabbits to wolves approaches $20/10$; e^{5t} dominates.

5 $d(v+w)/dt = dv/dt + dw/dt = (w - v) + (v - w) = 0$, so the total $v + w$ is constant. $A = \begin{bmatrix} -1 & 1 \\ 1 & -1 \end{bmatrix}$ has $\lambda_1 = 0$ and $\lambda_2 = -2$ with $\boldsymbol{x}_1 = \begin{bmatrix} 1 \\ 1 \end{bmatrix}$ and $\boldsymbol{x}_2 = \begin{bmatrix} 1 \\ -1 \end{bmatrix}$; $\begin{matrix} v(1) = 20 + 10e^{-2} \\ w(1) = 20 - 10e^{-2} \end{matrix}$.

8 $A = \begin{bmatrix} 0 & 1 \\ -9 & 6 \end{bmatrix}$ has trace 6, det 9, $\lambda = 3$ and 3 with only one independent eigenvector $(1, 3)$.

9 $my'' + by' + ky = 0$ is $\begin{bmatrix} m & 0 \\ 0 & 1 \end{bmatrix} \begin{bmatrix} y' \\ y \end{bmatrix}' = \begin{bmatrix} -b & -k \\ 1 & 0 \end{bmatrix} \begin{bmatrix} y' \\ y \end{bmatrix}$.

10 When A is skew-symmetric, $\|\boldsymbol{u}(t)\| = \|e^{At}\boldsymbol{u}(0)\| = \|\boldsymbol{u}(0)\|$. So e^{At} is an *orthogonal* matrix.

13 $\boldsymbol{u}_p = A^{-1}\boldsymbol{b} = 4$ and $u(t) = ce^{2t} + 4$; $\boldsymbol{u}_p = \begin{bmatrix} 4 \\ 2 \end{bmatrix}$ and $\boldsymbol{u}(t) = c_1 e^{2t} \begin{bmatrix} 1 \\ 0 \end{bmatrix} + c_2 e^{3t} \begin{bmatrix} 0 \\ 1 \end{bmatrix} + \begin{bmatrix} 4 \\ 2 \end{bmatrix}$.

14 Substituting $\boldsymbol{u} = e^{ct}\boldsymbol{v}$ gives $ce^{ct}\boldsymbol{v} = Ae^{ct}\boldsymbol{v} - e^{ct}\boldsymbol{b}$ or $(A - cI)\boldsymbol{v} = \boldsymbol{b}$ or $\boldsymbol{v} = (A - cI)^{-1}\boldsymbol{b} =$ particular solution. If c is an eigenvalue then $A - cI$ is not invertible.

18 The solution at time $t + T$ is also $e^{A(t+T)}\boldsymbol{u}(0)$. Thus e^{At} times e^{AT} equals $e^{A(t+T)}$.

19 $\begin{bmatrix} 1 & 1 \\ 0 & 0 \end{bmatrix} = \begin{bmatrix} 1 & 1 \\ 0 & -1 \end{bmatrix} \begin{bmatrix} 1 & 0 \\ 0 & 0 \end{bmatrix} \begin{bmatrix} 1 & 1 \\ 0 & -1 \end{bmatrix}$; $e^{At} = \begin{bmatrix} 1 & 1 \\ 0 & -1 \end{bmatrix} \begin{bmatrix} e^t & 0 \\ 0 & 1 \end{bmatrix} \begin{bmatrix} 1 & 1 \\ 0 & -1 \end{bmatrix} = \begin{bmatrix} e^t & e^t - 1 \\ 0 & 1 \end{bmatrix}$.

20 If $A^2 = A$ then $e^{At} = I + At + \frac{1}{2}At^2 + \frac{1}{6}At^3 + \cdots = I + (e^t - 1)A = \begin{bmatrix} 1 & 0 \\ 0 & 1 \end{bmatrix} + \begin{bmatrix} e^t - 1 & e^t - 1 \\ 0 & 0 \end{bmatrix}$.

22 $A = \begin{bmatrix} 1 & 1 \\ 0 & 3 \end{bmatrix} = \begin{bmatrix} 1 & 1 \\ 2 & 0 \end{bmatrix} \begin{bmatrix} 3 & 0 \\ 0 & 1 \end{bmatrix} \begin{bmatrix} 0 & \frac{1}{2} \\ 1 & -\frac{1}{2} \end{bmatrix}$, then $e^{At} = \begin{bmatrix} e^t & \frac{1}{2}(e^{3t} - e^t) \\ 0 & e^{3t} \end{bmatrix}$.

24 (a) The inverse of e^{At} is e^{-At} (b) If $A\boldsymbol{x} = \lambda\boldsymbol{x}$ then $e^{At}\boldsymbol{x} = e^{\lambda t}\boldsymbol{x}$ and $e^{\lambda t} \neq 0$.

25 $x(t) = e^{4t}$ and $y(t) = -e^{4t}$ is a growing solution. The correct matrix for the exchanged unknown $\boldsymbol{u} = (y, x)$ is $\begin{bmatrix} 2 & -2 \\ -4 & 0 \end{bmatrix}$ and it *does* have the same eigenvalues as the original matrix.

Problem Set 6.4, page 326

3 $\lambda = 0, 2, -1$ with unit eigenvectors $\pm(0, 1, -1)/\sqrt{2}$ and $\pm(2, 1, 1)/\sqrt{6}$ and $\pm(1, -1, -1)/\sqrt{3}$.

5 $Q = \dfrac{1}{3} \begin{bmatrix} 2 & 1 & 2 \\ 2 & -2 & -1 \\ -1 & -2 & 2 \end{bmatrix}$.

8 If $A^3 = 0$ then all $\lambda^3 = 0$ so all $\lambda = 0$ as in $A = \begin{bmatrix} 0 & 1 \\ 0 & 0 \end{bmatrix}$. If A is symmetric then $A^3 = Q\Lambda^3 Q^{\mathrm{T}} = 0$ gives $\Lambda = 0$ and the only symmetric possibility is $A = Q\,0\,Q^{\mathrm{T}} =$ zero matrix.

10 If $\boldsymbol{x}$ is not real then $\lambda = \boldsymbol{x}^{\mathrm{T}}A\boldsymbol{x}/\boldsymbol{x}^{\mathrm{T}}\boldsymbol{x}$ is *not* necessarily real. Can't assume real eigenvectors!

11 $\begin{bmatrix}3&1\\1&3\end{bmatrix}=2\begin{bmatrix}\frac12&-\frac12\\-\frac12&\frac12\end{bmatrix}+4\begin{bmatrix}\frac12&\frac12\\\frac12&\frac12\end{bmatrix}$; $\begin{bmatrix}9&12\\12&16\end{bmatrix}=0\begin{bmatrix}.64&-.48\\-.48&.36\end{bmatrix}+25\begin{bmatrix}.36&.48\\.48&.64\end{bmatrix}$

14 Skew-symmetric and orthogonal; $\lambda=i,\ i,\ -i,\ -i$ to have trace zero.

16 (a) If $Az=\lambda y$ and $A^{\mathrm T}y=\lambda z$ then $B[\,y;\ \ -z\,]=[\,-Az;\ \ A^{\mathrm T}y\,]=-\lambda[\,y;\ \ -z\,]$. So $-\lambda$ is also an eigenvalue of B. (b) $A^{\mathrm T}Az=A^{\mathrm T}(\lambda y)=\lambda^2 z$. The eigenvalues of $A^{\mathrm T}A$ are ≥ 0 (c) $\lambda=-1,\ -1,\ 1,\ 1$; $\boldsymbol{x}_1=(1,0,-1,0)$, $\boldsymbol{x}_2=(0,1,0,-1)$, $\boldsymbol{x}_3=(1,0,1,0)$, $\boldsymbol{x}_4=(0,1,0,1)$.

19 B has eigenvectors in $S=\begin{bmatrix}1&0&1\\0&1&0\\0&0&1+d\end{bmatrix}\to\begin{bmatrix}1&0&1\\0&1&0\\0&0&2\end{bmatrix}$; independent but not perpendicular.

21 (a) False. $A=\begin{bmatrix}1&2\\0&1\end{bmatrix}$ (b) True (c) True. $A^{-1}=Q\Lambda^{-1}Q^{\mathrm T}$ is also symmetric (d) False.

23 A and $A^{\mathrm T}$ have the same λ's but the *order* of the $\boldsymbol{x}$'s can change. $A=\begin{bmatrix}0&1\\-1&0\end{bmatrix}$ has $\lambda_1=i$ and $\lambda_2=-i$ with $\boldsymbol{x}_1=(1,i)$ for A but $\boldsymbol{x}_1=(1,-i)$ for $A^{\mathrm T}$.

24 A is invertible, orthogonal, permutation, diagonalizable, Markov; B is projection, diagonalizable, Markov. $QR,\ S\Lambda S^{-1},\ Q\Lambda Q^{\mathrm T}$ possible for A; $S\Lambda S^{-1}$ and $Q\Lambda Q^{\mathrm T}$ possible for B.

25 Symmetry gives $Q\Lambda Q^{\mathrm T}$ when $b=1$; repeated λ and no S when $b=-1$; singular if $b=0$.

26 Orthogonal and symmetric requires $|\lambda|=1$ and λ real, so every $\lambda=\pm1$. Then $A=\pm I$ or $A=Q\Lambda Q^{\mathrm T}=\begin{bmatrix}\cos\theta&-\sin\theta\\\sin\theta&\cos\theta\end{bmatrix}\begin{bmatrix}1&0\\0&-1\end{bmatrix}\begin{bmatrix}\cos\theta&\sin\theta\\-\sin\theta&\cos\theta\end{bmatrix}=\begin{bmatrix}\cos2\theta&\sin2\theta\\\sin2\theta&-\cos2\theta\end{bmatrix}=$ reflection.

28 The roots of $\lambda^2+b\lambda+c=0$ differ by $\sqrt{b^2-4c}$. For $\det(A+tB-\lambda I)$ we have $b=-3-8t$ and $c=2+16t-t^2$. The minimum of b^2-4c is $1/17$ at $t=2/17$. Then $\lambda_2-\lambda_1=1/\sqrt{17}$.

Problem Set 6.5, page 339

2 Positive definite for $-3<b<3$ $\begin{bmatrix}1&0\\b&1\end{bmatrix}\begin{bmatrix}1&b\\0&9-b^2\end{bmatrix}=\begin{bmatrix}1&0\\b&1\end{bmatrix}\begin{bmatrix}1&0\\0&9-b^2\end{bmatrix}\begin{bmatrix}1&b\\0&1\end{bmatrix}=LDL^{\mathrm T}$;

Positive definite for $c>8$ $\begin{bmatrix}1&0\\2&1\end{bmatrix}\begin{bmatrix}2&4\\0&c-8\end{bmatrix}=\begin{bmatrix}1&0\\2&1\end{bmatrix}\begin{bmatrix}2&0\\0&c-8\end{bmatrix}\begin{bmatrix}1&2\\0&1\end{bmatrix}=LDL^{\mathrm T}$.

3 $f(x,y)=x^2+4xy+9y^2=(x+2y)^2+5y^2$; $f(x,y)=x^2+6xy+9y^2=(x+3y)^2$.

6 $\boldsymbol{x}^{\mathrm T}A^{\mathrm T}A\boldsymbol{x}=(A\boldsymbol{x})^{\mathrm T}(A\boldsymbol{x})=0$ only if $A\boldsymbol{x}=\boldsymbol{0}$. Since A has independent columns this only happens when $\boldsymbol{x}=\boldsymbol{0}$.

8 $A=\begin{bmatrix}3&6\\6&16\end{bmatrix}=\begin{bmatrix}1&0\\2&1\end{bmatrix}\begin{bmatrix}3&0\\0&4\end{bmatrix}\begin{bmatrix}1&2\\0&1\end{bmatrix}$. Pivots outside squares, and L inside.

10 $A = \begin{bmatrix} 2 & -1 & 0 \\ -1 & 2 & -1 \\ 0 & -1 & 2 \end{bmatrix}$ has pivots $2, \frac{3}{2}, \frac{4}{3}$; $A = \begin{bmatrix} 2 & -1 & -1 \\ -1 & 2 & -1 \\ -1 & -1 & 2 \end{bmatrix}$ is singular; $A \begin{bmatrix} 1 \\ 1 \\ 1 \end{bmatrix} = \begin{bmatrix} 0 \\ 0 \\ 0 \end{bmatrix}$.

12 A is positive definite for $c > 1$; determinants $c, c^2 - 1, c^3 + 2 - 3c > 0$. B is never positive definite (determinants $d - 4$ and $-4d + 12$ are never both positive).

14 The eigenvalues of A^{-1} are positive because they are $1/\lambda(A)$. And the entries of A^{-1} pass the determinant tests. And $\boldsymbol{x}^{\mathrm{T}} A^{-1} \boldsymbol{x} = (A^{-1}\boldsymbol{x})^{\mathrm{T}} A (A^{-1}\boldsymbol{x}) > 0$ for all $\boldsymbol{x} \neq \boldsymbol{0}$.

17 If a_{jj} were smaller than all the eigenvalues, $A - a_{jj}I$ would have *positive* eigenvalues (so positive definite). But $A - a_{jj}I$ has a *zero* in the (j, j) position; impossible by Problem 16.

21 A is positive definite when $s > 8$; B is positive definite when $t > 5$ (check determinants).

22 $R = \frac{1}{\sqrt{2}} \begin{bmatrix} 1 & -1 \\ 1 & 1 \end{bmatrix} \begin{bmatrix} \sqrt{9} & \\ & \sqrt{1} \end{bmatrix} \frac{1}{\sqrt{2}} \begin{bmatrix} 1 & 1 \\ -1 & 1 \end{bmatrix} = \begin{bmatrix} 2 & 1 \\ 1 & 2 \end{bmatrix}$; $R = Q \begin{bmatrix} 4 & 0 \\ 0 & 2 \end{bmatrix} Q^{\mathrm{T}} = \begin{bmatrix} 3 & 1 \\ 1 & 3 \end{bmatrix}$.

24 The ellipse $x^2 + xy + y^2 = 1$ has axes with half-lengths $a = 1/\sqrt{\lambda_1} = \sqrt{2}$ and $b = \sqrt{2/3}$.

25 $A = \begin{bmatrix} 9 & 3 \\ 3 & 5 \end{bmatrix}$; $C = \begin{bmatrix} 2 & 0 \\ 4 & 3 \end{bmatrix}$.

29 $A_1 = \begin{bmatrix} 6x^2 & 2x \\ 2x & 2 \end{bmatrix}$ is positive definite if $x \neq 0$; $f_1 = (\frac{1}{2}x^2 + y)^2 = 0$ on the curve $\frac{1}{2}x^2 + y = 0$; $A_2 = \begin{bmatrix} 6x & 1 \\ 1 & 0 \end{bmatrix} = \begin{bmatrix} 6 & 1 \\ 1 & 0 \end{bmatrix}$ is indefinite and $(0, 1)$ is a saddle point.

31 If $c > 9$ the graph of z is a bowl, if $c < 9$ the graph has a saddle point. When $c = 9$ the graph of $z = (2x + 3y)^2$ is a trough staying at zero on the line $2x + 3y = 0$.

32 Orthogonal matrices, exponentials e^{At}, matrices with det $= 1$ are groups. Examples of subgroups are orthogonal matrices with det $= 1$, exponentials e^{An} for integer n.

Problem Set 6.6, page 349

1 $C = (MN)^{-1}A(MN)$ so if B is similar to A and C is similar to B, then A *is similar to* C.

6 Eight families of similar matrices: 6 matrices have $\lambda = 0, 1$; 3 matrices have $\lambda = 1, 1$ and 3 have $\lambda = 0, 0$ (two families each!); one has $\lambda = 1, -1$; one has $\lambda = 2, 0$; two have $\lambda = \frac{1}{2}(1 \pm \sqrt{5})$.

7 (a) $(M^{-1}AM)(M^{-1}\boldsymbol{x}) = M^{-1}(A\boldsymbol{x}) = M^{-1}\boldsymbol{0} = \boldsymbol{0}$ (b) The nullspaces of A and of $M^{-1}AM$ have the same *dimension*. Different vectors and different bases.

8 $\begin{bmatrix} 0 & 1 \\ 0 & 0 \end{bmatrix}$ and $\begin{bmatrix} 0 & 2 \\ 0 & 0 \end{bmatrix}$ have the same line of eigenvectors and the same eigenvalues 0, 0.

10 $J^2 = \begin{bmatrix} c^2 & 2c \\ 0 & c^2 \end{bmatrix}$, $J^3 = \begin{bmatrix} c^3 & 3c^2 \\ 0 & c^3 \end{bmatrix}$, $J^k = \begin{bmatrix} c^k & kc^{k-1} \\ 0 & c^k \end{bmatrix}$; $J^0 = I$, $J^{-1} = \begin{bmatrix} c^{-1} & -c^{-2} \\ 0 & c^{-1} \end{bmatrix}$.

13 (1) Choose M_i = reverse diagonal matrix to get $M_i^{-1}J_iM_i = M_i^{\mathrm{T}}$ in each block (2) M_0 has those blocks M_i on its block diagonal to get $M_0^{-1}JM_0 = J^{\mathrm{T}}$. (3) $A^{\mathrm{T}} = (M^{-1})^{\mathrm{T}}J^{\mathrm{T}}M^{\mathrm{T}}$ is $(M^{-1})^{\mathrm{T}}M_0^{-1}JM_0M^{\mathrm{T}} = (MM_0M^{\mathrm{T}})^{-1}A(MM_0M^{\mathrm{T}})$, and A^{T} is similar to A.

17 (a) True: One has $\lambda = 0$, the other doesn't (b) False. Diagonalize a nonsymmetric matrix and Λ is symmetric (c) False: $\begin{bmatrix} 0 & 1 \\ -1 & 0 \end{bmatrix}$ and $\begin{bmatrix} 0 & -1 \\ 1 & 0 \end{bmatrix}$ are similar (d) True: All eigenvalues of $A+I$ are increased by 1, so different from the eigenvalues of A.

18 $AB = B^{-1}(BA)B$ so AB is similar to BA. *Also* $AB\boldsymbol{x} = \lambda\boldsymbol{x}$ leads to $BA(B\boldsymbol{x}) = \lambda(B\boldsymbol{x})$.

19 Diagonals 6 by 6 and 4 by 4; AB has all the same eigenvalues as BA plus $6-4$ zeros.

Problem Set 6.7, page 360

2 (a) $AA^{\mathrm{T}} = \begin{bmatrix} 17 & 34 \\ 34 & 68 \end{bmatrix}$ has $\sigma_1^2 = 85$, $\boldsymbol{u}_1 = \begin{bmatrix} 1/\sqrt{5} \\ 2/\sqrt{5} \end{bmatrix}$, $\boldsymbol{u}_2 = \begin{bmatrix} 2/\sqrt{5} \\ -1/\sqrt{5} \end{bmatrix}$.

(b) $A\boldsymbol{v}_1 = \begin{bmatrix} 1 & 4 \\ 2 & 8 \end{bmatrix}\begin{bmatrix} 1/\sqrt{17} \\ 4/\sqrt{17} \end{bmatrix} = \begin{bmatrix} \sqrt{17} \\ 2\sqrt{17} \end{bmatrix} = \sqrt{85}\begin{bmatrix} 1/\sqrt{5} \\ 2/\sqrt{5} \end{bmatrix} = \sigma_1\boldsymbol{u}_1$.

4 $A^{\mathrm{T}}A = AA^{\mathrm{T}} = \begin{bmatrix} 2 & 1 \\ 1 & 1 \end{bmatrix}$ has eigenvalues $\sigma_1^2 = \dfrac{3+\sqrt{5}}{2}$ and $\sigma_2^2 = \dfrac{3-\sqrt{5}}{2}$.

Since $A = A^{\mathrm{T}}$ the eigenvectors of $A^{\mathrm{T}}A$ are the same as for A. Since $\lambda_2 = \frac{1-\sqrt{5}}{2}$ is *negative*, $\sigma_1 = \lambda_1$ but $\sigma_2 = -\lambda_2$. The eigenvectors are the same as in Section 6.2 for A, except for the effect of this minus sign:

$$\boldsymbol{u}_1 = \boldsymbol{v}_1 = \begin{bmatrix} \lambda_1/\sqrt{1+\lambda_1^2} \\ 1/\sqrt{1+\lambda_1^2} \end{bmatrix} \text{ and } \boldsymbol{u}_2 = -\boldsymbol{v}_2 = \begin{bmatrix} \lambda_2/\sqrt{1+\lambda_2^2} \\ 1/\sqrt{1+\lambda_2^2} \end{bmatrix}.$$

6 A proof that ***eigshow*** finds the SVD for 2 by 2 matrices. Starting at the orthogonal pair $\boldsymbol{V}_1 = (1,0)$, $\boldsymbol{V}_2 = (0,1)$ the demo finds $A\boldsymbol{V}_1$ and $A\boldsymbol{V}_2$ at angle θ. After a 90° turn by the mouse to $\boldsymbol{V}_2, -\boldsymbol{V}_1$ the demo finds $A\boldsymbol{V}_2$ and $-A\boldsymbol{V}_1$ at angle $\pi - \theta$. Somewhere between, the constantly orthogonal $\boldsymbol{v}_1, \boldsymbol{v}_2$ must have produced $A\boldsymbol{v}_1$ and $A\boldsymbol{v}_2$ at angle $\theta = \pi/2$. Those are the orthogonal directions for $\boldsymbol{u}_1$ and $\boldsymbol{u}_2$.

8 $A = UV^{\mathrm{T}}$ since all $\sigma_j = 1$.

14 The smallest change in A is to set its smallest singular value σ_2 to zero.

16 The singular values of $A+I$ are not σ_j+1. They come from eigenvalues of $(A+I)^{\mathrm{T}}(A+I)$.

Problem Set 7.1, page 367

4 (a) $S(T(\boldsymbol{v})) = \boldsymbol{v}$ (b) $S(T(\boldsymbol{v}_1) + T(\boldsymbol{v}_2)) = S(T(\boldsymbol{v}_1)) + S(T(\boldsymbol{v}_2))$.

5 Choose $\boldsymbol{v} = (1,1)$ and $\boldsymbol{w} = (-1,0)$. Then $T(\boldsymbol{v}) + T(\boldsymbol{w}) = \boldsymbol{v} + \boldsymbol{w}$ but $T(\boldsymbol{v}+\boldsymbol{w}) = (0,0)$.

7 (a) $T(T(\boldsymbol{v})) = \boldsymbol{v}$ (b) $T(T(\boldsymbol{v})) = \boldsymbol{v}+(2,2)$ (c) $T(T(\boldsymbol{v})) = -\boldsymbol{v}$ (d) $T(T(\boldsymbol{v})) = T(\boldsymbol{v})$.

10 (a) $T(1,0) = \mathbf{0}$ (b) $(0,0,1)$ is not in the range (c) $T(0,1) = \mathbf{0}$.

12 $T(\boldsymbol{v}) = (4,4)$; $(2,2)$; $(2,2)$; if $\boldsymbol{v} = (a,b) = b(1,1) + \frac{a-b}{2}(2,0)$ then $T(\boldsymbol{v}) = b(2,2) + (0,0)$.

16 No matrix A gives $A\begin{bmatrix} 0 & 0 \\ 1 & 0 \end{bmatrix} = \begin{bmatrix} 0 & 1 \\ 0 & 0 \end{bmatrix}$. To professors: The matrix space has dimension 4. Linear transformations come from 4 by 4 matrices. Those in Problems 13–15 were special.

17 (a) True (b) True (c) True (d) False.

20 $T(T^{-1}(M)) = M$ so $T^{-1}(M) = A^{-1}MB^{-1}$.

21 (a) Horizontal lines stay horizontal, vertical lines stay vertical (b) House squashes onto a line (c) Vertical lines stay vertical.

24 (a) $ad - bc = 0$ (b) $ad - bc > 0$ (c) $|ad - bc| = 1$. If vectors to two corners transform to themselves then by linearity $T = I$. (Fails if one corner is $(0,0)$.)

27 This emphasizes that circles are transformed to ellipses (figure in Section 6.7).

Problem Set 7.2, page 380

3 $A^2 = B$ when $T^2 = S$ and output basis = input basis.

6 $T(\boldsymbol{v}_1 + \boldsymbol{v}_2 + \boldsymbol{v}_3) = 2\boldsymbol{w}_1 + \boldsymbol{w}_2 + 2\boldsymbol{w}_3$; A times $(1,1,1)$ gives $(2,1,2)$.

7 $\boldsymbol{v} = c(\boldsymbol{v}_2 - \boldsymbol{v}_3)$ gives $T(\boldsymbol{v}) = \mathbf{0}$; nullspace is $(0, c, -c)$; solutions are $(1,0,0)$ + any $(0, c, -c)$.

9 We don't know $T(\boldsymbol{w})$ unless the $\boldsymbol{w}$'s are the same as the $\boldsymbol{v}$'s. In that case the matrix is A^2.

13 (c) is wrong because $\boldsymbol{w}_1$ is not generally in the input space.

15 (a) $\begin{bmatrix} 2 & 1 \\ 5 & 3 \end{bmatrix}$ (b) $\begin{bmatrix} 3 & -1 \\ -5 & 2 \end{bmatrix}$ = inverse of (a) (c) $A\begin{bmatrix} 2 \\ 6 \end{bmatrix}$ must be $2A\begin{bmatrix} 1 \\ 3 \end{bmatrix}$.

17 $MN = \begin{bmatrix} 1 & 0 \\ 1 & 2 \end{bmatrix}\begin{bmatrix} 2 & 1 \\ 5 & 3 \end{bmatrix}^{-1} = \begin{bmatrix} 3 & -1 \\ -7 & 3 \end{bmatrix}$.

19 $(a,b) = (\cos\theta, -\sin\theta)$. Minus sign from $Q^{-1} = Q^{\mathrm{T}}$.

21 $\boldsymbol{w}_2(x) = 1 - x^2$; $\boldsymbol{w}_3(x) = \frac{1}{2}(x^2 - x)$; $\boldsymbol{y} = 4\boldsymbol{w}_1 + 5\boldsymbol{w}_2 + 6\boldsymbol{w}_3$.

24 The matrix M with these nine entries must be invertible.

28 If T is not invertible, $T(\boldsymbol{v}_1), \ldots, T(\boldsymbol{v}_n)$ will not be a basis. We couldn't choose $\boldsymbol{w}_i = T(\boldsymbol{v}_i)$.

31 $S(T(\boldsymbol{v})) = (-1, 2)$ but $S(\boldsymbol{v}) = (-2, 1)$ and $T(S(\boldsymbol{v})) = (1, -2)$.

Problem Set 7.3, page 389

2 The last step writes 6, 6, 2, 2 as the overall average 4, 4, 4, 4 plus the difference 2, 2, -2, -2. Therefore $c_1 = 4$ and $c_2 = 2$ and $c_3 = 1$ and $c_4 = 1$.

3 The wavelet basis is (1, 1, 1, 1, 1, 1, 1, 1) and the long wavelet and two medium wavelets (1, 1,
$-1, -1, 0, 0, 0, 0)$ and $(0, 0, 0, 0, 1, 1, -1, -1)$ and 4 short wavelets with a single pair $1, -1$.

6 If $V\boldsymbol{b} = W\boldsymbol{c}$ then $\boldsymbol{b} = V^{-1}W\boldsymbol{c}$. The change of basis matrix is $V^{-1}W$.

7 The transpose of $WW^{-1} = I$ is $(W^{-1})^{\mathrm{T}}W^{\mathrm{T}} = I$. So the matrix W^{T} (which has the $\boldsymbol{w}$'s in its rows) is the inverse to the matrix that has the $\boldsymbol{w}^*$'s in its columns.

Problem Set 7.4, page 397

1 $A^{\mathrm{T}}A = \begin{bmatrix} 10 & 20 \\ 20 & 40 \end{bmatrix}$ has $\lambda = 50$ and 0, $\boldsymbol{v}_1 = \frac{1}{\sqrt{5}}\begin{bmatrix} 1 \\ 2 \end{bmatrix}$, $\boldsymbol{v}_2 = \frac{1}{\sqrt{5}}\begin{bmatrix} 2 \\ -1 \end{bmatrix}$; $\sigma_1 = \sqrt{50}$.

5 $A = QH = \frac{1}{\sqrt{50}}\begin{bmatrix} 7 & -1 \\ 1 & 7 \end{bmatrix} \frac{1}{\sqrt{50}}\begin{bmatrix} 10 & 20 \\ 20 & 40 \end{bmatrix}$. H is semidefinite because A is singular.

6 $A^+ = V\begin{bmatrix} 1/\sqrt{50} & 0 \\ 0 & 0 \end{bmatrix} U^{\mathrm{T}} = \frac{1}{50}\begin{bmatrix} 1 & 3 \\ 2 & 6 \end{bmatrix}$; $A^+A = \begin{bmatrix} .2 & .4 \\ .4 & .8 \end{bmatrix}$, $AA^+ = \begin{bmatrix} .1 & .3 \\ .3 & .9 \end{bmatrix}$.

9 $\begin{bmatrix} \sigma_1\boldsymbol{u}_1 & \sigma_2\boldsymbol{u}_2 \end{bmatrix}\begin{bmatrix} \boldsymbol{v}_1^{\mathrm{T}} \\ \boldsymbol{v}_2^{\mathrm{T}} \end{bmatrix} = \sigma_1\boldsymbol{u}_1\boldsymbol{v}_1^{\mathrm{T}} + \sigma_2\boldsymbol{u}_2\boldsymbol{v}_2^{\mathrm{T}}$. In general this is $\sigma_1\boldsymbol{u}_1\boldsymbol{v}_1^{\mathrm{T}} + \cdots + \sigma_r\boldsymbol{u}_r\boldsymbol{v}_r^{\mathrm{T}}$.

11 A^+ is A^{-1} because A is invertible.

13 $A = [\,1\,]\,[\,5 \;\; 0 \;\; 0\,]V^{\mathrm{T}}$ and $A^+ = V\begin{bmatrix} .2 \\ 0 \\ 0 \end{bmatrix}[\,1\,] = \begin{bmatrix} .12 \\ .16 \\ 0 \end{bmatrix}$; $AA^+ = [\,1\,]$;

$$A^+A = \begin{bmatrix} .36 & .48 & 0 \\ .48 & .64 & 0 \\ 0 & 0 & 0 \end{bmatrix}$$

15 If $\det A = 0$ then $\operatorname{rank}(A) < n$; thus $\operatorname{rank}(A^+) < n$ and $\det A^+ = 0$.

18 $\boldsymbol{x}^+$ in the row space of A is perpendicular to $\widehat{\boldsymbol{x}} - \boldsymbol{x}^+$ in the nullspace of $A^{\mathrm{T}}A =$ nullspace of A. The right triangle has $c^2 = a^2 + b^2$.

19 $AA^+\boldsymbol{p} = \boldsymbol{p}$, $AA^+\boldsymbol{e} = \boldsymbol{0}$, $A^+A\boldsymbol{x}_r = \boldsymbol{x}_r$, $A^+A\boldsymbol{x}_n = \boldsymbol{0}$.

21 L is determined by ℓ_{21}. Each eigenvector in S is determined by one number. The counts are $1+3$ for LU, $1+2+1$ for LDU, $1+3$ for QR, $1+2+1$ for $U\Sigma V^{\mathrm{T}}$, $2+2+0$ for $S\Lambda S^{-1}$.

24 Keep only the r by r invertible corner Σ_r of Σ (the rest is all zero). Then $A = U\Sigma V^{\mathrm{T}}$ has the required form $A = \widehat{U}M_1\Sigma_r M_2^{\mathrm{T}}\widehat{V}^{\mathrm{T}}$ with an invertible $M = M_1\Sigma_r M_2^{\mathrm{T}}$ in the middle.

Problem Set 8.1, page 410

3 The rows of the free-free matrix in equation (9) add to $[0\ \ 0\ \ 0]$ so the right side needs $f_1+f_2+f_3=0$. For $\boldsymbol{f}=(-1,0,1)$ elimination gives $c_2u_1-c_2u_2=-1, c_3u_2-c_3u_3=-1$, and $0=0$. Then $\boldsymbol{u}_{\text{particular}}=(-c_2^{-1}-c_3^{-1},-c_3^{-1},0)$. Add any multiple of $\boldsymbol{u}_{\text{nullspace}}=(1,1,1)$.

4 $\displaystyle\int -\frac{d}{dx}\left(c(x)\frac{du}{dx}\right)dx=\left[c(0)\frac{du}{dx}(0)-c(1)\frac{du}{dx}(1)\right]=0$ so we need $\displaystyle\int f(x)\,dx=0$.

6 Multiply $A_1^{\mathrm T}C_1A_1$ as columns of $A_1^{\mathrm T}$ times c's times rows of A_1. The first "element matrix" $c_1E_1=[1\ \ 0\ \ 0]^{\mathrm T}c_1[1\ \ 0\ \ 0]$ has c_1 in the top left corner.

8 The solution to $-u''=1$ with $u(0)=u(1)=0$ is $u(x)=\frac{1}{2}(x-x^2)$. At $x=\frac{1}{5},\frac{2}{5},\frac{3}{5},\frac{4}{5}$ this $u(x)$ equals $\boldsymbol{u}=2,3,3,2$ (discrete solution in Problem 7) times $(\Delta x)^2=1/25$.

11 Forward vs. backward differences for du/dx have a big effect on the discrete $\boldsymbol{u}$, because that term has the large coefficient 10 (and with 100 or 1000 we would have a real boundary layer = near discontinuity at $x=1$). The computed values are $\boldsymbol{u}=0, .01, .03, .04, .05, .06, .07, .11, 0$ versus $\boldsymbol{u}=0, .12, .24, .36, .46, .54, .55, .43, 0$.

The MATLAB code is $E=\mathrm{diag}(\mathrm{ones}(6,1),1)$; $K=64*(2*\mathrm{eye}(7)-E-E')$; $D=80*(E-\mathrm{eye}(7))$; $(K+D)\backslash\mathrm{ones}(7,1)$, $(K-D')\backslash\mathrm{ones}(7,1)$.

Problem Set 8.2, page 420

1 $A=\begin{bmatrix}-1&1&0\\-1&0&1\\0&-1&1\end{bmatrix}$; nullspace contains $\begin{bmatrix}c\\c\\c\end{bmatrix}$; $\begin{bmatrix}1\\0\\0\end{bmatrix}$ is not orthogonal to that nullspace.

2 $A^{\mathrm T}\boldsymbol{y}=\boldsymbol{0}$ for $\boldsymbol{y}=(1,-1,1)$; current $=1$ along edge 1, edge 3, back on edge 2 (full loop).

5 Kirchhoff's Current Law $A^{\mathrm T}\boldsymbol{y}=\boldsymbol{f}$ is solvable for $\boldsymbol{f}=(1,-1,0)$ and not solvable for $\boldsymbol{f}=(1,0,0)$; $\boldsymbol{f}$ must be orthogonal to $(1,1,1)$ in the nullspace.

6 $A^{\mathrm T}A\boldsymbol{x}=\begin{bmatrix}2&-1&-1\\-1&2&-1\\-1&-1&2\end{bmatrix}\boldsymbol{x}=\begin{bmatrix}3\\-3\\0\end{bmatrix}=\boldsymbol{f}$ produces $\boldsymbol{x}=\begin{bmatrix}1\\-1\\0\end{bmatrix}+\begin{bmatrix}c\\c\\c\end{bmatrix}$; potentials $1,-1,0$ and currents $-A\boldsymbol{x}=2,1,-1$; $\boldsymbol{f}$ sends 3 units into node 1 and out from node 2.

7 $A^{\mathrm T}\begin{bmatrix}1&&\\&2&\\&&2\end{bmatrix}A=\begin{bmatrix}3&-1&-2\\-1&3&-2\\-2&-2&4\end{bmatrix}$; $\boldsymbol{f}=\begin{bmatrix}1\\0\\-1\end{bmatrix}$ yields $\boldsymbol{x}=\begin{bmatrix}5/4\\1\\7/8\end{bmatrix}+\begin{bmatrix}c\\c\\c\end{bmatrix}$; potentials $\frac{5}{4}$, 1, $\frac{7}{8}$ and currents $-CA\boldsymbol{x}=\frac{1}{4},\frac{3}{4},\frac{1}{4}$.

9 Elimination on $A\boldsymbol{x}=\boldsymbol{b}$ always leads to $\boldsymbol{y}^{\mathrm T}\boldsymbol{b}=0$ which is $-b_1+b_2-b_3=0$ and $b_3-b_4+b_5=0$ ($\boldsymbol{y}$'s from Problem 8 in the left nullspace). This is Kirchhoff's Voltage Law around the loops.

11 $A^{\mathrm{T}}A = \begin{bmatrix} 2 & -1 & -1 & 0 \\ -1 & 3 & -1 & -1 \\ -1 & -1 & 3 & -1 \\ 0 & -1 & -1 & 2 \end{bmatrix}$ diagonal entry = number of edges into the node off-diagonal entry = -1 if nodes are connected.

13 $A^{\mathrm{T}}CAx = \begin{bmatrix} 4 & -2 & -2 & 0 \\ -2 & 8 & -3 & -3 \\ -2 & -3 & 8 & -3 \\ 0 & -3 & -3 & 6 \end{bmatrix} x = \begin{bmatrix} 1 \\ 0 \\ 0 \\ -1 \end{bmatrix}$ gives potentials $x = (\frac{5}{12}, \frac{1}{6}, \frac{1}{6}, 0)$ (grounded $x_4 = 0$ and solved 3 equations); $y = -CAx = (\frac{2}{3}, \frac{2}{3}, 0, \frac{1}{2}, \frac{1}{2})$.

17 (a) 8 independent columns (b) f must be orthogonal to the nullspace so $f_1 + \cdots + f_9 = 0$ (c) Each edge goes into 2 nodes, 12 edges make diagonal entries sum to 24.

Problem Set 8.3, page 428

2 $A = \begin{bmatrix} .6 & -1 \\ .4 & 1 \end{bmatrix} \begin{bmatrix} 1 & \\ & .75 \end{bmatrix} \begin{bmatrix} 1 & 1 \\ -.4 & .6 \end{bmatrix}$;

A^k approaches $\begin{bmatrix} .6 & -1 \\ .4 & -1 \end{bmatrix} \begin{bmatrix} 1 & 0 \\ 0 & 0 \end{bmatrix} \begin{bmatrix} 1 & 1 \\ -.4 & .6 \end{bmatrix} = \begin{bmatrix} .6 & .6 \\ .4 & .4 \end{bmatrix}$.

3 $\lambda = 1$ and $.8$, $x = (1, 0)$; $\lambda = 1$ and $-.8$, $x = (\frac{5}{9}, \frac{4}{9})$; $\lambda = 1, \frac{1}{4}$, and $\frac{1}{4}$, $x = (\frac{1}{3}, \frac{1}{3}, \frac{1}{3})$.

5 The steady state is $(0, 0, 1)$ = all dead.

6 If $Ax = \lambda x$, add components on both sides to find $s = \lambda s$. If $\lambda \neq 1$ the sum must be $s = 0$.

8 $(.5)^k \to 0$ gives $A^k \to A^\infty$; any $A = \begin{bmatrix} .6 + .4a & .6 - .6a \\ .4 - .4a & .4 + .6a \end{bmatrix}$ with $-\frac{2}{3} \leq a \leq 1$.

10 M^2 is still nonnegative; $[1 \ \cdots \ 1]M = [1 \ \cdots \ 1]$ so multiply by M to find $[1 \ \cdots \ 1]M^2 = [1 \ \cdots \ 1] \Rightarrow$ columns of M^2 add to 1.

11 $\lambda = 1$ and $a + d - 1$ from the trace; steady state is a multiple of $x_1 = (b, 1 - a)$.

13 B has $\lambda = 0$ and $-.5$ with $x_1 = (.3, .2)$ and $x_2 = (-1, 1)$; $e^{-.5t}$ approaches zero and the solution approaches $c_1 e^{0t} x_1 = c_1 x_1$.

15 The eigenvector is $x = (1, 1, 1)$ and $Ax = (.9, .9, .9)$.

18 $p = \begin{bmatrix} 8 \\ 6 \end{bmatrix}$ and $\begin{bmatrix} 130 \\ 32 \end{bmatrix}$; $I - \begin{bmatrix} .5 & 1 \\ .5 & 0 \end{bmatrix}$ has no inverse.

19 $\lambda = 1$ (Markov), 0 (singular), .2 (from trace). Steady state $(.3, .3, .4)$ and $(30, 30, 40)$.

20 *No*, A has an eigenvalue $\lambda = 1$ and $(I - A)^{-1}$ does not exist.

Problem Set 8.4, page 436

1 Feasible set = line segment from $(6, 0)$ to $(0, 3)$; minimum cost at $(6, 0)$, maximum at $(0, 3)$.

2 Feasible set is 4-sided with corners $(0, 0)$, $(6, 0)$, $(2, 2)$, $(0, 6)$. Minimize $2x - y$ at $(6, 0)$.

3 Only two corners $(4, 0, 0)$ and $(0, 2, 0)$; choose x_1 very negative, $x_2 = 0$, and $x_3 = x_1 - 4$.

4 From $(0, 0, 2)$ move to $\boldsymbol{x} = (0, 1, 1.5)$ with the constraint $x_1 + x_2 + 2x_3 = 4$. The new cost is $3(1) + 8(1.5) = \$15$ so $r = -1$ is the reduced cost. The simplex method also checks $\boldsymbol{x} = (1, 0, 1.5)$ with cost $5(1) + 8(1.5) = \$17$ so $r = 1$ (more expensive).

5 Cost = 20 at start $(4, 0, 0)$; keeping $x_1 + x_2 + 2x_3 = 4$ move to $(3, 1, 0)$ with cost 18 and $r = -2$; or move to $(2, 0, 1)$ with cost 17 and $r = -3$. Choose x_3 as entering variable and move to $(0, 0, 2)$ with cost 14. Another step to reach $(0, 4, 0)$ with minimum cost 12.

6 $\boldsymbol{c} = [\,3 \;\; 5 \;\; 7\,]$ has minimum cost 12 by the Ph.D. since $\boldsymbol{x} = (4, 0, 0)$ is minimizing. The dual problem maximizes $4y$ subject to $y \le 3$, $y \le 5$, $y \le 7$. Maximum = 12.

Problem Set 8.5, page 442

1 $\int_0^{2\pi} \cos(j + k)x \, dx = \left[\frac{\sin(j+k)x}{j+k}\right]_0^{2\pi} = 0$ and similarly $\int_0^{2\pi} \cos(j - k)x \, dx = 0$ (in the denominator notice $j - k \ne 0$). If $j = k$ then $\int_0^{2\pi} \cos^2 jx \, dx = \pi$.

4 $\int_{-1}^{1} (1)(x^3 - cx) \, dx = 0$ and $\int_{-1}^{1} (x^2 - \frac{1}{3})(x^3 - cx) \, dx = 0$ for all c (integral of an odd function). Choose c so that $\int_{-1}^{1} x(x^3 - cx) \, dx = [\frac{1}{5}x^5 - \frac{c}{3}x^3]_{-1}^{1} = \frac{2}{5} - c\frac{2}{3} = 0$. Then $c = \frac{3}{5}$.

5 The integrals lead to $a_1 = 0$, $b_1 = 4/\pi$, $b_2 = 0$.

6 From equation (3) the a_k are zero and $b_k = 4/\pi k$. The square wave has $\|f\|^2 = 2\pi$. Then equation (6) is $2\pi = \pi(16/\pi^2)(\frac{1}{1^2} + \frac{1}{3^2} + \frac{1}{5^2} + \cdots)$ so this infinite series equals $\pi^2/8$.

8 $\|\boldsymbol{v}\|^2 = 1 + \frac{1}{2} + \frac{1}{4} + \frac{1}{8} + \cdots = 2$ so $\|\boldsymbol{v}\| = \sqrt{2}$; $\|\boldsymbol{v}\|^2 = 1 + a^2 + a^4 + \cdots = 1/(1 - a^2)$ so $\|\boldsymbol{v}\| = 1/\sqrt{1 - a^2}$; $\int_0^{2\pi} (1 + 2\sin x + \sin^2 x) \, dx = 2\pi + 0 + \pi$ so $\|f\| = \sqrt{3\pi}$.

9 (a) $f(x) = \frac{1}{2} + \frac{1}{2}$ (square wave) so a's are $\frac{1}{2}$, 0, 0, . . ., and b's are $2/\pi$, 0, $-2/3\pi$, 0, $2/5\pi$, . . . (b) $a_0 = \int_0^{2\pi} x \, dx/2\pi = \pi$, other $a_k = 0$, $b_k = -2/k$.

11 $\cos^2 x = \frac{1}{2} + \frac{1}{2}\cos 2x$; $\cos(x + \frac{\pi}{3}) = \cos x \cos\frac{\pi}{3} - \sin x \sin\frac{\pi}{3} = \frac{1}{2}\cos x - \frac{\sqrt{3}}{2}\sin x$.

13 $dy/dx = \cos x$ has $y = y_p + y_n = \sin x + C$.

Problem Set 8.6, page 448

1 (x, y, z) has homogeneous coordinates $(x, y, z, 1)$ and also (cx, cy, cz, c) for any nonzero c.

4 $S = \begin{bmatrix} c & & & \\ & c & & \\ & & c & \\ & & & 1 \end{bmatrix}$, $ST = \begin{bmatrix} c & & & \\ & c & & \\ & & c & \\ 1 & 4 & 3 & 1 \end{bmatrix}$, $TS = \begin{bmatrix} c & & & \\ & c & & \\ & & c & \\ c & 4c & 3c & 1 \end{bmatrix}$, use $\boldsymbol{v}TS$.

5 $S = \begin{bmatrix} 1/8.5 & & \\ & 1/11 & \\ & & 1 \end{bmatrix}$ for a 1 by 1 square.

9 $\boldsymbol{n} = (\frac{2}{3}, \frac{2}{3}, \frac{1}{3})$ has $\|\boldsymbol{n}\| = 1$ and $P = I - \boldsymbol{n}\boldsymbol{n}^{\mathrm{T}} = \frac{1}{9}\begin{bmatrix} 5 & -4 & -2 \\ -4 & 5 & -2 \\ -2 & -2 & 8 \end{bmatrix}$.

10 Choose $(0, 0, 3)$ on the plane and multiply $T_-PT_+ = \frac{1}{9}\begin{bmatrix} 5 & -4 & -2 & 0 \\ -4 & 5 & -2 & 0 \\ -2 & -2 & 8 & 0 \\ 6 & 6 & 3 & 9 \end{bmatrix}$.

11 $(3, 3, 3)$ projects to $\frac{1}{3}(-1, -1, 4)$ and $(3, 3, 3, 1)$ projects to $(\frac{1}{3}, \frac{1}{3}, \frac{5}{3}, 1)$.

13 The projection of a cube is a hexagon.

14 $(3, 3, 3)(I - 2\boldsymbol{n}\boldsymbol{n}^{\mathrm{T}}) = (\frac{1}{3}, \frac{1}{3}, \frac{1}{3})\begin{bmatrix} 1 & -8 & -4 \\ -8 & 1 & -4 \\ -4 & -4 & 7 \end{bmatrix} = (-\frac{11}{3}, -\frac{11}{3}, -\frac{1}{3})$.

15 $(3, 3, 3, 1) \to (3, 3, 0, 1) \to (-\frac{7}{3}, -\frac{7}{3}, -\frac{8}{3}, 1) \to (-\frac{7}{3}, -\frac{7}{3}, \frac{1}{3}, 1)$.

17 Rescaled by $1/c$ because (x, y, z, c) is the same point as $(x/c, y/c, z/c, 1)$.

Problem Set 9.1, page 457

1 Without exchange, pivots .001 and 1000; with exchange, pivots 1 and -1. When the pivot is larger than the entries below it, ℓ_{ij} = entry/pivot has $|\ell_{ij}| \le 1$. $A = \begin{bmatrix} 1 & 1 & 1 \\ 0 & 1 & -1 \\ -1 & 1 & 1 \end{bmatrix}$.

4 The largest $\|\boldsymbol{x}\| = \|A^{-1}\boldsymbol{b}\|$ is $1/\lambda_{\min}$; the largest error is $10^{-16}/\lambda_{\min}$.

5 Each row of U has at most w entries. Then w multiplications to substitute components of $\boldsymbol{x}$ (already known from below) and divide by the pivot. Total for n rows is less than wn.

6 L, U, and R need $\frac{1}{2}n^2$ multiplications to solve a linear system. Q needs n^2 to multiply the right side by $Q^{-1} = Q^{\mathrm{T}}$. So QR takes 1.5 times longer than LU to reach $\boldsymbol{x}$.

7 On column j of I, back substitution needs $\frac{1}{2}j^2$ multiplications (only the j by j upper left block is involved). Then $\frac{1}{2}(1^2 + 2^2 + \cdots + n^2) \approx \frac{1}{2}(\frac{1}{3}n^3)$.

10 With 16-digit floating point arithmetic the errors $\|x - y_{\text{computed}}\|$ for $\varepsilon = 10^{-3}$, 10^{-6}, 10^{-9}, 10^{-12}, 10^{-15} are of order 10^{-16}, 10^{-11}, 10^{-7}, 10^{-4}, 10^{-3}.

11 $\cos\theta = 1/\sqrt{10}$, $\sin\theta = -3/\sqrt{10}$, $R = \frac{1}{\sqrt{10}}\begin{bmatrix} 1 & 3 \\ -3 & 1 \end{bmatrix}\begin{bmatrix} 1 & -1 \\ 3 & 5 \end{bmatrix} = \frac{1}{\sqrt{10}}\begin{bmatrix} 10 & 14 \\ 0 & 8 \end{bmatrix}$.

14 $Q_{ij}A$ uses $4n$ multiplications (2 for each entry in rows i and j). By factoring out $\cos\theta$, the entries 1 and $\pm\tan\theta$ need only $2n$ multiplications, which leads to $\frac{2}{3}n^3$ for QR.

Problem Set 9.2, page 463

1 $\|A\| = 2$, $c = 2/.5 = 4$; $\|A\| = 3$, $c = 3/1 = 3$; $\|A\| = 2+\sqrt{2}$, $c = (2+\sqrt{2})/(2-\sqrt{2}) = 5.83$.

3 For the first inequality replace x by Bx in $\|Ax\| \le \|A\|\|x\|$; the second inequality is just $\|Bx\| \le \|B\|\|x\|$. Then $\|AB\| = \max(\|ABx\|/\|x\|) \le \|A\|\|B\|$.

7 The triangle inequality gives $\|Ax + Bx\| \le \|Ax\| + \|Bx\|$. Divide by $\|x\|$ and take the maximum over all nonzero vectors to find $\|A + B\| \le \|A\| + \|B\|$.

8 If $Ax = \lambda x$ then $\|Ax\|/\|x\| = |\lambda|$ for that particular vector x. When we maximize the ratio over all vectors we get $\|A\| \ge |\lambda|$.

13 The residual $b - Ay = (10^{-7}, 0)$ is much smaller than $b - Az = (.0013, .0016)$. But z is much closer to the solution than y.

14 $\det A = 10^{-6}$ so $A^{-1} = \begin{bmatrix} 659{,}000 & -563{,}000 \\ -913{,}000 & 780{,}000 \end{bmatrix}$. Then $\|A\| > 1$, $\|A^{-1}\| > 10^6$, $c > 10^6$.

16 $x_1^2 + \cdots + x_n^2$ is not smaller than $\max(x_i^2)$ and not larger than $x_1^2 + \cdots + x_n^2 + 2|x_1||x_2| + \cdots = \|x\|_1^2$. Certainly $x_1^2 + \cdots + x_n^2 \le n \max(x_i^2)$ so $\|x\| \le \sqrt{n}\|x\|_\infty$. Choose $y_i = \operatorname{sign} x_i = \pm 1$ to get $x \cdot y = \|x\|_1$. By Schwarz this is at most $\|x\|\|y\| = \sqrt{n}\|x\|$. Choose $x = (1, 1, \ldots, 1)$ for $\sqrt{n}$.

Problem Set 9.3, page 473

2 If $Ax = \lambda x$ then $(I - A)x = (1 - \lambda)x$. Real eigenvalues of $B = I - A$ have $|1 - \lambda| < 1$ provided λ is between 0 and 2.

6 Jacobi has $S^{-1}T = \frac{1}{3}\begin{bmatrix} 0 & 1 \\ 1 & 0 \end{bmatrix}$ with $|\lambda|_{\max} = \frac{1}{3}$.

7 Gauss-Seidel has $S^{-1}T = \begin{bmatrix} 0 & \frac{1}{3} \\ 0 & \frac{1}{9} \end{bmatrix}$ with $|\lambda|_{\max} = \frac{1}{9} = (|\lambda|_{\max} \text{ for Jacobi})^2$.

9 Set the trace $2 - 2\omega + \frac{1}{4}\omega^2$ equal to $(\omega - 1) + (\omega - 1)$ to find $\omega_{\text{opt}} = 4(2 - \sqrt{3}) \approx 1.07$. The eigenvalues $\omega - 1$ are about .07.

15 The jth component of $A\boldsymbol{x}_1$ is $2\sin\frac{j\pi}{n+1} - \sin\frac{(j-1)\pi}{n+1} - \sin\frac{(j+1)\pi}{n+1}$. The last two terms, using $\sin(a+b) = \sin a\cos b + \cos a\sin b$, combine into $-2\sin\frac{j\pi}{n+1}\cos\frac{\pi}{n+1}$. The eigenvalue is $\lambda_1 = 2 - 2\cos\frac{\pi}{n+1}$.

17 $A^{-1} = \frac{1}{3}\begin{bmatrix}2 & 1\\ 1 & 2\end{bmatrix}$ gives $\boldsymbol{u}_0 = \begin{bmatrix}1\\0\end{bmatrix}$, $\boldsymbol{u}_1 = \frac{1}{3}\begin{bmatrix}2\\1\end{bmatrix}$, $\boldsymbol{u}_2 = \frac{1}{9}\begin{bmatrix}5\\4\end{bmatrix}$, $\boldsymbol{u}_3 = \frac{1}{27}\begin{bmatrix}14\\13\end{bmatrix} \to \begin{bmatrix}1\\1\end{bmatrix}$.

18 $R = Q^{\mathrm{T}}A = \begin{bmatrix}1 & \cos\theta\sin\theta\\ 0 & -\sin^2\theta\end{bmatrix}$ and $A_1 = RQ = \begin{bmatrix}\cos\theta(1+\sin^2\theta) & -\sin^3\theta\\ -\sin^3\theta & -\cos\theta\sin^2\theta\end{bmatrix}$.

20 If $A - cI = QR$ then $A_1 = RQ + cI = Q^{-1}(QR + cI)Q = Q^{-1}AQ$. No change in eigenvalues.

21 Multiply $A\boldsymbol{q}_j = b_{j-1}\boldsymbol{q}_{j-1} + a_j\boldsymbol{q}_j + b_j\boldsymbol{q}_{j+1}$ by $\boldsymbol{q}_j^{\mathrm{T}}$ to find $\boldsymbol{q}_j^{\mathrm{T}}A\boldsymbol{q}_j = a_j$ (because the $\boldsymbol{q}$'s are orthonormal). The matrix form (multiplying by columns) is $AQ = QT$ where T is *tridiagonal*. Its entries are the a's and b's.

23 If A is symmetric then $A_1 = Q^{-1}AQ = Q^{\mathrm{T}}AQ$ is also symmetric. $A_1 = RQ = R(QR)R^{-1} = RAR^{-1}$ has R and R^{-1} upper triangular, so A_1 cannot have nonzeros on a lower diagonal than A. If A is tridiagonal and symmetric then (by using symmetry for the upper part of A_1) the matrix $A_1 = RAR^{-1}$ is also tridiagonal.

27 From the last line of code, $\boldsymbol{q}_2$ is in the direction of $\boldsymbol{v} = A\boldsymbol{q}_1 - h_{11}\boldsymbol{q}_1 = A\boldsymbol{q}_1 - (\boldsymbol{q}_1^{\mathrm{T}}A\boldsymbol{q}_1)\boldsymbol{q}_1$. The dot product with $\boldsymbol{q}_1$ is zero. This is Gram-Schmidt with $A\boldsymbol{q}_1$ as the second input vector.

28 $\boldsymbol{r}_1 = \boldsymbol{b} - \alpha_1 A\boldsymbol{b} = \boldsymbol{b} - (\boldsymbol{b}^{\mathrm{T}}\boldsymbol{b}/\boldsymbol{b}^{\mathrm{T}}A\boldsymbol{b})A\boldsymbol{b}$ is orthogonal to $\boldsymbol{r}_0 = \boldsymbol{b}$: *the residuals* $\boldsymbol{r} = \boldsymbol{b} - A\boldsymbol{x}$ *are orthogonal at each step*. To show that $\boldsymbol{p}_1$ is orthogonal to $A\boldsymbol{p}_0 = A\boldsymbol{b}$, simplify $\boldsymbol{p}_1$ to $c\boldsymbol{P}_1$: $\boldsymbol{P}_1 = \|A\boldsymbol{b}\|^2\boldsymbol{b} - (\boldsymbol{b}^{\mathrm{T}}A\boldsymbol{b})A\boldsymbol{b}$ and $c = \boldsymbol{b}^{\mathrm{T}}\boldsymbol{b}/(\boldsymbol{b}^{\mathrm{T}}A\boldsymbol{b})^2$. Certainly $(A\boldsymbol{b})^{\mathrm{T}}\boldsymbol{P}_1 = 0$ because $A^{\mathrm{T}} = A$. (That simplification put α_1 into $\boldsymbol{p}_1 = \boldsymbol{b} - \alpha_1 A\boldsymbol{b} + (\boldsymbol{b}^{\mathrm{T}}\boldsymbol{b} - 2\alpha_1\boldsymbol{b}^{\mathrm{T}}A\boldsymbol{b} + \alpha_1^2\|A\boldsymbol{b}\|^2)\boldsymbol{b}/\boldsymbol{b}^{\mathrm{T}}\boldsymbol{b}$. For a good discussion see *Numerical Linear Algebra* by Trefethen and Bau.)

Problem Set 10.1, page 483

2 In polar form these are $\sqrt{5}e^{i\theta}$, $5e^{2i\theta}$, $\frac{1}{\sqrt{5}}e^{-i\theta}$, $\sqrt{5}$.

4 $|z \times w| = 6$, $|z + w| \le 5$, $|z/w| = \frac{2}{3}$, $|z - w| \le 5$.

5 $a + ib = \frac{\sqrt{3}}{2} + \frac{1}{2}i$, $\frac{1}{2} + \frac{\sqrt{3}}{2}i$, i, $-\frac{1}{2} + \frac{\sqrt{3}}{2}i$; $w^{12} = 1$.

9 $2+i$; $(2+i)(1+i) = 1+3i$; $e^{-i\pi/2} = -i$; $e^{-i\pi} = -1$; $\frac{1-i}{1+i} = -i$; $(-i)^{103} = (-i)^3 = i$.

10 $z + \bar{z}$ is real; $z - \bar{z}$ is pure imaginary; $z\bar{z}$ is positive; $z/\bar{z}$ has absolute value 1.

12 (a) When $a = b = d = 1$ the square root becomes $\sqrt{4c}$; λ is complex if $c < 0$ (b) $\lambda = 0$ and $\lambda = a + d$ when $ad = bc$ (c) the λ's can be real and different.

13 Complex λ's when $(a+d)^2 < 4(ad - bc)$; write $(a+d)^2 - 4(ad - bc)$ as $(a-d)^2 + 4bc$ which is positive when $bc > 0$.

14 $\det(P-\lambda I) = \lambda^4 - 1 = 0$ has $\lambda = 1, -1, i, -i$ with eigenvectors $(1,1,1,1)$ and $(1,-1,1,-1)$ and $(1,i,-1,-i)$ and $(1,-i,-1,i) =$ columns of Fourier matrix.

16 The block matrix has real eigenvalues; so $i\lambda$ is real and λ is pure imaginary.

18 $r = 1$, angle $\frac{\pi}{2} - \theta$; multiply by $e^{i\theta}$ to get $e^{i\pi/2} = i$.

21 $\cos 3\theta = \mathrm{Re}(\cos\theta + i\sin\theta)^3 = \cos^3\theta - 3\cos\theta\sin^2\theta$; $\sin 3\theta = \mathrm{Im}(\cos\theta + i\sin\theta)^3 = 3\cos^2\theta\sin\theta - \sin^3\theta$.

23 (a) e^i is at angle $\theta = 1$ on the unit circle; $|i^e| = 1^e = 1$ (c) There are infinitely many candidates $i^e = e^{i(\pi/2+2\pi n)e}$.

24 (a) Unit circle (b) Spiral in to $e^{-2\pi}$ (c) Circle continuing around to angle $\theta = 2\pi^2$.

Problem Set 10.2, page 492

3 $z =$ multiple of $(1+i, 1+i, -2)$; $Az = \mathbf{0}$ gives $z^{\mathrm{H}}A^{\mathrm{H}} = \mathbf{0}^{\mathrm{H}}$ so z (not $\bar{z}$!) is orthogonal to all columns of A^{H} (using complex inner product z^{H} times column).

4 The four fundamental subspaces are $\boldsymbol{C}(A)$, $\boldsymbol{N}(A)$, $\boldsymbol{C}(A^{\mathrm{H}})$, $\boldsymbol{N}(A^{\mathrm{H}})$.

5 (a) $(A^{\mathrm{H}}A)^{\mathrm{H}} = A^{\mathrm{H}}A^{\mathrm{HH}} = A^{\mathrm{H}}A$ again (b) If $A^{\mathrm{H}}Az = \mathbf{0}$ then $(z^{\mathrm{H}}A^{\mathrm{H}})(Az) = 0$. This is $\|Az\|^2 = 0$ so $Az = \mathbf{0}$. The nullspaces of A and $A^{\mathrm{H}}A$ are the ***same***. $A^{\mathrm{H}}A$ is invertible when $\boldsymbol{N}(A) = \{\mathbf{0}\}$.

6 (a) False: $A = \begin{bmatrix} 0 & 1 \\ -1 & 0 \end{bmatrix}$ (b) True: $-i$ is not an eigenvalue if $A = A^{\mathrm{H}}$ (c) False.

10 $(1,1,1)$, $(1, e^{2\pi i/3}, e^{4\pi i/3})$, $(1, e^{4\pi i/3}, e^{2\pi i/3})$ are orthogonal (complex inner product!) because P is an orthogonal matrix—and therefore unitary.

11 $C = \begin{bmatrix} 2 & 5 & 4 \\ 4 & 2 & 5 \\ 5 & 4 & 2 \end{bmatrix} = 2 + 5P + 4P^2$ has $\lambda = 2+5+4 = 11$, $2 + 5e^{2\pi i/3} + 4e^{4\pi i/3}$, $2 + 5e^{4\pi i/3} + 4e^{8\pi i/3}$.

13 The determinant is the product of the eigenvalues (all real).

15 $A = \frac{1}{\sqrt{3}}\begin{bmatrix} 1 & -1+i \\ 1+i & 1 \end{bmatrix}\begin{bmatrix} 2 & 0 \\ 0 & -1 \end{bmatrix}\frac{1}{\sqrt{3}}\begin{bmatrix} 1 & 1-i \\ -1-i & 1 \end{bmatrix}$.

18 $V = \frac{1}{L}\begin{bmatrix} 1+\sqrt{3} & -1+i \\ 1+i & 1+\sqrt{3} \end{bmatrix}\begin{bmatrix} 1 & 0 \\ 0 & -1 \end{bmatrix}\frac{1}{L}\begin{bmatrix} 1+\sqrt{3} & 1-i \\ -1-i & 1+\sqrt{3} \end{bmatrix}$ with $L^2 = 6+2\sqrt{3}$ has $|\lambda| = 1$. $V = V^{\mathrm{H}}$ gives real λ, trace zero gives $\lambda = 1, -1$.

19 The $\boldsymbol{v}$'s are columns of a unitary matrix U. Then $z = UU^{\mathrm{H}}z =$ (multiply by columns) $= \boldsymbol{v}_1(\boldsymbol{v}_1^{\mathrm{H}}z) + \cdots + \boldsymbol{v}_n(\boldsymbol{v}_n^{\mathrm{H}}z)$.

20 Don't multiply e^{-ix} times e^{ix}; conjugate the first, then $\int_0^{2\pi} e^{2ix}\,dx = [e^{2ix}/2i]_0^{2\pi} = 0$.

22 $R + iS = (R + iS)^{\mathrm{H}} = R^{\mathrm{T}} - iS^{\mathrm{T}}$; R is symmetric but S is skew-symmetric.

24 $[\,1\,]$ and $[\,-1\,]$; any $[\,e^{i\theta}\,]$; $\begin{bmatrix} a & b+ic \\ b-ic & d \end{bmatrix}$; $\begin{bmatrix} w & e^{i\phi}\overline{z} \\ -z & e^{i\phi}\overline{w} \end{bmatrix}$ with $|w|^2 + |z|^2 = 1$.

27 Unitary means $U^{\mathrm{H}}U = I$ or $(A^{\mathrm{T}} - iB^{\mathrm{T}})(A + iB) = (A^{\mathrm{T}}A + B^{\mathrm{T}}B) + i(A^{\mathrm{T}}B - B^{\mathrm{T}}A) = I$. Then $A^{\mathrm{T}}A + B^{\mathrm{T}}B = I$ and $A^{\mathrm{T}}B - B^{\mathrm{T}}A = 0$ which makes the block matrix orthogonal.

30 $A = \begin{bmatrix} 1-i & 1-i \\ -1 & 2 \end{bmatrix} \begin{bmatrix} 1 & 0 \\ 0 & 4 \end{bmatrix} \frac{1}{6} \begin{bmatrix} 2+2i & -2 \\ 1+i & 2 \end{bmatrix} = S\Lambda S^{-1}$.

Problem Set 10.3, page 500

8 $\boldsymbol{c} \to (1,1,1,1,0,0,0,0) \to (4,0,0,0,0,0,0,0) \to (4,0,0,0,4,0,0,0)$ which is $F_8\boldsymbol{c}$. The second vector becomes $(0,0,0,0,1,1,1,1) \to (0,0,0,0,4,0,0,0) \to (4,0,0,0,-4,0,0,0)$.

9 If $w^{64} = 1$ then w^2 is a 32nd root of 1 and $\sqrt{w}$ is a 128th root of 1.

13 $e_1 = c_0 + c_1 + c_2 + c_3$ and $e_2 = c_0 + c_1 i + c_2 i^2 + c_3 i^3$; E contains the four eigenvalues of C.

14 Eigenvalues $e_1 = 2 - 1 - 1 = 0$, $e_2 = 2 - i - i^3 = 2$, $e_3 = 2 - (-1) - (-1) = 4$, $e_4 = 2 - i^3 - i^9 = 2$. Check trace $0 + 2 + 4 + 2 = 8$.

15 Diagonal E needs n multiplications, Fourier matrix F and F^{-1} need $\frac{1}{2}n\log_2 n$ multiplications each by the **FFT**. Total much less than the ordinary n^2.

16 $(c_0 + c_2) + (c_1 + c_3)$; then $(c_0 - c_2) + i(c_1 - c_3)$; then $(c_0 + c_2) - (c_1 + c_3)$; then $(c_0 - c_2) - i(c_1 - c_3)$. These steps are the **FFT**!

A FINAL EXAM

***This was the final exam on December 17, 2002 in* MIT's linear algebra course 18.06**

1 The 4 by 6 matrix A has all 2's below the diagonal and elsewhere all 1's:

$$A = \begin{bmatrix} 1 & 1 & 1 & 1 & 1 & 1 \\ 2 & 1 & 1 & 1 & 1 & 1 \\ 2 & 2 & 1 & 1 & 1 & 1 \\ 2 & 2 & 2 & 1 & 1 & 1 \end{bmatrix}$$

(a) By elimination factor A into L (4 by 4) times U (4 by 6).

(b) Find the rank of A and a basis for its nullspace (the special solutions would be good).

2 Suppose you know that the 3 by 4 matrix A has the vector $\boldsymbol{s} = (2, 3, 1, 0)$ as a basis for its nullspace.

(a) What is the *rank* of A and the complete solution to $A\boldsymbol{x} = \boldsymbol{0}$?

(b) What is the exact row reduced echelon form R of A?

3 The following matrix is a *projection matrix*:

$$P = \frac{1}{21}\begin{bmatrix} 1 & 2 & -4 \\ 2 & 4 & -8 \\ -4 & -8 & 16 \end{bmatrix}.$$

(a) What subspace does P project onto?

(b) What is the *distance* from that subspace to $\boldsymbol{b} = (1, 1, 1)$?

(c) What are the three eigenvalues of P? Is P diagonalizable?

4 **(a)** Suppose the product of A and B is the zero matrix: $AB = 0$. Then the **(1)** space of A contains the **(2)** space of B. Also the **(3)** space of B contains the **(4)** space of A. Those blank words are

(1) ____________________ **(2)** ____________________

(3) ____________________ **(4)** ____________________

(b) Suppose that matrix A is 5 by 7 with rank r, and B is 7 by 9 of rank s. What are the dimensions of spaces **(1)** and **(2)**? From the fact that space **(1)** contains space **(2)**, what do you learn about $r+s$?

5 Suppose the 4 by 2 matrix Q has orthonormal columns.

(a) Find the least squares solution $\widehat{\boldsymbol{x}}$ to $Q\boldsymbol{x}=\boldsymbol{b}$.

(b) Explain why QQ^{T} is not positive definite.

(c) What are the (nonzero) singular values of Q, and why?

6 Let S be the subspace of $\mathbf{R}^3$ spanned by $\begin{bmatrix}1\\2\\2\end{bmatrix}$ and $\begin{bmatrix}5\\4\\-2\end{bmatrix}$.

(a) Find an orthonormal basis $\boldsymbol{q}_1, \boldsymbol{q}_2$ for S by Gram-Schmidt.

(b) Write down the 3 by 3 matrix P which projects vectors perpendicularly onto S.

(c) Show how the properties of P (what are they?) lead to the conclusion that $P\boldsymbol{b}$ is orthogonal to $\boldsymbol{b}-P\boldsymbol{b}$.

7 **(a)** If $\boldsymbol{v}_1, \boldsymbol{v}_2, \boldsymbol{v}_3$ form a basis for $\mathbf{R}^3$ then the matrix with those three columns is ______.

(b) If $\boldsymbol{v}_1, \boldsymbol{v}_2, \boldsymbol{v}_3, \boldsymbol{v}_4$ span $\mathbf{R}^3$, give all possible ranks for the matrix with those four columns. ____________________.

(c) If $\boldsymbol{q}_1, \boldsymbol{q}_2, \boldsymbol{q}_3$ form an orthonormal basis for $\mathbf{R}^3$, and T is the transformation that projects every vector $\boldsymbol{v}$ onto the plane of $\boldsymbol{q}_1$ and $\boldsymbol{q}_2$, what is the matrix for T in this basis? Explain.

8 Suppose the n by n matrix A_n has 3's along its main diagonal and 2's along the diagonal below and the $(1,n)$ position:

$$A_4=\begin{bmatrix}3&0&0&2\\2&3&0&0\\0&2&3&0\\0&0&2&3\end{bmatrix}.$$

Find by cofactors of row 1 or otherwise the determinant of A_4 and then the determinant of A_n for $n>4$.

9 There are six 3 by 3 permutation matrices P.

(a) What numbers can be the *determinant* of P? What numbers can be *pivots*?

(b) What numbers can be the *trace* of P? What *four numbers* can be eigenvalues of P?

MATRIX FACTORIZATIONS

1. $\mathbf{A = LU} = \begin{pmatrix}\text{lower triangular } L\\ \text{1's on the diagonal}\end{pmatrix}\begin{pmatrix}\text{upper triangular } U\\ \text{pivots on the diagonal}\end{pmatrix}$ *Section 2.6*

   Requirements: No row exchanges as Gaussian elimination reduces A to U.

2. $\mathbf{A = LDU} = \begin{pmatrix}\text{lower triangular } L\\ \text{1's on the diagonal}\end{pmatrix}\begin{pmatrix}\text{pivot matrix}\\ D \text{ is diagonal}\end{pmatrix}\begin{pmatrix}\text{upper triangular } U\\ \text{1's on the diagonal}\end{pmatrix}$

   Requirements: No row exchanges. The pivots in D are divided out to leave 1's in U. If A is symmetric then U is L^{T} and $A = LDL^{\mathrm{T}}$. *Section 2.6 and 2.7*

3. $\mathbf{PA = LU}$ (permutation matrix P to avoid zeros in the pivot positions).

   Requirements: A is invertible. Then P, L, U are invertible. P does the row exchanges in advance. Alternative: $A = L_1P_1U_1$. *Section 2.7*

4. $\mathbf{EA = R}$ (m by m invertible E) (any A) = rref(A).

   Requirements: None! *The reduced row echelon form* R has r pivot rows and pivot columns. The only nonzero in a pivot column is the unit pivot. The last $m - r$ rows of E are a basis for the left nullspace of A, and the first r columns of E^{-1} are a basis for the column space of A. *Sections 3.2–3.3.*

5. $\mathbf{A = CC^T}$ = (lower triangular matrix C) (transpose is upper triangular)

   Requirements: A is symmetric and positive definite (all n pivots in D are positive). This *Cholesky factorization* has $C = L\sqrt{D}$. *Section 6.5*

6. $\mathbf{A = QR}$ = (orthonormal columns in Q) (upper triangular R)

   Requirements: A has independent columns. Those are *orthogonalized* in Q by the Gram-Schmidt process. If A is square then $Q^{-1} = Q^{\mathrm{T}}$. *Section 4.4*

7. $\mathbf{A = S\Lambda S^{-1}}$ = (eigenvectors in S)(eigenvalues in Λ)(left eigenvectors in S^{-1}).

   Requirements: A must have n linearly independent eigenvectors. *Section 6.2*

8. $\mathbf{A = Q\Lambda Q^T}$ =(orthogonal matrix Q)(real eigenvalue matrix Λ)(Q^{T} is Q^{-1}).

   Requirements: A is *symmetric*. This is the Spectral Theorem. *Section 6.4*

9. $\mathbf{A = MJM^{-1}}$ = (generalized eigenvectors in M)(Jordan blocks in J)(M^{-1}).

Requirements: A is any square matrix. *Jordan form* J has a block for each independent eigenvector of A. Each block has one eigenvalue. *Section 6.6*

10. $$\mathbf{A = U\Sigma V^T} = \begin{pmatrix}\text{orthogonal}\\ U \text{ is } m\times m\end{pmatrix}\begin{pmatrix} m\times n \text{ singular value matrix}\\ \sigma_1,\ldots,\sigma_r \text{ on its diagonal}\end{pmatrix}\begin{pmatrix}\text{orthogonal}\\ V \text{ is } n\times n\end{pmatrix}.$$

Requirements: None. This *singular value decomposition* (SVD) has the eigenvectors of AA^{T} in U and of $A^{\mathrm{T}}A$ in V; $\sigma_i = \sqrt{\lambda_i(A^{\mathrm{T}}A)} = \sqrt{\lambda_i(AA^{\mathrm{T}})}$. *Sections 6.7 and 7.4*

11. $$\mathbf{A^+ = V\Sigma^+U^T} = \begin{pmatrix}\text{orthogonal}\\ n\times n\end{pmatrix}\begin{pmatrix} n\times m \text{ pseudoinverse of } \Sigma\\ 1/\sigma_1,\ldots,1/\sigma_r \text{ on diagonal}\end{pmatrix}\begin{pmatrix}\text{orthogonal}\\ m\times m\end{pmatrix}.$$

Requirements: None. The *pseudoinverse* has A^+A = projection onto row space of A and AA^+ = projection onto column space. The shortest least-squares solution to $A\mathbf{x} = \mathbf{b}$ is $\hat{\mathbf{x}} = A^+\mathbf{b}$. This solves $A^{\mathrm{T}}A\hat{\mathbf{x}} = A^{\mathrm{T}}\boldsymbol{b}$. *Section 7.4*

12. $\mathbf{A = QH}$ = (orthogonal matrix Q)(symmetric positive definite matrix H).

Requirements: A is invertible. This *polar decomposition* has $H^2 = A^{\mathrm{T}}A$. The factor H is semidefinite if A is singular. The reverse polar decomposition $A = KQ$ has $K^2 = AA^{\mathrm{T}}$. Both have $Q = UV^{\mathrm{T}}$ from the SVD. *Section 7.4*

13. $\mathbf{A = U\Lambda U^{-1}}$ = (unitary U)(eigenvalue matrix Λ)(U^{-1} which is $U^{\mathrm{H}} = \overline{U}^{\mathrm{T}}$).

Requirements: A is *normal*: $A^{\mathrm{H}}A = AA^{\mathrm{H}}$. Its orthonormal (and possibly complex) eigenvectors are the columns of U. Complex λ's unless $A = A^{\mathrm{H}}$. *Section 10.2*

14. $\mathbf{A = UTU^{-1}}$ = (unitary U)(triangular T with λ's on diagonal)($U^{-1} = U^{\mathrm{H}}$).

Requirements: *Schur triangularization* of any square A. There is a matrix U with orthonormal columns that makes $U^{-1}AU$ triangular. *Section 10.2*

15. $$\mathbf{F_n} = \begin{bmatrix} I & D\\ I & -D\end{bmatrix}\begin{bmatrix}\mathbf{F_{n/2}} & \\ & \mathbf{F_{n/2}}\end{bmatrix}\begin{bmatrix}\text{even-odd}\\ \text{permutation}\end{bmatrix} = \text{one step of the } \mathbf{FFT}.$$

Requirements: F_n = Fourier matrix with entries w^{jk} where $w^n = 1$. Then $\mathbf{F_n}\overline{\mathbf{F}}_\mathbf{n} = nI$. D has $1, w, w^2, \ldots$ on its diagonal. For $n = 2^l$ the *Fast Fourier Transform* has $\frac{1}{2}nl$ multiplications from l stages of D's. *Section 10.3*

CONCEPTUAL QUESTIONS FOR REVIEW

Chapter 1

1.1 Which vectors are linear combinations of $\boldsymbol{v} = (3, 1)$ and $\boldsymbol{w} = (4, 3)$?

1.2 Compare the dot product of $\boldsymbol{v} = (3, 1)$ and $\boldsymbol{w} = (4, 3)$ to the product of their lengths. Which is larger? Whose inequality?

1.3 What is the cosine of the angle between $\boldsymbol{v}$ and $\boldsymbol{w}$ in Question 1.2? What is the cosine of the angle between the x-axis and $\boldsymbol{v}$?

Chapter 2

2.1 Multiplying a matrix A times the column vector $\boldsymbol{x} = (2, -1)$ gives what combination of the columns of A? How many rows and columns in A?

2.2 If $A\boldsymbol{x} = \boldsymbol{b}$ then the vector $\boldsymbol{b}$ is a linear combination of what vectors from the matrix A? In vector space language, $\boldsymbol{b}$ lies in the _____ space of A.

2.3 If A is the 2 by 2 matrix $\left[\begin{smallmatrix} 2 & 1 \\ 6 & 6 \end{smallmatrix}\right]$ what are its pivots?

2.4 If A is the matrix $\left[\begin{smallmatrix} 0 & 1 \\ 1 & 1 \end{smallmatrix}\right]$ how does elimination proceed? What permutation matrix P is involved?

2.5 If A is the matrix $\left[\begin{smallmatrix} 2 & 1 \\ 6 & 3 \end{smallmatrix}\right]$ find $\boldsymbol{b}$ and $\boldsymbol{c}$ so that $A\boldsymbol{x} = \boldsymbol{b}$ has no solution and $A\boldsymbol{x} = \boldsymbol{c}$ has a solution.

2.6 What 3 by 3 matrix L adds 5 times row 2 to row 3 and then adds 2 times row 1 to row 2, when it multiplies a matrix with three rows?

2.7 What 3 by 3 matrix E subtracts 2 times row 1 from row 2 and then subtracts 5 times row 2 from row 3? How is E related to L in Question 2.6?

2.8 If A is 4 by 3 and B is 3 by 7, how many *row times column* products go into AB? How many *column times row* products go into AB? How many separate small multiplications are involved (the same for both)?

2.9 Suppose $A = \begin{bmatrix} \mathbf{I} & \mathbf{U} \\ \mathbf{0} & \mathbf{I} \end{bmatrix}$ is a matrix with 2 by 2 blocks. What is the inverse matrix?

2.10 How can you find the inverse of A by working with $[\,A \quad I\,]$? If you solve the n equations $A\boldsymbol{x}$ = columns of I then the solutions $\boldsymbol{x}$ are columns of ______.

2.11 How does elimination decide whether a square matrix A is invertible?

2.12 Suppose elimination takes A to U (upper triangular) by row operations with the multipliers in L (lower triangular). Why does the last row of A agree with the last row of L times U?

2.13 What is the factorization (from elimination with possible row exchanges) of any square invertible matrix?

2.14 What is the transpose of the inverse of AB?

2.15 How do you know that the inverse of a permutation matrix is a permutation matrix? How is it related to the transpose?

Chapter 3

3.1 What is the column space of an invertible n by n matrix? What is the nullspace of that matrix?

3.2 If every column of A is a multiple of the first column, what is the column space of A?

3.3 What are the two requirements for a set of vectors in $\mathbf{R}^n$ to be a subspace?

3.4 If the row reduced form R of a matrix A begins with a row of ones, how do you know that the other rows of R are zero and what is the nullspace?

3.5 Suppose the nullspace of A contains only the zero vector. What can you say about solutions to $A\boldsymbol{x} = \boldsymbol{b}$?

3.6 From the row reduced form R, how would you decide the rank of A?

3.7 Suppose column 4 of A is the sum of columns 1, 2, and 3. Find a vector in the nullspace.

3.8 Describe in words the complete solution to a linear system $A\boldsymbol{x} = \boldsymbol{b}$.

3.9 If $A\boldsymbol{x} = \boldsymbol{b}$ has exactly one solution for every $\boldsymbol{b}$, what can you say about A?

3.10 Give an example of vectors that span $\mathbf{R}^2$ but are not a basis for $\mathbf{R}^2$.

3.11 What is the dimension of the space of 4 by 4 symmetric matrices?

3.12 Describe the meaning of *basis* and *dimension* of a vector space.

3.13 Why is every row of A perpendicular to every vector in the nullspace?

3.14 How do you know that a column $\boldsymbol{u}$ times a row $\boldsymbol{v}^{\mathrm{T}}$ (both nonzero) has rank 1?

3.15 What are the dimensions of the four fundamental subspaces, if A is 6 by 3 with rank 2?

3.16 What is the row reduced form R of a 3 by 4 matrix of all 2's?

3.17 Describe a *pivot column* of A.

3.18 True? The vectors in the left nullspace of A have the form $A^{\mathrm{T}}\boldsymbol{y}$.

3.19 Why do the columns of every invertible matrix yield a basis?

Chapter 4

4.1 What does the word *complement* mean about orthogonal subspaces?

4.2 If $\boldsymbol{V}$ is a subspace of the 7-dimensional space $\mathbf{R}^7$, the dimensions of $\boldsymbol{V}$ and its orthogonal complement add to _____.

4.3 The projection of $\boldsymbol{b}$ onto the line through $\boldsymbol{a}$ is the vector _____.

4.4 The projection matrix onto the line through $\boldsymbol{a}$ is $P =$ _____.

4.5 The key equation to project $\boldsymbol{b}$ onto the column space of A is the *normal equation* _____.

4.6 The matrix $A^{\mathrm{T}}A$ is invertible when the columns of A are _____.

4.7 The least squares solution to $A\boldsymbol{x} = \boldsymbol{b}$ minimizes what error function?

4.8 What is the connection between the least squares solution of $A\boldsymbol{x} = \boldsymbol{b}$ and the idea of projection onto the column space?

4.9 If you graph the best straight line to a set of 10 data points, what shape is the matrix A and where does the projection $\boldsymbol{p}$ appear in the graph?

4.10 If the columns of Q are orthonormal, why is $Q^{\mathrm{T}}Q = I$?

4.11 What is the projection matrix P onto the columns of Q?

4.12 If Gram-Schmidt starts with the vectors $\boldsymbol{a} = (2, 0)$ and $\boldsymbol{b} = (1, 1)$, which two orthonormal vectors does it produce? If we keep $\boldsymbol{a} = (2, 0)$ does Gram-Schmidt always produce the same two orthonormal vectors?

4.13 True? Every permutation matrix is an orthogonal matrix.

4.14 The inverse of the orthogonal matrix Q is _____.

Chapter 5

5.1 What is the determinant of the matrix $-I$?

5.2 Explain how the determinant is a linear function of the first row.

5.3 How do you know that $\det A^{-1} = 1/\det A$?

5.4 If the pivots of A (with no row exchanges) are 2, 6, 6, what submatrices of A have known determinants?

5.5 Suppose the first row of A is 0, 0, 0, 3. What does the "big formula" for the determinant of A reduce to in this case?

5.6 Is the ordering (2, 5, 3, 4, 1) even or odd? What permutation matrix has what determinant, from your answer?

5.7 What is the cofactor C_{23} in the 3 by 3 elimination matrix E that subtracts 4 times row 1 from row 2? What entry of E^{-1} is revealed?

5.8 Explain the meaning of the cofactor formula for $\det A$ using column 1.

5.9 How does Cramer's Rule give the first component in the solution to $I\boldsymbol{x} = \boldsymbol{b}$?

5.10 If I combine the entries in row 2 with the cofactors from row 1, why is $a_{21}C_{11} + a_{22}C_{12} + a_{23}C_{13}$ automatically zero?

5.11 What is the connection between determinants and volumes?

5.12 Find the cross product of $\boldsymbol{u} = (0, 0, 1)$ and $\boldsymbol{v} = (0, 1, 0)$ and its direction.

5.13 If A is n by n, why is $\det(A - \lambda I)$ a polynomial in λ of degree n?

Chapter 6

6.1 What equation gives the eigenvalues of A without involving the eigenvectors? How would you then find the eigenvectors?

6.2 If A is singular what does this say about its eigenvalues?

6.3 If A times A equals $4A$, what numbers can be eigenvalues of A?

6.4 Find a real matrix that has no real eigenvalues or eigenvectors.

6.5 How can you find the sum and product of the eigenvalues directly from A?

6.6 What are the eigenvalues of the rank one matrix $[\,1\ \ 2\ \ 1\,]^{\mathrm{T}}[\,1\ \ 1\ \ 1\,]$?

6.7 Explain the diagonalization formula $A = S\Lambda S^{-1}$. Why is it true and when is it true?

6.8 What is the difference between the algebraic and geometric multiplicities of an eigenvalue of A? Which might be larger?

6.9 Explain why the trace of AB equals the trace of BA.

6.10 How do the eigenvectors of A help to solve $d\boldsymbol{u}/dt = A\boldsymbol{u}$?

6.11 How do the eigenvectors of A help to solve $\boldsymbol{u}_{k+1} = A\boldsymbol{u}_k$?

6.12 Define the matrix exponential e^A and its inverse and its square.

6.13 If A is symmetric, what is special about its eigenvectors? Do any other matrices have eigenvectors with this property?

6.14 What is the diagonalization formula when A is symmetric?

6.15 What does it mean to say that A is *positive definite*?

6.16 When is $B = A^{\mathrm{T}}A$ a positive definite matrix (A is real)?

6.17 If A is positive definite describe the surface $\boldsymbol{x}^{\mathrm{T}}A\boldsymbol{x} = 1$ in $\mathbf{R}^n$.

6.18 What does it mean for A and B to be *similar*? What is sure to be the same for A and B?

6.19 The 3 by 3 matrix with ones for $i \geq j$ has what Jordan form?

6.20 The SVD expresses A as a product of what three types of matrices?

6.21 How is the SVD for A linked to $A^{\mathrm{T}}A$?

Chapter 7

7.1 Define a linear transformation from $\mathbf{R}^3$ to $\mathbf{R}^2$ and give one example.

7.2 If the upper middle house on the cover of the book is the original, find something nonlinear in the transformations of the other eight houses.

7.3 If a linear transformation takes every vector in the input basis into the next basis vector (and the last into zero), what is its matrix?

7.4 Suppose we change from the standard basis (the columns of I) to the basis given by the columns of A (invertible matrix). What is the change of basis matrix M?

7.5 Suppose our new basis is formed from the eigenvectors of a matrix A. What matrix represents A in this new basis?

7.6 If A and B are the matrices representing linear transformations S and T on $\mathbf{R}^n$, what matrix represents the transformation from $\boldsymbol{v}$ to $S(T(\boldsymbol{v}))$?

7.7 Describe five important factorizations of a matrix A and explain when each of them succeeds (what conditions on A?).

GLOSSARY

Adjacency matrix of a graph. Square matrix with $a_{ij} = 1$ when there is an edge from node i to node j; otherwise $a_{ij} = 0$. $A = A^{\mathrm{T}}$ for an undirected graph.

Affine transformation $T(\boldsymbol{v}) = A\boldsymbol{v} + \boldsymbol{v}_0 =$ linear transformation plus shift.

Associative Law $(AB)C = A(BC)$. Parentheses can be removed to leave ABC.

Augmented matrix $[\,A \;\; \boldsymbol{b}\,]$. $A\boldsymbol{x} = \boldsymbol{b}$ is solvable when $\boldsymbol{b}$ is in the column space of A; then $[\,A \;\; \boldsymbol{b}\,]$ has the same rank as A. Elimination on $[\,A \;\; \boldsymbol{b}\,]$ keeps equations correct.

Back substitution. Upper triangular systems are solved in reverse order x_n to x_1.

Basis for $\boldsymbol{V}$. Independent vectors $\boldsymbol{v}_1, \ldots, \boldsymbol{v}_d$ whose linear combinations give every $\boldsymbol{v}$ in $\boldsymbol{V}$. A vector space has many bases!

Big formula for n **by** n **determinants**. Det(A) is a sum of $n!$ terms, one term for each permutation P of the columns. That term is the product $a_{1\alpha} \cdots a_{n\omega}$ down the diagonal of the reordered matrix, times $\det(P) = \pm 1$.

Block matrix. A matrix can be partitioned into matrix blocks, by cuts between rows and/or between columns. **Block multiplication** of AB is allowed if the block shapes permit (the columns of A and rows of B must be in matching blocks).

Cayley-Hamilton Theorem. $p(\lambda) = \det(A - \lambda I)$ has $p(A) =$ *zero matrix*.

Change of basis matrix M. The old basis vectors $\boldsymbol{v}_j$ are combinations $\sum m_{ij}\boldsymbol{w}_i$ of the new basis vectors. The coordinates of $c_1\boldsymbol{v}_1 + \cdots + c_n\boldsymbol{v}_n = d_1\boldsymbol{w}_1 + \cdots + d_n\boldsymbol{w}_n$ are related by $\boldsymbol{d} = M\boldsymbol{c}$. (For $n = 2$ set $\boldsymbol{v}_1 = m_{11}\boldsymbol{w}_1 + m_{21}\boldsymbol{w}_2$, $\boldsymbol{v}_2 = m_{12}\boldsymbol{w}_1 + m_{22}\boldsymbol{w}_2$.)

Characteristic equation $\det(A - \lambda I) = 0$. The n roots are the eigenvalues of A.

Cholesky factorization $A = CC^{\mathrm{T}} = (L\sqrt{D})(L\sqrt{D})^{\mathrm{T}}$ for positive definite A.

Circulant matrix C. Constant diagonals wrap around as in cyclic shift S. Every circulant is $c_0 I + c_1 S + \cdots + c_{n-1}S^{n-1}$. $C\boldsymbol{x} =$ **convolution** $\boldsymbol{c} * \boldsymbol{x}$. Eigenvectors in F.

Cofactor C_{ij}. Remove row i and column j; multiply the determinant by $(-1)^{i+j}$.

Column picture of $A\boldsymbol{x} = \boldsymbol{b}$. The vector $\boldsymbol{b}$ becomes a combination of the columns of A. The system is solvable only when $\boldsymbol{b}$ is in the column space $\boldsymbol{C}(A)$.

Column space $\boldsymbol{C}(A) =$ space of all combinations of the columns of A.

Commuting matrices $AB = BA$. If diagonalizable, they share n eigenvectors.

Companion matrix. Put $c_1, \ldots, c_n$ in row n and put $n - 1$ 1's along diagonal 1. Then $\det(A - \lambda I) = \pm(c_1 + c_2\lambda + c_3\lambda^2 + \cdots)$.

Complete solution $\boldsymbol{x} = \boldsymbol{x}_p + \boldsymbol{x}_n$ to $A\boldsymbol{x} = \boldsymbol{b}$. (Particular $\boldsymbol{x}_p$) + ($\boldsymbol{x}_n$ in nullspace).

Complex conjugate $\overline{z} = a - ib$ for any complex number $z = a + ib$. Then $z\overline{z} = |z|^2$.

Condition number $cond(A) = \kappa(A) = \|A\|\|A^{-1}\| = \sigma_{\max}/\sigma_{\min}$. In $A\boldsymbol{x} = \boldsymbol{b}$, the relative change $\|\delta\boldsymbol{x}\|/\|\boldsymbol{x}\|$ is less than $cond(A)$ times the relative change $\|\delta\boldsymbol{b}\|/\|\boldsymbol{b}\|$. Condition numbers measure the *sensitivity* of the output to change in the input.

Conjugate Gradient Method. A sequence of steps (end of Chapter 9) to solve positive definite $A\boldsymbol{x} = \boldsymbol{b}$ by minimizing $\frac{1}{2}\boldsymbol{x}^{\mathrm{T}}A\boldsymbol{x} - \boldsymbol{x}^{\mathrm{T}}\boldsymbol{b}$ over growing Krylov subspaces.

Covariance matrix Σ. When random variables x_i have mean = average value = 0, their covariances Σ_{ij} are the averages of $x_i x_j$. With means $\overline{x}_i$, the matrix Σ = mean of $(\boldsymbol{x} - \overline{\boldsymbol{x}})(\boldsymbol{x} - \overline{\boldsymbol{x}})^{\mathrm{T}}$ is positive (semi)definite; it is diagonal if the x_i are independent.

Cramer's Rule for $A\boldsymbol{x} = \boldsymbol{b}$. B_j has $\boldsymbol{b}$ replacing column j of A, and $x_j = |B_j|/|A|$.

Cross product $\boldsymbol{u} \times \boldsymbol{v}$ in $\mathbf{R}^3$. Vector perpendicular to $\boldsymbol{u}$ and $\boldsymbol{v}$, length $\|\boldsymbol{u}\|\|\boldsymbol{v}\||\sin\theta|$ = parallelogram area, computed as the "determinant" of $[\,\boldsymbol{i}\ \ \boldsymbol{j}\ \ \boldsymbol{k}\,;\ u_1\ \ u_2\ \ u_3\,;\ v_1\ \ v_2\ \ v_3\,]$.

Cyclic shift S. Permutation with $s_{21} = 1, s_{32} = 1, \ldots$, finally $s_{1n} = 1$. Its eigenvalues are nth roots $e^{2\pi ik/n}$ of 1; eigenvectors are columns of the Fourier matrix F.

Determinant $|A| = \det(A)$. Defined by $\det I = 1$, sign reversal for row exchange, and linearity in each row. Then $|A| = 0$ when A is singular. Also $|AB| = |A||B|$ and $|A^{-1}| = 1/|A|$ and $|A^{\mathrm{T}}| = |A|$. The big formula for $\det(A)$ has a sum of $n!$ terms, the cofactor formula uses determinants of size $n - 1$, volume of box = $|\det(A)|$.

Diagonal matrix D. $d_{ij} = 0$ if $i \neq j$. **Block-diagonal**: zero outside square blocks D_{ii}.

Diagonalizable matrix A. Must have n independent eigenvectors (in the columns of S; automatic with n different eigenvalues). Then $S^{-1}AS = \Lambda$ = eigenvalue matrix.

Diagonalization $\Lambda = S^{-1}AS$. Λ = eigenvalue matrix and S = eigenvector matrix. A must have n independent eigenvectors to make S invertible. All $A^k = S\Lambda^k S^{-1}$.

Dimension of vector space $\dim(\boldsymbol{V})$ = number of vectors in any basis for $\boldsymbol{V}$.

Distributive Law $A(B + C) = AB + AC$. Add then multiply, or multiply then add.

Dot product $\boldsymbol{x}^{\mathrm{T}}\boldsymbol{y} = x_1 y_1 + \cdots + x_n y_n$. Complex dot product is $\overline{\boldsymbol{x}}^{\mathrm{T}}\boldsymbol{y}$. Perpendicular vectors have zero dot product. $(AB)_{ij}$ = (row i of A)·(column j of B).

Echelon matrix U. The first nonzero entry (the pivot) in each row comes after the pivot in the previous row. All zero rows come last.

Eigenvalue λ **and eigenvector** $\boldsymbol{x}$. $A\boldsymbol{x} = \lambda\boldsymbol{x}$ with $\boldsymbol{x} \neq \boldsymbol{0}$ so $\det(A - \lambda I) = 0$.

Eigshow. Graphical 2 by 2 eigenvalues and singular values (MATLAB or Java).

Elimination. A sequence of row operations that reduces A to an upper triangular U or to the reduced form $R = \mathrm{rref}(A)$. Then $A = LU$ with multipliers ℓ_{ij} in L, or $PA = LU$ with row exchanges in P, or $EA = R$ with an invertible E.

Elimination matrix = Elementary matrix E_{ij}. The identity matrix with an extra $-\ell_{ij}$ in the i, j entry ($i \neq j$). Then $E_{ij}A$ subtracts ℓ_{ij} times row j of A from row i.

Ellipse (or ellipsoid) $x^{\mathrm{T}}Ax = 1$. A must be positive definite; the axes of the ellipse are eigenvectors of A, with lengths $1/\sqrt{\lambda}$. (For $\|x\| = 1$ the vectors $y = Ax$ lie on the ellipse $\|A^{-1}y\|^2 = y^{\mathrm{T}}(AA^{\mathrm{T}})^{-1}y = 1$ displayed by eigshow; axis lengths σ_i.)

Exponential $e^{At} = I + At + (At)^2/2! + \cdots$ has derivative Ae^{At}; $e^{At}u(0)$ solves $u' = Au$.

Factorization $A = LU$. If elimination takes A to U *without row exchanges*, then the lower triangular L with multipliers ℓ_{ij} (and $\ell_{ii} = 1$) brings U back to A.

Fast Fourier Transform (FFT). A factorization of the Fourier matrix F_n into $\ell = \log_2 n$ matrices S_i times a permutation. Each S_i needs only $n/2$ multiplications, so $F_n x$ and $F_n^{-1}c$ can be computed with $n\ell/2$ multiplications. Revolutionary.

Fibonacci numbers $0, 1, 1, 2, 3, 5, \ldots$ satisfy $F_n = F_{n-1} + F_{n-2} = (\lambda_1^n - \lambda_2^n)/(\lambda_1 - \lambda_2)$. Growth rate $\lambda_1 = (1+\sqrt{5})/2$ is the largest eigenvalue of the Fibonacci matrix $\left[\begin{smallmatrix}1&1\\1&0\end{smallmatrix}\right]$.

Four fundamental subspaces of $A = C(A), N(A), C(A^{\mathrm{T}}), N(A^{\mathrm{T}})$.

Fourier matrix F. Entries $F_{jk} = e^{2\pi ijk/n}$ give orthogonal columns $\overline{F}^{\mathrm{T}}F = nI$. Then $y = Fc$ is the (inverse) Discrete Fourier Transform $y_j = \sum c_k e^{2\pi ijk/n}$.

Free columns of A. Columns without pivots; combinations of earlier columns.

Free variable x_i. Column i has no pivot in elimination. We can give the $n-r$ free variables any values, then $Ax = b$ determines the r pivot variables (if solvable!).

Full column rank $r = n$. Independent columns, $N(A) = \{0\}$, no free variables.

Full row rank $r = m$. Independent rows, at least one solution to $Ax = b$, column space is all of $\mathbf{R}^m$. *Full rank* means full column rank or full row rank.

Fundamental Theorem. The nullspace $N(A)$ and row space $C(A^{\mathrm{T}})$ are orthogonal complements (perpendicular subspaces of $\mathbf{R}^n$ with dimensions r and $n-r$) from $Ax = 0$. Applied to A^{T}, the column space $C(A)$ is the orthogonal complement of $N(A^{\mathrm{T}})$.

Gauss-Jordan method. Invert A by row operations on $[\,A \;\; I\,]$ to reach $[\,I \;\; A^{-1}\,]$.

Gram-Schmidt orthogonalization $A = QR$. Independent columns in A, orthonormal columns in Q. Each column q_j of Q is a combination of the first j columns of A (and conversely, so R is upper triangular). Convention: $\mathrm{diag}(R) > 0$.

Graph G. Set of n nodes connected pairwise by m edges. A **complete graph** has all $n(n-1)/2$ edges between nodes. A **tree** has only $n-1$ edges and no closed loops. A **directed graph** has a direction arrow specified on each edge.

Hankel matrix H. Constant along each antidiagonal; h_{ij} depends on $i+j$.

Hermitian matrix $A^{\mathrm{H}} = \overline{A}^{\mathrm{T}} = A$. Complex analog of a symmetric matrix: $\overline{a_{ji}} = a_{ij}$.

Hessenberg matrix H. Triangular matrix with one extra nonzero adjacent diagonal.

Hilbert matrix hilb(n). Entries $H_{ij} = 1/(i+j-1) = \int_0^1 x^{i-1}x^{j-1}dx$. Positive definite but extremely small $\lambda_{\min}$ and large condition number.

Hypercube matrix P_L^2. Row $n+1$ counts corners, edges, faces, ... of a cube in $\mathbf{R}^n$.

Identity matrix I (or I_n). Diagonal entries $= 1$, off-diagonal entries $= 0$.

Incidence matrix of a directed graph. The m by n edge-node incidence matrix has a row for each edge (node i to node j), with entries -1 and 1 in columns i and j.

Indefinite matrix. A symmetric matrix with eigenvalues of both signs ($+$ and $-$).

Independent vectors $\boldsymbol{v}_1, \dots, \boldsymbol{v}_k$. No combination $c_1\boldsymbol{v}_1 + \cdots + c_k\boldsymbol{v}_k =$ zero vector unless all $c_i = 0$. If the $\boldsymbol{v}$'s are the columns of A, the only solution to $A\boldsymbol{x} = \boldsymbol{0}$ is $\boldsymbol{x} = \boldsymbol{0}$.

Inverse matrix A^{-1}. Square matrix with $A^{-1}A = I$ and $AA^{-1} = I$. No inverse if $\det A = 0$ and $\operatorname{rank}(A) < n$ and $A\boldsymbol{x} = \boldsymbol{0}$ for a nonzero vector $\boldsymbol{x}$. The inverses of AB and A^{T} are $B^{-1}A^{-1}$ and $(A^{-1})^{\mathrm{T}}$. Cofactor formula $(A^{-1})_{ij} = C_{ji}/\det A$.

Iterative method. A sequence of steps intended to approach the desired solution.

Jordan form $J = M^{-1}AM$. If A has s independent eigenvectors, its "generalized" eigenvector matrix M gives $J = \operatorname{diag}(J_1, \dots, J_s)$. The block J_k is $\lambda_k I_k + N_k$ where N_k has 1's on diagonal 1. Each block has one eigenvalue λ_k and one eigenvector $(1, 0, \dots, 0)$.

Kirchhoff's Laws. *Current law*: net current (in minus out) is zero at each node. *Voltage law*: Potential differences (voltage drops) add to zero around any closed loop.

Kronecker product (tensor product) $A \otimes B$. Blocks $a_{ij}B$, eigenvalues $\lambda_p(A)\lambda_q(B)$.

Krylov subspace $K_j(A, \boldsymbol{b})$. The subspace spanned by $\boldsymbol{b}, A\boldsymbol{b}, \dots, A^{j-1}\boldsymbol{b}$. Numerical methods approximate $A^{-1}\boldsymbol{b}$ by $\boldsymbol{x}_j$ with residual $\boldsymbol{b} - A\boldsymbol{x}_j$ in this subspace. A good basis for K_j requires only multiplication by A at each step.

Least squares solution $\widehat{\boldsymbol{x}}$. The vector $\widehat{\boldsymbol{x}}$ that minimizes the error $\|\boldsymbol{e}\|^2$ solves $A^{\mathrm{T}}A\widehat{\boldsymbol{x}} = A^{\mathrm{T}}\boldsymbol{b}$. Then $\boldsymbol{e} = \boldsymbol{b} - A\widehat{\boldsymbol{x}}$ is orthogonal to all columns of A.

Left inverse A^+. If A has full column rank n, then $A^+ = (A^{\mathrm{T}}A)^{-1}A^{\mathrm{T}}$ has $A^+A = I_n$.

Left nullspace $N(A^{\mathrm{T}})$. Nullspace of $A^{\mathrm{T}} =$ "left nullspace" of A because $\boldsymbol{y}^{\mathrm{T}}A = \boldsymbol{0}^{\mathrm{T}}$.

Length $\|\boldsymbol{x}\|$. Square root of $\boldsymbol{x}^{\mathrm{T}}\boldsymbol{x}$ (Pythagoras in n dimensions).

Linear combination $c\boldsymbol{v} + d\boldsymbol{w}$ or $\sum c_j\boldsymbol{v}_j$. Vector addition and scalar multiplication.

Linear transformation T. Each vector $\boldsymbol{v}$ in the input space transforms to $T(\boldsymbol{v})$ in the output space, and linearity requires $T(c\boldsymbol{v} + d\boldsymbol{w}) = c\,T(\boldsymbol{v}) + d\,T(\boldsymbol{w})$. Examples: Matrix multiplication $A\boldsymbol{v}$, differentiation in function space.

Linearly dependent $\boldsymbol{v}_1, \dots, \boldsymbol{v}_n$. A combination other than all $c_i = 0$ gives $\sum c_i\boldsymbol{v}_i = \boldsymbol{0}$.

Lucas numbers $L_n = 2, 1, 3, 4, \dots$ satisfy $L_n = L_{n-1} + L_{n-2} = \lambda_1^n + \lambda_2^n$, with eigenvalues $\lambda_1, \lambda_2 = (1 \pm \sqrt{5})/2$ of the Fibonacci matrix $\left[\begin{smallmatrix}1 & 1\\ 1 & 0\end{smallmatrix}\right]$. Compare $L_0 = 2$ with Fibonacci.

Markov matrix M. All $m_{ij} \geq 0$ and each column sum is 1. Largest eigenvalue $\lambda = 1$. If $m_{ij} > 0$, the columns of M^k approach the steady state eigenvector $M\boldsymbol{s} = \boldsymbol{s} > \boldsymbol{0}$.

Matrix multiplication AB. The i, j entry of AB is (row i of A)·(column j of B) = $\sum a_{ik}b_{kj}$. By columns: Column j of $AB = A$ times column j of B. By rows: row i of A multiplies B. Columns times rows: AB = sum of (column k)(row k). All these equivalent definitions come from the rule that AB times $\boldsymbol{x}$ equals A times $B\boldsymbol{x}$.

Minimal polynomial of A. The lowest degree polynomial with $m(A)$ = zero matrix. The roots of m are eigenvalues, and $m(\lambda)$ divides $\det(A - \lambda I)$.

Multiplication $A\boldsymbol{x} = x_1(\text{column } 1) + \cdots + x_n(\text{column } n)$ = combination of columns.

Multiplicities AM **and** GM. The algebraic multiplicity AM of an eigenvalue λ is the number of times λ appears as a root of $\det(A - \lambda I) = 0$. The geometric multiplicity GM is the number of independent eigenvectors (= dimension of the eigenspace for λ).

Multiplier ℓ_{ij}. The pivot row j is multiplied by ℓ_{ij} and subtracted from row i to eliminate the i, j entry: ℓ_{ij} = (entry to eliminate)/(jth pivot).

Network. A directed graph that has constants $c_1, \ldots, c_m$ associated with the edges.

Nilpotent matrix N. Some power of N is the zero matrix, $N^k = 0$. The only eigenvalue is $\lambda = 0$ (repeated n times). Examples: triangular matrices with zero diagonal.

Norm $\|A\|$ **of a matrix**. The "ℓ^2 norm" is the maximum ratio $\|A\boldsymbol{x}\|/\|\boldsymbol{x}\| = \sigma_{\max}$. Then $\|A\boldsymbol{x}\| \le \|A\|\|\boldsymbol{x}\|$ and $\|AB\| \le \|A\|\|B\|$ and $\|A + B\| \le \|A\| + \|B\|$. **Frobenius norm** $\|A\|_F^2 = \sum\sum a_{ij}^2$; ℓ^1 and ℓ^∞ norms are largest column and row sums of $|a_{ij}|$.

Normal equation $A^{\mathrm{T}}A\widehat{\boldsymbol{x}} = A^{\mathrm{T}}\boldsymbol{b}$. Gives the least squares solution to $A\boldsymbol{x} = \boldsymbol{b}$ if A has full rank n. The equation says that (columns of A)·$(\boldsymbol{b} - A\widehat{\boldsymbol{x}}) = 0$.

Normal matrix N. $NN^{\mathrm{T}} = N^{\mathrm{T}}N$, leads to orthonormal (complex) eigenvectors.

Nullspace $N(A)$ = Solutions to $A\boldsymbol{x} = \boldsymbol{0}$. Dimension $n - r$ = (# columns) − rank.

Nullspace matrix N. The columns of N are the $n - r$ special solutions to $A\boldsymbol{s} = \boldsymbol{0}$.

Orthogonal matrix Q. Square matrix with orthonormal columns, so $Q^{\mathrm{T}}Q = I$ implies $Q^{\mathrm{T}} = Q^{-1}$. Preserves length and angles, $\|Q\boldsymbol{x}\| = \|\boldsymbol{x}\|$ and $(Q\boldsymbol{x})^{\mathrm{T}}(Q\boldsymbol{y}) = \boldsymbol{x}^{\mathrm{T}}\boldsymbol{y}$. All $|\lambda| = 1$, with orthogonal eigenvectors. Examples: Rotation, reflection, permutation.

Orthogonal subspaces. Every $\boldsymbol{v}$ in $\boldsymbol{V}$ is orthogonal to every $\boldsymbol{w}$ in $\boldsymbol{W}$.

Orthonormal vectors $\boldsymbol{q}_1, \ldots, \boldsymbol{q}_n$. Dot products are $\boldsymbol{q}_i^{\mathrm{T}}\boldsymbol{q}_j = 0$ if $i \ne j$ and $\boldsymbol{q}_i^{\mathrm{T}}\boldsymbol{q}_i = 1$. The matrix Q with these orthonormal columns has $Q^{\mathrm{T}}Q = I$. If $m = n$ then $Q^{\mathrm{T}} = Q^{-1}$ and $\boldsymbol{q}_1, \ldots, \boldsymbol{q}_n$ is an **orthonormal basis** for $\mathbf{R}^n$: every $\boldsymbol{v} = \sum(\boldsymbol{v}^{\mathrm{T}}\boldsymbol{q}_j)\boldsymbol{q}_j$.

Outer product $\boldsymbol{u}\boldsymbol{v}^{\mathrm{T}}$ = column times row = rank one matrix.

Partial pivoting. In elimination, the jth pivot is chosen as the largest available entry (in absolute value) in column j. Then all multipliers have $|\ell_{ij}| \le 1$. Roundoff error is controlled (depending on the *condition number* of A).

Particular solution $\boldsymbol{x}_p$. Any solution to $A\boldsymbol{x} = \boldsymbol{b}$; often $\boldsymbol{x}_p$ has free variables = 0.

Pascal matrix P_S = pascal(n). The symmetric matrix with binomial entries $\binom{i+j-2}{i-1}$. $P_S = P_L P_U$ all contain Pascal's triangle with det = 1 (see index for more properties).

Permutation matrix P. There are $n!$ orders of $1, \ldots, n$; the $n!$ P's have the rows of I in those orders. PA puts the rows of A in the same order. P is a product of row exchanges P_{ij}; P is *even* or *odd* ($\det P = 1$ or -1) based on the number of exchanges.

Pivot columns of A. Columns that contain pivots after row reduction; not combinations of earlier columns. The pivot columns are a basis for the column space.

Pivot d. The diagonal entry (*first nonzero*) when a row is used in elimination.

Plane (or hyperplane) in $\mathbf{R}^n$. Solutions to $\boldsymbol{a}^{\mathrm{T}}\boldsymbol{x} = 0$ give the plane (dimension $n - 1$) perpendicular to $\boldsymbol{a} \neq \mathbf{0}$.

Polar decomposition $A = QH$. Orthogonal Q, positive (semi)definite H.

Positive definite matrix A. Symmetric matrix with positive eigenvalues and positive pivots. Definition: $\boldsymbol{x}^{\mathrm{T}}A\boldsymbol{x} > 0$ unless $\boldsymbol{x} = \mathbf{0}$.

Projection $\boldsymbol{p} = \boldsymbol{a}(\boldsymbol{a}^{\mathrm{T}}\boldsymbol{b}/\boldsymbol{a}^{\mathrm{T}}\boldsymbol{a})$ **onto the line through** $\boldsymbol{a}$. $P = \boldsymbol{a}\boldsymbol{a}^{\mathrm{T}}/\boldsymbol{a}^{\mathrm{T}}\boldsymbol{a}$ has rank 1.

Projection matrix P **onto subspace** $\boldsymbol{S}$. Projection $\boldsymbol{p} = P\boldsymbol{b}$ is the closest point to $\boldsymbol{b}$ in $\boldsymbol{S}$, error $\boldsymbol{e} = \boldsymbol{b} - P\boldsymbol{b}$ is perpendicular to $\boldsymbol{S}$. $P^2 = P = P^{\mathrm{T}}$, eigenvalues are 1 or 0, eigenvectors are in $\boldsymbol{S}$ or $\boldsymbol{S}^{\perp}$. If columns of A = basis for $\boldsymbol{S}$ then $P = A(A^{\mathrm{T}}A)^{-1}A^{\mathrm{T}}$.

Pseudoinverse A^+ **(Moore-Penrose inverse)**. The n by m matrix that "inverts" A from column space back to row space, with $\boldsymbol{N}(A^+) = \boldsymbol{N}(A^{\mathrm{T}})$. A^+A and AA^+ are the projection matrices onto the row space and column space. $\mathrm{Rank}(A^+) = \mathrm{rank}(A)$.

Random matrix rand(n) or randn(n). MATLAB creates a matrix with random entries, uniformly distributed on $[\,0 \quad 1\,]$ for rand and standard normal distribution for randn.

Rank one matrix $A = \boldsymbol{u}\boldsymbol{v}^{\mathrm{T}} \neq 0$. Column and row spaces = lines $c\boldsymbol{u}$ and $c\boldsymbol{v}$.

Rank $r(A)$ = number of pivots = dimension of column space = dimension of row space.

Rayleigh quotient $q(\boldsymbol{x}) = \boldsymbol{x}^{\mathrm{T}}A\boldsymbol{x}/\boldsymbol{x}^{\mathrm{T}}\boldsymbol{x}$ for symmetric A: $\lambda_{\min} \leq q(\boldsymbol{x}) \leq \lambda_{\max}$. Those extremes are reached at the eigenvectors $\boldsymbol{x}$ for $\lambda_{\min}(A)$ and $\lambda_{\max}(A)$.

Reduced row echelon form $R = \mathrm{rref}(A)$. Pivots = 1; zeros above and below pivots; r nonzero rows of R give a basis for the row space of A.

Reflection matrix $Q = I - 2\boldsymbol{u}\boldsymbol{u}^{\mathrm{T}}$. The unit vector $\boldsymbol{u}$ is reflected to $Q\boldsymbol{u} = -\boldsymbol{u}$. All vectors $\boldsymbol{x}$ in the plane mirror $\boldsymbol{u}^{\mathrm{T}}\boldsymbol{x} = 0$ are unchanged because $Q\boldsymbol{x} = \boldsymbol{x}$. The "Householder matrix" has $Q^{\mathrm{T}} = Q^{-1} = Q$.

Right inverse A^+. If A has full row rank m, then $A^+ = A^{\mathrm{T}}(AA^{\mathrm{T}})^{-1}$ has $AA^+ = I_m$.

Rotation matrix $R = \begin{bmatrix} \cos\theta & -\sin\theta \\ \sin\theta & \cos\theta \end{bmatrix}$ rotates the plane by θ and $R^{-1} = R^{\mathrm{T}}$ rotates back by $-\theta$. Orthogonal matrix, eigenvalues $e^{i\theta}$ and $e^{-i\theta}$, eigenvectors $(1, \pm i)$.

Row picture of $A\boldsymbol{x} = \boldsymbol{b}$. Each equation gives a plane in $\boldsymbol{R}^n$; planes intersect at $\boldsymbol{x}$.

Row space $\boldsymbol{C}(A^{\mathrm{T}})$ = all combinations of rows of A. Column vectors by convention.

Saddle point of $f(x_1, \ldots, x_n)$. A point where the first derivatives of f are zero and the second derivative matrix ($\partial^2 f/\partial x_i \partial x_j$ = **Hessian matrix**) is indefinite.

Schur complement $S = D - CA^{-1}B$. Appears in block elimination on $\left[\begin{smallmatrix} \mathbf{A} & \mathbf{B} \\ \mathbf{C} & \mathbf{D} \end{smallmatrix}\right]$.

Schwarz inequality $|\boldsymbol{v}\cdot\boldsymbol{w}| \le \|\boldsymbol{v}\|\,\|\boldsymbol{w}\|$.Then $|\boldsymbol{v}^{\mathrm{T}}A\boldsymbol{w}|^2 \le (\boldsymbol{v}^{\mathrm{T}}A\boldsymbol{v})(\boldsymbol{w}^{\mathrm{T}}A\boldsymbol{w})$ if $A = C^{\mathrm{T}}C$.

Semidefinite matrix A. (Positive) semidefinite means symmetric with $\boldsymbol{x}^{\mathrm{T}}A\boldsymbol{x} \ge 0$ for all vectors $\boldsymbol{x}$. Then all eigenvalues $\lambda \ge 0$; no negative pivots.

Similar matrices A **and** B. Every $B = M^{-1}AM$ has the same eigenvalues as A.

Simplex method for linear programming. The minimum cost vector $\boldsymbol{x}^*$ is found by moving from corner to lower cost corner along the edges of the feasible set (where the constraints $A\boldsymbol{x} = \boldsymbol{b}$ and $\boldsymbol{x} \ge \boldsymbol{0}$ are satisfied). Minimum cost at a corner!

Singular matrix A. A square matrix that has no inverse: $\det(A) = 0$.

Singular Value Decomposition (SVD) $A = U\Sigma V^{\mathrm{T}}$ = **(orthogonal** U**) times (diagonal** Σ**) times (orthogonal** V^{T}**)**. First r columns of U and V are orthonormal bases of $\boldsymbol{C}(A)$ and $\boldsymbol{C}(A^{\mathrm{T}})$ with $A\boldsymbol{v}_i = \sigma_i\boldsymbol{u}_i$ and singular value $\sigma_i > 0$. Last columns of U and V are orthonormal bases of the nullspaces of A^{T} and A.

Skew-symmetric matrix K. The transpose is $-K$, since $K_{ij} = -K_{ji}$. Eigenvalues are pure imaginary, eigenvectors are orthogonal, e^{Kt} is an orthogonal matrix.

Solvable system $A\boldsymbol{x} = \boldsymbol{b}$. The right side $\boldsymbol{b}$ is in the column space of A.

Spanning set $\boldsymbol{v}_1, \ldots, \boldsymbol{v}_m$ **for** $\boldsymbol{V}$. Every vector in $\boldsymbol{V}$ is a combination of $\boldsymbol{v}_1, \ldots, \boldsymbol{v}_m$.

Special solutions to $As = \boldsymbol{0}$. One free variable is $s_i = 1$, other free variables $= 0$.

Spectral theorem $A = Q\Lambda Q^{\mathrm{T}}$. Real symmetric A has real λ_i and orthonormal $\boldsymbol{q}_i$ with $A\boldsymbol{q}_i = \lambda_i\boldsymbol{q}_i$. In mechanics the $\boldsymbol{q}_i$ give the *principal axes*.

Spectrum of A = the set of eigenvalues $\{\lambda_1, \ldots, \lambda_n\}$. **Spectral radius** $= |\lambda_{\max}|$.

Standard basis for $\mathbf{R}^n$. Columns of n by n identity matrix (written $\boldsymbol{i}, \boldsymbol{j}, \boldsymbol{k}$ in $\mathbf{R}^3$).

Stiffness matrix K. If $\boldsymbol{x}$ gives the movements of the nodes in a discrete structure, $K\boldsymbol{x}$ gives the internal forces. Often $K = A^{\mathrm{T}}CA$ where C contains spring constants from Hooke's Law and $A\boldsymbol{x}$ = stretching (strains) from the movements $\boldsymbol{x}$.

Subspace $\boldsymbol{S}$ **of** $\boldsymbol{V}$. Any vector space inside $\boldsymbol{V}$, including $\boldsymbol{V}$ and $\boldsymbol{Z}$ = {zero vector}.

Sum $\boldsymbol{V}+\boldsymbol{W}$ **of subspaces**. Space of all ($\boldsymbol{v}$ in $\boldsymbol{V}$) + ($\boldsymbol{w}$ in $\boldsymbol{W}$). **Direct sum**: $\dim(\boldsymbol{V}+\boldsymbol{W}) = \dim \boldsymbol{V} + \dim \boldsymbol{W}$ when $\boldsymbol{V}$ and $\boldsymbol{W}$ share only the zero vector.

Symmetric factorizations $A = LDL^{\mathrm{T}}$ **and** $A = Q\Lambda Q^{\mathrm{T}}$. The number of positive pivots in D and positive eigenvalues in Λ is the same.

Symmetric matrix A. The transpose is $A^{\mathrm{T}} = A$, and $a_{ij} = a_{ji}$. A^{-1} is also symmetric. All matrices of the form $R^{\mathrm{T}}R$ and LDL^{T} and $Q\Lambda Q^{\mathrm{T}}$ are symmetric. Symmetric matrices have real eigenvalues in Λ and orthonormal eigenvectors in Q.

Toeplitz matrix T. Constant-diagonal matrix, so t_{ij} depends only on $j - i$. Toeplitz matrices represent linear time-invariant filters in signal processing.

Trace of A = sum of diagonal entries = sum of eigenvalues of A. $\operatorname{Tr} AB = \operatorname{Tr} BA$.

Transpose matrix A^{T}. Entries $A^{\mathrm{T}}_{ij} = A_{ji}$. A^{T} is n by m, $A^{\mathrm{T}}A$ is square, symmetric, positive semidefinite. The transposes of AB and A^{-1} are $B^{\mathrm{T}}A^{\mathrm{T}}$ and $(A^{\mathrm{T}})^{-1}$.

Triangle inequality $\|\boldsymbol{u}+\boldsymbol{v}\| \le \|\boldsymbol{u}\| + \|\boldsymbol{v}\|$. For matrix norms $\|A+B\| \le \|A\| + \|B\|$.

Tridiagonal matrix T: $t_{ij} = 0$ if $|i-j| > 1$. T^{-1} has rank 1 above and below diagonal.

Unitary matrix $U^{\mathrm{H}} = \overline{U}^{\mathrm{T}} = U^{-1}$. Orthonormal columns (complex analog of Q).

Vandermonde matrix V. $V\boldsymbol{c} = \boldsymbol{b}$ gives the polynomial $p(x) = c_0 + \cdots + c_{n-1}x^{n-1}$ with $p(x_i) = b_i$ at n points. $V_{ij} = (x_i)^{j-1}$ and $\det V =$ product of $(x_k - x_i)$ for $k > i$.

Vector $\boldsymbol{v}$ in $\mathbf{R}^n$. Sequence of n real numbers $\boldsymbol{v} = (v_1, \ldots, v_n) =$ point in $\mathbf{R}^n$.

Vector addition. $\boldsymbol{v} + \boldsymbol{w} = (v_1 + w_1, \ldots, v_n + w_n) =$ diagonal of parallelogram.

Vector space $\boldsymbol{V}$. Set of vectors such that all combinations $c\boldsymbol{v} + d\boldsymbol{w}$ remain in $\boldsymbol{V}$. Eight required rules are given in Section 3.1 for $c\boldsymbol{v} + d\boldsymbol{w}$.

Volume of box. The rows (or columns) of A generate a box with volume $|\det(A)|$.

Wavelets $w_{jk}(t)$ or vectors $\boldsymbol{w}_{jk}$. Stretch and shift the time axis to create $w_{jk}(t) = w_{00}(2^j t - k)$. Vectors from $\boldsymbol{w}_{00} = (1, 1, -1, -1)$ would be $(1, -1, 0, 0)$ and $(0, 0, 1, -1)$.

INDEX

MATLAB TEACHING CODES

cofactor	Compute the n by n matrix of cofactors.
cramer	Solve the system $A\boldsymbol{x} = \boldsymbol{b}$ by Cramer's Rule.
deter	Matrix determinant computed from the pivots in $PA = LU$.
eigen2	Eigenvalues, eigenvectors, and $\det(A - \lambda I)$ for 2 by 2 matrices.
eigshow	Graphical demonstration of eigenvalues and singular values.
eigval	Eigenvalues and their multiplicity as roots of $\det(A - \lambda I) = 0$.
eigvec	Compute as many linearly independent eigenvectors as possible.
elim	Reduction of A to row echelon form R by an invertible E.
findpiv	Find a pivot for Gaussian elimination (used by **plu**).
fourbase	Construct bases for all four fundamental subspaces.
grams	Gram-Schmidt orthogonalization of the columns of A.
house	2 by 12 matrix giving the corner coordinates of a house.
inverse	Matrix inverse (if it exists) by Gauss-Jordan elimination.
leftnull	Compute a basis for the left nullspace.
linefit	Plot the least squares fit to m given points by a line.
lsq	Least squares solution to $A\boldsymbol{x} = \boldsymbol{b}$ from $A^{\mathrm{T}}A\hat{\boldsymbol{x}} = A^{\mathrm{T}}\boldsymbol{b}$.
normal	Eigenvalues and orthonormal eigenvectors when $A^{\mathrm{T}}A = AA^{\mathrm{T}}$.
nulbasis	Matrix of special solutions to $A\boldsymbol{x} = \boldsymbol{0}$ (basis for nullspace).
orthcomp	Find a basis for the orthogonal complement of a subspace.
partic	Particular solution of $A\boldsymbol{x} = \boldsymbol{b}$, with all free variables zero.
plot2d	Two-dimensional plot for the house figures (cover and Section 7.1).
plu	Rectangular $PA = LU$ factorization with row exchanges.
poly2str	Express a polynomial as a string.
project	Project a vector $\boldsymbol{b}$ onto the column space of A.
projmat	Construct the projection matrix onto the column space of A.
randperm	Construct a random permutation.
rowbasis	Compute a basis for the row space from the pivot rows of R.
samespan	Test whether two matrices have the same column space.
signperm	Determinant of the permutation matrix with rows ordered by p.
slu	LU factorization of a square matrix using *no row exchanges*.
slv	Apply slu to solve the system $A\boldsymbol{x} = \boldsymbol{b}$ allowing no row exchanges.
splu	Square $PA = LU$ factorization *with row exchanges*.
splv	The solution to a square, invertible system $A\boldsymbol{x} = \boldsymbol{b}$.
symmeig	Compute the eigenvalues and eigenvectors of a symmetric matrix.
tridiag	Construct a tridiagonal matrix with constant diagonals a, b, c.

These Teaching Codes are directly available from the Linear Algebra Home Page:

http://web.mit.edu/18.06/www

They were written in MATLAB, and translated into Maple and Mathematica.

LINEAR ALGEBRA IN A NUTSHELL

((*A is n by n*))

Nonsingular	**Singular**
A is invertible	A is not invertible
The columns are independent	The columns are dependent
The rows are independent	The rows are dependent
The determinant is not zero	The determinant is zero
$A\boldsymbol{x}=\mathbf{0}$ has one solution $\boldsymbol{x}=\mathbf{0}$	$A\boldsymbol{x}=\mathbf{0}$ has infinitely many solutions
$A\boldsymbol{x}=\boldsymbol{b}$ has one solution $\boldsymbol{x}=A^{-1}\boldsymbol{b}$	$A\boldsymbol{x}=\boldsymbol{b}$ has no solution or infinitely many
A has n (nonzero) pivots	A has $r<n$ pivots
A has full rank $r=n$	A has rank $r<n$
The reduced row echelon form is $R=I$	R has at least one zero row
The column space is all of $\mathbf{R}^n$	The column space has dimension $r<n$
The row space is all of $\mathbf{R}^n$	The row space has dimension $r<n$
All eigenvalues are nonzero	Zero is an eigenvalue of A
$A^{\mathrm{T}}A$ is symmetric positive definite	$A^{\mathrm{T}}A$ is only semidefinite
A has n (positive) singular values	A has $r<n$ singular values

Each line of the singular column can be made quantitative using r.